PROJECT EVALUATION IN THE CHEMICAL PROCESS INDUSTRIES

PROJECT EVALUATION IN THE CHEMICAL PROCESS INDUSTRIES

J. Frank Valle-Riestra

Research Scientist
Dow Chemical U.S.A., Pittsburg, California

Lecturer in Chemical Engineering
University of California, Berkeley

McGraw-Hill, Inc.
New York St. Louis San Francisco Auckland Bogotá
Caracas Lisbon London Madrid Mexico City Milan
Montreal New Delhi San Juan Singapore
Sydney Tokyo Toronto

This book was set in Times Roman by Interactive Composition Corporation.
The editors were Kiran Verma and David A. Damstra;
the production supervisor was Charles Hess.
The drawings were done by Wellington Studios Ltd.
The cover was designed by Mark Wieboldt.

TP
155.5
V34
1983

PROJECT EVALUATION IN THE CHEMICAL PROCESS INDUSTRIES

Copyright © 1983 by McGraw-Hill, Inc. All rights reserved. Typeset in the United States of America. Except as permitted under the United States Copyright Act of 1976, no part of this publication may be reproduced or distributed in any form or by any means, or stored in a data base or retrieval system, without the prior written permission of the publisher.

9 10 11 12 13 14 BKMBKM 9 9 8 7 6 5 4

ISBN 0-07-066840-X

Library of Congress Cataloging in Publication Data

Valle-Riestra, J. Frank
   Project evaluation in the chemical process industries.

   Includes bibliographies and index.
   1. Chemical plants—Management. 2. Industrial project management—Evaluation. I. Title
TP155.5.V34 1983   658.4′04   82-20806
ISBN 0-07-066840-X

TO EDITH,
WHO HELPED A LOT.

CONTENTS

	PREFACE	xiii
1	**Introduction to Industrial Project Management**	**1**
1.1	NATURE OF THE INDUSTRIAL PROJECT	2
	The Industrial Environment / Industrial Functional Groups / The Corporate Organization / The Origin of Industrial Projects / Criteria of Project Success / The Role of Research and Development	
1.2	THE INDUSTRIAL PROJECT MANAGER: THE PROFESSIONAL CHEMICAL ENGINEER	26
	Starting a Professional Career in Industry / Advancement and Job Functions / Professional Development / Communications / Patents	
1.3	THE ECONOMIC BASIS OF THE INDUSTRIAL PROJECT	61
	The Scope of Economic Analysis in the Chemical Industry / The Nature of Money and Capital / Sources of Capital / The Cost of Capital	
	NOMENCLATURE / REFERENCES / PROBLEMS	
2	**The Mathematics of Finance**	**77**
2.1	THE MEASUREMENT OF INTEREST	78
	Classification by Compounding Method / Simple Interest / Discrete Compound Interest / Continuously Compounded Interest / Effective Interest / The Consideration of Inflation / Present and Future Worth	
2.2	THE PRESENT WORTH OF CASH FLOWS	86
	What Is Cash Flow? / Net Present Worth / The Continuous Discounting of Continuous Cash Flows / The Discrete Discounting of Discrete Cash Flows / The Continuous Discounting of Discrete Cash Flows	

	2.3	ANNUITIES	100

Proof by Induction / Periodic Payments / Perpetuity / Capitalized Cost / The Present Worth of Interest Payments

NOMENCLATURE / REFERENCES / PROBLEMS

	3	**Project Evaluation Systematics**	**122**
	3.1	PRINCIPLES OF PRELIMINARY PROJECT EVALUATION	122

Project Types: Similarities and Differences / Sequence of Preliminary Project Evaluation / Follow-up Evaluation

	3.2	MARKETING RESEARCH	126

The Scope of Marketing Research / Product Profile

	3.3	DEMAND PROJECTION	128

Trend Analysis / Price Elasticity of Demand / Other Aspects of Demand Projection / Market Penetration / Product and Plant Life Cycles / Long-Range Corporate Planning and Technological Forecasting

	3.4	PRICE PROJECTION	149

Pricing of Established Products / Criteria of Final Pricing Decision

	3.5	FLOW SHEET DEVELOPMENT METHODS IN COST ESTIMATION	152

Choice of Process / Block Flow Diagrams / Early-Stage Mass and Energy Balances / Process Flow Sheets / Some Frequent Flow Sheet Errors

NOMENCLATURE / REFERENCES / PROBLEMS

	4	**Equipment Design and Costing**	**187**
	4.1	EQUIPMENT SELECTION AND SIZING FOR PRELIMINARY COST ESTIMATES	188

The Heuristic Approach / Distillation / Heat Transfer / Liquid Pumping and Agitation / Storage and Surge Vessels / Other Items of Equipment / Safety Factors / Equipment Lists

	4.2	ESTIMATION OF PURCHASED COST OF EQUIPMENT	209

Definition of Purchased Cost / Methods of Presenting Cost Data / Sources of Equipment Cost Information / Estimation Methods in Absence of Specific Information

	4.3	EFFECT OF INFLATION UPON CAPITAL COSTS	223

Cost Indices / Limitations of Cost Indices

	4.4	COST OF EQUIPMENT INSTALLATION	227

Typical Installation Requirements / Installation Cost Estimation

4.5	RELIABILITY OF EQUIPMENT COST ESTIMATION Variability of Cost Data / Propagation of Errors NOMENCLATURE / REFERENCES / SPECIAL PROJECT / PROBLEMS	232
5	**The Direct Fixed Capital Investment**	**256**
5.1	THE ESTIMATION OF THE DIRECT FIXED CAPITAL Battery Limits and Auxiliary Facilities / The Definition of Direct Fixed Capital / The Elements of Direct Fixed Capital: The Method of Purchased Cost Factors / Alternate Methods of Fixed Capital Estimation / Functional Step Scoring	257
5.2	THE SEPARATE ESTIMATION OF AUXILIARY FACILITIES The Scope of Auxiliary Facilities / Auxiliary Facility Cost Correlations	285
5.3	THE ESTIMATION OF PIPING SYSTEMS The Design of Pipelines / Detailed Piping Estimates	290
5.4	RELIABILITY OF CAPITAL ESTIMATES NOMENCLATURE / REFERENCES / PROBLEMS	311
6	**Depreciation**	**334**
6.1	THE ECONOMIC IMPACT OF DEPRECIATION The Depreciation Concept / Depreciation and Its Impact upon Profit / Equipment and Plant Lifetimes / Salvage Value / Depreciation Methods and Tax Laws / Depreciation Accountancy / Expensed Capital	335
6.2	METHODS OF DETERMINING DEPRECIATION CHARGES Straight-Line (SL) Depreciation / Double-Declining-Balance (DDB) Depreciation / Composite DDB plus Straight-Line (DDB + SL) Depreciation / Sum-of-Years-Digits (SOYD) Depreciation / Preferred Depreciation Method NOMENCLATURE / REFERENCES / PROBLEMS	351
7	**The Cost of Manufacture**	**367**
7.1	THE ELEMENTS OF THE COST OF MANUFACTURE Definition and Boundaries / Computational Format / Stream Factors	368
7.2	RAW MATERIALS AND UTILITIES Standard Cost / Raw Material Costs / Utilities Demand and Costs	374
7.3	OPERATING LABOR Estimation of Labor Requirements / Labor Rates /	384

Fringe Benefits / The Cost of Supervision and
Clerical Help
7.4 REGULATED AND FIXED CHARGES 392
Maintenance / Process Analysis and Quality Control /
Waste Treatment / Operating Supplies / Fixed
Charges
7.5 THE TOTAL COST FOR SALE 397
Cost of Process-Oriented Technology / Bulk Cost:
Summary and Shortcut Methods / Cost of
Packaging
7.6 THE RESPONSIBILITY FOR ENVIRONMENTAL PROTECTION 400
NOMENCLATURE / REFERENCES / PROBLEMS

8 The Criterion of Economic Performance 421

8.1 THE CAPITAL FOR TRANSFER 422
Allocated Fixed Capital / Working Capital
8.2 RETURN ON INVESTMENT 428
General Expenses and Profit / Capital for Sale (CFS) /
Acceptable Return on Investment
8.3 PROFITABILITY UNDER VARIABLE CONDITIONS 435
Inflation and Profitability / Pricing for Acceptable
Profitability / Break-Even Charts / Economic
Optimization of Plant Size / Input Parameter
Uncertainties: Sensitivity Analysis
8.4 PROFITABILITY CRITERIA RELATED TO ROI 446
ROI and Variable Cash Flow / The Return on Average
Investment / Payout Time (Payback Period)
8.5 THE COSTS OF PRODUCT TRANSPORTATION 451
Rail Freight / Tank Trucks / Marine Transport
NOMENCLATURE / REFERENCES / PROBLEMS

9 Cash Flow Analysis 466

9.1 CASH FLOW CONCEPTS 467
The Cash Flow "Black Box" / Cash Flow Diagrams /
Tax Credits / Depreciation on Allocated Capital /
Equivalent Maximum Investment Period
9.2 NET PRESENT WORTH 474
Payout Time with Interest / The NPW Criterion
of Profitability
9.3 DISCOUNTED CASH FLOW 478
The DCRR Criterion of Profitability / The Projection
of Production Life Cycles / Time-Distributed
Investments / The Effect of Debt upon Discounted
Cash Flow / The Evaluation of Research Projects

9.4	RELATIVE MERIT OF PROFITABILITY CRITERIA The Low Equation / Critique of the Criteria	494
9.5	THE ANALYSIS OF RISK The Quantification of Risk / Risk Analysis Methodology NOMENCLATURE / REFERENCES / SPECIAL PROJECT / PROBLEMS	497

10	**The Analysis of Alternatives**	521
10.1	THE ANALYSIS OF EQUIPMENT ALTERNATIVES The Principle of Minimum Investment / The Differential Return on Investment / Capitalized Cost / Break-Even Analysis	522
10.2	THE ANALYSIS OF PROCESS AND INVESTMENT ALTERNATIVES The Ranking of Mutually Exclusive Projects / Differential Profitability Criteria / Present Worth Ratio Analysis / Capital Budgeting / Equipment Leasing / Economic Justification of Research Investments / Decision Trees	531
10.3	ECONOMIC OPTIMIZATION Economic Optimization in Plant Design / Optimization of Continuous Objective Functions / Optimization of Discrete Objective Functions / The Optimum Allocation of Resources	551
10.4	REPLACEMENT ANALYSIS Concepts / Methodology	567
10.5	PLANT MODIFICATION DECISIONS Equipment Debottlenecking Strategy / Plant Expansion Strategy / Termination of Operations NOMENCLATURE / REFERENCES / PROBLEMS	574

11	**Engineering Management of Construction Projects**	595
11.1	THE MANAGEMENT OF THE PROCESS DEVELOPMENT SEQUENCE The Substance of Process Development / The Organization of Pilot Plant Operations	596
11.2	COST CONTROL OF CONSTRUCTION PROJECTS The Requisites of a Successful Construction Project / The Organization of a Construction Project / The Significance of Cost Control / Cost Control Tools	613
11.3	TIME CONTROL OF CONSTRUCTION PROJECTS: SCHEDULING Bar Charts / CPM (Critical Path Method) / PERT (Program Evaluation and Review Technique)	626
11.4	SAFETY AND LOSS PREVENTION Preoperational Reviews / Fire and Explosion Index NOMENCLATURE / REFERENCES / PROBLEMS	642

12		**Corporate Performance Analysis**	**658**
	12.1	PERFORMANCE DOCUMENTATION	659
		The Annual Report / Balance Sheets / Income Statements / Footnotes	
	12.2	RATIO ANALYSIS	667
		The Significance of Ratio Analysis / Balance Sheet Ratios / Interstatement Ratios / The Return on Investment / The Internal Rate of Return	
	12.3	ANALYSIS OF PERFORMANCE OF SECURITIES	685
		Corporate Performance and Market Value of Securities / Book Value of Securities / Characterization of Security Performance / Leverage / Stock Performance History	
	12.4	PERFORMANCE OF THE CHEMICAL INDUSTRIES	692
		Some Corporate Statistics / Position of the Chemical Industry in the U.S. Economy	
		EPILOGUE / NOMENCLATURE / PROBLEMS	
		APPENDIXES	703
A		THE SUMMATION OF SERIES USING THE CALCULUS OF FINITE DIFFERENCES	703
	A.1	Finite Differences	703
	A.2	Factorial Notation	704
	A.3	Changing Polynomials into Factorials	705
	A.4	Finite Integration	706
	A.5	The Definite Integral	708
	A.6	Integration by Parts	710
	A.7	Method of Undetermined Coefficients and Functions	711
B		THE INTERCONVERSION OF ENGINEERING AND SI UNITS	715
		Fundamental SI Units	716
		Derived and Frequently Used Units	716
		Interconversion, FPS and SI Units	716
		Examples of Mixed Units, FPS System	718
		INDEX	719

PREFACE

We haven't the money, so we've got to think.

Ernest Lord Rutherford, 1871–1937

The subject matter of this book is the methodology used in the chemical process industries to evaluate the ultimate commercial feasibility of proposed new projects.

The text addresses itself to ways that engineers may blend principles drawn from diverse disciplines to evaluate and advance projects for which they are responsible. Economic, marketing, and managerial techniques are presented which, in combination with technological academic disciplines, impart the required skills for effective project evaluation and an early assumption of project managerial responsibilities.

Project evaluation may, indeed, be systematized, and this book represents such an effort. It is intended to meet the needs of chemical and other engineering students intent upon an industrial career with early project responsibilities; it is also intended for more-experienced practicing engineers who may wish to have an overview of project evaluation systematics. Hopefully, the subject matter will constitute a firm basis upon which both neophyte and experienced engineers will build their project evaluation skills.

The purpose of the book goes beyond mere description and systematization of the project evaluation theme. An attempt is made to expose the reader to the nature of the problems that are typical of the industrial environment. Such problems are often open-ended and unstructured, in contrast to the usually more highly structured problems typical of the academic environment, structured for the sake of pedagogical efficiency. The engineering student is given a wide-ranging indoctrination into methods of attacking such problems and is taught the techniques of problem *synthesis*, in contrast to the analytical approach which forms the background of the student's experience. The practicing engineer, on the other hand, will quickly recognize many of the situations described in the problem and exercise statements. The problems in this book are an important part of the pedagogical approach in that they not only illustrate but also expand and supplement material in the text.

The book is written in twelve chapters, but the encompassing design involves six sequential concepts:

1 The industrial environment. The purpose of this exposition is to give the student or neophyte engineer some understanding of the nature of the industrial workplace within which project evaluation is practiced.

2 The mathematical tools of project evaluation.

3 The evaluation process. This is the core material of the book, which includes:
Project definition
Investment analysis
Net income analysis
Project economic analysis
Evaluation of criteria of economic performance

4 The analysis of alternatives. Generalization of the core material to multicomponent systems.

5 Management of the developing project. The advancement of projects from the laboratory to operating commercial units.

6 Performance analysis of the chemical process industries. The CPI is evaluated as an ensemble of individual projects.

The book is *not* a source of design or economic data. Economic data in particular are rapidly rendered obsolete by the ravages of inflation and by changing technology. Whatever data are incorporated are used to illustrate methods and techniques; problems at the end of each chapter force reliance upon quantitative information available in the published literature. The reader is given guidance on how to locate desired information in the literature, but, again, the book is not an exhaustive source of literature references. A basic collection of the most reliable references is given for selected subject areas, along with a brief description of what the references contain and why they are useful; these collections are intended to constitute "starter collections" for neophyte engineers in industry.

The text is a distillation of the essence of lecture notes assembled for a graduate course, "Chemical Process Economics and Project Evaluation," taught during the past decade in the Chemical Engineering Department of the University of California, Berkeley. The establishment of such a course was originally recommended by the department's Industrial Advisory Board, and the author, on the strength of his industrial experience, was invited to organize it. The course has proved to be a popular one with graduate students, as well as senior students taking a process design course concurrently. The text is also suitable as a principal or supplemental resource book in allied courses encompassing process design, process synthesis, engineering economics, and industrial business management. Finally, professional engineers in the chemical process industries will find the subject matter so arranged as to make it particularly suitable for self-study.

The International System of Units (SI) is used as the basic one in the book, and the symbols in each chapter's Nomenclature are expressed in these units. However, to reflect common usage in the chemical process industries, a majority of examples and problems is characterized by engineering (FPS) units. Appendix B outlines some of the reasons for the mixed usage and contains information to facilitate interconversion. Units rarely constitute a problem in the text, since most of the concepts involved in the mathematical manipulations are simple ones, such as production rates in pounds per year.

ACKNOWLEDGMENTS

In 1974, when the aforementioned Industrial Advisory Board suggested the organization of a course on project evaluation, I was asked to give it a try by Prof. Scott Lynn, an old friend and erstwhile colleague at The Dow Chemical Company. Since my career at Dow offered more than sufficient opportunities to keep myself usefully occupied, my first stunned reaction was to say no. Scott tends to be a quiet and persistent fellow, and somehow he managed to drag me, kicking and screaming all the way, into the academic maelstrom—a plunge which, in retrospect, I consider to have been really rather pleasant. Over the past decade, I have, indeed, had the distinct pleasure of receiving the benefit of novel intellectual contacts with staff members of a superb chemical engineering department, and with an astonishingly stimulating group of fine students. These contacts have complemented the experiences I have absorbed as a member of the staff of Dow's Western Division Research Department, another superb organization with an outstanding record of intellectual and commercial achievements in the industrial world. I have been most fortunate to have lived in the best of two possible worlds, the industrial and the academic. I hope that through my gossamer presence in both I have done something that will benefit them mutually.

For all these opportunities I remain most grateful to Scott Lynn, and to Jud King, now Dean of the College of Chemistry at Berkeley. Jud took the bold step, as department chairman, of injecting the industrial presence, including mine, into the academic curriculum, and he was the first to encourage me to transform my lecture notes into book form. I must also gratefully acknowledge the encouragement, help, and understanding support of my colleagues at Dow, particularly Don Graham, Director of Research at Dow's Western Division, and Bob Steffanson, for many years my long-suffering supervisor. I cannot forget the inspiring example of Chuck Oldershaw, my chemical engineering mentor at Dow, who also devoted part of his career to academic work—the organization of a senior design course at Berkeley. When he used me as a guinea pig for his midterms and finals, I had no idea that some day I would be following in his footsteps.

The contents of this book have been shaped by the comments, criticisms, and suggestions of students, and some former students like George Tyson. A number of delightful persons contributed to the final manuscript: Lan To Pham, Deanna Smith, Jane Dillard—and Irene Marasco, a working tornado! Edith Valle-Riestra— my dear wife, Edith—cheerfully assumed all the mechanistic and style-modulating burdens that this neophyte author did not dream of, not even in his worst nightmares: proofreading, weeding out the most turgid of sentences, indexing, corrections of grammar and syntax, obtaining permissions, critique of style, and just general cajoling and handholding.

J. Frank Valle-Riestra

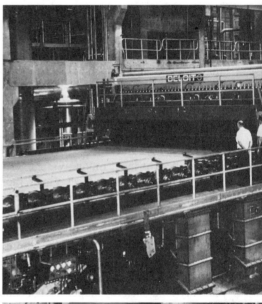

(left page) **Scenes from chemical process industries.**
Integrated chemical complex (*Dow Chemical USA*) Cement manufacture (*Portland Cement Association*)
Ore processing (*Mountain States R & D*) Paper manufacture (*St. Regis*)
Wine making (*Wine Institute*) Sugar refining (*C & H Sugar Co.*)

(right page) **Chemical engineering job functions** (*Dow Chemical USA*).

CHAPTER 1

INTRODUCTION TO INDUSTRIAL PROJECT MANAGEMENT

Engineering is the art of doing that well with one dollar, which any bungler can do with two after a fashion.

Arthur Mellen Wellington, 1847–1895*

The chemical engineer committed to a career in the petroleum or chemical process industry is, in due time, given the responsibility of managing an industrial project. Successful project management demands an intense focusing of the manager's technical and organizational skills acquired at various levels of his or her academic and industrial training. The demand to blend principles drawn from diverse disciplines, to "put it all together," poses an extraordinary challenge to the best-trained engineer. The shepherding of an idea from its inception, through stages of research and development to final plant construction and product commercialization, represents the crowning achievement of the chemical engineering methodology in the industrial milieu. In the final analysis, this is what chemical engineering is all about.

This intricate blending of disciplines into a grand design scheme is not an easy skill to acquire. The required sequence is subject to a certain amount of systematization; the system may certainly be described, but facility and confidence in carrying out the separate steps of the sequence come only with continued practice. The advanced chemical engineering student who wishes to obtain practice in project management is faced with still another obstacle, for without some familiar-

*"The Economic Theory of the Location of Railways," 6th ed., Wiley, New York, 1900.

ity and feeling for the nature of the industrial environment, the student attempts practice in the abstract, out of contact with the realities of the industrial work function. It would be futile to attempt to paint a clear picture of the industrial scene with a few words, to impart a feeling for the complex interplay of inputs experienced by the project manager in a petroleum refinery, in a pulp mill, in a chocolate factory. Perhaps a few initial descriptive brushstrokes will help the neophyte obtain an impressionistic image of a generalized industrial environment, an image which will hopefully gain some clarity as the various aspects of project management are presented.

1.1 NATURE OF THE INDUSTRIAL PROJECT

The Industrial Environment

The popular vision of the chemical or petroleum plant is replete with gleaming towers and vessels, often breathtaking in their overwhelming scale, set in a bewildering maze of pipes and support beams. This is, indeed, the reality of the industrial scene, the aspect of which does not change appreciably in the context of more specialized processing plants devoted to food preparation, paper manufacture, or the production of photographic film or artificial fibers. The photographs so frequently seen perhaps emphasize the neatness and orderliness, even an aura of technological esthetics, characteristic of most well-designed and well-maintained installations. The distant views may not reveal the flaws that also are reality—faded paint, rust deposits, hidden depositories of old equipment, and other evidence of the never-ending struggle with the forces of corrosion, mechanical fatigue, and obsolescence. Older installations may even project an aspect of gentle shabbiness, a patina of economic compromise; a well-managed plant, no matter how old, will nevertheless be characterized by scrupulously neat housekeeping.

The casual visitor to a chemical plant is immediately struck by the relative silence of the operations and the apparent absence of human activity. This deceptively peaceful scene contrasts sharply with the ear-splitting bustle of many other industrial operations, in steel mills, mines, and canneries. During normal operations the processes are run from centrally located control rooms; operators monitor instruments or computer outputs with little actual contact with the operating equipment. But operations are, more often than not, not entirely normal, and the hoped-for routine is repeatedly interrupted by many minor and a few major crises stemming from equipment breakdowns, unexpected operational upsets, and changing production schedules. Even during periods of relative operational stability the professional staff in the production plants is engaged in an endless search for improvements, review of past operations, and maintenance planning. The daily work function of production personnel in particular is characterized by a sense of urgency and intense activity, often far beyond normal working hours.

Another increasingly obvious aspect of modern processing plants is the absence of visible sources of atmospheric emission, an indication of the emphasis upon *environmental protection*. Particulate emission is rigorously controlled, and the discharge of chemical pollutants is subject to ever more rigid restrictions. Liquid waste control and water economy are manifested by large acreage devoted to cooling towers and biooxidation ponds. Concern for the environment is coupled with concern for the physical welfare of the employees, and evidence of *safety* programs permeates the industrial environment. The visible evidence is everywhere—in the eye and head protective gear of the employees; the pervading network of safety showers, sprinklers, and eye baths; the prominent warning signs and safety exhortations. In the well-managed plant and corporation, safety is a top priority, and the employees are taught to think individual and group safety in all their activities.

Occasionally, offices and laboratories are extremely well-appointed, even luxurious, particularly if regularly visited by customers or highly visible to the public. The more usual emphasis is upon functionality and economy rather than quality of appointments, and a great many chemical engineers work at desks and lab benches in an environment that would have to be characterized as Spartan. The work performed is quite often routine and not immediately exciting, something the neophyte scientist and engineer must learn to realize and accept. The basic process of synthesizing or analyzing a problem is intellectually very stimulating, but this process is necessarily interspersed with lengthy periods of data gathering, repetitive computation, and myriad other routine tasks. The particular excitement of a new discovery, a successfully completed project, or a consummated business arrangement occurs, after all, on rare occasion only—and yet, it is the prospect of this excitement, the prospect of creating with one's brain something of significant technological and social impact, which makes the chemical engineer's job ultimately so fascinating. It is this prospect that allows the engineer to tolerate periods of unscheduled overtime, lack of sleep, and even intense physical discomfort during critical phases of pilot plant operations or production plant start-ups.

The complex physical facilities of a chemical or petroleum plant are staffed by highly trained specialists who function in a curious atmosphere of well-coordinated cooperation and unspoken competitiveness. The successful blending of the efforts of specialists requires intense and careful planning, and the chemical engineer's working day is filled with a sequence of not-always-welcome meetings and conferences. Close teamwork is essential to the successful conclusion of projects, and yet each member of the project staff is also an individual who tries to make his or her contributions as "visible" as possible to managers, for these contributions are the rungs in the ladder of career advancement. Competitiveness and the search for peer approval and tangible rewards constitute the fact of life in both academic and industrial environments.

As we will see, the investment associated with process plant facilities is very large, indeed, and it should come as no surprise if we emphasize that

> The primary objective in the industrial environment is to make money.

This is not a moral judgment, but a simple statement of fact. Those of us who are neophyte engineers will more rapidly reach an understanding of the policies and motivation of the corporation or agency for which we work if we continuously remind ourselves of this primary objective. Many may protest that this point of view states the goals of a business enterprise in too crude a commercial form, that the primary objective of any business should be the satisfaction of society's needs. Perhaps as a broad overview of human endeavor this is correct, but the success of an individual enterprise must be measured in readily understood quantitative terms, terms that incorporate the enterprise's profitability. The emphasis upon profitability transcends the delineations of a particular economic system; it happens that profitability is most easily measured in terms of money in most politico-economic systems.

It also happens that chemical and petroleum corporations expend part of their profits in endeavors which do not appear to hold promise of commercial return. After all, such corporations finance theatrical performances on educational television; they give out scholarships to many university chemical engineering departments; they even support the activities of local Little League teams and United Crusade organizations. If each one of these activities were to be analyzed in detail, however, one would discover that each one had been motivated by a long-range goal of commercial advantage—in terms of a more attractive public image vis-à-vis commercial competitors, in terms of a more assured future supply of competent technical employees, in terms of a tax advantage. The employees of a corporation may, indeed, be dedicated to the preservation of the environment, but the primary motivation for the infusion of massive capital for pollution control equipment is the prospect of government-imposed economic penalties, not the love of nature. Actually, the profit incentive does not require a great deal of moralization; we all derive a living from it, and, more often than not, it coincides pretty well with the public interest.

The profit objective is possibly the most obvious differentiating characteristic between the academic and industrial environments, at least as far as the student is concerned (school administrators might wish to challenge this distinction). The young engineer in the first industrial job soon discovers major differences in the *goals* and in the *structure* of the two opposite environments.

The difference in *goals* is the obvious one between skill acquisition and skill application. The university student is faced with the primary task of acquiring skills that will be used in the course of an industrial career. At the graduate school level some practice in applying these skills is given to the student, but the primary motivation is still pedagogical. The process of acquiring new skills certainly spills over into the industrial environment, and continuing education is a never-ending aspect of a successful chemical engineering career. Nevertheless, the young engineer on first assignment quickly recognizes the change in goals—for now the acquired skills must be gathered and organized and applied to the solution of real problems. The acceptable solution is no longer just proof of skills successfully acquired; it is a matter of meeting the primary objective of making a profit.

Even more challenging to the neophyte engineer transferring from the academic to the industrial environment is the difference in *structure* of the two environments. The university environment tends to be highly structured. Courses are presented in logical sequence, and the problems are designed to facilitate the acquisition of skills, to sharpen up the process of *analysis*. The problem statements are well-organized, and because of limitations of time, all information and data required for solution are usually assembled as part of the problem statement. In contrast, the industrial environment is unstructured and open. The sequence of tasks is never well-defined and is to a considerable extent a matter of the engineer's own initiative. The engineer must develop skills in *synthesis,* in defining the very nature of the problems to be solved, in assembling and correlating the disparate elements of the project to be managed. Information and data must be laboriously dug up, and frequently they are nonexistent and must be generated as part of the project. The neophyte may at first feel uncomfortable in this apparent morass of uncertainty but soon learns to enjoy the challenge of a new form of creativity.

Still another uncomfortable aspect of the industrial scene is a vague sense of instability engendered by continuous change. To meet the challenge of the primary objective of making a profit, all functional groups are engaged in a struggle to change the status quo, to improve and modify facilities, to optimize the organizational structure, to discover new opportunities, to discard ideas, functions, and facilities considered obsolete. The process of change, characterized by personnel transfers, decisions to undertake new projects often of stunning magnitude and complexity, and equally stunning decisions to abort ongoing projects, is a part of the everyday experience of the practicing chemical engineer. Increasingly rapid change is an uncomfortable characteristic of our times.

Industrial Functional Groups

The work in the chemical process and allied industries is carried out by a number of functional groups, each with its own set of operational goals. A list of such functional groups may be found in Table 1.1.1; the list is intended to be representative of the process industries, although much the same kind of work function subdivision may be found in any industrial enterprise. At the same time, the list must not be considered as all-inclusive and exhaustive in its survey of work functions. Chemical engineers, by virtue of their training as nonspecific *problem solvers*, are often involved in the activities of each and every one of the functional groups shown.

The first five groups listed are generally considered to be the key functional groups in any industrial organization.

Research and development heads the list, since the R&D function is of particular importance to the chemical engineer entering industry. Often the practice is to hire graduating chemical engineers into research and development where, by virtue of the group's required interaction with all aspects of industrial operations, the neophyte engineer receives well-rounded industrial training. After a few years, an

TABLE 1.1.1
INDUSTRIAL FUNCTIONAL GROUPS

Research and development
Production (operations)
Engineering (including instrumentation and computations specialties)
Management and finance
Sales and marketing
   Technical service and development
   Planning and economic evaluation
   Traffic and distribution
   Specialties:
     Purchasing, raw materials procurement
     Power and energy
     Environmental control
     Corporate recruiting
     Legal and patents
     Maintenance
     Safety and health
     Facilities management
   Nontechnical functions:
     Accounting
     Payroll
     Personnel and labor relations
     Public relations
     Security
     Government affairs

---

individual with specific talents and career goals may transfer into the other functional groups or else may choose to continue in the R&D function. This practice is by no means universal, even in corporations that favor the practice, but it does make a lot of sense to use research and development as a training ground for staff personnel in an industry with such a strong technological base.

Production is the essence of any industry, the actual process of transforming raw materials into a marketable product. In a sense all the other functional groups are service organizations clustered around the production function, with the task of helping to make the process of producing continuous and profitable. The job of managing production demands an admixture, in equal parts, of technical, business, and people-organizational acumen. In the chemical process industry, production service and management is the core job for chemical engineers, and a wealth of job opportunities exists for young engineers, ranging from production engineering to the plant superintendent's job. Production is another frequent entry functional group for new graduates.

The *engineering* function has the objective of creating new production facilities as well as modifying and maintaining existing ones. Two broad classifications are recognized, process engineering and plant engineering. *Process engineering* has the task of transforming a process from the miniplant or pilot plant stage (usually an R&D responsibility) to a full-scale plant through the gamut of design, construction and start-up. *Plant engineering* personnel has the responsibility for the design, construction, and maintenance of structures and auxiliary facilities such as sewers and power distribution. Here chemical engineers work side by side with colleagues from a number of other engineering disciplines. In large industrial organizations the engineering department may incorporate separate and administratively autonomous groups devoted to special applications and skills in areas of instrumentation, computational operations, and vessel design and code enforcement. Occasionally the engineering functional group may include a *construction engineering* division that will assume the entire job of organizing the erection of new facilities, a task more often entrusted to independent contractors and construction engineering firms.

Line managerial duties are part and parcel of all the functional groups; the special *management and finance* function listing refers to the overall business decision-making process at the corporate level and includes decisions regarding the procurement, allocation, and management of financial resources. The ultimate decisions involving major corporate policy are made by the board of directors, but a large staff may be required to back up the decision-making process. The board is at the apex of a corporate managerial hierarchy which, along with its backup staffs, encompasses all aspects of the corporation's operations. Not surprisingly, more often than not chemical engineers rather than business or economics specialists are found at the corporate management level of the chemical process industries, since the decision-making process requires thorough grounding in the technical aspects of the business. Also not surprisingly, not many opportunities exist in this functional group for the starting chemical engineer, since many years of experience are generally required to gain the proper appreciation of the corporation's structure and functions.

The *sales and marketing* function operates at the interface between the corporation and its customers. Very frequently the "customer" of the chemical process industry is still another industrial organization that, in turn, uses the product as its own raw material. As an example, the manufacturer of polystyrene will market the product in bulk form such as beads, and the beads, in turn, are purchased by fabricators who employ extrusion and molding equipment to produce consumer products. In other instances the primary product may be sold through independent or subsidiary distributors; this is the case with petroleum products or industrial solvents. The job of the chemical sales agent goes far beyond the immediate step of consummating a sale—the salesperson must be a thoroughly competent technical professional who not only understands a broad range of technical properties of the product represented, but also must understand well the customer's operations and the way in which the product is to be used. The sales agent's job is that of a technical consultant, and the sales representative is also the primary agent of the corporation's *product stewardship* program—the acceptance of the responsibility that the product, when properly used, will not be deleterious to the environment or to the public safety and well-being.

Sales and marketing incorporates a broad range of tasks involving the distribution policy and pricing of existing products as well as the assessment of markets for new and proposed products. The functional group represents an excellent entrée for young chemical engineers whose career goals are in the business and managerial area, since the marketing experience offers a fine overview of the chemical business and the corporation's role in it.

The *technical service and development* (TS&D) group is the research arm of the sales force. Organizationally it may be part of the sales and marketing department, or else it may be a subdivision of research. The specific task of this functional group is to supply the sales representatives with the technological ammunition they need to sell their line of products. The nature of the TS&D function is such that it demands a great deal of flexibility and aggressive inventiveness from members of the staff. Projects may involve laboratory work to determine unknown physical properties of a product, or a search for improved formulations, or perhaps an investigation of corrosion rates of specific alloys in a boiling liquid product. TS&D may offer various service functions to the customer as an additional incentive for buying the parent corporation's product—analytical services, or perhaps an in-depth review of the customer's process to discover sources of avoidable product waste. Helpful technical service is a strong selling point in a competitive situation. Such service often entails joint visits with sales representatives to the customer's plant to consult on various processing problems.

One of the most challenging tasks often handled by TS&D is the development of new business opportunities for both established and new products. This undertaking may involve pilot plant or even full-scale plant work on the customer's plant site and consequently requires the thorough technical know-how of both the product and the customer's operations. A successful job of new business development demands not only technical skill but also a great deal of organizational, marketing, and business acumen. TS&D shares in the responsibility for product stewardship by providing the expertise to monitor the safety, health, and environmental impact of products after they leave the producing company's property.

A *planning and economic evaluation* group fulfills a number of important functions in a chemical processing complex. Using input from the sales force and production facilities, planners draw up production schedules and goals which serve as guides for raw material procurement, labor force scheduling, distribution planning—in fact, all production-dependent activities. The planning function, however, reaches out much further than the immediate physical plant; in effect, the planning group has the responsibility of charting the future course of the corporation. The successful planner is aware of the complexity of interactions between existing production units and must also remain continuously aware of external developments in politics, economics, and the consumer markets, developments which are likely to influence the future of the production complex. Proposals and ongoing projects are frequently reviewed by the planning and economic evaluation group, which obviously serves as an important advisory staff to corporate management. Economic evaluators must work closely with accountants to keep track of up-to-date standard costs incurred by existing operations, for these costs serve as basis for future projections and estimates.

It must be emphasized that even though most companies have economic evaluation departments, the job of monitoring the economic attractiveness of industrial projects still rests with the project manager. The economic evaluation group is there to help, but it is the project manager who knows the most about the project and is in the best position to evaluate its economic potential.

The focus of *traffic and distribution* is the finished product—how to get it to the customer in the form that the customer requires. The distribution function may involve the packaging of the product in containers ranging from tank cars for bulk liquids to boxes and cans for products destined directly for the consumer market. Distribution also encompasses transportation of the product to customers or redistribution agencies, a task that requires close coordination with truck, railroad, and marine transport systems. Some process industries (petroleum, for example) retain control over product distribution directly to the nonindustrial consumer, in which case the distribution function may match production in size and complexity. The nature of the distribution function dictates that it be closely coordinated with the activities of both production and sales. *Quality control*, the job of checking shipments to customers to make sure they meet all specifications, may be the responsibility of an independent department but is frequently a task entrusted to production.

Of the other specialties listed in Table 1.1.1, chemical engineers are most frequently employed in the first three. *Purchasing* agents have the responsibility of procuring for the corporation equipment, raw materials, and supplies of highest quality at lowest cost, within prescribed delivery time limitations. This goal often requires the negotiation of contracts with specific suppliers. A number of the process industries (petroleum, metals, paper) depend upon natural resources for their raw materials supply, and exploration and exploitation of the resources constitute a function often broader and more complex in scope than the raw materials processing function itself. The *power and energy* specialty encompasses the operation of steam boilers and power generation units, as well as the management of auxiliary facilities such as cross-country pipelines and power distribution systems. The tightening of energy resources has underlined the importance of another task of the power and energy specialist, the development of means of effective energy conservation. The *environmental control* specialty has the primary responsibility of monitoring operations to ascertain whether all federal and state regulations are satisfied. Staff members serve as consultants on environmental matters; however, the task of making sure that new projects incorporate appropriate design parameters to meet environmental standards rests with each project manager.

Most of the other technical functions listed utilize the services of engineers, scientists, and technologists other than chemical engineers, but the chemical engineers' know-how is often called upon. For example, many corporations use their technically employed chemical engineers to help with the college recruiting effort by interviewing chemical engineering job applicants and maintaining fruitful contacts with university chemical engineering departments. Even the nontechnical functions welcome the chemical engineer's input—after all, they are an extension of a technologically centered system.

The Corporate Organization

The activities, duties, and goals of some of the industrial functional groups have been described, and occasional allusions have been made to interrelationships between the groups, but so far nothing has been said of the system of organizing the functional groups to form the corporate entity. The organization of industrial corporations follows a pattern of a sort, but the details of the arrangement vary a great deal, for the management of each corporation believes that its particular organizational pattern is the best of all—or certainly the best-adapted to the particular nature of its corporation. Each organizational pattern is thus the result of a combination of historical precedent and current policy. When organizational diagrams of the various corporations constituting the chemical process industry are compared (Kern, 1958), one is struck by the diversity of organizational philosophy.

A representative organizational chart, freely adapted from Ireson and Grant (1955), is shown in Fig. 1.1.1. The diagram contains the graphical elements of all organizational charts—boxes representing various functions (not necessarily identifiable with the functional groups of Table 1.1.1) and connecting lines representing direct lines of authority. By convention, the highest authority is placed at the top of the diagram, and each level of boxes represents approximately equal levels of authority. Thus the corporation is run by a board of directors, but even the board has a "boss" that they are responsible to—the investors in the corporation, the

FIGURE 1.1.1
The corporate organizational chart. (*Adapted from Ireson and Grant, 1955, by permission of Pitman Learning Inc., original copyright holder.*)

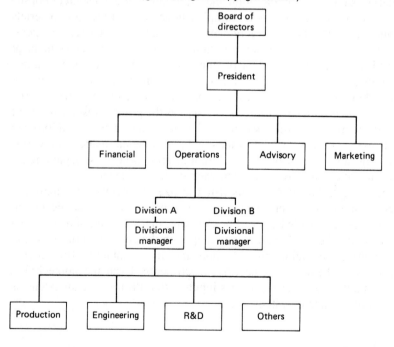

stockholders. Branching out below the board of directors is a managerial hierarchical system, with *staff* managers at the top of the system, *line* managers at the lower levels. The concept of staff and line positions stems from the military hierarchy; in the industrial organization, the distinction is not always clear-cut. Staff managers, in general, are responsible for decisions affecting overall policy of the corporate entity or its major administrative subdivisions, whereas line managers have supervisory responsibility in the various operational areas—production, research, marketing, and others.

The working organization of a large corporation is generally much more complex than that implied by the relatively ordered scheme of Fig. 1.1.1:

The organization may be characterized by *lines of multiple authority* (sometimes indicated on organizational charts by dotted lines), even though multiple authority runs counter to the often-voiced admonition that no individual or group have "more than one boss." For example, a divisional R&D director may report to the divisional manager as well as to the corporate research director. In situations such as this one, areas of authority must be clearly distinguished to avoid unproductive conflict.

The basic hierarchical network of Fig. 1.1.1 may be overlaid by other networks representing interactions generated by *committees, business teams,* and *technological centers*. Committees are generally charged with authority in particularly sensitive areas such as safety, emissions to the environment, or personnel policy. Business teams are committees charged with organizing all aspects of the business of a particular product or product grouping—research, production, marketing, pricing. Technological centers are responsible for centralization and appropriate dissemination of the best up-to-date corporate technological know-how in specific product areas; the centers may have their own budgets to support research programs. The authority entrusted to all such groups may complement or even counterbalance the hierarchical authority symbolized in the organizational chart, with important implications in areas of budgeting, the enforcement of directives, and marketing policy. The apposition of authority sets up stresses and push-pull situations with corresponding benefits of competitiveness and disadvantages of fractured authority.

In a system of overlying networks of authority,[1] major decisions are often a matter of hard-fought compromise. Divisional R&D budgeting and project assignment may be dictated by the often conflicting goals of the divisional manager (who wishes to optimize profits of the division) and of the product business team managers (who wish to optimize profits from manufacture of their own products). Automatic fire protection installations in a new plant may represent a compromise between the wishes of a loss prevention committee (charged with reducing corporate losses due to disasters) and the wishes of the plant manager (charged with keeping capital investment to a minimum). A considerable degree of political savoir faire is required by the successful project manager to compete in an environment of overlying networks of authority.

---

[1] Sometimes called a *matrix* organization.

12 CHAPTER 1: INTRODUCTION TO INDUSTRIAL PROJECT MANAGEMENT

Large corporations may be characterized by operating *divisions* (two are shown in Fig. 1.1.1) which are organized along either geographical or functional lines. Corporate management frequently encourages competition between geographical divisions (such as competition for the privilege of producing an attractive new product) to gain some of the economic advantages which accrue from any free competition. A successful corporation is, in many respects, an assembly of small companies, each one a corporate microcosm simultaneously supporting and yet competing with the others.

The hierarchical structure of Fig. 1.1.1 is repeated at various lower levels. For example, divisional research and development may be organized in the manner illustrated in Fig. 1.1.2. Again, the diagram is grossly simplified—the organization of R&D departments is greatly variable and may incorporate lines of multiple authority and overlying networks of responsibility. The TS&D function, for instance, may also have lines of administrative responsibility to the divisional director of marketing.

In Fig. 1.1.2, the category of *project managers* is indicated as part of the hierarchical structure. This is not always so; the category may be strictly project-oriented, and the title of project manager, if awarded in the first place, may last for the duration of the project only. In this case the project manager is chosen from the

FIGURE 1.1.2
Organizational chart of division R&D.

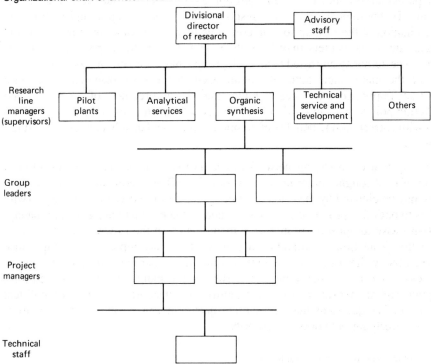

ranks of the technical staff. Of course, project managers need not be research people; they are just as likely to be taken from engineering or other departments, wherever the immediate project responsibility happens to lie. Project managers may assume a variety of titles; for example, the *manufacturing representative* is a project manager charged with supervision of the design, construction, and start-up of a production facility.

We will focus upon the tasks of the project manager in what follows, upon the basic engineering and economic tools that the successful project manager will need. Some aspects, particularly those referring to cost accounting, budget control, and scheduling procedures, will not be particularly emphasized; good expositions of these special procedures are available (Moder and Phillips, 1964; Ludwig, 1974).

The Origin of Industrial Projects

We may think of a project as a task with more or less well-established boundary conditions, and the sum total of the corporation's operations as an assembly of projects. Now, there are many projects which we might classify as *software* projects, such as the reorganization of a department or the task of writing an improved computer program for analyzing cooling tower performance; such software projects, of course, have an economic impact upon the corporation's operations, an impact often amenable to systematic analysis. However, we wish to focus upon *hardware* projects, and for our purposes we define such a project as *any proposal which will require investment of cash for new or modified processing facilities*. Project management within this definition involves the application of chemical engineering principles to specify the nature of the facilities and economic principles to ensure an acceptable return on the investment.

How do industrial projects originate? Here the situation is not much different from that in a typical university graduate engineering school. How do the teaching staff and graduate students select topics for research projects? First of all, it is necessary to establish a *need*, and *ideas* are then generated to meet that need. In an industrial organization, that need may be internal (high corrosion rates in an existing reactor feed line), or the perceived need may be external, an opportunity within the realm of capabilities of the corporation (demand for a species-specific insecticide, or a lead-free gasoline of high octane rating).

Who originates industrial projects? The answer here is that *any* of the functional groups in the corporation may be the originating source. For example, the sales engineer often perceives external needs as a result of customer contacts, and familiarity with the customer's operations may help generate ideas to meet those needs. The patent attorney may perceive the need to establish a better-defined range of product properties in attempting to shepherd successfully a patent application through the intricacies of the Patent Office. Production improvement ideas, more often than not, originate with production personnel. An important source of new ideas in the research environment stems from unexpected side discoveries made in the course of working on a specific project, discoveries that do not really impact upon the project at hand. Every research director hopes that staff members possess the gift of *serendipity*,[1] the ability to recognize the existence of an unexpected discovery and to recognize the applicability of the discovery to a particular need.

Examples such as these underscore the fact that, in the final analysis, projects originate with *individuals,* and that anyone, even the neophyte engineer, can, indeed, be the originator of an industrial project. Obviously, the concept of the project must be accepted by quite a number of individuals in positions of key responsibility before the machinery of carrying the project to its conclusion is set in motion. The key here is effective communication on the part of the originating individual, the *ability to sell ideas*. The most brilliant of discoveries or ideas is lost if buried in some unread report, if the attempt is not made to convince others, to kindle their enthusiasm, to prove the worth and applicability of the discovery.

Suppose, then, that the project is born, that it finds acceptance; how are project assignments made, who will handle the project? The personnel may come from one of several of the industrial functional groups, but if the project involves developmental work, research is usually given the job. In such cases the project manager reports to an administrative manager who, along with the research director, carries the responsibility of adequately staffing the project and obtaining proper budgeting from fund-dispensing sources. In some cases the research director (or the divisional manager in the case of projects handled by production personnel) may have a discretionary fund to support projects not funded from normal corporate sources. Occasionally the project originator may be asked to carry project management responsibilities during this developmental stage.

Criteria of Project Success

If the primary objective of a corporation is to make money, and if a corporation's operations may be characterized as an assembly of projects, it then follows that the objective of each new project should be, indeed, to make money, that the criterion of success is centered on economics. In fact, the objective of new projects is one of the following:

1 To find the best methods of new investment
2 To protect old investments

The eventual economic success of a new project may be gauged by means of various reasonably well-defined methods, much the same methods that are used to evaluate a completed project. During the early stages of a project, particularly one involving a new product or process, a number of noneconomic criteria must also be evaluated in conjunction with the economic criterion of success. These noneconomic criteria can, at least in principle, be quantitatively characterized by their economic impact, but during early stages of a project there is often insufficient information to account for these criteria quantitatively in an economic analysis. The projection of project success then becomes a matter of qualitative, subjective

---

[1] Etymologically derived from Horace Walpole's adaptation of the Persian folk tale, "The Three Princes of Serendip."

judgment, the consensus of experienced engineers and managers. Let us take a look at a few of the commonly considered noneconomic criteria.

Attractive Product Properties at a Good Price This criterion seems to be self-evident, and yet its implications are all too often overlooked. Nobody will argue that the customer must be offered a product at an attractive price, and the price structure is reflected in economic analyses at a very early stage. The economic impact of product properties is more difficult to gauge, and yet an overlooked flaw may scuttle a project that has apparently acceptable economics. A new food preservative, for instance, may display superior performance, acceptable safety, and low cost relative to competitive products. Yet the customer may exhibit reluctance to use it because of a tendency for it to cake in feeding devices or because of the slightly off-white color which you judged unimportant or perhaps even visually pleasing—nevertheless, the customer prefers the "clean white" of the competition.

Good Initial Markets Whether a product is new or established, investment for a new production facility must be justified on the basis of a strongly anticipated demand for the product. The projection of this demand is one of the end purposes of *marketing research*—the investigation of the likely time distribution of demand and the acceptable price structure of a product in the marketplace defined by a specific geographical area. Some reasonably firm concept of demand is essential even in very early stages of a project, and early-stage customer interviews and small-scale product tryouts may be required. Marketing research must establish good *initial* markets. Even though a new product may have attractive properties and a good price, initial markets may be small due to customer inertia, customer loyalty to competitive products, or even perfectly sound technological limitations. Resorcinol formaldehyde resins, for instance, may be used as adhesives in plywood manufacture in place of the more standard phenol formaldehyde (PF) resins. The resorcinol-based resins exhibit the distinct advantage of an ambient curing temperature, whereas PF resins require high temperature and steam-heated platens. As a result, simpler and cheaper equipment is required for forming plywood with the resorcinol-based resins. Regrettably, the argument of cheaper equipment will not guarantee good initial penetration into the PF resin market because existing plywood manufacturers already have the more expensive steam-heated equipment on hand, and they have insufficient incentive to deviate from current operating procedures with which they feel quite comfortable.

The promise of good initial markets is essential to assure those responsible for making the investment that a positive cash flow (i.e., revenues minus expenses) will occur at an early stage of plant operations. We will see that a dollar of profit is valued the highest during the earliest stages of plant life.

Growth Potential Only rarely is the production rate to meet initial market demands sufficient to meet the required long-range economic profitability of a new venture; a steady growth pattern in sales revenues is generally important for the continued economic health of the project. Any new production endeavor requires a pattern of growth to maintain the vigor of the total production effort; growth stimulates employees' motivation and draws the attention of corporate management, and the mutually generated enthusiasm breeds further commercial success. On the other hand, decline in demand for the product of an otherwise profitable plant leads to apprehension, lowering of morale, and downgrading by corporate management, phenomena which may well accelerate the process of decline. If the projected life cycle of a product indicates an eventual decline in demand, this fact should be well understood during the earliest possible stages of a project, and economic projections and subsequent commercial operations must account for the anticipated decline.

Raw Materials Position Good initial markets and a pattern of growing demand are prerequisites for successful marketing of a product, but situations exist which will block product manufacture and distribution in spite of a favorable demand. One such situation involves difficulties with the procurement of raw materials, and a *favorable raw materials position* is an early consideration in the assessment of chances of project success.

What does a good raw materials position imply? Preferably, the producing corporation wishes to have firm control over the supply of its starting materials—by being itself a manufacturer of the starting materials, or perhaps by owning or otherwise controlling natural resources, if these happen to be the source of the raw materials. If raw materials must be purchased from outside suppliers, a competitive supply situation is preferred so that the producer is not tied down to a single supplier, with corresponding penalties of noncompetition: price escalation, supply uncertainties due to labor strikes or poor shipping practices, quality variability. A limited choice of suppliers may be at least partly obviated by taking various steps to ensure a more reliable long-range supply—negotiation of a long-range contract or the stockpiling of raw materials.

Raw materials may be classified into categories reflecting the degree of synthetic chemical transformation incorporated into them (see, for example, Wei, Russell, and Swartzlander, 1979):

Natural resources ("ultimate" raw materials such as petroleum, coal, salt)
Commodity chemicals ("basic," or "heavy" chemicals such as sulfuric acid, caustic soda, chlorine)
Fine chemicals (intermediates derived from a combination of the above)

Processes utilizing fine chemicals as raw materials have the additional problem that the supply of the raw materials may well dry up if the supplier corporation decides, itself, to get into the manufacture of the product. For example, 2,4-tolylene diisocyanate (TDI), a widely used component in the production of polyurethane foams, may be prepared by the phosgenation of 2,4-tolyldiamine (TDA). No matter how attractive the economics may be for TDI production from TDA, if the latter must be purchased from an outside supplier, the haunting question will keep popping up, "If the project is all that good, why doesn't the supplier get into TDI production?" An acceptable raw materials position may demand that the TDI synthesis be extended back to commodity chemicals as starting materials, in this case, toluene and nitric acid.

Patent Protection Management is generally quite insistent that projects involving a new product or process enjoy the benefit of patent protection. Patent coverage gives the corporation holding the patent the advantage of exclusive rights to manufacture the product (or practice the process) for a period of 17 years following the patent issuance. As a practical matter, this means that a corporation enjoys the advantage of exclusive production for a prolonged period without threat of competition that could well reduce or destroy the project's anticipated economic attractiveness. Moreover, such a prolonged period of exclusive production gains the corporation a firm grasp upon the marketplace, and its dominance is likely to continue even after its patents have run out.

A corollary argument suggests that a project involving the production of a high-demand existing product not benefitting from patent protection is a somewhat risky proposition even if the economics are particularly attractive, for what is to stop an enterprising competitor from jumping into the business?

Established Know-How Any corporation which ventures into a project area that is not within the scope of its established know-how is undertaking a serious risk, indeed. Penetration into new areas of enterprise is facilitated by amalgamation or cooperative efforts with companies having experience in those areas, but in the absence of such help an irreversible step into the unknown may well turn out to be a disaster. Business opportunities in new areas may appear economically tempting, but lack of experience in such new areas often leads to errors of commission and omission which may soon turn the promise of economic success into a succession of disappointments. A corporation producing fine chemicals may find prospects of entering the area of, let us say, pharmaceutical production attractive, particularly if a promising new pharmaceutical product is under consideration—after all, the steps of the synthesis involved are in many ways analogous to those in synthesizing fine organic chemicals. Nevertheless, the corporation's managers would do well to recognize the problems facing them in conforming to acceptable manufacturing practices in an unfamiliar field, in penetrating what is to them an entirely new distribution network, in interfacing with an unfamiliar set of governmental regulating agencies, and in facing entirely new challenges in toxicological testing and product stewardship. The competitor already established in the pharmaceutical field has the advantage of familiarity with these sorts of problems, and the neophyte may become bogged down, only to find the marketplace dominated by competitive products.

Only a few of the noneconomic criteria have so far been mentioned, although they are perhaps the most significant ones:

1 Attractive product properties at a good price
2 Good initial markets
3 Growth potential

4 Raw materials position
5 Patent protection
6 Established know-how

Many others are important and should be considered; a more complete list is reviewed by Harris (1976).* The author illustrates how his employer (Monsanto) uses both economic and noneconomic criteria of success to construct a *product profile chart*, a graphical method which allows a rapid visual overview of the totality of criteria that characterize the product.

Construction of the chart is illustrated in Fig. 1.1.3. A grid is drawn in which each horizontal row represents a specific criterion of success (indicated by A, B, C, etc.). Four rating levels (represented by the columns) are assigned to each criterion—negative ratings (-1 for moderate, -2 for strong) if the condition represented by the criterion is considered detrimental to success, positive ratings if the condition is considered as contributing to success. For example, suppose that criterion A is "Patent protection"; the ratings might represent conditions as follows:

-2 No patent protection; product or process well known
-1 No patent protection; some proprietary process know-how
$+1$ Patent protection; some patent claims easily bypassed
$+2$ Strong patent position

If the product under study is not patented, but the process under consideration involves steps which represent proprietary know-how, then criterion A is given the rating of -1 by shading in the appropriate box, as shown in the illustration. Similarly if criterion B, say, "Market development requirements," is characterized by ratings such as:

-2 Extensive educational program
-1 Appreciable customer education
$+1$ Moderate customer resistance
$+2$ Ready customer acceptance

and a $+2$ rating is projected, the "plus" side of the row is shaded up to the $+2$ box as illustrated. Shading in two boxes for +2 (or -2) ratings gives greater visual impact.

Harris gives several application examples of the product profile chart. In Fig. 1.1.4, profiles based on knowledge available during the development stage are shown for two products, one of which turned out to be successful, the other of which failed. Differences in visual impact are readily apparent. Progress in the development of a product may be monitored by using the profile chart technique; an example, reproduced from Harris's paper, is shown in Fig. 1.1.5. During the earliest stage of development (stage I), insufficient information is available to characterize all criteria of success, but what information is available makes the project appear promising. The initial promise is reinforced as additional experience is accumulated during later stages of development (II). In stage III, reevaluation of

---

*Article originally appeared in *Chemical and Engineering News*.

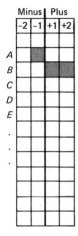

FIGURE 1.1.3
Product profile chart: construction method. (*Adapted with permission from Harris, 1976. Copyright 1976, American Chemical Society.*)

several criteria in the light of more refined information has rendered the profile considerably less attractive; as a result, serious consideration might have to be given to dropping the project.

Attempts have been made to quantify the estimation of project success using noneconomic criteria (Dean, 1968; Sarin, 1978). The procedure usually involves some method of "scoring"; each criterion is scored to reflect its rating level (in much the same way as Harris's rating method from -2 to $+2$), and in some procedures the score of each criterion is multiplied by a weighting factor reflecting the relative importance of a particular criterion. The sum of all the criteria scores is then the project score. Scores of various projects may be compared to focus upon the most promising ones, or else an individual project score may be compared with absolute values based upon past experience with projects of various proven success levels. Several corporations utilize their own favorite procedure, but no one procedure has gained general acceptance.

The Role of Research and Development

A great many projects, particularly those involving a new product or process, are handled by the R&D function, and a strong R&D effort is considered essential to the economic well-being of the chemical corporation (Wei, Russell, and Swartzlander, 1979).

An often-discussed classification of research is based upon its end objective; within this classification, research is considered to be *fundamental* or *applied*. Fundamental (basic) research is "uncommitted research, prompted by demonstrated curiosity and aimed primarily at the extension of the boundaries of knowledge" (American Council of Education, quoted in Heyel, 1959). Applied research, in contrast, is committed research, prompted by promise of commercial advantage and aimed primarily at the technological definition of a marketable product and the development of a process to make it. Within the context of the primary objective

I

	Minus		Plus	
	-2	-1	+1	+2
Financial Aspects				
Return on investment (before taxes)		■		
Estimated annual sales		■		
New fixed capital payout time		■		
Time to reach est. sales vol.				■
Research & Development Aspects				
Res. investment payout time			■	
Dev. investment payout time		■		
Research know-how		■		
Patent status		■		
Marketing & Product Aspects				
Similarity to present product lines			■	
Effect on present products			■	
Marketability to present customers			■	
Number of potential customers		■		
Suitability of present sales force			■	
Market stability		■		
Market trend		■		
Technical service		■		
Market development requirements		■		
Promotional requirements		■		
Product competition		■		
Product advantage			■	
Length of product life				■
Cyclical & seasonal demand	■			

II

	Minus		Plus	
	-2	-1	+1	+2
Financial Aspects				
Return on investment (before taxes)			■	
Estimated annual sales			■	
New fixed capital payout time			■	
Time to reach est. sales vol.				■
Research & Development Aspects				
Res. investment payout time			■	
Dev. investment payout time			■	
Research know-how		■		
Patent status		■		
Marketing & Product Aspects				
Similarity to present product lines			■	
Effect on present products			■	
Marketability to present customers			■	
Number of potential customers			■	
Suitability of present sales force			■	
Market stability			■	
Market trend			■	
Technical service			■	
Market development requirements			■	
Promotional requirements			■	
Product advantage			■	
Product competition		■		
Length of product life			■	
Cyclical & seasonal demand		■		

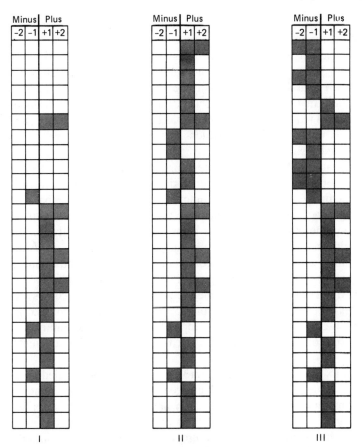

FIGURE 1.1.5
Product profile chart: a product development sequence. (*Adapted with permission from Harris, 1976. Copyright 1976, American Chemical Society.*)

FIGURE 1.1.4 (p. 20)
Product profile chart: profile of successful and unsuccessful products. I. Test of a chart's ability. Before the chart was put into use, it was tried out on several products previously put on the market by a company. Textile Preservative-B had been a market failure due primarily to poor profitability and heavy technical service and marketing demands. The chart would have emphasized these, and it could have saved much time and expense. II. Another test. Both charts on this page reflect only the information available during development of the products. Insecticide-N, unlike Textile Preservative-B, is a proven market success. Based on developmental data, the chart missed what actually happened only on annual sales and length of product life. Both were underestimated. (*Adapted with permission from Harris, 1976. Copyright 1976, American Chemical Society.*)

in the industrial environment, namely, to make a profit, it is not too surprising to find that the bulk of the research done in the chemical industry is the applied kind. Industry leaves fundamental research, for the most part, to the universities and research institutes, not because such research is not considered important, but rather because applied research can more easily be shown to result in a tangible return on the investment of time and money. This subdivision of research duties between the academic and industrial environments has been reinforced by time, and the fact is that each is doing the job for which it is best fitted by nature of its personnel, facilities, and general objectives.

The division of research duties is nevertheless characterized by a state of symbiosis, for one could not exist without the other. Industry depends upon academic research to provide the firm foundation of fundamental science required for backing up applied research. The academic environment, on the other hand, depends for its continued vitality upon the practical use of its ideas; without such application the academic world would, indeed, live in an intellectual vacuum. Industry *needs* the results of fundamental research to attain its commercial objectives, and financial support of universities by industry is a logical extension of the profit-making objective. Significantly, even though the academic staff obviously welcomes and encourages industrial support, it rejects the concept of *directed* research, that is, research directed by the source of financial support toward desired commercial objectives, for such form of control is not in the spirit of uncommitted fundamental research.

Clearly, the arbitrary division of research into fundamental and applied areas gives rise to a broad, ill-defined interface which feeds the ever-popular debate as to what, in fact, constitutes fundamental research and what constitutes applied research. Most research projects in one category exhibit nuances of the other classification, and the fact is that any research project, no matter how far into the applied category, will constitute a fascinating intellectual challenge if approached on the same basis of intellectual curiosity as the most fundamental of research problems.

As a matter of fact, a certain amount of fundamental research is done by the process industries, and some industrial leaders recommend that more such research be performed within their own domain (Heyel, 1959). Research of a more fundamental nature may be undertaken, for example, for the sake of establishing parameters governing the development of a particular product, or to strengthen a proprietary foundation of knowledge in an area of specialty. The manufacturer of artificial heart valves may well be compelled to engage in fundamental research on the rheology of human blood. In some organizations the term "basic research" is applied to *long-term* research, research that is directed toward commercial ends but, because of the anticipated complexity of a developmental program, is not expected to yield an economic return for many years. R&D programs leading to new biomedical devices; or fuel cell power sources; or artificial fibers; or any novel products which are projected as useful at some future time on the basis of technological forecasting—all of these are generally anticipated to last for many years, and many do involve fundamental research excursions, particularly during earlier stages of development. It is equally evident that applied research is done in conjunction with the more fundamentally oriented projects in universities, perhaps to perfect a specialized piece of equipment to measure fundamental properties, perhaps to perfect a chemical process as a means of teaching process research.

In the industrial environment, research may also be classified as *defensive* and *offensive*. This classification coincides with the objectives of new industrial projects; defensive research supports existing products and processes, whereas offensive research is directed toward the development of new products and processes. One of the challenging aspects of effective corporate research planning is how to strike the appropriate balance between defensive and offensive research. In the late 1970s, the balance in the U.S. chemical industry favored offensive research (Rosenzweig, 1978):

Offensive research	46.7% of R&D expenditures
Defensive research	39.2% of R&D expenditures
Environmental research	14.1% of R&D expenditures

American industry, the chemical process industries included, expends an enormous amount of money on R&D in well-justified anticipation of an attractive return for such a formidable investment. Five chemical companies were expected to spend over $100 million each for R&D in 1979 (Table 1.1.2), an average of about 3.25 percent of annual sales. For the total chemical industry, a good rule of thumb (based on performance during the seventies), is

$$\text{R\&D expenditure} = 3\% \text{ of gross sales}$$

Very few chemical corporations spend more than 4 percent of sales; some spend less than 1 percent, a reflection of management's differing attitudes toward research expenditures as an investment.

Whereas the chemical industry is strongly research-oriented, it falls far behind some other industries (aerospace, communications) in R&D expenditures as a per-

TABLE 1.1.2
PROJECTED 1979 R&D EXPENDITURES OF THE FIVE LARGEST U.S. CHEMICAL CORPORATIONS

Corporation	Worldwide R&D spending, millions of $	R&D spending as % of sales (1978)
Du Pont	403	3.6
Dow Chemical	265	3.4
Union Carbide	165	2.0
Monsanto	160	3.0
American Cyanamid	120	3.9

Source: Fallwell (1979).

TABLE 1.1.3
R&D EXPENDITURES, U.S. CHEMICAL AND ALLIED INDUSTRIES

	R&D expenditure, billions of $		
Industry	1965	1970	1975
Chemical	1.36	1.77	2.65
Petroleum	0.40	0.52	0.70
Food	0.16	0.24	0.32

Source: "Statistical Abstracts of the United States: 1977," U.S. Department of Commerce.

centage of sales. Such expenditures in the chemical and allied industries increased steadily during the sixties and seventies (Table 1.1.3) but at a rate not far different from inflation (about 7 percent/annum). In a sense, therefore, R&D expenditures have remained constant during those two decades, and, as a matter of fact, the number of professional engineers and scientists in chemical industry R&D has increased only very moderately in the seventies (Table 1.1.4).

The cost statistics may be manipulated to derive the cost per man-hour (or man-year) of research, a figure of considerable interest to research planners and project managers (see Table 1.1.4). If a man-year is taken as 2000 man-hours (a convenient approximation), then a reasonable research cost projection for 1980 was

$75,000 per R&D man-year
$37.50 per R&D man-hour

The anticipated escalation of this cost is 10 percent/annum. It must be understood that the cost is not the *salary* of the research worker; the total includes *all* overhead costs, even the pay of nonprofessional support people whose time is chargeable to research. It certainly behooves all research workers to keep the crushing costs of research in mind when scheduling their own work time.

TABLE 1.1.4
EMPLOYMENT OF PROFESSIONAL ENGINEERS AND SCIENTISTS IN THE CHEMICAL INDUSTRY

	Heavy chemicals industry only*			Total chemical industry†	
	1970	1975	1980 (est.)	1970	1975
Total no. employed	21,800	22,000	24,700	42,200	45,700
Cost, $/man-year	40,000	56,000	77,000	42,000	56,000

*From Fallwell (1979).
†From "Statistical Abstracts of the United States: 1977," U.S. Department of Commerce.

Corporate management understandably is interested in monitoring the *effectiveness* of the corporation's research effort. It turns out that this is not an easy thing to do, for the results of research are often difficult to translate into easily measured and understood quantitative terms. One frequently used quantitative criterion of research effectiveness is the percentage of sales represented by new products, say, products less than 5 years on the market. Defensive research efforts may be monitored by estimating how much such efforts save in production costs; the statistics on "dollars saved" are usually generated by the research workers themselves and perhaps tend to be exaggerated, although experienced research directors learn how to discount such self-aggrandizement efforts.

Many such efforts at monitoring the effectiveness of research, no matter how well organized, still leave in doubt the primary quantitative criterion—what is the rate of return on the R&D investment? We will see that capital investments involving production facilities may be evaluated using certain well-defined *profitability criteria;* these criteria may then be used to derive general rules of thumb which indicate how much return R&D expenditures *should* generate, and the ideal may then be compared with actual performance. One such general rule is:

> $1 of R&D investment should yield $2.50 of new production investment.

Rules of thumb for TS&D efforts include:

> $1 of TS&D investment should result in increased sales of an existing product of $1/year for 10 years.

Each research worker should use these rules to self-monitor contributions made to the total research effort. For example, the first rule suggests that each researcher should generate $75,000 × 2.5 = $187,500 of new production investment every year!

Another measure of research effectiveness is the *mortality rate* of new ideas (see, for example, Anonymous, 1968). Statistics from several sources are summed up in Table 1.1.5. The high mortality rate of research projects (only one out of eight projects in the intense development stage is commercially successful!) accounts for the heavy expense of research, and, as we will see, the successful project must bear the cost of the research failures. Over 6 years is required on the average to bring a product to commercial fruition, and 15-year development periods are not unknown.

R&D efforts thus exhibit a number of discouraging statistics, but in the chemical industry R&D continues to serve as the focus of a great number, perhaps the majority, of identifiable projects within our definition.

TABLE 1.1.5
THE MORTALITY RATE OF RESEARCH IDEAS

To obtain one successful new product:

540 ideas considered at research level
440 screened and eliminated
 92 selected for preliminary laboratory investigation
  8 sufficiently promising for further development
  7 dropped as unsalable or unprofitable up to semiworks
  1 survives and is placed in regular production

Average of 6 years and 2 months from original investigation to full-scale production

1.2 THE INDUSTRIAL PROJECT MANAGER: THE PROFESSIONAL CHEMICAL ENGINEER

Starting a Professional Career in Industry

We now turn our attention from the nature of the industrial project to the individuals who manage the projects, the professional chemical engineers, and we specifically focus upon the prospective employee who is about to receive a final university degree before taking the plunge into the industrial environment.

The question that occurs to many students, hopefully well before graduation, centers upon the choice between an industrial and an academic career. It is important that students monitor their interests, strengths, and weaknesses to arrive at a decision with which they can live comfortably. Some typical "indicators" to be monitored include:

Aptitude and fundamental interest in the mathematically oriented sciences
Aptitude and interests in laboratory work, research apparatus
Attitude toward mechanical work, mechanical details
Aptitude and attitudes toward scholarly research
Degree of comfort with and achievement level in applied sciences
Degree of interest in technological "fringe" areas (technical economics, organization of people, managerial science, business administration, marketing)
Attitude toward specialization
Experience and motivation to teach
Experience and attitude toward the social structures represented by academia and industry, respectively

Students frequently have only a vague idea about the nature of the professor's job, what academic life is really like; a few informal discussions will help to straighten out some common misconceptions. Similarly, most students have not had the experience to assess the "feel" of working in industry; summer jobs will prove to be of considerable help here.

Let us suppose, then, that you have decided upon a career in industry. How does one go about contacting the process industries to find a job? There are several possibilities:

Campus interviews of prospective graduates by corporate recruiters

Written or personal contacts based on "help wanted" ads in publications such as *Chemical Engineering Progress*

Placing classified ads in the "situation wanted" column of the same publications

Referrals by professional colleagues, professors

Contacts established through the efforts of professional placement agencies

Most students experience their first contact with a potential employer on campus, and we therefore focus here upon the *campus job interview*. Such interviews are generally arranged by the university placement office, and most major chemical corporations, as well as some smaller ones, have interviewers on campus for one or two days each year. The purpose of these preliminary interviews must be clearly understood—they are devices to establish an initial contact, to establish whether there is a significant degree of mutual interest which can, in turn, lead to more serious employment negotiations. The campus job interview is thus a screening device for both the student and the potential employer, and as such it has to be necessarily brief, to an almost uncomfortable degree. In spite of this imposed brevity and apparent superficiality, the initial interview is of great importance—to the corporation as a method of establishing a pool of identified candidates from which to select those best-suited to operate the business of the corporation in the future, and to the student as the first step in the molding of a personally satisfying lifetime career. If mutual interest is established, further interviews are subsequently arranged, usually as a result of a plant visit invitation extended by the interested corporation.

The format of the campus job interview is simple. The student and the corporate recruiter (rarely more than one) meet in a private room, generally for about 30 minutes—or perhaps a little less, for the recruiter needs to take a few minutes to jot down impressions of the preceding student between interviews. Within the narrow time span available, the recruiter must obtain a clear impression of the student's abilities, personality, and fitness to meet the company's needs, and the student, in turn, must obtain an impression of the kind of place the company is to work in, the nature of the job openings available, and the likelihood of attaining the goal of a satisfactory career. Each participant has perhaps 10 minutes to establish what the other wishes to know, obviously a task of considerable difficulty and one that requires some preparation. The success of the interview is certainly dependent upon the skills of the recruiter; chemical process industry corporations, with increasing frequency, employ their younger engineers and scientists as campus recruiters, for these people more easily establish quick rapport with the students. The success of the interview also depends a great deal upon how well the student is prepared, and we will therefore review a few of the criteria for such success. Many university placement offices display and distribute available brochures and organize student seminars on successful campus interviewing; additional information appears occasionally in journal articles (Anonymous, 1973) and even in books devoted to the subject (Billmeyer and Kelley, 1975).

Attitude Modulation   The campus job interview is so brief that, rightly or wrongly, first impressions count, and, redundant as it may seem, it is worth emphasizing that the student should approach the contact with the corporate recruiter with a strongly positive attitude. What does this really mean? It means that each interview should be looked at as the fundamental first step in building an industrial career, and all interviewees should be ready to sell themselves as top-notch potential employees in the most forceful and convincing manner possible. Approaching the interview in a casual manner, perhaps out of idle curiosity as to the nature of a particular corporation, will turn out to be a waste of time for all concerned. Personal appearance may not have much to do with technical ability, and yet a neat, nonostentatious appearance obviously helps in establishing a good first impression, in showing the recruiter that the interviewee does care. Similarly, it is important to speak up in a gracious, friendly, and natural manner, without pretensions or ostentation, and without extending forcefulness to the point of dominating the interchange. No student need approach a campus interview with apprehension or feeling of an impending ordeal; after all, in spite of its potential importance, the interview is just an exchange of information, and usually a genuinely interesting and even pleasant experience for both participants.

Knowledge of Company   The student bears responsibility of finding out ahead of time the pertinent information about the company represented by the recruiter. Such ground work can save a great deal of time and trouble; the student can pretty well establish a priori whether a particular company fits into a desired scheme of career goals and thus avoid a pointless interchange. If the student is acquainted with the basic facts about the company, the recruiter may devote the few available minutes to a more meaningful description of the work environment and career opportunities. University placement centers have industrial literature libraries in which factual material about many companies can be found, often in the form of brochures addressed to graduates in a specific academic discipline. Some corporations hold short group seminars during evening hours to give interested students a broader base of understanding of their nature.

Perception of Career Goals   The corporate recruiter may be interviewing for a specific job opening, although more often than not the job opportunity is general in nature—the corporation may be interested simply in acquiring the services of a few outstanding individuals who may be placed, for example, in research projects that appear to match their immediate talents. In either case, the recruiter needs to obtain some idea of the interviewee's career goals. Students should start to think seriously about these long before graduation, to investigate the functional structure of the chemical corporation, difficult as it may be from the academic vantage point, and to formulate their goals within the context of their understanding of that structure and the assessment of their own strengths, abilities, and particular likes.

It simply will not do to tell the recruiter that you wish "to work as a chemical engineer"—the recruiter already knows that. It stands to reason that each one of us has some ideas of what we would like to do with our life, and if we decide upon an industrial career, we owe it to ourselves to focus our goals, in a reasonably general manner. The student fascinated by work in areas of electrochemistry would obviously do well to concentrate on interviews with corporations with known electrochemical operations, and a reasonable career goal might be "process development work in the field of electrochemical engineering, with a long-range goal of production or business management in that area." At the same time, it is important not to get too specific, for overrestrictive career goals may seriously limit job opportunities. The electrochemical enthusiast in the example given would do well not to limit career goals to "laser beam interferometry in electrochemical boundary-layer investigation" in spite of a possibly strong background in this restricted area.

It should also be emphasized that a clear statement of career goals goes a long way toward making a permanent impression of the interviewee in the recruiter's mind. The recruiter often interviews as many as 15 potential employees in a single day, and it is amazing how easily impressions of that many people can dissolve away by the time the recruiter has the chance to write interview reports. Anything exceptional to jog the memory helps ("Oh yes—the electrochemical engineer").

Preparation of Key Questions The brevity of the campus interview demands the effective prior formulation and organization of key questions that reflect the student's concern about a future career. The thrust and phrasing of those questions should be such as to draw from the recruiter responses that will establish the "feel" of working in the corporation, a sense of the total working environment and how it affects perceived career and life-style goals. Concentrate upon areas of major impact, with questions such as:

How much freedom will I have to move between departments? If I do wish to transfer, how are such transfers initiated?

Does the corporation encourage publication?

What are the corporation's Equal Employment Opportunity (EEO) policies? Does it have any minorities or women on the board of directors? In top corporate staff positions?

What is the frequency of geographic transfers? If I prefer a specific geographic location, can I stay there without seriously jeopardizing my career?

Is there a ladder of advancement in a strictly technical career that corresponds to a ladder of advancement in management? Are technical and managerial careers rated equally in prestige, or is one or the other considered the more desirable? Are young people encouraged to launch management-oriented careers?

What career guidance programs does the corporation have (counseling, assertiveness training courses, sabbaticals, etc.)? Does the corporation support and subscribe to the Professional Guidelines for Chemical Engineers (AIChE)?

What kinds of new employee technical initiation and training programs does the company have?

You may not always get a "straight" answer to questions such as these, but the recruiter's responses and attitudes will tell you a lot about the company represented. At the same time, the questions you ask tell the recruiter a lot about you; needless to say, your questioning should be aboveboard, straightforward, sincere, with no traces of a hostile attitude!

There are some questions that should *not* be asked:

Items you can readily find in the company's brochures.

Questions of starting salary. Now, there is nothing immoral about wanting to know how much money you will make—after all, you are seeking a job to make a hopefully comfortable living—but the time to discuss salary figures is *after* you have decided the company is one of the select group which appears to fit best your career goals, and usually at the time you have been invited to visit the plant site for more in-depth interviews. Most corporations have competitive starting salaries anyway, and curiosity about monetary rewards during a screening interview may give the recruiter an unfavorable impression. On the other hand, there is nothing wrong with pursuing questions of salary policy—how, for example, does the corporation cope with *salary compression,* the deceleration of the salary advancement of older corporate employees relative to the more rapidly rising starting salaries?

Evidence of Quality of Performance There is an ongoing debate about the utility and validity of letter grades and whether grades should be dropped in favor of written evaluations of students by the teaching staff. The fact remains, however, that grades still give the best opportunity of making a rapid assessment of a student's performance. Many universities do not favor the use of transcripts during campus interviews, but the student should at least have available some evidence of the grade point average (GPA) attained at both undergraduate and graduate levels.

Written Résumé An effective written résumé is an *indispensable document* in any campus interview situation. Every chemical engineering student at the senior level or higher should have an up-to-date résumé ready for any eventuality, perhaps an entirely unanticipated opportunity to contact a potential employer.

There is no one single way of assembling a résumé, and résumés are as different as the persons they describe—and should be so. Nevertheless, it is a good idea in writing one to keep in mind its primary purpose and to stick to a few simple rules (Weismantel and Matley, 1974). First of all, a résumé is not a biography; it is a compact sales package, a device to catch the potential employer's eye, to sell the author's technical skills and total personality. Undergraduates in particular frequently complain that they do not have a sufficient record of technical accomplishments to "pad" a résumé, and yet, with the use of proper perspective, every student can assemble an impressive document, one that will do a good job of selling. A forceful résumé is possibly the most effective device for making a permanent favorable impression upon the recruiter's mind; it is something the recruiter can use to refresh the memory long after the interview, to get an overall impression of the interviewee's personality, provided, of course, that the résumé contains the stuff to jiggle one's memory!

What, then, should be in a résumé? A highly idealized example is given in Exhibit 1.2A; it is the purported résumé of an undergraduate, no doubt somewhat exaggerated (nobody could have been all that busy!), but intended to highlight some types of experiences and accomplishments which should at least be considered for inclusion. Note how the listed experiences emphasize aspects which the prospective employer is particularly likely to seek out—evidence of motivation, organizational skill, unusual technical competence, ability to work with others, leadership characteristics. The written material should include:

A statement of career objectives
A brief review of past performance: academic background, scholarships, awards, etc. (college level only, not high school)
Any work experience, no matter how humble!
Any unusual experiences or acquired skills that help define the writer's abilities
A brief personal portrait

The format of the résumé should be something akin to Exhibit 1.2A: one to two pages in length, structured with paragraph headings, each section a compromise between brevity and good written prose, an attempt to avoid a shopping list quality. Some experienced personnel managers would consider the example résumé much too long, and it is true that for purposes of a campus interview it could profitably be condensed into a one-page document, one that the recruiter could use for rapid scanning during the interview. Indeed, an argument may be made for the preparation of two résumés—a short one for use during the campus interview, the other as part of the total record made available to each recruiter.

Weismantel and Matley (1974) directed their advice on résumés primarily to employed engineers looking for a new job, but their simple rules are universally applicable and worth repeating:

Keep your résumé short ("If you persist in thinking you need more, you don't really understand the purpose of a résumé").
Keep it to the point—what are your objectives?
Be specific in outlining your accomplishments—avoid puffery.
Be different—do draw attention to yourself. Weismantel and Matley even suggest using colored paper, and an attached photograph is not at all a bad idea!

Note also that the résumé is conspicuous for its absence of exhortations and "sales pitches" to the prospective employer; these are in the province of the *letter of transmittal,* often a companion document to the résumé. A letter of transmittal is a cover letter accompanying a résumé sent directly to a prospective employer. Such a letter may be used to incorporate information that may be of specific interest to a particular employer, along with a sales pitch outlining how the writer's background would contribute to the corporation's operations, and pertinent data such as

EXHIBIT 1.2A

AN EXAMPLE OF A RÉSUMÉ

Dorothea G. ("Deegee") FERNANDEZ (213)-137-8808
215 S. LeRoy Avenue
Pasadena, CA 91125 Small photograph here

Career objectives

My immediate career goal is to obtain a broad base of experience in production work, with the eventual goal of assuming responsibility for line management of various aspects of the corporate business. I would welcome overseas assignment opportunities.

Academic background

1978 to present: California Institute of Technology course work leading to a B.S. degree in Chemical Engineering in June of 1982
Grade point average: 3.86/4.00

Honors and awards

Half scholarship, freshman year; California Scholastic Federation scholarship grant; Western Petroleum Company scholarship award
Tau Beta Pi
Two letters in competitive swimming; elected to student body activities honor society

Work experience

Summer 1981: Air Pollution Chemist, Los Angeles County Air Pollution Control District. My duties were to design and equip a small mobile laboratory for the collection and analysis of air samples using vapor phase chromatography, to establish more accurately the diurnal distribution of pollutants in the Los Angeles basin.
1981–1982: Teaching assistant, unit operations lab
1979 to present: Waitress, Faculty Club

Other technical experience, special skills

As part of a special project in analytical chemistry (sophomore year), I helped to develop a method for analyzing trace quantities of cadmium in power plant boiler blowdown water.

I was team leader of a project to measure the energy efficiency of the campus central power plant.

I designed and carried out an experiment to check quantitatively theories of price elasticity (term project, microeconomics).

I am an experienced Fortran programmer and have experience in operating a PDP-11 computer.

I have taken elective courses in Particulate Technology, Instrumental Analysis, Vapor Emission Control Technology, Engineering Economics, Microeconomics, Calculus of Finite Differences, Process Optimization.

I am fluent in Spanish and have a working knowledge of German and French.

Interests and activities

I have been active in student government affairs; I was treasurer of the sophomore class and the student body, social chairperson of the off-campus students' club, and legislative secretary of the Latin American League. I enjoy sports, particularly swimming and backpacking, and am interested in outdoor activities, nature study, and photography. I am a member of Sierra Club and have organized trail maintenance efforts. My hobbies include gourmet cooking and embroidery.

Geographical preferences

None

References

Furnished upon request.

---

date available for employment. Students interested in a particular company may wish to send a letter of this sort even on the heels of a campus interview as an immediate follow-up device. Such letters demonstrate interest and intent and may more readily lead to the next logical step in the hiring procedure, an invitation to visit the company in question.

To recapitulate, some of the requisites for a successful interview include:

1 Attitude modulation
2 Knowledge of company
3 Perception of career goals
4 Preparation of key questions
5 Evidence of quality of performance
6 Written résumé

The campus interview results, within a span of a few weeks, in one of two courses of action: a polite rejection or else an invitation to visit the company's facilities. An invitation represents a strong corporate interest in the prospective employee, but it certainly is no guarantee of a job offer. The format of the in-plant interaction varies a great deal, but typically it incorporates one or two days of interviews, each perhaps 1 hour long, with key individuals in various functional groups, as well as lab and plant tours, luncheons, and other opportunities for mutual evaluation. As with the campus interview, the whole procedure is a two-way proposition, and the prospective employee and the corporation have an excellent chance to learn a great deal about each other and thus come to a final understanding. The visit usually, but not always, ends with a detailed discussion with employee relations representatives. A definite job offer may accompany these discussions, but a formal offer is most often made by letter several days after the plant visit, to give time for employees who had interviewed the candidate to write their own evaluations and for various managers to meet and make the final decision whether to offer a position. Various specifics are agreed upon: most likely place to start work (based upon company needs as well as the young engineer's impressions from the interviews), starting date, starting salary, moving expenses. Rejections are almost always made through the medium of a formal letter, unless a mutual agreement is reached beforehand. In spite of the fact that the corporation bears the cost of all expenses associated with the visit, including travel, the candidate is in no way obligated and still has full freedom to accept or reject an offer within a reasonable time.

What sort of salary may the young engineer, in fact, anticipate? Available data are never very consistent, but the following numbers are reasonably typical. In Fig. 1.2.1 the trend in the *median starting salaries of chemical engineers* over almost two decades is shown. In 1978, the indicated annual starting salaries were:

B.S. $18,100/year
M.S. $19,400/year
Ph.D. $24,800/year

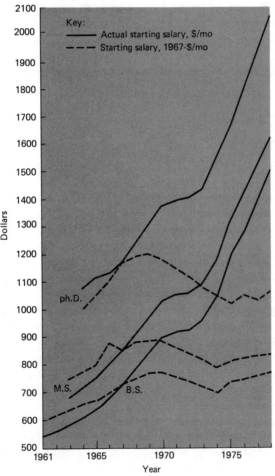

FIGURE 1.2.1
Median starting salaries of chemical engineers. (*From Matley and Ricci, 1979, by permission.*)

The most significant aspect of Fig. 1.2.1 is that since about 1968 the starting salary of B.S. chemical engineers, corrected for the inflated value of the dollar, has remained constant, and the starting salary of Ph.D.s has actually decreased! The trend of these inflation-corrected curves can be extrapolated to project starting salaries for many years to come.

How much of a *spread* in starting salaries is there? An assessment may be made by comparing median salaries with upper and lower deciles, that is, salary offer levels representing all but 10 percent of the top and bottom offers, respectively. In Table 1.2.1, the starting salary spread of *all* B.S. engineers (not just chemical engineers) is tabulated for one point in time (summer 1978). There certainly are differences; for example, the petroleum industry offers significantly higher starting salaries than the chemical industry, and both outstrip the rest of U.S. industry. And yet the spread is not really all that large; for the most part, corporations are competitive in their salary structure, and the student is well-advised to consider personal growth opportunities rather than small starting salary perturbations in choosing a place to work. The fact of the matter is that chemical engineering as a profession has many attractions, but the opportunity to get rich is usually not one of them, although chemical engineers are paid at a fairly high level in the professional spectrum. In 1975 the median income level of three professions (*"Statistical Abstracts of the United States"*) was

Doctors (self-employed)	$47,500
Chemical engineers	$21,500
Teachers (elementary, high school)	$11,700

The information on starting salaries suggests that Ph.D.s receive a great deal more money than B.S. engineers. It is instructive to examine the salary spread among various degree holders after, say, 20 years of industrial service to see whether the Ph.D. degree represents an economic advantage. The data in Table 1.2.2 were obtained from the 1978 AIChE Economic Survey Report; the basis of comparison is the year in which the B.S. degree was obtained. Even the initial large difference in starting salaries is reduced when viewed from the common vantage point of the B.S. year, and after 20 years or so the difference just about disappears. Clearly the more valid noneconomic criteria must be used in justifying academic work leading to a Ph.D. degree.

TABLE 1.2.1
STARTING SALARY SPREAD FOR B.S. ENGINEERS IN VARIOUS INDUSTRIES*

	$/year, August 1978		
	Chemical industry	Petroleum industry	All industry
Upper decile	19,100	21,700	18,800
Median	17,800	19,000	16,200
Lower decile	16,100	17,000	14,300

*Based on data of Mannon (1979).

TABLE 1.2.2
SALARY MEDIANS* OF CHEMICAL ENGINEERS BY YEAR OF BACHELOR'S DEGREE

B.S. year	B.S. degree	M.S. degree	Ph.D. degree
1973	20,030	20,330	23,600
1960	30,000	30,540	32,000
1959	32,640	30,000	34,300
1958	32,050	30,000	33,000

*Dollars per year, 1978.
Source: 1978 AIChE Economic Survey Report; used by permission.

We now turn our attention to the situation on the other side of the curtain that represents the entrance into the industrial environment. Suppose your efforts to penetrate the curtain are successful, and your first day on the job comes around—what happens next? As a newly hired chemical engineer you can wind up in any one of the many industrial functional groups, but statistically chances are good that you will start in research or development, since some companies use their R&D function as a training ground. The conversion of the neophyte engineer into an individual with a modicum of experience and insight may be done by assigning immediate project responsibility and depending upon "on-the-job" training. On the other hand, many corporations have training programs for neophyte engineers specifically designed to impart such experience in a more systematic manner than could be accomplished by haphazard project assignment. The program may not be anything more than a series of assignments to projects carefully screened for their particular educational value, or it may be a more formally organized *professional project program* (PPP; the program has different names throughout the industry).

PPP is typically a 1-year program consisting of three or four specially designed projects administered under the direction of experienced engineers. The projects are much better defined and simpler in scope than most industrial projects, straightforward enough to have a good chance of being completed within the allotted period of 3 to 4 months; usually they constitute a specific problem within the context of a broader project. The neophyte chemical engineer may choose from a list of projects, but acceptance of several consecutive projects sponsored by different functional groups is encouraged to give the neophyte the broadest possible basis of experience. PPP is particularly useful for those engineers who do not have firmly established career goals or who were not able to focus upon specific project responsibilities during plant visit interviews.

The tendency in the process industries is to give engineers immediate project responsibility involving significant technical and economic decisions. To allow a neophyte engineer to while away the time on some routine, insignificant task so as to "learn the ropes" just does not make economic sense in today's industrial climate. Young engineers welcome the opportunity to assume immediate significant responsibility, but the experience can be uncomfortable, particularly during the earliest stages of employment when the neophyte is faced with other difficult

EXHIBIT 1.2B

SOME TYPICAL INITIAL ASSIGNMENTS FOR CHEMICAL ENGINEERS IN INDUSTRY

After he was hired, Steve joined a development team working on a hollow-fiber gas separation device to provide in situ generated inert nitrogen pad gas for space vehicles and airplanes. Several small models with varying configuration had been built and tested, and Steve was assigned the job of designing the optimum prototype configuration. He started by using computer-based calculations to focus upon hypothetical fiber arrangements to meet imposed operating parameters. With the help of two laboratory technicians whose work he directed, he built and tested several versions of full-scale units, refining fabrication techniques along the way with mass production in mind. Evolution of the final design was influenced by consultation with potential customers in various parts of the country. Steve's part in the project ended with the completion of a formal report; 1 1/2 years after his first day on the job, he was offered the position of production engineer in a new production plant.

• • •

While interviewing during a plant visit, Dan was exposed to the problem of synthesizing a mining chemical from two organic raw materials, A and B. The first step in the synthesis is an addition reaction,

$$A + B \longrightarrow AB$$

This is followed by controlled addition of sufficient B to form an addition polymer of the average formula $AB_{4.5}$. Dan indicated an interest in developing a continuous process from the initial lab data available, and this was his first assignment after he had been hired. He decided to make a detailed study of reaction kinetics, but the first step was to develop reliable analytical methods for all intermediates as well as separation methods to obtain experimental quantities of intermediates such as AB_2. He found that the first step, the formation of AB, was very rapid; he hypothesized and experimentally confirmed that the kinetics of each succeeding step,

$$AB_n + B \longrightarrow AB_{n+1}$$

were identical, the rates being much slower than that of the initial step. After defining the pertinent kinetic parameters, Dan devised a computer program to calculate the reaction progress in continuous-coil reactors. He designed and built a small-scale continuous reactor based on the results of the computer analysis; this performed as expected, and Dan's program is in use by the engineering department to design a full-scale unit.

• • •

Peggy was given the problem of investigating the applicability of peeling fruit destined for production line canning by means of a process involving surface freezing of the epicarp (outer skin) followed by warm water thawing. Laboratory experimentation allowed her to define optimum conditions of temperature and time of exposure which gave an acceptable degree of separation of the epicarp without excessive loss of useful fruit pulp. She hypothesized that pulp lost during thawing was the pulp originally frozen; she derived generalized expressions for the depth of freezing and found that experiments confirmed her hypothesis. She was given the responsibility of designing and constructing a pilot plant to test the process with peaches. After several months pilot plant operations were discontinued pending the design and construction of a machine that would completely strip the loosened epicarp from the fruit.

• • •

For his first assignment in a professional project program, Bruce was asked to design a system for stripping volatile components (HC1, CO_2, and phenol) from batches of a nonvolatile resin. He did a few simple lab experiments and found that by pulling a moderate vacuum and slowly heating a batch to 100°C, the volatiles were reduced so as to meet specifications. After considering several alternatives for the full-scale system, Bruce decided to design a liquid jet eductor system, using 20% NaOH as motive fluid, that would combine the functions of a vacuum source and a scrubbing system to absorb the acidic volatiles. The design involved a number of key considerations such as the estimated solubility behavior of the mixed sodium salts, relative mass transfer and reaction rate in the eductor, and proper disposal of the phenol-containing spent caustic soda. Bruce summed up his recommended design in brief report form and proceeded to order the required equipment. He then moved to another project, and other engineers completed the construction and start-up of the system.

• • •

The day he was hired, John was assigned to a group working on the development of a new electrolytic cell to produce sodium chlorate. For the first month he worked exclusively on the mechanical assembly of experimental cells. Just as he was beginning to get discouraged at the lack of use of his chemical engineering skills, his boss assigned him full responsibility for designing an experimental program to evaluate optimum materials of construction. John incorporated his proposed program into the broader program of testing the operations of experimental cells, and the best materials of construction were pinpointed well ahead of the final cell design decision. After half a year John's responsibilities were extended to include supervision of round-the-clock operation of experimental cell units and correlation of data obtained. He contributed to an economic evaluation of the project which led to its indefinite suspension because of marginal profitability at projected business conditions.

• • •

Jane inherited her first project from a predecessor who transferred to another division. She was handed the responsibility of completing a partly assembled computer-controlled miniplant to test the life cycle time of a variety of hydrogenation catalysts. After completing the assembly, she found that nothing worked as anticipated, and she spent several months unscrambling faulty programs and incompatible logic sequences. At the height of her frustrations the system finally lined out properly, and 6 months after her starting day she was generating important information on relative catalyst properties and performance.

> • • •
>
> As part of his initial assignment in production, Gene was asked to check and review the calibration of all flow instruments in a plant producing a spectrum of flotation agents. He soon discovered that calibration records were lost or erroneous, that orifice plates were wrongly labeled or seriously corroded, and that in general flow measurements were so unreliable that a meaningful mass balance in the plant could not be made. Over a period of 6 months he systematically checked and recomputed the calibration of all flow instruments, replacing faulty DP cells, orifices, and instrument lines. His efforts resulted in improved product quality and better control of unit costs.

tasks—learning a thousand and one details about company operations, and learning to cope with the unstructured, open-ended problems typical of industry.

What are some typical initial assignments given to newly hired chemical engineers? Exhibit 1.2B describes a few cases related by a group of West Coast chemical engineers. Each case is characterized by aspects typical of initial assignments:

Chemical engineering skills are immediately utilized.

The new engineer works as part of a team.

The initial assignment is only a part of a broader project pattern and does not per se constitute some sort of revolutionary breakthrough, no matter how successful.

Advancement and Job Functions

Not too surprisingly, three out of four chemical engineers choosing an industrial career wind up in the chemical process and allied industries (CPI); the distribution of the profession by industrial sector is shown in Table 1.2.3. About half the

TABLE 1.2.3
DISTRIBUTION OF THE CHEMICAL ENGINEERING PROFESSION BY SECTOR

Industry	Percentage of the total population
Chemical process industries	65
Chemicals and allied products	46
Petroleum refining and related industry	12
Paper and allied products	2
Rubber	1
Stone, clay, and glass	1
Food processing	3
Other manufacturing	12
Nonmanufacturing	23

Source: 1981 AIChE Economic Survey Report; used by permission.

chemical engineers in industry work in research, process development, and allied functions (Table 1.2.4), and the newly hired professional may reasonably expect at least to start out in this functional group. The initial assignments listed in Exhibit 1.2B are mostly within the R&D province.

The industrial jobs of chemical engineers may be rather loosely classified as *technical* or *managerial*. Most jobs, of course, are a mixture of both; thus the jobs represented by the organizational chart in Fig. 1.1.1 are undeniably managerial but frequently require a great deal of technical expertise. Newly hired chemical engineers start in the technical area, but managerial responsibilities may become part of the job at an early stage, as illustrated by some of the initial assignments in Exhibit 1.2B. Traditionally, the career advancement of an individual has been monitored in terms of increasing managerial responsibilities. These responsibilities involve the administrative direction of *people* and the administrative control of *money,* and the degree of freedom to manipulate people and money is a measure of the *power* wielded by an individual; increasing power is equated with career advancement. In the traditional confrontation between the technologist and the manager, the manager's point of view has always prevailed since it is the manager who has the basis of power, and the technical job has correspondingly suffered from a lack of prestige; the managerial ranks have been filled by the more aggressive technologists moving into jobs where they can have a more effective say-so about the course of projects.

Career advancement therefore tends to be skewed toward managerial jobs, a situation of increasing concern in many corporations. The concern stems from the belief that a satisfactory career should not be wholly predicated upon managerial jobs, that the individual who continues to make valuable contributions to the technical side of the company's operations should be rewarded by satisfying career advancement criteria such as increased remuneration, peer recognition, enhanced power in directing projects, and the granting of prestigious privileges. Attempts

TABLE 1.2.4
JOB FUNCTION OF CHEMICAL ENGINEERS IN THE CHEMICAL INDUSTRY

	1967	1977
Development*	30%	47%
Research*	20%	
Design	10%	9%
Management	13%	12%
Operations	11%	11%
Sales	5%	6%
Economic evaluation	4%	5%
Others	7%	10%

*The R&D statistics incorporate all process development aspects (pilot plants, semiplants, etc.).
Source: AIChE Economic Survey Reports; used by permission.

have been made to answer the concern by establishing a *dual ladder system of advancement,* * and many companies try to attract high-quality technically oriented individuals into their ranks with the promise of this concept.

To understand the dual ladder system, it is best to describe first the nature of the ladder of advancement in industry. Each rung of the ladder is a *job category* with its own title, its generally defined limits of activity and responsibility, and its own salary range which is periodically adjusted to account for inflation. Various methods may be used to rank the job categories and to define the salary range; most of the methods are some variation of the *Hay points system*, a system which assigns a judiciously chosen point value to each one of several job aspects. The overall job ranking is a function of the Hay points total. The rated job aspects include items such as:

Accountability: number of people administratively supervised, degree of financial responsibility

Know-how requirements: special knowledge and training required for the job category

Problem-solving demands: exposure to nonroutine situations, demand for imaginative and creative solutions

The salary range within each category provides freedom for periodic routine and merit salary increases within the time span between promotions to a higher rung on the ladder.

The dual ladder system of advancement is based on the premise that parallel ladders may be established for managerial and technical careers and that a degree of *equivalence* can be established between the rungs of the side-by-side ladders. Why should this concept of equivalence be so important? Each ladder of advancement must, of course, represent wholly satisfactory career opportunities completely apart from the other in terms of attractive long-range goals at the top and an acceptable number of rungs on the way up. Nevertheless, the rungs of the technical ladder must inevitably be compared with those of the managerial ladder because of the traditional "visibility," prestige, and power basis of the managerial jobs; the sense of achievement in attaining a particular technical category is enhanced by being aware of the equivalent and traditionally more prestigious managerial job category.

True equivalence must, of course, be defined quantitatively. The obvious quantitative criterion of equivalence is salary, but this, in turn, is established on the basis of a quantitative evaluation of each job level responsibilities by means of some quantifying device such as the Hay points system. The understandable difficulty in constructing an equitable single advancement ladder by means of job level quantification is compounded when the concept of equivalence is used in constructing a dual ladder system. The problem is that different job aspects predominate depending upon whether a job is technical or managerial. For example, the aspect of accountability is dominant in managerial jobs, whereas know-how may be the key

---

*See Valle-Riestra (1980).

aspect of a technical job, and how is it possible to compare quantitatively accountability demands with know-how demands? These kinds of considerations are typical of the frequently encountered and challenging task of trying to quantify judgment factors with little or no quantitative basis. Nevertheless, in the case of devising a workable dual ladder system, the attempts have been at least partly successful, for workable and accepted systems do exist and are considered to be an important aspect of the career management of professional employees in many corporations.

An abbreviated and idealized example of a dual ladder system of advancement is illustrated in Figure 1.2.2; in this particular example, the dual ladder system is restricted to the R&D function, although corresponding dual ladders could be appended for engineering, production, and other major functions in the industrial environment. One difference is immediately apparent—the titles on the management ladder are *job*-oriented, whereas on the technical ladder they are *person*-oriented. Thus a "lab director" has a specific job to do, described in the title; on the other hand, the "senior research specialist" title does not imply any specific duties. This state of affairs is a reflection of the differences in the nature of the jobs and in the way individuals are promoted up the ladder. Managers are promoted into a higher job as a result of a new *assignment,* as part of the logical flow and evolution of managerial duties. Technologists are promoted as a result of a job well-performed, and the promotion is *reward* for performance.

Other aspects of the lack of equivalence between the dual ladders should be noted:

Technical	Management
Research fellow	
Research scientist II	Research director
Research scientist I	Lab director
Associate scientist II	Research manager II
Associate scientist I	Research manager I
Senior research specialist	Group leader II
Research specialist II	Group leader I
Research specialist I	Research supervisor
Senior research engineer	
Research engineer	
Engineer	

FIGURE 1.2.2
The dual ladder in research.

Disparity in power and prestige. The technologist may wield a great deal of influence on how money is spent and often controls the work progress of many individuals. But it is the manager who exercises *direct* control over money and people; after all, it is the manager who has the say-so about the technologist's promotion. Power stems from direct control, and power, in the popular view, merits prestige.

The atrophied technical ladder. It may happen that a corporation has a dual ladder system all right, with some rungs at equal level, but the technical ladder is much shorter. In this situation there is really no question of equivalence, for if there are no technical rungs to match the higher managerial positions, the successful and ambitious individual is driven to escape from the atrophied technical ladder into management.

Lack of two-way lateral movement. True equivalence would suggest that an individual could move from one ladder to the other in either direction. The fact of the matter is that lateral movement is almost invariably unidirectional, from the technical to the managerial ladder. One obstacle to movement in the other direction is the reluctance by managers to remove themselves, even temporarily, from the competitive race up the managerial ladder.

Disparity in stress exposure. Pressure engendered by the need to make difficult people and money decisions, decisions which are far from obvious or automatic, results in long working hours and often sleepless nights for managers. Technical people bear their own set of stress situations, but they are rarely as intense.

The dual ladder system is nothing more than a device for managing the careers of employees, and how well the system works out in practice is, to a considerable extent, a reflection of the attitudes inherent to the corporation, its life-style. A few such life-style attitudes which should be mentioned include the following.

The Whole-Job Concept This concept, which has a number of synonyms depending upon the corporation, refers to the attitude of allowing and encouraging each employee to assume any and all responsibilities that the individual feels are required for optimizing the effectiveness of job performance. It is an attempt to keep people from being pigeonholed into restricted job categories; by assuming functions that are normally not associated with a particular job, employees may substantially enhance the chances of success of their projects. For example, the research engineer may decide to broaden the work scope of a particular project by becoming involved in market research and customer contacts, or by organizing an after-hours course in microprocessors for technicians associated with the project, or by suggesting a streamlined procedure for project time accounting, or by any of a myriad other functions which could easily be left up to the boss or colleagues in other disciplines. Of course, the concept implies that the individual will not be discouraged and victimized by petty political considerations or unwarranted resentment. In this sort of an atmosphere, the equivalence of the dual ladders is much closer to reality.

Respect and Support of Professionalism Subprofessional standards refer to attitudes which do not give proper recognition to the professional status of chemical engineers, particularly those on the technical ladder. Equivalence is not likely to be realized in a corporation that tolerates the existence of subprofessional standards for its professional employees, for subprofessionalism pushes individuals from the technical toward the managerial ladder as a means of hoped-for escape from subprofessional practices. Subprofessional attitudes include reluctance to grant significant decision-making power to technical people and job assignments requiring skill levels far below those acquired as part of a professional education.

Advanced Career Guidance This subject merits special mention. "Career guidance" is a term the student starts to be aware of in high school. At this level it involves an exposition of the student to the facets of various possible careers and some attempt to arrange a curriculum to conform to a profession to which the student seems to be best-fitted. Career guidance in more advanced form may be encountered at the university level; in industry, *advanced career guidance* refers to attempts by more enlightened corporate management to shape the optimum lifelong career of each of its company's employees. Features such as those outlined in what follows are a manifestation of the corporation's dedication to creating for each employee a satisfactory career, one best fitted to the employee's desires and abilities, for the very good reason that such course of action results in the highest rate of return on the investment the corporation has in that employee. Let us then take a look at some of the features of a corporate advanced career guidance program.

Employee Rating Grading does not stop at the end of the last formal course taken in the university—it follows the chemical engineer throughout a life career. For each employee is graded by a superior at all stages of that career on the basis of total performance. The grades are usually not A's or F's, but they do represent four or five stages of performance characterized by designations such as "superior," "average," and so on. Moreover, each supervisor may be asked to "force-rank" people and fill each rating category with a prescribed percentage of the total number supervised. Promotions and raises, for example, are at least partly based on such grading.

The rating represented by grades can be a useful tool for career guidance, for a consistently high rating usually implies that the individual is on the right career track, and the result is rapid advancement up the promotion ladder. A consistently low rating spells trouble; the individual clearly has not been fitted into a proper job situation, and a radical change in career direction is called for. In extreme cases of consistently poor performance, dismissal, or at least a strong hint to establish a new career in a different company, may be called for. It is no great favor to mislead a poor performer and to let the individual drift in a morass of mediocrity. In the great majority of cases radical "career surgery" is not required, and rational career planning can be used to improve an individual's rating. Many a poor performer in, let us say, the R&D function has proved to be a great success in some other function.

Job Performance Reviews (JPRs)   The JPRs are regularly (usually annually) scheduled formal meetings between supervisor and employee for the purpose of apprising the employee of performance and general progress. They are a most important aspect of advanced career guidance, and as such are being adopted as standard policy by an increasing proportion of corporations. Since they represent a one-on-one confrontation, they are usually approached with a great deal of trepidation, by supervisor as well as employee.

And yet they need not be an unpleasant experience if they are recognized for what they are, not an occasion for criticism or downgrading but an occasion to exchange ideas and to take a constructive view of the direction of the employee's career. We all recognize that we are not perfect, that we have weaknesses in certain areas, and an understanding, factual, and friendly statement of perceived weaknesses and suggested remedies, balanced by proper emphasis of strong points, should be a welcome aspect of career guidance. The degree of success that this sort of confrontation engenders is certainly a function of the human relations skills of the supervisor, but the employee can contribute to that success immeasurably by adopting a positive, understanding, and receptive attitude. In fact, the JPR can lead to a more cordial and intimate relationship between supervisor and employee, and the sensitive supervisor will invite reciprocal evaluation, or insist upon "talking out" conflict situations.

Career Counseling   Understanding and sensitive supervisors can do a great deal toward successfully advancing the careers of those whose work they direct. Yet it is too much to expect that managers in the chemical industry, chosen on the basis of technical and business rather than human relations acumen, would always turn out to be competent career counselors, that they would possess the perception required to steer individuals along the optimum career track. The career guidance by supervision is often supplemented by that of professional career counselors to whom individuals can turn for advice. Often these individuals—industrial psychologists or staff members of employee relations departments—may be given influential say-so to facilitate transfers, if such a step is called for. These same individuals sometimes serve as ombudsmen to settle grievances or uncomfortable personality conflicts.

Formal Job Posting   Some companies give employees an opportunity to change jobs by letting them bid on specific posted jobs. This method of career modification is restricted to the lower rungs of the advancement ladders, but it represents a good opportunity for those who feel they need a change or new challenge. For example, the young engineer who feels trapped in the routines of the engineering department may use the bidding procedure to start a career in production.

Training Programs   In-house training programs span the range from occasionally short on-the-job surveys to elaborate course systems of universitylike diversity (Xerox Corporation, for example). Technical courses usually are designed to teach specialized skills not acquired as part of a regular university curriculum, skills that are considered important in context of the corporation's operations, although review courses may also be offered. An almost unlimited variety of courses and seminars may be made available to promote personal development, creativity, or managerial skills. Many companies encourage employees' enrollment in courses offered at nearby universities by paying part or all of the tuition fees. Chemical engineers may choose to obtain advanced degrees (such as M.B.A.) in this manner, either by attending night classes or occasionally by obtaining a leave of absence and attending regularly scheduled daytime classes.

These, then, are some of the aspects of advanced career guidance as practiced in the chemical process industry, aspects which are intended to facilitate and optimize professional employees' career progress. Clearly, career advancement depends upon the sum total of the personal characteristics of the employee, and some of these characteristics which are held to be particularly important will now be explored. Figure 1.2.3 illustrates the so-called *triangle of success,* a symbolic representation of the interplay of the most important personal characteristics.

The topmost apex of the triangle stands for *technical competence,* a characteristic of preponderant importance in chemical engineers, at the beginning of their careers in particular. The neophyte engineers are judged on the basis of their success in tackling technical problems, a process which involves a gamut of abilities in problem synthesis, in the organization of technical material, in facile application of learned theory, in writing and oral expression, and in stick-to-itiveness, the perseverance needed to pursue a stubborn problem to its ultimate resolution. Technical competence is always the mark of a successful engineer, but in advanced stages of the engineering career political adroitness, embedded in the bottom apices of Fig. 1.2.3, takes on increasing importance (Williams, 1979).

Personality refers to a great deal more than a pleasant smile. Of course, a warm, pleasing nature is an important asset in all human endeavor, but our meaning of the word "personality" encompasses more than that. For one thing, it implies the ability to deal with people, bosses and subordinates alike, to be sensitive to their attitudes and problems, to empathize. It involves the ability to adopt a positive attitude—not a mindless, Pollyanna viewpoint, but one based on a rational evaluation of the individual's environment, and a conscious decision to emphasize the positive aspects

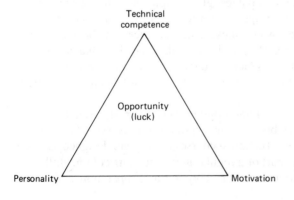

FIGURE 1.2.3
The triangle of success.

of the environment. Bad-mouthing the corporation or individual employees is a sterile release mechanism indeed, one that is bound to damage the reputation and chances for success of the unhappy perpetrator, and, really, nobody enjoys working with a habitual grouch.

Another important aspect of personality is the ability to perceive the political realities of a particular industrial environment and to adapt to them. We have alluded to the politics in industry, and we should perhaps briefly touch upon the significance. We have seen that the higher rungs on the managerial ladder represent an ever-increasing position of power, and the contest to attain these positions of power, as well as the very act of exercising that power, creates a structure of political influence, of political forces and counterforces. Technically minded individuals profess to abhor and despise politics, but they themselves tend to create political situations based on considerations of prestige; politics is an inevitable companion of human group endeavors. "Company politics" may assume a number of aspects. It may involve unwritten policies and beliefs of a powerful manager and strong pressures to conform to those policies in the context of the corporation's activities. It may involve efforts to impress and influence individuals who control the budgeting of specific projects or departments. Often it involves a silent, or not so silent, contest between conflicting managerial factions. Neophyte engineers luckily need not be too concerned with company politics, but they inevitably must expect to be involved in some politically oriented activities from the very start of their careers. For example, "dog and pony" shows—carefully orchestrated demonstrations of project progress for the benefit of budget-controlling managers—involve many a neophyte engineer. Progress up the promotion ladder sometimes involves a careful balancing of political and technical considerations; however, in all cases, personal and professional considerations of ethics must predominate.

The third apex of our triangle of success represents *motivation,* the sum total of a number of job-oriented attitudes. The successful engineer is motivated by an unfaltering desire to perform well and to be recognized for such efforts, a desire which is characterized by enthusiasm, drive, and aggressiveness, this latter not an attitude of hostility and abrasiveness but one of willingness to tackle new challenging situations and to accept the reality of competition. Proper motivation diminishes the need for close supervision since motivated engineers assume the responsibility of planning and carrying out the job sequence associated with their project. Motivation means dedication to the job, and this may mean a lot of overtime work when needed (although unnecessary overtime work just to impress a supervisor is foolish). Motivation means flexibility, the ability to adapt to changing characteristics of the job. The question often arises as to whether the desired flexibility should include the willingness to accept geographical transfers. The answer depends to a large extent upon the individual's career ambition. If a chemical engineer's life-style is such that a particular geographical area is strongly preferred, it certainly is possible to pursue a wholly satisfactory career in the preferred location, but it must be understood that the somewhat reduced flexibility does impose some career limitations. For example, the individual striving to reach top staff management positions must expect to transfer into higher-level jobs as they become available, and this may occur at widely separated geographical locations. Understandably, an unwillingness to make the geographical move would impose a serious limitation upon the career advancement sequence. Unlimited flexibility is perhaps a desirable theoretical attribute, but corporate management recognizes that most of us do have limitations imposed by various loyalties, family ties, and life concepts, and career advancement is a matter of a series of mutual compromises.

We have focused upon the apices of our triangle of success in Figure 1.2.3, but it should be noted that the area enclosed is appropriately marked as "Opportunity," with the subheading, "Luck." This is meant to show that the right kind of opportunities are the true binding material of a successful career, that without the opportunities and the lucky turns of events that spawn them, no career will be a success in spite of the strengths represented by the apices. Does this mean that an individual's career is, in the final analysis, just a matter of luck? Not really, because each one of us can help our career along by developing the ability and insight to recognize and take advantage of opportunities, in proverbial fashion, whenever they come along. Nevertheless, it must be conceded that simple luck does have a great deal to do with success in anybody's career.

Professional Development

We have seen how advanced career guidance, as practiced in the chemical process industry, contributes to the professional development of employees. In the final analysis, however, professional development is the *individual's* responsibility; the responsible chemical engineer does not depend upon management for guidance or encouragement but undertakes a number of activities which contribute to individual development. Let us take a look at what some of these activities are.

Professional Society Activities Membership in professional societies represents perhaps the most obvious way in which individuals can promote their own profession. Frequently corporate management does not particularly encourage participation in the activities of societies; even in the face of lack of managerial support, chemical engineers should not hesitate to contribute to the growth of their own profession by lending support to organizations which represent it—they should be proud of their profession and the accomplishments it represents. For chemical engineers, the core organization is the *American Institute of Chemical Engineers* (AIChE).

AIChE, an organization of some 60,000 chemical engineers, offers many activities typical of large professional societies. One such important activity is the promotion and support of the technological and scholarly advancement of the profession through the medium of technical publications and national meetings, which give chemical engineers the opportunity to disclose and discuss their work. Important as the promotion of technological progress may be, perhaps the most significant contribution of a society such as AIChE is the opportunity given to the individual to contact colleagues in the profession on a one-to-one basis, to keep abreast of what is going on in the chemical process industry as a whole and in its component corporations. Excellent opportunities for individual contact occur at the level of the *local sections* which encompass the membership in specific geographical areas. Local sections are the center of many autonomous activities ranging from periodic (usually monthly) business-technical meetings to regional symposia. Most universities with chemical engineering departments have AIChE *student chapters;* membership is strongly recommended to students for it constitutes an important introduction to professional activities.

Regrettably, a great many chemical engineers are not affiliated with AIChE, and others offer only nominal financial support as nonparticipating members. Such individuals fail to recognize the importance of the institute as a potential force in charting the course of their own profession and careers due to its influence upon governmental, industrial, and academic entities. Chemical engineers may, indeed, have a say-so about the nature of their profession and jobs by engaging in the only satisfactory professional society participation mode—an active form of participation on both the local and national level as officers, as committee members, as volunteers.

One of the ongoing key issues in AIChE focuses upon the formulation and implementation of the *AIChE Professional Guidelines* ("Guidelines to Professional Employment for Engineers and Scientists"). The guidelines constitute a system of commonsense rules, a guide to mutually satisfying relations between professional engineers and their employers. They spell out the obligations that employers have to recognize and respect the professional status of their employees, but they also emphasize the responsibilities that chemical engineers have as professionals toward their employers. A detailed outline of the guidelines may be found by reference to the articles in Exhibit 1.2C ("References on Professional Development").

The guidelines touch upon many of the issues we have examined during our discussion of advanced career guidance. For example, periodic performance reviews (JPRs) are identified as the proper basis for economic advancement. Employer obligations ranging from minimum pension standards to the avoidance of subprofessional level assignments are outlined—even the use of punched clocks is proscribed. Employees who change employers are properly reminded of their ethical obligation not to divulge a former employer's proprietary know-how. The current policy of the institute toward the guidelines is to consider them as minimum standards for professional employment, but not to publicize or "blacklist" nonconforming corporations. The opinion of the membership of AIChE is split regarding the blacklisting policy, and many chemical corporations are regrettably lukewarm toward the guidelines concept. Thus many conflicting attitudes need to be resolved, and chemical engineers at all levels should become involved in this important activity which has so many implications upon professionalism and basic job satisfaction.

Another key ongoing issue involves the *dynamic objectives of AIChE,* an attempt to define the goals and objectives of the profession. Again, the decisions regarding this issue may exert a profound influence upon the future of the profession and each job within its context, and members of the institute are strongly urged to participate

EXHIBIT 1.2C

REFERENCES ON PROFESSIONAL DEVELOPMENT

The following journals frequently have articles on subjects related to professional development:

AIChE Student Members Bulletin (ASMB)
Chemical Engineering Progress (CEP) (see "SCOPE")
Chemical Engineering (CE) ("You and Your Job")

Some specific references include the following:

Professional guidelines

CEP: April 1972, p. 23; March 1973, p. 15; April 1974, p. 30 (How Does the Young Engineer Feel?); August 1975, p. 127 (AIChE Opinion Survey on Professional Employment Guidelines); March 1976, p. 21.

Registration of engineers

CEP: March 1974 (Certification of Engineers); May, June, July, and October 1976. Constance, John D.: Getting Licensed in any State, *CE,* **83**(9): 115, Apr. 26, 1976.

Interviews and résumés

CE: **81**(24):164 (Nov. 11, 1974) (Is Your Résumé Junk Mail?, Guy Weismantel and Jay Matley)
ASMB, **14**(2) (Fall 1973) (Careers in Chemical Engineering: Interviewing and Résumés)

Project groups

ASMB, **14**:(1) (Spring 1973) (How to Have a Good Project Group, O. M. Fuller)

Communications

CEP: November 1974 (Joel Primack, The Nuclear Safety Controversy. A fascinating commentary on dissemination of technical information.)
CEP: February 1977 (The Communication Crisis)

Dynamic objectives for chemical engineering

CEP, April 1977, p. 41.

Ethics

CEP, July 1978, p. 13 (Multinationals and the Social Contract)
CEP, February 1965, p. 39 (AIChE Code of Ethics).
See also Kohn and Hughson (1980) in Chap. 1 references.

in the formulating and decision-making process. References in Exhibit 1.2C outline the nature and meaning of the objectives, which include the promotion of advanced career guidance (seriously neglected by some corporations); the formulation of an appropriate position on registration and certification (to be further discussed in a subsequent section of this chapter); and the promotion and encouragement of continuing education programs.

Chemical engineers newly out of school may become associate members; after 3 years of professional experience and with the recommendation of three professional references, full AIChE membership is available.

Chemical engineers often opt for membership in other professional societies. Many belong to the *American Chemical Society* (ACS), an organization of 120,000 chemists and engineers with an enviable record of publication in all the chemical sciences. *The American Institute of Chemists* (AIC) is a small (5000 members) but prestigious organization with the primary objective of promoting the professional status of chemists and chemical engineers. A number of professional societies serve as the focus for specialists from all the engineering and scientific disciplines; these include the *Institute of Food Technologists,* the *Combustion Institute,* the *American Association of Cost Engineers,* the *Electrochemical Society,* and many others. At the other end of the speciality spectrum are societies which promote the overall advancement of science or engineering, such as the *American Association for the Advancement of Science* and the *Society of Professional Engineers*. Chemical engineers who are successful in their profession often belong to three or four professional societies.

Engineering Registration and Certification The licensing and registration of chemical and other engineers is granted under the authority of state governments for the ostensible purpose of protecting the public against the dangers of malpractice by unqualified individuals. In some cases the need for a degree of governmental control through registration is obvious, as in the case of structural engineers whose design work directly affects the public welfare. In other cases the need for such control is not so obvious. For example, almost all chemical engineers are employed by private corporations which are perfectly capable of weeding out incompetents and quacks and thus serve as a protective barrier for the public's interest. The differences in the justification for engineering registration are reflected by two distinct categories of registration. Licensing under a *practice act* gives the registered engineer the right to practice a particular engineering discipline and prohibits practice by nonlicensed individuals. Licensing under a *title act* gives the registered engineer the right to use the title of Professional Engineer (P.E.) or the more restricted title within a particular discipline ("chemical engineer"), but practice by nonlicensed individuals is not prohibited. The latter form of registration might appear to be a particularly sterile gesture, but, as a matter of fact, there are some advantages. For example, final design drawings for chemical plants must be signed by a registered engineer; only a registered engineer may serve as an expert technical witness in court. Whether a particular engineering discipline is covered by a practice or title act depends upon the current law in a given state.

The existence of two categories, however, creates a problem. Some engineering groups have looked at the granting of registered status as a symbol of prestige, and perhaps rightfully so. They have gone so far as to promote forcefully the legislative passage of a practice act for their particular discipline. Since the practice act incorporates statements spelling out in some detail the limits of practice of the discipline in question, other engineering disciplines not covered by a practice act suddenly find themselves facing a situation wherein their own discipline overlaps the one covered by a practice act. In effect, the title act engineers find themselves forbidden to practice certain aspects of their profession. As an example, mechanical engineers may have heat exchanger design specified as within the realm of their practice, and if they are covered by a practice act, chemical engineers covered by a title act only are, strictly speaking, proscribed from designing heat exchangers. The result has been an increasing pressure by chemical engineers in many states to become registered under the provisions of a practice act, perhaps with an industrial exemption for those employed by large corporations.

Going into the early 1980s, the registration situation in many states is uncertain and confusing. The young chemical engineer may legitimately ask, "Why register?" There are some good reasons: practice act coverage for all chemical engineers may well become reality, and recent graduates might as well pass the written examinations while they have academic concepts still freshly in mind. After all, some benefits do accrue to the registered engineer, and registration may be properly considered as part of the individual's professional development.

The first step in the registration process is an examination which, if passed, establishes the applicant as an *Engineer in Training*. The EIT examination is 8 hours long and covers all phases of an undergraduate engineering curriculum; the same examination is given to applicants in all the engineering disciplines. It may be taken at the time of graduation (B.S. degree) or at any time thereafter. If the applicant does not pass, the examination may be taken again, as many times as desired. The final step involves a *Professional Engineer* examination, but certain prerequisites of experience must first be met, and the applicant must be recommended by a required number of professional colleagues. Typically, 4 years of "responsible charge" (professional engineering work) are required; prerequisites for each state are outlined in the references in Exhibit 1.2C. The P. E. examination is different for each discipline. It is 8 hours long, with 4 hours devoted to theory and 4 to engineering practice. Perhaps one-half of applicants fail the P. E. exam, but it may be repeated. An annual renewal fee is the only requirement to maintain an individual's registration status; in particular, certification that the engineer has kept up with developments in the engineering discipline is not currently required (1981).

To maintain their proficiency, chemical engineers need to keep up with up-to-date developments in their profession. A *certification* procedure is one whereby individuals periodically account for their efforts to develop themselves professionally. The accounting procedure involves a review of applicable activities undertaken by the applicant in continuing education, in professional society participation, and in publications and presentations; some sort of weighting procedure is then employed to establish a score which, if in excess of a preset minimum, merits the applicant a certificate of professional competence. If certification were to become more widely accepted, it would serve to identify those dedicated engineers who strive to be at the top both technically and professionally. It has been suggested that certification serve as basis for the renewal of registration licenses, a concept currently rejected by a majority of chemical engineers.[1] The American Institute of Chemists sponsors the *National Certification Commission in Chemistry and Chemical Engineering,* an organization which undertakes to certify individuals who meet certification prerequisites, with a renewal period of 3 years. Recipients are encouraged to use the title of Certified Chemical Engineer (C.Ch.E.).

In 1980, AIChE instituted a Professional Development Recognition Certificate program.

Continuing Education The profession of chemical engineering is an unusually dynamic one. Developments in theory, in new techniques and applications, in advanced equipment, and in related fields are all rapid, and engineers who are satisfied to sit tight on their academic background alone find themselves rapidly obsolete. Regrettably, the profession is full of individuals who have not heeded this warning and whose technical effectiveness has suffered correspondingly. The dedicated chemical engineer must be ready to continue to develop technically, to keep up-to-date by taking advantage of opportunities in continuing education.

There are many such opportunities, and some have been discussed in the context of advanced career guidance activities of individual corporations. Professional societies also underwrite continuing education courses. AIChE, for example, sponsors a wide variety of 1- and 2-day concentrated courses which are held just before or after national meetings (AIChE "Today Series," AIChE Management and Executive Seminars). Some sessions during the meetings themselves are in the form of new technology reviews. Advanced technical courses are offered by many universities, either during the evening or as correspondence courses, as well as by private continuing education organizations, such as the American Institute for Professional Education or the Center for Professional Advancement.

Motivated engineers need not depend on outside agencies to keep themselves up-to-date, however. A good way to brush up or to expand one's realm of knowledge is to read technical books at the individual's chosen pace; many books incorporate practice problems and exercises that serve a useful pedagogical purpose. All engineers should assign a minimum number of weekly hours to surveillance of current literature; many subscribe to journals such as *Chemical Engineering* and *AIChE Journal* and buy representative books to build up a personal reference library.

Bondi (1969) has used examples from his industrial experience to demonstrate that individuals who develop themselves professionally through continuing education may expect an exceptionally profitable return on their investment of time in terms of more rapid advancement and higher salary.

Civic Activities Professional development need not be all job-oriented. Chemical engineers enjoy eminent status in industry and in the society the industry serves, and they should strive to contribute to that society in all its aspects, to become leaders in their community. It is true that the experience they gain in such

---

[1]See reference on certification, Exhibit 1.2C.

activities serves them well in their job situation, but this should not really be the primary motivation; chemical engineers, just like any other responsible citizens, need to contribute some portion of their time and effort to community activities. Almost any interest may be satisfied in the context of civic affair participation, and chemical engineers are influential on school boards, in environmental organizations, in church groups, and in youth guidance organizations.

Considerations of Ethics Conformity to ethical standards is fundamental to what we think of as professional behavior. Decisions which affect personal ethics are an everyday part of the professional experience, and ethical problems are some of the most difficult ones facing the practicing chemical engineer throughout a career. Considerations of ethics permeate the basic relationships between employer and employee and serve to define the employee's responsibilities; some of these responsibilities have been codified and may be thus found in professional society codes of ethics. The AIChE Code of Ethics is reprinted in the reference in Exhibit 1.2C, although the focus of interest about issues of ethics has shifted toward the AIChE Professional Guidelines, which incorporate many of the provisions of the code.

Certain standards of acceptable human behavior are obvious and do not require any great philosophical rumination; for instance, you do not steal, lie, cheat, or betray your employer. But most personal ethical problems that arise as part of employment in the chemical process industry cannot be so easily resolved by reference to universal tenets of morality or professional codes; the final decision must be made on the basis of the engineer's own personal code of ethics. Some typical ethical problems that chemical engineers often must be ready to face are outlined in Exhibit 1.2D. Now, there are no "right" answers to the dilemmas which these situations present, and professionals must resolve them so as to optimize their own moral sensibilities and comfort. The resolution of such difficult situations may involve some pretty uncomfortable decisions, even termination of employment in extreme cases.

In general, if the corporate employer does something illegal or highly immoral, the engineer's proper course of action should be to face up to the subject and talk it over with superiors, preferably starting with the immediate supervisor. If nothing is done in spite of the engineer's best efforts, perhaps, after all, the company is not the kind of place where the engineer would wish to spend a career. In quite a number of cases an apparent ethical dilemma turns out to be a matter of misunderstanding, of poor communications, and this is one reason why it is so important to try to talk these situations out. As a matter of fact, the great majority of people do want to run their businesses legally and ethically. St. Clair (1979) points out, "The mainstream of American businesses that employ chemical engineers do not, and do not want to, twist a person's scientific or ethical standards to fit their profit motive."

Ethical controversies *must* be faced and resolved head on, never to be ignored or "swept under the rug." However, responsible engineers should always be ready to scrutinize their own ethical attitudes, not just the other fellow's. In this way the chemical engineer can best attain what should always be the first goal—the preservation and protection of ethical status and reputation.

EXHIBIT 1.2D

SOME TYPICAL PROBLEMS OF PERSONAL ETHICS

If faced with the following ethical dilemmas, what would *you* do?

1 In spite of your repeated reminders, your superiors have done nothing to eliminate the dumping of toxic wastes into a nearby river.

2 The hourly employees in your plant, many of them close friends, are about to go on strike. Your management intends to lock in all engineers and other salaried personnel to take over the hourly employees' jobs and keep the plant running.

3 You have been asked to interview applicants for a new professional-level position in your plant. You are told to reject automatically any applicant over 50, since this is an "unwritten policy" of the company.

4 Your company has signed a government contract to research and manufacture a new nerve gas for the military. Your political and moral principles are opposed to the use of such weapons.

5 After 20 years' experience with your company, you have gained unusual expertise in a highly specialized area of manufacture which has made your company successful. A competitor has offered you a job at a fabulous salary increase, with the understanding that you would not be required to divulge proprietary information but would be asked to teach your expertise.

6 Your direct supervisor is a successful technical innovator with a great deal of influence who can really help your career. However, he is an outspoken racist.

7 You have been offered a top managerial position in your company's installation in a foreign country where bribery of officials is standard business practice.

8 You have just received a $5000 cash bonus for a suggestion that has resulted in a substantial production improvement, but the idea was originally given to you by one of the operators on the graveyard shift.

9 The animals in your company's toxicological labs are not properly cared for, and experimental shortcuts result in cruel treatment. You are told that economic considerations do not allow for improvements, and you know that toxicological evaluation is vital to your company's business interests.

10 While entertaining the divisional personnel manager at dinner in your home, you find out that female engineers are paid 25 percent less than men in the same job. The manager defends the practice by pointing out that in this way more women are given the opportunity to enter professional ranks; moreover, since the company can afford to hire more women, its record looks particularly praiseworthy during annual Equal Employment Opportunity surveys by the federal government.

11 You believe strongly that "small is beautiful," but your employer is dedicated to rapid expansion of the business.

12 The boss asks you for a preliminary estimate of a project for the purpose of asking for supporting funds. You are informally requested to "make it look good," and your boss is notorious for not wanting to hear bad news, usually with the admonition to "be more positive."

13 One of your close friends has just been made manager of a production facility which, because of the nature of the process, has had a record of "close calls," toxic gas releases, and minor accidents. Only a few people, you among them, know that your friend is an alcoholic.

14 You have fought hard for years to have a "grass roots" plant built on a company-owned site near a small community. Construction has finally been started, and during excavation for foundations an old Indian cemetery is uncovered beneath the plant location.

15 Company business has taken you far away from home for the tenth day, and over a lonely dinner you would like to try out an $80 bottle of a 1963 Chambertin. No one at work has ever questioned your expense account.

16 You know that a number of years ago your company disposed of some 100 drums of wastes containing dangerously carcinogenic components through the services of a disposal firm that operates a burial site.

For other interesting cases, see Kohn and Hughson (1980).

Communications

The ability to express oneself well is a very important requirement of the successful engineer, for the proof of accomplishment must be communicated in some form so that the accomplishment may be ultimately usable in the industrial environment. The most brilliant piece of research, the most inventive production plan, the shrewdest evaluation of competitors' pricing policy, all are wasted unless effectively communicated so that the information can be used. Regrettably, many engineers are poor communicators, and their presentations, both oral and written, tend to be insufferably boring and, as a result, frequently misrepresented. Davidson (1974),* speaking of engineers, sadly points out that "to the public, we are single-minded solvers of industrial problems, indifferent to ecological or sociological consequences. To our management, we are detail-philes lost in a forest of data. To other engineers, we are perpetrators of ponderous, dull reports." He ascribes this negative image to communication failure: "Turgid prose larded with professional jargon, convoluted syntax, slavish devotion to detailed trivia, chronological instead of interpretational presentation, and more."

What, then is the answer to this sad state of affairs? The answer is no mystery; those engineers who enter the industrial environment and who lack the required background of communication skills had better start working hard to develop those skills, for they are needed badly in the course of personally satisfying professional development and career advancement. The problem of underdeveloped communication skills is no different from that of underdeveloped skills in any technical area; the solution is to face the problem head on, to take advantage of learning opportunities, and to practice. It is really not particularly surprising to find many neophyte engineers lacking communication skills, for so many university chemical engineering curricula concentrate on the technical requirements of the profession; some attempt is usually made to give practice in technical writing, but the humanities, so important to the development of a personal communication style, are often neglected.

*Reproduced by permission.

Oral communication is an integral part of each and every working day, and often a very substantial portion of the engineer's day is occupied by meetings and interactions with colleagues. Such interactions are most frequently quite informal—reports to the boss, instruction sessions with operators, group consultations—but they become more formal and structured as one progresses from oral progress reports to the previously mentioned, politically oriented dog and pony shows, and finally to formal technical presentations and speeches, both in-house and at professional meetings. Regardless of the degree of formality, any presentation should be properly prepared, and the speaker should at all times attempt to stay on "firm ground," to know exactly what he or she *wants* to say. Deviation from this self-evident rule results in stumbling speech, verbal gyrations, and a fuzzy result that nobody understands.

There are many available written guides to effective oral communication, and Davidson's (1974) collection of articles from *Chemical Engineering* magazine is particularly laudable. A few simple rules are worth emphasizing here. Try to deliver any talk in a natural, conversational tone, and do try to develop a strong voice, something that comes with practice. Under no circumstances read a written-out speech; in spite of your best efforts, it will sound stilted and artificial. You should know, and know well, what it is that you wish to talk about, and a brief outline on cards, perhaps with color-underlined key words to remind you of concepts, should be your only guide. Naturally, you must practice your presentation, perhaps by simply reviewing it a few times in your mind, or better yet by locking yourself up somewhere in front of a mirror and talking aloud. Another excellent way to lend smooth continuity to your presentation is to talk with the help of slides or overhead projections; the principal concepts can be outlined, and your job then is to tie the concepts together and to elaborate. Copies of your projections constitute a good handout for members of your audience.

Stage fright is an uncomfortable aspect of the more formal oral presentations, and, make no mistake, even the most experienced public speakers and actors have twinges of discomfort during the initial stages of their performance. Fortunately, this problem rapidly disappears once the speaker really gets going, but to avoid the danger of "freezing up," some speakers break the rule of not reading speeches and write out the first sentence or two of their talks. However, the only answer to the feeling of discomfort and anxiety which haunts the inexperienced speaker is practice, and the engineer dedicated to the attainment of oral communication skills should seek out opportunities to participate—in front of community groups, at Toastmaster Clubs, at company seminars, at professional society meetings.

Written communications represent a permanent record of the engineer's accomplishments and ideas. The basic medium for recording those accomplishments is the technical report. Each company, and each department in the company, has its own ideas as to how a proper report should be put together, but there are a number of common elements of structure and style which are first introduced to the engineering student in undergraduate laboratory courses. Graduate students have additional opportunities to sharpen up their report-writing skills, including the writing of a thesis required for advanced degrees. Neophyte engineers in the industrial environment may pick up valuable advice and help from report-writing manuals such as those of Kobe (1957) or Souther and White (1977).

Writing facility and the development of a readable personal style result from a great deal of reading, including good literature, and a great deal of writing practice. No matter how much experience the technical writer has, and no matter how small the writing task is, the same simple rules should be followed: make sure of exactly what it is you want to say, make an outline, and *plunge right in*. Do not sit and stew and stare off into space; once you force yourself to start, the job rapidly seems to get easier, and you can always go back to rewrite your awkward beginnings. In fact, you should check and, if necessary, rewrite your material several times, paying particular attention to syntax, grammar, and spelling. If you want to create an immediate poor impression of yourself and your work, leave in all those spelling and grammar errors. If these continue to be a problem to you, you had better think seriously about a night school brushup. Under no circumstances should you expect secretaries to do a repair job on your writing, but, with reports in particular, it is a good idea to have a colleague review your work.

There is an almost endless variety of written communications that the engineer must handle as part of project management. In addition to formal technical reports, progress reports must be submitted at frequent intervals to keep managers apprised of the current status of projects. Each working day brings new requirements for written memoranda, or business letters, or operating manual assembly. Occasionally, chemical engineers in the industrial environment do submit material for outside publication, but occasions to do this are less frequent than in the academic world because the results of industrial development work are considered proprietary. Publication is encouraged if the submitted material promotes the corporation's product, service expertise, or image in the world of technology, or if the work is performed under contract with a governmental agency.

Patents

Patents are still another form of written communication, and since projects sometimes result in patentable inventions, engineers are then asked to assemble information leading to the granting of a patent. Patent protection is one of the important noneconomic criteria of project success and as such is considered to be of utmost importance by corporations. Why should this be so? What is the importance of patents?

The patent is an agreement between an inventor and the government which gives the inventor the incentive to disclose the details of the invention "for the long-term benefit of society" (Maynard, 1978). The incentive consists of the inventor's right to exclude others from practicing the teachings of the patent for a period of 17 years from the date of issuance. One is tempted to say that the patent gives the inventor the exclusive right to practice its teachings; as a practical matter this is true, except if there is a "dominating" patent—a previous patent which incorporates teachings without which the more restricted teachings of the subsidiary patent could not be practiced.

If the inventor is an employee of a corporation, and the invention is made in the course of the employee's work for the corporation, the patent rights are transferred (assigned) to the corporation. New employees are usually asked to sign a document assigning their patent rights to the corporation on the day they are hired. This seems to be a reasonable requirement; after all, engineers are hired with the anticipation that their work will lead to the kind of success that a patent represents, and each patent is the culmination of a large investment of capital by the corporation. The corporation, in turn, agrees to pay for the assigned rights at the time the patent is filed—usually the magnificent sum of a symbolic $1, and corporation patent attorneys rarely decline the opportunity to hand over the dollar to the inventor as part of a tiny ceremony, laced with wry remarks. However, many corporations do have the policy of rewarding inventor employees with cash bonuses, and in any case patents are considered to be a prestigious measure of an employee's performance.

Since corporations are assigned the patent rights, it now becomes clearer why corporations consider patents to be so important. The right to exclude others from practicing the teachings of the patent means that the corporation enjoys the advantage of exclusive production for a prolonged period without threat of competition, that it may enjoy the use of the fruit of its research and development investment. Society enjoys long-term benefits in that the secrets of the invention are made public, and other entrepreneurs may use the know-how after 17 years.

There are several patent types, but those of primary interest to chemical engineers are:

1 Composition of matter patents. These cover new types of products, either chemical compounds (perhaps a new organosilicon heat-transfer fluid) or formulations (a novel paint stripper blend).

2 Process patents. These may range from unit operations and processes (a new thin-film-forming technique) to complete production processes (an improved process for synthesizing cyclohexyl chloride from toluene).

3 Machine patents. New processing equipment is included in this category, such as a novel molten salt electrolysis device or a crystal purifier employing sublimation techniques.

4 Article of manufacture patents. These are less commonly considered by chemical engineers; examples are filtration cartridges fabricated from a porous polymer, or plastic sandwich bags.

Many chemical products are marketed under their *trademark* names which are registered with the U.S. Patent Office. Some trademarks of particularly popular products (Lucite, Plexiglas, Teflon) tend to creep into everyday language, but most corporations prefer to keep the names of their products proprietary and insist that registration be acknowledged by proper identification, for example, Styrofoam®.

For an idea to be patentable, it must satisfy three basic elements of patentability:

1 It must be novel. For example, it cannot represent a combination of routine or well-known ideas.

2 It must be unobvious and "unexpected." The best kind of invention is the one that is based on an entirely unexpected or surprising occurrence in the laboratory or workshop.

3 It must be useful, and some attempt must be made to demonstrate its utility (to "bring it to practice"). "Paper" patents are unacceptable, and it is equally unacceptable to make a partial disclosure which fails to mention some key element of the invention, thus rendering the information useless.

Much of the tedious litigation that characterizes the process of advancing a patent from filing to issuance stems from disagreements between the inventor and the Patent Office as to whether the basic elements of patentability have been met.

Chemical engineers come in contact with patents in two ways: they read them and they write them. Reading the patent literature is an important element in keeping current in areas of individual expertise. Patents are not always easy or particularly enjoyable reading, and understanding them may be facilitated by reference to specialized treatises such as that of Maynard (1978). The format of patents is standard. Following introductory material such as title, inventor, assignee, number, and issuance date, the bulk of the text is devoted to the patent "teachings." These are arranged much like any technical report; the invention is introduced, background material is outlined—including issued patents which relate in some way to the subject of the invention—utility is emphasized, and examples may be given which demonstrate the nature of a process or the method of manufacturing a product. The patent text concludes with a series of "claims" which specify in a precise manner the scope of the invention. For example, if a process is the subject of the patent, one or more of the claims might specify the operating temperature. To broaden the coverage of the patent and to avoid possible circumvention, the operating temperature would normally be claimed as lying within a broad range, although a narrower "preferred" range might be specified if, indeed, the narrower range corresponds to more acceptable yield, or corrosion rate, or some other process result.

The language of patents ("patentese") is stilted and archaic-sounding to conform to traditional legalistic requirements, and it is the subject of many a joke or groan of impatience. Similarly, diagrams are drawn in a peculiarly archaic, nineteenth century style, in conformity to long-accepted practice. Many engineers downgrade patents as a source of information, and it is certainly true that a great deal of important know-how is missing in patent write-ups, some of it perhaps deliberately omitted (occasionally in violation of one of the elements of patentability). Nevertheless, patents serve as useful couriers of industrial progress, and even with their inevitable content limitations, they are an indispensable tool in product and process research and development.

The inventor who is employed by one of the corporations in the chemical process industry receives considerable help in writing, filing, and processing patents. Probably the first step in initiating a patent is a search of the patent literature to make sure that the idea has not already been patented or that there are no dominating patents. Particular care must be taken to identify closely related patents which may be cited by the Patent Office for interference. The inventor generally has the responsibility for the patent search. The invention is next written up in plain technical language, and an expert patent processor (possibly a patent attorney) translates the text into patentese; special care is taken to formulate proper claims, and the processor may recommend that the inventor undertake further experimental work to broaden the range of the claims. When completed, the patent is submitted to the Patent Office, and, more often than not, a prolonged tug-of-war ensues of "final rejections" by the government's attorneys, and appeals, rewriting, and refiling by the corporation, before final acceptance and issuance.

One of the patent rights granted to the inventor or assignee is that of *licensing* the patent know-how. Licensing agreements may be negotiated in several ways; the patent may be sold outright for a lump sum to an interested user, or else the user (or users) may be asked to pay *royalties,* either as a fixed fee per pound of product manufactured by means of the licensed process or as a small percentage of the profits.

In some instances corporations make a calculated decision not to patent an invention. For one thing, the invention may be judged to be relatively unimportant, and the fact of the matter is that patents are an expensive undertaking. Another reason for avoiding the patenting process is that the corporation prefers to keep the details of the invention secret. This may happen if it is felt that just the public knowledge of a particular detail may trigger potentially harmful competitive developments, or else the invention may be of such nature that, even though patented, it could be too easily circumvented or "pirated." The circumvention of patented know-how is a frequent preoccupation of the chemical engineer in industry, either in devising processes around the competition's patents or else in writing ironclad patents.

1.3 THE ECONOMIC BASIS OF THE INDUSTRIAL PROJECT

The Scope of Economic Analysis in the Chemical Industry

I don't know of any decision made against economics that turned out to be the right decision.

C. B. Branch
President, Dow Chemical USA

Economic considerations are dominant in charting the progress of an industrial project toward its ultimate fate—commercial success or abandonment. Many a technically successful project has been doomed by unfavorable economics; the most dazzling spectrum of favorable noneconomic criteria is meaningless unless it can be backed with a favorable projection of economic profitability. In broad terms, what then does the economic analysis of an industrial project entail?

Regardless of the size or scope of the project, any economic analysis is based upon two fundamental concepts:

1 The *capital investment,* a "one-time" expense required to transform the project idea into reality

2 The *project net income,* a time-continuous function that represents the funds generated by the project minus the operating costs that the project incurs

The project economic analysis involves estimating the magnitude of these concepts by techniques of various complexity and degree of accuracy, depending upon the nature of the result that is desired or justified by the state of current project know-how. The two estimated items are then combined in some fashion to generate a *criterion of economic performance*. In symbolic form,

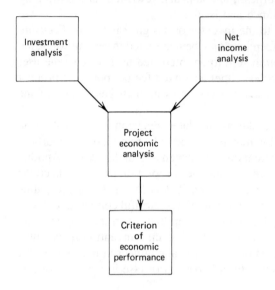

As a simple example of this procedure, let us consider the problem of choosing the insulation for a steam pipeline, a problem reasonably familiar to chemical engineering students. The idea here is to choose the appropriate insulation thickness. Now, there may be situations where the choice is made on the basis of some noneconomic criterion—perhaps the thickness of insulation is limited by spatial considerations, or else by the primary desire to avoid a burning injury to workers in the area. The usual choice, however, is based upon an economic analysis.

We start by making an investment analysis. Such analysis may include the choice of the type of insulation and weatherproofing, both of which are usually picked on the basis of some sort of economic benefit. Ultimately, a relationship between the insulation thickness and the investment for a specific job (or per 100 feet of pipeline) is developed; the investment includes not only the cost of the insulation but all the installation costs as well.

The next step is the income analysis. The funds generated by the project are equivalent to the heat saved by the insulation relative to the heat lost in the absence of insulation. Heat-transfer computations need to be performed for varying insulation thickness, and a reasonable cash value must be placed upon the heat (for instance, the cost of generating a unit amount of steam). Some time-continuous operating costs are associated with the use of the insulation; for example, the insulation will have to be properly maintained. As we will see, maintenance costs are usually estimated as a certain percentage of the initial investment per year.

An economic analysis may now be performed to arrive at a criterion of economic performance. The method is illustrated schematically in Figure 1.3.1. The annual value of the heat saved is plotted as a function of the insulation thickness, and the annual fixed charges are similarly plotted. Finally, the difference between the two, the net income, is computed and plotted. The criterion of economic performance in this instance is the *maximum net income,* and this then leads to the choice of the proper insulation thickness. Other, more sophisticated criteria of economic performance may be employed, as we will see, but the simple criterion illustrated here is very frequently used in the *economic optimization* of equipment design or operating schemes. Other well-known examples of the application of economic optimization are the sizing of pipelines for fluid flow and the choice of the optimum reflux ratio in distillation column design.

Actually, economic optimization may be considered to be a special case in an area of economic analysis known as the *analysis of alternatives* (in the case considered, the number of alternatives is infinite). Economic analysis in the chemical process industry may, in fact, be broadly classified into three categories:

1 *The economic analysis of individual production facilities (new).* The scope of the analysis may vary from an individual piece of equipment to a complete projected plant. The usual criterion of economic performance is one of many possible indices of profitability; that is, will the facility make an acceptable measure of profit?

2 *The analysis of alternatives.* Within this category, the principles of the economic analysis of individual production facilities are extended to the comparison of two or more alternatives, criterion of performance being the *highest* index of profitability. The analysis may consist of the economic optimization of a number of discrete or a continuous spectrum of equipment choices; it may involve the comparison of two or more processes to make the same product; or it may involve a decision of the best way to invest money, given several alternatives. A special

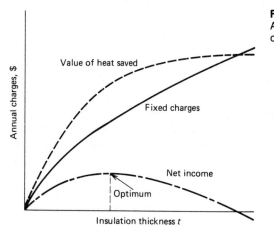

FIGURE 1.3.1
An example of economic analysis: optimum thickness of insulation.

case of the analysis of alternatives is *replacement analysis;* the problem here is to decide whether at some point a piece of equipment, or a whole production facility, should be replaced, maintained, or abandoned.

3 *The analysis of past or projected performance of operating entities already in existence.* The analysis may be applied to an existing operating plant, a corporation, or the whole chemical industry.

The degree of effort and care taken to estimate the capital investment and the projected income depends upon the stage of development of the project. In the earliest stages, rough and relatively rapid estimation methods are usually the only ones justified, but as the project advances toward the stage of final managerial approval and construction, more sophisticated and time-consuming procedures resulting in greater reliability are employed. Several such stages are recognized in making investment estimates (cost estimates); Roth (1979) lists four, Bauman (1964) five. The number is not really important; what is important is to recognize that estimates of increasing reliability are required throughout the lifetime of the project up to the final construction stage, and that, in fact, many more than four or five estimates may have to be made as the economic performance projection is "fine-tuned" during development stages. Following Roth, four approximate levels corresponding to stages of project development may be considered:

1 *Order-of-magnitude estimates.* These are made during the earliest possible stages of project development, just as soon as some reasonable ideas are obtained about likely processing schemes. They are usually based on very little information, perhaps some bench-scale laboratory data, backed by process projections of experienced chemists and engineers and analogy to other similar processes. And yet these estimates are of utmost importance, for even though quite rough, they will usually indicate whether a project is worth pursuing or not. Without them, many a hopeless project needlessly sucks up the corporation's development funds before the truth is ultimately recognized.

2 *Study estimates.* As the name suggests, these estimates are made throughout the middle development stages of the project; as more process information is obtained from the laboratory, from engineering studies, from miniplants and pilot plants, the estimates become increasingly sophisticated, detailed, and reliable. In fact, the results of economic analysis may be used to guide the development effort, perhaps to concentrate upon processing areas which are particularly costly. Regrettably, a great number of projects "bite the dust" during these middle stages as more detailed development and analysis reveal costly difficulties not anticipated earlier.

3 *Authorization estimates.* If the project continues to be a good business prospect, the time may come when the decision is made to proceed with the project implementation, usually plant construction. The capital that is to be invested must be authorized by the appropriate authority; in the case of large, costly production plants, the authority is usually the board of directors. The authorization request is accompanied by an authorization estimate, which is based on the best up-to-date process information and economic analysis. Even though all construction details of the proposed facility are not known at this point, a sufficiently accurate estimate of the required investment must be made so that the authorized funds will meet construction costs without any major "surprises." It is at this point that the estimators' skills are put to the "acid test."

4 *Project control estimates.* These are estimates based upon detailed drawings and equipment inquiries; as pointed out by Roth, their purpose is "to provide an accurate document against which to control expenditures" during construction.

Who is responsible for making these estimates? Basically, it is the chemical engineer with project management responsibilities. Depending upon the magnitude of the project, the responsible engineer will get help from various sources. In the earliest project stages, estimates may be made with the cooperation of bench-level chemists; during the middle stages, economic planning specialists make important study estimate contributions; project control estimates, particularly for large production plants, are handled by cost engineering specialists, but always with input from individuals with developmental experience, and with the project manager in control.

A complete spectrum of techniques is available for carrying out project economic analyses, ranging from educated guesses through an astounding assortment of shortcuts and approximations to a careful item-by-item cost accounting. As we explore these different techniques, the student may be puzzled as to which technique should be applied at a particular level of project development. There is little question that educated guesses belong in the order-of-magnitude estimates and that project control estimates will require quite a bit of item-by-item accounting. But what of the middle stages of project development?

The answer is that the engineer must learn to use proper *judgment* based on *experience,* and that the shortcut technique that best fits a particular situation must be used. For example, shortcut methods are used by cost specialists to estimate piping costs even for project control estimates; proper shortcut methods have been found by experience to be quite accurate, and the usual judgment is that the small error incurred is more than compensated by savings relative to a laborious item-by-item piping cost estimate. At any stage of development, the more time-consuming and accurate techniques should be used when the situation requires it and the backup information is known, but there is little point in wasting time with the same technique when only a rough answer is good enough. In what follows, many techniques will be explored without reference to just when and how the technique should be used; this will be left to the student's judgment.

Many of the shortcut methods that will be described involve the use of "factors" to determine the value of a cost category from that of some other category previously determined. For example, maintenance labor costs in a projected plant are often estimated as a certain percentage of previously estimated operating labor costs; the cost of instrumentation is taken as a particular multiple of the purchased equipment cost. The factors used to make these kinds of estimates are based upon numerous statistical correlations of the costs of completed plants, usually large production plants. Again, judgment must always be used in applying such shortcut factors. The author is aware of a particular case where the usual factor was used to estimate the cost of instruments in a pilot plant. To the complete embarrassment of the estimator, the actual instrument costs turned out to be about four times as high as estimated. Why? Because the usual purchased equipment-based factor is valid for large-scale production plants. In pilot plants, the small equipment results in smaller equipment costs, but *the instruments are not smaller*—a temperature-control loop costs the same in the pilot plant as in the large production plant, and the relative instrument cost is thus much higher. The estimator did not use proper judgment in choosing a shortcut method; perhaps a count of the total number of instrument loops and an average cost per loop would have yielded a better estimate.

Many large corporations develop their own methods and factors for making cost estimates, based upon the statistical correlation of their own past costs. The factors given in this text should therefore never be considered as absolute, rock-firm values; they are, however, representative and adequate for the purpose of illustrating the principles involved.

The Nature of Money and Capital

It is the purpose of the economic analysis of existing or proposed projects to measure their profitability. The yardstick which is ultimately used to measure the degree and magnitude of economic success is some unit of currency—the dollar, the pound, the peso, the yen. Our purpose here is not to discuss the macroeconomic nature of money; we simply accept the economists' judgment that the *function* of money is to serve as a unit of value and as a standard of deferred payment (Chandler and Goldfeld, 1977). Money is the universal exchange medium, but regrettably it has a poorly defined inherent value, a source of considerable confusion and frustration to the engineering economist. Among all the frustrations, a phenomenon which makes the definition of the value of money particularly difficult is that of *inflation*. The most carefully evaluated analysis of the capital investment required for a particular project is obsolete almost at the moment of completion, unless couched in such format as to account for the effect of inflation.

Attempts are continually made to define the effect of inflation upon the unit value of currency in a quantitative manner; one such evaluation that has received wide acceptance is the Consumer Price Index, published monthly by the U.S. Bureau of Labor Statistics. The index measures the average *change* in prices of a certain defined spectrum of consumer goods and services purchased in some 60 locations throughout the United States. The base year for the index is 1967, for which the index is arbitrarily taken as 100. In July of 1977 the index stood at 182.6; this means that the spectrum of consumer items considered increased in price by 82.6 percent since the base year. The index does not define the inherent value of the dollar, but it does indicate changes in the unit value.

A few values of the index are listed in Table 1.3.1. Not all the consumer categories incorporated into the index increase in price at the same rate relative to each other; for example, medical care (since 1967) has increased in cost much more rapidly than apparel. The cost of building chemical process facilities has increased

TABLE 1.3.1
A FEW SELECTED VALUES OF THE CONSUMER PRICE INDEX*
(1967 = 100)

1950	72.1	1965	94.5	1976	170.5
1955	80.2	1970	116.3	1978	195.4
1960	88.7	1975	161.2	1980	246.8

*From "Statistical Abstracts of the United States: 1979," U.S. Department of Commerce.

at its own particular rate, and some of the construction and equipment cost indices will be explored in a later chapter. To the engineering economist, the facts of inflation have severalfold significance:

1 The dollar cost of projected facilities is continually on the increase. Any delay in construction means that a higher dollar investment will be required to complete the project, although it is arguable whether any increase in inherent value is involved.

2 The operating costs associated with production in a completed facility will increase year by year, but the income for the selling of the product will also increase, hopefully in roughly the same proportion.

The average annual fractional inflation rate f may be related to the Consumer Price Index. Let R_{1-0} be the ratio of the indices for years 1 and 0. Then

$$R_{1-0} = 1 + f \tag{1.3.1}$$

by definition.

If f were, indeed, constant, then

$$R_{2-1} = 1 + f \tag{1.3.2}$$

where R_{2-1} is the ratio of indices for years 2 and 1. But

$$R_{2-0} = R_{2-1} \times R_{1-0}$$

so that

$$R_{2-0} = (1 + f)^2 \tag{1.3.3}$$

and by induction,

$$R_{n-0} = (1 + f)^n \tag{1.3.4}$$

or

$$f = R_{n-0}^{1/n} - 1 \tag{1.3.5}$$

Now, of course, **f** is *not* constant—it changes from year to year—but an *average* annual rate over a period of n years may be obtained from Eq. (1.3.5). For example, the equation may be used to obtain **f** between 1970 and 1975 from the data in Table 1.3.1:

$$\mathbf{f} = (161.2 \div 116.3)^{1/5} - 1 = 0.0675$$

that is, the average annual inflation rate in the United States during that period was 6.75 percent. The corresponding inflation rate in some other countries is shown in Table 1.3.2. The inflation rate in the United States compared favorably with that of other major industrial nations; during that same period, the rate varied from 0 percent in countries with rigidly controlled state economies to the catastrophic values experienced in some South American countries. The rate of inflation in the United States increased disturbingly at the end of the decade of the 1970s, with the dread specter of double-digit inflation becoming reality.

The problem of widely differing inflation rates in different countries is a particularly challenging one to the engineering economist with the task of evaluating overseas projects. Inflation indices for some of the major industrial countries appear frequently in the journal *Engineering Costs and Production Economics*.[1] It has been occasionally suggested that difficulties with currency exchange and inflation could be eased by using a more stable entity as a unit of value rather than some currency, perhaps a unit of energy (joule). The rapid rise of the value of energy relative to other commodities emphasizes the questionable validity of such an idea.

The term "capital" will be used, in a rather restricted sense, to designate the money that must be invested to bring a project to its desired state of completion. It is important to recognize that the capital associated with a project represents *all* the money invested, not just the money expended for the observable facilities ("direct" expenses). For example, if the project is a full-production plant, the capital investment includes expenses for the purchase of equipment, the wages of all construction workers, the cost of various ancillary facilities, contractors' fees, and many other items which will be more thoroughly investigated in subsequent chapters.

---

[1] Before 1980, name was *Engineering and Process Economics*.

TABLE 1.3.2
AVERAGE ANNUAL INFLATION RATE, %*
(1970–1975)

United States	6.75	Japan	11.46
Argentina	64.43	Peru	12.60
Canada	7.26	Switzerland	7.71
Chile	208.08	U.S.S.R.	0.00
France	8.88	United Kingdom	12.97
Germany	6.19	Zambia	7.26

*Recomputed from data in "Statistical Abstracts of the United States: 1979," U.S. Department of Commerce.

Sources of Capital

Where does a corporation obtain the capital required for a particular project? In a sense, the situation is not much different from that of an individual wishing to purchase, let us say, a TV set. The set can be purchased with the individual's savings, or else the money can be borrowed and repaid in a variety of ways. Corporations have available to them three principal sources of capital:

1 *Internal sources.* The most common among these are retained earnings.

2 *External financing.* Money may be borrowed in many ways, on either a short- or long-term basis. A common method of raising capital is through the sale of *bonds,* formal promissory notes issued by the corporation, which agrees to pay the buyer the full bond value at maturity and to pay a specified interest rate in the interval up to the maturity date.

3 *Issue of new stock.* The corporation may also raise money through the issue of new stock. The stock may be *preferred* stock, which guarantees the buyer a fixed dividend rate regardless of the current status of the business, and *common* stock, which yields a dividend rate that fluctuates with the corporation's business success. The stock is sold at some issuing price above the *par value,* and the proceeds accrue to the corporation. The *market value* of the stock is established by trading; this value is influenced by the economic performance of the corporation, the dividend rate it pays its stockholders, and perhaps open market trading of its own stock, but otherwise the corporation has no direct control over the market value. In the broadest sense, capital stock represents shares in the proprietary interest in the company, and the shares are represented by stock certificates issued by the corporation.

By far the largest proportion of capital is raised from internal sources, as is clear from Table 1.3.3. On the other hand, the amount of capital raised from new stock issue has, in recent decades, been quite small. There are several reasons for this, one being that in an inflationary period it is more attractive to borrow and to pay off borrowed money with cheap dollars. As for external long-term financing, there are wide differences in corporate policy. There are those corporations that have adopted the policy that long-term corporate financial stability depends upon a low *debt ratio* (ratio of debt to the sum of debt and stockholders' equity). Others feel that it is perfectly sound policy to borrow "up to the hilt," to obtain the capital to invest in projects which will help to dominate specific markets, and to pay off debts

TABLE 1.3.3
SOURCES OF CAPITAL, 15 MAJOR CHEMICAL COMPANIES*

	Percent distribution by year						
	1974	1975	1976	1977	1978	1979	1980
Internal sources	79.0	68.9	72.8	79.1	82.3	85.3	77.5
External financing	19.6	28.8	25.6	17.9	16.8	11.6	17.3
Stock	1.4	2.3	1.6	2.5	0.9	3.1	5.2

Chemical and Engineering News, with permission.

in cheap dollars during inflationary periods. The fact of the matter is that there are financially successful corporations in both ideological camps.

For any particular project in the planning stages, it is not at all clear where the capital to finance it will come from. If the capital were to be borrowed, the cost of the capital (interest) could logically be considered a legitimate operating expense once the project is operational. However, in view of the uncertainty as to the source of capital, we will adopt the policy in this text of *not including interest on invested capital as an operating expense*. This policy is not universally accepted for corporate financial analyses, and the neophyte engineer must establish what the employer's policy is. Interest on borrowed capital will be accounted for in analyzing overall corporate performance (Chap. 12).

The Cost of Capital

In the narrowest sense, interest is compensation for use of borrowed capital. The bank depositor receives interest because the bank, in turn, can use the money for other transactions; the depositor receives compensation for the use of the entrusted money which, of course, remains on nominal deposit, along with accumulated interest, at the depositor's disposition. The borrower receives an agreed-upon principal sum which may be handled as best seen fit; the principal must then be repaid according to some negotiated scheme, along with the interest upon any unpaid balance. The interest is frequently, but not always, expressed as an annual percentage (or fraction) of the amount of cash at the borrower's disposal during 1 year. Thus, for example, if $1000 were to be borrowed for just 1 year at a 10 percent interest rate, at the end of the year the borrower would be expected to pay back $1000, plus 10 percent, or $100, as interest. In the mathematics of finance, interest is expressed as a fraction ($i = 0.10$).

But capital, whether it is borrowed or not, has some inherent *time value* which is not necessarily the same as the going interest rate on borrowed money, for the capital can be invested in various ways to earn various equivalent "interest" rates. For example, the institution that made the 1-year loan of $1000 in the above example may have chosen to invest the $1000 in rare stamps instead, and with luck it could have realized $1800 by selling the stamps at the end of the year; the equivalent interest rate would have been 80 percent. The question is, "Why would the institution ever want to loan money at 10 percent when it could potentially realize 80 percent on stamp investments?" There are two basic reasons:

1 The institution generally prefers to invest its capital in ventures which fall within its range of expertise. The chemical corporation invests its capital in production facilities to make chemicals; it may decide to invest some capital in real estate because of favorable equivalent rates of return, but it had better have the expertise to do it.

2 Each investment has an element of risk associated with it. A banking institution may find the business of lending money relatively safe; it may consider philatelic investments needlessly risky (the stamps could depreciate to $800 in 1 year!).

The time value of money thus is not represented by some absolute interest rate; it is the rate of return that can be reasonably expected by an investor. The chemical corporation, for instance, would consider the time value of money to be equal to the rate of return on its overall investment. To the imprudent investor with savings stashed in a mattress, the time value of money during inflationary periods is negative.

Three categories of interest rates on capital must be distinguished in making economic analyses in the chemical process industry:

1 *Interest on borrowed capital.* The recognized standard is *the prime interest rate,* the rate charged by banks to preferred customers. In recent years the prime interest rate has fluctuated a great deal, as is evident from Table 1.3.4. At the end of the decade of the seventies, the prime interest rate reached the unprecedented figure of 20 percent and higher.

2 *The internal interest rate.* This is the rate at which the company's overall investments are earning money *after taxes.* The distinction, "after taxes," is very important; if the internal interest rate is to be compared with the prime interest rate on borrowed capital, certainly a reasonable undertaking, the prime interest rate must first be adjusted for tax payments. We will see presently how this can be done. We will also see in Chap. 12 how the internal interest rate can be computed from the corporation's annual reports.

3 *Prescribed interest rate of return, new investments.* The internal interest rate is the standard by which new investments are judged, but most new projects must hold promise of a rate of return higher than the internal interest rate. Just how much higher depends upon the perceived risk inherent in the project, and the prescribed rate represents the prerogative and reflects the judgment of the corporation's top management. The prescribed rate also depends upon the available supply of capital,

TABLE 1.3.4
PRIME INTEREST RATE AND INFLATION RATE: DECADE OF THE SEVENTIES*

Year	Average inflation rate, %	Average prime interest rate, %
1970–1971	4.30	7.91
1971–1972	3.30	5.72
1972–1973	6.23	5.25
1973–1974	10.97	8.03
1974–1975	9.14	10.81
1975–1976	5.77	7.86
1976–1977	6.63	6.84
1977–1978	7.66	6.83
1978–1979	11.26	9.06
1979–1980	13.52	12.67

*Based on "Statistical Abstracts of the United States," U.S. Department of Commerce.

and in periods of "tight money" (usually high interest rates), extraordinarily high rates of return may be demanded before a project is accepted.

In subsequent chapters we will talk a great deal about interest and rates of return, and the question may arise as to which of the above three interest rates is under consideration at any particular time. Strictly mathematical relationships are, of course, general; in specific situations, the applicable rate will be evident from the nature of the discussion.

We have seen what the cost of capital raised from external sources is, but what is the cost of capital raised from stock issues, and why has so little capital been raised in this manner in recent times? The answer lies in the tax policies of the government. *Interest* on borrowed capital (bonds, bank loans) is an operating cost, and the government allows it to be deducted (along with other operating costs) from sales revenue to arrive at a *profit before tax*, which is then taxed at a rate of approximately 50 percent (large corporations). *Dividends* on issued stock are *not* an allowable cost. What effect does all this have? Consider the following example:

	Project, borrowed capital	Project, capital from stock issue	Δ
Sales, say	100	100	
Costs, say	−70	−70	
Interest on $100 borrowed (at 10%)	−10	—	
Gross profit	20	30	+10
Tax (50%)	−10	−15	+ 5
Net profit	10	15	+ 5

In the column marked Δ, the differences between stock issue financing and bond financing are noted for key items. Note that by *borrowing* money, the tax is reduced by $5, and the actual interest, as far as the corporation is concerned, is

$$\$10 - \$5$$

or only $5 *after* taxes are paid; that is, the actual interest rate is 5 percent, and not the nominal 10 percent. The tax structure thus makes the cost of borrowing money quite attractive; the cost of capital raised from new stock issues is considerably more, about 10 percent/year. Thus, even though the net profit is *larger* when stock issue financing is used, a major part of that profit must go to the stockholders. The dividend rate is established by the corporation's directors each year, but it must be set high enough to make new stock issues an attractive investment.

NOMENCLATURE

A volumetric flow rate of aqueous phase in extraction, m³/h
c concentration of transferring species, kg·mol/m³; c_O, concentration in organic phase; c_A, concentration in aqueous phase

f   average annual fractional inflation rate
K distribution constant of transferring species in liquid-liquid extraction
n index symbol, year
O volumetric flow rate of organic phase in extraction, m^3/h
R_{j-k} ratio of Consumer Price Indices for years j and k

REFERENCES

Anonymous: "Management of New Products," Booz, Allen, and Hamilton, Management Consultants, 1968.

Anonymous: Careers in Chemical Engineering. Interviewing and Résumés, *AIChE Student Members Bull.*, **14**(2)(Fall 1973).

Bauman, H. Carl: "Fundamentals of Cost Engineering in the Chemical Industry," Reinhold, New York, 1964.

Billmeyer, F. W., Jr., and R. N. Kelley: "Entering Industry—A Guide for Young Professionals," Wiley, New York, 1975.

Bondi, A.: The Price and Rewards of Continuing Education, *Chem. Eng. Prog.*, **65**(10): 27(October 1969).

Chandler, Lester V., and Stephen M. Goldfeld: "The Economics of Money and Banking," 7th ed., Harper & Row, New York, 1977.

Davidson, Robert L.: "Effective Communication for Engineers," McGraw-Hill, New York, 1974.

Dean, Burton V.: "Evaluating, Selecting, and Controlling R&D Projects," American Management Association, Inc., 1968.

Eisenberg, A. E.,: "Effective Technical Communication," McGraw-Hill, New York, 1982.

Fallwell, William F.: 1979 R&D Budgets up 10% at Chemical Firms, *Chem. Eng. News*, p. 12 (Jan. 15, 1979).

Harris, John S.: New Product Profile Chart, *Chem. Technol.*, **6**:554, September 1976.

Heyel, Carl (ed.): "Handbook of Industrial Research Management," Reinhold, New York, 1959.

Ireson, William G., and Eugene L. Grant (eds.): "Handbook of Industrial Engineering and Management," Prentice-Hall, Englewood Cliffs, N.J., 1955.

Kern, Kenneth R. (ed.): "Corporate Diagrams and Administrative Personnel of the Chemical Industry," Chemical Economic Services, Princeton, N.J., 1958.

Kobe, Kenneth A.: "Chemical Engineering Reports," 4th ed., Interscience, New York, 1957.

Kohn, Philip M., and Roy V. Hughson: Perplexing Problems in Engineering Ethics, *Chem. Eng.*, **87**(9):96(May 5, 1980); (19):132(September 22, 1980).

Ludwig, Ernest E.: "Applied Project Management for the Process Industries," Gulf, Houston, 1974.

Mannon, James H.: US Engineers' Salaries: Gains Keep Inflation at Bay, *Chem. Eng.*, **86**(4):123(February 12, 1979).

Matley, Jay, and Larry J. Ricci: New Chemical Engineers: Too Many, Too Soon, *Chem. Eng.*, **86**(2):95(January 15, 1979).

Maynard, John T.: "Understanding Chemical Patents," American Chemical Society, Washington, D.C., 1978.

Moder, Joseph J., and Cecil R. Phillips: "Project Management with CPM and PERT," Reinhold, New York, 1964.

Rosenzweig, Mark D.: Cleanup Efforts Burden R&D Budgets, *Chem. Eng.*, **85**(19):54 (August 28, 1978).

Roth, Joanne E.: Controlling Construction Costs, *Chem. Eng.*, **86**(21):88(October 8, 1979).

Sarin, R. K., A. Sicherman, and K. Nair: Evaluating Proposals Using Decision Analysis, *IEEE Trans. Syst., Man Cybern.*, **SMC-8** (2):128(February 1978).

Souther, James W., and Myron L. White: "Technical Report Writing," 2d ed., Wiley-Interscience, New York, 1977.

St. Clair, J. B.: Ethics and Engineering Practice: A Two-Way Bridge, *Chem. Eng. Prog.*, **75**(8):27(August 1979).

Valle-Riestra, J. F.: The Dual Ladder System: A Critique from the Technical Side, *Chem. Eng. Prog.*, **76**(2):12(February 1980).

Wei, J., T. W. F. Russell, and M. W. Swartzlander: "The Structure of the Chemical Processing Industries," McGraw-Hill, New York, 1979.

Weismantel, Guy, and Jay Matley: Is Your Résumé Junk Mail?, *Chem. Eng.*, **81**(24):164 (November 11, 1974).

Williams, Howard O., Jr.: 'How' Will Help You Make It to the Top, *Chem. Eng.*, **86**(17):147 (August 13, 1979).

PROBLEMS

1.1 Write a personal résumé by using the guidelines in Exhibit 1.2A. If you already have a résumé, review it critically, expand if necessary, and bring it up-to-date.

1.2 Condense your personal résumé into a one-page document intended for use during a campus interview.

1.3 Obtain information from sources available to you about Celanese Corporation [for example, book on CPI by Wei, Russell, and Swartzlander (1979), or corporate brochures in the university placement office]. Write a letter of transmittal to the manager of corporate hiring in which you ask to be interviewed for a position, and emphasize some of your marketable qualifications; be sure that your letter demonstrates some recognition of the nature of Celanese Corporation.

1.4 Devise a method of your choice to generate a numerical rating of a project, based on noneconomic criteria. What are some of the strengths and weaknesses of your method? Does it properly reflect the sense of the sequence of profile charts in Figure 1.1.5?

1.5 Prepare a 10-minute classroom presentation outlining the nature and objectives of the American Association of Engineering Societies. Check back issues of *Chemical Engineering Progress* for reference material to get you started.

1.6 During the past year the common stock of a large pharmaceutical corporation has fluctuated between $25 and $30 a share, and the dividend rate was $1.50 per share for the year. The ratio of dividends to stock price has held constant for a number of years. The corporation wishes to raise new capital by either floating a new common stock issue at $25 per share or by selling bonds. What interest rate should the bonds carry to guarantee the same net profit on the newly raised capital after all interest or dividend payments?

1.7 Obtain a process patent of your choice that had been issued in 1980. Write a short report-form summary that "translates" the teachings and claims into usable *design* information.

1.8 From the data in Table 1.3.1, make a reasoned projection of the Consumer Price Index in 1994. Comment on the reliability of your projection. What would you anticipate the median starting salary of M.S. chemical engineers to be at that time?

1.9 Use a method analogous to the product profile chart of Harris to establish a graphical profile of two universities you know well. Establish your own rating criteria, and make a comparison of the two.

1.10 Investigate the current status of engineering registration and certification in your state or in a foreign country of your choice. Write a two-page summary of your findings.

1.11 Review the meaning of the operators Σ and Π. Derive an expression for the fractional inflation rate **f**, averaged over n years, in terms of the known values of the annual rates f_i for each of the n years. Use your result to compute the value of **f** for the 10-year period shown in Table 1.3.4. Also calculate **f** for the period 1970–1975; does this value agree with one calculated from Eq. (1.3.5) and the data in Table 1.3.1?

1.12 List some of the ethical problems that may be encountered in the university. Organize a classroom discussion of these problem areas.

1.13 2268 liters/h flow of an aqueous solution of metal M^{2+} is to be extracted with an organic, water-insoluble carrier in a Karr column (reciprocating perforated plates). The equilibrium between the organic and aqueous phases is given by the expression

$$K_{O-A} = c_O/c_A = 20.0$$

where c is in g · mol/liter.

The initial concentration of M^{2+} in the aqueous phase is 0.1 g · mol/liter. The desired concentration of the organic extract is 1.5 g · mol/liter, and the stripped organic phase is recycled at a level of 0.1 g · mol/liter.

Based on the cost data supplied and a criterion of economic performance of your choice, recommend how many actual reciprocating plates should be used in the Karr column.
Data:

Five reciprocating plates are required per theoretical extraction stage.

The differential cost (complete, installed) of each reciprocating plate is $2000.

The differential operating cost of the unit is 24 percent of the differential installed cost per year ("fixed" charges).

At this point of the process, the metal is valued at 1.4 cents/g · mol.

Stripping costs of the organic phase are independent of the amount of metal extracted when the recovery from the aqueous phase exceeds 50 percent.

1.14 A petroleum company wishes to invest $20 million for expansion of existing refining capacity. Management has agreed that 50 percent of the investment capital is to be derived from its own capital reserves. Two schemes have been suggested for raising the balance of the capital:

Scheme *A*: 25 percent from bonds at 8 percent interest
25 percent from preferred stock at 5 percent equivalent interest
Scheme *B*: 50 percent from preferred stock at 5 percent

If management wishes to maximize growth of capital reserves, which scheme would you recommend? Combined federal and state income tax rate = 50 percent.

1.15 The prime interest rate may be considered a criterion of the time value of money, the percent increase in total value each year. Using the method of Prob. 1.11 (solve Prob. 1.11 if you have not already done so), obtain an average value of the prime interest rate over the 10-year period in Table 1.3.4. The result cannot be considered as corresponding to an increase in total value, since it does not account for inflation. How would you correct the result to account for inflation? If you invested your savings at the prime interest rate, would your accumulated capital increase its purchasing power?

1.16 A divisional R&D department has 193 professional employees at the end of 1980, with a historical growth of 3 percent/year during the seventies. The research director is asked to submit a preliminary operating budget for 1985. What do you think it should be? How much new capital investment should the 1985 R&D effort generate?

CHAPTER 2

THE MATHEMATICS OF FINANCE

Economists love to draw curves.

Martin Gardner[1]

Before we can undertake the economic analysis of projects, we must first examine the tools which we will use to accomplish the task. The necessary mathematical tools are not difficult to learn; in fact, most chemical engineering students are exposed to them by the end of the second year of their undergraduate curriculum as part of their study of mathematical analysis and the calculus. The mechanics, then, may be mastered rather rapidly, but even more important is the acquisition of an appreciation of the significance of the concepts that the mathematical manipulations represent. The mathematically sophisticated may be unsatisfied by the superficial manner in which some of the mathematical subject matter is presented. There are, however, good reasons for this. The material represents a mix of selected subjects, chosen for their specific utility from a number of identifiable mathematical disciplines, and a logical and rigorous development is not really practical. Moreover, the objective is not to learn mathematical theory or to develop deductive acuity, but rather to demonstrate how accepted methodology may be applied to the specific goal of economic analysis.

[1]*Scientific American*, vol. 245, no. 6, December 1981.

2.1 THE MEASUREMENT OF INTEREST

Classification by Compounding Method

In the first chapter we explored some extended definitions of the term "interest rate." It was pointed out that basically interest is the cost of borrowing capital, but we saw that the income generated by investing money in any manner can be expressed as equivalent interest. Moreover, we saw that the interest rate is often expressed as a fraction of the investment (or loan outstanding) per year.

Let us agree, for the moment, to restrict our discussion of interest to the financial manipulation of depositing a certain sum of money in a banking institution. We refer to the sum with which the financial transaction is started as the *principal* (mathematical symbol P). The principal is allowed to accumulate interest charges to produce the *accumulated sum* S. If the interest rate $\mathbf{i}$ is defined as a fraction of the principal per year, then at the end of the first year,

$$S_1 = P(1 + \mathbf{i}) \tag{2.1.1}$$

We may now examine the classification of interest rates based on the concept of *compounding*. With the same definition of $\mathbf{i}$, the accumulated sum S_1 at the end of the first year, if it is allowed to remain on deposit for still another year, draws interest on the *total* S_1—both the principal P and the first-year interest $P\mathbf{i}$. Thus

$$S_2 = S_1(1 + \mathbf{i}) \tag{2.1.2}$$

and if the two equations are combined,

$$S_2 = P(1 + \mathbf{i})^2 \tag{2.1.3}$$

The interest drawn during the first year is

$$I_1 = S_1 - P = P\mathbf{i} \tag{2.1.4}$$

The interest drawn during the second year is

$$\begin{aligned} I_2 &= S_2 - S_1 \\ &= P(1 + 2\mathbf{i} + \mathbf{i}^2) - P(1 + \mathbf{i}) \\ &= [P(1 + \mathbf{i})]\mathbf{i} \end{aligned} \tag{2.1.5}$$

That is, the second-year interest is drawn on both the principal and the accumulated first-year interest; the interest is *compounded,* and clearly $I_2 > I_1$.

The following classification of interest rates is based on the specific manner of compounding:

1 Simple interest
2 Discrete compound interest
3 Continuously compounded interest

Each will be explained in turn.

Simple Interest

Simple interest, as the name implies, is not compounded; the annual interest is on the principal amount only and is the same for every year the principal continues to be on deposit. We use the symbol i_s for simple interest, and, for any one year, the interest drawn is

$$I = Pi_s \tag{2.1.6}$$

so that n years after the deposit is made,

$$S = P(1 + ni_s) \tag{2.1.7}$$

Simple interest is occasionally paid on short-maturity-period securities such as U.S. Treasury bills.

Discrete Compound Interest

Discrete compound interest refers to interest that is compounded at regular (discrete) intervals. If the intervals are 1 year, the interest is termed *discrete annual compound interest;* as we will see, all other classifications of interest rate are compared with this interest classification as the comparison basis. We reserve the boldface symbol **i** for discrete annual compound interest.

Continuing the sequence represented by Eq.(2.1.1) and (2.1.3),

$$S_3 = S_2(1 + \mathbf{i}) = P(1 + \mathbf{i})^3 \tag{2.1.8}$$

and by extension to n years,

$$\boxed{S = P(1 + \mathbf{i})^n}^* \tag{2.1.9}$$

The student may wish to show (Prob. 2.3) that

$$I_n = P(1 + \mathbf{i})^{n-1}\mathbf{i} \tag{2.1.10}$$

*The notation S_n is dropped in favor of S when the subscript is redundant.

Example 2.1.1

What discrete annual compound interest rate will result in a 10-year doubling of money? Referring to Eq. (2.1.9), $S/P = 2$; therefore

$$i = 2^{1/10} - 1 = 0.0718 \quad \text{or} \quad 7.18\%$$

The compounding of interest need not be annual; any regular interval may be used. For example, quarterly compounding of interest (i.e., every 3 months) is often encountered; monthly and even daily compounding are proudly advertised by banking institutions. The equations for the accumulated sum and the interest drawn are similar in format to Eq. (2.1.9) and (2.1.10), but modifications must be made to account for certain differences. We reserve the symbol n for the number of annual intervals and use m to indicate other specified discrete intervals, such as months. The symbol i_m is the discrete compound interest rate *for the time interval of m;* for example, if m stands for number of months, then i_m is the discrete monthly compound interest rate. The working equations then become

$$S = P(1 + i_m)^m \tag{2.1.11}$$
$$I_m = P(1 + i_m)^{m-1} i_m \tag{2.1.12}$$

i_m, however, is rarely quoted as such; that is, banking institutions do not quote quarterly or monthly interest rates. The custom is to use the *nominal* annual interest rate i, where

$$i = i_m \cdot p \tag{2.1.13}$$

and p is the number of intervals under consideration in 1 year.

Example 2.1.2

A deposit of $1000 draws interest compounded quarterly at a nominal interest rate of 10 percent. What is the accumulated sum at the end of 1 year?
Here $p = 4$, and $i_m = 0.10/4 = 0.025$. Therefore, for $m = 4$,

$$S = 1000(1.025)^4 = \$1103.81$$

In the example, note that S is not the same as it would be with 10 percent discrete annual compound interest rate; i.e.,

$$i \neq i_m \cdot p \quad \text{(Why?)}$$

Continuously Compounded Interest

As the time interval for discrete compound interest is allowed to approach zero, the interest approaches the condition of *continuous compounding*. For continuously compounded interest (or "continuous interest") i_m no longer has any meaning, and the rate is expressed as the nominal annual interest rate i.

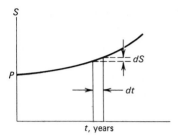

At any differential section of the tS curve,

$$dS = Si\, dt \qquad (2.1.14)$$

$$\int_P^S \frac{dS}{S} = i \int_0^n dt \qquad (2.1.15)$$

$$\ln \frac{S}{P} = in \qquad (2.1.16)$$

or

$$\boxed{S = Pe^{in}} \qquad (2.1.17)$$

$$I_n = S_n - S_{n-1}$$
$$= P(e^{in} - e^{i(n-1)})$$

whence

$$I_n = Pe^{in} \frac{e^i - 1}{e^i} \qquad (2.1.18)$$

Example 2.1.3
What is the accumulated sum after 1 year for a $1000 principal compounded continuously at a nominal annual interest rate of 10 percent?

$$S = 1000e^{0.1} = \$1105.17$$

The interest yield is 1.3 percent higher than with quarterly compounding at the same nominal interest rate.

Effective Interest

Continuous interest has been accepted by most economic analysts as the standard for the economic analysis of projects, but discrete annual compound interest continues to serve as the basis for comparison. There are reasons for this; some are a matter of long-standing tradition and public "feel" for the concept, others relate to the perception of loan repayments, which are by necessity discrete. The economic analyst must learn to use either tool with equal facility. Nevertheless, it is useful to express nominal interest rates (for continuous or nonannual discrete compounding) as *effective* interest rates, that is, discrete annual compound interest rates **i** that would yield the same S at the end of each year.

Let us start by examining the effective interest rate which is equivalent to the nominal interest rate for nonannual discrete compounding.

Consider S_n, the accumulated sum at the end of n years. From the definition of p, the total number of nonannual intervals at the end of n years is $m = pn$. From Eq. (2.1.9),

$$S_n = P(1 + \mathbf{i})^n \qquad (2.1.19)$$

Substituting (2.1.13) into (2.1.11),

$$S_n = P\left(1 + \frac{i}{p}\right)^{pn} \qquad (2.1.20)$$

Equating (2.1.19) and (2.1.20),

$$\boxed{\mathbf{i} = \left(1 + \frac{i}{p}\right)^p - 1} \qquad (2.1.21)$$

As for the effective interest equivalent of the nominal rate for continuous compounding,

$$S_n = P(1 + \mathbf{i})^n = Pe^{in}$$

$$\boxed{\mathbf{i} = e^i - 1} \qquad (2.1.22)$$

Example 2.1.4

What is the effective interest rate for the two cases in Examples 2.1.2 and 2.1.3?
For the quarterly compounding case,

$$\mathbf{i} = \left(1 + \frac{0.10}{4}\right)^4 - 1 = 0.1038 \quad \text{or} \quad 10.38\%$$

TABLE 2.1.1
THE FAMILY OF INTEREST RATES

Classification	Equivalent nominal annual interest rate	Equivalent effective annual interest rate
Simple, i_s	i_s	See Prob. 2.2
Discrete compound, i_m	$i = i_m p$	**i** [Eq. (2.1.21)]
Discrete annual compound, **i**	**i**	**i**
Continuously compounded, i	i	**i** [Eq. (2.1.22)]

For the continuous compounding case,

$$\mathbf{i} = e^{0.10} - 1 = 0.1052 \quad \text{or} \quad 10.52\%$$

It may be well at this point to review our collection of little i's; this is done in Table 2.1.1. In subsequent discussions, unless otherwise stated, we will deal with either discrete annual or continuous interest; the distinction will be made clear with the use of the appropriate symbol: the boldface **i** for discrete annual compound interest, the italicized i for continuous interest.

The Consideration of Inflation

In Chap. 1 we saw that inflation may be characterized by an average annual fractional inflation rate **f**. The value of **f** generally changes from year to year, but an average value may be used to represent the inflationary trend over a period of several years.

Now, we have seen that deposition of a principal P accumulates at the end of 1 year the sum S_1:

$$S_1 = P(1 + \mathbf{i}) \tag{2.1.1}$$

The problem is that S_1 in the above equation should be measured in currency that has the same value as the currency used to measure P; that is, the unit of monetary measurement should be current[1] or *constant-value* money (Rose, 1976). If the money is withdrawn at the end of the year, however, the currency so received is *inflated;* in terms of current money, its actual value is

$$S_1' = \frac{S_1}{1 + \mathbf{f}} \tag{2.1.23}$$

If the above two equations are combined,

---

[1]Inflation terminology raises problems in semantics. In this book, the term "current" (or "uninflated") money refers to money *valued at the zero-time reference point* (usually the present time). In contradistinction, Rose (1976) uses the term for inflated money, in the sense of current (or cocurrent) value at *any* time.

$$S_1' = P\frac{1+i}{1+f} \tag{2.1.24}$$

and by extending the above arguments,

$$S' = P\left(\frac{1+i}{1+f}\right)^n \tag{2.1.25}$$

Equation (2.1.25) expresses the compound interest growth of principal in terms of constant-value money. The relationship clarifies the hard fact that true growth occurs only if $i > f$. Regrettably, this has not been the case with bank deposits in recent years. What appears to be an attractive 7.5 percent long-term certificate of deposit account in a savings and loan association is nothing but money lost during periods of double-digit inflation.

We will repeatedly return to considerations of inflation, particularly in discussing how inflation affects the economic evaluation of projects made in "current" dollars.

Present and Future Worth

We started this chapter by agreeing to restrict our discussion of interest to the financial manipulation of depositing a certain sum of money in a banking institution. The definitions of the various kinds of interest and the derived mathematical expressions do not depend upon the financial manipulation, however; they are completely general. For instance, they apply equally well to the financial manipulation of borrowing money; P, the principal, then becomes the amount of the loan, and S, the accumulated sum, becomes the total owed at the end of a specified period (in the absence of intermediate repayments). Thus if a corporation borrows $1 million at $i = 10$ percent, the balance outstanding at the end of 1 year is

$$1.0 \times 10^6 \, e^{0.1 \times 1} \quad \text{or} \quad \$1.105 \times 10^6$$

Moreover, all the derived expressions hold for the extended meaning of interest implied by the various classifications in Chap. 1. That is, we can speak of an *investment* P which results in a *return* S at the end of n years, and we can look at the return as if P were invested at an interest rate, as in Eq. (2.1.9):

$$S = P(1 + i)^n \tag{2.1.9}$$

Example 2.1.5

A capital investment of $1000 results in a net return (before taxes) of $1500 after 1 year. What is the interest rate i on the investment?

$$1500 = 1000(1 + i)$$

and

$$i = 0.50 \quad \text{or} \quad 50\%$$

This would be an exorbitantly high interest rate on borrowed money, but a perfectly possible interest rate on a high-risk investment.

But the relationship in Eq. (2.1.9) has even broader implications, for it expresses in mathematical form the very concept of the *time value of money*. Within this context, P becomes the *present worth* of some future income or expenditure S. The concept of the time value and present worth of money is a key one in the economic evaluation of projects, one that we will refer to repeatedly. Of course, the question immediately arises as to what value of **i** (or *i*) is to be used in assessing the present worth. The answer is that it depends upon the nature of the financial problem under discussion. For example, if the discussion involves a projected new investment, **i** may be the prescribed interest rate of return mentioned in Chap. 1. The bank depositor concerned about the present worth of an accumulated sum 5 years hence would use a value of **i** corresponding to the interest rates on deposits advertised by the banking institution. (This usually is well below the prime interest rate. Why?) The previously mentioned unfortunate individual who prefers to stuff savings into a mattress may safely use the value $i = 0$ for computing the present worth of the mattress stuffing.

This last example serves to underscore an important characteristic of the present worth concept—it does *not* account for inflation. Present worth is based on a firm projected value of interest, not on an uncertain projection of inflation rates. The present worth is thus expressed in terms of noncurrent dollars, and we will see how this fact is accounted for in making our project economic evaluations. If desired, the present worth may be expressed in current dollars by using a modification of Eq. (2.1.25):

$$P' = S \frac{1}{[(1 + i)(1 + f)]^n} \tag{2.1.26}$$

where the primed quantities are expressed in current dollars.

Exercise 2.1.1

Equations (2.1.25) and (2.1.26) are somewhat different in form. Explain the difference, and derive Eq. (2.1.26).

Example 2.1.6

A new project is expected to yield a rate of return on investment of **i** = 15 percent. What is the present worth of income of $1000 five years from now?

$$P = S(1 + i)^{-n} = 1000(1.15)^{-5} = \$497.18$$

Note that the present worth is less than the future sum, and the higher the anticipated rate of return, the smaller it is. We will see that the present worth concept is a powerful one in comparing the worthiness of investments, and the following little example serves as a preview.

Example 2.1.7

The same investment may be made in two alternate projects: one will yield an income of $1000 ten years from now, the other $2500 twenty years from now. If the time value of money to the investor is $i = 0.10$, which project results in the larger present worth?

First project: $\quad P = Se^{-in} = 1000e^{-10 \times 0.1} = \$367.88$
Second project: $\quad P = 2500e^{-20 \times 0.1} = \$338.34$

We can also speak of the *future worth* of an investment, or expenditure, or income; the future worth is identified with the term S in the compound interest equations. Thus if money is worth 10 percent ($i = 0.10$), the future worth of $1000 ten years from now is $P(1 + i)^n = 1000(1.1)^{10} = \$2593.74$. Sometimes the present worth and future worth concepts can be confused, and the true significance must be kept in mind. For example, we can speak of the present worth of a past expenditure, but in this case the present worth is identified with the term S (why?).

Example 2.1.8

What is the present worth of $100,000 invested 2 years ago into an R&D project? Use $i = 0.14$.

$$S = 100{,}000\,(1.14)^2 = \$129{,}960$$

With the extended definition of interest rate discussed in this section, it will be recognized that a negative interest rate may exist; this occurs if $S < P$.

The engineering economics literature frequently refers to the $(1 + i)^n$ portion of Eq. (2.1.9) as the "discrete single-payment compound-amount factor," and the reciprocal $(1 + i)^{-n}$ as the "discrete single-payment present-worth factor." Similar factors are defined for the case of continuous compounding. Many handbooks and textbooks are replete with tabulated values of these factors, but in this age of the hand-held computer this seems a bit superfluous.

2.2 THE PRESENT WORTH OF CASH FLOWS

What is Cash Flow?

The history of a project is characterized by a series of *cash flows* into and out of the project, the movement of hard cash across the boundary of the "black box" which represents the project. Outflowing cash may be the investment itself, out-of-pocket everyday expenses, or tax payments incurred by profits from the project;

incoming cash is generally the revenue from selling the product generated by the project. For the sake of consistency, cash outflow is considered as negative, cash inflow as positive. Sometimes when the term "cash flow" is used, the *net* cash flow is implied, i.e., the incoming cash with expenses and taxes already subtracted. This is certainly the meaning when we hear corporate management speak of methods to "generate more cash flow."

All cash flows associated with a project have some present worth, based upon a specified time value of money. Just as is the case with the compounding process itself, cash flows can be *discrete* or *continuous,* or, at least, this is an easy way of classifying them. In fact, the cash flows associated with a project are usually something in between; they occur in fits and starts throughout the year, not as a continuous uniform ooze, and almost certainly not as a single halcyon event at the end of the year. Nevertheless, this is the way cash flows are characterized—the discrete cash flow occurs at the *end* of each year following time zero (until project termination), at the *beginning* of each year preceding time zero (if applicable), and precisely at time zero, the basis of the "present worth" (which, incidentally, need not be the same thing as "right now"; zero time may be set anywhere on the time coordinate). Continuous cash flow is generated in the form of a continuous function with no time discontinuities. We use the symbol C_k to indicate discrete cash flow at the end of year k, and $\overline{C}_t$ (with an overbar) to indicate continuous cash flow, represented as an instantaneous annual rate at time t.

We have already looked at both discrete and continuous compounding of a single discrete cash flow, Eqs. (2.1.9) and (2.1.17), respectively. The equations, as written, yield the future worth S of a zero-time cash flow P (or C_0). The reciprocal process of computing the present worth P of a future cash flow C_n ($= S$) (this process is also termed *discounting*) can be easily deduced from the same equations.

But a project is more than a single cash flow occurrence, and we will explore methods of computing the present worth of any multiplicity of cash flows. Three combinations of cash flow and discounting (or compounding) may be recognized:

1 Continuous discounting of continuous cash flows
2 Discrete discounting of discrete cash flows
3 Continuous discounting of discrete cash flows

Exercise 2.2.1

The discrete discounting of continuous cash flows is not listed. Point out why such listing would be redundant.

A combination of discrete and continuous compounding is normally not considered within the scope of a single project.

Net Present Worth

Each individual cash flow associated with a project, whether positive or negative, may be converted to its present worth. The arithmetic sum of all the present worths is the *net present worth* (NPW) of the project (also called the *net present value*, NPV). A few examples will serve to illustrate the significance of the concept.

Consider a project with the following projected cash flows:

Time	Cash flow, $
Zero time	− 1000 (initial investment)
End of year 1	+ 500 ⎫
End of year 2	+ 400 ⎬ net positive cash flows: net income after all expenses and taxes paid
End of year 3	+ 300 ⎭
Project terminated	

If money did not have time value ($i = 0$), the NPW would be $-1000 + (500 + 400 + 300) = +\$200$. Clearly, the project does make money, but we wish to ask ourselves whether the money could be better invested, and to do this we must see whether the project satisfies some minimum criterion of rate of return. For example, we might ask whether a better return could be realized by lending the money for a period of 3 years at a before-tax interest rate **i** of 14 percent. Or else we might establish whether the investment matches management's criterion of minimum rate of return of $i = 10$ percent (after taxes).

To check the latter proposition, we determine the NPW at a time value of money of 10 percent. We do this by applying Eq. (2.1.9) to each cash flow to discount it to zero time. Thus

$$\text{NPW} = -1000 + 500(1.1)^{-1} + 400(1.1)^{-2} + 300(1.1)^{-3}$$
$$= +\$10.52$$

The net present worth is larger than zero. What does this mean? If money is indeed worth 10 percent, then the project has a favorable NPW; the project is *better* than if the initial $1000 were invested for 3 years at a 10 percent rate of return.

If the NPW of a project is just zero, then the cash flow spectrum *just matches* the expected time value of money. A trivial example of this is the financial transaction of borrowing money and then paying it off at 10 percent interest. From the borrower's point of view, the time value of money is, indeed, 10 percent, and the NPW of the transaction to the borrower is zero. Similarly, if management stipulates a 15 percent rate of return minimum on new investments and the NPW of projected cash flows of a new project is zero, then the requirements of management are just met. It is useful to keep in mind that a zero NPW is not "bad," that it is not a basis for judging the project to be "worthless"; it simply means that the criterion of the time value of money, whatever it may be, has been met, but only barely, and a positive NPW is certainly preferable, less subject to the vagaries of risk and forecasting uncertainties.

As a further extension of the NPW concept, if the NPW of a contemplated or existing project is negative, the cash flow spectrum does not satisfy the stipulated criterion of rate of return.

Example 2.2.1

Two years ago a sum of $100,000 was budgeted to support research into the optimization of the operating conditions in an existing plant. The research has been completed, and it is expected that the work will result in annual savings of $47,000 for the next 3 years, at which time the plant will be completely rebuilt. For $i = 10$ percent, was the project justified?

$$\text{NPW} = -100,000(1.10)^2 + 47,000[(1.10)^{-1} + (1.10)^{-2} + (1.10)^{-3}]$$
$$= -\$2875$$

In this particular case, the NPW is negative, and so the project was not economically justified. Note that the NPW concept is an important *criterion of economic performance* which we discussed in the first chapter. Just like many other such criteria, this one is dependent upon the time value of money that is acceptable in a particular case.

Example 2.2.2

At what time value of money (i) will the project in Example 2.2.1 be justified?
For this case, we choose i so that NPW = 0. Thus

$$0 = -100,000(1 + i)^2 + 47,000[(1 + i)^{-1} + (1 + i)^{-2} + (1 + i)^{-3}]$$

By trial and error, $i = 8.94$ percent.

Example 2.2.3

Suppose the $100,000 for research support in Example 2.2.1 had been invested only 1 year ago and the research were now completed. Would the project be justified at $i = 10$ percent?
In this particular case,

$$\text{NPW} = -100,000(1.10) + 47,000[(1.10)^{-1} + (1.10)^{-2} + (1.10)^{-3}]$$
$$= +\$6882$$

and the project would, indeed, be justified.

It has been pointed out that present worth need not be identified with right now—zero time can be set anywhere on the time axis.

Example 2.2.4

In Example 2.2.3 what was the NPW at the time the investment was made?
The problem is essentially one of discounting the previously computed NPW back 1 year (why?). Therefore

$$\text{NPW}_{-1} = 6882(1.10)^{-1} = \$6256$$

This example illustrates the principle that the sign of NPW is the same no matter where on the time axis zero is chosen; that is, the success or failure of a project is not a function of the chosen time basis (why?).

It should perhaps be emphasized again that care must be taken in comparing the NPW of a project by basing the computation on, say, the corporation's "internal" interest rate and again on, let us say, the prime interest rate. It must be remembered that the internal interest rate is based on net income (after taxes), whereas the prime interest rate is given on a before-tax basis. It will not do to reject an internal interest rate–based investment at 10 percent by arguing that the money could be more profitably loaned at 15 percent; the interest payments from the loan are taxed at about 50 percent!

The Continuous Discounting of Continuous Cash Flows

We now turn our attention to the evaluation of the present worth of sequences of cash flows and start with the continuous discounting of continuous cash flows. In deriving Eq. (2.1.17), we found that the *discrete* quantities P and S could be related by continuous compounding. If we modify our symbols and let C represent a discrete cash flow, Eq. (2.1.17) may be rearranged to

$$P = Ce^{-it} \qquad (2.2.1)$$

where t now represents the time at which C occurs. Now, if the cash flow is, in fact, continuous, let it be represented by the continuous function

$$\overline{C}_t = \phi(t) \qquad (2.2.2)$$

where, as previously mentioned, $\overline{C}_t$ is in the form of an annual rate.

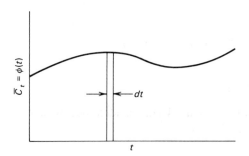

At some specific time t, during the differential time interval dt, a differential cash flow dC occurs, where

$$dC = \phi(t)\, dt \qquad (2.2.3)$$

But dC may be considered a discrete event, and its differential present worth dP is obtained from Eq. (2.2.1):

$$dP = dCe^{-it} \tag{2.2.4}$$

Combining (2.2.3) and (2.2.4),

$$dP = \phi(t)e^{-it} \, dt \tag{2.2.5}$$

This expression may now be integrated over the period of n years of continuous cash flow to yield

$$\boxed{P = \int_0^n \phi(t)e^{-it} \, dt} \tag{2.2.6}$$

Equation (2.2.6) is the basic equation of continuous discounting of continuous cash flows. Of course, the zero time for calculating present worth need not coincide with the beginning of the cash flow period. This difference could be accommodated by modifying the integral indices to n_1 and n_2, but this is not necessary; as we have seen, P may be, in turn, discounted to any place on the time axis.

Exercise 2.2.2

Derive an expression analogous to Eq. (2.2.6) for the continuous *compounding* of continuous cash flows to the end of year n.

Example 2.2.5

Integrate Eq. (2.2.6) for the case of uniform constant cash flow, $\phi(t) = \overline{C}$.

$$P = \int_0^n \overline{C} e^{-it} \, dt = \overline{C} \left. \left(-\frac{1}{i} e^{-it} \right) \right|_0^n$$

whence

$$P = \frac{\overline{C}}{i} \frac{e^{in} - 1}{e^{in}} \tag{2.2.7}$$

Exercise 2.2.3

Derive an expression analogous to Eq. (2.2.7) for the present worth of uniform constant cash flow $\overline{C}$ occurring for the past n years.

In Eq. (2.2.7), $\overline{C}$ is an annual cash flow rate. The total cash flow $C = \overline{C}n$, so that

$$P = \frac{C}{in} \frac{e^{in} - 1}{e^{in}} \tag{2.2.8}$$

The factor

$$\mathcal{F}_D = \frac{1}{in} \frac{e^{in} - 1}{e^{in}} \tag{2.2.9}$$

which, when multiplied by the total cash flow C gives its present worth, is called the *discount factor;* its form, of course, depends upon the nature of $\phi(t)$. Similarly, a *compounding factor* is one which, when multiplied by C, yields the accumulated sum S.

Exercise 2.2.4

What is the form of the compounding factor $\mathcal{F}_C$ for the case of uniform constant cash flow?

Exercise 2.2.5

Derive an expression for the present worth of a step function with uniform constant cash flow $\overline{C}_1$ between 0 and n_1, and $\overline{C}_2$ between n_1 and n_2.

Example 2.2.6

A cash flow increases uniformly over n years as shown in the diagram; the total amount is C. What is the expression for the present worth?

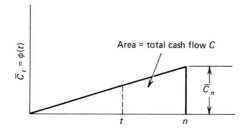

$$\text{Area } C = \tfrac{1}{2}n\overline{C}_n$$

or

$$\overline{C}_n = \frac{2C}{n} \tag{2.2.10}$$

At any time $t \leq n$,

$$\overline{C}_t = \frac{t}{n} \overline{C}_n = C \frac{2t}{n^2} \tag{2.2.11}$$

$$\begin{aligned} P &= \int_0^n \overline{C}_t e^{-it} \, dt \\ &= \frac{2C}{n^2} \int_0^n t e^{-it} \, dt \\ &= \frac{2C}{n^2} \left. \frac{-e^{-it}}{i^2} (it + 1) \right|_0^n \end{aligned}$$

or

$$P = \frac{2C}{(ni)^2} [1 - (in + 1) e^{-in}] \tag{2.2.12}$$

Exercise 2.2.6

What is the present worth of a ramp cash flow extending over a period of T years and ending n years from now? The total cash flow is C.

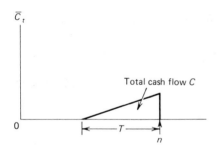

Ans.: $P = \dfrac{2C}{(Ti)^2} [e^{iT} - (1 + iT)] e^{-in}$

Exercise 2.2.7

What is the present worth of a decreasing ramp cash flow shown in the diagram?

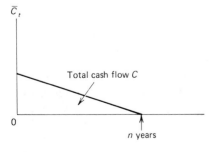

Ans.: $P = \dfrac{2C}{in} \left[1 - \dfrac{1}{in} (1 - e^{-in}) \right]$

It may well be asked, "What is the point of exercises such as those above?" What is the significance of a ramped cash flow distribution, or one perhaps even more complex?

It happens that one of the challenges of project evaluation is the prediction of the cash flow distribution that the project is likely to generate. In the absence of the necessary information, the assumption is often made that the cash flow will be constant,

$$\phi(t) = \overline{C}$$

even though this situation is rarely met in practice. Whenever possible, the projection of cash flow distribution is based upon some sort of systematic investigation of the market potential. The result of such an investigation may be an assessment of the most likely year-by-year cash flow throughout the projected lifetime of the project; in many cases the function $\phi(t)$ cannot be expressed in analytical form, and Eq. (2.2.6) must be evaluated numerically by using techniques well known to chemical engineering students (graphical integration, Simpson's rule, etc.).

The prediction of the form of $\phi(t)$ is within the scope of *product life-cycle theory*, a subject we will return to when we explore subjects such as marketing research. A number of product life cycles and corresponding cash flow distributions are recognized in the literature of engineering economics. The up-ramp and down-ramp distributions in previous examples are not a very likely model for project-generated cash flows, but a combination certainly is—the *triangular* distribution, Fig. 2.2.1a. The *Gompertz* curve (Fig. 2.2.1b)(Holland et al., 1974) is not often used as a predictive tool in the marketing of chemicals, but it is useful in the evaluation of sales trends of consumer products. The curve qualitatively predicts the growth profile of a new product: an early slow growth rate, a period of accelerated growth, and a sustained maturity stage. Mathematically, the Gompertz curve takes the form

$$\ln \frac{\overline{C}_t}{A_1} = -A_2 e^{-A_3 t} \qquad (2.2.13)$$

where the A's are appropriate constants.

Another S-shaped curve is the *logistic* curve (Holland et al., 1974) defined by the equation

$$\overline{C}_t = \frac{B_1}{1 + B_2 e^{-B_3 t}} \qquad (2.2.14)$$

and shown in Fig. 2.2.1c; the B's are constants.

The S-shaped curves may represent the period of cash flow growth and maturity, but they do not properly reflect the later stages of product life characterized by decline. The complete life cycle of cash flows from a given project is sometimes approximated by sections of the *normal distribution* curve (Valle-Riestra, 1979),

SECTION 2.2: THE PRESENT WORTH OF CASH FLOWS

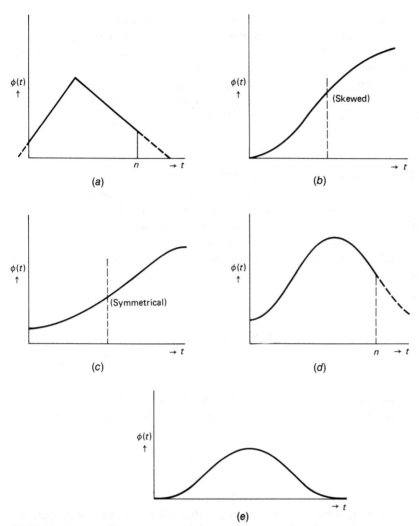

FIGURE 2.2.1
Some common cash flow distributions. (a) Triangular distribution. (b) The Gompertz curve. (c) Logistic curve. (d) Truncated normal distribution curve. (e) Weibull function.

Fig. 2.2.1d, or by the *Weibull function* (Beenhakker, 1975), Fig. 2.2.1e:

$$\overline{C}_t = \frac{A_4}{A_5}\left(\frac{t}{A_5}\right)^{A_4-1} \exp - \left(\frac{t}{A_5}\right)^{A_4} \qquad (2.2.15)$$

where, as before, the A's are appropriate constants.

The evaluation of the present worth of even some of the simpler cash flow distributions in Fig. 2.2.1 can be quite an analytical challenge. Beenhakker (1975) uses the Laplace transform to render the mathematical task more manageable, but, except for the simplest cases, the resulting expressions grow to formidable proportions and are too subject to the analyst's error. The present worth of some distributions such as the normal distribution cannot be determined analytically (Valle-Riestra, 1979), and numerical methods must be used. This is probably the best approach to use for all but the simplest cash flow distributions anyway, and packaged programs are available for modern hand-held computers to facilitate the task.

The Discrete Discounting of Discrete Cash Flows

Whereas the modeling of projects with continuous discounting of continuous cash flows is gaining increasing acceptance, the discrete case is still preferred by many engineering economists and serves as a basis of comparison. A general expression for the present worth of a series of discrete cash flows C_k may be written down on the basis of Eq. (2.1.9) and some of the examples presented as part of the discussion of net present worth:

$$\boxed{P = \sum_{k=1}^{n} C_k (1 + \mathbf{i})^{-k}} \qquad (2.2.16)$$

To evaluate Eq. (2.2.16), year-by-year numerical values of C_k must be known or else C_k must be expressible as a general term in k. For example, C_k might take the form

$$C_k = A \frac{k!}{2^k}$$

This may now be substituted into Eq. (2.2.16), and P is evaluated by methods to be explored momentarily. By convention, C_k is taken as a discrete event at the end of a future year k, or at the *start* of a *past* year k. Occasionally a project or financial manipulation involves cash flows at the *beginning* of a future year k, but this case is easily handled by a slight modification of Eq. (2.2.16) (for instance, cash flow C_k may be considered as taking place at the end of year $k - 1$).

Exercise 2.2.8

Write an expression analogous to Eq. (2.2.16) for the future worth S, at the end of year n, of n cash flows C_k occurring at the beginning of each year k.

For several simple cash flow distributions, the series represented by Eq. (2.2.16) is an arithmetic or geometric progression. An *arithmetic progression* (see, for example, Jolley, 1961) is a sequence such that each term a_k differs from the previous one a_{k-1} by a common difference d:

$$a_k - a_{k-1} = d \tag{2.2.17}$$

The sum Σ_1^n of the series is given by

$$\sum_1^n = \frac{n}{2}[2a_1 + (n-1)d] \tag{2.2.18}$$

A *geometric progression* is characterized by a common ratio r between a_k and the previous term a_{k-1}:

$$\frac{a_k}{a_{k-1}} = r \tag{2.2.19}$$

and the sum is given by

$$\sum_1^n = a_1 \frac{r^n - 1}{r - 1} \qquad r \neq 1 \tag{2.2.20}$$

In evaluating the sum of a series, the analyst should always be on the lookout whether the series is, in fact, a geometric progression.

Example 2.2.7

Evaluate P for the case of constant annual cash flow $C_k = R$.
Substituting,

$$P = R \sum_1^n (1+i)^{-k}$$

This is a geometric progression:

$$a_k = (1+i)^{-k} \qquad a_{k-1} = (1+i)^{-k+1}$$

$$r = \frac{a_k}{a_{k-1}} = \frac{(1+i)^{-k}}{(1+i)^{-k+1}} = (1+i)^{-1}$$

$$a_1 = (1+i)^{-1}$$

$$\sum_1^n = (1+i)^{-1} \frac{[(1+i)^{-1}]^n - 1}{(1+i)^{-1} - 1}$$

Therefore

$$P = \frac{R}{i} \frac{(1 + i)^n - 1}{(1 + i)^n} \qquad (2.2.21)$$

For this case, what is $\mathcal{F}_D$? The total cash flow $C = Rn$. Therefore

$$P = \frac{C}{ni} \frac{(1 + i)^n - 1}{(1 + i)^n}$$

and, by definition,

$$\mathcal{F}_D = \frac{(1 + i)^n - 1}{ni(1 + i)^n} \qquad (2.2.22)$$

Example 2.2.8

What is P for a series of cash flows starting with $C_1 = R$ and $C_k = rC_{k-1}$?
Here $C_k = Rr^{k-1}$. Therefore

$$P = R \sum_{1}^{n} r^{k-1}(1 + i)^{-k} \qquad (2.2.23)$$

This is a geometric progression.

Exercise 2.2.9

Sum the series in Eq. (2.2.23).

Series which are not arithmetic or geometric progressions can sometimes be summed by using the techniques of the *calculus of finite differences*. For those students who are not familiar with this calculus, a brief exposition is given in Appendix A. The subject is treated more thoroughly in an excellent small book by Richardson (1954), and the summation of series forms the subject of a chapter in a book by Hamming (1962).

Example 2.2.9

Evaluate $\sum_{1}^{5} C_k(1 + i)^{-k}$ for $i = 0.10$, $C_k = 1000\,k^2$.
In Appendix A we find that

$$\sum_{1}^{n} x^2 a^{-x} = \frac{a^{-n}}{1 - a}(n + 1)^2 - \frac{2a^{-n}}{(1 - a)^3}[(n + 1)(1 - a) - 1]$$

$$- \frac{a^{-n}}{(1 - a)^2} - \frac{1}{1 - a} - \frac{2a}{(1 - a)^3} + \frac{1}{(1 - a)^2}$$

Substitute $x \equiv k$, $a = 1 + i = 1.10$, $n = 5$:

$$\text{Sum} = \$37,427.96$$

In cases such as the above example, it is much less laborious to evaluate the annual C_k's and then solve for the sum arithmetically. Unless the cash flow spectrum is very simple or a compelling reason exists, numerical integration of discrete cash flows is recommended.

Example 2.2.10

Obtain the sum in Example 2.2.9 by direct calculation.

k	C_k	$C_k(1.1)^{-k}$
1	1,000	909.09
2	4,000	3,305.79
3	9,000	6,761.83
4	16,000	10,928.22
5	25,000	15,523.03
Total		$37,427.96

The Continuous Discounting of Discrete Cash Flows

The continuous discounting of discrete cash flows may seem like mixing apples and oranges, and yet it is commonplace. For example, discrete bank deposits are often compounded continuously; tax payments are discrete events which may be continuously discounted as part of a project economic analysis.

The continuous discounting of a discrete financial event is represented by Eq. (2.2.1); for $t = k$ years, this is written

$$P = C_k e^{-ik} \quad (2.2.24)$$

A sequence of cash flows leads to a present worth equation of the form

$$P = \sum_{k=1}^{n} C_k e^{-ik} \quad (2.2.25)$$

Mathematically, the solution of Eq. (2.2.25) is entirely analogous to the solution of Eq. (2.2.16) for the discrete discounting of discrete cash flows.

Example 2.2.11

Evaluate P for the case of constant annual cash flow $C_k = R$.

$$P = R \sum_{k=1}^{n} e^{-ik}$$

This is a geometric progression.

$$a_k = e^{-ik} \qquad a_{k-1} = e^{-i(k-1)}$$

$$r = \frac{a_k}{a_{k-1}} = \frac{e^{-ik}}{e^{-ik+i}} = e^{-i}$$

$$a_1 = e^{-i}$$

$$\sum_{1}^{n} = e^{-i} \frac{e^{-in} - 1}{e^{-i} - 1}$$

Therefore

$$P = R \frac{e^{in} - 1}{(e^i - 1) e^{in}} \qquad (2.2.26)$$

and since $C = Rn$,

$$\mathscr{F}_D = \frac{e^{in} - 1}{n(e^i - 1)e^{in}} \qquad (2.2.27)$$

Exercise 2.2.10

Study Eqs. (2.1.22), (2.2.22), and (2.2.27), and suggest a rule of thumb to convert expressions for discrete compounding of discrete cash flows to the corresponding expressions for the continuous compounding of discrete cash flows.

2.3 ANNUITIES

Proof by Induction

A powerful method of verifying theorems or formulas that are believed but not definitely known to be true is *proof by induction*. The method can be used only with formulas that depend upon a variable which assumes positive integral values, and as such it finds wide application in the analysis of staged processes as well as economic analysis involving regularly spaced discrete cash flows.

The proof consists of two parts. In the first part, the formula is shown to be valid for some small integral value of the variable. In fact, the usual procedure is to deduce the format of the general formula from the format of the results for three or four of the lowest values of the variable. The second part of the proof is to show that if the formula can be assumed to be valid for a value of the variable equal to k, it is valid for $k + 1$. Since k can assume any value, including the lowest values previously considered, the validity of the formula is proved.

The method is best illustrated by example.

Example 2.3.1

An aqueous solution of volume A, having an initial concentration of transferring species c_{in} mol per unit volume, is to be extracted with n equal portions of organic extractant, each of volume O. The two solvents are entirely nonsoluble in each other. The distribution of the transferring species is given by

$$\frac{c_{O,k}}{c_{A,k}} = K$$

where $c_{O,k}$ is the concentration in the organic phase after the kth extraction, and $c_{A,k}$ is the residual concentration in the aqueous phase. The value of the transferring species is T dollars/mol.

Using the method of induction, derive an expression for the value of the species recovered in the total organic extract after n extractions.

SOLUTION

$$\text{Total species in} = Ac_{in}$$
$$\text{Species remaining in aqueous phase after } n \text{ extractions} = Ac_{A,n}$$

Therefore

$$\text{Amount recovered after } n \text{ extractions} = A(c_{in} - c_{A,n})$$

and

$$\text{Value} = AT(c_{in} - c_{A,n}) \tag{2.3.1}$$

The problem therefore resolves itself to finding the general formula of $c_{A,n}$.
For $n = 1$:

Mass balance:
$$Ac_{A,1} + Oc_{O,1} = Ac_{in}$$

Distribution:
$$c_{O,1} = Kc_{A,1}$$

Therefore
$$c_{A,1} = c_{in}\frac{A}{A + OK}$$

For $n = 2$:

Mass balance:
$$Ac_{A,2} + Oc_{O,2} = Ac_{A,1}$$
$$= Ac_{in}\frac{A}{A + OK}$$

Distribution:
$$c_{O,2} = Kc_{A,2}$$

Therefore
$$c_{A,2} = c_{in}\left(\frac{A}{A + OK}\right)^2$$

For $n = 3$:

Mass balance:
$$Ac_{A,3} + Oc_{O,3} = Ac_{A,2}$$
$$= Ac_{in}\left(\frac{A}{A+OK}\right)^2$$

Distribution:
$$c_{O,3} = Kc_{A,3}$$

Therefore
$$c_{A,3} = c_{in}\left(\frac{A}{A+OK}\right)^3$$

By induction,
$$c_{A,n} = c_{in}\left(\frac{A}{A+OK}\right)^n \qquad (2.3.2)$$

This constitutes the first part of the proof. For the second part, we assume that

$$c_{A,j} = c_{in}\left(\frac{A}{A+OK}\right)^j$$

For $n = j + 1$:

Mass balance:
$$Ac_{A,j+1} + Oc_{O,j+1} = Ac_{A,j}$$
$$= Ac_{in}\left(\frac{A}{A+OK}\right)^j$$

Distribution:
$$c_{O,j+1} = Kc_{A,j+1}$$

Whence

$$c_{A,j+1} = c_{in}\left(\frac{A}{A+OK}\right)^{j+1}$$

and the second condition of the proof is satisfied. Thus

$$\text{Value} = ATc_{in}\left[1 - \left(\frac{A}{A+OK}\right)^n\right] \qquad (2.3.3)$$

We will find the method of induction useful in deducing certain relationships involving the concept of annuities.

Periodic Payments

In a restricted sense, an *annuity* is an annual income derived from an initial investment or principal. (Occasionally the principal itself is referred to as an annuity.) In a broader sense, the term "annuity" refers to any periodic payment (usually annual) used to repay a loan, deplete a principal, or accumulate a sum. Specifically, three types of financial manipulations involving periodic payments may be recognized:

1 The repayment of a loan, whereby each payment is a combination of a charge for interest and partial repayment of the loan principal
2 The true annuity, whereby a periodic amount is withdrawn from a principal which is allowed to accumulate interest until it is reduced to zero
3 Periodic deposits which are allowed to accumulate interest up to a maturity date ("sinking fund")

It is important for the student to understand and to feel confident that *all three financial manipulations are mathematically equivalent*. For example, it is a matter of common sense that, given the same principal P, the same interest rate **i**, and the same number of annual payments n, the periodic payments R will be exactly the same whether they are loan repayments or a true annuity. For if this were not so, it should then be possible to borrow P, redeposit it as an annuity investment, and then pay off the loan with the annuities and perhaps come out ahead, clearly an absurdity. Similarly, if the same periodic deposits R accumulate a sum S, S must be the future worth of P; otherwise it should be possible to borrow P, deposit the payments R rather than use them to repay P, and, at the end of n years, repay the loan out of S and hopefully have something left over. Both examples are a matter of getting something for nothing.

In any of the three financial manipulations mentioned, the payments need not be equal, and they need not be regularly spaced. In the most common situation, however, the annuities are equally spaced equal payments R. In that case, the manipulations are completely defined by three parameters:

The interest rate **i** (or i; that is, the kind of interest must be specified)
The number of equal annual payments n (or number of equally spaced nonannual payments m)
The principal (or its future worth S)

The equivalence of the three financial manipulations becomes even more evident when each is considered as a sequence of cash flows. In Eq. (2.2.21) we saw that the present worth of a series of equal cash flows R (discretely discounted) was

$$P = \frac{R}{i} \frac{(1 + i)^n - 1}{(1 + i)^n}$$

If these cash flows are periodic payments on a loan, then their present worth P is identified with the amount of the original loan. If the cash flows are annuities, then P is identified with the original principal on deposit. Therefore, to calculate what the equal annual payments are, it is only necessary to rearrange Eq. (2.2.21):

$$R = Pi \frac{(1 + i)^n}{(1 + i)^n - 1} \qquad (2.3.4)$$

Note that three parameters (P, i, and n) completely define the value of R.

Example 2.3.2

A recently hired chemical engineer wishes to buy a $120,000 house. $20,000 is available as down payment, and $100,000 may be borrowed at a 15 percent interest rate (*discrete annual compounding*) over 25 years. What would the engineer's monthly payments be?

From Eq. (2.1.21), the *nominal annual* interest rate i is

$$i = \left(1 + \frac{i}{p}\right)^p - 1$$

and, for $p = 12$ months/year,

$$0.15 = \left(1 + \frac{i}{12}\right)^{12} - 1 \qquad i = 0.1406$$

$$i_m = \frac{i}{p} = 0.01171$$

Equation (2.3.4) can be written for m and i_m:

$$R = Pi_m \frac{(1 + i_m)^m}{(1 + i_m)^m - 1}$$

$$m = 25 \times 12 = 300$$

$$R = 100,000 \times 0.01171 \frac{(1.01171)^{300}}{(1.01171)^{300} - 1}$$

$$= \$1207.74/\text{month}$$

How does this compare with the anticipated monthly starting salary for B.S. chemical engineers in 1981?

For the case of the financial manipulation involving periodic deposits which are allowed to accumulate, Eq. (2.3.4) is best modified by replacing P with its future worth, the accumulated sum S; using Eq. (2.1.9),

$$S = P(1 + i)^n$$

so that

$$R = Si \frac{1}{(1 + i)^n - 1} \qquad (2.3.5)$$

For the case of continuous compounding of discrete payments, the equations corresponding to (2.3.4) and (2.3.5) are

$$R = P(e^i - 1) \frac{e^{in}}{e^{in} - 1} \qquad (2.3.6)$$

and

$$R = S(e^i - 1) \frac{1}{e^{in} - 1} \qquad (2.3.7)$$

Exercise 2.3.1

Derive Eq. (2.3.7).

Exercise 2.3.2

Derive equations corresponding to (2.3.4) and (2.3.5) for the case of the continuous compounding of continuous payments with an annual rate of $\overline{R}$.

The answers to the last exercise are

$$\overline{R} = Pi \frac{e^{in}}{e^{in} - 1} \qquad (2.3.8)$$

and

$$\overline{R} = Si(e^{in} - 1)^{-1} \qquad (2.3.9)$$

For the computation of annuities, discrete compounding of discrete payments is generally used.

For nonuniform payments, if the payment interrelationship may be expressed analytically, the payment sequence may be computed from Eq. (2.2.16). For example, suppose the first payment is R_1 and payments will be such that succeeding payments always have the ratio r; that is,

$$\frac{R_{k+1}}{R_k} = r$$

This particular cash flow (or payment) situation was treated in a previous example, and we saw [Eq. (2.2.23)] that

$$R_1 = \frac{P}{\sum\limits_{1}^{n} r^{k-1}(1+\mathbf{i})^{-k}}$$

It must again be emphasized that Eqs. (2.3.4) and (2.3.5) are based on payments R made at the *end* of year k. In Eq. (2.3.5), which represents the accumulation S of annual deposits R, it may certainly happen that the deposits will be made at the beginning of each year, with the first deposit at time zero.[1] The student should be able to attack this problem in several ways.

Exercise 2.3.3

Derive an equation analogous to (2.3.5) for the case of deposits made at the beginning of the year, starting at time zero, with the last deposit made at the beginning of year n. S is the accumulated sum of the end of year n. Start with Eq. (2.2.16), substituting R for C_k and modifying the summation indices to 0 and $n-1$ (why?).

One may also argue as follows:

In Eq. (2.3.5), the sum S at the end of year n is the result of deposits R at the end of years 1, 2, 3, ..., n. This is the same thing as deposits at the beginning of years 2, 3, 4, ..., $n+1$. What would be the sum S^* if we kept the same payment sequence but (1) eliminated R at the beginning of year $n+1$ and (2) added R at the beginning of year 1?

Future worth of R at $n+1$ = exactly R
Future worth of R at time zero = $R(1+\mathbf{i})^n$

Thus

$$S^* = S - R + R(1+\mathbf{i})^n$$

Substituting for S in Eq. (2.3.5), we get

$$R = \frac{S^*\mathbf{i}}{1+\mathbf{i}} \frac{1}{(1+\mathbf{i})^n - 1}$$

that is, for the same R, S^* will be different, or, to obtain the same S, the required R are somewhat smaller than indicated in Eq. (2.3.5).

---

[1] Time zero is the beginning of year 1.

Exercise 2.3.4

By comparing the equation obtained with (2.3.5), can you think of a much simpler argument which will let you convert expressions based on end-of-the-year payments to those based on beginning-of-the-year payments?

Perpetuity

A *perpetuity* is an annuity paid out over an indefinite length of time. Specifically, the term refers to periodic payments generated by a principal amount large enough so that the periodic payments will continue indefinitely. In order that this process be perpetual, the interest, and interest alone, between payments must just match each payment.

Suppose payments are made n years apart. Then

$$P(1 + i)^n - P = R \qquad (2.3.10)$$

and for the specific case of $n = 1$,

$$R = Pi \qquad (2.3.11)$$

Example 2.3.3

A corporation wishes to establish a retirement fund. It is believed that proper management of the fund will guarantee a continuous compounding interest rate of 10 percent. How large a fund is required to guarantee an average annual retirement income of $20,000 if, on the average, 3000 retirees participate?

From Eq. (2.1.22),

$$i = e^i - 1 = e^{0.10} - 1 = 0.1052$$
$$P = \frac{R}{i} = \frac{20{,}000}{0.1052} = 190{,}167$$
$$\text{Total} = 190{,}167 \times 3000 = \$570 \times 10^6$$

Exercise 2.3.5

In the above example the income is assumed to be paid as a once-a-year lump sum. For the fund as a whole, the retirement income payments may better be considered as a continuous cash flow. How would you handle the problem in that case? [*Hint*: Start with Eq. (2.1.14), and find $dS/dt = \overline{R}$ at $t = 0$.]

Capitalized Cost

The concept of perpetuity may be used to compute the *capitalized cost* of an item of equipment. The capitalized cost is the initial principal required to guarantee that funds will always be on hand to replace an item of equipment that has worn out, for an indefinite period of time. For example, a particular pump might have a usable lifetime of $m = 3$ years. If the replacement cost of the pump is C and the initial principal is S, then the balance, $P = S - C$, has 3 years to be compounded at an interest rate **i** back to the initial principal sum S. In this way, the process can be repeated indefinitely. It is the sum S that is termed the capitalized cost of the pump.

What is the purpose of estimating the capitalized cost of equipment? Does one really set aside a potful of money just to perpetuate the replacement of the equipment? Certainly this is not done literally, although one may wish to think of some small portion of the corporation's total investment, drawing the "internal" interest rate **i,** as being devoted to this purpose. The capitalized-cost concept, however, is a powerful one in the analysis of alternatives (Chap. 10); we will see, for example, how a choice between two possible pumps for the same job may be made on the basis of capitalized costs. For the present, we restrict our discussion to methods of computing the capitalized cost.

For the case where the replacement cost C of the equipment is just equal to W, the purchased cost, Eq. (2.3.10) can be rewritten as

$$P(1 + i)^m - P = W$$

or

$$P = \frac{W}{(1 + i)^m - 1} \quad (2.3.12)$$

The capitalized cost S is

$$S = P + W$$

or

$$\boxed{S = W \frac{(1 + i)^m}{(1 + i)^m - 1}} \quad (2.3.13)$$

where the term

$$\mathscr{F}_K = \frac{(1 + i)^m}{(1 + i)^m - 1} \quad (2.3.14)$$

is sometimes referred to as the *capitalized-cost factor*.

The replacement cost C may include not only the purchased cost W; it may also account for installation costs and the salvage value of the worn item of equipment. Installation cost may be taken as a percentage of the purchased cost (say, bW), and similarly the *salvage value* may be related to the purchased cost (say, sW). In this case W in Eq. (2.3.13) is replaced by C:

$$C = W(1 + b - s) \tag{2.3.15}$$

However, to compute the capitalized cost correctly, we must recognize that *no salvage value sW is obtained at time zero;* this term must therefore be added to the right-hand side, and a more general version of Eq. (2.3.13) is

$$S = W[\mathscr{F}_K(1 + b - s) + s] \tag{2.3.16}$$

A capitalized cost may also be computed for the case of a finite projected *plant life* of n years; we assume that

$$n = km$$

where m is the equipment life, and k is the number of times the equipment must be purchased during the lifetime of the plant. The problem will be attacked using the method of induction. A graphical representation is shown in Fig. 2.3.1.

Each interval in Fig. 2.3.1 represents a span of m years, the equipment lifetime; there are k such intervals. The subscripts refer to the number of the interval; thus the first interval starts at time zero, and the subscripts refer to the quantities S and P at the beginning of each interval. For the moment, installation costs and salvage value will be neglected, so that the replacement cost C is just equal to W, the purchased cost. With this simplification, S_1 in Fig. 2.3.1 is the capitalized cost, and at the start of each interval,

$$P_N = S_N - W \qquad \text{(interval } N\text{)}$$

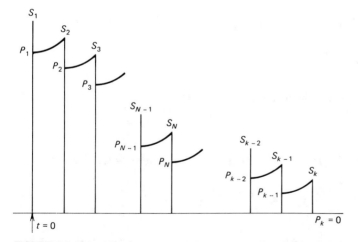

FIGURE 2.3.1
A graphical representation of the cash flow position, capitalized cost computation for finite plant life.

After the initial purchase, P_1 is compounded at the interest $\mathbf{i}$ for m years to S_2. At this point another purchase is made, and the process is repeated until the last period k. Since no additional principal is required,

$$P_k = 0$$

For any interval such as N,

$$S_N = P_{N-1}(1 + \mathbf{i})^m$$
$$P_N = S_N - W$$

Now, $P_k = 0$, $S_k = W$. Therefore

$$P_{k-1} = S_k(1 + \mathbf{i})^{-m}$$
$$= W(1 + \mathbf{i})^{-m}$$
$$S_{k-1} = P_{k-1} + W = W(1 + \mathbf{i})^{-m} + W$$
$$= W[1 + (1 + \mathbf{i})^{-m}]$$

Continuing,

$$P_{k-2} = S_{k-1}(1 + \mathbf{i})^{-m} = W[1 + (1 + \mathbf{i})^{-m}](1 + \mathbf{i})^{-m}$$
$$= W[(1 + \mathbf{i})^{-m} + (1 + \mathbf{i})^{-2m}]$$
$$S_{k-2} = P_{k-2} + W$$
$$= W[1 + (1 + \mathbf{i})^{-m} + (1 + \mathbf{i})^{-2m}]$$

By induction,

$$S_1 = W[1 + (1 + \mathbf{i})^{-m} + (1 + \mathbf{i})^{-2m} + \cdots + (1 + \mathbf{i})^{-(k-1)m}]$$
$$= W \sum_{0}^{k-1} (1 + \mathbf{i})^{-Nm}$$

But $\sum_{N=0}^{k-1} (1 + \mathbf{i})^{-Nm}$ is a geometric series with ratio

$$r = (1 + \mathbf{i})^{-m}$$

and the sum is $a[(r^k - 1)/(r - 1)]$, where a is the first term $= 1$. Therefore

$$S_1 = W \frac{(1 + \mathbf{i})^{-km} - 1}{(1 + \mathbf{i})^{-m} - 1} \tag{2.3.17}$$

The second part of the proof by induction is left as an exercise to the student.

By analogy to Eq. (2.3.16), the capitalized cost S_1 with installation costs and salvage value is given by

$$S_1 = W(1 + b - s)\frac{(1 + i)^{-km} - 1}{(1 + i)^{-m} - 1} + sW \qquad (2.3.18)$$

It may be argued quite legitimately that the capitalized cost should also account for inflation, since W is not a constant but will increase with time. By analogy to Eq. (2.1.25), one might expect the expression for capitalized cost, corrected for inflation, to be

$$S_1 = W_1(1 + b - s)\frac{\left(\dfrac{1 + i}{1 + f}\right)^{-km} - 1}{\left(\dfrac{1 + i}{1 + f}\right)^{-m} - 1} + sW_1 \qquad (2.3.19)$$

The student is asked to prove that this expression is correct (Prob. 2.10).

Exercise 2.3.6

Is it possible to set up a perpetuity so that annual payments keep up with projected inflation?

The Present Worth of Interest Payments

We saw in Eq. (2.3.4) that a borrowed principal P may be repaid with the periodic payments R extending over n periods:

$$R = Pi\frac{(1 + i)^n}{(1 + i)^n - 1} \qquad (2.3.4)$$

The quantity R incorporates both the interest payment for the period in question and a partial repayment on principal; the relative amounts of the two are different for each period. Since, as we saw in Chap. 1, interest payments are tax-deductible, it is useful to know how much interest any particular payment R_k incorporates. Moreover, for purposes of some economic analyses, it may be useful to compute the present worth of the interest payments. It should be perhaps emphasized at this point that in popular usage the periodic payments R are sometimes referred to as interest payments; this term is distinctly erroneous and must be avoided.

To obtain a general expression for the amount of interest incorporated in payment R_k, use the method of induction.

For year 1:   Balance at start $= P$
              Balance at end $= P(1 + i) - R$
              $I_1 = Pi$

112 CHAPTER 2: THE MATHEMATICS OF FINANCE

For year 2:

$$\text{Balance at start} \equiv \text{balance at end of year 1}$$
$$= P(1 + i) - R$$
$$\text{Balance at end} = [P(1 + i) - R](1 + i) - R$$
$$= P(1 + i)^2 - R[1 + (1 + i)]$$
$$I_2 = [P(1 + i) - R]i \quad \text{(balance at start} \times i\text{)}$$

For year 3:

$$\text{Balance at end} = \{[P(1 + i) - R](1 + i) - R\}(1 + i) - R$$
$$= P(1 + i)^3 - R[1 + (1 + i) + (1 + i)^2]$$
$$I_3 = \{P(1 + i)^2 - R[1 + (1 + i)]\}i$$

For year 4: $\quad I_4 = \{P(1 + i)^3 - R[1 + (1 + i) + (1 + i)^2]\}i$

By induction,

$$I_k = \{P(1 + i)^{k-1} - R[1 + (1 + i) + (1 + i)^2 + \cdots + (1 + i)^{k-2}]\}i$$
$$= \left[P(1 + i)^{k-1} - R \sum_{0}^{k-2}(1 + i)^j\right]i$$
$$= \left[P(1 + i)^{k-1} - R \frac{(1 + i)^{k-1} - 1}{i}\right]i$$

or

$$\boxed{I_k = R - (R - Pi)(1 + i)^{k-1}} \quad (2.3.20)$$

To prove the second part of the proof by induction, we note that, based on Eq. (2.3.20), the balance at the start of year k must be

$$\frac{1}{i} I_k = \frac{1}{i}[R - (R - Pi)(1 + i)^{k-1}]$$

Thus

$$\text{Balance at end of year } k = \frac{1}{i}[R - (R - Pi)(1 + i)^{k-1}](1 + i) - R$$
$$= \frac{R}{i} - (R - Pi)\frac{(1 + i)^k}{i}$$

Whence $\quad I_{k+1} = R - (R - Pi)(1 + i)^k \quad$ Q.E.D.

Exercise 2.3.7

Show that Eq. (2.3.20) correctly predicts the relationship

$$\sum_{1}^{n} I_k = nR - P \qquad (2.3.21)$$

What is the significance of Eq. (2.3.21)?

The present worth of the interest portion of all periodic payments is given by

$$\sum PW = (Pi - R) \sum_{1}^{n} (1 + i)^{k-1}(1 + j)^{-k} + R \sum_{1}^{n} (1 + j)^{-k} \qquad (2.3.22)$$

Note that the rate at which the interest payments are discounted (j) need not be the same as the interest rate i on the loan (why?).

Equation (2.3.22) is readily integrated to

$$\sum PW = \frac{Pi}{(1 + i)^n - 1} \left\{ \frac{1 - \left(\frac{1 + i}{1 + j}\right)^n}{i - j} + \frac{(1 + i)^n[(1 + j)^n - 1]}{j(1 + j)^n} \right\} \qquad (2.3.23)$$

NOMENCLATURE

$a_1, a_2, \ldots$	terms in mathematical progression
A	volumetric flow rate of aqueous phase in extraction, m³/h
$A_1, A_2, \ldots$	constants
b	installation cost factor
$B_1, B_2, \ldots$	constants
c	concentration of transferring species, kg·mol/m³; c_O, concentration in organic phase; c_A, concentration in aqueous phase
C	total cash flow; replacement cost, $
C_k	discrete cash flow at end of year k, $
C_t	annual cash flow rate at time t, $/year
d	common difference, arithmetic progression
e	base of natural logarithms
f	average annual inflation rate, fractional
$\mathscr{F}_C$	compounding factor
$\mathscr{F}_D$	discount factor, Eq. (2.2.9)
$\mathscr{F}_K$	capitalized cost factor, Eq. (2.3.14)
i	nominal annual interest rate, fractional
i_m	discrete compound interest rate for time interval m (for example, monthly interest rate)
i_s	simple interest rate
$\mathbf{i}$	discrete annual compound interest rate; effective interest rate
I	annual interest payment, $
j	indexing symbol

j	symbol for discrete annual compound interest rate used to distinguish borrowing rates and investment return rates
k	indexing symbol
K	distribution constant of transferring species in liquid-liquid extraction
m	indexing symbol for time interval other than years
n	indexing symbol for time in years
N	indexing symbol for multiyear periods
O	volumetric flow rate of organic phase in extraction, m³/h
p	number of time intervals (indexed by m) in 1 year (for example, if m represents weeks, $p = 52$); also plant capacity, kg/year
$\bar{p}$	production rate, kg/year
P	principal; present worth; investment, $
P'	present worth in current dollars
q	ratio of initial production to maximum sustained plant capacity
r	ratio of consecutive terms in a geometric progression; ratio of consecutive discrete periodic payments
R	annuity; periodic payment, $
$\bar{R}$	continuous periodic payment, uniform over period
s	salvage value factor; sales price, $/kg
S	accumulated sum; future worth; return on investment; capitalized cost
S'	accumulated sum in constant-value currency
t	time, years
T	value of transferring species in extraction, $/(kg·mol)
W	purchased cost of equipment, $
α	superproductivity factor
Δ	difference between consecutive payments, gradient uniform series
ϕ	a function
Σ	summation symbol

Notes

1 Meaning of subscripts (j, k, 0, 1, 2, . . .) must be deduced from text.
2 NPW = net present worth

REFERENCES

Almond, Brian, and Donald S. Remer: Present Worth Analysis of Capital Projects Using a Polynomial Cash Flow Model, *Eng. Costs Prod. Econ.*, **5**:33 (1980).

Beenhakker, H. L.: Sensitivity Analysis of the Present Value of a Project, *Eng. Econ.*, **20** (2):123 (1975).

Hamming, R. W.: "Numerical Methods for Scientists and Engineers," McGraw-Hill, New York, 1962.

Holland, F. A., F. A. Watson, and J. K. Wilkinson: "Introduction to Process Economics," Wiley, London, 1974.

Jolley, L. B. W.: "Summation of Series," 2d ed., Dover, New York, 1961.

Newnan, Donald G.: "Engineering Economic Analysis," Engineering Press, San Jose, Calif., 1976.

Richardson, C. H.: "An Introduction to the Calculus of Finite Differences," Van Nostrand, New York, 1954.

Rose L. M.: "Engineering Investment Decisions: Planning Under Uncertainty," Elsevier, Amsterdam, 1976.

Valle-Riestra, J. F.: The Evaluation of the Present Worth of Normally Distributed Cash Flows, *Eng. Process Econ.*, **4**:37 (1979).

Woods, Donald R.: "Financial Decision Making in the Process Industry," Prentice-Hall, Englewood Cliffs, N.J., 1975.

PROBLEMS

2.1 Some engineering economics texts (Woods, 1975; Newnan, 1976) use a peculiar type of line chart to illustrate problems in the mathematics of finance. Investigate the nature of these charts, show how some of the concepts in this chapter may be illustrated, and give your own evaluation of the value of these charts in facilitating the understanding of the nature of the relationships involved.

2.2 What is the equivalent effective annual interest rate for the simple interest rate i_s?

2.3 Use the method of induction to obtain Eq. (2.1.10), the interest earned during some year n by undisturbed principal P.

2.4 Convince yourself by using the method of induction that Eq. (2.3.4) does, indeed, represent the annual payment required to pay off a loan P in n years.

2.5 An undergraduate 4-year education in a private university may cost $8000/year* for tuition and books. How much additional average annual income (before taxes) must such an education guarantee if money is worth 20 percent (discrete annual compounding) and a typical career lasts 40 years beyond graduation? Include in the computations a 4-year loss of income averaging $12,000/year, income not earned because of the time required in the university.

2.6 For the case of nonuniform payments on a loan, with

$$\frac{R_{k+1}}{R_k} = r$$

find an expression for the sum total of all the payments.

2.7 A small plastics fabricator wishes to purchase an extrusion molding machine for $100,000 and must borrow money for the purchase. He expects his product sales to be slow during the first year and to pick up subsequently. He therefore arranges with the bank to pay off the loan on a "sliding scale" which results in the lowest payment at the end of the first year.

Annually compounded interest rate = 10 percent
Loan to be paid off at the end of 5 years in five annual payments
Each payment to be just twice the previous one

Compute and tabulate the sum of all payments made to the bank for three cases:

a The sliding-scale arrangement described
b Five equal annual payments
c Single payment at the end of 5 years

*1980 dollars.

Rework the three cases and compute the "current" dollar value of the sum of all payments for an average inflation factor of 5 percent per annum.

What do you conclude is the effect of inflation upon the real value of the various payment schemes?

2.8 What is the discretely discounted present worth of discrete cash flows given by the distribution function

$$C_k = Ak$$

Use the calculus of finite differences. For this case, what is the discount factor $\mathcal{F}_D$?

2.9 One possible way to pay off a loan P is as a series of periodic payments, each differing from the previous one by the fixed amount Δ,

$$R, R + \Delta, R + 2\Delta, \ldots$$

Show the required relationship between R and Δ for fixed P, i, and n ("gradient uniform series").

2.10 Use the method of induction to derive Eq. (2.3.19), which gives the value of the finite plant life capitalized cost with inflation accounted for.

2.11 Use Simpson's rule to evaluate the present worth of the continuously discounted continuous cash flow given by the functional relationship

$$\phi(t) = \sinh(1 + 0.1t) + \cosh(1 + 0.1t)$$

for $n = 17$ years, $i = 0.0875$. t is in years.

2.12 Integrate Eq. (2.3.22) to obtain Eq. (2.3.23), the present worth of the interest portion of periodic payments. What form does the result assume for the special case of $\mathbf{i} = \mathbf{j}$?

2.13 In an example illustrating the concept of a perpetuity, we saw that a $570 million retirement fund, managed to yield 10 percent, would provide an average of 3000 retirees with a $20,000 annual income. What would the retirement fund have to be to provide automatic 5 percent annual inflation increases?

2.14 Because of rapid corrosive attack, a pump in a large production plant must be replaced every 4 years. What is the theoretical amount of capital that must be available ("capitalized cost") at "zero time" (new plant) so that the pump may continue to be replaced until final plant shutdown, 20 years from start-up?

Original purchase cost (new plant): $5000
Cost of installation: 50 percent of purchase cost
Interest rate: 10 percent, discrete annual compounding
Annual pump cost inflation: 5 percent
Salvage value at end of each 4-year period: 10 percent of purchase cost at the time of salvaging

In practice, such a capital pool is not held in reserve for the sole purpose of perpetuating a piece of equipment. Why?

2.15 A corporation wishes to raise capital by selling sinking fund debentures through the agency of a syndicate of investment bankers who agree to market and underwrite the securities for a commission fee of 3 percent of the nominal capital. *Debentures* are bonds unsecured by any specific property; the investment banker guarantees to sell the entire issue at the stated price, and the corporation is relieved of the risk of not being able to sell the issue at the stated price. The debentures mature and are due at face value after a stated number of years. The *sinking fund* refers to the method the corporation wishes to employ to accumulate the redemption value of the bonds at maturity; a separate fund is established into which the corporation makes annual deposits so that the accrued sum will just match the total redemption value.

The nominal capital the corporation wishes to raise is $300 million, with 300,000 debentures, each with a 30-year maturity value of $1000. The debentures are priced to sell for $997, and an annual interest payment of $78.88 (7 7/8 percent) will accrue to the owner. The sinking fund will earn interest at 8 1/4 percent, discrete annual compounding. The investment bankers agree to accept their commission in five equal annual payments at 10 percent interest, with the first payment to be made at the end of the year during which the securities are issued.

a From the point of view of the *buyer*, what is the *yield to maturity* of the debentures; i.e., at what interest rate is the money earned? (*Hint:* Investment equals present worth at maturity plus present worth of interest payments.)

b Compute the annual payments that the corporation will be required to make for bond interest, sinking fund payments, and commission payments.

c From the point of view of the *corporation,* what rate of interest is paid on the capital raised?

2.16 Derive Eq. (2.3.17), the capitalized cost for finite plant life, by considering the capitalized cost as an annuity with periodic payments at the *beginning* of each period of m years. Start by modifying Eq. (2.3.4) for the case of beginning-of-year payments and the compound interest i_m for m-year periods. Note that

$$n = mk$$

2.17 What is the expression for the continuously discounted present worth of a truncated ramp continuous cash flow shown in the diagram? The total amount of cash is C.

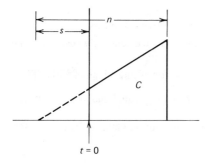

2.18 Use the result of Prob. 2.17 to compute the present worth of cash flow having a triangular distribution:

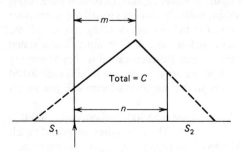

For the computation, use

$$C = \$5 \times 10^6 \qquad S_1 = 3 \text{ years}$$
$$n = 17 \text{ years} \qquad S_2 = 7 \text{ years}$$
$$m = 10 \text{ years} \qquad i = 14\%$$

2.19 An investment of $\$10 \times 10^6$ has resulted in the following annual cash flows up to project termination at the end of year 10:

Year	Cash flow, $	Year	Cash flow, $
1	0.012×10^6	6	1.998×10^6
2	0.872×10^6	7	3.581×10^6
3	1.652×10^6	8	4.290×10^6
4	1.371×10^6	9	2.010×10^6
5	0.858×10^6	10	0.107×10^6

In retrospect, was the investment justified on the basis of an expectation of $i = 12$ percent?

2.20 Use a method of your choice to compute the present worth (at $i = 9\ 1/2$ percent) of the discrete cash flows given by

$$C_k = 10,000 \ln (k + 5)$$

where k is the year number, and the cash flows will extend over 24 years.

2.21 A total cash flow C occurs uniformly over a period of T years ending n years from the present. What is the discount factor for continuous discounting?

2.22 The financial officer of a corporation has two options for raising a loan of $50 million:

Option 1: Bank loan at 12 percent effective interest rate, equal monthly payments, 20-year loan

Option 2: Sale of bonds at $996 sales value, $1000 redemption value in 20 years, $100 annual "interest" payments on each bond

Which option should the officer recommend?

2.23 The projected production of a pharmaceutical is expected to follow a logistic curve [Eq. (2.2.14)] with parameters

$$B_1 = 1.00 \times 10^6 \qquad B_2 = 3.0 \qquad B_3 = 1.0$$

t is expressed in years. What is the continuously discounted present worth of the cash flows if the pharmaceutical plant is expected to last just 10 years? Use $i = 0.15$.

2.24 The production life cycle of a product with a limited market life due to anticipated obsolescence may follow a normal (gaussian) distribution. The present worth of normally distributed cash flows cannot be computed analytically, but results of numerical discounting procedures are available (Valle-Riestra, 1979). Use the discussion and charts in the reference to compute the present worth of the sales income of a product with a symmetrical normally distributed life cycle; the following parameters describe the life cycle:

$i = 0.20$ (use discrete discounting of discrete cash flows)
$n = 15$ years (plant life)
p_R (rated plant capacity) $= 30 \times 10^6$ kg/year
p_{max} (maximum sustained plant capacity, or "superproductivity") $= p_R (1 + \alpha)$, with $\alpha = 0.10$
q (ratio of production during first, or last, year to p_{max}): 0.50
Selling price: $1/kg

2.25 During the lifetime of a particular plant, assume that sales revenues may be approximated by a portion of a sine wave. Derive an expression for the discount factor for the total revenue C, an integral of an instantaneous cash flow that initially starts at zero (at plant start-up), may be represented by the positive half of the sine wave, and decreases back to zero after n years. The instantaneous cash flow rate $\overline{C}$ may be represented as

$$\overline{C} = Q \sin \frac{t}{n} \pi$$

where Q is a constant to be determined, and t is expressed in years. Use continuous discounting with nominal interest rate i.

What is the value of $\mathcal{F}_D$ for $i = 0.10$, $n = 10$?

2.26 Repeat the derivation of Prob. 2.25 for the case where the total revenue C is represented by the area under a portion of the positive half of a sine wave, starting with fraction A of the maximum production P at time zero, and fraction B of P after n years:

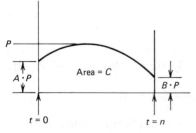

Does the derived expression reduce to that in Prob. 2.25 for $A = B = 0$?
Compute $\mathcal{F}_D$ for $i = 0.10$, $n = 10$, $A = 0.20$, $B = 0.60$.

2.27 A capital investment of $\$1.0 \times 10^6$ (at time zero) is expected to generate cash flows (after taxes) of $\$0.25 \times 10^6$/year for 10 years, starting the first year. Following completion of the investment, it is found that market factors will force delay in the start-up of the facility of 2 years. If a 10-year project life is still anticipated following start-up, what annual cash flows must be realized to maintain the same project NPW at $\mathbf{i} = 14$ percent?

2.28 Suppose the production of a plant follows the symmetrical triangular distribution shown; each annual production is characterized as an average value $p_1, p_2, p_3, \ldots$ as illustrated. Because of symmetry, $p_1 = p_n$, and p_1 is taken as

$$p_1 = p_n = qp_{\max}$$

that is, as a fraction q of the maximum plant production (represented by the *apex* of the triangle).

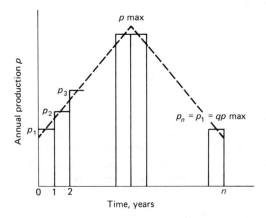

Using the calculus of finite differences, derive an expression for the discretely discounted present value of the sales revenue, based on the sales price s per unit of production. The derived expression should have the functional relationship

$$PV = \phi(p_{\max}, s, q, \mathbf{i}, n)$$

Assume n is even. See also Valle-Riestra (1979).

2.29 The production of an epoxy resin has decayed exponentially according to the relationship

$$\bar{p}_t = Ae^{-Bt}$$

where $\bar{p}_t$ is the production *rate* at time t. Simultaneously, the price has increased linearly,

$$s \text{ (price per unit of production)} = s_0(1 + gt)$$

What is the expression for the present worth of sales revenues from $t = 0$ to $t = n$? Use continuous discounting of continuous cash flows.

2.30 Almond and Remer (1980) have made a present worth analysis of discretely discounted polynomial cash flows. Use the approximations suggested in the article to evaluate the present worth of cash flows represented by

$$C_k = 10{,}000\,(0.9k^3 + 8k^2 - 15k + 18)$$

from $k = 1$ to $k = 10$ years, with $\mathbf{i} = 0.10$.

CHAPTER **3**

PROJECT EVALUATION SYSTEMATICS

We are now ready to turn our attention to the principal theme, that of project evaluation. The first two chapters were in the nature of a prelude, an introduction that hopefully contributed to the student's perception of the nature of the industrial environment, and that additionally laid down a mathematical foundation for some of the evaluation techniques which will follow. In this chapter we return to the concept, initially explored during our cursory review of the economic basis of industrial projects (Sec. 1.3), of project economic analysis as a blending of investment and income analyses pointing toward the examination of a criterion of economic performance. We will examine the preferred sequencing of such an economic analysis as a first step leading to a more detailed exposition of methodology in subsequent chapters. We will also examine methods of assembling background information required for the proper definition of the nature of the project. Specifically, we will explore ways of making product pricing and demand projections by using the techniques of marketing research, and we will also examine some of the ways of defining a processing scheme that can serve as a basis for subsequent investment and income analyses.

The differences among various project types will be pointed out, but the exposition of project evaluation systematics will emphasize those aspects which are common to all projects.

3.1 PRINCIPLES OF PRELIMINARY PROJECT EVALUATION

Project Types: Similarities and Differences

No two projects are alike. Each one differs in nature and scope, and each one presents its own set of challenges. And yet, as we attempt to evaluate projects to establish their economic promise, we find that the mechanism of evaluation has a number of elements, as well as sequencing of elements, most of which are common to each project regardless of the project's nature.

Even one particular project changes in nature in the course of project development. For example, the overall process to manufacture a particular product, spanning the gamut from raw material preparation to final product purification, may change in such a way as to be virtually unrecognizable during its progress from the laboratory bench to the pilot plant. We begin by focusing upon project evaluation at the preliminary stage of development. By *preliminary* stage we refer to the stage of a project when it is still in the laboratory, the product preparation appears to be technically feasible, and the product itself holds promise of more than casual acceptance in the world of commerce. We will find that these preliminary-stage techniques and sequences can be used throughout the lifetime of a project, throughout the various stages of project development and "fine-tuning," perhaps with some refinements. We emphasized in the first chapter that the decision as to when refinements and more thorough evaluation techniques are to be used is primarily a matter of *judgment* and *experience*. We should perhaps emphasize again that project evaluation must be attempted at the earliest possible stages of development, just as soon as some processing scheme can be visualized, to avoid catastrophic development expenditures on a project with a dismal economic prospect.

The nature and scope of a project are distinctly influenced by the history of the product in the corporation, and a product-oriented project classification is a useful one:

Existing products in company. These projects involve new plants to manufacture an existing company product on an existing site (for example, to expand production capacity), new plants in new locations, and plants to produce an existing product but utilizing a new process.

Existing products but new to company. Even though a great deal of information may be available on the existing products, the very fact that they are new to the company increases the complexity of this type of project.

Entirely new products. These projects involve the greatest number of unknowns and are therefore the most difficult to handle.

Limited-scope projects. There are, of course, those projects which do not really fit into any of the previous categories, principally because of their limited scope. These include (1) existing plant changes and improvements and (2) pilot plant and semiplant projects.

Sequence of Preliminary Project Evaluation

The following steps describe the normal sequence used during preliminary project evaluation. As the project progresses toward more advanced stages of development, one or another of the sequential items may receive special emphasis, but the *total* sequence should be kept up-to-date; it applies to projects up to the stage of final authorization, and perhaps beyond.

1 Product profile. Assembly of data and information that describe the nature of the product.

2 Marketing research. Answer to the fundamental question, "How much can we sell at what price?"

3 Process choice. For a new product, this step may be more a matter of process definition rather than process choice.

4 Flow sheet synthesis, material and energy balances. In preliminary stages, flow sheet synthesis may involve a great deal of speculation.

5 Equipment choice and design. Shortcut methods or rules of thumb are particularly useful in preliminary stages.

6 Plant investment estimate. Estimate of the capital required to build a plant to meet the likely demand.

7 Manufacturing cost estimate. Projection of how much it will cost to produce the product in the specified plant.

8 Profitability evaluation. Project profitability based upon a chosen criterion of economic performance.

Some of the sequence items will constitute the subject matter of separate chapters.

Follow-up Evaluation

The product and process development which characterize successive stages of a project do not occur as quantum jumps but rather as a developmental continuum. Nevertheless, discrete states of advancement may be approximately identified and described on the basis of firmness of process, degree of backup toxicological work, scale-up status, and many other criteria. Often the stages of development are formally described and numerically identified, although the methodology differs among corporations. One may speak, for example, about a stage 0 or a stage 1 project, the latter corresponding perhaps to the preliminary stage previously described. Projects may also be identified by reference to the status of the corresponding capital investment estimate. In Chap. 1, four estimate categories were identified. These estimate categories are sometimes characterized as "class levels" IV (order-of-magnitude) down to I (project control) (Maples and Hyland, 1980).

A project development sequence may be described in terms of a few developmental landmarks:

- *Laboratory development work*
    Definition of desirable product characteristics.
    Measurement of physical properties.
    Reaction definition, usually batch reaction, often in glass equipment.
    Separation studies in laboratory equipment.
    Purification studies.
    Satellite studies: toxicology, product performance, safe handling, waste disposal.

- *Miniplant*

    At this stage, process is moved into larger-scale equipment, but still on "bench scale."

    More continuous operation, attempt to integrate several process steps.

    Equipment may still be glass, but materials of construction studies initiated.

    Initiation of studies to check effect of recycle streams, long residence time in recovery units, etc.

    Some process control studies.

- *Pilot plant*

    The purpose of the pilot plant is to study problems of scale-up.

    Plant prototype equipment first used; performance study of special equipment difficult to study on small scale (centrifuge; plug flow reactor).

    Process need not be fully integrated; may focus on specific process steps.

    Waste disposal studies.

    Prototype process control.

- *Semiplant*

    The semiplant manufactures product for *market development* studies.

    Management usually prefers that pilot plant and semiplant functions be combined for reasons of economy, even though this tends to deemphasize needed scale-up work.

    Semiplants may span the gamut from a few items of portable equipment for the manufacture of needed product batches to fully integrated, highly automated production units not much smaller than projected full-scale plants.

- *Market development*

    The purpose of market development is to lay the groundwork for subsequent sales of the product. Key activities include distribution of trial quantities for customers' processes, and definition of product formulation properties to meet customers' specific needs. Adequate market development usually entails semiplant operation.

- *Authorization study*

    The authorization study leads up to a specific request to the board of directors to authorize the expenditure of capital needed to build a full-scale plant. At this stage, the process and product properties must be essentially completely defined to allow detailed plant design.

- *Plant construction*

    Once the moneys for plant construction are authorized, the detailed plant design can proceed, followed by equipment procurement and the construction process itself.

- *Plant start-up*

    Plant start-up, which may typically last 4 months, is usually undertaken with the help of the research and engineering functions. Start-up is the last step in the sequence characterizing a project within our definition.

It is important to recognize that no project is likely to be characterized by all the landmarks mentioned. There are occasions, for example, when full-scale plants may be built on the basis of miniplant data only; that is, the pilot plant stage is bypassed. If an existing product is involved, a semiplant is usually not required, and market development needs are often less demanding than those for a new product. Some of the activities mentioned will certainly overlap others. Laboratory development work, for example, may continue up to the authorization study; market development should preferably be initiated during very early stages of project advancement; and pilot plant work may extend well into the plant construction period to better define the mechanical performance of equipment components such as pump seals.

Regardless of the exact developmental sequence, follow-up evaluation, the fine-tuning of the project economics, must never stop. As much as any factor, the projected profitability guides the fate of the project.

3.2 MARKETING RESEARCH

The Scope of Marketing Research

In the context of project evaluation requirements, the purpose of marketing research is perfectly straightforward: how much of a particular product can be sold, and at what price? The quantity that can reasonably be sold is of interest because it establishes the size of the plant that is to be used for the investment analysis. The likely sales price is the key item required for the income analysis.

The methodology of marketing research is far from systematic. Basically, anything, any idea that yields the answers required, is used. Those ideas that have worked particularly well in the past are remembered more fondly and perhaps become part of the more formalized expositions of the scope and techniques of marketing research (see, for example, Pacifico and Williams, 1967). The responsibility for marketing research within the corporation is often rather loosely assigned, and whereas some companies have centralized commercial development departments, the tendency in recent years has been to transfer that responsibility to business teams which subdivide marketing research responsibilities among several functional groups through project budgetary control (Giragosian, 1978).

Particularly in the earliest stages of project evaluation, a great deal of the marketing research must be done by individuals who are members of the functional group handling the project; this usually includes the project manager. Since most projects involving process and product development are handled by the R&D function, it is that function that most often is given the job of early-stage marketing research. More advanced techniques of marketing research are entrusted to specialists in functional groups such as sales and marketing. The emphasis in the discussion to follow will be on early-stage techniques, with some elaboration of the data and results that may be expected from more advanced techniques.

The difficulty of the task of establishing demand and acceptable pricing is a function of the product history. Established products may be evaluated in a reasonably systematic way; particularly if the products are commodity chemicals, useful statistical information is available on production, demand, and pricing. The task becomes more difficult if the project involves an existing "fine" chemical. An entirely new product is the most difficult to evaluate, and various survey techniques are usually required to obtain the proper marketing perspective:

Detailed customer contact. Field calls by sales representatives to establish potential customer demand, price acceptance.

Analogous product analysis. Investigation of the marketing history of analogous products—forecasting by "historical analogy" (Pacifico and Williams, 1967).

"Leading indicator" and "user" analysis. Investigation of the projected growth of the main users of the new product. *Leading indicators* are any statistical data from which product demand and growth may be projected. For example, statistics on consumer trends in cotton apparel may be used to project demand for pesticides to control cotton crop pests.

Market development. Potential customers may have to be convinced of the value of a new product by giving them pilot amounts of the product to try out in their own operations.

Product Profile

The first step in marketing research (and in the overall evaluation sequence) is the assembly of information to formulate a product profile. It stands to reason that if an effort is to be expended to evaluate the profitability of manufacturing a product, all available know-how about the nature of the product should be assembled first.

At all stages of project development, consideration should be given to obtaining information, as complete as possible, in the following categories:

- *Product specifications*

    Purity: Nature and maximum allowable content of impurities. Specifications are frequently end-use dependent.

    Physical form: Phase behavior; limitation upon form imposed by customers' requirements—crystal form, particle size, color, stray odors, stickiness. Some properties may be defined in terms of specific testing methods. Distribution packaging may be specified—bulk, drums, fiberpacks, cylinders, etc.

    Formulation: Some products are blends of compounds, or solutions. Many require additives such as preservatives, antioxidants, anticaking agents, perfumes, buffers, dyes.

- *Product properties*

    Physical properties: These need not be exhaustively listed. The dielectric constant of liquids is of no great interest in the context of a project evaluation, unless, of course, the liquid is meant to be used as a transformer fluid. Properties of interest are those that generally characterize the product in light of its end uses.

    Toxicity (see Steere, 1967). *Short-range effects:* Skin exposure, eye exposure. Inhalation hazard: Threshold limit value (TLV) (allowable working space atmosphere concentration). Ingestion hazard. *Long-range effects:* Carcinogenicity, chronic toxicity, teratogenicity.

Safe handling (see Steere, 1967) Flammability (flammable range in air, flash point, ignition point). National Fire Protection Association (NFPA) ratings: health-fire-reactivity. Shock sensitivity. Results of calorimetric screening to test thermal stability—differential thermal analysis (DTA), Accelerated Rate Calorimetry (ARC). Dusting hazard (explosions). Protective equipment required for safe handling. Special handling procedures, transportation restrictions.
- *Product uses*
  Chemical profile: Summary of useful chemical reactions.
  Use areas: Existing uses, as well as proposals for new uses; check with TS&D function.

Many of the items that contribute to a finished product profile are not known during preliminary stages of development, but their early definition is most important. The nature of the product influences the course of marketing research, the choice of process, and, in fact, all the subsequent steps of project evaluation. For example, if HCl is the by-product of a chlorination process, an early decision must be made whether to produce an anhydrous HCl gas to be compressed and marketed in cylinders, or an aqueous solution. The decision profoundly affects the recovery and purification scheme for the HCl, storage and distribution, marketing decisions, and perhaps even profitability.

Note that our sequence of preliminary project evaluation said nothing about the noneconomic criteria of project success mentioned in Chap. 1. The consideration of noneconomic criteria spans the total project evaluation effort and cannot really be fitted into any particular slot in the sequence. The time to start is at the time when the initial attempts to assemble a product profile are made. Note that many of the items in the product profile constitute the required information for the construction of a product profile chart (Figs. 1.1.4 and 1.1.5).

3.3 DEMAND PROJECTION

Trend Analysis

Predicting the future is a difficult and risky business at best, and it certainly cannot be done without examining the past. To project the future demand for a chemical product, providing the product is one that is already part of the world of commerce, then historical production statistics can be of immense value. For existing products, marketing research involves the assembly of those production statistics which hopefully best fit the nature of the demand projection required; that is, the statistical data should be corporate, regional, national, or worldwide, depending on the kind of projection needed.

Corporate production data constitute proprietary information not readily available to the public, but certainly available to corporate marketing researchers. Regional data (for example, product demand in the southeastern United States) are extremely difficult to come by, although corporate marketing and sales departments try to assemble such data for the corporation's own planning and sales projections.

Excellent estimates are available for national and worldwide production; some useful sources are listed in Exhibit 3.3A. Note that in the discussion so far, it has been implied that future *demand* may be projected from historical *production* data. The assumption is implicit that past production has just matched demand within the historical price structure and that the price structure will remain approximately the same.

The principle involved in projecting demand from past production data is the *principle of historical continuity* (Park, 1973), which suggests that an established historical trend may be projected into the future unless there is a good reason to believe otherwise. With this principle in mind, the projection task then involves assembling the data, fitting the data into one of several possible trend curve formats, and finally extrapolating the curve. The assumption is that periodic production data do follow some general trend in spite of temporary upsets, that they are not random.

Analytical expressions used for the correlation of historical production data fall into the following categories:

1 Polynomial correlations. The general format is

EXHIBIT 3.3A

HISTORICAL PRODUCTION DATA: SOURCES

1 Faith, W. L., D. B. Keyes, and R. L. Clark: "Industrial Chemicals," 4th ed., Wiley, New York, 1975. About 150 chemical processes described briefly, including production statistics, block flow diagrams, uses, pricing, etc. New editions appear about every 10 years.

2 Kirk–Othmer, "Encyclopedia of Chemical Technology," 3d ed., McGraw-Hill, New York, 1978. This vast encyclopedic work has production data shown under specific chemical headings. So far new editions have appeared in less than 10 years.

3 Stanford Research Institute: "Chemical Economics Handbook," with monthly "Manual of Current Indicators," Menlo Park, Calif. An invaluable resource, continuously updated, with production statistics, process description, producers, pricing, uses, international trade. Available only to subscribing industrial customers. A 6-foot shelf of books.

4 Meegan, Mary K. (ed.): "The Kline Guide to the Chemical Industry," 3d ed., Charles H. Kline & Co., Inc., Fairfield, N.J., 1977. A more modest compilation of chemical industry statistics.

5 "Chemical Statistics Handbook," Manufacturing Chemists' Association, Washington, D.C. Annual.

6 "Chemical Statistics Directory," U.S. Department of Commerce. For detailed research, a directory outlining where information on specific products can be obtained.

7 The journal *PROMPT* (*Chemical Market Abstract*) has a great deal of current information in abstract form, with references.

$$p = a + bt + ct^2 + \cdots \tag{3.3.1}$$

where p is the production (or demand) during year t measured from some base year, and $a, b, c, \ldots$ are constants. The simplest form is one that reflects linear growth of demand with time:

$$p = a + bt \tag{3.3.2}$$

2 Exponential correlations. The general format is

$$\log p = a + bt + ct^2 + \cdots \tag{3.3.3}$$

The simple exponential

$$\boxed{\log p = a + bt} \tag{3.3.4}$$

is used very frequently to correlate production statistics exhibiting steady growth; expressions with terms higher in t are used to correlate data exhibiting a maximum and decline.

3 Modified-exponential correlations. These include two analytical functions previously mentioned—the Gompertz curve,

$$\log p = a - br^t \tag{3.3.5}$$

and the logistic curve,

$$p = (a + br^t)^{-1} \tag{3.3.6}$$

with $r < 1$. The modified-exponential curves are S-shaped (see Fig. 2.2.1). They reflect a growth pattern characterized by slow initial growth, a rapid intermediate growth, and a slow approach toward market saturation. They are used most frequently to correlate the demand for consumer products rather than for chemical commodities.

Of the various correlation methods, the simple exponential in Eq. (3.3.4) has proved to be the most generally applicable historically. The equation reflects the relationship

$$\text{Rate of growth of production} \propto \text{demand}$$

that is, $\quad\dfrac{dp}{dt} = k_1 p$

whence $\quad\ln p = k_1 t + k_2$

where $\quad k_2 = \ln p \text{ at } t = 0$

Equation (3.3.4) follows. If production data conform to this equation, then a plot on semilog paper is a straight line. A great number of chemical products exhibited an astoundingly steady semilog growth behavior during the fifties and sixties, the "Golden Age" of chemical industry growth. It almost made one believe that the proportionality between production growth and demand finally constituted that one economic "law" that proved reliable; then came the seventies and eighties. Nevertheless, production data continue to be shown on semilog plots.

An example of a successful semilog correlation is afforded by production data for sorbitol, Fig. 3.3.1. Interestingly enough, the trend of the fifties and sixties seems to have survived the year 1975, a serious recession year in the chemical industry. Note that the curve marked "sales" lies below that marked "production" since some production is "captive"; i.e., the producer uses up the product internally.

If the data for a particular product exhibit a steady growth pattern, an attempt should be made to fit the data to one of the two principal linear equation forms, (3.3.2) or (3.3.4). The best fit can be estimated visually by plotting on both cartesian and semilog coordinates. For purposes of projection, the trend of the most recent data only needs to be established. The best line through the data can be computed by using the techniques of linear regression analysis (method of least squares); the forecast of future demand then involves nothing more than extension of the best fit line, and if so desired, 90 percent confidence limits on the forecast may be computed (Mapstone, 1973). The fact of the matter is, however, that unless the data scatter badly, a surprisingly good linear fit may be made by "eyeballing" the best fit line.

FIGURE 3.3.1
U.S. production of sorbitol. (*Adapted from "Chemical Economics Handbook," Stanford Research Institute, Menlo Park, CA., by permission.*)

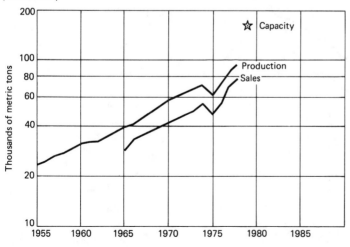

Clearly, the act of demand projection from the historical trend must be an informed one; the strictly mechanical extrapolation must be tempered with some understanding of the likely future commercial pressures upon the product. After all, many products do reach a stage of decline and obsolescence because of competition from a superior product, changes in consumption patterns, or unanticipated environmental problems. As illustrated in Fig. 3.3.2, the demand prognosticator in 1970 would have felt well justified to use the fitted line to predict the 1977 demand for dichlorodifluoromethane as 310,000 metric tons. The error of such a projection is quite apparent, and the 1970 prognosticator had no way of knowing about the upcoming controversy regarding the effect of dichlorodifluoromethane in aerosol can propellants upon the stratospheric ozone shield. A 1973 prognosticator, on the other hand, would have done well to consider what effect the current newspaper reports on dangers to the ozone layer would have upon the future use of fluorocarbons in spray cans.

Special techniques are required to fit data to nonlinear expressions. Mapstone (1973), for example, shows how the Gompertz curve may be fitted to applicable data.

FIGURE 3.3.2
U.S. production of fluorocarbons. (*Adapted from "Chemical Economics Handbook," Stanford Research Institute, Menlo Park, CA, by permission.*)

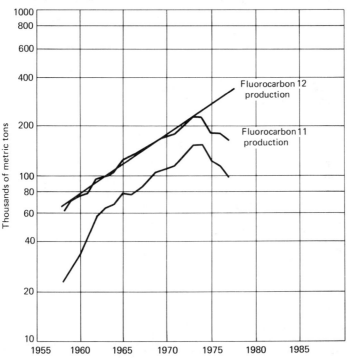

Price Elasticity of Demand

No matter whether a product is an existing one or a new one in the world of commerce, it is a matter of common experience that the demand for it is price-dependent. Corporate management that wishes to market a new product, or perhaps expand the sales of an existing one, must have some a priori idea of the *price elasticity of demand* of the product. Ideally, one would prefer to have available a curve analogous to Fig. 3.3.3, which shows just how much of a given product could be sold at any particular sales price. Unfortunately, a curve of this kind is extremely difficult to come by, and most projects do not have the benefit of such a quantitative aid. It is possible to generate such curves, but the effort is time-consuming and expensive, and the reliability of the result is by no means assured. The data required to generate such a curve are usually obtained by corporate sales representatives and purchasing personnel. The effort involves contact with existing and potential customers, usually at the purchasing or production planning level, by means of personal interviews, questionnaires, and other survey techniques. Giragosian (1967) gives an in-depth presentation of techniques needed to establish price elasticity, including "field call" techniques and effective reporting.

The price elasticity of demand may be defined quantitatively from a consumer demand curve such as that in Fig. 3.3.3 (see Wei, Russell, and Swartzlander, 1979). It is the percent change in demand for a product divided by the percent change in sales price causing the change in demand (the negative sign accounts for the normal negative slope):

$$E = -\frac{dp/p}{ds/s}$$

or

$$E = -\frac{s}{p}\frac{dp}{ds} \quad (3.3.7)$$

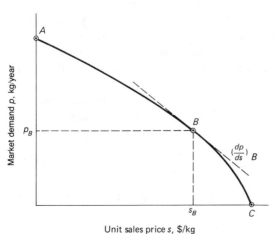

FIGURE 3.3.3
Market demand versus sales price curve.

Suppose point B in Fig. 3.3.3 represents the existing price-demand structure of a commercial product. Its price elasticity is

$$E_B = -\frac{s_B}{p_B}\left(\frac{dp}{ds}\right)_B$$

Normally the price elasticity is different at each point of the demand-price curve.

Exercise 3.3.1

Derive the analytical expression for a demand-price curve that exhibits a constant price elasticity over its full range. What is the expression for the elasticity E?

Even though quantitative data for characterizing price elasticity are rarely available, the elasticity of a product is frequently expressed in qualitative terms. *Unitary* elasticity is the term ascribed to a product for which a small percent price increase is perceived to result in roughly the same percent demand decrease. An *inelastic* product is one for which $E < 1$; that is, the price has a relatively small effect upon demand. This is true of chemical commodities which are "necessities of life," such as gasoline. *Elastic* products are those for which $E > 1$, and the price has a very important effect upon demand. Some luxury products, or products with latent competition, fall into this category. An example might be plastic food wraps, which compete with aluminum foil, wax paper, and rigid plastic containers.

Justification for the effort required to generate a market demand-price curve is usually not present until a project is in fairly advanced stages of development. For existing products, a rough approximation may be generated during early stages of development, but even that effort is not easy. Referring to Fig. 3.3.3, three points, A, B, and C, may be established. Point B represents the existing demand, how much of the product is currently sold at the prevailing price (although, as we will see, the actual selling price may be different than the listed price). Point C represents the price at which the *customers'* profit would be reduced to zero. The price can be calculated from the knowledge of the customers' process (or processes) and economics of that process, certainly not an easy task but not an impossible one. Point A represents the demand if the product were to be given away free; for the present rough purpose, this demand is the equivalent total capacity of all potential customers' plants.

Even after the three points have been generated, the question remains as to the best curve to be drawn through them. In the absence of restricting criteria, a hyperbolic correlation of the form

$$s = a_1 + a_2 p^{a_3} \tag{3.3.8}$$

is easy to apply; the three points are just sufficient to determine the constants a_1, a_2, and a_3.

Demand-price curves ("consumer demand" curves) such as the one in Fig. 3.3.3

may be used to compute an economically based optimum production plant size; the method will be presented in Chap. 8.

Other Aspects of Demand Projection

In our discussion of demand projection up to this point, we have shown that the demand for established products may be mechanically projected from historical production data, and we have said that it would be nice to have demand-price curves, but that these are hard to come by. This is not much to go on, particularly in preliminary stages of project evaluation, and obviously additional techniques must be employed to check the validity of the mechanically projected demand for established products, or to establish the demand for new products in the first place.

An important part of the marketing research procedure involving an established product is a survey of existing producers. Exhibit 3.3B shows some useful references which list producing plants and their locations as well as other current product statistics. Information in these references includes items such as total production capacity, the range of plant sizes, the geographical distribution of plants, and many other items which may be judiciously applied to the job of establishing a desirable capacity of a projected plant. Exhibit 3.3C shows the contents of a typical "Chemical Profile" published by the Schnell Publishing Company (see Exhibit 3.3B); the wealth of information needed for marketing research is apparent.

One of the most powerful methods of forecasting the demand of both established and new products involves *leading indicator* analysis. The concept is obvious—if one has information on the size and projected growth of the "users" of the product

---

EXHIBIT 3.3B

SURVEY OF CURRENT PRODUCTION: SOURCES

The sources in Exhibit 3.3A are useful. In addition, the following are well worth consulting:

1 "Chemical Profiles," Schnell Publishing Co., Inc., New York. Thumbnail sketches of principal chemicals—production, pricing, producers, uses, future prospects. This is a loose-leaf publication kept up-to-date by subscription. It is a companion to the weekly journal *Chemical Marketing Reporter;* a different "Chemical Profile" appears in it each week.

2 U.S. International Trade Commission: "Synthetic Organic Chemicals," U.S. Government Printing Office. An annual publication with producers and current production statistics.

3 "Directory of Chemical Producers—USA," Stanford Research Institute, Menlo Park, Calif. Annual publication.

4 *Chemical Engineering* magazine devotes part of one issue each year to a "Construction Alert," a summary of chemical production plants in preconstruction or construction status.

EXHIBIT 3.3C

THE CONTENTS OF A TYPICAL "CHEMICAL PROFILE"
(Copyright Schnell Publishing Company, Inc.)

All profiles consist of the following sections:

    Supply (producing companies and capacities)
    Demand
    Growth
    Price
    Uses
    Strength
    Weakness
    Outlook

The following examples are intended to convey the gist of material found in the various sections. Data on *sorbitol* will be used for the examples (data as of January 1, 1979).

1 Supply: producers and capacity

Producer	Capacity*
Ethichem Corp., Carlstadt, N.J.	5
ICI United States, Atlas Point, Del.	125
Lonza, Mapleton, Ill.	70
Hoffmann-La Roche, Belvedere, N.J.	45
Merck, Danville, Pa.	22
Pfizer, Groton, Conn.	80
Total	347

*Millions of pounds annually, 70% basis, including sorbitol not isolated during the manufacture of ascorbic acid and its salts. Hoffmann-La Roche consumes all its sorbitol captively in the manufacture of vitamin C. Lonza and Merck sell all their sorbitol. Remaining producers have captive requirements and also sell to the merchant market. ICI and Pfizer also make crystalline product: Pfizer's Groton plant can produce about 50 million pounds of this material a year.

2 Demand and growth projection

Current demand: 280 million pounds
Historical growth (1968–1978): 6.5% per year
Projected growth (1979–1983): 6.5% per year
Projected demand, 1983: 383 million pounds

3 Pricing

Current price: 34¢ per pound of 70% solution, tank cars, FOB works.

4 Uses

Confections and food, 32%; toothpaste and toiletries, 20%; ascorbic acid and salts,

18%; surfactants, 14%; pharmaceuticals and cosmetics, 8%; exports and miscellaneous, 8%.

5 Strength, weakness, and outlook

Typical comments might be:

"Outlets are growing well, but confections, particularly sugarless gums and hard candies, have increased their share of sorbitol consumption significantly and are expanding at about 12–15% a year."

"There is overcapacity in overseas markets, and exports have not shown consistency nor growth, but have averaged about 20 million pounds a year, including material sent to Puerto Rico."

"Sorbitol is a good growth product, with several attractive qualities and a good diverse base of markets. New capacity has recently been added and older plants shaped up, thus easing the tight supply situation that prevailed in the mid-1970's."

---

or other "leading indicators," one has a basis for projecting the demand and growth of the product itself. If the product has only a single use, then the total *potential* demand for the product is uniquely determined by the size and growth pattern of the user. An example of this type of one-on-one relationship is illustrated by the example in Fig. 3.3.4; this shows the annual production history of synthetic elastomers and the corresponding demand for rubber processing chemicals (antioxidants, accelerators, and modifiers). Note how the demand for the chemicals follows the elastomer production; the slopes of the growth lines are the same, and the major dips and surges are faithfully reproduced. The total production of elastomers thus establishes the demand for the processing chemicals, and the demand growth estimate can be tied into projections of the growth rate of synthetic elastomers.

In the more usual case, more than one leading indicator is involved in the projection. For example, the growth pattern of degreasing solvents is tied to the projected manufacturing activity of the automobile industry, the aircraft industry, the household appliance industry, and any other industry based on the mass production of machined parts. The leading indicators need not be immediate areas of use; population growth is often used as an indicator of the growth of established consumer products. In fact, in our discussion here we have used a rather broad interpretation of the term "leading indicator." Economists generally use the term to indicate some national political or economic trend which is likely to have a long-range impact upon a product. Examples of leading indicators in this sense are interest rates, or stock prices, or average family incomes. Such indicators may well be important in advanced stages of project evaluation but are only rarely considered in the earlier stages.

User and leading indicator statistics may be located in many of the references in Exhibits 3.3A and 3.3B; a few additional references are given in Exhibit 3.3D. During early project stages it is useful to undertake "back of the envelope" user analyses which may be extremely rough, but at least they establish the order of magnitude of product demand, and unattractive business prospects may be eliminated more promptly.

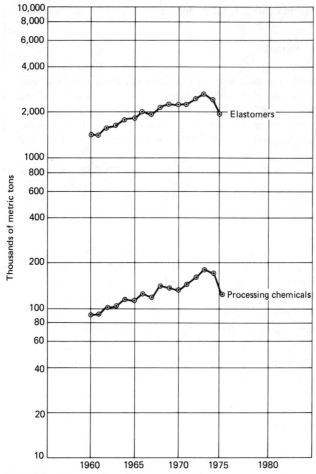

FIGURE 3.3.4
U.S. production of synthetic elastomers and rubber-processing chemicals. (*Data from "Chemical Economics Handbook," Stanford Research Institute, Menlo Park, CA., by permission.*)

Example 3.3.1

A new product is tied to the U.S. production of beer. Develop a rough estimate of beer production.
SOLUTION

$$\text{Population of United States} \approx 225 \times 10^6$$

Assume 60 percent of the population drinks beer, or 135×10^6. Assume the average per capita consumption = six-pack per week, or

> **EXHIBIT 3.3D**
>
> LEADING INDICATORS: SOURCES
>
> **1** U.S. Department of Commerce: "U.S. Industrial Outlook for 200 Industries." One- and four-year projections. Annual.
> **2** "Statistical Abstracts of the United States," U.S. Department of Commerce. Annual.
> **3** Charles H. Kline & Co., Inc., Fairfield, N.J., publish marketing research guides for various process industries. Examples of titles are "Paper and Pulp Industry" and "Plastics Industry."
> **4** A large number of industry-specific publications are in existence. In the food industry, for example, a useful annual publication is the "Almanac of the Canning, Freezing, Preserving Industries," Edward E. Judge & Sons, Inc., Westminster, Md..

$$6 \times 12 = 72 \text{ oz} = 0.562 \text{ gal}$$
$$52 \times 0.562 = 29.2 \text{ gal/year per capita}$$

Thus

$$29.2 \times 135 \times 10^6 = 3.94 \times 10^9 \approx 4 \text{ billion gal/year}$$

From "Statistical Abstracts of the United States," 1978 U.S. consumption was 5.5 billion gal/year.

For preliminary purposes, the accuracy of an estimate typified by the example is good enough. At least the business prospect is not based on 4×10^6 gal or 4×10^{10} gal.

It should be pretty evident at this point that the task of demand projection is far from systematic, that a great deal of insight and good judgment and determined data digging are required. Not only positive but negative indicators must be considered—the effect of competitive products, governmental regulations and restrictions, possible environmental effects of the product, and many others. Since Fluorocarbon 12 is one of the prime users of carbon tetrachloride (CCl_4 is one of the starting materials), the individual forecasting future demand for the chlorinated methanes in 1974 would have done well to consider the effect of prospective restrictions of fluorocarbon use upon prospects for CCl_4. The production of CCl_4, shown in Fig. 3.3.5, shows, in fact, much the same decline after 1974 as Fluorocarbon 12 in Fig. 3.3.2.

Forecasting by historical analogy is still another method of applying leading indicator statistics. An example, originally published by Van Arnum (1964), is shown in Fig. 3.3.6. The idea is to project the growth of a product by drawing a growth curve similar to that of a similar product on an offset time basis. In the example shown, a 13-year forecast of the demand growth of polyester fibers was

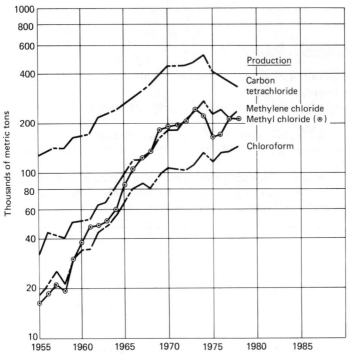

FIGURE 3.3.5
U.S. production of chlorinated methanes. (*Adapted from "Chemical Economics Hnadbook," Stanford Research Institute, Menlo Park, CA., by permission.*)

made in 1964 based on the growth of nylon fibers during the preceding two decades. Based on present-day knowledge, the growth curve, bold as it must have appeared at the time, has turned out to be quite conservative. Van Arnum could not have foreseen the explosive demand increase for polyesters in the 1965–1975 era. Note that the actual growth is linear on the usual semilog plot.

Market Penetration

The estimated product demand may actually be there, or it may be only the *potential* demand. The actual demand is one that is communicated by customers as a part of field studies, or it may be the result of extrapolation of current sales data. Potential demand, on the other hand, is one that is computed for a product assuming complete acceptance in user applications. In either case, the problem remains of making a reasonable estimate of *market penetration*. An even more difficult task is that of making *regional* demand and penetration estimates.

Regional considerations are important because a plant may be built to service a particular geographical region. Why should this be so important? For one thing, the

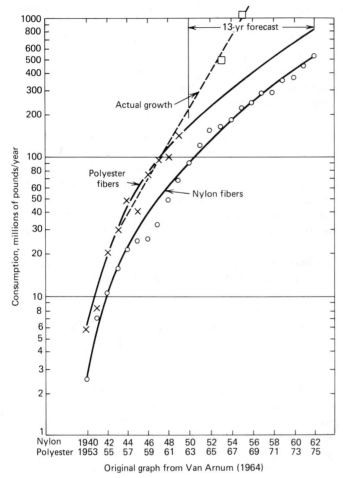

FIGURE 3.3.6
Forecasting by "historical analogy." (*Van Arnum, 1964; by permission, AIChE.*)

users of the product under consideration may be concentrated in a particular region; for instance, textile mills are concentrated in the southeast, and a plant to produce a new fiber antistatic agent may well be located in this region. Transportation costs result in a considerable advantage to regional plants satisfying regional demand. Railroad freight costs of 4 to 5 cents/lb (early eighties) for shipping bulk chemicals from Texas to the Pacific northwest result in an appreciable cost advantage to commodity chemical producers in the west satisfying western markets. Regional producers in general are able to give better service to their local customers, and even "regional loyalty" plays a role—local customers are eager to help themselves by promoting the regional economy through support of regional producers.

Regional production competition may be readily ascertained from the references in Exhibit 3.3B. For example, it is immediately obvious (Exhibit 3.3C) that sorbitol producers are, for the most part, clustered along the eastern seaboard. However, regional production is not the same thing as regional demand; the clustering does not mean that, for example, all food processing (a principal user of sorbitol) is located in the east, and a regional plant might well be justified to serve western food processors. On the other hand, the location of still another new sorbitol plant in New Jersey might be questioned.

There may be various reasons for the kind of geographical clustering observed with sorbitol plants, in spite of geographically diffuse demand. Probably the most obvious one is that certain areas have been historically developed as chemical manufacturing areas and have many of the advantages associated with optimum plant siting—good transportation, an experienced labor force, readily accessible raw materials, more accommodating political institutions, and so forth. Moreover, successful chemical producers have developed their own *integrated complex* in one or more key locations—an assembly of diverse chemical producing plants tied together by a system of auxiliary facilities such as steam plants, railroad loading facilities, employee services, and administrative centers. There are compelling reasons to build new plants as an extension of such an integrated complex. *Grass roots* plants—isolated producing units with their own auxiliaries, built on virgin sites—are relatively expensive and must be justified by extraordinarily compelling regional demand reasons. The selection of optimum plant locations is a whole separate discipline with its own requirements of expertise (see, for instance, Landau, 1966).

As was the case with overall demand, regional demand may often be established by reference to statistics on user areas and leading indicators. Caustic soda, for example, is used for the removal of the outer skin layer (epicarp) of a number of fruits and vegetables as part of the commercial canning process. Regional demand for 50% NaOH destined for the canning markets can be established from statistics on canned peaches (southeast), hominy (south), or peeled machine-harvested tomatoes (central California). Corporate marketing departments make an effort to assemble good regional demand statistics for their own products, based on field inquiries by sales personnel. If product demand is geographically diffuse and primarily population-dependent, then population distribution is a good indicator of regional demand. As an example, some 15 percent of the population of the United States lives in the western states[1] (west of the Rocky Mountains); it is reasonable to assume that some 15 percent of the demand for dry-cleaning solvents will be centered in the west, barring secondary effects of weather upon type of apparel.

The likely market penetration of new products, regional or nationwide, depends upon a large number of factors (see Giragosian, 1978); each product is unique and must be judged on its own merits. Obviously, a safe and effective new cancer chemotherapeutic agent can be expected to approach rapidly 100 percent market penetration. But even otherwise excellent products may find market penetration

---

[1]From "Statistical Abstracts of the United States."

slow going. A new, unusually effective broadleaf herbicide for railroad and transmission line rights of way may experience disappointingly slow market acceptance, and why should this be so? Among many possible reasons, the slowness may simply be due to customer inertia to change, or loyalty to existing suppliers of the old herbicide, or reluctance to modify existing spraying equipment, or fear of the unexpected environmental effects of a more powerful agent. It is usually best to limit the pioneer plant to a relatively low percentage of the potential market, perhaps 10 to 20 percent, and expand when demand justifies it.

On the other hand, the tendency currently is to build very large plants to satisfy demand growth for large-volume commodity chemicals. Plants to satisfy 35 to 50 percent of the projected regional demand growth during the coming decade are quite reasonable. Regardless of demand analysis, however, the selected plant size or size range must satisfy other important criteria:

1 An acceptable criterion of economic performance. Economic acceptability is the ultimate criterion. It may well happen that projected demand analysis will establish a plant size so small that production costs would be uneconomically high, and production would not be profitable with the anticipated price structure.

2 Criteria of corporate operational policy. The size range may violate such criteria as availability of internally generated raw materials, long-range energy utilization policies, or availability of investment capital.

Example 3.3.2

The year is 1973, and a decision is to be made whether a plant to manufacture perchloroethylene on a western site would be a good business prospect. The existence of a chlorine-producing integrated complex in the west is assumed. 1980 is to be taken as the key projection year.

SOLUTION Based on the material presented in this chapter and the references listed in the exhibits, the procedure for projecting a demand-based plant size might be about as follows:

Historical U.S. production data up to 1973 are shown in Fig. 3.3.7. Using an eyeballed straight line to extrapolate the data, the 1980 production of C_2Cl_4 is judged to be 1640 million lb/year.

Differences between production and consumption (due to exports and imports) will be neglected.

Corporate data indicate that, in 1973, 10 percent of U.S. consumption was in the west. Some 1973 western trends:

1 Current use almost all in dry cleaning.

2 Apparel use indicates retreat from polyesters (do not require dry cleaning).

3 Hydrocarbon dry-cleaning agents (Stoddard solvent) being phased out due to air pollution problems.

4 Fluorocarbon dry-cleaning agents exhibit very slow market penetration due to high costs.

5 Under Rule 66 (California state air pollution regulation), C_2Cl_4 is an "exempt" solvent because of proven low tendency toward smog formation. Phasing out of trichlo-

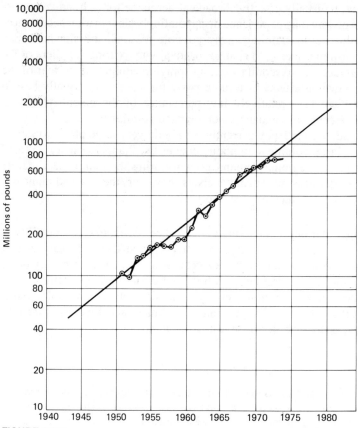

FIGURE 3.3.7
U.S. production of perchloroethylene. (*Adapted from "Chemical Economics Handbook," Stanford Research Institute, Menlo Park, CA., by permission.*)

roethylene from vapor-degreasing applications expected to create new markets for perchloroethylene.

Some limitations:

1 Perchloroethylene process coproduces CCl_4, and markets must be found for coproduct (raw material for booming fluorocarbons).

2 Chlorinated solvents under "cloud of suspicion" due to association with chlorinated hydrocarbon pesticides alleged to cause environmental problems.

The balance is well on the side of an optimistic projection, and in 1980 consumption in the west is expected to match the proportion of population, or 15 percent. Therefore

$$1980 \text{ west coast demand} = 246 \text{ million lb}$$

In 1973, the only listed western producer of C_2Cl_4 (Exhibit 3.3B) had a stated capacity of 20 million lb/year (although capacity was listed as "extremely flexible"). The nearest producers were in Texas; the listed price at the time was 10 cents/lb, and freight charges for railroad shipment from Texas to the west coast averaged 3 cents/lb, an appreciable differential.

This information indicates that 50 percent penetration of the 1980 market is reasonable; however, so that plant will not be oversized for lower markets before that date, choose a plant for 40 percent of the 1980 west coast market, or 100 million lb/year.

This, then, might be chosen as a basis for subsequent economic evaluation of the project.

In retrospect, how good a business prospect would the plant in the above example have been? The answer is given in Fig. 3.3.8, which gives U.S. production figures for C_2Cl_4 five years after those of Fig. 3.3.7. The fact of the matter is that, far from growing in the linear fashion assumed, C_2Cl_4 production experienced a drastic "flattening out" in the seventies. In view of the optimistic predictions in the example, what happened? There were two key considerations:

1 It must be understood that *all* C_2Cl_4 sold into the dry-cleaning industry eventually winds up in the environment. Even though C_2Cl_4 may decompose harmlessly in the atmosphere to CO_2 and chlorides, the sum total of environmental, energy, and cost considerations has forced the dry-cleaning industry to tighten up on its

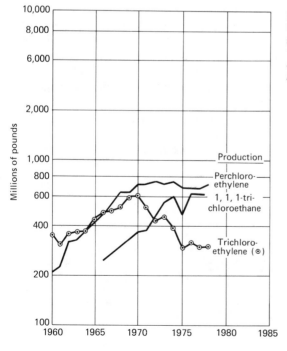

FIGURE 3.3.8
U.S. production of chlorinated solvents. (*Adapted from "Chemical Economics Handbook," Stanford Research Institute, Menlo Park, CA., by permission.*)

emission control. Vents on dry-cleaning machinery have been provided with efficient condensers and even charcoal adsorbers, and greater care has been taken to remove solvent residues from apparel before removal from the machinery. These steps have obviously reduced the demand for fresh perchloroethylene.

2 The anticipated penetration of perchloroethylene into the vapor-degreasing market has not materialized. This can be seen in Fig. 3.3.8; the slack due to phasing out of trichloroethylene has been taken up by 1,1,1-trichloroethane, since existing vapor degreasers could be more easily converted to its use rather than to C_2Cl_4.

The result is that U.S. demand at the end of the seventies was about 700 million lb annually, and the West Coast demand of 15 percent would barely exceed 100 million lb. Recent data show that the one existing West Coast producer increased output to 40 million lb, and it is clear that a new 100-million-lb plant would have had a hard struggle. Unfortunately, such is the fate of many plants in the process industry; to predict the future is, indeed, a risky business.

Product and Plant Life Cycles

There are those commodity chemicals, such as chlorine, caustic soda, or ethylene oxide, which are closely tied to the gross national product and have historically exhibited steady growth. Extrapolation of historical data is a reasonably reliable operation. There are other products, perhaps specialties and formulations such as pharmaceuticals or metal polishes, which exhibit a characteristic life cycle and lifetime. The volume of demand typically follows some such pattern as shown in Fig. 3.3.9. Region A represents slow market introduction, penetration, and acceptance. Region B is a period of rapid growth as the product proves its worth and gains acceptance. Region C is a period of maturity; the potential demand is saturated, and growth, if any, is moderate, perhaps in conformity with the GNP. The final region D is a period of decline due to product obsolescence and competition from alternatives. The point T represents the lifetime of the product.

The pattern in Fig. 3.3.9 is, of course, different for each product, although

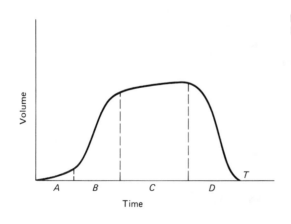

FIGURE 3.3.9
Product life-cycle chart.

groups of products (detergent formulations, herbicides) may exhibit some similarity. In any case, the prediction of the shape for a new product is clearly very difficult, and the unpredictability must be accommodated by appropriately high values of the criterion of economic performance chosen to characterize the production economics.

For many existing products the life cycle pattern may not be obvious due to the long time span of the various regions. Region A plus the first part of region B, and parts of region C, may appear as straight lines on the usual semilog plots over a period of many years. It is interesting to conjecture whether the flattening out of the perchloroethylene production in Fig. 3.3.8 corresponds to the maturity region C.

So far we have been examining the *product* life-cycle concept; however, it is important to distinguish between product life cycles and *plant* life cycles. If a product is made in a unique plant, then the two cycles coincide. The introductory region A is hopefully accommodated by a semiplant. The plant life cycle may then be roughly projected in terms of an estimated product lifetime T and the production rate at maturity, i.e., production to meet estimated market penetration. For purposes of economic evaluation, the production volume-time relationship may be approximated as constant annual production at capacity, or perhaps more realistically by some of the distribution patterns in Fig. 2.2.1, such as the truncated normal distribution curve.

If the plant is not unique, regardless of whether the product is new or established, the plant life cycle is still similar to the pattern of Fig. 3.3.9. This is because much the same marketing patterns shape the curve. For example, slow market penetration as well as unanticipated start-up problems may reduce production in the new plant. Decline in plant production may be caused by product, process, and plant obsolescence.

The plant life is, in general, completely different from the product life. Moreover, plant life is different from average *plant equipment* life. In the chemical industry, the average life span of equipment is about 11 years (Park, 1973). Quite often the plant life span is longer than that; plant life does not end as equipment wears out but is continually regenerated through equipment replacement. If, due to decline in demand, plant life is shorter than equipment life, the terminated plant may have considerable salvage value.

During preliminary stages of project evaluation, projection of plant life-cycle curves is obviously difficult, and the following simple rules are used:

1 Unless there are other good reasons, *estimate plant life to be 10 years* (i.e., average equipment life).

2 Postulate *constant annual production at rated capacity*.

Clearly, these simple rules can raise problems unless managers making key project decisions are aware of their use. The evaluator is obligated to undertake a *sensitivity analysis* to gauge the effect of less than full-scale production, and marketing analysts must make the effort to establish a more reliable time-volume relationship during intermediate project stages.

Long-Range Corporate Planning and Technological Forecasting

Demand projection extends beyond the immediate scope of a specific project evaluation. The continued economic health of a corportation depends upon the judicious projection of future demand trends for its full spectrum of products. Moreover, technological trends need to be monitored to focus upon new products likely to meet changed societal needs in the decades to come; this is the function of the intriguing discipline of *technological forecasting* (see Olenzak, 1972). Without such long-range planning, the corporation is unable to initiate those new ventures and changes that will allow it to keep up with the times.

Such long-range projections are obviously difficult and risky—perhaps not so much so in the case of commodity chemicals which are permanently established in the economy, but risky nonetheless. In the case of integrated complexes, long-range projections are further complicated by the restrictive interrelationship of certain products. This sort of situation is typified by chloralkali complexes, illustrated schematically in Fig. 3.3.10. Regardless of what market projections might be, attempts to meet them will be restricted, for example, by the interrelationship between the products of brine electrolysis. It is not possible to make chlorine without making caustic soda. A certain degree of flexibility is available among some of the products shown. The ratio of CCl_4 to C_2Cl_4 may be varied over a considerable range, but the HCl product is then inevitably locked into the ratio selected. Clearly, the long-range growth of the various products shown must inevitably be closely interwoven.

FIGURE 3.3.10
A chloralkali complex.

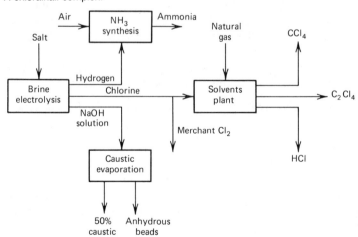

Exercise 3.3.2

Derive a relationship between the kilograms of HCl produced per kilogram of a CCl_4–C_2Cl_4 mix as a function of the CCl_4 weight fraction. Use the formula CH_4 for natural gas.

The complexity of multiproduct interrelationships is formidable, indeed, in industries such as petroleum refining. Long-range projections of the product spectrum must account for the effects of energy and environmental crises, the demands of international politics, and economic forces. Interdiscipline teams are used to analyze future areas of industrial activity, regional economic trends, and perceived societal needs to make reasonable technological forecasts for the guidance of corporate management. It is such guidance that is needed to start the plastics manufacturer on a course of research into biodegradable beverage containers or to indicate to the petroleum refiner whether entry into the geothermal energy area is justified.

3.4 PRICE PROJECTION

Pricing of Established Products

We now turn our attention to the other side of the marketing research coin, the projection of product price. As we have seen, the product price is closely tied to product volume, although the exact relationship is usually difficult to establish. In the case of existing commercial products, the obvious basis for the economic evaluation of business prospects is the current listed price.

Perhaps the best source of current prices of chemical products is the weekly journal, *Chemical Marketing Reporter*. The listings shown must be understood in the context of a few basic rules and facts:

1 The quotations reflect the list prices of producers accounting for at least 80 percent of annual U.S. production.

2 For many products, two prices are listed; these represent the range of quotations from producers in different localities.

3 The quoted prices may or may not reflect freight charges; this is indicated next to each quotation. For example, "FOB Works" indicates free on board at the producing site; i.e., the quotation includes charges for delivering the product aboard the carrier at the manufacturing plant site. Some prices are for products delivered at the customer's site. In the absence of any indication, it is understood that the quotation is for large lots FOB New York.

4 Prices quoted are a function of the type of packaging (bulk, drums, bags), product volume (car lots or less), and product quality (technical, USP, others). All these factors must be carefully evaluated in picking a likely sales price for economic evaluation.

Railroad freight charges for bulk chemicals may be an appreciable fraction of the sales price, and a new producing plant in a consuming, nonproducing region may have a very real marketing advantage. In an attempt to compete with this regional advantage, more distant producers will sometimes "swallow" at least part of the freight charges and quote "freight-equalized" prices—that is, prices which are competitive with those of a product manufactured at a nearby site with a correspondingly lower freight burden.

The price history of established products may be obtained from a number of sources, most notably the "Chemical Economics Handbook" published by the Stanford Research Institute. Unlike production statistics, price statistics do not follow a time-consistent pattern and cannot be extrapolated, as is readily apparent from Fig. 3.4.1, the sales price history of phenol. By 1980, the price had risen above the scale shown, to 36 cents/lb. The startling post-1973 rise was due to skyrocketing costs of energy and other inflationary pressures. Between 1950 and 1970, phenol prices experienced a steady decline in spite of moderate inflation, a reflection of the stiff competition in the phenol business and the decrease in costs of production as the result of technological improvements.[1]

The list prices, of course, do not immediately reflect the pricing strategy of individual corporations. In the case of commodity chemicals in particular, actual sales prices may be different to each customer, a matter of individual contract negotiations and guarantees involving product specifications and quality variance.

---

[1] See discussion of "experience curves" in Wei, Russell, and Swartzlander (1979).

FIGURE 3.4.1
U.S. list price, synthetic phenol. (*Composite data from Chemical Marketing Reporter and "Chemical Economics Handbook," Stanford Research Institute, Menlo Park, CA.*)

The unit sales value of the product may consequently be appreciably lower than the listed price. Pricing strategy may involve techniques such as *penetration* pricing, the purposeful use of initially low prices for an established product exhibiting some price elasticity. *Skimming* is the exact opposite, a strategy of artificially high initial prices for new products that are perhaps unique, or at least have little competition. Giragosian (1978) describes many aspects of such pricing strategies, and Doerr (1980) emphasizes the importance of focusing upon an overall marketing strategy.

Pricing strategies are outside the province of early-stage project evaluation. The rule is to *use the current list prices* for establishing project economics. If the project cannot meet acceptable criteria of economic performance under this rule, all the pricing strategy in the world is not likely to do much good.

Criteria of Final Pricing Decision

The economic criterion, the criterion of acceptable economic performance, is, with very few exceptions, the dominant one in setting prices in the chemical process industry. Corporate management has the responsibility of choosing the profitability index type and magnitude that it believes best fits its own business organization and current economic realities. For each project, however, the final choice must be tempered with other criteria of commercial acceptability of the product.

We have seen, for example, that the economics of production of established products is limited by the existing price structure. That structure may not be compatible with the price necessary to meet the required profitability. Intense competition and oversupply have a depressing effect upon the price of many commodity chemicals, and producers who wish to penetrate such competitive markets and still realize acceptable profit margins must sometimes build very large plants, indeed, to gain the advantage of large-scale lowering of unit costs (i.e., costs of manufacturing one kilogram of product). If a relationship is perceived between price and demand, this relationship may also be incompatible with the simultaneous choice of production level based on market projections and price based on acceptable margin of profit. A final marketing research decision is often the result of a great deal of balancing and compromising.

Pricing considerations up to and beyond the primary criterion of acceptable economic performance must be used to arrive at a reasonable sales price of new products. If the new product is competitive with other similar products already on the market, the price must be such as to offer the customer some obvious economic advantage. A new specific herbicide for weed control of wheat fields may be five times as effective per kilo as the nearest competitor, but the price per kilo cannot be five times as high, for, all other things being equal, no economic benefit accrues to the customer. Holding the price down to, say, twice that of the competition may, however, be a strong incentive to switch.

Even if a product is unique, its price range can often be narrowed by analyzing the extent of economic benefit to the customer. For example, a new compound which helps plants to better assimilate nitrogen from fertilizers may double the fertilizer effectiveness (half the required fertilizer application rate per acre). The gross benefit to the farmer is 50 percent of the fertilizer cost, but again the price of the new compound cannot absorb the total benefit, for the farmer would have no incentive to use it.

3.5 FLOW SHEET DEVELOPMENT METHODS IN COST ESTIMATION

Choice of Process

The project evaluation sequence now progresses to process definition. If there is a choice among processes to make a particular product, this is the time to make it.

Even if the project involves a new product, process choices may exist. For established products, several process choices are almost always available. The references in Exhibit 3.5A are particularly useful as sources of process information. The procedure of establishing a choice involves decisions based upon a number of criteria:

EXHIBIT 3.5A

PROCESS DESCRIPTION SOURCES

1 Stanford Research Institute: "Process Economics Program" (PEP Reports), Menlo Park, Calif. A vast library, each volume covering a single product in admirable detail—complete process descriptions, economics, and production statistics. Available only to subscribing industrial customers.

2 Stanford Research Institute: "Chemical Economics Handbook," with monthly "Manual of Current Indicators." See Ref. 3, Exhibit 3.3A.

3 Kirk–Othmer: "Encyclopedia of Chemical Technology," 3d ed., McGraw-Hill, New York, 1978. A gold mine of process description.

4 Faith, W. L., D. B. Keyes, and R. L. Clark: "Industrial Chemicals," 4th ed., Wiley, New York, 1975. A valuable source—about 150 chemical processes described briefly, including block flow diagrams, uses, economics.

5 Shreve, R. N., and J. Brink: "Chemical Process Industries," 4th ed., McGraw-Hill, New York, 1977. Great detail—industries described in "blocks." Only general background for preliminary estimates.

6 Weissermel, K., and H. Arpe: "Industrial Organic Chemistry," Verlag Chemie GmbH, Weinheim and New York, 1978. Excellent overview of reaction schemes, favored processes, production statistics in the organic chemical industry—emphasis is primarily upon European practice.

7 Chem Systems International Ltd.: "Chemical Process Economics," Parts I and II, London, March 1981. Excellent summaries of over 125 chemical processes. Each item includes a process description, simple flow sheet, and economic data in U.S. dollars for a Benelux location. Obtainable only by subscription.

8 McKetta, John J., and William A. Cunningham: "Encyclopedia of Chemical Processing and Design," Marcel Dekker, Inc., New York. Well-presented, up-to-date process description entries. Full set of volumes has not so far been published.

Economics. The criterion of the most favorable economics is an obvious one, but the student may protest at this point that the goal of the evaluation procedure is to establish the process economics in the first place; how, then, can comparative process economics be used to make a decision at the beginning of the evaluation sequence? In some cases, preliminary information may be available in sources such as those in Exhibit 3.5A. In other instances, the process alternatives may exhibit rather obvious economic impact differences. For example, if the process choice involves one of two alternate catalysts, one of which demands pure methane as feed whereas the other can use natural gas, the latter is likely to offer a considerable economic advantage. On the other hand, a process choice may involve the alternate use of air or oxygen. The temptation is to avoid the use of oxygen and the associated expense of an oxygen plant, but that expense may be more than balanced out by reduction in equipment size due to elimination of large volumetric throughputs of nitrogen. In many cases where the economic choice is not obvious from the start, parallel economic evaluation procedures may have to be undertaken for the process alternatives, at least during the preliminary stages of project development.

Simplicity. If there is a choice, choose the process that is simple, that involves standard and well-known unit operations and processes, even if the more complex procedure promises to result in cost savings. This is particularly true for unproved or only partly developed processes. Tube polymerization may result in some substantial cost advantages over polymerizations carried out in a series of stirred reactors, but unless the former has been tried or is otherwise known to be a practical approach, it is best to choose the more expensive alternative for the initial evaluation work. In this way a conservative estimate is obtained, and unpleasant surprises are avoided in case the more expensive alternative is later found to be necessary, and uneconomical.

Raw materials. The preferred process is one which utilizes raw materials which are readily available and over which the corporation preferably exerts some form of control. The importance of a favorable raw materials position has been already emphasized. A producer who wishes to manufacture ethylene and has a reliable hydrocarbons position is more likely to select a process involving hydrocarbon cracking rather than ethanol dehydration.

By-products or coproducts. By-products may occasionally make the economics of a process look particularly attractive. Nevertheless, if there is a choice, pick the process that produces a minimum of by-products, for eventually these are as likely as not to create a marketing headache. Keep it simple!

State of know-how. If at all possible, stick with processes that are commercially proven in preference to those that have limited developmental backup.

Patent coverage. The preferred process is one that is backed by patents issued to the producing corporation. The choice of a process patented by others and requiring the payment of royalties is justified only by extraordinarily compelling reasons. *Turnkey plants* (standard plants which can be purchased as ready-to-go contractor-built units) to manufacture intermediates, such as methanol, for internal consumption may involve patented processes and royalty payments.

Energy usage. Energy shortages and high costs point toward the process with smallest energy usage, all other things being equal.

Wastes and environmental considerations. Skyrocketing costs of ultimate waste disposal make this criterion of process choice an unusually important one. Restrictions upon emissions, both voluntary and imposed, result in difficult choices regarding ultimate disposal. Tar wastes may be burned, but it is necessary to clean up the products of combustion, acid gases such as SO_2 or HCl. These may be absorbed in a caustic solution, but we are faced with the disposal of an aqueous salt solution. If the solution is evaporated in solar ponds, there is the matter of disposal of the solids. This sort of endless chain can impose enormous economic penalties upon the process that creates unwanted wastes.

Safety aspects. It is just as well to stay away from those processes that involve safety and loss prevention hazards such as unstable intermediates—peroxides, acetylides, azo and nitroso compounds, and many others; highly flammable solvents; or unusually toxic compounds. If there is no choice, there are ways of accommodating such hazards, but the economic penalty is high, and the element of doubt and suspicion remains to haunt the producer.

An excellent example of process comparison is given by Brownstein (1975).

Note that many of the process choice criteria are similar to the noneconomic process criteria considered in Chap. 1, and the product profile chart is a useful comparison tool.

If a new product is the subject of the economic evaluation procedure, the concern is not so much process choice as process definition. The chemical engineer can usefully contribute to this definition even during the earliest stages of process development by working along with the research chemists, guiding their choice of process conditions and reaction solvent carriers, helping with the definition of kinetic parameters, suggesting the best approaches to product recovery and purification.

Block Flow Diagrams

Once the process is chosen or defined, the next step is to arrange the known information into a basic process sequence in written form. The follow-up procedure is a critical one—the translation of the written information into a diagram that represents a network of interrelated operations, the integrated process. This undertaking of *process synthesis* is one of the most challenging tasks the chemical engineer can assume, one that demands the utmost in creativity and innovation. Process synthesis is an evolutionary operation; the very step of creating a diagram that symbolizes the process suggests changes and modifications and new ideas useful in guiding subsequent process development efforts. The evolutionary aspect is a long-range one that carries over into advanced stages of process development.

Process synthesis is an intuitive skill, an "art," but one that can be practiced and developed. Chemical processes do have some common systematic aspects; in fact, most processes can be generalized in the manner illustrated in Fig. 3.5.1. The core of the process is the reaction, or at least some form of physical transformation. The raw materials may require various types of conditioning (comminution, purification, phase transformation) before introduction into the reaction zone. Reaction

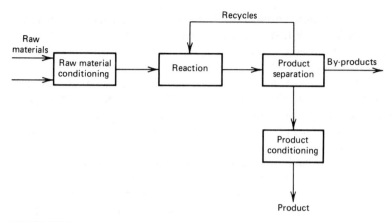

FIGURE 3.5.1
A prototypic chemical process.

products must be separated into recycle streams (unreacted raw materials, solvent carriers), by-products and waste products, and the desired product, which itself may require further conditioning (purification, formulation, packaging). This chemical process prototype represents the skeletal structure of the most complex processes, some of which may be made up of a sequence of such prototypes (the product of one "block process" is the feed to the next one).

The simplest diagram to represent a process, and one which should be drawn as a first step, is the *block flow diagram*. Some of the symbolic features of such a diagram are illustrated in Fig. 3.5.2. Each block represents a unit operation ("Distillation"), a process step ("Color removal"), or perhaps even an item of equipment ("Sulfonator"). Arrowed lines connecting the blocks represent the principal process streams, which may be labeled for clarity ("Extract," "Raffinate"). Terminal lines (i.e., lines not connecting two blocks but those that start at a block and go "off the paper," or vice versa) represent feeds and products; these *must* be properly labeled. It is one of the purposes of the block flow diagram to force the designer to think

FIGURE 3.5.2
Block flow diagram symbolism.

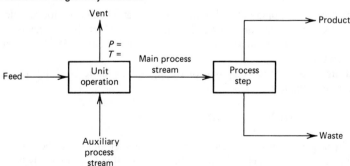

about the significance of these terminal lines. If the line represents a raw material, where does it come from? Will storage and transfer facilities be required? Is raw material conditioning required that should be represented by further blocks? A vent line must be similarly questioned—is an atmospheric vent acceptable, or should additional blocks to represent vent scrubbing be shown? In this way the designer is forced to develop a diagram that is complete, for failure to incorporate required process steps will result in serious errors in the subsequent economic analysis. Any dangling off-the-paper lines, particularly those representing wastes, must be clearly identified as to ultimate disposal, and the indicated disposal facilities (biooxidation ponds, thermal oxidizers, land disposal sites) must either be existing ones or else must be accounted for separately during the economic analysis.

Key operating conditions, such as unusual pressures or temperatures, may also be shown on the block flow diagram, as well as indication whether some steps are to be batch operations. In chemical engineering practice, continuous processing steps are greatly preferred by reason of economy and simplicity; nevertheless, some operations may have to be chosen as batch steps because of their nature or because insufficient information is available to visualize a continuous operation.

Example 3.5.1

Investigate the possible production of monoethylamine using alkylation of ammonia:

$$CH_3CH_2Cl + NH_3 \longrightarrow CH_3CH_2NH_2 \cdot HCl$$

Draw a *block flow diagram* for the process, which involves the reaction of a 20:1 NH_3–ethyl chloride mixture at 600 psig and 225°C for 1 min. The NH_3 is added to the reaction mix as a 2:1 H_2O–NH_3 aqueous solution, and the mixture is homogeneous at reaction conditions.

Per pass, 99% C_2H_5Cl is reacted; 60% ethyl chloride forms the primary amine $CH_3CH_2NH_2$, 30% forms the secondary amine $(CH_3CH_2)_2NH$, and 10% forms the tertiary amine.

The reaction product is cooled and mixed with 50% NaOH sufficient to convert all HCl to NaCl, including HCl from the hydrochloride salts of the amines. The reaction

$$NH_4Cl + NaOH \longrightarrow NH_3 + NaCl + H_2O$$

is forced to completion by stripping out NH_3 (and unreacted ethyl chloride) for recycle. The amines remain dissolved in the aqueous NaCl solution, from which they are removed by distillation. (Boiling points: primary amine 17°C, secondary amine 56°C, tertiary amine 90°C, no azeotropes.)

The unwanted higher amines are disposed of by high-temperature cracking on the bricks of a regenerative furnace:

$$(C_2H_5)_2NH \longrightarrow NH_3 + 4C + 4H_2$$

The ammonia-hydrogen off-gas is recycled into the process; clean H_2 may be vented. The carbon remaining on the bricks is periodically burnt off to provide the cracking heat.

Raw materials for the process are liquid ethyl chloride, compressed NH_3 gas, and 50% NaOH.

SOLUTION A block flow diagram is shown in Fig. 3.5.3. Note how formulation of the diagram has led to some process modifications and improvements not considered in the written information:

Heat interchange of reactor feed and products, an energy conservation device
Use of water first used for vent scrubbing as feedwater to ammonia absorber

Note also that waste streams such as the aqueous sodium chloride are not only properly identified, but clear indication is given that additional processing is probably necessary. This is a "red flag" to draw attention to the fact that proper disposal will have to be resolved before a meaningful economic analysis of the project can be accomplished. Ultimate disposal need not involve an economic penalty; the vented hydrogen, for example, may represent an important economic credit if used as a supplementary fuel in a furnace or steam plant.

Note the relationship between the ethylamine process block flow diagram and the prototype process in Fig. 3.5.1, a relationship that should be kept in mind as part of the process synthesis procedure. The absorber represents the "raw material conditioning," and the reactor-heat interchange-neutralizer combination is the reaction system. "Product separation" incorporates the stripper and columns. The cracking furnace and scrubber represent another sequential process involving a by-product of the first process.

FIGURE 3.5.3
Ethylamine process. Block flow diagram.

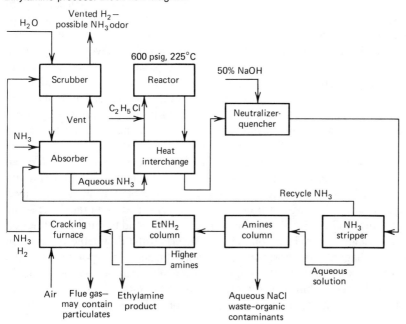

Exercise 3.5.1

Suppose the tertiary amine is found to be an additional marketable product. Indicate how the process could be modified to recover the tertiary amine with no secondary amine production.

The question of the systematization of process synthesis has been addressed by King (1974). He finds that certain logical elements of the synthesis procedure may, indeed, be identified, and that a considerable degree of systematization, perhaps even automation, may be incorporated in each element to the extent that computers may be used as an aid in process definition. Nevertheless, the overall grand design remains a matter of intuition, innovative skill, and inspiration which no degree of systematization can replace.

Early-Stage Mass and Energy Balances

There is no chemical process that can be properly evaluated without a mass and energy balance. The balances define the block flow diagram process quantitatively and form the basis for subsequent preliminary equipment sizing. The very act of making a mass and energy balance suggests needed process modifications which may well be reflected by changes on the block flow diagram.

In early stages of process development, the information required to make balances may be very limited. This obviously raises problems, for it is not possible to do a partial or incomplete mass balance. It is necessary to define chemical reactions, including side reactions, as soon as possible, for these define the mass balance around the reaction system of the process prototype in Fig. 3.5.1. The mass balance around the other elements of the process prototype is a matter of the desired separations; these are generally fixed by the requirements of the process. Assumptions based on intuition and experience must often be made; for instance, the products of the distillation of multicomponent mixtures often have to be specified with no a priori assurance that the spectrum of separations can be practically attained. Clearly, the procedure of developing a mass balance is evolutionary, something that is true of all elements of the project evaluation sequence. In the early stages of process development, the mass balance focus is upon the principal process components, with gradually increasing emphasis upon the nature and behavior of minor components.

The results of a process mass balance are incorporated in a *flow table*, an important developmental tool that, the first time around, may be based upon and may serve as an adjunct to the block flow diagram. Flow table formats depend upon project requirements, the stage of process development, and corporate lore; some features are shown in Fig. 3.5.4. Stream numbers correspond to numbered lines on the block flow diagram (or, later, on the flow sheet). If the flow table is based on the block flow diagram, the corresponding mass balance incorporates only the principal process streams; at a later stage, the flow table may reflect more "finely tuned" mass balances within one or more blocks of the block flow diagram.

Componer⁺ flows may be shown as either mass or mol units per hour. The flows

Stream No. Component	1	2	3
Total lb/h			
Total mol/h			
Temp, °C			
Pressure, psig			
Density, lb/ft^3			
Flow, gpm (acfm)			

FIGURE 3.5.4
Elements of a flow table.

may be based upon projected plant capacity, although if this is not yet firmly determined, the flows may be normalized (for example, based on 100 mol/h of product). The flow table should also show the principal properties of each stream: temperature, density, and volumetric flow.

Many thermal and thermodynamic properties are not available during early project stages; nevertheless, the calculation of the heats of reaction is a must. Reaction heats are computed from the heats of formation of reactants and products; data are available for a great many compounds in dozens of handbooks and specialized compilations, but information on organic compounds is limited. Stull et al. (1969) is a good source of existing data; thermal and thermodynamic properties can be estimated by using methods such as those reviewed by Reid et al. (1977). An important method of estimating reaction heat, certainly quite acceptable in early process development stages, involves estimation on the basis of structural analogy. The underlying assumption is that the heat of a reaction is primarily determined by the bond energies of chemical bonds broken and formed as part of that reaction, and that atomic groupings which do not directly participate in the reaction have only a minor effect upon the reaction heat (i.e., upon the participating bond energies). Since thermodynamic data are available for relatively simple organic compounds with a given reactive grouping, reaction heats involving more complex molecules with the same reactive grouping may be estimated.

Example 3.5.2

Estimate the heat of reaction for the alkylation of ammonia with α-picolyl chloride at 298 K:

$$\text{(pyridyl)}-CH_2-Cl(g) + 2NH_3(g) \longrightarrow \text{(pyridyl)}-CH_2-NH_2(g) + NH_4Cl(s)$$

SOLUTION   Handbooks do not list the heats of formation of picolyl chloride or picolylamine, but an analogous reaction is

$$C_2H_5Cl(g) + 2NH_3(g) \longrightarrow C_2H_5NH_2(g) + NH_4Cl(s)$$

The handbook values for ΔH_f° are

$C_2H_5Cl(g)$	-25.1 kcal/(g·mol)
$NH_3(g)$	-11.04 kcal/(g·mol)
$C_2H_5NH_2(g)$	-11.6 kcal/(g·mol)
$NH_4Cl(s)$	-75.38 kcal/(g·mol)

$$\Delta H_R = -11.6 - 75.38 - (-25.1) - 2(-11.04) = -39.8 \text{ kcal/(g·mol)}$$

Exercise 3.5.2

Using handbook values (for example, Lange's "Handbook of Chemistry")*, estimate the above heat of reaction from the reaction

$$CH_3Cl(g) + 2NH_3(g) \longrightarrow CH_3NH_2(g) + NH_4Cl(s)$$

In your opinion, are the two answers close enough for preliminary evaluation purposes? Why do you think there is a difference?

Process Flow Sheets

Process flow sheets are an extension, an elaboration of block flow diagrams. Their purpose is not only to illustrate the basic process relationships but also to define the equipment needed to carry out the process steps and to indicate need for utilities and peripheral facilities. The process flow sheet serves as an index for generating equipment lists and equipment costs needed for a reliable investment analysis. As such, the flow sheet must be complete; *all* new equipment associated with a proposed process must be shown, or at least identified in sufficient detail to ensure proper economic consideration. Failure to incorporate needed equipment into a flow sheet results in a sequence of interdependent errors leading up to the most common of the economic evaluation failures—an underestimate of investment requirements.

Process flow sheets may be generated directly from the block flow diagrams by selecting the most likely equipment needed to perform the function represented by each box on the block flow diagram. The blocks of equipment are then arranged into a process train; of course, a certain amount of coordination is needed to fit these blocks together, an evolutionary procedure in itself. In fact, many items of equipment cannot be completely defined "the first time around." For example, if the block flow diagram box indicates liquid-liquid extraction, the first process flow sheet (and the one possibly used for the initial economic evaluation) might show

---

*12th ed., McGraw-Hill, New York, 1979.

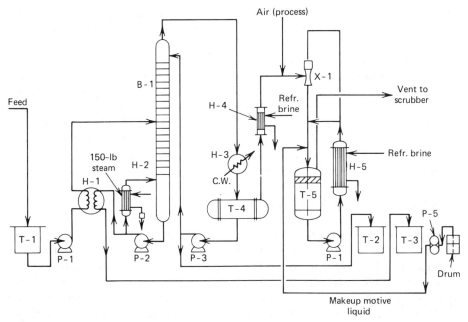

FIGURE 3.5.5
Distillation equipment.

an extraction column, perhaps a Rushton column. Subsequent design might indicate that mixer-settler units would be more advantageous, and the subsequent phase flow sheet would then show, say, two stages of mixer-settlers.

To illustrate the kinds of decisions that need to be made to generate the initial process flow sheet, consider the "translation" of a block flow diagram box labeled "Distillation." The equipment that could at least potentially be associated with this unit operation is shown in Fig. 3.5.5. The "standard" equipment associated with a distillation step is

B-1 Distillation column
H-2 Reboiler
H-3 Overhead condenser
T-4 Distillate accumulator

Even here choices must be made which may be modified by subsequent design calculations. A plate column is shown, but a packed column may eventually be chosen. The reboiler is a forced circulation exchanger (the bottoms stream is recirculated with P-2), but a free circulation reboiler (calandria) may be preferred.

Other items of equipment are optional or depend upon the nature of the distillation operation. The interchanger H-1 is a potential energy-saving device, but it might not be incorporated if its cost does not justify the energy saved or if the bottoms stream might freeze in such a device. These kinds of decisions are part of the advanced-phase evolutionary process. The reflux pump P-3 is required if H-3 and T-4 cannot be located in such a way as to take advantage of gravity flow. Some plant operators prefer to use pumped reflux and to have H-3 and T-4 at ground level for easy maintenance; others prefer overhead condensers and accumulators to be integral parts of the column hardware.

A good part of the equipment illustrated in Fig. 3.5.5 is associated with vacuum distillation and would not be required with operation above atmospheric pressure. The system shown represents a liquid eductor as the vacuum source; other possibilities are a vacuum pump or a steam ejector system. The proper choice is largely a matter of processing conditions and energy economy. Whether the storage tanks T-1, T-2, and T-3 are required is a matter of coordinating the distillation step with other process steps. Surge capacity may, indeed, be desirable, but it is not unusual to have the product of one operation feed the next operation in a sequence directly.

A few aspects of flow sheet construction methodology merit emphasis:

Equipment Pictographs Over the years, more or less generally accepted pictographic symbols have evolved for the more common items of equipment, and attempts have been made to collect the most widely used versions (Austin, 1979). A few such symbols are illustrated in Fig. 3.5.6; even these common ones, however, are far from standardized, and each industrial entity seems to have its own favorites. The pictographs do serve the function of facilitating visualization of equipment interaction, and the flow sheet designers are urged to create their own descriptive symbols for unusual equipment. Do not use boxes; save those for block flow diagrams.

Equipment Labeling Equipment labels are even less standardized, and each corporation does, indeed, have its own system. A reasonable labeling code is reproduced in Exhibit 3.5B.

Valves and Instruments Valves and instruments normally need not be shown on process flow sheets unless they help in visualizing process operations. For example, in Fig. 3.5.5, a control valve and a pressure controller symbol might be shown on the air line adjacent to eductor X-1 to explain the possibly puzzling reason for the air line, namely, pressure control in the distillation system:

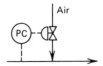

Utilities and Auxiliary Facilities In general, it is not necessary to show on process flow sheets equipment that is associated with existing utilities and auxiliary facilities within the integrated complex. Thus in Fig. 3.5.5 the 150 psig steam (one of the standard generated steam pressures) required for H-2 is shown as a utility stream with no part of the generating equipment shown, for this is not new equipment—it is already available. Similarly, the condensate from H-2 is shown passing

SECTION 3.5: FLOW SHEET DEVELOPMENT METHODS IN COST ESTIMATION

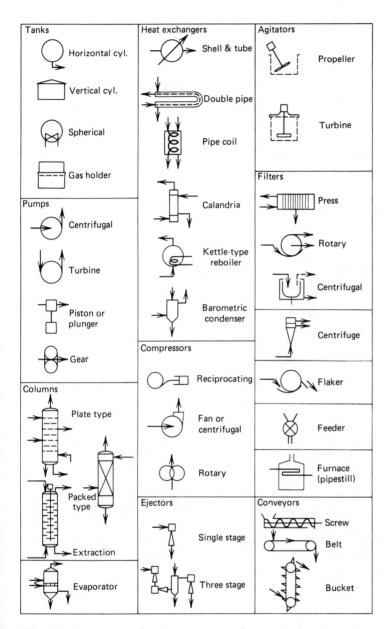

FIGURE 3.5.6
Some typical equipment pictographs.

> **EXHIBIT 3.5B**
>
> FLOW SHEET LABELING OF EQUIPMENT
>
> Each installed item will be designated with a letter prefix for type identification, a serial number, and a letter suffix to indicate parallel or spare service.
>   The letter prefixes used to identify the various types of equipment are as follows:
>
>   (A) Agitators, portable type or removable. If agitator is an integral part of the equipment, do not assign it a separate number.
>   (B) Distillation columns, quench columns, scrubbing columns, acid drying towers, absorption columns, stripping columns, extraction columns
>   (C) Compressors, blowers, fans, vacuum pumps, jet pumps
>   (D) Driers, drying ovens, spray driers, drum driers, flakers, dehydrators, desiccators
>   (E) Evaporators
>   (F) Filters, centrifuges, screens
>   (H) Heat exchangers
>   (L) Lifts, hoists, chain falls, jib cranes, conveyors, elevators
>   (P) Pumps
>   (R) Reactors, reaction vessels, converters; in general, any piece of equipment used to contain materials during chemical reaction
>   (T) Tanks, drums, receivers, hoppers, boxes, separators, decanters, vent bottles
>   (U) Furnaces, burners, direct-fired heaters, boilers, flare stacks, inert-gas generators
>   (X) Miscellaneous equipment, blenders, cyclones, multiclones, scales, tables, grinders, pulverizers, rotary valves, feeders
>
> The equipment shall be numbered from "1" up in each classification. In many cases, it will be desirable to set aside blocks of numbers to identify equipment in various process steps, operating areas, or similar service. A master list of assigned numbers should be maintained by the various plants or by the engineers.
>
> Several systems for identifying parallel and standby items can be used. In general, equivalent items in parallel trains or which operate together in parallel service shall have the same number with letter suffixes "A," "B," "C," etc. The same system can be used when one of the units is in standby service. Alternately, a common spare for duplicate parallel units can be identified with the addition prefix "S." A common spare for two units in dissimilar service should have the same number as the operating unit in the more severe service, with a suitable suffix.

through a steam trap and off the paper, but existing condensate collecting and recycling facilities are available. For a grass roots plant, the distillation flow sheet utilities might be shown in the same manner, but separate flow sheets would be necessary for the required new steam-generating facilities (although these might be estimated as a "package" unit rather than separately designed).

In the same way, cooling water for H-3 is shown without associated equipment: water tower or feed pump (which serves many cooling water "users"), cooling

SECTION 3.5: FLOW SHEET DEVELOPMENT METHODS IN COST ESTIMATION **165**

towers, and so forth. On the other hand, the flow sheet indicates the need for other auxiliary facilities, and the project evaluator is forced to ask questions about these. For instance, the use of a refrigerated-brine coolant is indicated in H-4, the accumulator vent condenser, and in H-5, the eductor motive liquid cooler. Is refrigerated brine available for general plant use in the integrated complex? If not, a refrigerated-brine system must be designed for use in the contemplated plant, and a separate flow sheet should be developed for purposes of economic analysis, unless packaged systems can be purchased. The vent from T-5, which may contain vapors of the motive liquid, is shown as going to a scrubber. Is such a scrubber available? If not, a plant-dedicated scrubbing system must be included in the overall process flow sheet.

Process Information It is always helpful to incorporate some process information into a flow sheet to facilitate understanding of the overall process. The more advanced the stage of development, the more information is likely to be shown on the flow sheet, and advanced-stage flow sheets may consist of dozens of sequential drawings. Some commonly appearing items of explanation are:

Equipment names ("accumulator vent condenser")
Basic equipment specifications (vessel volumes, exchanger heat-transfer duties)
Intermediate stream labels ("reflux")
Extreme or unusual operating conditions (temperature, pressure)

Example 3.5.3

An existing facility produces an adipic acid product in a crystalline powder form. A small but potentially profitable market exists for a repurified product in denser form than the powder. It has been proposed that a small plant addition be constructed incorporating the following steps:

Redissolve powder in hot water (25 wt %, 100°C).
Cool solution to 30°C (1.5 wt % solubility).
Filter resulting crystals on a rotary vacuum filter.
Wash crystals on filter with cold water; recycle combined wash water and mother liquor back into existing process.
Remove water below 2 wt % by melting crystals and heating to 170°C in a jacketed kettle (adipic acid melting point=150°C).
Flake molten product on double-drum flaker.
Bag flakes in 50-lb bags.
Make a flow sheet sketch incorporating *all* new equipment required for the addition. Assume all auxiliary facilities (utilities, process water, warehouse, etc.) are available.
SOLUTION We start by drawing a block flow diagram, Fig. 3.5.7, a relatively straightforward procedure. We note, however, that vapor from the evaporation step must be properly disposed of, particularly since it is likely to contain some adipic acid. A logical disposal step suggests itself—condense the vapor, and combine the condensate with the mother liquor returned into the existing process. The additional step eliminates an environmental nuisance and results in an economic benefit in the form of recovered adipic acid.

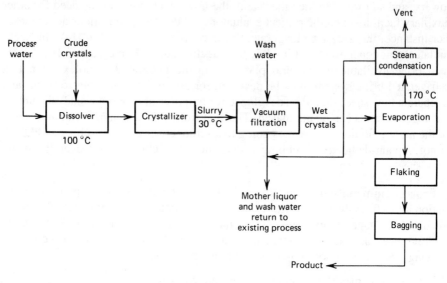

FIGURE 3.5.7
Adipic acid purification: block flow diagram.

The process flow sheet, Fig. 3.5.8, is derived from the block flow diagram. Interstep coordination is typified by the "crystallizer" block. In the absence of more detailed information on crystallization, a Swenson-Walker crystallizer has been selected, a water-cooled jacketed trough with an internal screw that scrapes the forming crystals from the cold wall and moves them forward. This is not always the cheapest device for crystallization, but, among other advantages, it avoids the nuisance of crystal adherence to heat-transfer surfaces. A mini flow sheet representing the crystallization step might include feed and product storage; coordination with the other steps eliminates the need for a feed tank, for the dissolver fills that need. A feed pump is required, but the filter vat serves as a perfectly adequate slurry receiver.

Note that an attempt has been made to be complete by including conveying equipment, all rotary vacuum filter auxiliaries, and a polishing filter for the dissolved crude crystals.

Exercise 3.5.3

In Fig. 3.5.7, can you see some ways of attaining economy of water usage? Could some of the mother liquor be used for dissolving the crude crystals?

In advanced stages of plant design, another category of flow sheets is developed, the *piping and instrumentation diagrams* (P&IDs). In these diagrams, *all* process pipelines are shown and specified, including all valves, insulations, tracing, orifices, and even reducers, as well as *all* instruments, including instrument lines,

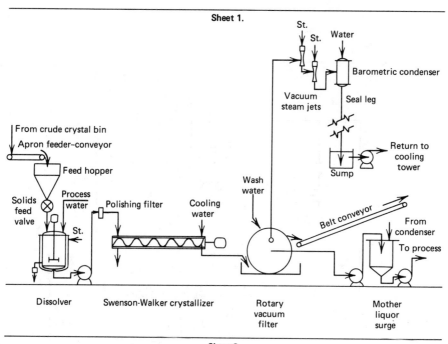
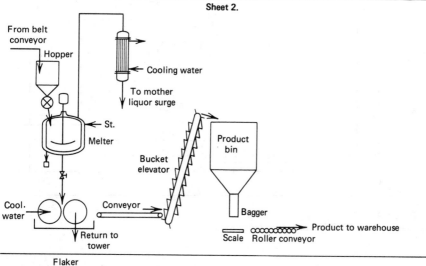

FIGURE 3.5.8
Adipic acid purification flow sheet.

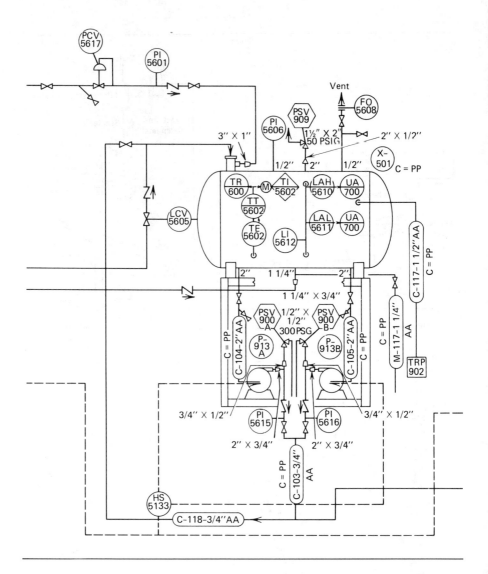

FIGURE 3.5.9
Small section of a P&ID (boiler feedwater deaerator).

relief devices, computer connections—in fact, any aspect of piping and instrumentation. An example is shown in Fig. 3.5.9, a P&ID of a boiler feedwater deaerator.

The P&IDs serve as detailed "blueprints" of the plant to be built, as sources for detailed piping and instrumentation costing, and as references for the construction of piping isometrics, instrument installation diagrams, and plant construction models.

Some Frequent Flow Sheet Errors

The principal sins associated with flow sheets used for economic evaluation are sins of omission. Of all the sins, the worst one is the process line arrow pointing to the edge of the paper with no firm definition or identification shown. Other frequently omitted items include:

Raw materials and product storage tanks. These must be sized to conform to reasonable volumes of deliveries and shipments. For example, raw material storage volume may be dictated by a combination of daily process requirements, volume of railroad tank cars making the delivery, and a reasonable frequency of delivery.

Product check tanks or bins. Products that have imposed specifications should first be stored in check tanks pending lot analysis. If specifications are met, the product can be transferred into final storage; if not, the material is transferred back for reprocessing or blending.

Surge tanks. Surge capacity is needed between various process steps to allow for temporary shutdown of operating blocks for maintenance without disrupting continuous operation of the balance of the plant. For example, large distillation columns are often difficult to start up, and if there is a sequence of several columns, some intercolumn surge capacity will be most welcome by plant operators, sophisticated control systems notwithstanding. The choice of surge tank location and capacity is a matter of careful process analysis and downtime forecasting; for preliminary estimation, surge capacity of 4 or 8 hours is often assumed. Some progress has been made toward mathematical systematization of optimum surge tank policy (Henley and Hoshino, 1977).

Duplicated equipment. Certain items of critical equipment subject to particularly severe service should be duplicated and piped up in parallel for alternative use. Examples are principal process pumps and critical compressors; cartridge filters and other items of equipment operated on alternate cycles must, of course, also be duplicated.

Equipment accessories and auxiliaries. Quite often these are not shown on the flow sheet, with serious effects upon the subsequent economic analysis. For example, a vacuum source must be associated with the use of a vacuum rotary filter, as in Fig. 3.5.8.

Start-up equipment. Start-up problems should be recognized early in process development and equipment provided to cope with them. If a vapor phase reactor is to be operated at high temperature and the reaction kinetics depend upon that temperature, how will that temperature be attained in the first place?

Emergency items. Suppose even a slight possibility exists that a reaction may undergo a thermal runaway, rendering the contents explosive. What equipment, such as quench pots for the dumped contents, is needed?

Provisions for treating purges and side cuts. Any closed loop stream is likely to accumulate undesirable impurities, and even if the nature of such impurities is not known in early process development stages, some provisions for purging and purge treatment should be made.

Utilities not available for general use in the integrated complex. Typical ones are Dowtherm and other hot heat-transfer media, refrigeration systems, and inert gas systems. These systems should be shown on separate flow sheets, or else indication should be given that they can be purchased as package units.

Waste stream treatment   Many years ago flow sheets were replete with the little symbols

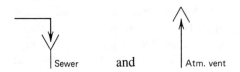

These are still legitimate ultimate disposal methods, provided that the waste streams involved are compatible. It is the responsibility of chemical engineers to make sure that such streams are, indeed, compatible and that proper waste treatment has been provided. The flow sheet must reflect this top-priority concern for the environment, and if specific waste treatment facilities are not available in the integrated complex, they must be provided as part of the flow sheet and as part of the investment for the proposed plant.

The modern integrated complex must be thought of as an essentially closed system as far as wastes are concerned. The ultimate internal disposal of the many kinds of waste streams—vapors, solids, aqueous streams, nonaqueous streams—poses an extraordinary challenge to creative chemical engineers. The exportation of wastes beyond the boundaries of the complex must be viewed as a manifestation of failure to deal effectively with internal problems of disposal.

NOMENCLATURE

a, b, c	constants
E	price elasticity (dimensionless)
ΔH_f°	standard heat of formation at 25°C, kcal/(g · mol)
ΔH_R	heat of reaction, kcal/(g · mol)
k	a constant (such as $k_1, k_2, \ldots$)
p	annual production, kg
r	a constant
s	unit sales price, $/kg
t	time, years

REFERENCES

Austin, D. G.: "Chemical Engineering Drawing Symbols," George Goodwin, Ltd., London; Wiley, New York, 1979.

Brownstein, A.M.: Economics of Ethylene Glycol Processes, *Chem. Eng. Prog.*, **71**(9):72 (September 1975).

Dobie, John B., Phillip S. Parsons, and Robert G. Curley: Systems for Handling and Utilizing Rice Straw, *Trans. ASAE,* **16**(3):533(1973).
Doerr, J. S.: Marketing of Commodity Chemicals, *Chem. Eng. Prog.*, **76**(6):27 (1980).
Giragosian, Newman H.: "Chemical Marketing Research," Reinhold, New York, 1967.
——(ed.): "Successful Product and Business Development," Marcel Dekker, Inc., New York, 1978.
Gouw, T. H.: "Guide to Modern Methods of Instrumental Analysis," Wiley, New York, 1972.
Grover, P.: A Waste Stream Management System, *Chem. Eng. Prog.*, **73**(12):71 (1977).
Henley, Ernest J., and Hiromitsu Hoshino: Effect of Storage Tanks on Plant Availability, *Ind. Eng. Chem., Fundam.*, **16**(4):439 (November 1977).
King, C. Judson: Understanding and Conceiving Chemical Processes, *AIChE Monogr. Ser. No. 8*, (1974).
Landau, Ralph: "The Chemical Plant: From Process Selection to Commercial Operation," Reinhold, New York, 1966.
Maples, R E., and M. J. Hyland: What Is Involved in Major Venture Financing?, *Chem. Eng. Prog.*, **76**(1):24 (1980).
Mapstone, George E.: Forecasting for Sales and Production, *Chem. Eng.*, **80**:126 (May 14, 1973).
Olenzak, Albert T.: Technological Forecasting: A Pragmatic Approach, *Chem. Eng. Prog.*, **68**(6):27 (June 1972).
Pacifico, Carl, and Roger Williams, Jr.: Marketing and Marketing Research, in "Kirk–Othmer Encyclopedia of Chemical Technology," 2d ed., vol. 13, p. 66, Wiley, New York, 1967.
Park, William R.: "Cost Engineering Analysis," Wiley, New York, 1973.
Peters, Max S., and Klaus D. Timmerhaus: "Plant Design and Economics for Chemical Engineers," McGraw-Hill, New York, 1980.
Reid, Robert C., John M. Prausnitz, and Thomas K. Sherwood: "The Properties of Gases and Liquids," 3rd ed., McGraw-Hill, New York, 1977.
Steere, Norman V. (ed.): "Handbook of Laboratory Safety," The Chemical Rubber Co., Cleveland, Ohio, 1967.
Stull, Daniel R., Edgar F. Westrum, Jr., and Gerard C. Sinke: "The Chemical Thermodynamics of Organic Compounds," Wiley, New York, 1969.
Valle-Riestra, J. F.: The Evaluation of the Present Worth of Normally Distributed Cash Flows, *Eng. Process Econ.*, **4**:37 (1979).
Van Arnum, K. J.: Measuring and Forecasting Markets, *Chem. Eng. Prog.*, **60**(12):18 (December 1964).
Weast, Robert C., and Samuel M. Selby: "Handbook of Tables for Mathematics," 3rd ed., The Chemical Rubber Co., Cleveland, Ohio, 1967.
Wei, J., T. W. F. Russell, and M. W. Swartzlander: "The Structure of the Chemical Process Industries: Function and Economics," McGraw-Hill, New York, 1979.

PROBLEMS

3.1 Prepare a product profile on one of the following established commercial products:
  a Maleic anhydride, commercial grade
  b Glycerine, USP
  c Salicylic acid, USP

3.2 What is the threshold limit value (TLV) of a vapor? How is it measured? What are the TLV values for H_2S, HCN, phosgene, methylene chloride, octane, DDT? For information, see "Threshold Limit Values for Chemical Substances" published annually by the American Conference of Governmental Industrial Hygienists, Cincinnati, Ohio. See also Steere (1967).

3.3 How is the TLV determined for a mixture of vapors? A laboratory atmosphere contains 3 ppm HCl, 6 ppm H_2S, and 10 ppm nitric oxide. Does it meet the recommended TLV? See Prob. 3.2 for references.

3.4 Another product property that can usefully be included as part of the product profile is the National Fire Protection Association rating. Describe the NFPA ratings and the corresponding hazard identification system. Look up NFPA ratings of the following chemicals:

50% NaOH
Hydrogen fluoride
Benzene
Pyridine

Guess, or make up your own, ratings of the following:

Aqueous nitric acid
NaCl brine
Gasoline
Sodium metal
Vinegar

Information on NFPA ratings may be obtained from the periodically published booklet, "Hazardous Chemicals Data," National Fire Protection Association, Boston, Mass. A useful companion booklet by the same organization is the "Manual of Hazardous Chemical Reactions." Steere (1967) has some NFPA ratings.

3.5 Dangerous thermal instability of chemical products may be monitored by means of calorimetric screening systems such as differential thermal analysis (DTA) and accelerated rate calorimetry (ARC). What are these systems, and what kind of information do they give? For information, check books on "Thermal Analysis" (for example, Gouw, 1972) or technological encyclopedias under the headings "Thermal analysis" or "Calorimetry."

3.6 The national production of an organic intermediate was 110 million lb in 1972 and 220 million lb in 1979. What is the likely production in 1985?

3.7 The rapid growth of demand for polyester fibers is shown in Fig. 3.3.6. The production subsequent to the data shown is as follows (in millions of pounds per year):

1968	1081
1970	1465
1972	2328
1975	2995
1976	3340

Predict the 1980 demand, and check the validity of your prediction against available 1980 data.

3.8 The following data represent the production history of a poultry feed supplement (in millions of kilograms per year):

Year	Production	Year	Production
1965	9.0	1974	18.5
1966	10.7	1975	15.0
1967	11.0	1976	20.9
1968	11.0	1977	22.0
1969	10.5	1978	24.0
1970	12.5	1979	28.0
1971	18.5	1980	40.0
1972	17.0	1981	35.5
1973	18.0	1982	35.3

a Plot the data on semilog paper, and predict the 1985 production by drawing the best eyeball straight line.

b Perform a least-squares regression analysis to predict the 1985 production. How do the two answers compare?

3.9 In Prob. 3.8b, what are the 90 percent confidence limits of the predicted production? For a rapid review of the required methodology, look over the discussion on "Regression Analysis" in Peters and Timmerhaus (1980) or the article by Mapstone (1973).

3.10 The data points in Prob. 3.8 exhibit a certain amount of fluctuation. The process of extrapolation is sometimes aided by computing and plotting a *moving average;* such a plot suppresses the amplitude of fluctuations and results in a more obvious trend (see also discussion in Mapstone, 1973). Starting with 1970, calculate the 5-year average production (that is, 1966 to 1970 for the first point), and plot this moving average versus year (1970, etc.) on semilog paper. Extrapolate a best-fit eyeball line to find the 1985 moving average, and back-calculate the predicted 1985 production.

3.11 It is interesting to conjecture whether the flattening of the perchloroethylene growth curve in Fig. 3.3.8 is due to saturation of existing markets and whether the growth pattern can be approximated by the Gompertz curve, Eq. (2.2.13). Starting with the year 1955, fit the production (not sales) data in Figs. 3.3.7 and 3.3.8 to the Gompertz curve format given by

$$\log_e(\text{production} \times 10^{-6} \text{ lb/yr}) \equiv \ln P \equiv y = a - br^t$$

where t is the time in years (for 1955, $t = 1$), and a, b, and r are constants to be determined. As a rough approximation, use the three-point method (Mapstone, 1973). Proceed as follows:

a Arrange the data between 1955 and 1978 (24 years) into three consecutive groups of eight. Find the arithmetic average $\bar{y}$ of each group, and associate this with the median year $\bar{t}$ of each group (example: $\bar{t} = 4.5$ for the first group).

b Substitute these values into the given format to yield three equations which can be solved simultaneously to give values of a, b, and r.

c Plot the resulting curve on rectangular coordinate paper as production (not the log) versus year, and reproduce the actual data. Do you consider the fit to be satisfactory?

d Use the derived parameter values to predict perchloroethylene production in 1980 and 1985. How well did you do? Consult current production data.

3.12 The following comments are intended to represent a grossly simplified and abbreviated example of the results of a marketing field survey to establish a consumer demand curve. See if you can generate such a curve from the information given.

The product in question is a formulated resin bead that the industrial customer fabricates into a film which is distributed and sold directly to the consumer. There are three suppliers and five principal industrial customers. The following table gives the current consumption and average price paid by each individual customer:

Customer	Consumption, million lb/yr	Price, ¢/lb
A	3.5	32.5
B	3.0	35.0
C	1.5	34.0
D	1.5	36.5
E	0.5	35.0

Price differences are the result of contract negotiation skills and differences in product specifications.

The field survey report sums up interviews with each customer's purchasing manager:

A "The cost of raw materials is only a small part of our operating costs; price changes over a relatively wide range would probably have little effect upon our operations and sales. Now, you folks have done a good job for us, you have a good product and your pricing is fair, but if you are thinking of raising your prices, we'll just have to go with the cheapest supplier."

B "Our profit margin is real tight, and we face stiff competition from similar consumer products. Each extra penny we pay for your beads will have to be charged to our customers. Let's say you double your price; we will have to hike the price of our film correspondingly, and I would guess we would lose a quarter of our customers."

C The purchasing manager was extremely irritated by our questions and at first refused to discuss pricing, claiming we were out to raise prices to drive him out of business. He finally calmed down and admitted that raw materials costs were actually a small proportion of his total costs. He confirmed that competition among products serving a similar function was stiff and precariously balanced. He thought that if the product price contribution of raw materials was taken out entirely, there would be little effect upon demand—"maybe 20 percent more."

D The agent here was very understanding and cooperative, willing to share his company's operating costs and pricing strategy. He is very worried about inflation. "If the cost of our beads triples—and let me tell you, that's not far-fetched, you people yourselves raised your prices 50 percent in the last 3 years—you can just bet we're out of business."

E "I am a small operator, and I'm barely hanging on. If I have to increase my prices, I'll drive my customers to buy competitive products. Now, of course, a price decrease would be great, but I don't think it would have much effect on my sales, even if I passed on all of my savings. If my bead cost went down to 20 cents/lb, I doubt we could pick up one more customer for every ten we have. And as for price increases, I don't even want to think about that—we could swallow 5, maybe 10 cents a pound, but, believe me, if my cost goes up to a dollar a pound, we are in deep trouble."

3.13 The graph in Fig. P3.1 represents the projected demand curve for a new fadeproof dye for blue denim. The manufacturer wishes to maximize revenues.

a Show that maximum revenue coincides with unitary price elasticity of demand for any demand curve.

b Make a plot of annual revenue versus price. What production level would you recommend?

c The product is price inelastic at low prices. At the highest prices shown, the demand again becomes less sensitive to price. Does this mean that the product again becomes inelastic?

3.14 a Tolylene diisocyanate is a key raw material in the manufacture of polyurethane foams. A Texas producer competes successfully in the West Coast markets by rail shipping 50 million lb/year and selling to customers at the prevailing rate of 55 cents/lb (and "swallowing" the freight rate); no further market penetration has been possible at this price. Of course, larger West Coast market penetration would be possible by lowering the price; in fact, it is estimated that the total West Coast demand of 225 million lb could be captured by lowering the price to 25 cents/lb, the point at which competition would be forced to shut down their plants. On the other hand, it is believed that forcing the price up to $1/lb would reduce West Coast demand to zero, for foam producers could not compete with other regions and products.

From the above data, derive an expression for a price-demand curve as it applies to the Texas producer. Use the hyperbolic format of Eq. (3.3.8).

b The Texas producing company wishes to build a large plant on the West Coast. The cost of manufacture of TDI is 32 cents/lb (note that this is more than the "shutdown" selling price!); in large plants, this cost may be taken as independent of size. If the company wishes

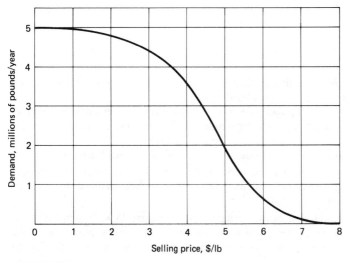

FIGURE P3.1
Blue denim dye demand curve.

to maximize profit after tax, how much should it produce in the new plant, and what should be the selling price?

3.15 Make a rough estimate of the passenger vehicle gasoline consumption in the United States. If gasohol containing 5 vol % ethanol were to be used to satisfy this demand, what annual volume of ethanol would be involved? What proportion of current ethanol production does this represent?

3.16 Confirm the validity of the gasoline consumption estimate in Prob. 3.15 by reference to the "Statistical Abstracts of the United States." Leaded gasoline used to contain about 3 g tetraethyl lead/gal. What is the order of magnitude of decrease in demand for lead as a result of the switch to unleaded gasoline in the 1970s? Establish evidence that this has had a serious effect upon the market for lead chemicals. What other development in the last several decades has had a disastrous impact upon lead chemicals?

3.17 Trichloroisocyanuric acid,

$$
\begin{array}{c}
\text{O} \\
\parallel \\
\text{Cl}-\text{N} \diagup \overset{\text{C}}{} \diagdown \text{N}-\text{Cl} \\
| | \\
\text{O}=\text{C} \diagdown \diagup \text{C}=\text{O} \\
\text{N} \\
| \\
\text{Cl}
\end{array}
$$

sells for $1.25/lb (1980). It is used as a disinfectant in residential swimming pools by virtue of the hydrolysis reaction

$$> \text{N}-\text{Cl} + \text{H}_2\text{O} \longrightarrow > \text{N}-\text{H} + \text{HOCl}$$

analogous to

$$\text{Cl}_2 + \text{H}_2\text{O} \longrightarrow \text{HCl} + \text{HOCl}$$

The HOCl is thought to be the active antimicrobial species. Two of the three chlorines on the cyanuric acid molecule may be considered available.

The trichloroisocyanuric acid powder must be added to each pool in sufficient quantity to maintain an equivalent level of 1 ppm (by weight) of Cl_2; the consumption rate (average over warm and cold seasons) is equivalent to 2 ppm Cl_2/day.

Estimate the potential annual sales in the residential swimming pool market.

3.18 Using back-of-the-envelope methods, estimate the annual U.S. consumption of rubber in passenger automobile tires, and calculate the corresponding demand for rubber-processing chemicals from Fig. 3.3.4.

3.19 A new ink formulation for throwaway felt point pens contains 25 wt % of methyl ethyl ketone (MEK). Inspect a typical felt point pen, and make a reasonable projection of the probable consumption of MEK assuming a 10 percent market penetration in the United States.

3.20 Epoxy resin undercoatings for passenger automobiles are an important corrosion-resistance feature, particularly in geographical areas with cold winters where salts may be used for controlling highway ice. A new water-suspended resin formulation, selling for $3/lb of contained resin, promises to exhibit exceptional resistance in this application. Make a rough estimate of the maximum potential annual sales should the underside surfaces of new U.S.-manufactured passenger cars be coated with a 1-mm layer of resin having a density roughly equal to that of water.

3.21 Example 3.3.2 incorporated the projection of the demand for perchloroethylene on the West Coast. A 100 million lb/year C_2Cl_4 plant was recommended, but we saw that the 1980 demand projection was regrettably high.

Carbon tetrachloride is coproduced with C_2Cl_4 (see Fig. 3.3.10); within certain limitations, the ratio of the coproduced products is at the operator's discretion. The question is whether some of the overcapacity that a 100-million-lb C_2Cl_4 plant would represent could be absorbed by switching to more CCl_4 production. Use the data in Fig. 3.3.5 to project West Coast demand for CCl_4 in 1980. "Chemical Profiles" (Exhibit 3.3B) lists the capacity of the only existing West Coast producer at 80 million lb/year.

Would switching to CCl_4 production in an overdesigned C_2Cl_4 plant be a viable option in 1980?

3.22 Use the technique of the perchloroethylene example (Example 3.3.2) and the data in Fig. 3.3.1 and Exhibit 3.3C to establish a reasonable capacity of a sorbitol plant proposed for West Coast construction in 1982. The West Coast has a sizable concentration of food-processing plants.

3.23 The characteristic flavor of beer is imparted by a group of terpenelike compounds called *humulones* and *lupulones*, which are extracted from hops during the brewing process. It is also possible to extract the flavoring agents from hops with methylene chloride; the viscous extract obtained following solvent evaporation may be added to the beer directly during the brewing process, thus avoiding the problem of handling the cellulosic hops wastes in the brewery. From the data given and assumption of reasonable market penetration, estimate how much methylene chloride could potentially be sold to the brewing industry.

Hops used in direct brewing: 120 g/hectoliter beer. CH_2Cl_2 losses, hops extraction: 75 lb/ton hops.

Is this a potentially important solvent business?

The 1978 U.S. consumption of beer was 5.5 billion gal/year. Methylene chloride production data are given in Fig. 3.3.5.

3.24 Another application of methylene chloride in the food-processing industry is in the decaffeination of coffee. The green coffee beans are steamed and then contacted with liquid CH_2Cl_2 which extracts the alkaloid from the bean without coextracting appreciable quantities of coffee flavors. All traces of the solvent are eliminated during subsequent steaming and roasting.

The CH_2Cl_2 consumption in this process is a function of the care taken to reduce fugitive vapor losses; a solvent loss equivalent to 5 wt % of coffee processed may be assumed. About 15 percent of the coffee consumed in the United States is decaffeinated instant coffee. Estimate the total coffee consumption, and decide whether decaffeination represents an important solvent market.

3.25 Activated charcoal is extremely effective in eliminating odors from air, and it does not lose its effectiveness even after adsorbing 10 to 20 percent of its weight of common odoriferous agents. Sketch a small device that could be sold to households for deodorizing refrigerators. Assume the device could be used for 1 year, after which it would be thrown away. How much do you think the consumer would be willing to pay for such a device?

Make a rough estimate of the market potential and market penetration and the gross sales potential.

3.26 If a plant is constructed to fill the growing demand for an established product, the annual production is not likely to start and continue at plant rated capacity. A more likely scenario is very low production during the first year (the start-up period) and gradual production increase up to rated capacity after several years, with rated capacity production after that. This kind of trend is illustrated by the following:

Draw a graph of the projected production of a plant during the first 10 years of life.

Rated capacity: 75 million lb/year.
Total market demand at time zero (start-up): 100 million lb/year.
Total market demand doubles every 5 years.
Production from existing suppliers is 100 million lb/year at time zero but decreases 10 percent every 5 years due to plant attrition.
Market penetration is 50 percent of new demand above production level of suppliers existing at time zero.

3.27 A production plant has a lifetime of 10 years and a rated capacity of P. Its production life cycle may be approximated by a symmetrical section of the normal distribution curve, with initial and final production equal to $\frac{1}{2} P$ and maximum production just equal to P.

a Compute the equivalent uniform annual production rate (as a fraction of P) that would result in the same grand total production output over the 10-year life span. Probability density and cumulative distribution functions for normal probability distribution are tabulated in handbooks of mathematics such as Weast and Selby (1967).

b The product sells for s dollars per unit (in current dollars). What equivalent annual production rate (as a fraction of P) would yield the same present worth of sales revenues at 20 percent time value of money (discrete annual discounting)? See also Valle-Riestra (1979).

3.28 A chloralkali complex such as the one illustrated in Fig. 3.3.10 has an annual capacity of 500 million lb Cl_2 produced in the electrolytic cells. The following is a long-range demand forecast (in millions of pounds per year):

50% aq NaOH	465	(100% NaOH basis)
NaOH anhydrous beads	155	
Anhydrous ammonia	50	(primarily consumed in integrated complex)
Merchant chlorine	50	(compressed gas)
CCl_4	65	
C_2Cl_4	165	
HCl (anhydrous)	200	

Is this long-range forecast compatible with the capacity and nature of the complex? What additional processing steps could be undertaken to increase the flexibility in choosing a product spectrum?

Let us suppose the relative importance of the business represented by each product is as follows:

50% NaOH
Merchant Cl_2
C_2Cl_4
NaOH beads
CCl_4

Ammonia
HCl

With this list as a guide and on the basis of process modifications of your choice, recommend a more realistic production plan, but one not based on an increase in cell capacity.

3.29 Use data in "Chemical Profiles" (Exhibit 3.3B) and other references available to you to generate a plot of sales price versus total U.S. annual production for a number of chemicals. Be sure to include a full spectrum of product types—heavy inorganics, metals, organic intermediates, pharmaceuticals. Do you detect any correlation? Do you think your plot would be a useful guidance tool? Can you, for instance, define "exclusion areas" which could serve to give warning of an unrealistic pricing structure?

3.30 Economic analyses of processes utilizing biomass and agricultural wastes frequently suffer by virtue of underestimation of the cost of collecting, transporting, and storing the raw materials in a central processing location. Using the data of Dobie et al. (1973), estimate a reasonable current cost per ton of rice straw which will be used as feed into a methanol production facility.

3.31 At the beginning of the eighties, some 10 million tons of machine-harvested tomatoes were grown annually in the United States. The crop yield was about 20 tons/acre, and the field value (truck loaded) was about $100/ton.

A new maturing agent is about to be introduced on the market which allows all the tomatoes to ripen on the vine at the same time, and which eliminates "yellow and green shoulders," a condition of localized lack of ripeness which may result in rejection of the fruit. It is claimed that an application rate of 150 lb/acre each season will result in a 27 percent increase in acceptable fruit. The agent is applied as an aqueous solution by several aerial sprays at a total cost of $50/acre. The agent costs $1/lb to manufacture. What might be a reasonable sales price range, based on the benefit to the farmer?

3.32 In Prob. 3.20 mention was made of an effective epoxy resin undercoating for automobiles. An automobile manufacturer finds that his total cost for undercoating is $300. The car buyer is offered the undercoating as an optional extra for $500, with the claim (but no guarantee) that the undercoating will "add at least a year to the life of your car."

Assuming the claim is true, and assuming that the net effect to the buyer will be a delay in the purchase of a new $8000 car (current dollars) from the end of the fifth year to the end of the sixth year, determine whether the buyer is justified in obtaining the undercoating. Use a present worth analysis at 20 percent discrete annual discounting.

3.33 Use source material such as the references in Exhibit 3.5A to investigate processes for the manufacture of propionaldehyde. Use some of the process choice criteria in Sec. 3.5 to decide upon the most likely process candidate.

3.34 Grover (1977) describes an aqueous waste stream management system applicable to a specific production unit or a whole integrated complex. Draw a block flow diagram which shows the overall movement of the various water supply streams (river water, rain, etc.) and their ultimate disposal.

3.35 Peters and Timmerhaus (1980) describe a process for the manufacture of a sodium dodecylbenzene sulfonate detergent as follows (by permission, McGraw-Hill Book Co.).

This process involves reaction of dodecene with benzene in the presence of aluminum chloride catalyst; fractionation of the resulting crude mixture to recover the desired boiling range of dodecylbenzene; sulfonation of the dodecylbenzene and subsequent neutralization of the sulfonic acid with caustic soda; blending the resulting slurry with chemical "builders"; and drying.

Dodecene is charged into a reaction vessel containing benzene and aluminum chloride. The reaction mixture is agitated and cooled to maintain the reaction temperature at about 115°F maximum. An excess of benzene is used to suppress the formation of by-products. Aluminum chloride requirement is 5 to 10 wt % of dodecene.

After removal of aluminum chloride sludge by settling, the reaction mixture is fractionated to recover excess benzene (which is recycled to the reaction vessel), a light alkylaryl hydrocarbon (which is recycled to the reaction), dodecylbenzene, and a heavy alkylaryl hydrocarbon (waste product).

Sulfonation of the dodecylbenzene may be carried out continuously or batchwise under a variety of operating conditions using oleum (usually 20% SO_3). The optimum sulfonation temperature is usually in the range of 100 to 140°F depending on the strength of acid employed, mechanical design of the equipment, etc. Removal of spent sulfuric acid from the sulfonic acid is facilitated by adding water to reduce the sulfuric acid strength to about 78%; the spent H_2SO_4 and sulfonic acid form two liquid phases separated by settling. This dilution prior to neutralization results in a final neutralized slurry having approximately 85% active agent based on the solids. The inert material in the final product is essentially Na_2SO_4.

The sulfonic acid is neutralized with 20 to 50% caustic soda solution to a pH of 8 at a temperature of about 125°F; an aqueous slurry is formed. Chemical "builders" such as trisodium phosphate, tetrasodium pyrophosphate, sodium silicate, sodium chloride, sodium sulfate, carboxymethyl cellulose, etc., are added to enhance the detersive, wetting, or other desired properties in the finished product. A bead product is obtained by spray drying.

The basic reactions which occur in the process are the following:

Alkylation:

$$C_6H_6 + C_{12}H_{24} \xrightarrow{AlCl_3} C_6H_5 \cdot C_{12}H_{25}$$

Sulfonation:

$$C_6H_5 \cdot C_{12}H_{25} + H_2SO_4 \longrightarrow C_{12}H_{25} \cdot C_6H_4 \cdot SO_3H + H_2O$$

Neutralization:

$$C_{12}H_{25} \cdot C_6H_4 \cdot SO_3H + NaOH \longrightarrow C_{12}H_{25} \cdot C_6H_4 \cdot SO_3Na + H_2O$$

A literature search indicates that yields of 85 to 95 percent have been obtained in the alkylation step, while yields for the sulfonation process are substantially 100 percent, and yields for the neutralization step are always 95 percent or greater. All three steps are exothermic and require some form of jacketed cooling around the stirred reactor to maintain isothermal reaction temperatures.

Draw a block flow diagram of the process, and make sure you identify streams that may require further treatment, waste streams, and other edge-of-the-paper streams.

3.36 A pure form of sodium benzoate is used as an antibacterial food and beverage preservative. Make a cursory investigation of the properties of sodium benzoate and benzoic acid available in standard handbooks, and devise a process for the production of USP sodium benzoate involving the neutralization of benzoic acid with caustic soda. Draw a block flow diagram of the process. Assume that a food-grade 50% NaOH is available but that a technical grade of benzoic acid will have to be purified in a manner similar to that used for adipic acid purification in Figs. 3.5.7 and 3.5.8. The water-soluble impurities will have to be eliminated in an environmentally acceptable manner.

3.37 Liquid bromine is recovered from natural brines of Michigan, which typically contain 1300 ppm Br_2 and 26% total solids (chlorides of sodium, calcium, and magnesium). The brines are forwarded to the process directly from deep wells, and the bromine-free brine is stored in ponds prior to treatment in salt recovery units.

The brine is preheated to just below the boiling point and charged into a stripping tower where it is reacted with a countercurrent stream of chlorine:

$$CaBr_2 + Cl_2 \longrightarrow Br_2 + CaCl_2$$

The chlorine is introduced as a vapor, obtained by vaporization of a liquid Cl_2 stream pumped from chlorine storage facilities. A small amount of 98% H_2SO_4 is introduced into the top of the tower for proper pH control. Steam is introduced into the bottom of the tower well below the chlorine; the steam acts to strip the liberated Br_2 from solution and keeps the product brine chlorine-free.

The bromine-steam mixture leaving the top of the tower contains small amounts of chlorine; the mixture is condensed and decanted. In the decanter, a lower layer of crude bromine and an upper layer of bromine water separate. The aqueous layer is air-stripped in a blowing-out tower, and the debrominated water is evaporated in a steam-heated evaporator to regenerate the low-pressure stripping steam for the brine stripping tower.

The air-bromine vapor from the blowing-out tower is passed into an absorption tower filled with water-irrigated iron filings to form aqueous ferrous bromide:

$$Fe + Br_2 \longrightarrow FeBr_2$$

The clean air is vented, and the concentrated bromide solution is treated with Cl_2 vapor in a vented packed tower to liberate more Br_2, which is only slightly soluble in the simultaneously formed ferrous chloride:

$$FeBr_2 + Cl_2 \longrightarrow Br_2 + FeCl_2$$

The ferrous chloride–liquid Br_2 mixture is passed into a settler where the crude bromine settles to the bottom. The combined crude bromine streams are purified by distillation in two columns. Chlorine is removed in the overhead of the first column and forms part of the Cl_2 feed to the ferrous chloride tower. In the second tower, Br_2 is separated as an overhead stream from small quantities of water and heavier compounds; this bottom stream is returned to the top of the brine stripping tower. The purified Br_2 product is condensed and stored for tank car loading.

The ferrous chloride solution is converted to a marketable ferric chloride solution in another chlorination tower filled with iron filings:

$$FeCl_2 + 3Fe + 5Cl_2 \longrightarrow 4FeCl_3$$

Excess Cl_2 from this operation (along with traces of liberated Br_2) is used as part of Cl_2 feed to the brine stripping column.

Draw a neat *block flow diagram* of the process. Identify key process streams and waste streams which might need additional attention.

3.38 Catalytic reforming is typified by the conversion of *n*-heptane to toluene:

$$n - C_7H_{16} \longrightarrow C_6H_5CH_3 + 4H_2$$

Reactions such as this one take place over catalysts at temperatures around 1000 K. Estimate the heat of reaction at 1000 K by using a number of estimation methods, such as those outlined in Reid et al. (1977), and compare your results with the heat calculated from accepted thermodynamic parameters (listed in appendix A of Reid).

3.39 Could the heat of alkylation of ammonia with α-picolyl chloride be approximated with heat of formation data of the compounds in the following reaction:

$$\text{C}_5\text{H}_4\text{N-CH}_2\text{-Cl}(g) + 2\text{NH}_3(g) \longrightarrow \text{C}_5\text{H}_4\text{N-CH}_2\text{-NH}_2(g) + \text{NH}_4\text{Cl}(s)$$

Calculate the reaction heat at 298 K by using accepted heats of formation, and compare the result with that in Example 3.5.2. Why is the above reaction not a good analog?

3.40 Not all chloralkali complexes (Fig. 3.3.10) have an associated ammonia plant, and in some such complexes hydrogen has been in the past vented to the atmosphere. This is clearly a case for energy conservation. Calculate the approximate percentage of the theoretical energy required to electrolyze a 5.0 M NaCl solution that could be recovered by burning the generated hydrogen.

3.41 Rapid preliminary estimation of heat-transfer requirements is illustrated by the ethylamine process summed up in Fig. 3.5.3. Data required for these preliminary calculations may be found in standard handbooks (Perry, Lange).

a The reaction forming monoethylamine may be written as

$$\text{CH}_3\text{CH}_2\text{Cl}(l) + \text{NH}_3(aq) \longrightarrow \text{CH}_3\text{CH}_2\text{NH}_2 \cdot \text{HCl}(aq)$$

or

$$\text{CH}_3\text{CH}_2\text{Cl}(l) + 2\text{NH}_3(aq) \longrightarrow \text{CH}_3\text{CH}_2\text{NH}_2(aq) + \text{NH}_4\text{Cl}(aq)$$

Show that the heat of reaction is approximately the same regardless of which reaction predominates.

b Show that the heat of reaction for the formation of the higher amines, such as

$$\text{CH}_3\text{CH}_2\text{Cl}(l) + \text{CH}_3\text{CH}_2\text{NH}_2(aq) + \text{NH}_3(aq) \longrightarrow (\text{CH}_3\text{CH}_2)_2\text{NH}(aq) + \text{NH}_4\text{Cl}(aq)$$

is about the same as the heat of reaction of the first step, the monoethylamine formation. Data are not available to make the comparison using the ethyl compounds, but data for methyl compounds are available. Compare the heats of

$$\text{CH}_3\text{Cl}(l) + 2\text{NH}_3(aq) \longrightarrow \text{CH}_3\text{NH}_2(aq) + \text{NH}_4\text{Cl}(aq)$$

and

$$\text{CH}_3\text{Cl}(l) + \text{CH}_3\text{NH}_2(aq) + \text{NH}_3(aq) \longrightarrow (\text{CH}_3)_2\text{NH}(aq) + \text{NH}_4\text{Cl}(aq)$$

c Calculate the total heat of reaction at 225°C generated by reacting 10.0 lb · mol/h of ethyl chloride. Neglect the effect of temperature upon heat of reaction.

d If the total heat of reaction is absorbed by the liquid feed to the reactor (relative quantities are specified in Example 3.5.1), what must be the temperature of the feed entering the reactor?

e The aqueous NH_3 (and ethyl chloride) are fed to the heat interchanger at 30°C. What is the heat-transfer rate in the interchanger?

f Assuming reactor feed and product have the same heat capacity, what is the log-mean temperature difference in the interchanger?

3.42 The wet adipic acid cake feed to the melter in Fig. 3.5.8 contains 60 wt % H_2O. At 170°C the vapor pressure of adipic acid is 2 mmHg. What percentage of the adipic acid will wind up in the vapor out of the melter? Is there any danger of plugging up the condenser above the melter with solids? What can be done about it?

3.43 You have been asked to undertake an energy conservation investigation of a vacuum distillation operation with a well-defined feed and narrow overhead product specifications. Outline some of the areas of investigative concern which could lead to energy savings.

3.44 A volatile aromatic compound RH is to be chlorinated in the liquid state to form the much less volatile monochloro compound, RCl, on the surface of a solid pelletized catalyst. RCl, in turn, can be further chlorinated in like manner to form polychlorinated compounds. Devise and sketch an equipment system that will minimize the formation of the polychlorinated species by providing rapid removal of the RCl formed from the reaction zone, as well as a large excess of RH.

3.45 "Translate" one or more of the following "blocks" from a block flow diagram into a corresponding flow sheet section, in much the same way that Fig. 3.5.5 was generated from a box marked "Distillation."

    a "Liquid-liquid extraction" (show two stages of mixer-settlers)
    b "Evaporation" (show double effect forward-feed system)
    c "Vacuum crystallization" (aqueous solution of inorganic salt)
    d "Crushing and grinding" (metallic ore, 4-in rock to 100-mesh powder)
    e "Solvent desiccation" (water-insoluble solvent dried with molecular sieves)
    f "Vent scrubber" (5% NaOH used to remove traces of HF from vented air)

3.46 From the process descriptions and block flow diagrams shown in the book of Faith, Keyes, and Clark (see Ref. 4, Exhibit 3.5A), construct a process flow sheet for one of the following processes. Be sure to identify all process streams that may need additional processing.

    a Diethyl ether from 95% ethanol
    b Boric acid from borax
    c 35% aq HCl from H_2 and Cl_2 (H_2 and Cl_2 available directly from an existing chloralkali complex)
    d USP hexamethylenetetramine

3.47 Utilization of the spent acid from the manufacture of sodium dodecylbenzene sulfonate (see Prob. 3.35) poses a serious problem. It has been suggested that the spent 78% acid be blended with 60% oleum to produce the required 20% oleum for this process as well as other sulfonation processes.

As a preliminary step in evaluating this proposal, make a mass and energy balance and draw an equipment flow sheet for a blending scheme of your choice. (You need not size the equipment.) Be sure to show all required equipment. Assume the small amount of detergent sludge in the spent acid need not be separated. 50,000 lb/day of 78% H_2SO_4 are generated. What is the total daily production of 20% oleum?

Some useful data include:

Enthalpy at 70°F:
    78% acid   −104 Btu/lb
    20% oleum   +94 Btu/lb
    60% oleum   +280 Btu/lb
Specific heat of 20% oleum: 0.34 Btu/(lb · °F)
    Vapor pressure at 100°F:
    20% oleum   0.1 psia (290°F bp)
    60% oleum   3.8 psia
Note: % oleum refers to wt % free SO_3 in SO_3–H_2SO_4 mix.

3.48 The final detergent slurry obtained in the process described in Prob. 3.35 is spray-dried with hot gases obtained from natural gas combustion. The exhaust from the spray drying system consists of 500 lb · mol/h of dry flue gas, 140 lb · mol/h of water vapor, and traces of detergent, all at 215°C. The disposal of the water, either as a vapor discharge to the atmosphere or as a condensate, poses an environmental problem. Suggest how this environmental problem might be reduced, and draw a flow sheet of your suggested method.

3.49 Figure P3.2 is an "index" flow sheet submitted as a basis for the economic evaluation of a process to produce 25 tons/day of sodium dodecylbenzene sulfonate detergent (see Prob. 3.35). An estimate based on the flow sheet submitted is almost certain to be too low, for important pieces of process equipment have not been properly accounted for, a common error in preliminary estimates.

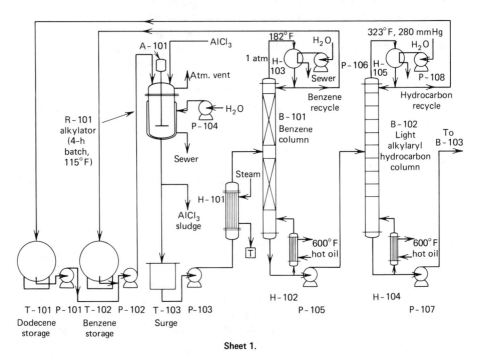

FIGURE P3.2
Sodium dodecylbenzene sulfonate flow sheet.

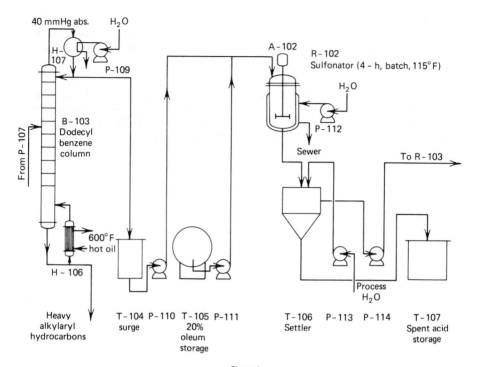

Sheet 2.

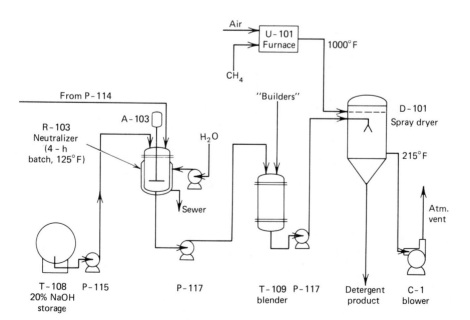

Sheet 3.

Make a list of errors which you can find in the flow sheet. Pay particular attention to disposal of waste streams and to storage requirements. You may find that important items of processing equipment have been omitted; on the other hand, some equipment may be redundant. In each case, make a brief suggestion as to how the error may be corrected and what type of additional equipment may be needed.

3.50 In the ethylamine process illustrated in Fig. 3.5.3, the high reactor temperature is maintained by virtue of the heat of reaction as well as feed preheat by hot reaction products. During process start-up, however, the preheat source is not available. Describe or sketch out equipment that may be required to effect a smooth process start-up.

CHAPTER 4

EQUIPMENT DESIGN AND COSTING

The Law of Optimum Sloppiness: *For any problem, there is an optimum amount of sloppiness we can use to solve the problem.*

Corollary: *There are occasions when we must be sloppy or imprecise in our calculations, and there are times when we must be precise. The essence of engineering is to be only as complicated as you have to be, but you must also be able to get as complicated as the problem demands.*

Donald R. Woods (1975)

This quotation from Woods, reproduced by permission, serves to underscore the importance of adjusting the effort of designing equipment to fit the requirements at any particular stage of project development. The term "sloppiness" is perhaps a little too strong; even during preliminary project stages we would prefer to have equipment design specifications that are correct and accurate, not sloppy. Nevertheless, during earlier stages of development, process characteristics are usually not firmly enough defined to justify a great deal of effort in making a careful, detailed design, and similarly the time is not there to obtain firm cost quotations from equipment vendors. We therefore rely upon approximations and rule-of-thumb shortcuts which are not really sloppy but are rather the distillation of a great deal of experience and judgment; in this way we arrive rather rapidly at equipment design specifications and costs which are, hopefully, good enough for the purpose at hand. The equipment design and costing methodology is gradually refined as part of the more advanced stages of project development.

188 CHAPTER 4: EQUIPMENT DESIGN AND COSTING

Given the flow sheet and mass and energy balances discussed in Chap. 3, we are, then, ready to size the equipment we have selected and to arrive at an estimate of the total cost to purchase the equipment. This total purchased equipment cost is the fundamental financial entity required for an investment analysis, and as such it must naturally be based upon a *complete* flow sheet. That is why completeness was so strongly emphasized as part of the flow sheet synthesis discussion in the last chapter; any omissions lead to a low purchased equipment cost total, and the error is propagated throughout the investment analysis. Rapid equipment design and costing methods also result in errors, but we will see that these kinds of errors tend to cancel out.

4.1 EQUIPMENT SELECTION AND SIZING FOR PRELIMINARY COST ESTIMATES

The Heuristic Approach

By the *heuristic* approach we mean a process design procedure which uses well-known, sometimes unprovable, rules of thumb to arrive at a rapid and reliable specification of equipment size or arrangement. Such shortcut methods are an invaluable and powerful component of the process synthesis itself; they are indispensable during preliminary stages of economic evaluation because they save time and they lead to a reasonably quantitative characterization of equipment size even in the absence of sufficient process data. One example of the heuristic approach is given by King (1980, reproduced by permission) as part of his discussion of distillation sequence selection; four "rules" are stated as:

1 Separations where the relative volatility of the key components is close to unity should be performed in the absence of nonkey components.

2 Sequences which remove the components one by one in column overheads should be favored.

3 Sequences which give a more nearly equimolal division of the feed between the distillate and bottoms product should be favored.

4 Separations involving very high specified recovery fractions should be reserved until last in a sequence.

These four rules, which are not necessarily mutually consistent, are the result of informed judgment and a great deal of experience. It is possible that their substance could be arrived at by repeated computations involving a specific multicomponent mixture that is to be separated. Even with highly sophisticated computer technology, the optimization task could be formidable, indeed, and a priori application of the heuristics greatly simplifies the task at hand.

In much the same way, the essence of informed judgment and experience can be expressed in the form of rules of thumb useful in the rapid preliminary design of items of equipment. Such preliminary design may, in fact, serve as a useful guide to more detailed design in advanced stages of development. An obvious prerequisite in this approach is that, at least the "first time around," the equipment

SECTION 4.1: EQUIPMENT SELECTION AND SIZING FOR PRELIMINARY COST ESTIMATES **189**

selected to perform a desired operation be simple and supported by a large amount of performance data and experience. It may very well happen, for example, that two viscous liquids may be acceptably blended in a centrifugal pump before introduction into a heated plug-flow reactor. The fact of the matter is that there is no reliable heuristic, no rule of thumb, that relates the pump energy and configuration to the blending requirements. It is preferable in the absence of more detailed experimental information to provide for a small blending tank ahead of the reactor for which design rules are available. Another approach might be to provide for a static mixer at the pump discharge; the pump itself may then be sized to satisfy the pumping requirements alone, rather than the blending requirements as well.

For the purposes of a preliminary cost estimate, only a few key characteristics of each item of equipment need be defined; cost correlations are given in terms of these key characteristics, which include:

1 *Equipment classification* (if applicable). In the case of heat exchangers, for example, the design specifications would call for a "shell-and-tube exchanger," or "panel coil," or "vertical-tube calandria," or one of many other possible heat-exchanger configurations.

2 *Size criterion,* which consists of one or more numbers which uniquely characterize the size of the item of equipment for costing purposes. Shell-and-tube-exchanger sizes are uniquely characterized by the heat-transfer area, based upon the tube external area. Plate columns are characterized by diameter, total height, and number of plates. The point is that preliminary costing may be accomplished with reasonable reliability without resorting to detailed design that would, in the case of the shell-and-tube exchanger, involve tube length, number of tube passes, tube layout, shell diameter, baffling, expansion joints, and many other size and service-type specifications.

3 *Extreme operating conditions.* For preliminary costing purposes, these should be listed as part of the equipment specifications. Frequently, cost correlations show what effect high pressure, for example, might have on the cost of equipment such as a shell-and-tube exchanger relative to the cost of a low-pressure exchanger. Again, detailed design to accommodate the high pressure is not required.

4 *Materials of construction.* This characteristic is very important in establishing equipment cost and may even dictate the choice of equipment type. The project manager must come to grips with the choice of likely materials of construction during very early stages of project development. Perry and Chilton's "Chemical Engineers' Handbook" (1973)* has charts and tables which facilitate the choice of construction materials for a wide range of chemical services.

Rules of thumb and rapid design methods focus upon the definition of these characteristics, particularly the size criterion. Exhibit 4.1A lists some references which are particularly valuable as sources of these rapid design methods; practicing chemical engineers soon develop their own file of their favorite methods. In the

---

*Hereafter referred to as "Perry."

> **EXHIBIT 4.1A**
>
> RAPID DESIGN METHODS: SOURCES
>
> **1** Happel, John, and Donald G. Jordan: "Chemical Process Economics," 2d Ed., Marcel Dekker, Inc., New York, 1975. Appendix C is a useful compendium of rapid approximation methods.
> **2** Aerstin, Frank, and Gary Street: "Applied Chemical Process Design," Plenum, New York, 1978. A treasure trove of rapid design methods, with step-by-step instructions.
> **3** Peters, Max S., and Klaus D. Timmerhaus: "Plant Design and Economics for Chemical Engineers," 3d Ed., McGraw-Hill, New York, 1980. A more detailed treatise with a wealth of information.
> **4** Jordan, Donald G.: Some Rapid Approximation Methods for the Design of Chemical Equipment, in "Process Economics," AIChE Today Series, American Institute of Chemical Engineers, New York. Many rules of thumb suggested.
> **5** *Chemical Engineering* magazine publishes many rapid design methods; consult the annual indices.
> **6** Baasel, William D.: "Preliminary Chemical Engineering Plant Design," Elsevier, New York, 1980 (4th printing). Rules of thumb throughout the text; many worked examples.

sections that follow, the heuristic approach is illustrated for a few key items of equipment. The purpose is to illustrate the method, not to be exhaustive.

Distillation

Let us start by taking a look at some of the approaches to rapid preliminary design in the area of distillation. The first step is to do something to establish vapor-liquid equilibrium (VLE) data. Some useful references for locating experimental data for systems previously studied are listed in Exhibit 4.1B. The first reference listed is particularly noteworthy as a source of literature references (not data) for a wide range of binary, ternary, and multicomponent systems. If the system under consideration has not been previously studied, the simplest recourse is to assume ideal behavior represented by Raoult's law,

$$p_i = x_i P_i^\circ \qquad (4.1.1)$$

and Dalton's law,

$$p_i = P y_i \qquad (4.1.2)$$

The VLE relationship for each component in the ideal mixture is given by the combination of these two equations,

EXHIBIT 4.1B

VAPOR-LIQUID EQUILIBRIA: REFERENCES

   1 Hála, Pick, Fried, and Vilím: "Vapor-Liquid Equilibrium," 2d Ed., Pergamon, Oxford, 1967.
   2 Horsley: Azeotropic Data, *Adv. Chem. Ser.* (6) (1952), plus decennial supplements.
   3 Chu, Wang, Levy, and Paul: "Vapor-Liquid Equilibrium Data," Edwards, Ann Arbor, Mich., 1956.
   4 Hála, Wichterle, Polák, and Boublík: "Vapor-Liquid Equilibrium Data at Normal Pressures," Pergamon, Oxford, 1968.
   5 Hirata, Ohe, and Nagahama: "Computer Aided Data Book of Vapor-Liquid Equilibria," Kodansha Limited/Elsevier, Tokyo/Amsterdam, 1975.
   6 Wichterle, Linek, and Hála (eds.): "Vapor-Liquid Equilibrium Data Bibliography," Elsevier, New York, 1973; Suppl. I, 1976; Suppl. II, 1979.

$$\frac{y_i}{x_i} = \frac{P_i^\circ}{P} = K_i \qquad (4.1.3)$$

Even if the system under consideration turns out to be quite nonideal, the ideal behavior assumption results in equipment cost projections within acceptable limits of error. This may not be true, however, if the nonideality results in azeotropic behavior, and the engineer must be on the lookout for systems which are likely to display such behavior. The existence of azeotropes may require the use of multiple column assemblies to accomplish a desired separation.

Azeotropic behavior may be predicted by considering the hydrogen-bonding tendency of the liquids in the mixture; a few simple rules are given, for example, in the article on Azeotropes in the McKetta–Cunningham "Encyclopedia of Chemical Processing and Design."* These rules predict (and correctly so) that chloroform and propanethiol form almost ideal mixtures, whereas 2-methoxyethanol and heptane should form an azeotrope (they do). If azeotropic behavior is indicated, it is best to attempt to develop an approximate VLE curve in spite of the extra effort required; methods to do this for unknown mixtures are given in Horsley (Ref. 2, Exhibit 4.1B), and additional references are given by Perry.

Sparingly intersoluble species form heterogeneous (two liquid phases) azeotropes. If solubility and pure-component vapor pressure data are available, the construction of a VLE diagram is straightforward (Prob. 4.3). Consider, for example, a system of water plus a hydrocarbon such as pentane. Pentane is very

*Marcel Dekker, Inc., New York/Basel, 1977, vol. 4.

sparingly soluble in water, and water is very sparingly soluble in pentane; in the region where the two liquid phases coexist, each phase is close to being a pure liquid. Thus, the partial pressure of water over the two liquid phases will be approximately the vapor pressure of pure water, and the partial pressure of pentane the vapor pressure of pure pentane. It follows that, at any total pressure, the temperature of the heterogeneous azeotrope will be that at which the sum of the pure-component vapor pressures equals the total pressure. Once this has been established, the composition of the equilibrium vapor can be obtained from Dalton's law. The equilibrium curves in the small single-liquid-phase regions may be calculated from Henry's law,

$$p_i = \mathcal{H} x_i \qquad (4.1.4)$$

Heterogeneous azeotrope equilibrium diagrams are particularly useful in the design of drying columns.

The prediction of VLE of petroleum fractions has been reviewed by Sim and Daubert (1980).

The next step in the quick design procedure is to calculate the minimum number of theoretical plates required for the desired separation. For binary systems, or systems that can be approximated as a binary system, the McCabe-Thiele graphical method (with constant molar overflow) is favored. The equilibrium curve is constructed directly from experimental data, or else, if ideality is assumed, the procedure is to develop an analytical expression in terms of α_{1-2}, the relative volatility of the more volatile component (no. 1). By convention, the equilibrium curve (that is, xy diagram) is drawn for the more volatile component. The relative volatility is defined as

$$\alpha_{1-2} = \frac{y_1 x_2}{y_2 x_1} \qquad (4.1.5)$$

Substituting Eq. (4.1.3) for each component,

$$\boxed{\alpha_{1-2} = \frac{P_1^\circ / P}{P_2^\circ / P} = \frac{P_1^\circ}{P_2^\circ}} \qquad (4.1.6)$$

that is, for ideal binary mixtures, the relative volatility at any particular temperature is simply the ratio of vapor pressures. Now, the temperature varies throughout the column, and so does the relative volatility, although the normal value change between reboiler temperature and overhead condensing temperature is small. A reasonably accurate design is afforded by generating an equilibrium curve based on

$$\overline{\alpha}_{1-2}$$

the arithmetic average of α's at reboiler and overhead condensing temperatures. The analytical expression for the equilibrium curve for the more volatile component (no.1) can now be developed from Eq. (4.1.5); since $x_1 + x_2 = 1$ and $y_1 + y_2 = 1$,

$$\bar{\alpha}_{1-2} = \frac{y_1(1 - x_1)}{(1 - y_1)x_1}$$

and rearranging,

$$y_1 = \frac{\bar{\alpha}_{1-2} x_1}{1 + (\bar{\alpha}_{1-2} - 1)x_1} \qquad (4.1.7)$$

The ideal mixture equilibrium curve for the McCabe-Thiele method can be plotted by computing y_1's for the full range of x_1's. The minimum number of theoretical plates can now be stepped off directly. Moreover, the minimum external reflux ratio may be calculated from the measured slope of the minimum internal reflux line (for initial rapid design, take $q = 1$ for slope of the q line; i.e., saturated liquid feed). The slope of the internal reflux line in the rectifying section, L_R/V_R, is related to the external reflux ratio R/D by

$$\frac{R}{D} = \frac{1}{(L_R/V_R)^{-1} - 1} \qquad (4.1.8)$$

The design number of theoretical plates can be determined graphically on the basis of the observation that the economically optimum reflux ratio is often 1.2 to 1.4 times the minimum external reflux ratio, or roughly

$$\frac{R}{D} \approx 1.3 \left(\frac{R}{D}\right)_{min} \qquad (4.1.9)$$

The proper corresponding L_R/V_R can be computed from Eq. (4.1.8) and plotted to allow the stepping off of the optimum number of theoretical plates. The procedure is so simple and rapid that one is tempted to characterize even a multicomponent system as a binary system. In some cases this can be done by literally neglecting very light and very heavy components. Very light components (high volatility relative to the light key) can be treated as inert vapors which simply serve to reduce the vapor concentration of distilling species in the rectifying section. On the other hand, very heavy components (low volatility relative to the heavy key) can be treated as "rocks" with no vapor pressure which serve to reduce the liquid-phase concentration of the distilling species, and drop to the bottom of the column. However, a rapid shortcut calculation of optimum plates for multicomponent mixtures (see, in particular, Ref. 2, Exhibit 4.1A) is really only slightly more time-consuming than the McCabe-Thiele method.

An often-used approximation is that the optimum number of theoretical plates is just twice the minimum number of plates.

To convert theoretical plates to actual plates to be used in a distillation tower, a plate efficiency must be estimated. For preliminary costing purposes, an efficiency of 50 percent may be assumed. A more accurate estimate may be obtained from the O'Connell correlation (Perry, for example) of computed plate efficiency of commercial columns versus the product of the relative volatility of the keys and the viscosity at the feed plate. It is important to remember that the reboiler is roughly equivalent to one theoretical stage; thus one theoretical plate must be subtracted from the total before correcting for plate efficiency and computing the actual number of plates. Similarly, one theoretical stage must be subtracted if the overhead condenser is a partial condenser.

The plate column diameter is based upon the maximum superficial vapor velocity in the column. The allowable velocity for optimum performance may be calculated readily from correlations given in many standard chemical engineering texts, or else more simply from the rule of thumb:

Choose vapor velocity = 2 ft/s for atmospheric distillation or moderate pressures.

Choose vapor velocity = 6 ft/s for vacuum distillation.

Choose sieve trays up to 6 ft in diameter, valve trays for larger diameters.

To estimate overall column height, use 2-ft plate spacing; add about 4 ft for gas disengagement at top, 6 ft for bottoms liquid level and reboiler return at bottom (2-ft-diameter basis; adjust for smaller or larger columns; also account for accidental dumping of all liquid on trays to the bottom).

Packed columns may be used for distillation up to about 4-ft diameters; poor liquid distribution becomes a problem in larger columns. Packed columns are in general less expensive than plate columns, but they do have some disadvantages, such as danger of plugging with solids or tars. Rapid design of packed columns may be realized with the aid of a few simple rules:

Choose packing diameter about one-tenth of column diameter.

Limit packed height to no more than about six column diameters before the liquid is redistributed.

Base column diameter on 70 percent of flooding velocity, using flooding velocity correlations in Perry or other references in Exhibit 4.1A.

Calculate HETP (column height equivalent to a theoretical plate) from empirical equations (Perry) or from rules of thumb listed by Aerstin and Street (Ref. 2, Exhibit 4.1A):

$$\text{HETP} = \begin{cases} 2\tfrac{1}{2} \text{ ft} & \text{for 2-in packing} \\ 2 \text{ ft} & \text{for } 1\tfrac{1}{2}\text{-in packing} \\ 1\tfrac{1}{2} \text{ ft} & \text{for 1-in packing} \end{cases}$$

Considerably lower values of HETP can be obtained with woven-wire packings.
Note that columns are normally mounted on "skirts" about 6 ft above ground level; this allows easy piping up of the bottoms stream and ready access for maintenance.

Example 4.1.1

Make a quick preliminary design (for costing purposes) of a sieve plate column required for the first-step drying of a mixture of γ-picoline and water. The following flow table data are available (all flows in lb·mol/h):

	Feed	Distillate	Bottoms
γ-picoline	200.0	12.0	188.0
Water	50.0	48.0	2.0

SOLUTION Assume distillation will be performed at atmospheric pressure.
Hála et al. (Ref. 1, Exhibit 4.1B) list three sources for γ-picoline–water VLE data. Data at 1 atm, computed from information in Andon et al. (1957), are reproduced in Fig. 4.1.1; an azeotrope occurs at $x_{H_2O} = 0.89$.
From flow table data,

$$x_F = 0.20 \qquad x_D = 0.80 \qquad x_B = 0.0105$$

In Fig. 4.1.2, the McCabe-Thiele method is used to determine the minimum number of theoretical plates and the minimum internal reflux, assuming feed at its bubble point.

$$N_{min} \approx 2.7$$

By direct slope measurement,

$$\left(\frac{L}{V}\right)_{min} = 0.191$$

From Eq. (4.1.8),

$$\left(\frac{R}{D}\right)_{min} = 0.237$$

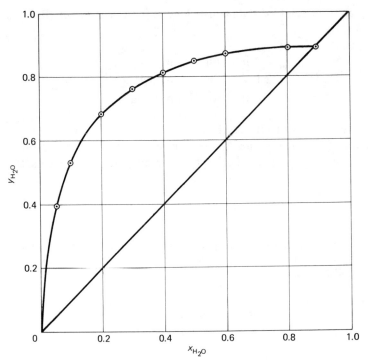

FIGURE 4.1.1
Equilibrium diagram, system H_2O–γ-picoline, $P = 760$ mmHg.

Take $R/D = 1.4(R/D)_{min} = 0.332$ (higher value chosen due to pinch point). From Eq. (4.1.8),

$$\frac{L}{V} = 0.249$$

In Fig. 4.1.3, the $L/V = 0.249$ reflux line is plotted, and the McCabe-Thiele method leads to

$$N_{opt} \approx 5.8$$

Note that $N_{opt} \approx 2N_{min}$; this would have been a satisfactory assumption.
At feed plate, using Eq. (4.1.5) and Fig. 4.1.1,

$$\alpha_{1-2} = \frac{0.685 \times 0.8}{0.2 \times 0.315} = 8.70$$

Handbooks do not list the viscosity of γ-picoline; assume viscosity is the same as that of pyridine, which is the same as water. Feed plate temperature will be somewhere

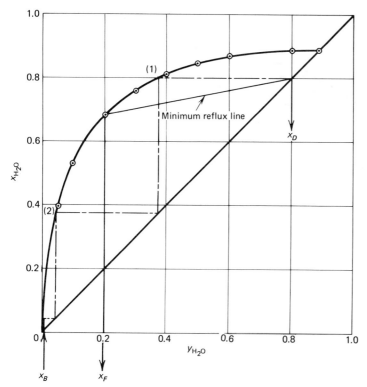

FIGURE 4.1.2
Equilibrium diagram, system H_2O–γ-picoline, $P = 760$ mmHg. Minimum reflux, minimum plates.

between boiling points of pure components (100 and 143°C); at these temperatures, liquid water viscosity ranges from 0.28 to 0.20 centipoise (cP).

Take $$\mu = 0.24 \text{ cP}$$

and $$\alpha\mu = 8.70 \times 0.24 \approx 2.1$$

From the O'Connell correlation, plate efficiency ≈ 40 percent (50 percent assumption would result in small error).

Accounting for reboiler and assuming total condenser overhead, actual number of plates is

$$N = \frac{5.8 - 1}{0.40} = 12$$

As a safety factor (see discussion later in Sec. 4.1), *recommend 15 plates.*

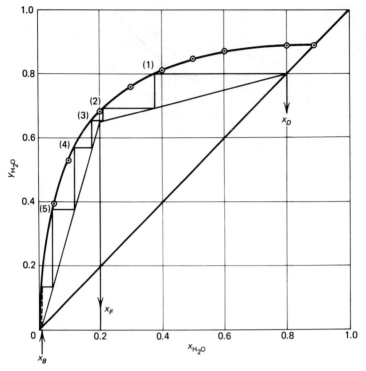

FIGURE 4.1.3
Equilibrium diagram, system H_2O–γ-picoline, $P = 760$ mmHg. Optimum plates.

With feed at bubble point, vapor flow in column is constant; vapor velocity will be highest in bottom of column where temperature (boiling point of γ-picoline) is about 143°C. Assuming perfect gas behavior,

$$M \text{ (molecular weight of } \gamma\text{-picoline)} = 93.13$$

$$\rho_v = \frac{PM}{\tilde{R}T} = 14.7 \times \frac{93.13}{10.73} \times \frac{1}{1.8(273 + 143)}$$
$$= 0.1704 \text{ lb/ft}^3$$

Since $R/D = 0.332$ and $D = 60$ lb · mol/h (flow table),

$$R = L \text{ in rectification section} = 19.9 \text{ lb} \cdot \text{mol/h}$$

But $L/V = 0.249$; $V = {19.9}/{0.249} = 80.0$ lb · mol/h. Thus

$$\mathring{Q}_V = 80 \times \frac{93.13}{0.1704} \times 3600 = 12.15 \text{ ft}^3/\text{s}$$

SECTION 4.1: EQUIPMENT SELECTION AND SIZING FOR PRELIMINARY COST ESTIMATES

Take allowable vapor velocity = 2 ft/s (flooding velocity correlations indicate that 3 ft/s would be satisfactory).

$$\text{Cross-sectional area} = 12.15/2.0 = 6.08 \text{ ft}^2$$
$$\text{Diameter} = 33.4 \text{ in}$$

Choose the next logical diameter (i.e., do not choose diameters that are "odd"), say, recommend 36-in (3-ft) column.

Use 2-ft plate spacing, plus 10 ft; column height is

$$(15 - 1)(2) + 10 = 38 \text{ ft}$$

Summary: Choose a 3-ft column, 38 ft high, with 15 sieve plates.

Exercise 4.1.1

Suppose the feed to the column in Example 4.1.1 is actually at 40°C. What effect would this have upon the number of plates? Take the feed boiling point as 112°C; the heat of vaporization of γ-picoline is 90 cal/g, and its heat capacity is 0.45 cal/(g · °C). In your opinion, would the assumption of $q = 1$ result in a serious cost estimating error?

Exercise 4.1.2

As a result of temporary upsets, the column in Example 4.1.1 could dump the liquid content on the trays into the column bottom, and the bottoms stream could be temporarily contaminated with large quantities of water. Can you suggest modifications that would avoid this potential problem?

Exercise 4.1.3

The minimum number of theoretical plates can also be calculated by using the Fenske equation:

$$N_{min} = \frac{\log\{[x_1/(1 - x_1)]_D[(1 - x_1)/x_1]_B\}}{\log \bar{\alpha}_{1-2}} \qquad (4.1.10)$$

Take $\bar{\alpha}_{1-2}$ as α_{1-2} at the feed plate; how well does the answer using Eq. (4.1.10) agree with the graphical solution in Example 4.1.1?

Heat Transfer

The basic equation that is used for the rapid preliminary design of heat-transfer equipment is the familiar one

$$A = \frac{Q}{U \Delta T_{LM}} \tag{4.1.11}$$

The heat-transfer area A is the primary costing criterion of heat-transfer equipment. If shell-and-tube exchangers are under consideration, A is the area of the outside surface of the tubes in contact with the shell fluid. Q is the heat load, calculated as part of the flow sheet energy balance.

For preliminary costing purposes, the overall heat-transfer coefficient U is normally not computed. A value typical of the equipment type and service is selected from tables of representative heat-transfer coefficients found, for example, in Perry, or Refs. 2 and 3, Exhibit 4.1A. The practicing engineer learns to recognize the range of most likely values of U for various services and is sensitive to the effect of unusual operating conditions upon those values of U. Perry shows a range of U of 80 to 200 Btu/(h · ft^2 · °F) for condensing volatile hydrocarbons with water; the experienced engineer will reduce that value to as little as 10 if the hydrocarbons are diluted with large quantities of noncondensable gases. Organic solvent-water exchangers (shell and tube) are operated with U values in the range of 50 to 150; the same liquids in agitated, jacketed glass-lined reactors exhibit a U of perhaps 30 (here glass thermal resistance limits the U).

The log-mean temperature difference ΔT_{LM} is normally calculated for the case of simple countercurrent flow; for shell-and-tube exchangers, this implies single-pass flow on both shell and tube sides. ΔT_{LM} is calculated from the expression

$$\Delta T_{LM} = \frac{\Delta T_1 - \Delta T_2}{\ln(\Delta T_1/\Delta T_2)} \tag{4.1.12}$$

where ΔT_1 is the difference between hot and cold fluid temperatures at one end of the exchanger, and ΔT_2 is the difference at the other end. The following rules of thumb are followed:

1 Do not use less than 15°F "approach" temperatures; i.e., neither ΔT_1 or ΔT_2 should be less than 15°F. (Small ΔT's require uneconomically large exchangers for small heat duties.)

2 Watch out for thermal "pinch points" of 15°F or less; these can occur, for example, in condensers that also act as desuperheaters and subcoolers, as shown in Fig. 4.1.4. Even for rapid preliminary sizing it is a good idea to sketch out a temperature-enthalpy diagram such as Fig. 4.1.4.

3 If water is used for cooling, design for no more than a 15°F temperature rise. Be conservative and use a cooling water temperature of 80°F; if process fluid cooling to much below 100°F is required, chilled brine may have to be used.

SECTION 4.1: EQUIPMENT SELECTION AND SIZING FOR PRELIMINARY COST ESTIMATES

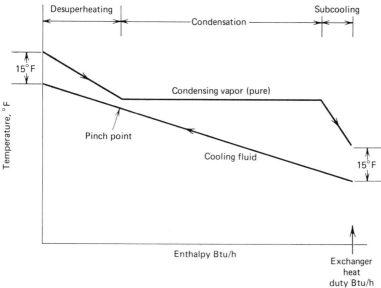

FIGURE 4.1.4
Temperature-enthalpy diagram, condenser.

The total area of condensers with heat duty typified by the situation in Fig. 4.1.4 can be estimated by adding up the separate areas for desuperheating, condensing, and subcooling. This works reasonably well even though in practice the three regions are not sharply defined and, for example, some condensation does take place in the desuperheater.

Preliminary design of reboilers is usually based upon maximum allowable heat flux. For free circulation reboilers, the usual design heat flux is about 12,000 Btu/(h · ft^2) with organic liquids, up to 30,000 Btu/(h·ft^2) with aqueous solutions.

Whereas very large shell-and-tube exchangers may be fabricated, the practical limit, at least for preliminary design purposes, is about 5000 ft^2 per unit. If larger heat-transfer areas are required, multiple units should be considered.

Example 4.1.2

Make a rapid preliminary sizing design of the total condenser required for the γ-picoline column in Example 4.1.1.

SOLUTION
Distillate $D = 60.0$ lb · mol/h (80 mol % water). From Example 4.1.1, $R/D = 0.332$. Therefore

$$\text{reflux } R = 19.9 \text{ lb} \cdot \text{mol/h}$$

and total to be condensed is

$$60.0 + 19.9 = 79.9 \text{ lb} \cdot \text{mol/h}$$

Initial condensing vapor (80 mol % water) condenses at 104°C (from data of Andon et al., 1957); the final condensing vapor is the azeotrope that condenses at 93°C. For rough sizing purposes, neglect the shape of the condensate temperature-enthalpy curve and take the condensing temperature as 93°C, a conservative step.

Cooling water is adequate for condensing. Take water in at 80°F, water out at 95°F.

For 93°C (199°F), $\Delta T_1 = 199 - 95 = 104°F$
$\Delta T_2 = 119°F$
$\Delta T_{LM} = 111°F$

Note that subcooling is neglected; if exchanger overdesign does result in some subcooling, so much the better.

Estimate heat duty from atmospheric latent heats of the pure components:

$$\text{Water} = 17{,}500 \text{ Btu/(lb} \cdot \text{mol)}$$
$$\gamma\text{-picoline} = 16{,}900 \text{ Btu/(lb} \cdot \text{mol)} \quad \text{(handbook values)}$$
$$Q = 79.9(0.80 \times 17{,}500 + 0.20 \times 16{,}900)$$
$$= 1.389 \times 10^6 \text{ Btu/h}$$

Perry (table 10-10) indicates U of 40 to 80 in similar service. Take

$$U_0 = 60 \text{ Btu/h} \cdot \text{ft}^2 \cdot °F$$

From Eq. (4.1.11),

$$A = \frac{1.389 \times 10^6}{60 \times 111} = 208 \text{ ft}^2$$

With a 25 percent overdesign factor (we will come back to these factors), specify

$$A = 250 \text{ ft}^2$$

Exercise 4.1.4

The effect of fouling must be considered early in the design of exchangers. The U value in Example 4.1.2 is based on an overall fouling coefficient of 200 Btu/(h · ft² · °F). Suppose the known effect of polymerizable impurities in the overhead reduced the recommended overall fouling coefficient to 100 Btu/(h·ft²·°F). What effect would this have upon the design?

Liquid Pumping and Agitation

Liquid pump size is usually characterized by the horsepower rating of the motor drive. The first task is to decide upon the pump type which best fits the demands of pumping rate, pumping head, and fluid corrosivity and which satisfies other possible restrictions such as those upon losses of dangerous fluids through seals. The choice is facilitated by reference to pump selection charts reproduced in Perry. The drive horsepower is calculated from the expression

$$\text{hp} = \frac{\dot{Q}\rho'H}{33{,}000\varepsilon} \qquad (4.1.13)$$

In this dimensional equation, $\dot{Q}$ is the required liquid pumping rate in gallons per minute (gpm), and ρ' is the liquid density in pounds per gallon. The head H, expressed in feet of liquid, must be estimated as a sum of static and dynamic heads. The static head may be estimated from knowledge of the elevation and operating pressure of equipment the pump serves. The dynamic head will, of course, depend upon the piping, valve, and instrument design. For preliminary purposes, in the absence of more specific knowledge, use a liquid head equivalent to a very liberal pipeline frictional pressure drop of 5 psi, and use a 5-psi drop across control valves.

ε is the combined fractional efficiency of the pump-motor unit, and for preliminary costing it is taken as 0.5. Adjust horsepower calculated from Eq. (4.1.13) to the next higher integral horsepower rating commonly used for electric motors (i.e., do not specify a 17.7-hp pump and drive—make it 20 hp). If the answer is below 1 hp, specify a 1-hp pump-motor for large plant service; fractional horsepower motors are single-phase devices with sparking hazards.

Motor-driven agitators are also characterized by the motor horsepower rating. A useful rule of thumb for specifying the required horsepower for agitated vessels is in terms of horsepower per unit volume:

Use 1 hp/1000 gal for blending.
Use 1 hp/250 gal for normal reactions, jacket cooling.
Use 1 hp/50 gal for multiphase reactions with extreme mass-transfer barriers, rapidly settling catalysts.

The last represents very violent agitation, indeed, and if the vessel were not closed, the contents would walk right out of the vessel.

Reactor volumes are set by considerations of kinetics. For extraction mixers, a 5- to 10-min residence time is often assumed for preliminary sizing. For blending and liquid-liquid dispersion, consideration should be given to low-cost, low-maintenance static mixers.

In the absence of compelling reasons, agitated vessel design should conform to the standard geometry illustrated in Fig. 4.1.5. The impeller is a flat six-blade turbine.

Example 4.1.3

Specify a pump for the γ-picoline column reflux of Example 4.1.1. Assume the accumulator will be essentially at ground level ("pumped reflux," see Fig. 3.5.5).

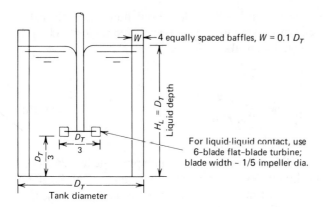

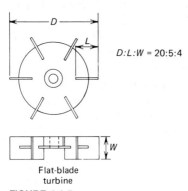

FIGURE 4.1.5
Standard agitation geometry.

SOLUTION   From Example 4.1.2 the reflux rate is 19.9 lb · mol/h. The material is 80 mol % water;

$$\text{Molecular weight} = 0.80 \times 18.0 + 0.20 \times 93.1 = 33.0$$

Thus   Reflux rate = $33.0 \times 19.9 = 657.1$ lb/h

Liquid density (assuming additive volumes, handbook values at room temperature) is

$$\text{Density} = \frac{33.0}{0.80 \times {}^{18}\!/_{1.00} + 0.20 \times {}^{93.1}\!/_{0.957}}$$
$$= 0.975 \text{ g/cc} = 8.13 \text{ lb/gal} \quad (\rho')$$

Therefore

$$\overset{\circ}{Q} = \frac{657.1}{60 \times 8.13} = 1.35 \text{ gpm}$$

This is quite a small reflux flow; for the sake of operational flexibility, design for a maximum of several times this value; say, 5 gpm.

Column is 38 ft high; when mounted on a skirt, column top will be higher than that, although static head will be slightly diminished by liquid level in the accumulator, several feet above the pump level. Choice of a static head of 45 ft seems reasonable.

Pipeline from pump to top of column is likely to be short and simple. Design for a reflux rate control valve, but total pressure drop for pipe and valve of 5 psi appears to be adequate. That is,

$$\text{Dynamic head} = \frac{5 \times 144}{62.4 \times 0.975}$$
$$= 11.8 \text{ ft}$$

Take $H = 60$ ft. From Eq. (4.1.13),

$$\text{hp} = \frac{5 \times 8.13 \times 60}{33{,}000 \times 0.5}$$
$$= 0.15$$

Therefore choose 1-hp drives.

A pumping rate of 5 gpm with a 60-ft head is a typical single-stage centrifugal pump application (Perry, fig. 6-3). Reflux pumps are critical items of equipment and should be spared.

Recommend *twin 1-hp centrifugal pumps.*

Exercise 4.1.5

A standard $1 \times 1\,\tfrac{1}{2}$ in centrifugal pump with an 8-in impeller will deliver 5 gpm against a head of 80 ft at 1750 rpm. Can this pump be used for the service in Example 4.1.3?

Storage and Surge Vessels

The sizing of storage vessels for raw materials is a matter of the commonsense evaluation of the effect of a number of variables, including:

1 Maximum size of delivery conveyance. Railroad tank cars are available in sizes up to 30,000-gal jumbos and larger. Tank trucks can deliver up to about 7500 gal.

2 Frequency of delivery. If, on the average, a volume equivalent to a tank car is required every 2 days, what should be the delivery schedule—one car every other day? Two cars every 4 days? Ten cars every 3 weeks? The final decision is a matter of an economic balance involving the cost of capital, the costs of unloading, railroad charges, and savings on large orders.

3 Need for stockpiling. Clearly, it would be intolerable to have to shut down a plant because of raw material delivery failure. Nevertheless, delays do occur, at the source and in transit, and major delivery interruptions due to strikes and natural disasters are not uncommon. Many manufacturers protect themselves against this eventuality by maintaining emergency stocks of key raw materials. Processors who depend upon seasonal supplies (for instance, in the food-processing industry) may also require large stockpiling facilities.

Much the same kinds of considerations enter into the sizing of product storage facilities. For preliminary estimates of new production plants, one month's storage of both raw materials and products is assumed. In most cases, this is a conservative estimate.

The desirability of adequate surge capacity has been mentioned in context of the discussion of common flow sheet errors (Chap. 3). If inadequate information is available to estimate likely surge requirements, a capacity equivalent to 4 to 8 hours' throughput is usually estimated. This provides for sufficient downtime of block units for many routine maintenance jobs: cleaning of plugged strainers, replacement of faulty instruments, realignment of a compressor drive. If a plant consists of sequential units typified by Fig. 3.5.1, several days' interunit surge may be provided.

Other Items of Equipment

Manufacturers' brochures and handbooks such as Perry are useful in providing rapid preliminary design methods and comparative performance data of many specialized items of equipment such as conveyors, rotary tray dryers, and continuous centrifuges. Often a quick survey of performance data and judicious interpolation will result in selection of a standard size for which reliable cost data are available.

Example 4.1.4

A 300 gpm (gallons per minute) wastewater stream contains 1 wt % of a toxic slime which must be removed before the water is discharged to waste ponds. The slime is to be removed as a mud containing 90 wt % water. Specify the approximate size of a helical conveyor centrifuge to accomplish the required sedimentation.

SOLUTION   The production of "solids" is

$$\frac{300 \times 8.34 \times 60 \times 0.01}{0.10 \times 2000} = 7.5 \text{ tons/h}$$

Performance characteristics of helical conveyor centrifuges are summed up in Perry (see table, page 207).

By interpolation, specify *40-in bowl, 100-hp motor drive*.

This is, of course, only a rough approximation of the actual requirement. Note that, conservatively, an oversized unit has been chosen, although it is quite possible that once data have been developed on settling rates and required clarification, more

Bowl dia., in	Speed, rpm	Centrifugal force, g's	gpm liquid	tons/h solids	Motor hp
6	8000	5500	To 20	0.03–0.25	5
14	4000	3180	75	0.5–1.5	20
18	3500	3130	50	0.5–1.5	15
25	3000	3190	250	2.5–12	150
32	1800	1470	250	3–10	60
40	1600	1450	375	10–18	100
54	1000	770	750	20–60	150

Source: Perry and Chilton (1973); reproduced by permission, McGraw-Hill Book Co.

careful design procedures would show that the item as specified could prove to be inadequate.

Safety Factors

In some of the examples of the methodology of rapid preliminary design, safety factors were applied to the answers obtained to arrive at a size somewhat larger than that computed from first principles. In other cases, whenever there was the choice between two standard sizes, one somewhat smaller and the other somewhat larger than the size computed, the larger size was chosen. What is the rationale for the apparently single-minded tendency toward overdesign?

Even in advanced stages of process design, safety factors are applied to design calculations. There are a number of reasons for this:

1 Most design methods have associated with them a degree of inaccuracy or error, and safety factors are applied to compensate for these, to make sure the specified item of equipment will, indeed, work as expected.

2 Most equipment is designed for steady-state operations, or perhaps for analyzable start-up and shutdown conditions. During actual operations, however, unanticipated upsets do occur which sometimes result in flow surges which could well overtax equipment designed "close to the vest." The problem is met at least part of the way by equipment overdesign. An example of the need for such overdesign is a column for scrubbing toxic components out of process vessel vents. Normally the volume of gases released from vessels operating at steady state might be small, and the practice is to design for the worst conceivable flow surge—perhaps an emergency release from a pressure relief device. In fact, the practice is to design beyond this worst conceivable surge, using a safety factor hopefully based upon some experience, for the risk of a toxic vapor "breakthrough" is unacceptable.

3 Perhaps the most controversial reason for overdesign is the unspoken desire to have some built-in overcapacity ("superproductivity"). The process designer has the clear responsibility to minimize costs, and yet it is difficult to forget instances of praise and relief when a plant, in response to strong demand, manages a superproductivity performance of 10 or 20 percent, without the necessity of expensive "debottlenecking."

TABLE 4.1.1
A TYPICAL EQUIPMENT LIST

No.	Description	Type of exchanger	Heat transfer, Btu/h	Transfer area, ft²	Transfer medium and amount	Material of construction	Bare cost, $
H-1	Reactor overhead condenser	Shell-and-tube condenser	1,046,000	300	Cooling water, 208 gpm	316 SS tubes, steel shell	
H-2	Steam condenser	Shell-and-tube condenser	271,000	40	Cooling water, 60 gpm	Copper tubes, steel shell	
H-3	Hydrolysis stream exchanger	Shell-and-tube exchanger; 1000 psig streams	540,000	102		316 L SS tubes and shell	
H-4	Evaporator condenser	Shell-and-tube condenser	1,920,000	127	Cooling water, 250 gpm	316 SS tubes, steel shell	
H-5	Aqueous-phase evaporator reboiler	Forced circulation vertical shell-and-tube	1,940,000	100	150-lb steam, 1940 lb/h	316 SS tubes, steel shell	
H-6	By-product crystallizer	Scraped wall cooler, 5-hp drive	59,000	230 (Three 25-ft sections, 2-ft dia.)	Cooling water, 12 gpm	316 SS	
H-7	B-3 condenser	Spiral condenser-accumulator	478,000	110	Cooling water, 100 gpm	316 SS	
H-8	B-3 reboiler	Shell-and-tube thermosiphon reboiler	478,000	25	DTA, 50 gpm	Titanium tubes, steel shell	
H-9	B-4 condenser	Shell-and-tube condenser	384,000	80	Boiling H₂O, 400 lb/h steam	316 SS	
H-10	B-4 reboiler	Shell-and-tube thermosiphon reboiler	384,000	40	DTA, 40 gpm	Titanium tubes, steel shell	
H-11	Product flaker	Chilled scraped drum roll	20,000	3-ft roll: 40	Cooling water, 4 gpm	316 SS	
						Total	

Peters and Timmerhaus (Ref. 3, Exhibit 4.1A) recommend the use of safety factors of 10 to 20 percent. Engineering departments of chemical corporations usually develop their own schedule of factors. The magnitude of the factors also depends upon the job at hand; well-tested processes based on experience in many plants can be designed with safety factors approaching zero.

Equipment Lists

Once all items of process equipment have been sized, the information is assembled in the form of equipment lists, in anticipation of the follow-up job of cost estimation. Each list incorporates items of the same general classification (that is, equipment with common letter label designations). An example of a heat-exchanger equipment list is shown in Table 4.1.1. The column headings common to all lists include:

Equipment label
Description and type of equipment
Criterion of size (for costing purposes; heat-transfer area in the case of heat exchangers)
Extreme operating conditions
Utility requirements (coolants, power, etc.)
Materials of construction
Purchased ("bare") cost

The format facilitates the assembly of cost data and a rapid survey of total demand for various utilities.

4.2 ESTIMATION OF PURCHASED COST OF EQUIPMENT

Definition of Purchased Cost

The total purchased cost of the equipment shown on a complete process flow sheet is the basis of perhaps the most frequently employed method to predict capital investment—the method of purchased cost factors—which we will examine in the next chapter. The cost of individual items of equipment can, of course, be obtained by contacting vendors, but this procedure is sufficiently time-consuming to prevent its use for preliminary stage estimates. Adequately accurate data may be obtained from a number of sources, including published purchased cost data correlations, and a rather rapid estimate may be assembled using these sources.

The concept of purchased cost of equipment would seem to be perfectly straightforward, but there are some complications. The purchased cost is not the same thing as the total of the costs incurred in receiving a crated, disassembled piece of equipment on the loading dock near the new plant site; i.e., the *delivered* cost. The delivered cost includes not only the vendor's selling price but all sales taxes and freight and insurance charges as well. Cost correlations in the published literature are not always clearly defined as to the status of these taxes and freight charges.

Most reliable correlations, such as those of Guthrie (1969), specifically identify purchased costs as FOB[1] costs at the *vendor's* loading dock. If the only cost available is the vendor's selling price (the FOB cost), delivered cost may be estimated by adding approximately

> 10% of selling price for sales tax and freight

Another problem with published cost data is improper identification as to whether the costs are for "bare" or installed equipment. Installed equipment is equipment that is set up in its proper location on the construction site, ready to be connected to piping and instrumentation. As we will see, the costs of installation are a substantial fraction of the purchased (or base) equipment cost, and the confusion of bare and installed costs can lead to serious error. Cost data which are not properly identified as either purchased costs or installed costs should be rejected.

Some items of equipment, particularly very large vessels and towers, may be partly fabricated "in the field." Tall tray columns, for example, may be shop-fabricated in sections, shipped to the construction site, and welded just before erection. The point is that it may be difficult to separate the purchased cost from the costs of installation, and cost correlations must be carefully surveyed to ascertain the nature of the correlation. It will not do to mix up purchased and installed costs.

Methods of Presenting Cost Data

The most common method of purchased cost data correlation is a log-log plot of cost versus size (or performance) criterion. In Fig. 4.2.1 the cost of reciprocating positive-displacement pumps is shown as a plot of this type (Pikulik and Diaz, 1977). The advantage of log-log plots is that cost data frequently result in a straight line, or very nearly so (the cynical will point out that this is a good way of squeezing together and quasi-linearizing any set of data).

In Fig. 4.2.1, a number of important characteristics should be noted, characteristics that should be incorporated as part of any reliable cost correlation:

1 The legend states clearly what the cost is for; as an example, it is explicitly pointed out that the cost does *not* include a motor, and this then will have to be separately estimated. Many pump cost correlations do include a motor.

2 The size criterion should be carefully noted; in this case, it is the delivery capacity in gallons per minute. Other pumps (particularly centrifugals) are often correlated on the basis of drive horsepower.

3 The cost is specified as purchased (FOB) cost, and the time basis is given.

---
[1]Free on board.

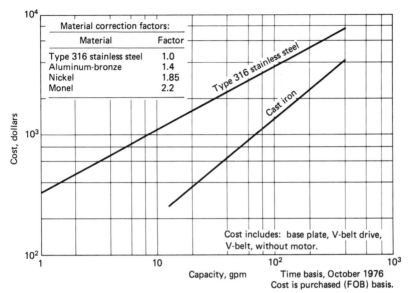

FIGURE 4.2.1
Costs for reciprocating (positive-displacement) pumps. (*Pikulik and Diaz, 1977, reproduced by permission of Chemical Engineering magazine.*)

4 The logarithmic grid is reasonably clear. Some correlations show the 1.5 and even 2.5 grid lines without clear identification, a considerable opportunity for error.

Correlations such as those in Fig. 4.2.1 must be understood for what they are. Usually the cost data of several manufacturers are included, and the best line is drawn through points that may, indeed, exhibit a considerable amount of scatter (some correlations show the range of reported costs). In addition, the data are often *list* prices, whereas many corporations, as a result of various business agreements with vendors, pay less than list prices for some items of equipment. This tends to make the use of such correlations slightly conservative, a perhaps welcome "cushion" in preliminary estimates.

Another variation of the log-log plot concept is shown in Fig. 4.2.2. Several characteristics of pressure vessels (50 psi design) are accommodated—diameter, length, and orientation. Note the carefully articulated legend, which makes it clear, for example, that the plot can be used to generate cost data for towers (as well as tanks) with a normal number of nozzles and manholes, but that additional data are required for estimating the cost of "internals" (trays, or packing, or lining).

Cost correlations are also available for standard assemblies of several items of equipment, or packaged units. The cost of absorption liquid chillers is shown in Fig. 4.2.3 as a function of capacity expressed in tons of refrigeration. Such units consist of storage vessels, heat exchangers, and internal pumps; it is important to know what items of equipment are *not* included (the chilled water circulating pump, for example) so that the missing items can be accounted for on the flow sheet and properly costed.

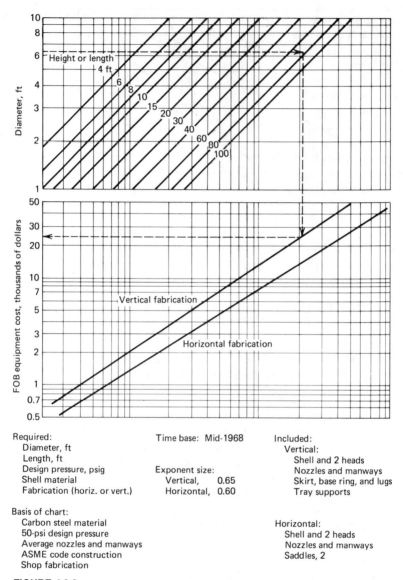

FIGURE 4.2.2
Purchased cost of pressure vessels. (*Guthrie, 1969, reproduced by permission of Chemical Engineering magazine.*)

The *component buildup method* is sometimes useful for estimating shop-fabricated tanks and towers. In this method the total weight of the metal to be fabricated is estimated by separately designing and computing the weight of the shell, heads, nozzles, manholes, skirts, and supports. The cost of fabricating the vessel can then be estimated from graphical correlations or equations of the form

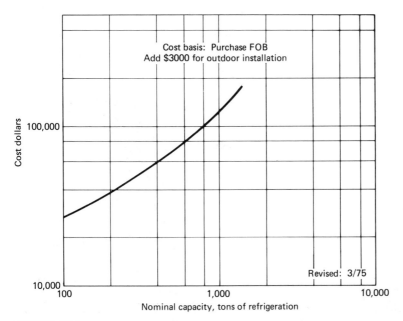

FIGURE 4.2.3
Refrigeration: absorption liquid chillers. 9 psig steam, 44°F chilled water, 85°F cooling water. (*Dow Chemical USA.*)

(Peters and Timmerhaus, Ref. 3, Exhibit 4.1A; 1979 basis)

$$\text{Fabrication cost, \$/lb} = 50w^{-0.34} \qquad (4.2.1)$$

where w is the weight (lb) of the vessel. To this total are added costs of other components, such as trays or shell linings. In fact, the component buildup method is used to broaden the scope of the graphical correlation in Fig. 4.2.2; cost data on additional vessel components are shown in Fig. 4.2.4.

This latter figure illustrates another useful procedure which greatly expands the scope of graphical cost correlations, the use of *adjustment factors*. These are factors which, when multiplied into the base cost of an item of equipment, account for different materials of construction, stronger design for high pressure, even equipment design variations. For example, material adjustment factors are listed in Figs. 4.2.1 and 4.2.4; tray type and tray spacing factors are listed below the tray costs in Fig. 4.2.4. Material adjustment factors are equipment-type-dependent and not interchangeable.

Exercise 4.2.1

Monel plate costs about 10 times as much as steel on a per pound basis. Yet the adjustment factor for monel relative to steel is 6.34 for vessels (Fig. 4.2.4), about 4.00 for pumps. Can you rationalize these differences?

Linings	Thickness, in	M & L, $/ft²
Acid brick	3	3.80
	4	5.50
	6	8.25
Firebrick	4½	7.16
	9	10.79
Rubber	3/16	4.37
	1/4	4.75
Refractory	2	7.50
	4	10.52
Gunite	2	3.20
	4	4.55
Chemical lead	5 lb	6.25
	10	7.13
	15	8.86

Process vessel cost, $ = Base cost × F_m × F_p
Adjustment factors:

Shell material	F_m Clad	Solid	Pressure factor, psi	F_p
Carbon steel	1.00	1.00	Up to 50	1.00
Stainless 316	2.25	3.67	100	1.05
Monel	3.89	6.34	200	1.15
Titanium	4.23	7.89	300	1.20
			400	1.35
			500	1.45
			600	1.60
			700	1.80
			800	1.90
			900	2.30
			1000	2.50

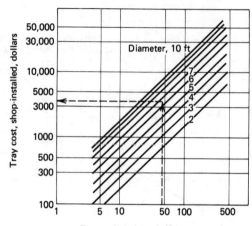

Required:
 Tray stack height, ft
 Tray diameter, ft
 Tray spacing, in
 Tray type
 Material

Time base: mid-1968

Exponent size: 1.0

Included:
 Trays (as specified)
 Supports
 All fittings
 Shop fabrication
 Shop installation

Tray cost, $ = Base cost ($F_s + F_i + F_m$)
Adjustment factors:

Tray spacing, in	F_s	Tray type	F_i^*	Tray material	F_m^*
24	1.0	Grid (no downcomer)	0.0	Carbon steel	0.0
18	1.4	Plate	0.0	Stainless	1.7
12	2.2	Sieve	0.0	Monel	8.9
		Trough or valve	0.4		
		Bubble cap	1.8		
		Koch Kascade	3.9		

*If these factors are used individually, add 1.00 to the above values.

FIGURE 4.2.4
Add-on costs of typical vessel components. (*Guthrie, 1969, reproduced by permission of Chemical Engineering magazine.*)

Example 4.2.1

Estimate the 1968 cost of a monel-clad distillation column with forty-five 4-ft-diameter monel valve trays, 24-in spacing, designed to operate at 200 psig.

SOLUTION

 Column height: (45 − 1) × 2 + ~12 ft top and bottom sections ≈ 100 ft, 4-ft dia
 Base cost of fabricated vessel (Fig. 4.2.2, vertical fabrication): $19,000
 Adjustment factor, monel-clad vessel (Fig. 4.2.4): 3.89

Material-adjusted vessel = 3.89 × $19,000 = $73,900
Pressure factor (200 psig) = 1.15
1.15 × 73,900 = $85,000
Tray stack height = 44 × 2 = 88 ft
Base cost = $3000

With adjustment factors,

Tray cost = 3000(1.0 + 0.4 + 8.9)
= $30,900
1968 purchased cost = 85,000 + 30,900
= $115,900
≈ $120,000

Exercise 4.2.2

If you do not already know, find out the meaning of monel, cladding, valve trays, skirts, and vessel manholes.

Occasionally cost data are presented in the literature in the form of an equation,

$$\text{Cost} = aS^n \quad (4.2.2)$$

where a = characteristic constant
S = size of capacity criterion
n = *scale-up*, or *capacity exponent*

Exercise 4.2.3

If an equation having the format of (4.2.2) applies, what shape will it have when plotted on log-log paper? What is the capacity exponent for type 316 stainless steel positive-displacement pumps (Fig. 4.2.1)?

Sources of Equipment Cost Information

A number of cost data sources, both public and proprietary, is available to the estimator:

Books Some of the books which contain a reasonably broad selection of equipment cost correlations are listed in Exhibit 4.2A.

Journals The journals *Chemical Engineering* and *Hydrocarbon Processing* are particularly useful and prolific in publishing current cost data. Refer to annual indices for listings of equipment covered. Special mention should be made of a series of unusually thorough and careful cost data reviews that have been appearing

> **EXHIBIT 4.2A**
>
> **A SHORT BIBLIOGRAPHY OF EQUIPMENT COST SOURCES**
>
> **1** Peters, Max S., and Klaus D. Timmerhaus: "Plant Design and Economics for Chemical Engineers," 3d ed., McGraw-Hill, New York, 1980. A thorough coverage of equipment costs in graphical format, plus an exhaustive bibliography of cost sources.
>
> **2** "Modern Cost Engineering: Methods and Data," edited by staff of *Chemical Engineering*, McGraw-Hill, New York, 1979. A compendium of cost-related articles appearing in *Chemical Engineering* magazine in the seventies.
>
> **3** Popper, H.: "Modern Cost-Engineering Techniques," McGraw-Hill, New York, 1970. A similar compendium of cost-related articles from issues of *Chemical Engineering* in the sixties.
>
> **4** Guthrie, Kenneth M.: "Process Plant Estimating, Evaluation, and Control," Craftsman Book Co., 1974. The "module" concept designed for more detailed estimates, but a lot of good information on equipment costs. Leans toward refinery practice.
>
> **5** Perry, Robert H., and Cecil H. Chilton: "Chemical Engineers' Handbook," 5th ed., McGraw-Hill, New York, 1973. Equipment cost data are scattered throughout.
>
> **6** Chauvel, Alain, et al.: "Manual of Economic Analysis of Chemical Processing," McGraw-Hill, New York, 1981. A fine collection of 1975 equipment costs.
>
> **7** Hall, Richard S., Jay Matley, and Kenneth J. McNaughton: Current Costs of Process Equipment, *Chem. Eng.*, **89**(7):80 (Apr. 5, 1982). A large collection of graphical cost correlations brought up to January 1982 levels.

in *The Canadian Journal of Chemical Engineering,* authored by Professor Woods and colleagues at McMaster University.

Sales Representatives Even though vendors are generally not contacted for quotations on standard equipment for the purpose of making preliminary estimates, sales representatives are obviously a reliable source of costs, and they are glad to quote on an informal basis. A special project is outlined at the end of this chapter to give the student some initial experience with sales representative contacts. The purchasing departments of corporations keep up-to-date files on vendors and serve the function of establishing contacts between the corporate employees and the vendors' sales representatives. Formal cost quotations are obtained by sending out written purchase inquiries, often to several competing vendors.

Manufacturers' Brochures and Catalogs A number of years ago manufacturers' catalogs were a good source of cost data, but the pressures of inflation have dried up this source. Cost information is now obsolete almost the day that it is printed, and few manufacturers wish to commit themselves to a firmly stated price, or one that would be escalated on the basis of an anticipated inflation rate. Nevertheless, price lists can still be occasionally found, particularly in catalogs listing

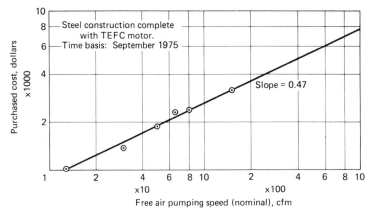

FIGURE 4.2.5
Purchased cost of liquid ring vacuum pumps. (*Cost data taken, by permission, from price lists, Kinney Vacuum Co., Canton, MO.*)

small mechanical equipment, hardware, or spare parts, and engineers can broaden the scope of their personal cost data collections by developing catalog-based information as in Fig. 4.2.5.

In-Company Cost Books Economic evaluation and production planning departments in the larger corporations assemble cost data based on their own plant construction experience into "Cost Estimation" books. Information in analytical form is frequently stored in computer bulk-memory storage devices for use in standard computer-generated economic evaluations. In industrial practice, these cost data sources are possibly the most commonly used ones.

Other In-Company Data Direct purchasing evidence is an obvious source of reliable information, the principal input to the cost books (or computer storage) mentioned. Corporate purchasing departments keep a record of *purchase orders* on file; an example is shown in Fig. 4.2.6. Note that the purchase order contains all the key information items required (indicated by the arrows): complete equipment description, price, FOB information, sales tax information, and time basis. Still another source of in-company cost data should not be neglected, the equipment "card file" (or its computer memory storage analog) in operating plants, which incorporates a great deal of information on each item of installed equipment, including the purchase cost.

Estimation Methods in Absence of Specific Information

Not infrequently the estimator is faced with the task of costing out an item of equipment for which no cost data can be located. An instance of this is the situation whereby the cost is known for a similar piece of equipment, but one of the wrong

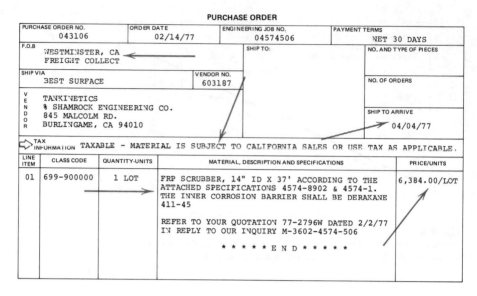

FIGURE 4.2.6
Some useful information in a purchase order. (*Shamrock Engineering Co., Burlingame, CA.*)

size or performance capability. In this case a procedure known as the *six-tenths factor rule* is often employed. It is based on the observation that, on the average, cost data on a log-log plot can be approximated by a straight line represented by Eq. (4.2.2), and the scale-up exponent n assumes a value of about 0.6. Suppose, then, we know the cost A of equipment having size criterion S_A; the cost of a similar piece of equipment with size criterion S_B can be computed from the expression

$$\frac{\text{Cost } B}{\text{Cost } A} = \left(\frac{S_B}{S_A}\right)^{0.6} \tag{4.2.3}$$

It should be emphasized that the 0.6 value of the capacity exponent is a convenient average. The actual value is somewhat different for different equipment classifications, and some of these values have been computed and tabulated (Peters and Timmerhaus, Ref. 1, Exhibit 4.2A). For projects of limited scope (few items of equipment), the tabulated values are preferred, if available.

Exercise 4.2.4

The capacity exponent for liquid ring vacuum pumps (Fig. 4.2.5) is 0.47. Suppose the cost of a 50 cfm pump were known to be $1880; how much of an error (percent) would be incurred by costing a 150 cfm pump using the six-tenths factor?

The actual value of the capacity exponent is also dependent upon the size criterion used. Valve trays characterized by diameter have a cost capacity exponent of 1.20; to use the six-tenths factor rule properly, some judgment must be used in choosing the proper size criterion.

Exercise 4.2.5

What do you think is the significance of the high value of the capacity exponent for valve trays?

Caution, and that somewhat nebulous concept of judgment, must be used in applying the six-tenths factor rule to extrapolate a single data point to much larger or smaller sizes. The cost of a 100-ft^2 exchanger can be extrapolated to 500 ft^2 with reasonable confidence, but beware of extrapolating down to, say, 10 ft^2. The 100- and 500-ft^2 items are both medium-sized exchangers with no large variation in the material-labor ratio of costs. A 10-ft^2 exchanger is quite small, and the relative amount of expensive labor required is high; such exchangers therefore tend to be relatively expensive. In essence we are saying that no single value of capacity exponent can cover the full range of equipment sizes, and it is risky to extrapolate even log-log plots much beyond the size range for which the cost data were obtained in the first place.

Exercise 4.2.6

Derive a reasonable value for the capacity exponent for absorption liquid chillers (Fig. 4.2.3), and use this to estimate the cost of a 20-ton unit. What is your estimate of the reliability of the computed value?

In those cases where the equipment is unusual or "custom-made" and no cost data can be located, the estimator must use ingenuity and a bit of informed guesswork to obtain a preliminary cost estimate. One method is *estimation by analogy;* the estimator takes advantage of similarities in design features to derive a cost from the known costs of similar equipment items. Cost data on packed distillation towers may be extended in this way to apply to extraction towers, or perhaps even evaporator bodies. Adjustment factors, even though specific to particular equipment classifications, may be applied judiciously to similar items; a monel adjustment factor of 4.0 for centrifugal pumps is likely to be quite acceptable for positive-displacement pumps or even vacuum pumps.

The idea of the component build-up method can also be used to generate cost data for more exotic items of equipment. The following example illustrates the method.

Example 4.2.2

Estimate the purchased cost of a pulsed column to extract uranium from a sulfuric acid leach solution into perchloroethylene containing a dissolved complexing agent. The leach solution (sp gr = 1.18) flows at a rate of 10 gpm and is to be extracted counter-

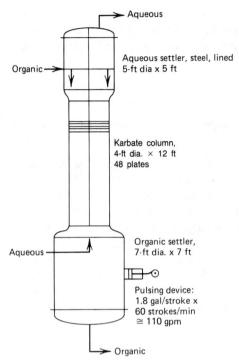

FIGURE 4.2.7
Schematic diagram of a pulsed column (results of Example 4.2.2 computations are shown).

currently at a phase volumetric feed ratio of 2:1 organic to aqueous. To attain the desired degree of transfer, it is estimated that 12 theoretical stages are required. Specific gravity of organic phase = 1.68.

The general features of a pulsed column are illustrated in Fig. 4.2.7. The column is a particularly compact device for multistage extraction. The heavy phase is introduced through dispersal nozzles into the top settling chamber, and the light phase is similarly introduced into the bottom chamber. Either phase may be the continuous one in the extraction column between the two chambers, depending upon whether the phase interface is maintained in the top chamber (as shown) or the bottom chamber. A pulsing device forces the mixed phases up and down through the narrowly spaced perforated plates in the column. On the average, the heavy phase moves downward, the light phase upward under the influence of gravity. Partial settling occurs in the chambers between the plates, particularly at the extreme up or down positions of each pulse. The top and bottom chambers act as settlers to disengage the phases.

Materials of construction:

Column and plates: Karbate (impervious graphite)
Settling sections: Epoxy-lined steel
Pulsing device: Hastelloy C

Use the following rules of thumb for sizing:

Column throughput: 2000 lb/(h · ft^2) of *combined* phases
Pulse amplitude: ¼ in (in column)
Pulse frequency: 60 min^{-1}
Plate spacing: 3 in* (Karbate); plates drilled, ⅛-in holes
Stage efficiency per plate: 25 percent
Settling velocity in settlers: Continuous-phase superficial velocity = 0.5 gal/(min · ft^2)
Settler height–diameter ratio = 1.0

Use the following approximations:

Fabrication cost of 1-in Karbate column shell (density = 145 lb/ft^3) is 30 percent higher on a per pound basis than carbon steel, including stiffening rings.
No nozzles on column; plates are mounted on a central rod support (i.e., plates are not attached to shell).
Karbate trays cost 50 percent of equivalent stainless steel trays.
Cost of epoxy lining (seven coats): $5/ft^2.
Estimate the cost using a component build-up method; i.e., estimate column, settlers, and pulsing device separately, add and adjust to accommodate assembly costs.

SOLUTION
(1) *Column design*:

$$\text{Leach solution flow} = 10 \times 8.34 \times 1.18 \times 60$$
$$= 5900 \text{ lb/h}$$
$$\text{Organic phase flow} = 20 \times 8.34 \times 1.68 \times 60$$
$$= 16,800 \text{ lb/h}$$
$$\text{Combined phase flow} = 22,700 \text{ lb/h}$$
$$\text{Cross-sectional area of column} = 22,700/2000 = 11.37 \text{ ft}^2$$
$$\text{Diameter} = 3.80 \text{ ft} \approx 4 \text{ ft}$$
$$\text{No. of plates} = 12/0.25 = 48$$
$$\text{Column height} = \frac{(48 - 1)(3)}{12} \approx 12 \text{ ft}$$

Settler design:
Upper settler, aqueous phase continuous:

$$A_x = 10/0.5 = 20 \text{ ft}^2$$
$$\text{Diameter} = 5.05 \text{ ft} \quad \text{say, 5 ft dia.} \times 5 \text{ ft}$$

Lower settler, organic phase continuous:

$$A_x = 20/0.5 = 40 \text{ ft}^2$$
$$\text{Diameter} = 7.12 \text{ ft} \quad \text{say, 7 ft dia.} \times 7 \text{ ft}$$

*Center line to center line.

Pulsing device design:

$$\text{Volume pulsed per stroke} = \tfrac{1}{4} \times 11.37 \times \tfrac{1}{12} \times 7.48$$
$$= 1.77 \text{ gal}$$
$$\text{Strokes per min} = 60$$

Thus

$$\text{Equivalent throughput per min} = 106.3 \text{ gpm}$$

(2) *Costs of separate items*: Use January 1979 data in Peters and Timmerhaus (Ref. 1 Exhibit 4.2A).

Column fabrication cost:

$$\text{Weight of Karbate shell} = 4.0 \times \pi \times \tfrac{1}{12} \times 12.0 \times 145$$
$$= 1822 \text{ lb}$$

Add two 5-ft OD flanges:

$$2\pi(5^2 - 4^2) \times \tfrac{1}{12} \times 145 = 683 \text{ lb}$$
$$\text{Total weight} = 2505 \text{ lb}$$

Based on Eq. (4.2.1), fabrication cost per pound is

$$1.30 \times 50(2505)^{-0.34} \approx \$4.50/\text{lb}$$

Thus

$$\text{Shell cost} = 2505 \times 4.50 \approx \$11{,}300$$
$$\text{Tray cost, stainless steel sieve trays} = \$400$$

Thus cost of 48 drilled Karbate trays is

$$48 \times 400 \times 0.50 = \$9600$$
$$\text{Total column cost} = \$20{,}900$$

Cost of settlers:

$$\text{Top settler volume} = 735 \text{ gal}$$
$$\text{Cost for steel vessel} = \$3200$$
$$\text{Bottom settler volume} = 2020 \text{ gal}$$
$$\text{Cost} = \$5500$$

Approximate coating area (assume column opening just about balances extra coating area for nozzles and manholes):

Top settler:

$$5\pi \times 5 + 2 \times \frac{\pi}{4} \times 25 = 118 \text{ ft}^2$$

Bottom settler: 232 ft^2

$$\text{Coating cost} = 5.00(118 + 232) = \$1750$$
$$\text{Purchased cost of settlers} = \$10{,}500$$

Cost of pulsing device: The pulsing device is assumed to be analogous to a low-head diaphragm pump. From Woods et al. (1979), purchased cost of stainless steel pump and drive is $1600 (110 gpm). The adjustment factor to Hastelloy C is about 1.5. Therefore

$$\text{Cost} = 1.5 \times 1600 = \$2400$$

(3) *Total*:

$$\begin{array}{r} 20{,}900 \\ 10{,}500 \\ \underline{2{,}400} \\ \$33{,}800 \end{array}$$

Add 10 percent for bolts, tie rods, other assembly items; call it $37,000 (in 1979).

4.3 EFFECT OF INFLATION UPON CAPITAL COSTS

Cost Indices

The cost of any particular item of equipment is estimated for some particular point in time. Inflationary and other economic factors result in frequent changes (usually, but not always, increases) in that cost with the passage of time. Clearly, if different equipment costs have different time bases, a method is needed to convert the sum of all the costs to a common time basis. Usually the preferred time basis is the present, although frequently some cost projection into the future may be required.

The method commonly used to alter the time basis of equipment costs or total plant costs involves cost indices. We seek index values for any two years m and n, I_m and I_n, so that the costs of a particular piece of equipment can be related in the same proportion,

$$\boxed{C_n = C_m \frac{I_n}{I_m}} \qquad (4.3.1)$$

Here the C's are the costs (for instance, purchased costs of a particular pump); if C_m is known for the year m, we then have a way of predicting the cost for year n. Clearly m is usually some year in the past, but n may be in the past, it may be the current year, and it may even be in the future.

Now, to find an index that will fit any and all items of equipment is quite an order. We have already discussed such an index in Chap. 1, the Consumer Price

Index. We saw that such an index is the result of a great deal of statistical balancing; in the case of the CPI, the U.S. Bureau of Labor Statistics evaluates the prices of a defined spectrum of consumer products and services purchased in about 60 locations throughout the country. A similar index is generated for industrial equipment, the *Marshall and Swift (M&S) Equipment Cost Index*. It is based on much the same kind of statistical balancing that is used to generate the CPI, except that industrial equipment and services rather than consumer goods and services are evaluated. The evaluation involves a fixed spectrum of equipment classifications (such as heat exchangers) purchased by a fixed spectrum of industries (about 70 percent in the chemical process and petroleum industries); equipment installation costs are also accounted for, so that the M&S index reflects changes in construction labor costs as well. Strictly speaking, the M&S index does not apply to the equipment *purchased* cost, but it is used in this way, and justifiably so if the purchased cost is used to generate total capital investment numbers by using the method of purchased cost factors (Chap. 5).

The arbitrary base for the M&S index is 100 for the year 1926. The base value, however, is not particularly significant, and each index seems to have a different base year (CPI has a base of 100 for 1967). The important item is the ratio R_{j-k} of the index values I_j and I_k:

$$C_n = C_m R_{n-m} \quad (4.3.2)$$

In Example 4.2.1, the 1968 cost of a 4-ft diameter by 100-ft monel distillation column was estimated as $120,000. The 1968 M&S index was 273. In 1980 the M&S index was 650;

$$R_{1980-1968} = {}^{650}\!/_{273} = 2.38$$

and the estimated 1980 cost is

$$120,000 \times 2.38 = \$286,000$$

Current values of the M&S index (and values for the preceding decade) appear routinely in each issue of *Chemical Engineering* magazine (page marked "Economic Indicators"). Older values are listed in the references in Exhibit 4.2A.

An index such as the M&S index reflects more than inflationary pressures; technological improvements are also accounted for as part of the data-gathering methodology. This is certainly true of another commonly used index, the *Chemical Engineering (CE) Plant Cost Index*. Input data for this index, which is used to time-adjust the fixed capital investment of whole plants, include items such as the cost of engineering design; such costs are certainly affected by changes in computer technology, for example.

How do some of these indices compare? A comparison for the years 1965 to 1980 is shown in Table 4.3.1 for CPI, M&S, and CE indices, with 1965 arbitrarily

TABLE 4.3.1
RELATIVE VALUES OF THREE COST INDICES
(Basis: 1965 = 1.0)

Year	Consumer Price Index	M&S index	CE plant index
1965	1.000	1.000	1.000
1967	1.058	1.073	1.058
1970	1.231	1.237	1.210
1975	1.706	1.807	1.754
1976	1.804	1.926	1.848
1978	2.068	2.224	2.105
1980	2.612	2.651	2.481

chosen as 1.0. It turns out that the CPI is a reasonably good approximation of chemical plant cost trends; the M&S index runs a little ahead of CPI.

Exercise 4.3.1

What has been the average annual increase in the cost of equipment between 1965 and 1980?

Future trends can be predicted from calculations such as Exercise 4.3.1, or from regression analysis, recommended by Mascio (1979). Mascio analyzed 1971–1978 data and predicted that the percentage increase in the M&S index above the value at the beginning of 1971 (315.2) was given by

$$\mathcal{P} = -12.08 + 10.87T' \qquad (4.3.3)$$

where T' is the time (years) since the start of 1971. His equation predicts a value of 594 for March 1980; the average increase calculated from Exercise 4.3.1 and used to extrapolate the 1971 index results in a value of 575. The actual index was 640.

It is customary to submit capital estimates with the time basis clearly stated and the explicit warning, "After this date, escalate estimated cost by 1 percent per month." During the early eighties this turned out to be a realistic piece of advice.

Other cost indices exist and are used in specific applications; Mascio (1979) lists several. Plant cost indices for a number of the world's industrialized countries are published in the journal *Engineering Costs and Production Economics* (before 1980, name was *Engineering and Process Economics*).

Limitations of Cost Indices

The very nature of the cost index concept suggests what its limitations are going to be. After all, it is a great deal to ask that such a simple concept accommodate the long-range cost variation of a bewildering array of technologically sophisticated equipment. The concept does take into account technological improvements within the narrow spectrum of common equipment that is monitored as part of the index generation; yet it is unlikely to account for the effect of technological breakthroughs in specific areas. Titanium, tantalum, and zirconium equipment was an expensive curiosity only a couple of decades ago; improvements in metallurgy and fabrication techniques have greatly reduced the relative cost of such equipment, in contradistinction to the cost index trend. Obsolescence of certain items of equipment; regional variations in fabrication costs; supplier-user business agreements—these are just some of the items which contribute to major perturbations from the cost index average. For these reasons the scaling up of the cost of a specific piece of equipment may be a risky business, whereas the scaling up of the cost of an *assembly* of equipment works quite well.

In view of some of the limitations that have been alluded to, a reasonable approach to older cost data is the following:

Data up to 10 years old are acceptable for cost index adjustment.

Data between 10 and 20 years old should be treated with caution; data for standard items such as centrifugal pumps and electric motors are more likely to be acceptable.

FIGURE 4.3.1
Fabrication cost of distillation columns in Europe. (*From Stallworthy, 1980; by permission, Process Economics International.*)

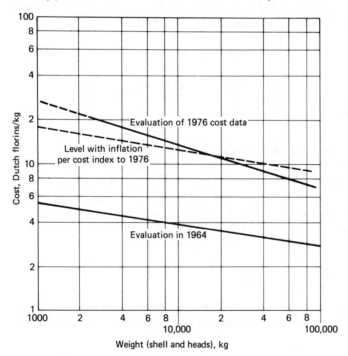

TABLE 4.3.2
COMPARISON OF 1949 AND 1979 EQUIPMENT COSTS

Item	1949 cost, $*	1949 cost scaled to 1979 (M&S index)	1979 cost, $†
Centrifugal pump, Worthite construction, 200 gpm, 100-ft head, no motor	613	2,190	2,300
Heat exchanger, 500 ft², fixed tube sheet, 304 stainless tubes and tube sheets	3,000	10,700	10,500
Two-stage reciprocating compressor, 1000 scfm, 125-psig discharge, steel, with drive	14,300	51,100	61,000
Spray dryer, 18-ft drying chamber, 8000 lb/h water evaporation, 316 stainless steel	78,000	280,000	320,000
Centrifuge, carbon steel suspended basket, 20-in basket, 20-hp drive	14,200	50,700	30,000

*Zimmerman and Lavine (1950).
†Peters and Timmerhaus (Ref. 1, Exhibit 4.2A).

Data more than 20 years old are frequently obsolete.

Stallworthy (1980) points out an interesting case of the deviation of costs of a specific type of equipment from the cost index trend. The cost of distillation columns in 1964 and 1976 is shown in Fig. 4.3.1 (Dutch currency). During the 12-year period, small columns became relatively more expensive, large columns cheaper; i.e., the capacity exponent changed value.

Occasionally old data on standard equipment can be scaled up with surprisingly good results. In Table 4.3.2 some 1949 costs (M&S = 157) from Zimmerman and Lavine (1950) are scaled up and compared with costs in Peters and Timmerhaus (Ref. 1, Exhibit 4.2A) (M&S = 561; 1979). For preliminary costing purposes the projection over a 30-year period is acceptable. Only the centrifuge projection tends to be high, probably due to intervening improvements in fabrication techniques and automatic controls.

4.4 COST OF EQUIPMENT INSTALLATION

Typical Installation Requirements

We must be very careful in talking about equipment installation, for the term means something different to different individuals. Some engineers, for example, think of the installed cost of an item of equipment as that portion of the total plant capital investment assignable to that equipment. In fact, this is a perfectly legitimate concept, one that is the basis of the *modular* method of cost estimation proposed by Guthrie (1969). For our present purposes, we will think of an installed item as equipment which has been set in place in its proper location on the construction site, ready to be integrated with other items through piping, instrument loops, utility distribution systems, and other tie-ins. This narrower definition of installation is perhaps best illustrated by an example.

Suppose we have purchased a pump that is to be part of a new processing plant. What do we get for our money? In the usual case, we receive a crated item aboard a truck or railroad car (assuming that we have paid our transportation bills). We are now ready to install the pump. What is involved in the installation process? A great deal; for instance:

Unloading, temporary warehousing, transportation to local construction site (all laced with a considerable amount of paperwork)
Unloading and assembly; inspection for structural integrity, presence of all parts
Construction of concrete pad or other mounting platform
Procurement of motor drive (motor may come with pump, both mounted on a common base, shafts already coupled); bolting down of unit, installation of drive coupling
Installation of power supply (local switches, local conduits); wiring of motor
Installation of pump seal flushing fluid system (if required), including rotameters and throttling valves
Painting, and insulation, if required
Pre–start-up checkout: shaft lineup, check on direction of rotation (which will be wrong in about 50 percent of the cases)

We now have an installed pump, and yet much more remains to be done—the pump must be connected to other equipment with piping, instrumental controls must be installed, a switch room with switch gear must be constructed; in fact, all the jobs that will integrate the pump with the rest of the plant remain to be done. We will be concerned about what these jobs are and how much they will cost in the next chapter.

Some very large equipment is at least partly field-fabricated; that is, some of the preinstallation assembly jobs are completed on the plant construction site, and it may be a matter of some nit-picking as to which job is encompassed by the purchased cost and which is part of the installation. Large towers may be shop-fabricated in sections, transported to the site, and joined by welding in a horizontal position before erection; this is part of the "bare" equipment assembly. Large API storage tanks are assembled in situ from shop-fabricated plate sections; bare equipment fabrication and equipment installation are certainly difficult to separate in such cases. Nevertheless, even those large API tanks have a quoted purchased price, which includes the welding assembly labor. Installation, on the other hand, encompasses costs associated with construction of pads and foundations, construction of ladders and service platforms (all of which are not included in the purchased cost), and insulation and painting.

Installation Cost Estimation

The detailed estimation of the cost of installing an item of equipment (within our definition of installation) is often difficult. It involves a step-by-step analysis of the installation procedure, with labor and material estimates for each step. For this reason even advanced-stage estimates are frequently based upon installation cost factors which are a certain percentage of the equipment purchased cost. Such factors are a distillation of a great deal of experience; some of these factors have been published and can be found in the references listed in Exhibit 4.2A, but most corporations prefer the use of factors based upon their own background of experience.

The estimation of cost categories (such as installation costs) as a certain percentage of the equipment purchased cost is the substance of the *method of purchased cost factors,* which we will examine in more detail in Chap. 5. We will discuss the numerical values of these factors, although it must be recognized that each organization has a little different idea as to what these factors should be, and the values that will be given are best looked upon as being "representative." It should also be mentioned that some published methods analogous to the method of purchased cost factors are reportedly based upon equipment *delivered* costs, although the difference between purchased (FOB) and delivered costs is seldom properly emphasized. We have seen that this difference amounts to about 10 percent and is ascribable to sales taxes, freight, and insurance. In "factored estimate" methods based on purchased costs, these charges are generally incorporated under the estimated "indirect" costs associated with the construction of a new facility.

If the cost of installation is to be taken as a certain factored percentage of the equipment purchased (or delivered) cost, it stands to reason that the factor will be a function of the equipment type, the materials of construction (expensive materials do not necessarily increase costs of installation), and the degree of preinstallation in the fabricator's shop (for example, assembly of skid-mounted units). Indeed, pump installation costs may range from 25 to 60 percent of the purchased cost (Peters and Timmerhaus, Ref. 1, Exhibit 4.2A); the high value is normal for steel centrifugals, which, by virtue of quantity production, tend to have a relatively low purchased cost, whereas the low value applies to expensive alloy pumps. Limited-objective projects involving a small number of equipment items are better handled by using individual item installation factors, if available.

On the other hand, a single factor based upon the sum total of all equipment costs may be used for estimating the installation costs in a projected plant. A single factor is commonly used for making order-of-magnitude and study estimates; this estimation method is reasonably reliable, and is often used even for advanced-stage estimates. The factor method of estimating the installed cost is just the first step in the method of purchased cost factors, leading up to an estimate of the plant direct fixed capital investment; as the name implies, the factors are based on the equipment *purchased* cost. A representative factor for the cost of installation is 43 percent of purchased equipment cost:

$$\text{Total installed equipment cost} = 1.43 \Sigma (\text{purchased costs}) \quad (4.4.1)$$

The 43 percent factor is a rough average for plants handling both solids and fluids. In plants handling fluids exclusively, the installation factor is higher, about 47 percent.

In many instances, it may be necessary to estimate the man-hours of labor and the labor costs involved in installing an item of equipment. This is certainly an important part of project control estimates, but the project manager of limited-scope projects in production units or pilot plants may be called upon to make a quick and reasonably reliable projection. An often-used rule of thumb is that about *three-quarters of the installation cost is for labor*. Construction labor costs are very high, indeed; in 1979, typical West Coast contract labor cost per man-hour was

$$\boxed{\$38-\$45 \text{ per man-hour}}$$

(This cost includes all "fringe benefits," contractor's overhead, and contractor's profit.)

Occasionally the estimated man-hours required for "installing" equipment may be found in the published literature; an example is shown in Fig. 4.4.1. The installation time shown, however, does not include the time required for constructing concrete foundations, steel supports, electrical connections, and many other jobs which fall within our definition of equipment installation. The time refers essentially to the job of erection, of "setting the equipment in place." The man-hours for erection must not be confused with the total installation labor in our sense of the word; nevertheless, information such as that in Fig. 4.4.1 is useful, particularly in the case of limited-objective projects involving production plant changes and pilot plant installations. Estimators working on project control estimates use data such as those in Fig. 4.4.1, along with similar labor correlations for concrete and steel structures and other construction items, to generate total construction labor costs.

Example 4.4.1

A 500-gal glass-lined jacketed reactor is to be installed on an existing slab in a pilot plant area. What is the likely *delivered* cost of the reactor in 1980 (M&S = 650), and what is the likely labor cost for setting the reactor in place?

SOLUTION From Fig. 4.4.1 (M&S = 238.8),

$$\text{Purchased cost} = \$9200$$
$$\text{1980 purchased cost} = {}^{650}/_{238.8} (9200) = \$25{,}000$$

Assuming 10 percent delivery costs,

$$\text{1980 delivered cost} = 1.10 \times 25{,}000 = \$27{,}500$$

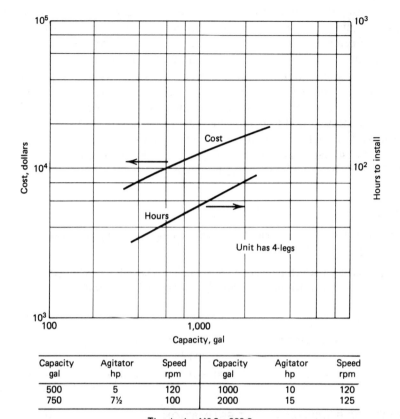

FIGURE 4.4.1
Cost and erection labor, glass-lined reactors. (*Mills, 1964; reproduced by permission of Chemical Engineering magazine.*)

From graph

$$\text{``Hours to install''} = 38$$

Assuming $40/man-hour cost in 1979 and 10 percent annual labor cost increase, labor cost of setting reactor in place is

$$1.10 \times 40 \times 38 = \$1670$$

Note that the labor cost in the above example is only about 7 percent of the purchased cost. This is far below the total installation labor costs that we would calculate from Eq. (4.4.1) and the three-quarters labor rule of thumb. This is because no account has been taken, for example, of labor required to build the supporting concrete slab. The total installation cost for glass-lined reactors, nevertheless, is well below the 43 percent average of Eq. (4.4.1); they are relatively expensive items of equipment, and they are delivered largely preassembled. The 43 percent factor (or 47 percent for all-fluid processing plants) should be used only for a collection of equipment items with a wide spectrum of functions.

Occasionally the cost of equipment such as evaporators and dryers is reported in the cost literature as *installed cost*. Clearly, extreme care must be taken to look over the descriptive legend on graphical cost correlations so as not to get purchased and installed costs confused, and to make sure that "installed cost" is used in the same sense that we have defined. Total equipment purchased cost is used as the basis for "factored" estimates, and the best procedure is to divide any literature installed costs by 1.43 before adding to the purchased cost total.

Example 4.4.2

Estimate the installation costs in a plant for which the following equipment cost totals have been obtained:

Total of equipment for which purchased costs data are available: $1,500,000
Total of equipment for which installed cost data are available: $500,000

SOLUTION

$$\text{Purchased cost} = 500{,}000/1.43 + 1{,}500{,}000$$
$$= \$1{,}850{,}000$$
$$\text{Installation cost} = 1{,}850{,}000 \times 0.43 = \$800{,}000$$

4.5 RELIABILITY OF EQUIPMENT COST ESTIMATION

Variability of Cost Data

It may have occurred to the student that the preliminary costing of the equipment for a proposed process rests upon a somewhat shaky foundation, a foundation of assumptions, approximations, and uncertainties. Since the total of equipment costs serves as the takeoff point for the estimation of the direct fixed capital, how can one expect to obtain a reliable estimate from individual cost components of uncertain reliability?

Let us examine first the question of the variability of the cost data for an otherwise well-defined item of equipment. Cost correlations are usually based upon compilations of actual moneys expended for equipment purchased within the various authors' spheres of experience. One would therefore expect to find differences in reported costs for a specific piece of equipment, differences ascribable to the multiplicity of fabricators and distributors, or to secondary design variations, or to regional variations in business. The most reliable cost correlations are based upon sufficient data to average out some of these variations, but the likelihood remains that the actual expenditure for a purchased item of equipment will not be the same as the estimated cost. For most items, there really is no such thing as a "firm" cost, and published correlations must be viewed as being merely "representative."

TABLE 4.5.1
REPORTED PURCHASED COST OF 500-gal GLASS-LINED REACTOR

Source	Year	M&S index	Reported bare cost, $	Cost scaled to M&S = 561 (end of 1978)
1	1954	184	7,260	22,140
2	1961	237	4,600	(10,890)
3	1964	239	9,100	21,380
4	1967	263	11,830	25,230
5	1968	256	9,600	21,040
6	1970	303	18,460	(34,170)
7	1976	472	29,000	(34,470)
8	1979	561	23,000	23,000

Sources: 1. Aries and Newton (1955); 2. Bauman (1964); 3. Mills (1964); 4. Derrick (1967); 5. Peters and Timmerhaus (Ref. 1, 2d ed., Exhibit 4.2A); 6. Happel and Jordan (1975); 7. Hoerner (1976); 8. Peters and Timmerhaus (Ref. 1, 3d ed., Exhibit 4.2A).

As an example of the reality of the situation, consider the cost data in Table 4.5.1. The costs shown are reported values for a 500-gal glass-lined steel reactor, rated for 50 to 100 psi internal pressure, with jacket and agitator. Not all the data are "original"; that is, some of the sources simply take the reported costs from previous references and update them, or present them in more convenient form. Variability may be introduced as part of this process [for example, an equation of the form of (4.2.2) may be used to approximate a graphical correlation]. The costs are all updated to the end of 1978 by using the M&S index; this procedure may also introduce variability due to uncertainties introduced with the cost index concept. The table incorporates some pretty ancient data, but cost index updating does not reveal any obvious chronological trend.

If all eight sources are considered, the range of the updated costs is surprising: $10,890 to $34,470. The surprise stems from the fact that historically there has been only one major manufacturer of glass-lined reactors in the United States, and only in the last few years has significant European competition made its mark; one would think that the cost information would reflect rather firm quotations. If the cost data are considered to be a random population, the following statistical parameters can be calculated:

$$\text{Mean (most probable value)} = \$24,040$$
$$\text{Standard deviation, } S(x) = \$7620$$
$$\text{Standard deviation of the mean, } S(\bar{x}) = \$2690$$

We can establish the 95 percent confidence limits of the most probable value by using the t test.

For $(8 - 1)$ degrees of freedom, $t_{0.025} = 2.365$, and the limits therefore are

$$\bar{x} \pm t \cdot S(\bar{x}) = \$24,040 \pm 6,360$$

that is, the confidence limits are ±26.5 percent of the most probable value. Even though in this case we have an unusually high number of cost sources available, considerable uncertainty is associated with the computed average value.

Inspection of the data in Table 4.5.1 reveals that three data points (enclosed in parentheses) are suspect. If these are ignored, the statistical parameters become

Mean, $\bar{x}$ $22,560
$S(x)$ $1,670
$S(\bar{x})$ $750

and the 95 percent confidence limits are

$$\$22{,}560 \pm 2{,}080 \quad (\pm 9.2\%)$$

a much more acceptable indication of variability.

Of course, cost data differ from many other statistical ensembles in that the true value eventually becomes known—the actual purchased cost of the item of equipment when the plant is built. For the set of data in Table 4.5.1, a comparison standard is available; at the end of 1978, the manufacturer's quoted price for the 500-gal vessel was (author's files)

$$\$25{,}142 \quad \text{(actual cost)}$$

Assuming that the actual cost is the true mean of the infinite set of data from which Table 4.5.1 was sampled (a debatable point), we see that the actual cost is well within the 95 percent confidence limits of the eight-point set but outside the limits of the supposedly improved five-point set!

Clearly, it is a very good idea to check and double-check cost data in the literature whenever possible, even for the purpose of order-of-magnitude estimates. Regrettably, most of the time the estimator does not have access to as broad a range of values as shown in Table 4.5.1; one or two cost values are the norm. This means that an estimate could very well be based upon, say, source 7 in the table, $34,470. In fact, based on the data in the table, chances are even that a single cost source would supply a value below $18,900 or above $29,100 (±21 percent).

The variability of reported cost data is enhanced by the fact that cost correlations tend to be a little "simplistic"; for the sake of compactness, secondary design considerations are often ignored, and yet these may have a substantial influence upon the purchased cost. For example, the pressure rating of the *jackets* of glass-lined steel reactors is an important parameter. A 1000-gal agitated reactor (100 psi internal rating) with a jacket rated at 90 psi has a quoted price (M&S = 655) of $43,800; a 150 psi jacket would increase the price by some $12,000 (27 percent).

Figure 4.5.1 is a collection of 1973 data (normalized to M&S = 300) for rotary sliding vane, cast iron pumps. The scatter of the data may be ascribed not only to the multiplicity of fabricators and distributors but to differences in design of the basic pump item as well. A 100 gpm pump has a predicted cost (least-squares fit) of $830, with a reported range of $440 to $1400 (−47 percent, +67 percent).

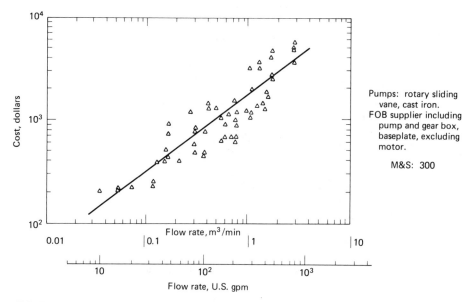

FIGURE 4.5.1
Cost of rotary sliding vane pumps. (*Woods et al., 1979; reproduced by permission of The Canadian Journal of Chemical Engineering.*)

Another contributing factor to the uncertainty of cost data used in estimates is the random error generated in reading the data. Eighteen chemical engineering students were asked to determine the cost of a 1000-gal glass-lined steel reactor from Fig. 4.5.2a. The chart is not easy to read, and the frequency distribution of the reported results is shown in Fig. 4.5.2b with a superimposed eyeballed bell-shaped curve. The mean of the results is $13,700, close to what one would reasonably pick from Fig. 4.5.2a ($14,000); the standard deviation is $1300.

If we further keep in mind the tentative nature of the flow sheets generated during the earlier stages of process development and the approximations used in preliminary equipment design, we come to realize that the error range in the cost of any particular item of equipment may be uncomfortably large. In fact, the best we can hope for is to zero in on the order-of-magnitude cost of each item; as a rough guess, we expect an error range of perhaps ±50 percent.

Propagation of Errors

Regardless of the precise method that is used to generate an estimate of the required capital to construct a proposed new facility, the usual starting point is the sum total of all equipment purchased (or delivered) costs. In view of the large error range that we have projected for the cost of any individual item of equipment, is the estimation effort doomed in advance to being nothing more than an order-of-magnitude guess?

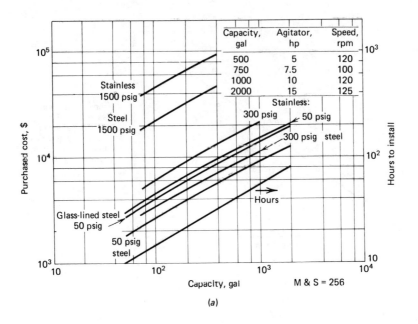

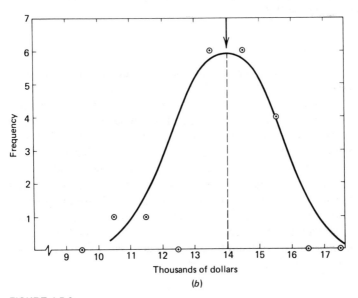

FIGURE 4.5.2
Random errors generated in reading cost graphs. (a) Cost and installation time of jacketed and agitated reactors. (*From Peters and Timmerhaus, Ref. 1, 2d ed., Exhibit 4.2A; by permission, McGraw-Hill Book Company*). (b) Frequency distribution of reported values.

SECTION 4.5: RELIABILITY OF EQUIPMENT COST ESTIMATION

Let us represent the sum total of costs as

$$Y = \sum_n y_i \quad (4.5.1)$$

where Y is the total, and y_i is the cost of each of n items.

If ΔY represents the maximum possible error in Y, it follows from the linear nature of (4.5.1) that

$$|\Delta Y| = \sum_n |\Delta y_i| \quad (4.5.2)$$

where Δy_i is now the maximum error in y_i.

The error may be expressed as a percentage (or better yet, a fraction) of the base value:

$$|\Delta y_i| = f_i y_i \quad (4.5.3)$$

and

$$|\Delta Y| = FY \quad (4.5.4)$$

If the four equations are combined, we find that

$$F = \frac{\sum_n f_i y_i}{\sum_n y_i} \quad (4.5.5)$$

In the special case that all f_i are the same, $F = f$. Consider an ensemble of equipment for which all f_i are 0.50, a value that we have previously conjectured. The conclusion we reach is that the maximum error in the total cost Y, our basis for estimating the capital, is ± 50 percent, not a particularly comforting thought.

Our conclusion does not mean, however, that we will always be saddled with such a giant error; except in cases of incredibly bad luck, the Δy_i's will tend to compensate, since they can have opposing signs.

For the sake of argument, let us consider each y_i to be based on a sample from a normally distributed population. Focus first on a very simple case of Eq. (4.5.1),

$$Y = y_1 + y_2 \quad (4.5.6)$$

Here the equipment ensemble consists of just two items; the total Y is the sum of two costs y_1 and y_2, each cost being a *single* value selected by an estimator from the cost data population—actually, the immediately available sample. Now, estimator A will arrive at the result $(Y)_A$,

$$(Y)_A = (y_1)_A + (y_2)_A$$

whereas estimator B will get

$$(Y)_B = (y_1)_B + (y_2)_B$$

where $(y_1)_A$ may or may not be the same value as $(y_1)_B$.

Following the development in Box et al. (1978), we note that

$$\bar{Y} = \bar{y}_1 + \bar{y}_2 \qquad (4.5.7)$$

that is, the mean of all the sums generated from all the estimates is just equal to the sum of the mean costs of each item of equipment. For M separate estimates,

$$\bar{y}_1 = \frac{(y_1)_A + (y_1)_B + \cdots + (y_1)_M}{M} \qquad (4.5.8)$$

We can now write, based on the above equations,

$$(Y)_A - \bar{Y} = [(y_1)_A - \bar{y}_1] + [(y_2)_A - \bar{y}_2]$$
$$(Y)_B - \bar{Y} = [(y_1)_B - \bar{y}_1] + [(y_2)_B - \bar{y}_2]$$
$$\cdots\cdots\cdots\cdots\cdots\cdots\cdots\cdots\cdots\cdots\cdots\cdots$$
$$(Y)_K - \bar{Y} = [(y_1)_K - \bar{y}_1] + [(y_2)_K - \bar{y}_2] \qquad (4.5.9)$$

Squaring (4.5.9),

$$[(Y)_K - \bar{Y}]^2 = [(y_1)_K - \bar{y}_1]^2 + (y_2)_K - \bar{y}_2]^2 + 2[(y_1)_K - \bar{y}_1][(y_2)_K - \bar{y}_2]$$

Summing over M,

$$\sum^M [(Y)_K - \bar{Y}]^2 = \sum^M [(y_1)_K - \bar{y}_1]^2 + \sum^M [(y_2)_K - \bar{y}_2]^2 + 2 \sum^M [(y_1)_K - \bar{y}_1]$$
$$[(y_2)_K - \bar{y}_2] \quad (4.5.10)$$

If we now divide Eq. (4.5.10) by $M - 1$, we note that

$$\frac{\Sigma_M [(Y)_K - \bar{Y}]^2}{M - 1}$$

is the definition of the variance of Y, $S^2(Y)$. Similarly, the first two terms on the right-hand side become $S^2(y_1)$ and $S^2(y_2)$, the variances of the y_1 and y_2 sets used by the estimators. The last term becomes the covariance of y_1 and y_2; if y_1 and y_2 sets are completely independent (and there is little reason to believe otherwise), then the covariance is zero. We arrive at the relationship

$$S^2(Y) = S^2(y_1) + S^2(y_2) \qquad (4.5.11)$$

The relationship can be readily broadened to

$$S^2(Y) = \sum_{i=1}^{n} S^2(y_i) \qquad (4.5.12)$$

Exercise 4.5.1

Generalize the above derivation to show that Eq. (4.5.12) follows.

Exercise 4.5.2

What is meant by saying that the sets $y_1, y_2, \ldots, y_n$ are independent? Give a reasoned argument why

$$\text{cov}(y_i, y_j)_K \approx 0$$

if the sample sets y_i and y_j are independent.

It must be emphasized that the variances $S^2(y_i)$ refer to the particular set of y_i's selected in the course of M estimates; the variances need not be the same as the variances of the samples from which the sets were chosen, unless $M \to \infty$.

What, then, is the significance of Eq. (4.5.12)? For any set y_i, the standard deviation $S(y_i)$ is a measure of the scatter of the data about the mean value $\bar{y}_i$; it is a "handle" on the magnitude of the error that a single cost value might be expected to incorporate. For example, if we are willing to settle upon 95 percent confidence limits, with reasonably large samples, the limits are approximately

$$\bar{y} \pm 2S(y_i)$$

Equation (4.5.12), rewritten as

$$S(Y) = \left[\sum_{i=1}^{n} S^2(y_i) \right]^{1/2} \qquad (4.5.13)$$

relates the expected error in the equipment cost total to the error spread of the costs of individual items of equipment. If we express the standard deviation as a fraction of the mean value and use the 95 percent confidence limit approximation,

$$\frac{2S(Y)}{\bar{Y}} = F \quad \text{and} \quad \frac{2S(y_i)}{\bar{y}_i} = f_i$$

(4.5.13) becomes

$$F = \frac{[\sum_{i=1}^{n} (f_i \bar{y}_i)^2]^{1/2}}{\bar{Y}} \quad (4.5.14)$$

Suppose, now, that we agree that the individual equipment cost data are as bad as we have indicated, that all f_i are equal to ± 0.50. Equation (4.5.14) then simplifies to

$$F = 0.5 \frac{[\sum_{i=1}^{n} \bar{y}_i^2]^{1/2}}{\bar{Y}} \quad (4.5.15)$$

As a simple example of the significance of Eq. (4.5.15), suppose we have an ensemble of 120 items of equipment (all different) which all have the identical mean sample cost $\bar{y}_i$. Then

$$\bar{Y} = 120 \bar{y}_i$$

and

$$\left[\sum_{i=1}^{120} \bar{y}_i^2 \right]^{1/2} = \sqrt{120}\, \bar{y}_i$$

and, from (4.5.15),

$$F = \frac{0.5}{\sqrt{120}} = 0.0456 \approx 4.6 \text{ percent}$$

In qualitative terms, the result of this example indicates that, given a sufficiently large equipment ensemble typical of new process plants, the equipment cost total, albeit based upon very rough individual item costs, turns out to be quite acceptable. This is one reason why order-of-magnitude and study estimates of capital investment turn out to be reasonably good, a useful tool for assessing project profitability. We will return to the question of ultimate capital estimate accuracy and reliability in the next chapter.

Does the relationship expressed by Eqs. (4.5.13) and (4.5.15) suggest that we need not be particularly careful in pinning down the costs of individual items of equipment? Of course not. Perhaps the statistical realities tend to be somewhat forgiving, but in the final analysis, "you get what you pay for"—sloppy procedures beget sloppy results. Provided the time is available, the conscientious estimator consults several cost sources for each item of equipment to reduce the margin of error to a practical minimum.

Exercise 4.5.3

In the above example, suppose all 120 items are exactly the same piece of equipment. What is F?

NOMENCLATURE

a	a characteristic constant [Eq. (4.2.2)]
A	area, m^2
C	cost, $
D	distillate flow rate, kmol/h; also, diameter, m
f	ratio, error to mean value, individual equipment costs
F	ratio, error to mean value, total equipment cost
H	fluid head, ft [in Eq. (4.1.13)]
$\mathcal{H}$	Henry's law constant
I	index value
K	equilibrium constant
L	liquid flow rate in column, kmol/h; also, turbine impeller blade length, m
M	molecular weight; also, counting index for number of separate estimates
n	capacity (scale-up) exponent
N	number of plates or stages
p	partial pressure, Pa
P	total pressure, Pa
$P°$	vapor pressure of pure component, Pa
$\mathcal{P}$	percentage increase in index [Eq. (4.3.3)]
Q	heat transfer rate, W
$\mathring{Q}$	fluid flow rate [gpm in Eq. (4.1.13)]
R	reflux flow rate, kmol/h
R_{j-k}	ratio of index values for years j and k
$\check{R}$	perfect gas law constant, 8313.0 J/(kmol·K)
S	size or capacity criterion of equipment
$S(x)$	standard deviation; also $S(\bar{x})$, standard deviation of mean; $S^2(x)$, variance, etc.
T	temperature, K
T'	time, years
U	overall heat transfer coefficient, W/(m^2·K)
V	vapor flow rate in column, kmol/h
w	weight, lb [in Eq. (4.2.1)]
W	width of turbine impeller blade, m
x	mol fraction in liquid
y	mol fraction in vapor
y_i	cost of ith equipment item
Y	total cost of ensemble of equipment
$\bar{y}_i, \bar{Y}$	mean values of y_i, Y
α	relative volatility; specifically, α_{1-2} is the volatility of component 1 relative to component 2, and $\bar{\alpha}_{1-2}$ is the average α_{1-2} in a column
Δ	difference, as in ΔT
ε	fractional efficiency of pump-motor unit
ρ	density, kg/m^3; ρ' in Eq. (4.1.13), lb/gal

Subscripts

$A, B, C, \ldots$	sequential items
B, D, F, R, V	in bottoms, distillate, feed, reflux, vapor, respectively
$\ldots, i, j, k, \ldots, m, n$	sequential items

1, 2, 3, . . .   numbered components; numbered flow sections
x cross-sectional, as in A_x
0   based on outside area, as in U_0
T tank, as in D_T
LM   Log-mean, as in ΔT_{LM}
min   minimum
opt   optimum

Abbreviations

$\text{cov}(y_i, y_j)$ covariance of sample sets i and j
gpm   gallons per minute
HETP   height equivalent to a theoretical plate
hp   horsepower rating of pump motor
rpm   revolutions per minute
VLE   vapor-liquid equilibrium

REFERENCES

Andon, R. J. L., J. D. Cox, and E. F. G. Herington: Phase Relationships in the Pyridine Series, *Trans. Faraday Soc.*, **53**:410(1957).

Aries, Robert S., and Robert D. Newton: "Chemical Engineering Cost Estimation," McGraw-Hill, New York, 1955.

Bauman, Carl H.: "Fundamentals of Cost Engineering in the Chemical Industry," Reinhold, New York, 1964.

Box, George E. P., William G. Hunter, and J. Stuart Hunter: "Statistics for Experimenters," Wiley, New York, 1978.

Denzler, Rudolph E.: Blower and Fan Costs, *Chem. Eng.*, **59**:120 (October 1952).

Derrick, George C.: Estimating the Cost of Jacketed, Agitated, and Baffled Reactors, *Chem. Eng.*, **74**:272(Oct. 9, 1967).

Guthrie, Kenneth M.: Capital Cost Estimating, *Chem. Eng.*, **76**:114 (Mar. 24, 1969).

Happel, John, and Donald Jordan: "Chemical Process Economics," 2d ed., Marcel Dekker, Inc., New York, 1975.

Hoerner, George M., Jr.: Nomograph Updates Process Equipment Costs, *Chem. Eng.*, **83**:141 (May 24, 1976).

King, C. Judson: "Separation Processes," 2d ed., McGraw-Hill, New York, 1980.

Lundeen, R. W., and W. G. Clark: How much for Rubber-Lined Vessels?, *Chem. Eng.*, **62**:191 (March 1955).

Oldershaw, C. F., L. Simenson, T. Brown and F. Radcliffe: Absorption and Purification of Hydrogen Chloride from the Chlorination of Hydrocarbons, *Chem. Eng. Prog.*, **43**(7):371 (July 1947).

Mascio, Nicholas E.: Predict Costs Reliably Via Regression Analysis, *Chem. Eng.*, **86**:115 (Feb. 12, 1979).

Mills, H. E.: Costs of Process Equipment, *Chem. Eng.*, **71**:133 (Mar. 16, 1964).

Perry, Robert H., and Cecil H. Chilton: "Chemical Engineers' Handbook," 5th ed., McGraw-Hill, New York, 1973.

Pikulik, Arkadie, and Hector E. Diaz: Cost Estimating for Major Process Equipment, *Chem. Eng.*, **84**:106 (Oct. 10, 1977).

Sim, William J., and Thomas E. Daubert: Prediction of Vapor-Liquid Equilibria of Undefined Mixtures, *Ind. Eng. Chem., Process Des. Dev.,* **19**(3):386 (1980).
Stallworthy, E. A.: Cost Inflation - The Effect of Time, *Proc. Econ. Int.,* **1**(2):29 (Winter 1979/1980).
Valle-Riestra, J. F.: Method of Removing Solvent Residues from Aqueous Paper Pulp, U.S. Patent 4,048,007, Sept. 13, 1977.
Van Winkle, M., and W. G. Todd: Optimum Fractionation Design by Simple Graphical Methods, *Chem. Eng.,* **78**:136 (Sept. 20, 1971).
Woods, D. R.: "Financial Decision Making in the Process Industry," Prentice-Hall, Englewood Cliffs, N.J., 1975.
———, Susan J. Anderson, and Suzanne L. Norman: Evaluation of Capital Cost Data: Gas Moving Equipment, *Can. J. Chem. Eng.,* **56**(4):413 (August 1978).
———, Suzanne L. Norman, and Susan J. Anderson: Evaluation of Capital Cost Data: Liquid Moving Equipment, *Can. J. Chem. Eng.,* **57**:385 (August 1979).
Zimmerman, O. T., and Irvin Lavine: Chemical Engineering Costs, Industrial Research Service, Dover, N.H., 1950.

SPECIAL PROJECT

The purpose of the project is to give the student some initial experience with equipment vendor contacts. Select one or two categories from Table SP4.1 and proceed as follows:

1 Search out phone numbers of one or two regional vendors, using resource material in Table SP4.2, local phone directories, and any material available at the university.
2 Telephone vendor and ask for sales engineer.
3 Identify yourself, explain why you are calling (university course project).
4 Ask for quotation on specific item; use listed scenario, or make up your own application.
5 Ask for brochures, descriptive literature (for your own files).
6 Write a brief report (one page) outlining what you asked for and what information you obtained, including purchased cost. How does the quoted cost compare with cost given in available references?

Most vendors will be delighted to help you—you are their future customer!

PROBLEMS

4.1 Using resource material in Perry and your own background of knowledge, select two or three materials of construction for each of the listed services, and choose the one you think would be the cheapest.
  a Phase separation vessel: perchloroethylene (C_2Cl_4) and 60% aq H_2SO_4 (cold)
  b Centrifugal pump: 20% aq $CaCl_2$ with suspended sand particles
  c Heat exchanger (shell-and-tube): hot ethylene glycol, seawater coolant
  d Helical ribbon mixer: tomato paste and onion puree (both pH = 4.5)
  e Sieve tray distillation column: atmospheric separation of phenol and toluene (some moisture)

TABLE SP4.1
EQUIPMENT APPLICATIONS LIST

Item		Typical application
1.	Moyno pump	Meter 22°Bé HCl in range of 0 to 20 gpm, 50 psig discharge pressure. Rubber-lined construction, variable-speed drive.
2.	Gyratory crusher	Crush 50 tons/h of dolomite from $2\frac{1}{2}$- to $\frac{1}{2}$-in size.
3.	Rotary vacuum filter	Filter 20 gpm of a 0.5 wt % slurry of gelatinous $Al(OH)_3$ in 20% solution of $CaCl_2$ in water.
4.	Conveyor dryer	Dry 5 tons/h of alfalfa pellets, $\frac{1}{2}$-in dia. $\times$ $1\frac{1}{2}$ in, from 40 to 5% moisture.
5.	Centrifugal pump and drive	50 gpm of 98% H_2SO_4, 150-ft head.
6.	Extraction column, glass, Rushton type	Contact 5 gpm of aqueous phase with 5 gpm of kerosene, eight stages.
7.	Fiberglass storage tank	1000 gal, 25% LiCl solution; 2-in bottom nozzle, integral cover with manhole. Four other nozzles.
8.	Crystallizer	500 lb/h of $-20/+50$ mesh KCl crystals from saturated solution at 80°C.
9.	Wiped-film evaporator	Devolatilize 2 gpm of molten paraffin wax from 5 wt % hexane to 20 ppm.
10.	Glass centrifugal pump	10 gpm of an emulsion of perchloroethylene (2 vol %) in 20% phosphoric acid, 50-ft head.
11.	Spray dryer	Spray-dry a 40% sodium acetate solution to give 500 lb/h of 20-μm powder particles.
12.	Refrigeration unit	25-ton unit, 0°F refrigeration level (to refrigerate circulating $CaCl_2$ brine).
13.	Glass-lined reactor	1000 gal, agitated, jacketed (50 psig internal rating, 100 psig on jacket).
14.	Chain conveyor	Elevate 25 tons/h of $\frac{1}{2}$-in dolomite from 1000-ton bin into process vessel, 75-ft elevation, 50 ft horizontal displacement.
15.	Impervious graphite heat exchanger	Interchange two 5 gpm streams of sulfuryl fluoride, SO_2F_2; require 12.5 ft^2 transfer area (10 psig on one side).
16.	Centrifuge (horizontal suspended basket)	Filter and through wash 1000 lb/h of benzoic acid needle crystals (50% moisture) from 5% aqueous slurry. 316 stainless steel.
17.	Lapp pulsafeeder	10 gpm of hydrocarbon tar, 100°C, 1000 cP, 50-ft head.
18.	Centrifugal compressor, electric drive	1000 scfm of H_2, from 5 to 500 psig. Steel construction.
19.	Filtration cartridge	50 gpm of aqueous waste containing 0.01 wt % of $CaCO_3$ sludge.
20.	U-tube, removable bundle heat exchanger (two-pass)	Preheat phenol from 150 to 250°F with 150 psig steam; estimate 150 ft^2 area requirement. All surfaces in contact with phenol 316 stainless steel.

PROBLEMS

TABLE SP4.2
RESOURCE MATERIAL, NAMES AND SALES OFFICES OF EQUIPMENT VENDORS

1. "Chemical Engineering Catalog," Penton/IPC, Cleveland. Annual.
2. "Hydrocarbon Processing Catalog," Gulf, Houston. Annual.
3. "Chemical Engineering Equipment Buyers' Guide," published annually by *Chemical Engineering* magazine.
4. "Thomas Register," Thomas Publishing, New York. Annual.

f 20,000-gal storage tank: cold-mixed sulfuric-nitric acids (about 10% aqueous concentration)

g Basket centrifuge: NaCl crystals in 50% aq NaOH

4.2 Cost correlations are based upon some capacity or size criterion. Fixed-tube heat exchangers, for example, are correlated on the basis of tube external area; considerations of materials of construction, pressure ratings, etc., are accommodated by adjustment factors. What unique capacity criteria do you think might apply to the following types of equipment? (Check Perry for description if you are not acquainted with the equipment.)

Jaw crusher
Reciprocating compressor
Distillation column, Koch trays
Apron conveyor
Electrolytic cell
Rotary kiln
Tramp iron magnetic separator
Electrostatic precipitator
Evaporative crystallizer
Fourdrinier (paper-forming) machine
Steam jet ejector
Hollow fiber dialyzer

4.3 Construct a vapor-liquid equilibrium (VLE) diagram at one atmosphere for the system water-tetrachloroethylene, which forms a heterogeneous azeotrope. Use the vapor pressure and solubility data in Figs. P4.1 and P4.2. The liquid density of C_2Cl_4 may be taken as constant at 1.48 g/cm^3.

What is the boiling temperature of the azeotrope?

After you have determined the limits of the heterogeneous region, the equilibrium lines in the two narrow homogeneous regions may be approximated as straight lines drawn through the origin. For very sparingly intersoluble species, this is a good enough, and somewhat conservative, approximation. What assumption regarding the Henry's law constant $\mathcal{H}$ is implicit in this approximation?

The H_2O–C_2Cl_4 system is so sparingly intersoluble that a normal xy diagram cannot be clearly drawn. Expand the liquid mol fraction scale in the homogeneous regions.

4.4 The water–hydrogen chloride system forms a high-boiling azeotrope. The effect of pressure upon VLE is shown in Fig. P4.3 (adapted from Oldershaw et al., 1947). Design a process for obtaining liquid anhydrous HCl from a 10% aq HCl product out of a dilute HCl vapor scrubber.

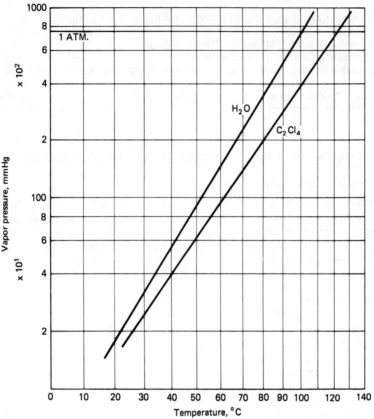

FIGURE P4.1
Vapor pressure of water and perchloroethylene. (*Perry and Chilton, 1973.*)

HCl has an atmospheric boiling point of $-85°C$. Cold liquid HCl reflux may be obtained by compressing anhydrous HCl vapor, condensing to a liquid by using cooling water or chilled brine, and partly flashing the liquid.

For quick preliminary calculations, assume that the latent heat of vaporization of HCl is 190 Btu/lb and the specific heats 0.50 Btu/(lb · °F) for the liquid, 0.18 Btu/(lb · °F) for the vapor. HCl vapor pressures above 1 atm are given in Perry.

Describe your process with the aid of a block flow diagram.

4.5 A bubble-point liquid of the following composition is to be separated by distillation:

Component	Mol fraction	
Propylene	0.2158	
Propane	0.1817	(light key)
Butadiene	0.2010	(heavy key)
Butane	0.2312	
Pentane	0.1703	

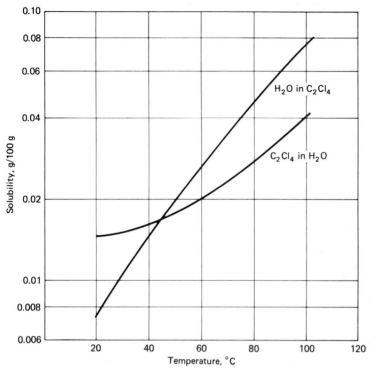

FIGURE P4.2
Mutual solubility, H_2O–C_2Cl_4 system. [Data from McGovern, *Ind. Eng. Chem.*, **35**:1230 (1943).]

The column is to be operated at 400 psia, and the desired split is 99 percent of the light key in the distillate, 99 percent of the heavy key in the bottoms. The average relative volatility of the keys is

$$\alpha_{LK-HK} = 2.05$$

Estimate the optimum number of theoretical plates, N_{opt}.

A solution to this problem is outlined in Aerstin and Street (Ref. 2, Exhibit 4.1A) who, in turn, adapted the problem from Van Winkle and Todd (1971). They use a shortcut method based upon graphical correlations and find

$$N_{opt} = 24.5$$

An even faster solution is afforded by using the rules of thumb outlined in this chapter. Assume components above the light key will wind up completely in the distillate; those below the heavy key will wind up in the bottoms. Calculate N_{min} from the Fenske equation [Eq. (4.1.10)], and use the rough relationship between N_{opt} and N_{min}.

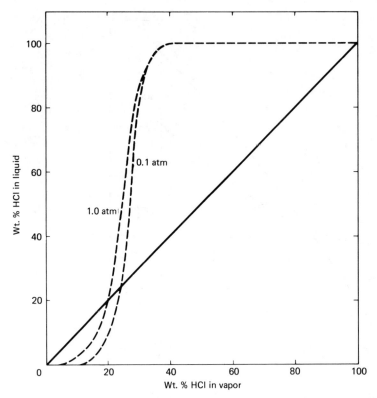

FIGURE P4.3
VLE of HCl–H$_2$O system. [Oldershaw et al. (1947).]

Assuming the computation of Aerstin and Street is the basis for buying the column, about what percent error in the estimate of column cost would be incurred by using your answer?

4.6 Make a rapid preliminary design of a packed column to accommodate the separation specified in Example 4.1.1.

4.7 100 gpm (gallons per minute) of a hot brine (60°C, sp gr = 1.08) contains 800 ppm (parts per million by weight) of iodine (as the species I$_2$). The iodine is to be extracted into CCl$_4$ in a countercurrent mixer-settler unit (CCl$_4$ sp gr at 60°C = 1.52).

The species I$_2$ is distributed between the CCl$_4$ and brine layers according to the relationship

$$\frac{c \text{ in CCl}_4}{c \text{ in brine}} = \text{constant} = 86 \text{ (at 60°C)}$$

where the concentration c is in grams per liter.

It is required that 95 percent of the I$_2$ be removed from the brine, and the desired I$_2$ concentration in the extract is 25 g/liter. The solvent is supplied to the extraction train iodine-free; volume expansion of CCl$_4$ due to dissolved I$_2$ may be neglected. The brine and CCl$_4$ may be considered to be mutually insoluble.

The internal phase ratio in the mixer may be the same as the feed phase ratio. The settler may be designed using the rule of thumb that a typical settling rate to minimize entrainment is 0.5 gpm/ft^2 of settling area, based on the more rapidly flowing phase.

Design a mixer-settler system to accomplish the extraction. Sketch a flow sheet of the system (see also Prob. 3.45a), and show instrumentation and/or piping arrangement that will make the system workable.

In some cases the maintenance of an internal phase ratio in the mixers equal to the feed phase ratio results in a problem with stable emulsions. This may sometimes be eliminated by "inverting" the internal phase ratio; in the present case, for example, the organic phase might be the continuous phase in the mixer if the organic-aqueous internal ratio were maintained at, say, 2.0. How could this be accomplished? What effect would this phase inversion have upon the equipment design?

4.8 In Prob. 3.47, a process of producing 20% oleum by blending 78% spent acid and 60% oleum was suggested. For the case considered, the net 20% oleum product amounts to about 8.7 gpm. If feeds to the blender are at 70°F and the oleum product is limited to 100°F, about 640,000 Btu/h must be removed.

Show that this heat duty could not be handled in a jacketed, agitated glass-lined blender vessel cooled with water from a cooling tower (at 80°F).

4.9 (Adapted from Happel and Jordan, 1975, Marcel Dekker, Inc.) A topic of much interest in present-day process engineering is the use of air-cooled heat exchangers instead of water-cooled exchangers. Consider the use of air-cooled heat exchangers to condense 10,000 lb/h of benzene in the overhead from an atmospheric distillation column. Assume a negligible amount of liquid subcooling, and estimate the purchase cost of

    a A water-cooled condenser of conventional design

    b An air-cooled condenser

Use data available in the 5th edition of Perry (1969 cost basis).

Which appears to be more expensive? Is the greater expense balanced by other advantages?

4.10 A 1% consistency paper pulp (1 wt % of dry fibers) must be elevated a net of 10 ft into a headbox above a paper-forming machine. The paper mill is rated at 500 tons/day of dry paper product. What is the approximate rating (hp) of the pump drive?

4.11 In the absence of specific data, a rule of thumb for filtration rates is:

5 gpm/ft^2 or more of filtrate, filters like sand
1 gpm/ft^2, average
0.1 gpm/ft^2, poor filtration, Al(OH)$_3$ sludge

The above are based on passing mother liquor through a 1-in cake on a laboratory glass frit at essentially full vacuum in the filtration flask.

Make a preliminary design of a rotary drum vacuum filter to process 500 gpm of a slurry containing 10 wt % of solids. The slurry exhibits "average" filtration behavior (filter aid is not required); the mother liquor has a density of 68.6 lb/ft^3, the solid has a true density of 93.6 lb/ft^3, and the filter cake has a porosity of 50 percent.

For quick preliminary design of rotary drum vacuum filters, the following assumptions are often made:

Neglect filter medium resistance
Consider cake to be incompressible
Maximum available pressure drop across cake = 2000 lb/ft^2 (0.945 atm)

Rotational speed of drum = 1 rpm (revolutions per minute)
Drum submergence = 25 percent (total filtering area basis)

Use typical filter sizes listed in Perry to pick one for the job.

The student may wish to review constant-pressure filtration theory before tackling this problem.

4.12 50,000 pounds of crushed limestone (maximum 1-in lumps) are required daily in a large trickle-bed reactor. The limestone is to be fed into the reactor once per day over a period of 1h. The crushed rock (bulk density 90 lb/ft^3) is delivered by truck and dumped from a ramp into a concrete bin.

From listed items in Perry, pick a centrifugal discharge bucket elevator size to load the reactor, and estimate the required power consumption. The chute discharge from the bin is 8 ft below ground level, and the reactor loading chute is 60 ft above ground level.

4.13 For every piece of equipment in a processing plant there is a rational basis for design. The rationale, however, is not always clear or understood, and design engineers are frequently puzzled by apparently indeterminate design criteria. An example is the problem of designing atmospheric pressure vertical cylindrical tanks for the storage of a specified volume of liquid—what diameter-height ratio is one to choose?

Assume that the metal wall thickness is determined by requirements of stiffness and is independent of diameter. What design criterion would you choose, and what diameter-height ratio would the criterion lead to? Can you think of other design or economic considerations which would modify your conclusion?

4.14 A process has been proposed to produce 1 ton/h of a metal selling for $1/lb. The process involves a high-pressure leaching of the powdered ore directly with a complexing agent dissolved in an organic carrier (density = 56.2 lb/ft^3). For every pound of metal produced, 250 lb of the extractant must be recycled. The leaching system operates at 200 psig, and the recycled extractant must be pumped in from atmospheric pressure storage vessels.

Some concern has been expressed that the huge extractant recycle might impose an undue economic burden, that the cost of pumping might be too high relative to the value of product.

a Estimate the current purchased cost of positive-displacement pumps and motors to recycle the extractant. It is proposed that four pump-motor units be purchased, three to operate at any one time and the fourth to stand by. Because of the corrosive nature of the extractant, nickel pumps must be used (Fig. 4.2.1 may be used for pump cost data). The motors must be explosionproof; cost data are available in the references listed in Exhibit 4.2A.

b If ac power costs 3 cents/kWh, what is the pumping cost (power only) per pound of product? Is this likely to be an undue economic burden?

c Suppose the rich extract could be discharged from the high-pressure leaching equipment through turbines that would result in recovery of 75 percent of the useful power input to the pumps. If the turbines cost roughly the same as the pump-motors and last just 2 years, will the purchased cost of the turbines be repaid by the power saved? What is your recommendation?

4.15 Estimate the current purchased cost of the column designed in Example 4.1.1, using data in Figs. 4.2.2 and 4.2.4 (M&S = 273). Assume use of carbon steel shell, stainless steel sieve trays.

4.16 Using pressure vessel design methods summed up in Peters and Timmerhaus (Ref. 1, Exhibit 4.2A), recommend the shell thickness for the column in Example 4.1.1. Assume 80 percent welded joint efficiency and a corrosion allowance of 1/16 in (corrosion is expected to be minor for carbon steel). Use the component buildup method in the same reference to estimate the purchased cost; estimate the shell weight, assume a reasonable number of nozzles and manholes, and provide for shop installation of the stainless steel sieve trays.

How does the estimated current cost compare with that in Prob. 4.15?

4.17 (Adapted from Aries and Newton, 1955, by special permission.) A simple crystallization unit consists of the following process equipment:

- 3 1000-gal agitated steel tanks
- 1 200-ft^2 cast iron plate and frame filter
- 1 1000-ft^2 cast iron horizontal-tube evaporator, Cu tubes
- 1 300-ft steel Swenson-Walker crystallizer
- 1 200-ft^2 steel atmospheric tray dryer
- 2 2000-gal steel storage tanks
- 1 42-in steel batch top-suspended centrifugal
- 12 15 gpm cast iron centrifugal pumps with 1-hp motors, 30-ft head

Using cost data available to you, estimate the current total purchased equipment cost.

4.18 The 1974 purchased cost of an all-carbon-steel heat exchanger with 100 ft^2 of heating surface was $1680. The capacity exponent of similar exchangers is 0.60 between 100 and 400 ft^2, 0.81 between 400 and 2000 ft^2. What is the estimated current purchased cost of a 1000-ft^2 exchanger?

4.19 Make a plot of the M&S index for your own files. Do your own least-squares fitting of the data (see Mascio, 1979) to predict what the index value will be 5 years from the present.

4.20 Estimate the total purchased cost of equipment, illustrated in Fig. 3.5.8, to manufacture 1 ton/h of pure adipic acid. All equipment in contact with liquid streams containing adipic acid is to be fabricated from type 316 stainless steel or rubber; equipment in contact with the dry powder may be type 304 stainless steel.

Use shortcut methods to arrive at approximate sizes of the equipment. The references in Exhibit 4.1A and Perry's handbook may prove helpful. Exhibit 4.2A lists readily available cost data sources.

4.21 The recycling of wastepaper may involve extraction of aqueous pulp suspensions with insoluble solvents such as tetrachloroethylene (perchloroethylene, C_2Cl_4) to remove paper contaminants ("pernicious contraries") such as waxes and glues. Following countercurrent extraction and solvent separation, some solvent remains occluded on the fibers. This may be removed by countercurrent air stripping in a sieve plate column, provided the consistency (i.e., the weight percent of fibers in the suspension) is sufficiently low.

The effectiveness of pulp stripping on each plate may be represented with the concept of the Murphree vapor efficiency E_m, given by

$$E_m = \frac{y - y_0}{y^* - y_0} \quad (P4.1)$$

where y_0 = mol fraction of solvent in air entering a particular plate
y = mol fraction in air leaving the plate
y^* = mol fraction if the gas leaving the plate were in equilibrium with the liquid on the plate

Valle-Riestra (1977) determined that E_m varied with the weight percent of occluded (separate liquid phase) C_2Cl_4 in 0.5% consistency pulp as follows (20°C):

Wt % C_2Cl_4	E_m, %
1.60	39.4
0.941	10.2
0.688	6.33
0.418	2.91
0.312	2.15
0.205	2.18
0.118	1.53
0.062	0.92

a Based upon the format of Eq. (P4.1), construct a "pseudo-equilibrium" xy diagram that may be used for a McCabe-Thiele graphical construction (you may wish to read the discussion of the Murphree vapor efficiency in King, 1980). Note that for a separate C_2Cl_4 phase in the pulp suspension, the true equilibrium line on the xy diagram is a horizontal line. At 20°C, the vapor pressure of C_2Cl_4 is 14.7 mmHg, and its solubility in water is 147 ppm (parts per million by weight). The diagram is to be computed for atmospheric pressure stripping.

b Choose what seems to you to be a reasonable stripping air rate, construct an operating line on the xy diagram, and determine graphically the number of actual plates required to strip out 95 percent of C_2Cl_4, present at a level of 1.5 percent in the liquid feed.

c Design a column to handle 100 gpm (gallons per minute) of feed, based on an allowable superficial gas velocity of 5 ft/s. Determine the current purchased cost of a steel column by using cost data in this chapter.

4.22 The cost of a long-tube vertical (LTV) evaporator [see, for example, fig. 11-16(f), Perry, 5th ed.] may be approximated, using the component buildup method, as a heat-exchanger cost plus a tank cost. Using this method, estimate the cost of an all-steel LTV evaporator to evaporate 20,000 lb/h of water at atmospheric pressure. LTV evaporators may be designed for a heat flux of 12,000 Btu/(h · ft^2) and a disengaging vapor velocity in the tank of 2 ft/s. Include a reasonable additional cost for joining the items into a single unit, tie rods, flanged joints, etc.

Compare your computed current cost with the cost index adjusted value given in the same edition of Perry (fig. 11-21, selling price).

4.23 What average annual European equipment inflation index can you derive from Fig. 4.3.1, and how does that compare with values derived from the M&S index (for example, Table 4.3.1)?

4.24 An estimated fabricated cost of an electrolytic cell to produce 2.4 tons/day of Cl_2 gas is required. The Cl_2 will be generated by electrolysis of an HCl–NaCl solution; at the anode,

$$Cl^- = \tfrac{1}{2} Cl_2 + e^- \qquad (P4.2)$$

whereas at the cathode,

$$H^+ + e^- = \tfrac{1}{2} H_2 \qquad (P4.3)$$

The proposed cell design is shown in Fig. P4.4. The design constitutes a *bipolar cell* consisting of N compartments, each compartment a separate cell with its own brine feed and product outlets. The compartments are internally divided by a diaphragm which keeps the generated gases separated but allows current flow and ion diffusion. The compartments are separated by 2-in graphite electrode plates; each plate acts as an anode for one compartment and cathode for the adjacent one. The voltage drop across the whole cell is NE, where E is the voltage drop across each compartment.

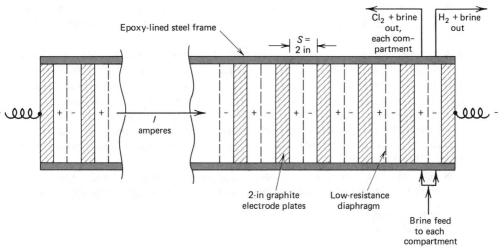

FIGURE P4.4
Bipolar chlorine cell.

a For a bipolar cell, typically

$$E = E_D + \bar{\iota}\rho S \tag{P4.4}$$

where E_D = apparent minimum decomposition voltage
$\bar{\iota}$ = current density, A/cm^2
ρ = specific resistance of the electrolyte, $\Omega \cdot$ cm
S = compartment width, cm

Calculate the number of compartments N for the following conditions:

$$E_D = 3.00 \text{ V}$$
$$\bar{\iota} = 0.1 \text{ A/cm}^2$$
$$\rho = 5.91 \text{ } \Omega \cdot \text{cm}$$
$$S = 2.0 \text{ in}$$

Total dc voltage available = 220 V.

b What is the required total direct current I? (Note that I generates the same amount of Cl_2 in each compartment.)
c Assuming a square cell cross section, what graphite plate dimensions are indicated?
d Very little information is available in the literature on cell costs—cell design is just

too variable. However, the bipolar cell may be constructed much like a plate and frame filter, and by analogy, an order-of-magnitude fabrication cost may be estimated from available costs of such filters. Data in Perry suggest a cost of $110/ft^2 of filtration area for plastic-lined steel (M&S = 650), large units. (Note that in filters, *both* sides of the frame are active filtration areas!) Use this figure to estimate the cost of the cell, but add a charge of $5/lb for material cost and special fabrication of the graphite (density = 135 lb/ft^3). Cost of copper bus bars will not be included.

4.25 Compute the value of the capacity exponent for agitated, jacketed glass-lined steel reactors from Fig. 4.5.2a. From the data in Table 4.5.1, estimate the current cost of a 2500-gal reactor.

4.26 Describe how the component buildup method could be used to estimate the purchased cost of an Oldshue-Rushton extractor. The extractor is described in Perry's handbook.

4.27 Woods et al. (1978) have made an exhaustive investigation of the FOB costs of gas-moving equipment and list recommended values "normalized" to an M&S index of 300. How do 25-year old data, such as those of Denzler (1952), compare on the "normalized" basis? Make the comparison for several common items of equipment: fans, centrifugal blowers, rotary lobe compressors, sliding vane compressors. (The Denzler data are for M&S = 180.) What is your conclusion?

4.28 In some cases, certain items of process equipment are associated with a grouping of satellite items. For example, a distillation column is normally associated with the following:

Overhead condenser
Overhead accumulator
Reboiler
Feed and product pumps
Vacuum pump (vacuum distillation)

This is, of course, the minimum; Fig. 3.5.5 demonstrates that additional or modified equipment may well be part of the distillation system.

Such a "normal" minimum association suggests a possible cost estimating shortcut—an "adjustment" factor which, when multiplied by the cost of the central item (distillation column in this case), accounts for the cost of the satellite items.

a Estimate the total cost of a vacuum distillation system to separate an equimolar feed of a binary component liquid into an overhead and a bottoms stream, each 98 mol % pure. Use the following data:

Operating pressure, top of column: 100 mmHg absolute pressure
Total feed: 100 mol/h
Pressure drop through column: 50 mmHg (use valve trays)
Overhead condensing temperature: 80°C
Bottoms boiling temperature: 120°C
Molecular weights: light component, 180; heavy component, 120
Average α: 2.50
Plate efficiency: 50 percent
Heat of vaporization of either component: 100 Btu/lb
Air leakage into system under vacuum: 10 lb/h
Vapor pressure of overhead at 35°C: 20 mmHg
Design for minimum equipment listed above; cost out for mild steel

b Estimate the total purchased cost of column and minimum satellite items for the column in Example 4.1.1. Use type 316 stainless steel as material of construction (except, of course, mild steel shells on exchangers). See also Example 4.1.2 and Prob. 4.15.

c Repeat part (b) for the case of a packed column (see Prob. 4.6), ceramic packing.

d Make a rough estimate of the variation of the cost of the column of part (a) alone with the relative volatility α (between, say, 1.20 and 10.0).

e On the basis of the above calculations, what is your conclusion regarding the possibility of developing a reasonably reliable adjustment factor?

4.29 A large refinery-petrochemical complex is scheduled to use 10,000 gal 50% aq NaOH per day. The material will be supplied in tank cars. Estimate the current bare cost of storage facilities for the caustic soda.

4.30 An alternative to the duplication of critical items of equipment is the purchase, installation, and simultaneous operation of two parallel items rated at just one-half full capacity. In this way, if one item must be shut down for maintenance, the other still allows production throughput at one-half capacity. On the average, what percent saving in purchased equipment cost is realized by using this approach over that involving duplication of full-scale equipment, with one item on standby?

4.31 A 10,000-gal horizontal rubber-lined storage tank (ASME Code, up to 25 psig rating) incurred a shop fabrication cost on the Gulf Coast of $26,820 (M&S index = 650).

a What cost would be predicted from the data of Guthrie (1969) (reproduced in Figs. 4.2.2 and 4.2.4)?

b Much of the published information on rubber-lined equipment costs is quite old. What costs would be predicted on the basis of this older information, adjusted to an M&S index of 650? Some possible sources for consultation include Mills (1964), Aries and Newton (1955), and Lundeen and Clark (1955).

4.32 Make a statistical analysis of the estimates, made by a number of students, of the adipic acid process in Prob. 4.20. Ignore estimates that are clearly "out of the ball park," presumably due to fundamental misinterpretations; for the balance, determine the range, the mean value, and the 95 percent confidence limits.

4.33 Estimate the mean total purchased equipment cost and the 95 percent confidence limits for a plant consisting of the following equipment assembly:

   4 items A
   1 item B
   3 items C
   7 items D
   11 items E
   2 items F
   25 items G

The following are the available cost data for each item, adjusted to a common cost index basis; each cost is from a different literature source.

Item A	Item B	Item C	Item D	Item E	Item F	Item G
17,500	140,000	51,800	12,500	6400	95,000	1250
22,300	110,000	73,700	10,700	5900	85,000	1510
19,800	155,000	58,900	17,100	6500		1460
27,100	103,000	61,600	13,000	6000		1220
15,900		61,600	9,100	6100		1370
		60,900		2300		1740
				6300		1540
						1410

CHAPTER 5

THE DIRECT FIXED CAPITAL INVESTMENT

We are now well on the way toward the first objective of a project economic analysis, namely, the definition and analysis of the project's capital investment. It has been implied that the total cost of the process equipment purchased to satisfy the demands of the project is the fundamental building block of such definition and analysis. Indeed, we will learn how to integrate the individual items of process equipment into an operating plant, how to "flesh out" the space in between those items with the various components of the plant construction process, and how to create an economic counterpart to the physical plant, constructed on the foundation of equipment costs.

We will focus upon the capital that is required to construct the facilities which are directly involved in the manufacture of the products which constitute the end objective of the project. This capital will be referred to as the *direct fixed capital* (abbreviated DFC), and we will presently define the expenditures that it encompasses, and the meaning of the modifying adjectives "direct" and "fixed." Perhaps to our dismay, we will discover that a project involves other capital expenditures for a spectrum of satellite facilities not directly a part of the physical plant. After all, a new plant, even a plant expansion or improvement of an existing plant, must bear some of the capital burden of satellite facilities which support them. For the present time, however, we will concentrate on DFC.

We will explore methods of estimating DFC, ranging from quick and rough guesses to more elaborate and time-consuming methods, some of which are used even in making advanced-stage detailed estimates. In general, the various methods correspond in complexity to the stages of project development—but, again, the final decision as to when a particular method should or should not be used is a matter of experience and judgment. Hopefully the material in this chapter will give the student background material to make such judgments intelligently.

One more point must be reemphasized, perhaps at the risk of overstating it. Some of the approximate methods which will be described involve the use of factors for estimating the cost of various construction items (the so-called *factored* estimates). For example, the cost of piping may be estimated by multiplying the total equipment purchased cost by a specific factor. Now, such factors are based upon the judicious analysis of many plant construction expenditure records, but there is widespread disagreement as to what the magnitude of those factors should be, and even what range of expenditures they really cover. We will give representative values for factors, such as the piping factor mentioned, which may be expected to generate unkind comments from experienced estimators, who will object to the opulence or penuriousness of the stated values, as the case may be. To repeat, however, the values are representative, and our purpose is to illustrate the various methods, to get the neophyte engineer acquainted with them. Many corporations have developed their own favorite factors, and the new employee will quickly learn what they are.

5.1 THE ESTIMATION OF THE DIRECT FIXED CAPITAL

Battery Limits and Auxiliary Facilities

A new process plant about to be built may be part of an *integrated complex,* or else it may be a *grass roots* plant. We have defined an integrated complex as a production site which incorporates several separate production facilities tied together by satellite process-oriented service facilities called *auxiliary facilities*. A steam plant is a typical auxiliary facility, and in an integrated complex one steam plant serves several production plants, perhaps all the plants on site. A steam plant may well be required for a grass roots plant, a plant constructed on a "virgin" site, standing all by itself (hence the politically originated term "grass roots," removed from the centers of political power). On a grass roots site the steam plant uniquely serves the one production facility; it becomes an integral part of that production facility.

In an integrated complex, the imaginary limits which define the equipment directly associated with a specific process are called the *battery limits* of the plant. The auxiliary facilities are service facilities vital to the operation of the plant. If they are wholly dedicated to the plant (such as a refrigerated brine facility, for example), they are a part of the plant's battery limits. If the auxiliary facility is shared by several plants (such as the previously mentioned steam plant), then that auxiliary facility is outside the plant's battery limits (*offsite* facility).

The concept is illustrated in Fig. 5.1.1. The battery limits enclose a specific production unit, for instance, the plant which incorporates the process to produce monoethylamine, described in Example 3.5.1 and illustrated in the block flow diagram, Fig. 3.5.3. The production unit is surrounded by satellite auxiliary facilities, vital to the operation of the monoethylamine plant as well as other plants within the integrated complex.

CHAPTER 5: THE DIRECT FIXED CAPITAL INVESTMENT

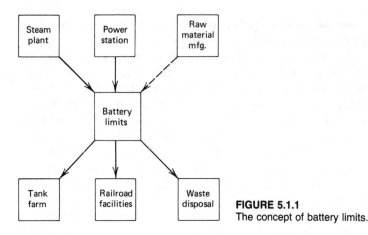

FIGURE 5.1.1
The concept of battery limits.

Some of the auxiliary facilities contribute their services to operations within the battery limits. This is indicated in Fig. 5.1.1 by arrows pointing into the battery limits box. Utilities such as steam and ac power are typical of such input services. A special case is provided by raw material manufacturing facilities (indicated by a dashed arrow). It may happen that one (or more) of the raw materials required for operations within the battery limits is manufactured in another plant in the integrated complex, a plant which also supplies merchant markets and may, indeed, service still other plants in the complex. For instance, in the case of the monoethylamine process, the supporting raw material manufacturing facility may be an ammonia plant, which supplies only a fraction of its output to monoethylamine production. A separate production plant can hardly be considered a satellite service facility and is therefore not included in the "auxiliary facilities" category (the reason for the dashed arrow). Nevertheless, it certainly serves a vital supporting function, and we will examine methods of accounting for separate raw material manufacturing facilities in the economic evaluation of the battery limits plant.

There are other auxiliary facilities which service the various outputs of the battery limits plant; these are indicated in Fig. 5.1.1 by arrows pointing away from the battery limits box. The separate identification of facilities such as a tank farm or waste disposal units outside the battery limits implies that those facilities are shared by other plants in the integrated complex. For instance, the tank farm may be located adjacent to marine loading facilities, and the storage tanks in the tank farm may be used to store a variety of liquid products, depending upon the schedule of shipments. However, if any tanks in the tank farm are wholly dedicated to the product of the battery limits plant, then they themselves are within the battery limits. In fact, product storage tanks constitute a part of the process flow sheet equipment and are not an auxiliary facility, unless shared by other plants.

The tank farm example serves to emphasize that the battery limits are not necessarily geographically contiguous. Raw material storage tanks may be located in a tank farm which is a considerable distance away from the production unit, but if those storage tanks are wholly dedicated, they are considered to be inside the battery limits.

The Definition of Direct Fixed Capital

Capital that is invested into production facilities may be broadly categorized as:

1 Fixed capital
  a Direct
  b Allocated
2 Working capital

An initial understanding of the significance of these categories may be gained by the analogy to one of a chain of department stores. The permanently fixed facilities—the store building, the shelves, the escalators—constitute the direct fixed capital. We say "direct" because the capital is associated directly with the fixed facilities in a particular store. The working capital is represented by all the items on the shelves that are to be sold; these items, of course, also represent an investment, but an investment which is presumably recoverable. Finally, there are service facilities, such as a central distribution warehouse, which are shared by several stores. A proportional share of the fixed capital represented by the warehouse is allocated to our one store and added to the other capital investments to give a grand total of capital which is associated with that particular store.

Fixed capital, then, is defined as capital invested in equipment and other permanently fixed facilities, that is, facilities that are not portable, temporary, or consumable. It is a characteristic of fixed capital[1] facilities that they have a limited lifetime; after a certain number of years, they wear out or become obsolete and must be replaced if the life of the production unit is to be continued. Some examples of fixed capital are:

Process equipment
Warehouse
Railroad spur line
Sewer
Laboratory building

Investment in portable, temporary, or consumable materials is termed the *working capital;* it is a characteristic of working capital that the investment is recoverable, at least in principle. Examples of working capital are:

Raw materials stored in warehouse
Maintenance shop tools and supplies
Laboratory stockroom chemicals
Spare parts for pump seals

---

[1] By custom, the word "capital" refers both to the investment and the facilities which the investment produces.

We will focus in this chapter upon the estimation of fixed capital, and specifically the direct fixed capital (DFC). The DFC refers to the capital investment in facilities which are located within the battery limits of the production unit;[1] this includes new auxiliary facilities which are wholly dedicated to this unit. It should perhaps be reemphasized that the DFC encompasses all expenditures required to complete the unit within the battery limits, including plant "indirect" costs such as contractor fees; this will become clearer when we discuss the elements of DFC.

Allocated fixed capital will be considered in depth in Chap. 8. For present purposes, it may be looked upon as a proportional share of the fixed capital investment in auxiliary facilities such as those shown in Fig. 5.1.1. Thus the capital investment in a steam plant, for example, is allocated between the units the steam plant services in proportion to the fractional amount of steam consumed by each unit. The concept is illustrated schematically in Fig. 5.1.2. We will see that a project must show a favorable return not only on the DFC invested but upon allocated fixed capital and working capital as well.

Regrettably, the terminology of investment analysis is not well-defined, perhaps because it is not an exact science, such as thermodynamics. What has been defined as direct fixed capital is sometimes referred to in the literature as the fixed capital investment, an incorrect term since allocated capital is also fixed. Occasionally the term used for DFC is the total capital investment, which is decidedly wrong, since the total capital investment includes the working capital.

[1] For this reason, the DFC is sometimes called *battery limits investment*.

FIGURE 5.1.2
Concept of allocated fixed capital.

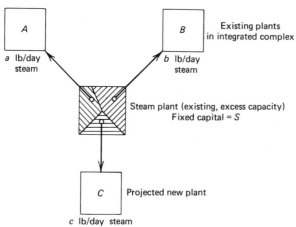

The Elements of Direct Fixed Capital: The Method of Purchased Cost Factors

We now examine the elements of the direct fixed capital, all the construction categories required to convert a collection of delivered equipment into an organized, operating production plant. We will combine the description of the elements themselves with a description of the method of purchased cost factors. This method, as previously advertised, consists of assigning to each element a judiciously chosen factor which, when multiplied by the total purchased equipment cost, gives the estimated cost of bringing that element to reality as part of the overall construction process. The DFC then is the sum of the separate element costs. The appropriate factor range will be given as part of the discussion of each element; the method is summed up in shortened form for ready reference in Exhibit 5.1A.

The first set of elements contributes to the *total plant direct cost* (TPDC), also known as the *physical* cost; it is essentially the cost of assembling the plant hardware.

We start with the total equipment purchased cost (PC), sometimes referred to as *bare* cost.

Equipment Installation This DFC element was outlined in the last chapter; it refers to the in-place erection of equipment at the new plant site, and it includes cost of foundations, slabs, supports, and local equipment services. The cost factor for the average chemical plant (both solid and fluid processing) is 43 percent of PC; in plants processing fluids exclusively, the factor tends to be a little higher, about 47 percent of PC.

As is true of the other elements of the TPDC, the associated factor incorporates both material and labor costs. Often the subdivision between material and labor is documented in the cost literature; we saw, for example, that some 75 percent of the installation cost is for labor. Such subdivisions are particularly useful for the quick estimation of labor requirements when the project is in the preconstruction stages.

Process Piping This element incorporates not only process fluid piping that connects the principal items of equipment but piping for connecting various items to main utility headers, and vent piping as well. Included are all manual and control valves, piping supports, piping insulation, tracing and "winterizing" installations (to keep process fluids and utilities from freezing), relief valves, check valves, and other myriad items associated with equipment piping.

Next to the equipment itself, piping is often the most expensive element of DFC, and some care must be taken in choosing the appropriate factor. The estimator has a range of factors to choose from, and different ranges are available, depending upon whether the proposed plant will process solids, mixed solids and fluids, or fluids exclusively. An ore crushing and grinding facility is an example of a solids processing plant (although some fluids may be involved, say, in wet grinding). An antifreeze canning facility involves both liquid handling (storage, pumping, proportioning) and solids handling (the cans themselves). A wastewater liquid-liquid

262 CHAPTER 5: THE DIRECT FIXED CAPITAL INVESTMENT

EXHIBIT 5.1A

ELEMENTS OF DIRECT FIXED CAPITAL (DFC) COST ESTIMATE:
METHOD OF PURCHASED COST FACTORS

A. Total plant direct cost (TPDC) (physical cost)

1 Equipment purchased cost (PC) (bare cost)	$1.00 \times$ PC
2 Installation	43% PC; all fluid processing 47% PC
3 Process piping	Solid: 10–20% PC
	Solid-fluid: 14–43% PC
	Fluid: 43–86% PC
4 Instrumentation	No auto control: 15% PC
	Auto control: 30% PC
	Computer control: 40% PC
5 Insulation	Low-temp. plants: 8% PC
6 Electrical facilities	11% PC
7 Buildings	6–45% PC, fluid processing
	15–70% PC, solids processing
8 Yard improvement	16% PC
9 Auxiliary facilities	Existing: 0%; minor new: 15%
	Major additions: 15–40% PC
	Completely new (grass roots) 40–140%
	Σ = TPDC

B. Total plant indirect cost (TPIC)

10 Engineering	⎫ 25% TPDC (normal)
11 Construction	⎬ 30–35% TPDC: small jobs
	⎭ 40% TPDC: high labor jobs
	Σ = TPIC

C. Total plant cost (TPC)   TPC = TPDC + TPIC

D. Direct fixed capital (DFC)

12 Contractor's fee	5–8% TPC
13 Contingency	5% TPC: established commercial project
	10% TPC: process subject to change
	20% TPC: speculative process
14 Size factor	Contingency should be adjusted as follows:
	$\times$ 0.80: huge commercial plant
	$\times$ 1.00: standard plant
	$\times$ 1.3: experimental units, pilot plants, small additions
	Σ (12 + 13)

DFC = TPC + Σ (12 + 13)

extraction facility is a typical fluid processing unit (again, some solids handling may well be involved, say, precoating of polishing filters). Most plants can be located approximately somewhere along this spectrum.

Typical factor ranges are:

Solids processing	10–20% of PC
Mixed solid-fluid processing	14–43% of PC
Fluid processing	43–86% of PC

Not surprisingly, solids processing plants have much lower associated piping costs.

As an example of the choice to be made within a given range, consider a plant processing primarily liquids. If the process is a simple batch process, the piping cost is likely to correspond to the low end of the range, 43 percent of PC. On the other hand, continuous ion exchange or liquid-liquid extraction plants, with their "Christmas tree" arrangements of complex piping, may easily attain the other end of the range, 86 percent. A majority of processes in the chemical process industry tend to lie on the fluid-processing end of the spectrum—even unit operations such as filtration involve a great deal of fluid processing—and a piping cost of 60 percent of PC is about average.

The piping factor is *not* a function of the costliness of the materials of construction. This is because the same expensive materials tend to be used for both processing equipment and the associated piping, and the adjustment factor is the same for both.

The neophyte estimator may feel quite uncomfortable about the responsibility of picking a number from such a broad range of possibilities (10 to 86 percent), with no real quantitative methodology as a basis. Nevertheless, for large plants this judgmental method, which, it must be remembered, is based upon a great deal of a priori cost experience, yields estimates within ±10 percent of the correct value. It is true that these factors are based upon cost data for large processing units, and some suspicion is justified when they are applied toward limited-scope projects such as existing plant modifications. In case of doubt, it is best to make a rough piping design for such limited-scope modifications and then use some of the more-detailed piping estimation methods outlined in this chapter.

Instrumentation Instrumentation includes transmitters and controllers, with all required wiring and tubing for installation; field and control room terminal panels; alarms and annunciators; indicating instruments, both in field and control room; on-stream analyzers; control computers and local data-processing units; and control room display graphics.

The cost of instrumentation depends upon the degree of automatic control built into the process. With little automatic control, instrumentation cost can be as low as 15 percent of PC; the factor is closer to 30 percent for full automatic control, and computer control in large plants increases the factor to 40 percent of PC or more.

Some caution must be exercised in using the recommended factors in estimates for limited-scope projects. The factors are ordinarily applicable to plant expansion

264 CHAPTER 5: THE DIRECT FIXED CAPITAL INVESTMENT

or modification projects and to pilot plants which are sufficiently experimental in nature to preclude the use of a high degree of automatic control. On the other hand, smaller-scale semiplants for the production of market research quantities of a chemical commodity are often highly automated, a "testing ground" for full-scale plant automation, should a full-scale plant be eventually built. The cost of instrumentation for such semiplants is often higher than indicated by the recommended factors, particularly with computer-controlled units.

Lipták (1970) refined the purchased cost factor method for instrumentation by accounting for both control complexity and project size. The method is illustrated in Fig. 5.1.3. The basic factor for continuous processes is obtained from the graph as a function of the total plant cost, calculated for an M&S index of 285. For preliminary purposes, the total plant cost may be taken as five times the total equipment purchased cost (we will learn about the background of this approxima-

FIGURE 5.1.3
Instrumentation cost factor. (*Lipták, 1970; reproduced by permission, Chemical Engineering magazine.*)

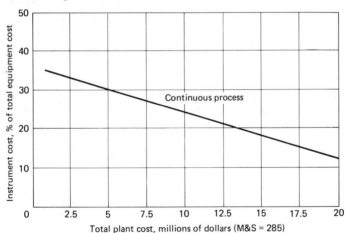

Correction factors to reflect instrumentation philosophy

Features	Factor F
Localized control	−0.20
Pneumatic instrumentation	0.00
Centralized control	0.00
Sample analysis performed in laboratory	0.00
General-purpose process area	0.00
Explosionproof process area	+0.10
Graphic panel display	+0.10
Special alloys required for pipeline items	+0.15
Sample analysis by on-line analyzers	+0.20
Electronic instrumentation	+0.20
Limited-scope optimizer computer included	+0.25
All loops on computer control	+0.45

tion later in this chapter). The basic factor F_0 is then corrected by other factors reflecting the "instrumentation philosophy," according to the expression

$$F = F_0(1 + F_1 + F_2 + \cdots)$$

For example, suppose the project under study is a semiplant with a total purchased equipment cost (M&S = 700) of $1,000,000. The plant will be located in an explosionproof process area, will use electronic (rather than pneumatic) instrumentation, and all control loops will be on computer control. The total plant cost is $\sim 5 \times 1.0 \times 10^6$, or $\sim \$5,000,000$ (M&S = 700); for M&S = 285, the plant cost is $2,040,000.

From Fig. 5.1.3, $F_0 = 33.5$ percent, and

$$F = 33.5(1 + 0.1 + 0.2 + 0.45) = 58.6\%$$

a figure considerably higher than the top of the range indicated in Exhibit 5.1A.

A few rule-of-thumb numbers are also worthy of consideration (all 1980 figures):

Small control computer (PDP-11) to handle up to 200 control modules: $50,000

If computer analytical control and operating discipline (for instance, start-up procedure programming) are desired, each will require a separate $50,000 computer as a minimum

Cost of control computer software development: $500 per control valve

Approximate cost of computer control: $5000 per control valve, *not* counting cost of valve

Lipták tabulates the cost of a large variety of instrumentation items.

Insulation The cost of insulation and painting is normally included in the recommended factors for equipment installation and piping. In low-temperature plants, however, insulation may become an unusually high cost burden; it is not uncommon for cryogenic process equipment to be literally buried in loose insulation contained in fabricated metal boxes. An insulation surcharge of 8 percent of PC is recommended for such plants.

There are also those processing plants which require an unusual degree of electrical or steam tracing, careful application of heat-transfer cements, and other unusual measures to avoid the freezing of high-melting process fluids. The costs of these apparently minor embellishments can mount alarmingly to embarrass the cavalier estimator; these should be accounted for by using a high piping cost factor, or else by considering the cost to be an extra "insulation" cost, as in the case of cryogenic processing plants.

Electrical Facilities These include battery limits substations and transmission lines, motor switch gear and control centers, emergency power supplies, wiring and conduit, bus bars, and area lighting. Separate equipment estimation is required for electrolytic installations.

The average for electrical facilities is about 11 percent of PC.

Buildings The processing areas of modern chemical plants are, for the most part, open structures. A typical process "tower" consists of a concrete slab ground floor, a fireproofed steel I-beam framework, and open checker-work decks at 15-ft vertical intervals. The equipment layout is confined primarily to such a process tower (layouts are not usually required for preliminary estimates). The deck area and tower height are determined by considerations of logical proximity of adjacent items of equipment, access convenience for maintenance (such as space for pulling out heat-exchanger tubes for cleaning), and height requirements (for example, barometric condensers, or gravity-feed bins). Process towers are rarely more than five decks high; particularly tall columns may extend far above the top deck, with individual column ladders and catwalks for emergency access. Pumps and heavy rotating equipment are preferably confined to the ground floor; in fact, operators prefer that any equipment that is likely to require frequent maintenance be located on the ground floor to avoid a lot of stair and ladder climbing. The location of large-volume containers of hazardous fluids on upper decks is normally discouraged; nobody cherishes the thought of cascades of flammable or toxic liquids from accidental vessel breaks.

Open structures are favored even in the coldest and least hospitable climates, in spite of the obvious discomfort and even potential safety risk (icy surfaces) to operating personnel. Enclosed structures pose a much more serious hazard of accumulation of flammable or toxic vapors.

The category of "buildings" includes process towers, subsidiary concrete slabs, stairways and catwalks (not equipment-specific), control rooms, and other battery limits buildings—change rooms, lunch rooms, furnished offices, and warehouses. The recommended factors incorporate costs for nonelectric building services (heat, potable water, sanitary sewers) as well as a variety of safety related items—safety showers, blast walls, fire sprinklers, fire monitors.

The recommended building factors also cover a wide span; here the determining characteristics are the scope of the project and whether or not the projected plant is a grass roots one:

Fluid processing   6–45% of PC
Solids processing   15–70% of PC

Use the low figures for plant expansions or modifications (in some cases, limited-scope projects may not involve building modifications at all). The upper limit of the range is recommended for grass roots plants. The average for integrated new large plants which are primarily fluid processing units is 25 percent.

Solids processing units generally require more protection from the weather elements, and charges for buildings and structures are correspondingly larger. Consider, for example, plants to extract oil from cottonseeds. Huge piles of the raw material must be stored up during the cotton harvesting and ginning season, and these must be protected from moisture by means of elaborate covers and roofs to avoid fungal infections and moisture-induced spontaneous combustion.

Yard Improvement Some aspects of yard improvement represent the very first steps of the plant construction process, excavation and site grading. Roads, fences, railroad spur lines, fire hydrants, and paved areas for parking or storage are some other items in this category, which averages about 16 percent of PC.

Land is not considered part of this cost, since the cost of land is recoverable. Direct fixed capital items are characterized by limited life, but the land is there and usable long after the production unit has outlived its usefulness.

Auxiliary Facilities Of all the direct plant cost categories, this is perhaps the most difficult to delineate properly. Auxiliary facilities have been defined as satellite process-oriented service facilities vital to the proper operation of the battery limits plant. In an integrated complex, some of these auxiliary facilities, such as a steam plant, may already exist as part of the complex, before the production facility under study has been built. Such a facility is shared by a number of production units (as illustrated in Fig. 5.1.2), and if the plant under study is also scheduled to share that facility, no direct expense is involved; the auxiliary facility is already there, although, as already implied, the plant under study must bear the burden of a certain amount of allocated ("backup") capital. This, however, is not the direct cost of auxiliary facilities that we are concerned with here.

The direct cost of concern is for those auxiliary facilities which must be *newly* built to make proper operation of the plant under study possible. This may mean auxiliary facilities which are wholly dedicated to the new plant and are within the plant battery limits. It may mean an expansion of offsite facilities to accommodate the new plant (installation of economizers in the steam plant, or an additional lagoon in existing biooxidation facilities). In a grass roots plant, the cost category incorporates *all* the auxiliary facilities. Clearly, an early definition of auxiliary facility requirements is necessary to focus upon the proper cost factors to be used.

Now, as was so strongly emphasized in Chap. 3, wholly dedicated auxiliary facilities should be incorporated in the process flow sheet and estimated as part of the process equipment. For example, a caustic scrubber for all the plant vent discharges is a waste disposal auxiliary facility which should constitute part of the flow sheet. There are many auxiliary facilities, however, which can be estimated as installed, *turnkey* (i.e., ready to operate) packaged units (see, for instance, the cost of absorption liquid chillers, Fig. 4.2.3), and detailing on a preliminary flow sheet is not necessary. How are these alternative approaches handled in a factored estimate?

The various possibilities may be illustrated by a hot oil supply system (for instance, Dowtherm), required for high-temperature-level heat-transfer equipment (i.e., above steam temperatures). A typical system consists of a direct-fired oil heating furnace, an oil head tank, and pumps to circulate the oil through the furnace and out to the "users." One way to handle this auxiliary facility is to incorporate the various equipment items in the plant equipment list and thus to estimate the facility as part of the overall plant. Another way is to find cost data on packaged hot oil systems; this cost then appears as part of the "auxiliary facilities" category in the factored estimate, but is *not* incorporated in the purchased cost factor used to estimate other less well defined auxiliary facilities. We will return to the estimation of separate auxiliary facilities later in this chapter.

Either of the above methods is perfectly acceptable. The third method is to lump the hot oil system in with other auxiliary facilities and to choose an appropriate factor which, when multiplied by the equipment purchased cost, will give the cost of all new auxiliary facilities not separately estimated. This method requires the least effort, but the result is subject to the greatest uncertainty.

Auxiliary facilities include a number of categories:

Process-dedicated facilities inside of battery limits, such as refrigeration systems, hot oil systems, inert gas generators, dedicated cooling towers, solid waste furnaces

Offsite utilities such as steam plants, power plants, substations, water treatment plants, instrument air compressors

Dedicated or shared waste treatment facilities

Dedicated or shared distribution facilities (railroad spur lines, truck loading stations, docking facilities, and shared storage facilities and warehouses)

The recommended purchased cost factors are one of the following:

0 percent for small improvements or additions in existing plants, other limited-scope projects. Minor modifications rarely result in any appreciable change in auxiliary facilities.

15 percent for new plant requiring minor auxiliary facility additions. Even if existing facilities are all adequate, some cost is incurred by new sewers, utility lines, outside lines (i.e., pipelines outside of battery limits), waste disposal facilities.

15 to 40 percent for major new additions to service a new facility in an integrated complex.

40 to 140 percent for grass roots plants.

For the grass roots plant in particular the magnitude of the requirement for auxiliary facilities makes the estimation with purchased cost factors suspect, and it is clearly better to identify the required facilities and estimate them separately. No matter how careful the estimator may be to identify all the requirements, 15 percent of PC should be included in the estimate for some of the "hidden" costs, such as the aforementioned sewer lines.

The total plant direct cost (physical cost) is the sum of the nine physical elements, each characterized by a factor F_i:

$$\text{TPDC} = 0.01(\text{PC}) \sum_{i=1}^{9} F_i \qquad (5.1.1)$$

We next turn our attention to some of the indirect costs which, nevertheless, are a logical part of the DFC. The total plant indirect cost refers to costs incurred by engineering and construction. Engineering costs refer to expenses associated with all the usual engineering functions:

Preparation of "design books" that document the total process. Some of this may involve R&D personnel, additional laboratory or pilot plant work to "nail down" final design criteria, physical properties.

"Chemical engineering" design of equipment. This phase of design refers to process-oriented specifications, for example, actual number of plates for distillation columns.

"Mechanical engineering" design of equipment. This is the "nuts and bolts" engineering phase—design of distillation column shell thickness, support structures.

Preparation of specification sheets for process equipment, instruments, auxiliaries, etc.

Design of control logic, computer software.

Preparation of drawings and models—P&IDs, piping diagrams, layouts, instrument loop diagrams, three-dimensional models.

Civil and electrical engineering work preparation—structures, motor control centers, yard improvements.

Purchasing functions—preparation of inquiries, equipment purchasing, expediting, control of payments (liaison with accounts payable department), negotiation of raw material contracts, procurement of supplies and spares.

Project management in the usual sense of the word—scheduling, cost control, construction supervision, inspection. (See Chap. 11 for additional detail.) Detailed cost estimation is required for cost control.

Plant start-up costs. Both engineering and R&D personnel are used in the start-up phase, which usually lasts 4 months to a year, to help solve unanticipated problems. Production personnel are responsible for preparing instruction books and for training operators.

Many large corporations have their own construction divisions which take on the responsibility of organizing the plant construction, and often perform many of the engineering functions as well. On the other hand, independent construction companies are also available to take on the construction task and, if desired, the engineering task. In either case, subcontractors are usually used to perform work in specific areas—pipe fitting, or electrical installation, or erection of structures. Construction costs are associated with the expenses of organizing the total construction effort; they do *not* include the cost of construction labor, which is already included in the elements of TPDC. What, then, do construction expenses entail?

There are, in fact, many indirect expenses associated with the construction effort and organization. For example, we have already mentioned taxes and freight charges for equipment delivery. As another example, consider the various costs engendered by the efforts of each subcontractor; planning the "attack"—the procedure for accomplishing a specific portion of the construction work in conformity and harmony with other simultaneous jobs—is in itself a time-consuming task demanding a great deal of experience and skill. Other indirect costs include the wages of construction foremen, expenses of erecting temporary buildings (onsite construction offices), procurement and rental of tools and heavy machinery, and materials and labor for start-up. The construction effort extends into the plant start-up period, and resources must be available to effect repairs, changes, and improvements during this difficult part of the plant's life span.

Engineering and construction are usually estimated jointly, since some phases of those two elements are hard to separate and are often handled by the same organization. The normal estimation factor for these indirect costs is 25 percent of TPDC. The factor, however, should be increased to 30 or 35 percent for smaller-scale projects (existing plant improvements, pilot plants), as high as 40 percent for labor-intensive construction efforts (for example, plants requiring an unusual degree of pipe tracing or start-up labor). The total plant cost (TPC as indicated in Exhibit 5.1A) is then

$$\text{TPC} = \text{TPDC} + \text{TPIC} \qquad (5.1.2)$$

The TPC is still not the DFC we seek. One obvious expense that must be added is the contractor's fee (the contractor's profit), which varies from about 5 to 8 percent of the total plant cost (TPC). Even if a corporation does its own construction, this item should be a part of the factored estimate, since the construction division is expected to show a profit. Nevertheless, this rather appreciable expense is one reason why some of the larger chemical and petroleum corporations have formed their own construction divisions.

The final elements in the method of purchased cost factors are contingency factors (items 13 and 14 in Exhibit 5.1A). In spite of appearances, the contingency factor is more than a "botch factor," although one must confess that there is an element of that. The fact of the matter is that experience has shown that the more speculative a process is, the more likely it is that key elements, later found to be important, have been overlooked or not anticipated during the project's early stages. The contingency factor then represents an attempt to compensate for those missing elements. However, even advanced-stage estimates will include a contingency to account for unexpected problems during construction—strikes, delays, unusually high price escalations, and so forth.

Typical contingency factors (as a percentage of total plant cost) are

5% of TPC for established commercial processes
10% of TPC for processes subject to change
20% of TPC for speculative processes

The contingency factors should be adjusted by a size factor as follows:

Multiply by 0.80 for very large commercial plants.
Multiply by 1.00 for standard plants.
Multiply by 1.30 for experimental units, pilot plants, small additions to existing plants.

The size factor reflects the common experience that very large plants tend to be overestimated, small-scope projects tend to be underestimated.

The direct fixed capital, then, is the sum of the total plant cost, contractor's fee, and contingency (Exhibit 5.1A).

Alternate Methods of Fixed Capital Estimation

The method of purchased cost factors is just one of several methods commonly used for the estimation of DFC. The various available methods differ considerably in complexity and the effort required to carry them through; the choice is a matter primarily of the stage of the project development and the justification of time expenditure to obtain greater reliability. The method of purchased cost factors is perhaps unique in that it is commonly used in all stages of project development, from the earliest stages up to preauthorization estimates.

Advanced-stage estimates for the purpose of budgetary control of the construction effort are detailed, item-by-item affairs assembled by cost estimation specialists. Even at this stage, however, shortcut methods based upon statistical correlations are often used to ease the burden of detailed estimation. An example of such a method is illustrated in Fig. 5.1.4, a correlation of the cost of various kinds of buildings. The costs shown are analogous to the unit cost (dollars per square foot) quoted by residential architects to their clients. If based upon plentiful data, such correlations can prove to be very reliable, and they do save a great deal of labor. Sometimes they are useful to chemical engineers who may need to estimate items (such as buildings) for limited-scope projects.

Example 5.1.1

A production plant superintendent wishes to add a small machine shop, 30 × 75 ft, to the plant's facilities. What is the likely cost (M&S = 700) of the shop equipped with standard equipment?

From Fig. 5.1.4, the base cost for 2250 ft^2 is $62,000.

In 1976, the M&S index was 472. The adjustment factor for equipped shops is 2.0. Therefore

$$\text{Cost} = 62,000 \times \frac{700}{472} \times 2.0 = \$184,000$$

It is also possible to estimate the total cost of a building such as that in Example 5.1.1 by detailed consideration of the costs of foundations, walls, roofing, utilities, individual machine tools, and so forth. Cost data are available for these items, but the inexperienced estimator is discouraged from trying this detailed approach—it is too easy to leave something out!

At the other end of the spectrum of methods of fixed capital estimation are shortcut methods which are more rapid, and usually less reliable, than the method of purchased cost factors. For preliminary purposes, however, they may be perfectly acceptable.

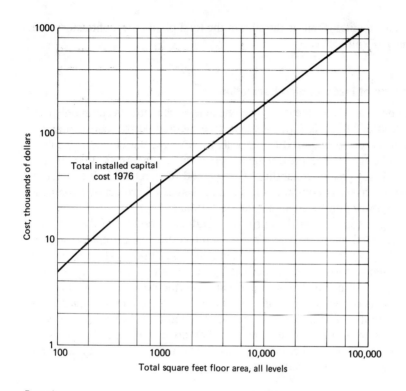

FIGURE 5.1.4
Cost of buildings (general). *(Dow Chemical USA.)*

Lang Factors In the method of purchased cost factors, a range of values of the factor for each element of the DFC is suggested; in many cases the suggested value depends upon whether the plant under consideration is one that will involve fluids alone, solids alone, or something in between. It should be possible to add up all the average factors for, say, an all-fluids-handling plant, and to arrive at an overall factor which, when multiplied into the total equipment purchased cost, will yield an estimate of the DFC. This grossly simplified approach was suggested many years ago by Lang (1948), and the factors that he recommended, or modified values thereof, have come to be known as *Lang factors*.

Lang originally proposed factors which were to be multiplied by the delivered equipment cost. If 10 percent of the purchased cost is allowed for the various delivery costs, the modified Lang factors become

> 5.2 × PC for all-fluid plants
> 4.0 × PC for mixed fluids-solids plants
> 3.4 × PC for all-solids plants

These factors give rise to the rough rule of thumb that the DFC is about five times the equipment purchased cost for plants processing primarily fluids. This is really a rather uncomfortable fact of life, one which managers bent upon discovering a novel way to save money occasionally try to ignore, but always to no avail.

The simplicity of the Lang factor concept constitutes a temptation to bypass the method of purchased cost factors, but this is not recommended. If sufficient background knowledge about the proposed plant is available, more realistic results are obtained by estimating the full set of purchased cost factors.

Power Factors A logical consequence of the six-tenths factor rule for characterizing the relationship between equipment capacity and cost is that a similar relationship should hold for the DFC of specific plants, provided the Lang factor is independent of capacity. In point of fact, the capacity exponent for plants, on the average, turns out to be closer to 0.7. Thus, if the DFC for a plant to produce a specified annual amount of a given product is known, a method is available for estimating the DFC of a smaller or larger plant to manufacture the same product (using the same process):

$$\text{DFC}_B = \text{DFC}_A \left(\frac{r_B}{r_A}\right)^{0.7} \tag{5.1.3}$$

A certain amount of discretion must be observed in using Eq. (5.1.3). As a rough guideline, the relationship is valid in the range of one-quarter to four times the capacity of the base plant. In the range of very large plants (say, 1/2 billion lb/annum), the exponent increases to 0.8 or more. Small plants, on the other hand, may be characterized by very small exponents, perhaps 0.3 to 0.5. This happens because, as equipment gets small, the cost becomes almost independent of size. It would be a serious error, indeed, if a 100-million-lb plant were to be scaled down by a factor of 100 by using the 0.7 power factor. Another embarrassing error arises from estimating the cost of scaled-down semiplants or pilot plants. This is often required after a particular size has been estimated, and management decides that the cost is too high and asks for the anticipated cost of a plant, say, one-half the original size. The truth of the matter is that the cost of the smaller unit is often very little less, contrary to what would be computed from Eq. (5.1.3).

Exercise 5.1.1

For extremely large plants the capacity exponent approaches a value of 1.0. Why should this be so?

Eq. (5.1.3) is sometimes characterized as the *seven-tenths factor rule*. The actual value of the power factor depends upon the nature of the particular plant; some typical values are shown in Table 5.1.1. These were computed by Guthrie (1970) and were based upon DFC data for existing plants.

TABLE 5.1.1
DFC PARAMETERS FOR PROCESSING PLANTS

Product	1980 DFC (battery limits only) (M&S = 650) for 50,000 tons/yr plant, millions of $	Capacity exponent	1980 turnover ratio 50,000 tons/yr plant
	Group 1		
Ammonium sulfate	1.5	0.73	2.0
Sulfuric acid	1.5	0.65	2.0
Ethylene dichloride	1.9		7.3
	Group 2		
Phosphoric acid	3.8	0.60	6.5
Urea	5.9	0.70	1.6
Ethanol	6.7	0.73	4.2
Mixed glycols	6.8	0.75	5.9
	Group 3		
$CCl_4 + C_2Cl_4$	10.8		2.1
Hydrogen peroxide	11.3	0.75	4.5
Trichloroethylene	12.3		2.2
Phthalic anhydride	12.4	0.70	3.2
Acetone	14.0	0.45	2.2
	Group 4		
Phenol	24.1	0.75	1.6
Acrylic acid	26.4		1.6
Vinyl acetate	30.4	0.65	1.1
Maleic anhydride	42.9		1.2
	Group 5		
Butadiene	73.3	0.68	0.4
Caprolactam	76.9		1.1

Source: Guthrie (1970); reproduced by special permission.

Of course, to apply the seven-tenths factor rule, information must be available on the DFC for at least one size of plant. Such evaluations appear frequently in the current cost literature, most notably in the Process Economics Project reports distributed to industrial customers by the Stanford Research Institute. The journal *Process Economics International* publishes DFC costs for plants constructed in Europe; an example is shown in Fig. 5.1.5 (vinyl chloride monomer and polymer).

In Table 5.1.1 the battery limits DFC for several plants, all producing 100 million annual pounds of product, has been computed from various sources. The plants have been categorized into five groups, each group being characterized by the same order of magnitude of DFC. Whereas chemical plants do vary greatly in the DFC per unit of production, and even though the spectrum of costs is, of course, continuous, it is curious that it is possible to form cost-similar groupings as demonstrated in the table. Quite obviously, a great deal more equipment, and more expensive equipment, is required to produce 1 ton of caprolactam than 1 ton of ammonium sulfate. But what is it that distinguishes the two processes in concept, and is it possible in some way to place the plant for a proposed new product into one of the groups a priori, using some distinguishing characteristic? For if this could be done, we would have still another way of projecting the DFC of a proposed plant. We will see that the functional step scoring method is just such an attempt to characterize a plant; it is a method that does not require preliminary equipment identification and sizing.

The information in Table 5.1.1 is occasionally reported as *investment per unit capacity J*, and it has been suggested that an approximate DFC can be estimated from the expression

$$DFC = Jr \qquad (5.1.4)$$

FIGURE 5.1.5
VCM and PVC (vinyl chloride monomer and polyvinylchloride polymer). Battery limits capital cost, mid-1978, Benelux location. *(Process Economics International, vol. I, no. 2, Winter 1979/80. Reproduced by permission, Chem Systems Inc., New York.)*

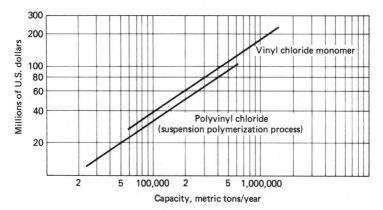

where J is expressed, say, in dollars per annual short ton of production, and r is the annual production tonnage. However, Eq. (5.1.4) is a very rough approximation; it holds true only for that value of r at which J has been determined. At all other values of r it is in conflict with the concept of Eq. (5.1.3), the seven-tenths factor rule. Nevertheless, investment per unit capacity is often quoted in the cost literature. In Table 5.1.1, J varies from $30 per annual ton for sulfuric acid to $1540 for caprolactam.

Quite obviously, care must be taken in using published DFC information to identify which process is covered by that information. For example, ethanol (group 2 in Table 5.1.1) is manufactured by several different processes; the data in the table happen to be based upon hydration of ethylene.

Exercise 5.1.2

Suppose the investment per unit capacity of a specific process plant is determined to be J_0 at an annual production rate of r_0 tons. How could Eq. (5.1.4) be modified to reflect more properly the variation of DFC with plant capacity?

Exercise 5.1.3

Guthrie (1970) gives the DFC for hydrochloric acid plants involving the absorption of HCl gas into water. The cost for a 10,000-ton plant (M&S = 273) is $2.0 million. The cost of a hydrofluoric acid plant, having the same capacity and using the same process (water absorption), is $2.5 million. Give some reasons for the higher HF plant investment.

Turnover Ratio A rough indication of the profitability of a plant is the turnover ratio, defined as the number of times that the DFC is "turned over" (i.e., is returned as gross sales) per annum:

$$T = \frac{sr}{\text{DFC}} \qquad (5.1.5)$$

where s = sales price per unit of product
r = annual capacity (in the same units)
T = turnover ratio

Some 1980 values of the turnover ratio are listed in Table 5.1.1. A number of years ago an often-proposed rule of thumb was that the average turnover ratio for the chemical process industry was 1.0. This is no longer true, for the pricing structure of chemical products has changed a great deal in the sixties and seventies, and for the most part the value of T has increased by a factor of 2 or more. Table 5.1.2 compares the turnover ratios for large plants (50,000 tons/year) computed in 1955 (Aries and Newton, 1955) and 1980 (from Table 5.1.1).

TABLE 5.1.2
TURNOVER RATIO TRENDS

	Turnover ratio	
Product	1955*	1980†
Butadiene	0.21	0.44
Sulfuric acid	0.54	2.00
Phenol	0.74	1.58
Ethanol	1.01	4.24
Mixed glycols	1.11	5.86
Phosphoric acid	1.70	6.52
Urea	2.36	1.58
Ammonium sulfate	5.55	2.03

*From Aries and Newton (1955), by permission.
†From Table 5.1.1.

If the turnover ratios in Table 5.1.2 (and in other sources) are averaged, it appears that the average between 1955 and 1980 has increased by about 1.75. If, indeed, the industrywide average in 1955 was 1.0, a 1980 average T of 1.75 is indicated. This value can then be substituted into Eq. (5.1.5) to obtain a very rough approximation of the plant DFC for a product that is to be sold at a price level of s.

Example 5.1.2

What should have been the 1980 selling price of phenol, produced at a rate of 50,000 tons/year, to reflect the normal turnover ratio for the chemical industry?
SOLUTION From Table 5.1.1, the DFC for a 50,000-ton phenol plant is $24,100,000. Substituting T = 1.75 into Eq. (5.1.5),

$$s = \frac{1.75 \times 24{,}100{,}000}{50{,}000} \approx \$844/\text{ton or } 42.2¢/\text{lb}$$

The listed price in the *Chemical Marketing Reporter* for mid-1980 was 38 cents/lb.

Exercise 5.1.4

What is the relationship between the turnover ratio T and production capacity r for a fixed selling price s?

Reference must also be made to a method of DFC estimation that rivals the method of purchased cost factors in popularity and, in fact, is thought to give more reliable results (Baasel, 1980), although at the expense of more effort. This is the *modular factor* approach that has been fine-honed by Guthrie (1969). In this approach, the purchased cost of each item of equipment is estimated as before. The equipment is then apportioned into "modules," such as, for instance, a shell-and-tube exchanger module; each module is then multiplied by a "bare module factor" which brings the cost of the module up to its share of the plant total DFC. The modular factors are thus equivalent to Lang factors for each item of equipment.

An example of one of Guthrie's modules is given in Fig. 5.1.6; this one is for shell-and-tube exchangers. The FOB cost of each exchanger is estimated from the graph and adjustment factors; the total of the exchangers is then multiplied by a factor which depends upon the dollar magnitude of the module. For example, if the total purchased cost of the exchangers were to be $500,000, this would be classified as a 3C module, and the modular factor is 3.14. Thus the DFC contribution of shell-and-tube exchangers would be

$$3.14 \times 500,000 = \$1,570,000$$

Note the amount of detail contained in the tables; these are very useful, for example, in making detailed estimates of insulation costs for the whole plant, or direct field labor requirements. Note also that the modular factors are considerably smaller than the normal Lang factors (5.2 for all-fluid plants); this is because some of the items reflected in the normal Lang factors are covered by separate modules—buildings, site development, auxiliary facilities, etc.

Estimators who wish to use the modular factor approach are urged to survey Guthrie's presentation very carefully to avoid errors of omission. A similar method developed by W. E. Hand is outlined in Peters and Timmerhaus (1980).

Functional Step Scoring

Returning to Table 5.1.1, we again note the curious fact that ethanol and glycol plants, for example, appear to cost about the same, whereas a phenol plant costs about four times as much as a urea plant having the same capacity. Are there some general process characteristics that may be examined which would place a plant for a new product into a certain cost category, say, group 4 in Table 5.1.1? This question has been addressed repeatedly in the published cost estimation literature, and several distinctive procedures have been proposed. The procedures usually demand some degree of process definition, but they do not require individual equipment specification, a considerable saving of time and effort. At the same time, the accuracy of the estimates derived from such simplified procedures cannot be expected to be high. Nevertheless, for preliminary stage estimates such shortcut procedures may be perfectly adequate. They are good enough to tie down the order of magnitude of the investment, and they do allow a reasonable assessment of operating costs which are investment-based.

The process characteristic which forms the basis of many of such shortcut procedures is the number of identifiable process steps classified by their "functionality"—distillation, reaction, filtration, and so forth. The idea is to "keep score" of the number of such functional steps, and hence such procedures are sometimes titled *functional step scoring*. The method is based upon a concept originally

SECTION 5.1: THE ESTIMATION OF THE DIRECT FIXED CAPITAL

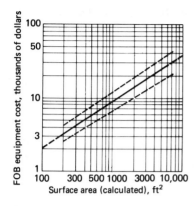

Required:
Surface area, ft²
Design type
Tube, shell material
Design pressure
Design temperature

Time base:
mid-1968

Size exponent:
0.65

Included:
Complete fabrication

Basis of chart:
Floating head
Carbon steel construction
Design pressure, 150 psi

Field installation modules

Module	3A	3B	3C	3D	3E
Base dollar magnitude, $100,000	Up to 2	2 to 4	4 to 6	6 to 8	8 to 10
Equipment FOB cost, E	100.0	100.0	100.0	100.0	100.0
Piping	45.6	45.1	44.7	44.4	44.3
Concrete	5.1	5.0	5.0	5.0	5.0
Steel	3.1	3.0	3.0	3.0	3.0
Instruments	10.2	10.1	10.0	9.9	9.8
Electrical	2.0	2.0	2.0	2.0	2.0
Insulation	4.9	4.8	4.7	4.7	4.7
Paint	0.5	0.5	0.5	0.5	0.5
Field materials, m	71.4	70.5	69.9	69.5	69.3
Direct material, E + m = M	171.4	170.5	169.9	169.5	169.3
Material erection	55.4	54.7	54.2	53.9	53.8
Equipment setting	7.6	6.5	5.9	5.5	5.2
Direct field labor, L	63.0	61.2	60.1	59.4	59.0
Direct M & L cost	234.4	231.7	230.0	228.9	228.3
Freight, insurance, taxes	8.0	8.0	8.0	8.0	8.0
Indirect cost	86.7	78.8	75.9	75.5	73.0
Bare module cost	329.1	318.5	313.9	312.4	309.5
L/M ratios	0.37	0.36	0.35	0.35	0.35
Material factor, E + M	1.71	1.70	1.70	1.69	1.69
Direct cost factor, M & L	2.34	2.32	2.30	2.29	2.28
Indirect factor	0.37	0.34	0.33	0.33	0.32
Module factor (norm)	3.29	3.18	3.14	3.12	3.09

Note: All data are based on 100 for equipment, E.
Dollar magnitudes are based on carbon steel.

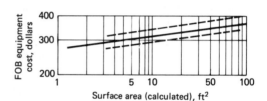

Double-pipe exchanger costs (for process requirements less than 100 ft².)

Exchanger cost, $ = base cost $(F_d + F_p) \times F_m$

Adjustment factors:

Design Type	F_d	Design Pressure, psi	F_p*
Kettle, reboiler	1.35	Up to 150	0.00
Floating head	1.00	300	0.10
U tube	0.85	400	0.25
Fixed tube sheet	0.80	800	0.52
		1000	0.55

*If these factors are used individually, add 1.00 to these values.

Adjustment factors:
Material: CS/CS = 1.0, CS/SS = 1.85
Pressure: up to 600 psi 1.00
 900 1.10
 1000 1.25

Module factors:
Field installation 1.35
Module factor (norm) 1.83

Shell/tube materials, F_M:

Surface area, Ft²	CS/CS	CS/Brass	CS/Mo	CS/SS	SS/SS	CS/Monel	Monel/Monel	CS/Ti	Ti/Ti
Up to 100	1.00	1.05	1.60	1.54	2.50	2.00	3.20	4.10	10.28
100 to 500	1.00	1.10	1.75	1.78	3.10	2.30	3.50	5.20	10.60
500 to 1,000	1.00	1.15	1.82	2.25	3.26	2.50	3.65	6.15	10.75
1,000 to 5,000	1.00	1.30	2.15	2.81	3.75	3.10	4.25	8.95	13.05
5,000 to 10,000	1.00	1.52	2.50	3.52	4.50	3.75	4.95	11.70	16.60

Shell-and-tube Exchangers
M&S = 273

FIGURE 5.1.6
An example of Guthrie's (1969) modules. *(Guthrie, 1969; reproduced by permission, Chemical Engineering magazine.)*

developed by Zevnik and Buchanan (1963), who postulated that the DFC of plants *processing primarily fluids* could be correlated with the following parameters alone:

1. The plant production capacity
2. The number of functional steps in the process
3. The severity of operating conditions
4. An appropriate cost index

The authors succeeded in correlating known plant costs with the equation

$$\text{DFC} = 1.33 \, NC \, \frac{I}{I_0} \qquad (5.1.6)$$

where N = number of functional steps
C = direct cost per functional step
I = cost index
I_0 = cost index value at the time the authors correlated their cost data: ENR (*Engineering News Record*) index = 300 (1939 ≡ 100) or M&S index = 185

The factor 1.33 is an allowance for auxiliary facilities and indirect costs.

The direct cost per functional step, C, is a function of the production capacity and a parameter which reflects the severity of the operating conditions, the so-called *complexity factor*. The recommended correlation is given in Fig. 5.1.7. The complexity factor is given by

$$K = 2 \times 10^{F_t + F_p + F_a} \qquad (5.1.7)$$

where F_t, F_p, and F_a are factors which reflect the severity of temperature, pressure, and alloy construction requirements, respectively. The values of these factors are shown in Fig. 5.1.8. F_t and F_p are selected for the one temperature (or pressure) in the process which maximizes the factor; note, for example, that both low and high temperatures increase F_t. The alloy factor F_a may be interpolated from the table provided to conform to a particular alloy mix in the proposed plant.

Since DFC is directly proportional to N, the proper identification of the number of functional steps is obviously critical. In some cases the delimitation of a functional step is not difficult. A distillation step, for example, incorporates equipment such as is shown in Fig. 3.5.5. It is true that some distillation units will have more equipment than others, but errors ascribable to such second-order differences will presumably be of a lower order of magnitude and perhaps somewhat compensating. In a distillation train, each column and associated equipment constitute a functional step (i.e., the "separation of A from the feed"). In the same functional sense, all equipment associated with a vacuum filter would meet the process function of "filtration" and would constitute a single functional step. We see that "functional steps" coincide pretty well with the "blocks" that one might expect to find on a block flow diagram—except that such blocks might incorporate several distillation steps, as an example.

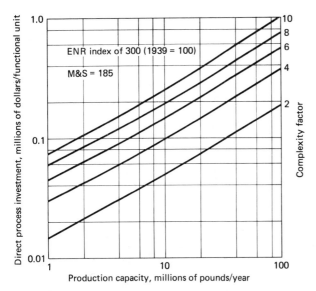

FIGURE 5.1.7
Direct cost of functional steps. *(Zevnik and Buchanan, 1963; reproduced by permission, Chemical Engineering Progress.)*

On the other hand, what of the heat interchanger that may be located between two distinct functional units—does it, in itself, constitute a functional step ("heat interchange")? Zevnik and Buchanan think not, unless the process involves a number of such heat-transfer items, in which case they consider the ensemble of exchangers to constitute a "functional unit." Similarly, dedicated raw materials and product storage facilities are considered to be equivalent to a functional step ("storage"). Clearly, there is no such thing as an unequivocal enumeration of functional steps; perhaps some examples will serve to illustrate the approach used by Zevnik and Buchanan.

The Andrussow process for hydrogen cyanide is illustrated in Fig. 5.1.9. For this case, the authors estimated N to be 8:

1 Feed preparation
2 HCN converter—heat recuperation
3 Excess ammonia removal
4 HCN absorption (in water)
5 HCN stripping from water
6 HCN purification by distillation
7 Storage
8 Waste disposal

Let us examine the philosophy underlying this categorization. The attempt has clearly been made to lump equipment into functional units of about the same degree

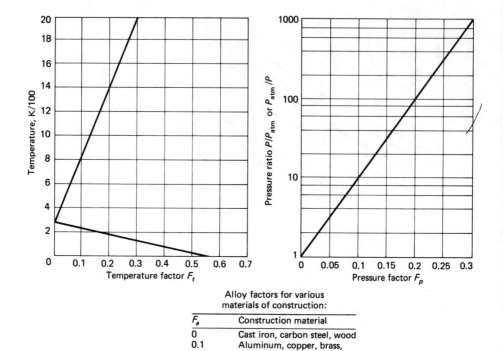

FIGURE 5.1.8
Severity factors. *(Zevnik and Buchanan, 1963; reproduced by permission, Chemical Engineering Progress.)*

of equipment complexity. For instance, the process is one operated at only slightly above atmospheric pressure; therefore ammonia vaporizer, air blower, and filters are all modest items that can reasonably be lumped under "feed preparation." If the process were operated at 1000 psig, the air compressor and ammonia vaporizer could well be considered as separate functional steps. Note also the waste disposal is justifiably characterized as a separate functional step, since a considerable aqueous purge is involved. Some heat interchange is involved, but the exchangers are considered to be part of other functional units, and "heat exchange" is not a separate step.

Another example is given in Fig. 5.1.10, a process for the production of phthalic anhydride from naphthalene. The functional units identified by Zevnik and Buchanan are listed in the legend. How believable is the conclusion that $N = 8$? Well, let us say that there are some reasons for skepticism. Why are "storage" and "product shipment" separate functional steps? Why are the product condensers two

SECTION 5.1: THE ESTIMATION OF THE DIRECT FIXED CAPITAL   **283**

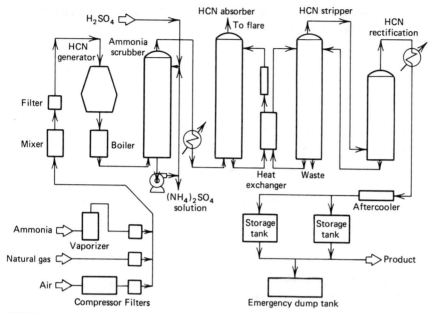

FIGURE 5.1.9
Hydrogen cyanide process (Andrussow process). *(Zevnik and Buchanan, 1963; reproduced by permission, Chemical Engineering Progress.)*

FIGURE 5.1.10
Phthalic anhydride process. Functional steps are (1) air compressor, (2) vaporizer, (3) reactor, (4) condenser (product), (5) condenser (product), (6) refining column, (7) storage, and (8) raw materials receipts and product shipments. *(Zevnik and Buchanan, 1963; reproduced by permission, Chemical Engineering Progress.)*

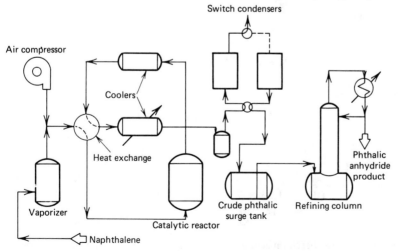

steps? Is the air blower justifiably considered to be a whole functional unit? A more reasonable step sequence might be:

1 Feed preparation
2 Coolers and interchangers (here an important component of process)
3 Catalytic reactor
4 Product condensers
5 Product refining
6 Product flaking (not shown on flow sheet!)
7 Storage and distribution

The modified sequence yields $N = 7$. Thus, in spite of possible disagreements, the final value of N is not all that much different. Zevnik and Buchanan admit with refreshing insight that they may have been influenced by unconscious bias to select factors more likely to give the "right" answers.

Example 5.1.3

Estimate the DFC (M&S = 650) of a 50,000-ton phthalic anhydride plant using the process in Fig. 5.1.10. Use the following information:

Most extreme pressure: 1/3 atm absolute
Extreme temperature: 750 K
Materials of construction: 316 stainless steel, Hastelloy

$$\frac{p_{atm}}{p} = 3$$

From Fig. 5.1.8,

$$F_p = 0.05 \qquad F_t = 0.09$$

Interpolate table,

$$F_a = 0.25$$
$$K = 2(10)^{0.39} = 4.9 \approx 5$$

From Fig. 5.1.7,

$$C = 0.44 \times 10^6$$
$$N = 7$$

From Eq. (5.1.6),

$$\text{DFC} = 1.33 \times 7 \times 0.44 \times 10^6 (650/185)$$
$$= \$14.4 \times 10^6$$

In Table 5.1.1, the value based on actual plant costs is 12.4×10^6; however, this does not include any offsite auxiliary facilities. If the phthalic anhydride process were a new one, the functional step scoring estimate would be quite acceptable for the preliminary evaluation of the new process.

Baasel (1980) found that the method of Zevnik and Buchanan overestimated the DFC of an isopropanol plant by about 40 percent. Cevidalli and Zaidman (1980) modified Eq. (5.1.6) to account for processes with dilute aqueous feeds; they found the method to be quite reliable on the basis of actual plant costs in Israel. A modified version of functional step scoring has been recently proposed by Viola (1981).

5.2 THE SEPARATE ESTIMATION OF AUXILIARY FACILITIES

The Scope of Auxiliary Facilities

It has been pointed out as a part of the discussion of the elements of DFC that auxiliary facilities are peripheral installations vital to the operations of a particular process. There are, of course, other facilities which may be terribly important to the proper operation of production facilities, but they do not lie within the meaning of the narrower definition. The plant cafeteria and the payroll offices may both be essential to smooth operations, but they are not process-oriented facilities in the sense of power substations or warehouses. We will return to the consideration of non-process-oriented support installations in Chap. 8.

Some typical auxiliary facilities which are normally shared by a number of production units in an integrated complex are listed in Exhibit 5.2A. In some cases such facilities may be built as units dedicated to a particular plant under consideration, and one of the early-stage decisions that must be made is whether an auxiliary facility is to be dedicated, or whether an existing facility is to be shared, or perhaps expanded, if need be. The tendency in recent years has been to place the responsibility for certain auxiliary facility operations upon each individual plant, that is, to build smaller dedicated facilities, particularly when there is a risk of cross-contamination in jointly used facilities. Cooling towers, for instance, are built to serve the cooling water load of individual plants, in spite of the size economy afforded by large integrated units. The dedicated units isolate the plant superintendent's responsibility for water contamination (due to leaking heat-exchanger tubes, for example), for proper blowdown (circulating stream purge) and water treatment, for acceptable water economy; failure in any of these areas imposes a penalty upon the responsible manager's operations rather than the whole integrated complex. Similarly, waste disposal furnaces are now often dedicated items.

In the case of some of the utilities, the investment in the auxiliary facility depends upon whether the utility is purchased or locally generated. Alternating current power, for example, is normally purchased, and the investment incorporates the required substations and transmission lines. In many instances, however, required power is partially or totally generated within the integrated complex, often as an energy-saving device in conjunction with the disposal of combustible wastes. Such a facility represents additional DFC, although the sum total of the charges for power is hopefully less than if the power were purchased. Nitrogen is usually generated in inert gas generators; if an air distillation plant is located close to the

---

EXHIBIT 5.2A

TYPICAL AUXILIARY FACILITIES

A. Commonly available utilities in integrated complex
Steam, 150 or 400 psig
Power, ac
Process water
Cooling water
Compressed air
Instrument air (dry)
Natural gas

B. Occasionally available utilities in integrated complex
Nitrogen (for inert gas padding)
Power, dc
Dowtherm (or other hot oil system)
Condensate, or deionized water, or reverse-osmosis water
Refrigeration (chilled brine)

C. Waste disposal systems for general use
Process wastewater treatment
Evaporation ponds
Sewers and sewage treatment
Biooxidation ponds
Disposal furnaces
Land burial sites

D. Distribution systems
Ship docking and loading facilities
Railroad sidings
Tank truck loading facilities
Shared warehouses and storage tanks

---

integrated complex, purchase of nitrogen is likely to be more economical, and the only DFC expenditure is for a pipeline.

If auxiliary facilities are to be individually estimated, some criterion of preliminary sizing must be chosen. For the common utilities, such as steam, power, and cooling water, the demand may be estimated from individual equipment demands shown in equipment lists (see Table 4.1.1). Another 10 percent should be added to the subtotals for power (for lighting, electric appliances) and steam (building heating). The total demand is then used to judge whether existing facilities in the complex are adequate to handle the additional load or whether expansion is required. The criteria for sizing facilities other than those supplying the common utilities may usually be established even during early project stages by judicious analysis of process requirements and a little common sense.

Example 5.2.1

An oxygen-sensitive liquid pharmaceutical intermediate is to be subjected to further chemical treatment in 1000-gal well-agitated reactors. There are six batch reactors operating on a staggered schedule, and each batch takes 12 h to complete. Before the intermediate is introduced into any one reactor, air introduced during cleanup between batches must be purged with a stream of dry nitrogen containing 10 ppm of O_2 (by volume) until the oxygen residual is reduced to 100 ppm. What is the daily nitrogen demand in standard cubic meters, assuming good gas mixing in the vessels?

SOLUTION

```
       N₂ in              gas out
      ────────→  ┌──────┐  ────────→
                 │ V, c │
       Q̇, c_in   └──────┘  Q̇, c
```

From the continuity concept

$$\text{Rate of } O_2 \text{ in} - \text{rate of } O_2 \text{ out} = \text{accumulation}$$

$$\dot{Q}(c_{in} - c) = V \frac{dc}{dt}$$

Rearranging,

$$\frac{\dot{Q}}{V} dt = \frac{dc}{c_{in} - c}$$

Integrating,

$$\frac{\dot{Q}}{V} t = -\ln(c_{in} - c) + K$$

When $t = 0$, $c = c_0$. Thus

$$K = \ln(c_{in} - c_0)$$

and

$$\frac{\dot{Q}}{V} t = \ln \frac{c_{in} - c_0}{c_{in} - c} \tag{5.2.1}$$

Total Gas in $= \dot{Q}t$, and

$$\dot{Q}t = V \ln \frac{c_{in} - c_0}{c_{in} - c} \tag{5.2.2}$$

For the case stated in the problem,

$$V = 1000 \text{ gal} \equiv 3.785 \text{ m}^3$$
$$c_0 = 210{,}000 \text{ ppm (approximate } O_2 \text{ in air)}$$
$$c_{in} = 10 \text{ ppm}$$
$$c = 100 \text{ ppm}$$

Therefore

$$\text{Total purge } N_2 \text{ per vessel} = 3.785 \ln \frac{10 - 210{,}000}{10 - 100}$$
$$= 29.35 \text{ m}^3$$

On the average, six vessels must be purged every 12 h, or 12 vessels per day.

Assume purging is done at 25°C, atmospheric pressure. A standard cubic meter is measured at 0°C, atmospheric pressure. Therefore

$$\text{Daily nitrogen demand} = 29.35 \times 12 \times \frac{273}{273 + 25} = 323 \text{ m}^3$$

Exercise 5.2.1

An antifreeze canning facility has been proposed for an integrated complex. The peak production will be 20,000 gal/day in 1-gal plastic containers and 10,000 gal/day in quart cans. Estimate the cost (M&S = 700) of a dock-high warehouse (use Fig. 5.1.4) that will hold up to 1 month's production. Assume product will be boxed and that boxes will be stacked on pallets. Make reasoned estimates of stacking height (20-ft ceilings), passageway requirements for forklift trucks, open areas at docking gates, etc.

Auxiliary Facility Cost Correlations

Many auxiliary facilities are complex units, in their way as complex as the primary production units they support. An estimate starting with item-by-item equipment design could be as time-consuming as the primary plant estimate itself. Fortunately, cost correlations are available which make the job of estimating the facilities separately relatively straightforward. We have seen an example of such a cost correlation in Fig. 5.1.4, which shows the cost of various kinds of buildings. The cost of steam plants in Fig. 5.2.1 is another example. The correlation gives the base cost of a field-erected steam plant, plus a bare module factor of 1.96 which converts the base purchased cost to the steam plant's contribution to the DFC. Remember, however, that it is necessary to make a separate estimate of a facility such as a steam plant only if the facility is to be a dedicated one, such as would be the case in a grass roots installation. If an existing facility in an integrated complex is to be shared, the capital cost burden is accounted for by means of the allocated capital concept.

Many correlations such as that in Fig. 5.2.1 are available for utilities, water treatment, and waste disposal systems. Evaporation ponds for the disposal of aqueous wastes are particularly popular in the more arid regions of the country. If the gross evaporation rate in an area substantially exceeds the annual precipitation,

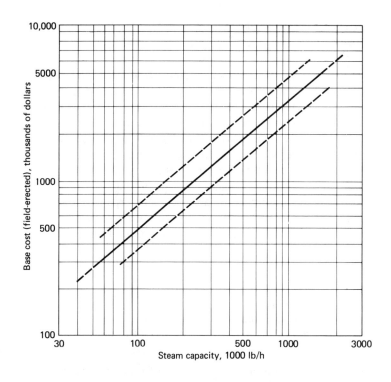

Required:
 Generating capacity, lb/h
 Steam pressure, psig
 Superheat, °F

Basis of chart:
 Saturated steam
 Above 100,000 lb/h cap.

Time base: mid-1968

Size exponent: 0.80

Included:
 All materials; FD fans, instruments, controls, burners, soot blowers, etc.
 Feedwater deaerator
 Chem. injection system
 Structural steel and platforms
 Gas and oil firing
 Stack
 Field erection
 Subcontractor indirects

Boiler erected cost, \$ - base cost $(F_p + F_s)$

Adjustment factors:

Steam pressure, psi	F_p	Superheat, °F	F_s*
Up to 400	1.00	Sat.	0.00
500	1.02	100	0.07
600	1.06	200	0.13
700	1.15	250	0.16
1000	1.30	300	0.20

*If these factors are used individually, add 1.00 to the above values.

Installation (prime contractor work only):	
Field installation (M & L)	1.44
Bare module factor	1.96
L/M ratio	0.16
M&S = 273	

FIGURE 5.2.1
Field-erected boiler units. *(Guthrie, 1969; reproduced by permission, Chemical Engineering magazine.)*

evaporation ponds are applicable for the treatment of aqueous solutions and suspension of inorganic and organic solutes, food wastes, colloidal mineral dispersions, and other similar aqueous streams. This method of disposal does demand proper attention to evaporation pond construction; class I ponds have compacted crushed rock walls and bottoms with incorporated plastic sheeting to minimize penetration of the aqueous waste into the underlying strata. Such ponds are quite expensive—about $125,000/acre (1978), not counting the land cost. Proper pH control and other chemical treatment are important to avoid the escape of toxic gases from the pond surface. Under prolonged exposure to sun and air, many organic species are slowly oxidized to inorganic residues. Of course, after a number of years, the ponds are filled up with residual sludges; these must then be bulldozed out and disposed of by land burial.

Exercise 5.2.2

An ore treatment plant in El Paso, Texas, generates 10 gal/min of an aqueous waste stream. What is the 1978 cost of evaporation ponds to handle this stream? The mean annual pond evaporation rate in El Paso is 72 in, and the mean annual precipitation is 8 in.

There are occasions when the preliminary evaluation of a project requires the separate estimation of the DFC required to manufacture a particular raw material. This may occur if the raw material cannot be purchased, or if there is reason to believe that integrated manufacture of the raw material would exhibit an economic advantage. Such raw material production plants can be considered as auxiliary facilities, and approximate DFC correlations such as Fig. 5.1.5 are most welcome and valuable in assessing the overall economic attractiveness of the core process. Another example is given in Fig. 5.2.2, the DFC required for oxygen plants. Such a correlation is valuable, for example, as part of an initial assessment of whether air or oxygen should be used as the oxidant in a proposed process.[1]

Clearly, if supporting raw materials production units are a part of the auxiliary facilities, the method of purchased cost factors is wholly inadequate to estimate the auxiliary facilities, and separate estimation as outlined above is mandatory. In fact, the separate estimation of any of the auxiliary facilities will result in a more accurate or realistic estimate; nevertheless, if the required new facilities in support of a new project are modest, the factored estimate method is perfectly adequate for preliminary purposes.

5.3 THE ESTIMATION OF PIPING SYSTEMS

The Design of Pipelines

The important role of piping in the overall design and cost of a chemical-processing plant was emphasized during the discussion of the elements of DFC. The purchased

---

[1] A more up-to-date report on the cost of oxygen plants is available in the Chem Systems International "Chemical Process Economics" (see Exhibit 3.5A).

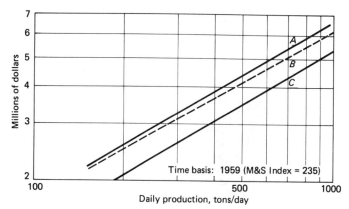

FIGURE 5.2.2
Capital requirements for oxygen plants. Curve A: split cycle plant: 450 psig discharge; 99.5% pure oxygen. Curve B: low-pressure plant: 0.5 to 5 psig discharge; compressed to 450 psig; 95% pure oxygen. Curve C: low-pressure plant: 0.5 to 5 psig discharge; 95% pure oxygen. *(Katell and Faber, 1959; reproduced by permission, Chemical Engineering magazine.)*

cost factor for piping is quite large, and yet the estimation of piping costs using the method of purchased cost factors is commonly done, with a satisfactory degree of accuracy for project evaluation purposes. For limited-scope projects such as plant modifications, however, the use of factors results in some uncertainties, and preliminary piping design, along with a more detailed estimation, is preferred. Preconstruction estimates of large projects also require detailed piping design and estimation, but these tasks are performed by specialists, whereas chemical engineering generalists may be called upon to perform the same tasks at a limited-scope level.

One of the first things that the neophyte chemical engineer must learn is something about the bewildering array of pipes, tubing, flanges, valves, and fittings that constitute the reality of piping systems. New developments in piping components are frequent and commonplace. We often hear about revolutionary developments in, let us say, automatic process control, but seldom do we hear mentioned revolutionary developments in piping systems. And yet a development such as the ball valve (Fig. 5.3.1) is truly revolutionary, for its adoption within the last two decades as a standard element of piping systems has decidedly simplified the operation and maintenance of those systems. Ball valve passages offer an unobstructed flow path in the open position with a correspondingly reduced tendency to "plug up," and Teflon seats in particular result in free movement of the ball without "freezing."

Chemical engineers, by virtue of their training, are particularly well-qualified to design and analyze piping systems; it is with some justification that they are occasionally referred to as "plumbers with a degree." Beginners may, indeed, have some difficulty in understanding how the learned principles of fluid flow are applied to the design of practical systems; success in the design endeavor comes with practice and exposure to the myriad details that piping design entails. An element

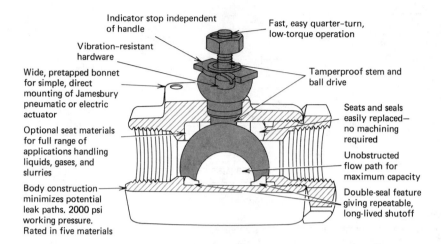

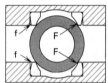

Floating Ball — Concept of flexible-lip seat plus floating ball allows upstream pressure to force ball into downstream seat for tight shutoff capability. A "double-seal" is created as lip on upstream seat is held against ball by line pressure.

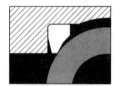

Unique Seat Design — Patented flexible-lip seats exert continuous sealing pressure, automatically compensate for wear and for changes in temperature and pressure, and provide extremely high cycle life.

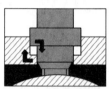

Stem Sealing — "Corner sealing", the compression of low-friction TFE box rings, eliminates stem leakage by avoiding straight-line leakage paths. Stem seal adjustment, if required, is done simply by tightening bonnet screws or bolts.

FIGURE 5.3.1
Jamesbury ball valve. *(From Bulletin 211, Jamesbury Corp., Worcester, Mass.; reproduced by permission.)*

of systematization does, fortunately, tie the diverse world of piping together; process piping, for example, in spite of the wide variability in construction materials and methods of joining, is for the most part based upon the familiar steel pipe dimensions prescribed by the American Standards Association. Three kinds of joints are commonly used:

1 Screwed joints (usually in smaller pipe sizes up to 1 1/2 in).
2 Welded joints.
3 Flanged joints. These are more expensive and more subject to leakage along gasketed surfaces, but they are often used when frequent disconnection is anticipated (for cleanout; at valves; equipment nozzle connections) or when no other joining method is available, as with lined pipe (Fig. 5.3.2).

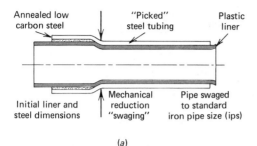

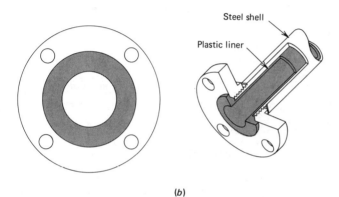

FIGURE 5.3.2
Flanged joints, lined pipe. (a) Swaging pipe to finished size. (b) Molded raised face joint. *(From brochure, "Field Installation of Dow Plastic Lined Piping Products," Dow Chemical USA, Midland, Mich.; by permission.)*

Many organizations develop their own *piping standards* which systematize the design of piping systems in specific service situations. An example of a piping standard is given in Fig. 5.3.3. This happens to be a "general service" standard, although specific applications are indicated as WOG (water-oil-gas) and saturated steam. Note how systematization eases the designer's task; materials are specified (steel pipe,[1] malleable iron fittings, brass valves), as are the joints (for example, flanged joints for 2-in valves and larger), pipe schedules, and valves. The specification number (AA in this case) may be shown on the line number flags on P&IDs, so that the nature of the pipeline in question is immediately identified.

Pipeline design involves much more than connecting equipment *A* with equipment *B*. For example, there is an endless variety of piping configurations, most of

---

[1] Standard carbon steel pipe is an extruded, seamless product specified by ASTM Standard A-53. Malleable iron fittings (2.5% C, 1% Si, 0.5% Mn) are rated at 150 or 300 lb; they are inexpensive relative to forged steel fittings, which are rated for much higher pressure (3000 to 6000 lb). The terms "black iron" or "black steel" are often heard; these are used to emphasize the difference from galvanized products, easily distinguished by their luster and lighter color.

| PIPING STANDARD | SPEC. NO. | AA |

MATERIAL AND SERVICE STEEL - MALLEABLE IRON - BRASS   GENERAL SERVICE

PRESSURE RANGE WOG (NON SHOCK) ½"-1½" - 200#, 2"-12" - 175#, 14"-24" - 150#

PRESSURE RANGE SAT. STEAM ½"-1½" - 144#, 2"-16" - 125#, 18"-24" - 100#

TEMPERATURE RANGE -20° F TO +410° F (300° F FOR BALL, 350° F FOR PLUG VALVES)

\multicolumn{4}{c}{PIPE}			
SIZE	½" THROUGH 1½"	2" THROUGH 10"	12" and LARGER
SPECS.	SCH. 80 A-53 SMLS.	SCH. 40 A-53 SMLS.	0.375 WALL A-53 SMLS.
JOINTS	SCREWED	WELDED	WELDED
COMPOUND	KEYTITE	—	—
\multicolumn{4}{c}{FITTINGS E - ELLS, T - TEES, R - REDUCERS, X - CROSSES, C - CAPS}			
SIZE	½" THROUGH 1½"	\multicolumn{2}{c}{2" and LARGER}	
E, T, R, X, C	300# MI SCREWED A-197	\multicolumn{2}{c}{STD. WT. SMLS. WELDING A-234}	
BUSHINGS	6000# FS SCREWED A-105-11	—	
COUPLINGS	3000# FS SCREWED A-105-11	—	
UNIONS	3000# FS SCREWED A-105-11	—	
PLUGS	3000# FS SCREWED A-105-11	—	
FLANGES	\multicolumn{3}{c}{150# FORGED STEEL SLIP-ON or WELDING NECK A-181-1}		
FACING	\multicolumn{3}{c}{STANDARD RAISED FACE}		
GASKETS	\multicolumn{3}{c}{1/16" GARLOCK #7819}		
BOLTS	\multicolumn{3}{c}{A-307B HEX HEAD BOLTS WITH HEAVY HEX NUTS}		
BOLTS COMPOUND	\multicolumn{3}{c}{GRAPHITE AND OIL}		
\multicolumn{4}{c}{VALVES, ETC.}			
SIZE	½" THROUGH 1½" (SCR.)	\multicolumn{2}{c}{2" and LARGER (FLGD.)}	
GATE	LUNKENHEIMER #2150	LUNK. #1430 (TO 24")	
GLOBE	LUNKENHEIMER #1021PS	LUNK. #1123 (TO 10")	
PLUG	ROCKWELL #114	ROCK. #115 and #143	
CHECK	LUNKENHEIMER #554Y	LUNK. #1790 (TO 12")	
NEEDLE	DRAGON #10F01		
BALL	SALISBURY #350	JAMES. #D150F22-36TT (TO 16")	
MISC.			
STRAINER - BODY	250# CAST IRON SCREWED	125# CAST IRON FLANGED	
STRAINER - SCREEN	STANDARD MONEL OR BRASS	STANDARD MONEL OR BRASS	

REMARKS

1. DO NOT USE THIS STANDARD FOR TOXIC FLUIDS.
2. WITH THE EXCEPTIONS OF TANK VALVES, BALL VALVES, AND FLANGED STRAINERS. THIS STANDARD CAN BE USED FOR FLAMMABLE FLUIDS IN SIZES ½" THROUGH 4".
3. THE MAXIMUM PRESSURE FOR GLOBE VALVES IN SIZES 6" THROUGH 10" IS 125#.
4. DO NOT USE THIS STANDARD FOR FLAMMABLE FLUIDS OVER 75 PSIG OR 150°F.
5. STEAM RATINGS FOR FLANGED GATE VALVES REDUCES THEIR TEMPERATURE RATING TO 353°F IN SIZES 2" TO 16" AND TO 338°F IN SIZES 18" TO 24".

FIGURE 5.3.3
An example of a piping standard.

which are not codified by means of something like a piping standard, but which nevertheless are based on years of operational experience and are commonly used. A few examples of such configurations serve to illustrate the reason for the complexity of piping systems (see Fig. 5.3.4).

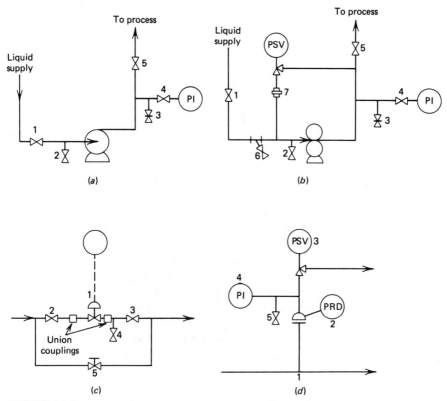

FIGURE 5.3.4
Some typical piping configurations. (a) Centrifugal pump. (b) Positive-displacement pump. (c) Control valve. (d) Relief device system.

Centrifugal Pump Configuration   (See Fig. 5.3.4a.) A field-mounted pressure gauge on pump discharges is standard, for it gives quick visual indication whether the pump is working properly. Block valve 4 (i.e., a shutoff valve which may be a gate valve or, more popularly, a ball valve) is provided to permit servicing of the gauge. The tee and valve 3 are often provided for convenient sampling of the higher-pressure stream at the pump discharge. Valve 3 is preferably a needle valve (to avoid a potentially unsafe sudden eruption of a high-pressure stream).

Exercise 5.3.1

What is another piping configuration that would allow sampling of the pump discharge without the danger of a high-pressure ejection of liquid?

Block valves 1 and 5 are provided so that the pump may be isolated from the rest of the process in case it must be opened or removed for servicing. Drain valve 2 on the pump suction, normally the low point in the pump circuit, allows removal of liquid before servicing. It is a common rule of thumb to provide drain valves at all low points in piping systems to facilitate removal of trapped liquids.

Exercise 5.3.2

To drain liquid from the pump before servicing, which valves must be open?

Positive-Displacement Pump Configuration (See Fig. 5.3.4b.) Many of the features of the positive-displacement pump configuration are the same as those associated with centrifugal pumps. An additional requirement is a strainer (item 6) which removes stray solids from the liquid supply; many positive-displacement pumps, such as gear pumps, may "freeze up" with even a trace of solids. Positive-displacement pumps by their very nature can develop excessively high discharge pressures if they happen to be deadheaded (as would occur, for instance, if valve 5 were to be closed inadvertently during pump operation). Many such pumps are protected from high-pressure damage by internal flow relief devices or mechanical shear pins. Nevertheless, a relief system such as the one illustrated is frequently incorporated into the piping configuration. Pump discharge and suction are short-circuited through a pressure relief valve (pressure safety valve, PSV) set at the desired upper limit of the pressure difference across the pump. Note that to fabricate the illustrated bypass using screwed fittings, at least one union coupling (7) must be used (normally, these are not shown on P&IDs). The frequent use of union couplings in screwed pipe networks is important to facilitate pipeline disassembly for maintenance.

Control Valve Configuration (See Fig. 5.3.4c.) Control valves (1) are frequently provided with block valves (2 and 3) to permit in-place maintenance or removal; a drain valve 4 may also be provided. A globe valve 5 on a bypass line allows the process to be operated on manual control during periods of maintenance. Piping layout must be such that manually operated valves are easily accessible.

Relief Device Configuration (See Fig. 5.3.4d.) If overpressure relief is required at 1 (usually to protect a vessel or other item of equipment from catastrophic failure), a pressure safety valve (3) is often provided. The valve discharge is then directed into scrubbers, or large containment vessels, or flares, or just into the atmosphere—the choice depends upon the nature of the material vented. PSVs have the distinct advantage that they reclose once the pressure has dropped below a set point, so that the now-relieved system is again isolated and operable, and yet the valves are again ready to open upon demand. A serious disadvantage is that the valves are exposed to prolonged contact with process fluids; the result may be corrosion, or solids deposition, which may affect the proper operability of the valves. Occasionally it will prove worthwhile to install the PSV in tandem with a bursting disk (pressure rupture disk, PRD, 2). PRDs consist of disks (metals, graphite) designed to burst above a certain pressure difference; they serve the same relief function as PSVs, but they have the disadvantage of not reisolating the relieved system once it is below the set overpressure; the total contents of a vessel may well spew out once the disk has ruptured. In the tandem arrangement, the PRD protects the PSV from contact with corrosive process fluids; if the PRD bursts, however, the PSV is immediately activated, but closes once the high pressure has been relieved. The PSV is often placed on an elbow beyond the PRD to avoid damage from directly impacting pieces of the bursting disk.

One may argue that if process fluids affect the PSV, they may also damage and change the effective bursting pressure of the PRD. Such damage would be indicated initially by pinhole leaks through the PRD, and if the pressure of 1 is normally even slightly above 1 atm, the leak can be detected by means of the "telltale" pressure gauge 4 (or an automatic alarm on the control room panel board).

An important rule is that no block valve is allowed anywhere on a relief line. A three-way ("Trans Flow") valve may be installed between point 1 and the PRD 2 to permit switching to a duplicate tandem relief system if one system has been activated or has been found to be faulty. In this way the relief system may be properly maintained with uninterrupted operation and undiminished relief protection.

Exercise 5.3.3

Suppose a small amount of an inert gas such as nitrogen is permitted in a process stream. Sketch a system that would protect a pressure safety valve from process fluid contact using the inert gas.

Exercise 5.3.4

A vapor lock may occur in an unvented vertical pipe loop whenever liquid flows through such a loop at low enough velocity, so that any entrained gas bubbles have the opportunity to become disengaged and to form a gas pocket in the top part of the loop. Vapor locks may seriously reduce the liquid flow capacity. How and why? Assuming that a vertical loop is unavoidable, can you think of a device that would prevent gas accumulation?

The *sizing* of pipelines is a matter of choosing the appropriate sizing criterion for the job on hand. The following are some of the criteria and associated shortcut methods commonly used.

Economic Pipe Diameter Criterion This is perhaps the most commonly used criterion of pipe sizing; a balance is made between the cost of the pipe ("fixed" charges and cost of capital investment increase with pipe diameter) and the cost of pumping (which decreases with larger pipe diameter). The balance represents one of the few cases of optimization within the province of chemical engineering that can be readily developed by using the techniques of simple calculus, and that is quite generally applicable. As such, it is a favorite subject in beginning treatises on process optimization, and several good reviews are readily available—in Perry's handbook and in plant design textbooks (Peters and Timmerhaus, 1980; Happel and

Jordan, 1975). Nomographs are included as part of the reviews which allow rapid sizing of whole piping systems. A few limitations of the nomographs, however, must be kept in mind:

The charts apply to commercial steel pipe. For more expensive materials of construction, use the next standard pipe size below the size indicated.

The usual nomographs (Perry) apply to turbulent flow only; however, similar nomographs are available for laminar flow (Peters and Timmerhaus, 1980), which may well occur, for example, in the flow of high-viscosity crudes.

Maximum Allowable Pressure Drop Criterion It may happen that limitations on allowable pressure drop override economic optimization considerations. There are many such instances: gravity flow, flow of high-pressure steam, flow in vacuum systems. Standard fluid flow design procedures can be used to size pipe for a specified pressure drop, although a trial-and-error calculation is usually involved. Again, nomographs are available to simplify the task ("pipe flow chart," Perry's handbook), but care must be taken to recognize the limitations; the charts are for turbulent flow, and for surface roughness equivalent to commercial steel pipe.

The equivalent pipe length of valves and fittings must, of course, be accounted for in estimating the total length of the pipe run in which the pressure drop is to be limited. For preliminary purposes, it is usually more than adequate to assume that the equivalent length of the fittings is one-third of the straight-run pipe length.

Even if the pressure drop is the limiting design criterion, an economic pipe diameter estimation should be performed. If it turns out that the economic pipe diameter is *larger* than the diameter based on allowable pressure drop, the larger diameter should be chosen; the pressure drop will be less than the specified maximum.

Exercise 5.3.5

Why is it usually important to limit the pipeline pressure drop of high-pressure steam?

Pipe Velocity Criterion Occasionally the criterion of pipe sizing is the allowable velocity of the fluid (usually liquid) in the pipe, limited by considerations of excessive corrosion or erosion of protective films. Kern (1974) has recommended the maximum velocities listed in Table 5.3.1 for a few typical systems.

"Typical" pipeline velocities are sometimes used as a pipe-sizing criterion. Numbers such as those in Table 5.3.1 simply reflect what is "normally done," and may or may not represent the economic optimum. The only thing that can be said in favor of this approach is that it is fast and simple. A rough rule of thumb for pipeline velocities is:

1 ft/s for viscous liquids
10 ft/s for nonviscous liquids
100 ft/s for gases

TABLE 5.3.1
PIPELINE VELOCITIES

Maximum velocities to prevent erosion or corrosion of pipe wall, ft/s	
Liquid in carbon-steel pipe:	
Phenolic water	3
Concentrated sulfuric acid	4
Cooling-tower water	12
Salt water	6
Calcium chloride brine	8
Caustic soda (>5%)	4
Aqueous amine (mono- or diethanolamine)	10
Wet phenolic vapor	60
Liquid in plastic or rubber-lined pipe	10

Typical velocities in gas and vapor lines, ft/s

Nominal pipe size, in	Saturated steam or saturated vapor	Superheated steam, superheated vapor, or gas	
	Low pressure	Medium pressure	High pressure
2 or less	45–100	40–80	30–60
3–4	50–110	45–90	35–70
6	60–120	50–120	45–90
8–10	65–125	80–160	65–125
12–14	70–130	100–190	80–145
16–18	75–135	110–210	90–160
20	80–140	120–220	100–170

Typical liquid velocities in steel pipelines, ft/s

Liquid and line	Nominal pipe sizes, in		
	2 or less	3 to 10	10 to 20
Water:			
Pump suction	1–2	2–4	3–6
Pump discharge (long)	2–3	3–5	4–7
Discharge leads (short)	4–9	5–12	8–14
Boiler feed	4–9	5–12	8–14
Drains	3–4	3–5	
Sloped sewer		3–5	4–7
Hydrocarbon liquids (normal viscosities):			
Pump suction	1.5–2.5	2–4	3–6
Discharge header (long)	2.5–3.5	3–5	4–7
Discharge leads (short)	4–9	5–12	8–15
Drains	3–4	3–5	
Viscous oils:			
Pump suction:			
Medium viscosity		1.5–3	2.5–5
Tar and fuel oils		0.4–0.75	0.5–1
Discharge (short)		3–5	4–6
Drains	1	1.5–3	

Source: Kern (1974); reproduced by permission of *Chemical Engineering*.

300 CHAPTER 5: THE DIRECT FIXED CAPITAL INVESTMENT

Net Positive Suction Head Criterion The NPSH criterion translates into the maximum allowable pressure drop criterion. The net positive suction head is defined as the difference between the pressure at the suction side of pumps (usually centrifugal pumps) and the vapor pressure of the pumped liquid at the operating temperature. The NPSH is usually expressed as a head in feet of the liquid pumped. A minimum value of NPSH is specified for each pump and flow rate; such values often appear on the manufacturer's pump performance curves (Fig. 5.3.5). Since the NPSH criterion fixes the required pump suction pressure, the suction side piping must be sized so that the total pressure drop will not reduce the suction pressure below the required value.

The point-by-point thermodynamic path of a liquid moving into the vanes of a centrifugal pump impeller is complex. Depending upon specific mechanical design, pump throughput, and impeller rotational speed, the accelerating fluid is subjected to a variable drop in absolute pressure. If that pressure drops below the vapor pressure, the liquid flashes, and the pump is subjected to potentially destructive cavitation—at the very least, it will not work very well. If the suction pressure satisfies the NPSH criterion, the problem does not arise. Potential flashing is most commonly a problem with liquids that are not far from their bubble point temperature at the prevailing operating pressure.

Example 5.3.1

A pump is to be selected to pump 250 gpm of hot water. The required static head that must be developed is 100 ft. In addition, it is estimated that a friction head loss of 10

FIGURE 5.3.5
Centrifugal pump performance curves. ("Goulds Pump Manual," Goulds Pumps, Inc., Seneca Falls, N.Y. Reproduced by permission.)

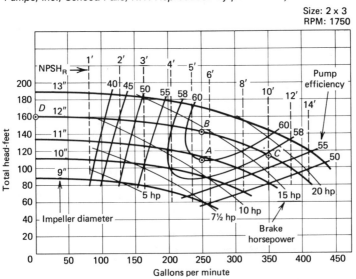

ft will occur in the pipeline at the 250 gpm flow (not counting pressure drop through a control valve). The flow will be controlled with a control valve on the pump discharge. The required range of flow is 0 to ~350 gpm. If the pump with the performance curves in Fig. 5.3.5 is selected, specify:

a Recommended impeller diameter
b Required NPSH for piping design purposes
c Required motor horsepower

SOLUTION The design operating point (250 gpm, 110-ft dynamic head) is indicated by point A in Fig. 5.3.5. The required delivery could be obtained with an 11-in impeller, but no pressure drop would be available for the control valve operation. If a 12-in impeller is selected, the head developed at 250 gpm is 140 ft (point B), and $140 - 110 = 30$ ft is available for pressure drop through the control valve. This is more than adequate; therefore *recommend a 12-in impeller*.

At the maximum flow of 350 gpm (point C), the total dynamic head developed will be about 116 ft. Since frictional head in turbulent flow varies at about the 1.75 power of flow (do you agree?), the 350 gpm flow will be attained approximately with the control valve fully open. The required NPSH is *10 ft at 350 gpm, 5.6 ft at 250 gpm*.

The suction line should be designed to satisfy NPSH at the maximum flow, and the design should be also checked out at the design flow.

The maximum pressure drop across the control valve occurs at zero flow (point D); it is $160 - 100$ (100-ft static head, zero friction head), or 60 ft.

The maximum power is required at point C. The pump efficiency is 58 percent; taking the density of the hot water as 8.2 lb/gal and using Eq. (4.1.13), we calculate

$$\text{bhp} = \frac{350 \times 8.2 \times 116}{33{,}000 \times 0.58} = 17.4$$

Choose a *20-hp motor*.

Exercise 5.3.6

Does the NPSH depend upon the liquid that is pumped?

Piping design involves a great deal more than the rather straightforward sizing based upon the above criteria. There are many very interesting flow situations that test the ingenuity and experience of the chemical engineer—critical flow, two-phase flow, nonnewtonian flow, flow distributors, unsteady-state flow. Rules of thumb exist that facilitate the design of piping for these less conventional flow situations. The complexities of flow distribution in manifolds, for example, have been analyzed in considerable detail (Acrivos et al., 1959); in most cases, the design of manifolds can be simplified, at little economic sacrifice, by sizing the main manifold header in such a way that the pressure drop across the manifold branches will be substantially the same:

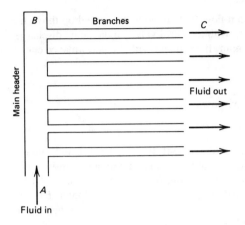

The header must be sized so that the pressure drop from A to B will be small enough to make the pressure drop from the header to the discharge at C the same for all branches. If the pressure drop from B to C is very small, the header sizing may still be a problem. One way to avoid this problem is to put restricting orifices at the entrance to each branch; this artificially increases the BC pressure drop but renders the AB pressure drop less distribution dominant. This artifice is occasionally used to improve liquid distribution in forced circulation reboilers. (See Perry's handbook for additional manifold design rules of thumb.)

Insulation represents another important component of piping systems. Strictly speaking, an economic insulation thickness exists for each insulation application (Fig. 1.3.1), and publications are available (Govan and Grabiner, 1973) which tabulate the economic thickness for a wide variety of parameters. As a practical matter, condensed tables are available (Perry's handbook, for example) which approximate the economic optimum and limit the selection of preformed insulation shapes to a few standard thicknesses; with the aid of such tables, the selection process is rapid. Insulation is required for other applications besides heat loss prevention; in some cases, personnel protection from hot surface contact is the only criterion of insulation selection, and in cold geographical areas "winterizing" may be necessary—insulation protection to avoid freezing of fluids (for example, water to safety showers and eye baths). In outdoor areas, in particular, insulation must be protected from the weather by means of the judicious use of canvas or sheet metal protective jackets.

If high-melting process fluids are handled in pipelines, insulation alone may not be adequate to keep the fluids from freezing. The pipelines may have to be "traced" by using electrical resistance heating strips or else steam tubing that is wrapped around pipe and fittings in the manner illustrated in Fig. 5.3.6. Proper thermal contact between the steam tracing and pipelines is provided by thermally conducting cements. The few examples in the illustration demonstrate the complexity of steam tracing and the reason for its high expense.

Other aspects of piping system design include the specification of supports and central pipeways. The appropriate layout of pipelines—to assure adequate accessi-

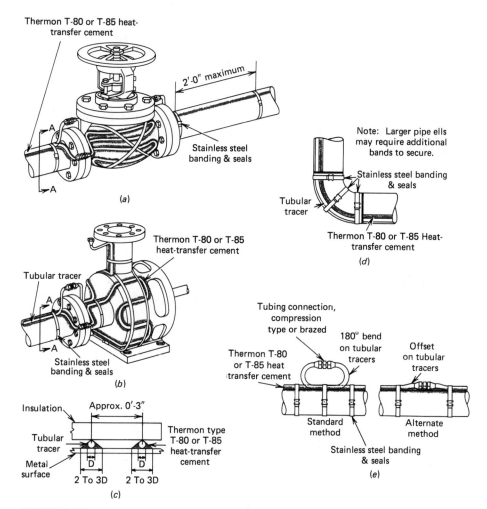

FIGURE 5.3.6
Examples of steam tracing. (a) Typical installation on valve and piping; (b) typical installation on pump and piping; (c) typical cross section of heat-transfer cement on pumps, valves, and other irregular surfaces; (d) typical installation of tubular tracer and heat-transfer cement on pipe elbow; (e) typical connection of tubular tracers on process piping. *(Thermon Manufacturing Co., San Marcos, Tex. Reproduced by permission.)*

bility, to eliminate dangerous stresses, to avoid embarrassing bumping or tripping hazards—is facilitated with the help of adequate piping diagrams. Computer-aided pipeline design and the construction of scale models of complete plants are both increasingly popular aspects of advanced-stage projects. For limited-scope projects, isometric drawings such as the illustration in Fig. 5.3.7 are adequate for both costing and construction guidance to pipe fitters. Such diagrams are easily drawn to scale, particularly if isometric graph paper is available. Common symbols include those for the various kinds of pipe joints:

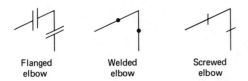

Flanged elbow   Welded elbow   Screwed elbow

Other symbols reproduce, more or less faithfully, the many types of valves and fittings that constitute piping systems. Thus, for example, a ball valve is represented by

FIGURE 5.3.7
An isometric piping diagram.

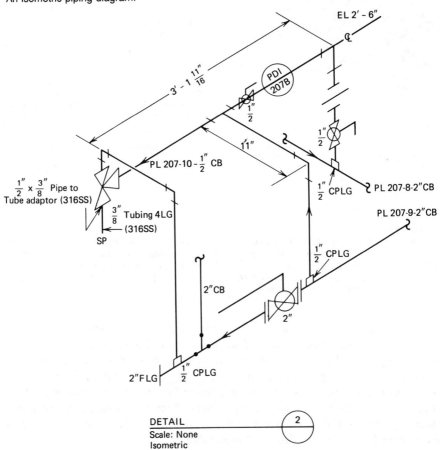

Detailed Piping Estimates

The accurate preconstruction cost estimation of piping systems for battery limits plants is a tedious, nit-picking job that is best left to specialists who are well-versed in the pitfalls of their profession. The demands of project evaluation are most often adequately satisfied by using the equipment purchased cost factor method. This method, however, is not really applicable to some limited-scope projects. The piping for a substantial semiplant may be estimated in this fashion, but what of the minor modification in an existing plant, for example? If the project involves, let us say, a pipe connection between two vessels, and the decision to proceed with the project must await a reliable cost projection, how is the cost of the pipe connection, in fact, estimated?

The first step, of course, is the design of the pipeline, and the proper documentation of the design with P&IDs, isometrics, and other detailed sketches. The cost estimation then involves *takeoff* from the piping diagrams—an enumeration, item by item, of the pipeline components, followed by one of several ways of estimating the purchased cost of the components and the total labor cost of joining the components into the desired unit.

The basic "installed" cost of a piping system, then, consists of the two principal cost items:

Purchased cost of piping materials
Installation labor cost

To obtain the total investment required, some additional costs must be accounted for:

Insulation
Painting
Pressure testing, weld examination
Tracing
Hangers and pipeways
Trenching (for underground lines)

Still another cost component to be considered is the so-called *indirect cost*, analogous to the plant indirect costs (Exhibit 5.1A):

Sales tax and freight
Construction overhead
Contractor engineering
Contractor fee

Methods are available to estimate indirect costs, but care must be taken not to duplicate such costs if detailed pipeline estimation is used in conjunction with the method of purchased cost factors. Total plant indirect cost factors incorporate piping system indirect costs.

Some useful piping cost references are listed in Exhibit 5.3A.

The purchased cost of piping materials may be estimated, of course, by generating a list using the takeoff procedure from piping diagrams and computing the cost from price lists. The small price list excerpt in Fig. 5.3.8 demonstrates the burden of making such a detailed estimate, even with the help of computer-stored information. The method is fine for very simple piping jobs, but for more complex systems various shortcuts are used. One shortcut method is illustrated in Fig. 5.3.9, a graphical cost correlation. An *average* cost of fittings is shown, so that it is now necessary to enumerate just the total number of fittings of each size, rather than each fitting type separately. Similarly, the average cost of valves is shown (although this average does not include control valves).

Correlations are available (for instance, Guthrie, Ref. 1, Exhibit 5.3A) for estimating the installation labor ("direct labor") cost of piping systems, as man-hours (or dollars) per foot of pipe, per fitting, and so forth. Again, shortcut methods may be used to ease the burden of labor cost estimation. Several of these methods are outlined by Bauman (Ref. 4, Exhibit 5.3A); we will focus here on just one of those methods, originally proposed by Clark (1957), the diameter-inch method. The basis of the method is the observation that installation labor depends primarily upon the number of connections that must be made, rather than the length of piping between the connections. Clark defines diameter-inches as the product of nominal diameter and number of connections (screwed or welded). For example, three welded connections to join lengths of 8-in pipe are equivalent to 3×8, or 24 dia.-in. The diameter-inches for each pipe size are added up and multiplied by the appropriate labor factor from Table 5.3.2. The sum of all these products represents the number of man-hours of direct labor required to build the pipeline. An example from Clark's article will serve to illustrate the method.

EXHIBIT 5.3A

SOME USEFUL PIPING COST REFERENCES

1 Guthrie, Kenneth: Costs of Liquid-Handling Systems, *Chem. Eng.*, **76**:201 (Apr. 14, 1969). Perhaps the most generally useful compendium; cost indices must be used to scale up costs.

2 Peters, Max S., and Klaus D. Timmerhaus: "Plant Design and Economics for Chemical Engineers," 3d ed., McGraw-Hill, New York, 1980. An up-to-date coverage of costs in graphical format.

3 Aries, Robert S., and Robert D. Newton: "Chemical Engineering Cost Estimation," McGraw-Hill, New York, 1955. An older reference which contains factors and methods that are still applicable.

4 Bauman, H. Carl: "Fundamentals of Cost Engineering in the Chemical Industry," Reinhold, New York, 1964. A good exposition of estimation shortcuts.

5 Yamartino, James: Installed Cost of Corrosion-Resistant Piping, *Chem. Eng.*, **85**:138 (Nov. 20, 1978). Useful for obtaining estimates for nonsteel piping systems.

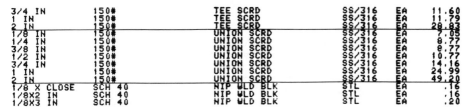

3/4 IN	150#	TEE SCRD	SS/316	EA	11.60
1 IN	150#	TEE SCRD	SS/316	EA	11.79
2 IN	150#	TEE SCRD	SS/316	EA	28.83
1/8 IN	150#	UNION SCRD	SS/316	EA	7.05
1/4 IN	150#	UNION SCRD	SS/316	EA	8.77
3/8 IN	150#	UNION SCRD	SS/316	EA	8.77
1/2 IN	150#	UNION SCRD	SS/316	EA	10.77
3/4 IN	150#	UNION SCRD	SS/316	EA	14.16
1 IN	150#	UNION SCRD	SS/316	EA	24.99
2 IN	150#	UNION SCRD	SS/316	EA	49.20
1/8 X CLOSE	SCH 40	NIP WLD BLK	STL	EA	.16
1/8X2 IN	SCH 40	NIP WLD BLK	STL	EA	.16
1/8X3 IN	SCH 40	NIP WLD BLK	STL	EA	.20

FIGURE 5.3.8
An excerpt from a fittings price list (1980 cost of type 316 stainless steel screwed union couplings, 150-lb rating).

FIGURE **5.3.9**
Piping cost correlation. (1) Pipe, (2) fitting, (3) valve. *(Aries and Newton, Ref. 3, Exhibit 5.3A. Reproduced by permission of McGraw-Hill Book Co., Inc.)*

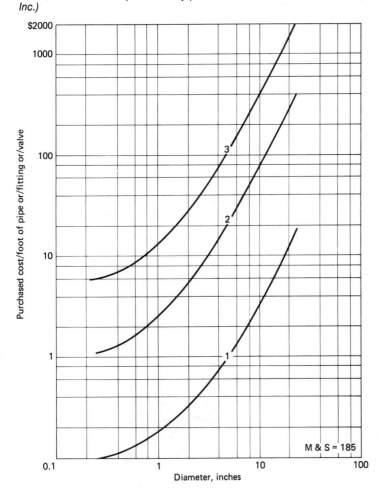

TABLE 5.3.2
CLARK'S LABOR FACTORS

	Pipe layout, cutting, welding, threading, and erection		
	Man-hours/dia.-in		
Pipe size	Carbon steel Sch. 40	Stainless steel Sch. 5 and 10	Saran-lined
$\frac{1}{2}$ through $1\frac{1}{2}$ in	1.30	1.60	1.50
2 and 3 in	1.00	1.30	1.50
4 through 8 in	0.90	1.25	1.40

Flanged valves and fittings; handling and bolting up	
Pipe size	Man-hours/dia.-in per end
$\frac{1}{2}$ through $1\frac{1}{2}$ in	0.40
2 and 3 in	0.35
4 in	0.35
6 and 8 in	0.30

Source: Clark (1957); by permission of *Chemical Engineering*.

Example 5.3.2*

Estimate the man-hours of direct labor required to make the connections between flanges A and B in Fig. 5.3.10. A side connection is also required to an existing 1-in pipe, as shown; a 1-in coupling will be welded to a hole drilled through the wall of the 3-in pipe at C.

SOLUTION The various items are taken off the engineering drawing and tabulated in pipe size categories, so that the diameter-inches may be quickly computed:

*Reproduced by permission of *Chemical Engineering* magazine.

FIGURE 5.3.10
Pipeline isometric drawing for Example 5.3.2. *(Clark, 1957; reproduced by permission, Chemical Engineering magazine.)*

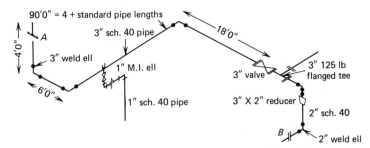

	No.	Connections	Dia.-in.
1-in-dia. ells	2	4	4
1-in-dia. union	1	2	2
1-in-dia. valve	1	2	2
1-in-dia. coupling	1	2	2
			10
2-in-dia. ell	1	2	4
2-in-dia. reducer	1	1	2
2-in-dia. flange	1	1	2
			8
3-in-dia. pipe	118 ft	4	12
3-in ell	4	8	24
3-in flanges	4	4	12
3-in reducer	1	1	3
			51
Bolting up 3-in valve	1	2	6
Bolting up 3-in tee	1	2	6
			12

The labor factors are obtained from Table 5.3.2:

	Dia.-in.		Labor factor		Man-hours
	10	×	1.30	=	13
	8	×	1.00	=	8
	51	×	1.00	=	51
	12	×	0.35	=	4.20
Total man-hours to fabricate and erect					76.20

The *labor rate* to be used to calculate the cost of piping systems must be carefully considered. If the average hourly basic pay rate is used to compute the direct labor cost, separate estimation of indirect costs must be made; this is the approach chosen by Guthrie (Ref. 1, Exhibit 5.3A). In mid-1979, the average hourly rate was about $12/h. However, contract labor rates (including all indirect costs) at that time were $38 to $45/man-hour (West Coast).

Example 5.3.3

Estimate the total installed cost (1979, M&S index = 580) of the piping system in Fig. 5.3.10, fabricated from black steel. Use $40/man-hour for contracted labor.

SOLUTION   Use Fig. 5.3.9 for material costs and Example 5.3.2 for material takeoff:

118 ft of 3-in pipe at 50 cents/ft	$59.00
Fittings (including flanges):	
1-in pipe: 5 at $2.60 each	13.00
2-in pipe: 3 at $5.30 each	15.90
3-in pipe: 9 at $9.00 each	81.00
2-in valve	12.00
3-in valve	48.00
Flanges on 3-in valve	27.00
Total	$255.90

Use M&S index to scale up to 1979:

$$580/185 \times 255.90 \approx \$800 \text{ for materials}$$

[Item-by-item estimation from Guthrie (Ref. 1, Exhibit 5.3A), cost-adjusted to 1979, gives $840; Fig. 5.3.9 is a good approximation.]

$$\text{Man-hours required (Example 5.3.2)} = 76.20$$

Thus

$$\text{Estimated installed cost} = 76.20 \times 40 + 800 \approx \$3850$$

Thus even a simple connection such as that illustrated in Fig. 5.3.10 can be very expensive, indeed. The estimate in Example 5.3.3 does not include insulation, painting, pipe hangers, and other items that may be required for a specific pipeline. Guthrie lists the costs of such items. He also recommends that direct labor and materials be escalated by 10 percent to account for errors in the engineering drawing takeoff procedure. The cost of pipe supports and hangers is often accounted for by adding 10 to 15 percent to the cost of the installed pipeline.

Example 5.3.4

Suppose the pipeline in Fig. 5.3.10 is to be steam-traced, insulated, painted, and hydrostatically tested. What additional cost, over and above that in Example 5.3.3, would be incurred? Include an estimate for pipe supports.

SOLUTION The pipeline in question is relatively short, and 10 percent of installed cost should be sufficient for supports. Thus

$$\text{Cost of supports (material and labor, including overhead)} = 385 \approx \$400$$

Perry's handbook suggests that 1½ in of insulation is optimum for steam temperatures. Guthrie gives the following costs for 3-in pipe items:

$1\frac{1}{2}$-in insulation: $5/ft
Painting: 45¢/ft
Testing: $1.30/ft
Tracing: $1/ft

Take total length about 130 ft, and scale up by using the M&S index (303 in Guthrie's work, 580 in mid-1979). Therefore

Additional cost = $400 + {}^{580}\!/_{303} \times 130(5.00 + 0.45 + 1.30 + 2.00)$
$\approx \$2600$

Yamartino (Ref. 5, Exhibit 5.3A) has calculated convenient cost ratios of the installed cost of pipelines fabricated from materials other than steel relative to the installed cost of steel lines. For example, for a typical complex piping system, he finds that the ratio for Schedule 40 nickel is 3.51 for 2-in pipelines, 5.21 for 4-in pipelines. The anticipated value for a 3-in line is thus about 4.36, and the installed cost of the pipeline in Example 5.3.3, fabricated from nickel, would be

$$3850 \times 4.36 \approx \$16,800$$

The extra items in Example 5.3.4 will cost the same.

5.4 RELIABILITY OF CAPITAL ESTIMATES

How reliable do capital estimates turn out to be? How closely do estimates at various stages of project development agree with the actual DFC of the installation, once it has been completed? Do estimates meet their objective of giving the decision makers a sound foundation upon which to base their business decisions?

Not very surprisingly, the deviation of estimates from actual costs depends upon the stage of the project development and the amount of effort that had gone into the estimate. The result of a study by Bauman (Ref. 4, Exhibit 5.3A) is reproduced in Fig. 5.4.1. The lines represent the range of the deviations at various stages of project development. Some 48 projects are involved in Bauman's study—30 plants, 9 auxiliary facilities, and 9 laboratories—ranging in DFC (1964 dollars) from $100,000 to $38,000,000.

Two characteristics are noticeable in Fig. 5.4.1:

1 The range of the deviations decreases in context of more-advanced projects.
2 The distribution is skewed; order-of-magnitude estimates tend to be too low.

The skewness may be ascribed to the fact that the project is insufficiently defined during early stages of development, and too many items important to the process are overlooked. This is why the completeness of preliminary flow sheets has been so strongly emphasized. The skewness disappears as the projects become better-defined, and the range of the deviations decreases as a result of the more-concentrated effort during later project stages. It is likely that the range could be rendered narrower during earlier stages by expending sufficient effort, but this does not appear to be justified. The fact of the matter is that estimates, not excluding the preliminary kind, turn out to be surprisingly good in view of the large number of uncertainties and guesses involved. The reason is much the same as was the case

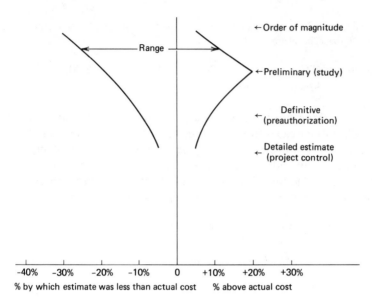

FIGURE 5.4.1
Deviation of estimates from actual DFC. *(Bauman, Ref. 4, Exhibit 5.3A. Copyright ©1964 by Van Nostrand Reinhold Company. Adapted by permission of the publisher.)*

with estimated purchased equipment cost totals (Sec. 4.5): the errors tend to be compensating.

In spite of Bauman's favorable assessment, there are, of course, some very bad estimates, and there is a latent tendency in the chemical industry to view early-stage estimates with distrust, even contempt. Reasons for such distrust may be counteracted only by a corporate effort to eliminate them, and many companies have departments devoted to a type of retrospective evaluation of estimates represented by Fig. 5.4.1. An attempt is made to assess the reliability of in-company estimates, to pinpoint reasons for failures, and to generate more reliable data and factors for future estimates.

Where do estimates go wrong? The most common causes include the following:

Lack of completeness. This is perhaps the most notorious of the causes, and it can assume various forms. The incomplete flow sheet has been mentioned as a prime cause. Another variation is the omission of one of the elements of the DFC—because "we don't need it," or "it is already there." Such omissions are regrettably commonplace whenever projects are in the process of being "ramrodded" by managers with overpoliticized motivation. If a process tower to house a new small plant already exists, it does not come free; its book value must be included in assessing the criterion of economic performance (although it certainly is useful to point out to the decision makers that the required new investment will be reduced). Moreover, the existing tower is likely to require considerable modification to accommodate the new equipment; this is often an unanticipated part of the

DFC. Attempts to close one's eye to the need for facilities and auxiliaries, to make estimates "lean" and attractive, are, more often than not, an invitation to trouble—trouble in the form of requests for "variances" (i.e., more money) to the board of directors during construction; trouble in the form of high operating costs when the plant is on-stream.

Poor project definition. This is actually a variation on the theme of incompleteness. Its manifestation is a large variation in the results of two or more estimators who are trying to estimate the cost of the same thing. Often it turns out that the reason for the large differences is that they are not really talking about the same thing. Indeed, it is important that the nature of the investment subject to analysis be properly defined. Is the result the direct fixed capital? Or is it just the total plant direct cost (TPDC)? Or is it, perhaps, the capital for transfer, or the capital for sale (terms we will further explore in Chap. 8)? We find that published economic analyses of projects of current interest (for instance, biomass conversion to gasohol) are often contradictory, not necessarily because they are wrong, but because the scope of the analyses is insufficiently defined and, unbeknownst to the reader, the analyses are mutually incompatible. Kermode and Jones (1980) succeeded in reconciling three apparently contradictory estimates of coal conversion to methanol (involving both investment and operational cost analysis) by forcing the estimates to a common basis.

Lack of managerial dedication. Again this category is a variation on incompleteness. It does sometimes happen that a manager is unwilling to pay for the effort required to come up with a reliable estimate to match the stage of project development. The "put something together, but don't spend too much time" approach is rarely justified; managers get "what they pay for." It is true that a back-of-the-envelope estimate can be made in a couple of hours, but considerably more time is required for more-advanced estimates. The time required for estimates is difficult to distinguish from time required for the process design itself. The two objectives are sought jointly as part of the preliminary project evaluation, and Bauman indicates that "study" estimates may require 150 to 250 man-hours, figures supported by the present author's experience. Thus even study estimates are expensive; at the 1980 R&D rate of $37.50/man-hour, the indicated cost is $6000 to $10,000. Managers must be ready to support such costs to adequately promote promising projects.

Inflation. An estimate is obsolete from the moment it has been completed. It is vitally important that the completion date of an estimate be prominently identified so that the projected costs are not "frozen in time," a most unfortunate error. It is wise to emphasize the inflation rate as part of the results; in the early eighties, the DFC requirement for a plant escalated at just about 1 percent/month.

Geographical cost variations. Published cost data are rarely identified as to regional origin, and yet some regional cost variation may be expected. The error incurred by this factor is not terribly important within the continental United States. If plant construction costs on the Gulf Coast (Houston) are taken as unity, regional factors vary from 0.89 in the southeast to 1.12 in New York (Guthrie, 1974). More serious errors may be incurred by projecting U.S. costs to construction overseas.

The reconciliation of construction costs in the United States and abroad is a fascinating and highly specialized aspect of cost estimation. The basic estimation principles that have been outlined, however, are universally applicable.

NOMENCLATURE

- c concentration, parts per million (ppm) by volume
- C direct cost per functional step, $
- F instrument or adjustment factor
- I index value of cost index
- J investment per unit capacity, $/annual production (kg)
- K complexity factor; also integration constant
- N number of functional steps
- $\overset{\circ}{Q}$ volumetric flow rate, m³/h
- r annual plant production capacity, kg/year
- s sales price per unit of production, $/kg
- t time, h
- V volume, m³
- T turnover ratio (dimensionless)

Subscripts

- 0, in   base value, incoming value, as in c_0, c_{in}
- t, p, a temperature, pressure, alloy designation, as in F_p
- 1, 2, 3, . . .   separate factor designations

Abbreviations

- bhp   brake horsepower
- DFC   direct fixed capital
- gpm   gallons per minute
- NPSH   net positive suction head
- PC   total equipment purchased cost
- PI   pressure indicator
- PRD   pressure relief device
- PSV   pressure safety valve
- TPC   total plant cost
- TPDC   total plant direct cost
- TPIC   total plant indirect cost

REFERENCES

Acrivos, A., B. D. Babcock, and R. L. Pigford: Flow Distributions in Manifolds, *Chem. Eng. Sci.*, **10**:112 (1959).

Allen, D. H., and R. C. Page: Revised Technique for Predesign Cost Estimating, *Chem. Eng.*, **82**:142 (Mar. 3, 1975).

API (American Petroleum Institute): "Recommended Practice for the Design and Installation of Pressure-Relieving Systems in Refineries," Bulletin API RP 520, 4th ed., December 1976.

Aries, Robert S., and Robert D. Newton: "Chemical Engineering Cost Estimation," McGraw-Hill, New York, 1955.
Baasel, William D.: "Preliminary Chemical Engineering Plant Design," Elsevier North-Holland, New York, 4th printing, 1980.
Bosworth, D. A.: Installed Costs of Outside Piping, *Chem. Eng.*, **75:**132 (Mar. 25, 1968).
Cevidalli, Guido, and Beno Zaidman: Evaluate Research Projects Rapidly, *Chem. Eng.*, **87:**145 (July 14, 1980).
Clark, W. G.: Accurate Way to Estimate Pipe Costs, *Chem. Eng.*, **64**(7):243 (1957).
Fang, Cheng-Shen: The Cost of Shredding Municipal Solid Waste, *Chem. Eng.*, **87:**151 (Apr. 21, 1980).
Govan, F. A., and J. W. Grabiner: "How to Determine Economic Thickness of Thermal Insulation," Thermal Insulation Manufacturers Association, Mt. Kisco, N. Y., 1973.
Guthrie, Kenneth M.: Capital Cost Estimating, *Chem. Eng.*, **76:**114 (Mar. 24, 1969).
———: Capital and Operating Costs for 54 Chemical Processes, *Chem. Eng.*, **77:**140 (June 15, 1970).
———: "Process Plant Estimating, Evaluation, and Control," Craftsman Book Co., Solana Beach, Calif., 1974.
Gutierrez, Alfonso, and Scott Lynn: Minimum Critical Velocity for One-Phase Flow of Liquids, *Ind. Eng. Chem., Process Des. Dev.*, **8:**486 (October 1969).
Happel, John, and Donald G. Jordan: "Chemical Process Economics," 2d ed., Marcel Dekker, Inc., New York, 1975.
Katell, Sidney, and John H. Faber: What Does Tonnage Oxygen Cost?, *Chem. Eng.*, **66:**107 (June 29, 1959).
Kermode, R. I., and J. E. Jones, Jr.: A Comparison of Independent Product Cost Estimates; The Cost of Methanol from Coal, *Eng. Costs Prod. Econ.*, **5:**143 (1980).
Kern, Robert: Useful Properties of Fluids for Piping Design, *Chem. Eng.*, **81:**58 (Dec. 23, 1974).
Kohn, Philip M.: CE Cost Indexes Maintain 13-Year Ascent, *Chem. Eng.*, **85:**189 (May 8, 1978).
Lang, Hans J.: Simplified Approach to Preliminary Cost Estimates, *Chem. Eng.*, **55**(6):112 (1948).
Lipták, Béla G.: Costs of Process Instruments, *Chem. Eng.*, **77:**60 (Sept. 7, 1970).
Lockhart, R. W., and R. C. Martinelli: Proposed Correlation of Data for Isothermal Two-Phase, Two-Component Flow in Pipes, *Chem. Eng. Prog.*, **45:**39 (1949).
Peters, Max S., and Klaus D. Timmerhaus: "Plant Design and Economics for Chemical Engineers," 3d ed., McGraw-Hill, New York, 1980.
Simpson, Larry L., Sizing Piping for Process Plants, *Chem. Eng.*, **75:**192 (June 17, 1968).
Viola, J. L., Jr.: Estimate Capital Costs via a New, Shortcut Method, *Chem. Eng.*, **88:**80 (Apr. 6, 1981).
Zevnik, F. C., and R. L. Buchanan: Generalized Correlation of Process Investment, *Chem. Eng. Prog.*, **59**(2):70 (1963).

PROBLEMS

5.1 You have been asked to prepare a capital estimate for a seawater desalination plant in Key West, Florida. The estimated total equipment purchased cost is $12,000,000 (M&S index = 720). Because of tight capital, two alternative approaches to instrumentation have been proposed:

Alternative 1:
Pneumatic instrumentation
Laboratory sample analysis
Computer control for key process variables only
Centralized control with limited graphics
Alternative 2:
Electronic instrumentation
Online automatic analyzers
All loops on computer control
Complete graphic panel display

What is the likely difference in the total plant direct fixed capital investment for the two alternatives? Use M&S = 720.

5.2 The Marshall and Swift Equipment Cost Index has been used implicitly in Chap. 5 to scale up the DFC of complete plants. What is the justification? In developing Eq. (5.1.6), Zevnik and Buchanan (1963) suggested the use of the *Engineering News Record* (ENR) Index as a basis. What is the ENR index, and can it be reasonably used for scaling up plant DFC costs? How do relative values of the ENR index compare with those of other indices listed in Table 4.3.1? (See Kohn, 1978.)

5.3 Use the results of Prob. 4.20 to estimate the DFC of the adipic acid purification plant in Fig. 3.5.8. Assume all auxiliary facilities, including cooling water and steam, are available. Use the method of purchased cost factors; assume that the plant will be enclosed in a concrete block building, and adjust your factors to account for warehousing space.

5.4 Dried alfalfa is processed for use in cattle feedlots by grinding, blending with nutrition supplements, and pelletizing in extrusion machines into $1/2 \times 1$ in pellets. The nutritional value of the alfalfa can be considerably enhanced by treating the ground material with caustic soda (NaOH); the treatment hydrolyzes some cellulosic fractions to more readily available sugars.

The treatment consists of adding aqueous NaOH through several recessed nozzles into a 20-ft-long screw conveyor carrying the ground alfalfa from the grinder to the pelletizer feed hopper. The recommended level of addition of NaOH is 4 lb anhydrous NaOH/100 lb feed material. Caustic soda is shipped as a 50% aqueous solution by weight; however, it must be diluted before addition to the alfalfa to bring the *total* moisture content of the treated product to 17% of the finished product weight. This final moisture content is required to impart proper cohesiveness and lubricity in the pelletizer. The feed material already has 10 wt % water. The diluted NaOH must be added at less than 100°F to avoid undesirable side reactions.

You have been asked to make a rapid preliminary estimate of the DFC required to incorporate a caustic treatment facility into an existing alfalfa pelletizing mill in a remote location in the Texas Panhandle. The facility can handle up to 10 tons dry alfalfa/h; however, the instantaneous rate fluctuates a great deal. Cooling water at 60 psi is available onsite; even in summer it does not exceed 80°F. A railroad siding exists that will accommodate 10,000-gal tank cars of 50% NaOH. Power supply is no problem.

Proceed as follows:

a Draw a *neat* flow sheet of a processing scheme of your choice; assign identification numbers to the equipment and identify equipment functions and process streams. You need not show instrumentation. The screw conveyor must be provided as a new piece of equipment, but it can be easily installed between the existing grinder and the pelletizer feeder (which need not be pictured on your flow sheet).

b Using the data in Figs. P5.1 to P5.3, calculate

Maximum daily requirement of 50% NaOH (gal).
Additional water requirement (gal/h) for dilution. *Note:* Cooling water can be used as process water.
Cooling requirement (Btu/h) for NaOH dilution.

c From information in the references in Exhibit 4.2A, size and obtain cost of screw conveyor. Alfalfa has about one-quarter the bulk density of grain.

d Size balance of equipment you think is needed. Use reported heat-transfer coefficients in Perry's handbook.

FIGURE P5.1
Specific heats of caustic soda solutions. *(The Dow Chemical Co. Caustic Soda Handbook, 1962. Reproduced by permission.)*

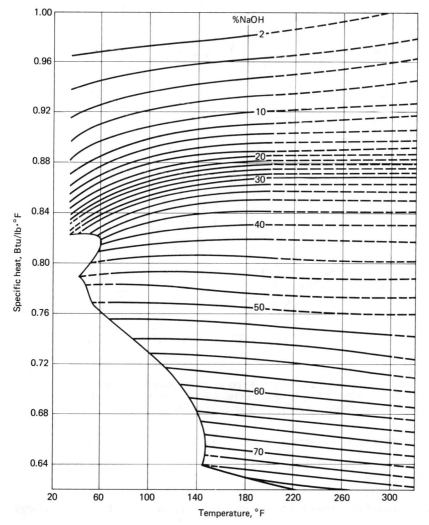

318 CHAPTER 5: THE DIRECT FIXED CAPITAL INVESTMENT

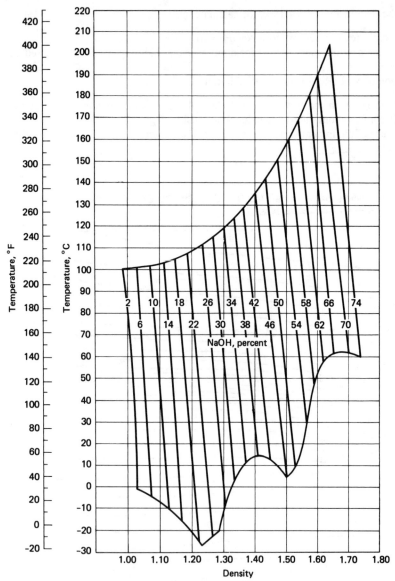

FIGURE P5.2
Density of caustic soda solutions at various temperatures (based on tables from International Critical Tables, vol. 3, p. 79 and extrapolation). *(The Dow Chemical Co. Caustic Soda Handbook, 1962. Reproduced by permission.)*

e Determine costs from sources available to you and the following data:

Heat exchangers (all steel, M&S = 500):
    10 ft² $2000
    50 ft² $5200
However, nickel tubes should be used in exchangers of your choice; the adjustment factor

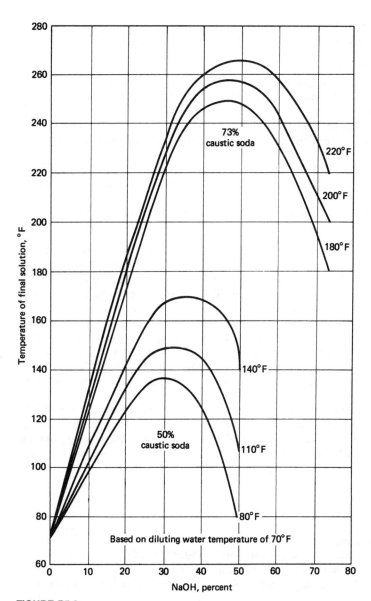

FIGURE P5.3
Dilution temperatures of caustic soda solutions. *(The Dow Chemical Co. Caustic Soda Handbook, 1962. Reproduced by permission.)*

for such tubes, in the size range given above, is 2.5 for the total exchanger purchased cost. Use Hastelloy C centrifugal pumps; open motors are OK.

f Adjust costs to M&S index of 650 and obtain rough DFC by using Lang factors.

5.5 In Prob. 4.17 the current total purchased equipment cost for a simple crystallization unit was obtained. Estimate the DFC of a production unit subject to the following conditions:

Grass roots plant in Minnesota, rural area.

Processing area will be constructed on an inclined site so that solids may be forwarded by gravity.

The process is well established; batch processing.

Since no explosive vapor hazard is involved and winters are extremely cold, processing area will be enclosed in substantial buildings. Nonprocess buildings will include a small office building, a warehouse, and operator facilities.

No central control will be used. Field instruments will be used for process control by operators, and analyses will be performed in modest laboratory facilities in the processing area.

There will be no steam; all heating will be with natural gas or electrical power. Clean subsurface water is available year-round. Aqueous wastes are not toxic and will be disposed of in deep wells.

Estimate factors on the basis of the discussion associated with Exhibit 5.1A.

5.6 A hot vapor mixture of anhydrous benzene sulfonic acid and SO_2 is passed through a series of condensers in an existing plant to condense out the sulfonic acid. Benzene sulfonic acid melts at 65°C, and to avoid freezing in the condensers, tempered water at 70°C is supplied to the condensers as cooling fluid. The use of water, however, has been occasionally disastrous; leaky condenser tubes have resulted in water penetration into the process stream, with locally catastrophic corrosion. It has been suggested that tempered perchloroethylene (C_2Cl_4) be used as a cooling fluid instead of tempered water. Any condenser tube leakage would result in process stream contamination with C_2Cl_4, but the chlorinated solvent is chemically inert and, in an emergency, separable by distillation. Since C_2Cl_4 is anhydrous and noncorrosive, leaks would not result in the disastrous corrosion currently experienced.

Make a preliminary estimate of the likely fixed capital investment required to provide a supply of warm perchloroethylene heat-transfer fluid to the condensers in the existing process plant. This new system should be looked at as an addition to the existing plant. In normal operation the C_2Cl_4 will be *heated up* in the existing exchangers and will therefore have to be cooled for recycle.

Specifications:

1 Required temperature level of C_2Cl_4 delivered to existing condensers: 70°C.

2 Maximum heat-transfer demand in condensers: 250,000 Btu/h.

3 Available cooling water temperature in plant: 80°F (summer).

4 Demand fluctuations may be expected, as well as prolonged shutdown periods and start-ups.

5 Mild steel is a satisfactory material of construction.

Layout:

1 A concrete slab is available for required equipment on the ground level adjacent to the plant.

2 Holdup in the shells of existing condensers is 220 gal of any cooling medium.

3 All condensers are located 35 ft above grade.

4 The supply and return pipes between the condensers and the new system will be 175 ft long each, standard steel pipe.
5 100 psig steam is available in the plant (337°F).

Physical properties of C_2Cl_4:

Specific gravity: 1.542
Viscosity: 0.47 cP
Specific heat: 0.215 Btu/(lb · °F)
Thermal conductivity: 0.061 Btu/(h · ft^2 · °F · ft)
Vapor pressure at 70°C: 140 mmHg
Atmospheric boiling point: 121°C

Other information:

1 All auxiliary facilities are available onsite.
2 Explosionproof motor drives are required.
3 Cooling water return to cooling towers must not exceed 95°F.
4 Recent vendor quotations on all mild-steel shell-and-tube exchangers, fixed tube sheets, are given in Prob. 5.4.

Required:

1 Flow sheet of new installation, excluding instruments and controls. (Existing condensers already are provided with temperature-regulated control valves controlling flow of cooling medium.)
2 Equipment list summing up size and purchased cost (M&S = 650).
3 Estimate of direct fixed capital investment.

5.7 A 396 gpm waste process water stream saturated at 80°F with perchloroethylene (C_2Cl_4, 150 ppm by weight) is presently forwarded to aeration ponds. This procedure represents an air pollution problem as well as a considerable loss of valuable solvent, and you have been asked to make a rapid preliminary estimate of the investment (DFC) in a proposed facility that would air-strip the water down to a 1 ppm residual and recover the C_2Cl_4 from the stripping air by charcoal adsorption.

The proposed facility would incorporate a sieve-tray steel stripping tower with stainless steel trays. Preliminary calculations indicate that six actual trays would result in a C_2Cl_4 residue well below the required 1 ppm. Sufficient air is supplied with a blower to give 2000 ppm (by weight) of C_2Cl_4 in the air effluent.

The tower air effluent is compressed in a booster blower and fed to one of two charcoal adsorption beds where the C_2Cl_4 is quantitatively removed from the vented air. The adsorbers are switched every 24 h; while one operates on the adsorption cycle, the other is eluted with atmospheric steam. The adsorbed C_2Cl_4 is eluted quantitatively, and the C_2Cl_4–steam vapor mixture is condensed. The immiscible liquid phases are decanted; the light water phase is returned by gravity back to the top of the stripping tower, and the C_2Cl_4 phase is returned by gravity to process.

The waste must be pumped to the top of the stripper, but the bottom liquid effluent flows to ponds by gravity.

a Make a *neat* flow sheet of the process and label all equipment and discharge streams. Do not bother showing all piping detail around the twin adsorbers—show one on adsorption cycle, the other on elution cycle. You need not show instrumentation.

b Compute a reasonable stripping tower size and estimate the fabricated cost from references in Exhibit 4.2A. Assume tower will be fabricated from ½-in steel.

c With 2000 ppm of C_2Cl_4 in the feed air, the charcoal beds will adsorb 0.4 lb C_2Cl_4/lb charcoal (density 30 lb/ft^3, cost $1/lb). The superficial gas velocity during adsorption is 2 ft/s, and the 24-h adsorption cycle ends just before breakthrough. Compute reasonable vessel dimensions for the adsorbers, and the charcoal bed depth.

d Compute the purchased cost of the balance of the required equipment, all-steel construction:

Condenser: 200 ft^2, fixed tube sheet.
Steel vessels: Add 50 percent to cost of adsorber vessels for bed supports, cooling coils. Decanter is 1 ft dia. × 2 ft.
Blowers (3 in of water pressure differential): Supply 2½-hp open motors.
Centrifugal pumps: Use 316 stainless fittings.
Pumps should be spared, but not blowers.

e Estimate the direct fixed capital, adjusted to an M&S index of 750.

5.8 Using the method of purchased cost factors, derive values for the Lang factors for the following cases:

a Fertilizer blending plant, standard size, all-solids handling. Automated proportioning and weighing of bagged product. Product is not a standard one, but solids-blending procedure is well-established. Process completely enclosed in process building, with large warehouses and railroad loading facilities required. Feed materials are partly produced in integrated complex, partly brought in by hopper cars and stored in bins.

b Simple crystallization unit in Prob. 5.5.

c Experimental semiplant, small scale [about $1 million (1981) of purchased equipment], continuous all-liquid and vapor process. Expensive alloy construction. Centralized process control with all loops on computer control, on-stream analyzers, graphic display. Class I, Division 1 electrical classification (explosionproof motors and instruments). Open structure in integrated complex with complete facilities, including offices and change rooms, already existing; however, small control room is required. Process is reasonably firm, but some changes may be anticipated.

d Giant commercial plant involving a complex, currently speculative process to produce a commodity chemical in high demand. Except for a few filtration steps, process involves fluids only. Significant parts of process involve very low temperature, or else high-melting liquids requiring pipe tracing. Modern, up-to-date process control in centralized control rooms with computers and advanced graphic display. Open structure; electrical classification accounts for possible release of flammable liquids and vapors. Giant size will require expansion of existing utilities, construction of waste disposal facilities.

5.9 Recently built phenol plants (cumene process) have shown a fixed capital–annual production ratio of $550/annual short ton. What should be the selling price to reflect the normal turnover ratio for the chemical industry? Assuming that the investment per unit capacity is a 1979 figure, what is the anticipated current selling price, and how does it compare with the quoted price in the *Chemical Marketing Reporter*?

5.10 Search cost data sources available to you to determine the current DFC investment for the following plant auxiliaries:

a 50 ton/day oxygen plant (>99 percent, high pressure).
b 100 ton/day turnkey methanol plant.
c 0.1 m^3/s inert gas generator (low-pressure nitrogen at ordinary temperatures).

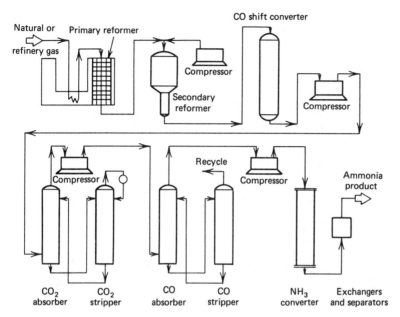

FIGURE P5.4
Kellogg ammonia process flow sheet. *(Zevnik and Buchanan, 1963; reproduced by permission, Chemical Engineering Progress.)*

5.11 Make your own estimate of the number of "functional units" associated with the Kellogg ammonia process (Fig. P5.4) and the Wulff acetylene process (Fig. P5.5). Zevnik and Buchanan (1963) have estimated 11 and 12 units, respectively; refer to the original article for their justification.

5.12 Use the functional step scoring procedure to estimate the DFC of a 30 million lb/year monoethylamine plant described in Example 3.5.1 and Fig. 3.5.3. Assume that type 316 stainless steel is the predominant alloy.

5.13 One difficulty with the functional step scoring procedure of Zevnik and Buchanan (1963) is that for rather simple processes an "error" in estimating functional steps, which is equivalent to just one step, may be an appreciable percentage of the total. This difficulty is illustrated by the adipic acid purification process shown in Figs. 3.5.7 and 3.5.8. How many functional steps do you estimate? Using this result, estimate the DFC of a facility to process 1 short ton/h of adipic acid. For additional data, see Example 3.5.3 and the problem statements, Probs. 4.20 and 5.3. How does your result compare with the result of the more-detailed analysis in Prob. 5.3?

5.14 Allen and Page (1975) have devised what they claim is a more reliable DFC projection method than the functional unit scoring method. Their "scoring" is based upon the concept of "main plant items" and requires a flow sheet, although equipment shown does not have to be sized. Review the discussion in the original publication and estimate the DFC of a 1 ton/h adipic acid purification plant. Use Fig. 3.5.8 as the basic flow sheet; other data sources are summed up in the statement of Prob. 5.13. How does the result of the Allen-Page method compare with the other adipic acid plant estimates in Probs. 5.3 and 5.13?

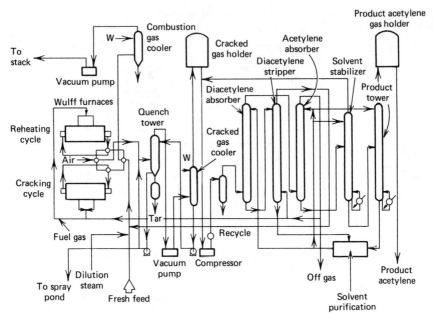

FIGURE P5.5
Wulff acetylene process flow sheet. *(Zevnik and Buchanan, 1963; reproduced by permission, Chemical Engineering Progress.)*

5.15 Make a rough estimate of the installed capital invested in the laboratory room (or office) in which you have your desk. Estimate separately "nonstandard" equipment (computers, spectroscopes, microtomes, etc.).

5.16 A process for producing pure adipic acid crystals is illustrated in the flow sheet, Fig. 3.5.8. The equipment was sized as part of the problem requirement, Prob. 4.20. Provided that you have previously sized the equipment, make a preliminary equipment *layout* of the 1 ton/h plant. The process is to be housed in a single- (ground) level concrete block building. The walls of the building may be higher than the standard 14 ft; however, unusually high or bulky equipment should be kept out of the building confines in an open structure. The layout may be facilitated by using paper cutouts representing equipment plan views drawn to scale, and moving the cutouts on a sheet of paper until a satisfactory arrangement is obtained. In deciding upon optimum arrangement, be sure to take into account the need for short pipeline runs to connect sequential equipment, convenient operator access space, passageways for personnel, and access space for heavy equipment during construction or for subsequent maintenance (although permanently installed moving hoists may often be used for the same purpose). Remember that equipment such as the Swenson-Walker crystallizers may be assembled in place.

Estimate the required area of the building and the open structure, and estimate the current cost from Fig. 5.1.4. For building walls other than 14 ft, adjust the cost by using the suggested scaling factor

$$\left(\frac{H}{14}\right)^{0.53}$$

Also estimate the cost of a grade-level warehouse to house up to 1 month's production in 50-lb bags. Use the criteria outlined in Exercise 5.2.1 to decide upon the size needed.

Assume that the process buildings and the warehouse are the only buildings required for the new purification plant annex. What is the total investment in buildings?

The data in Fig. 5.1.4 represent *all* direct and indirect costs, including contractor's fees. In the method of purchased cost factors, the "buildings" factor represents the direct costs only. Estimate the direct cost alone of the buildings, using reasonable purchased cost factors. Compare this cost with the total equipment purchased cost (Prob. 4.20); what is the purchased cost factor for the buildings?

5.17 For the reactor ensemble considered in Example 5.2.1, what percentage of the computed daily nitrogen demand could be saved by modifying the purge procedure as follows:

Evacuate each reactor to 0.1 atm absolute pressure.
Repressure up to atmospheric pressure using the nitrogen.
Repeat until oxygen level is below 100 ppm.

5.18 A projected 500 million lb/year butylene oxide plant will incorporate a new process involving direct oxidation of butylene in an inert liquid solvent. Either air or oxygen may be used as the oxidant. Oxygen has two distinct advantages: the reaction may be run at low pressure (5 psig), and the product recovery train is much smaller in size because large volumes of nitrogen introduced with the air need not be processed. It is estimated that about one-half the cost-identified equipment in a plant using air would be reduced in size by a factor of 4 if 95% oxygen were used instead.

The estimated DFC for an air-based plant is $150/annual short ton of butylene oxide (M&S = 700). Will the DFC saving realized by using oxygen more than balance out the required oxygen plant DFC? Assume 80% of the O_2 supplied winds up in the butylene oxide molecule.

5.19 A new catalytic reforming unit will use cooling water for a cooling and condensing demand of 75 million Btu/h. Estimate the DFC investment required for a dedicated cooling tower installation. Use the method of Guthrie (1969); see Fig. P5.6.

5.20 Municipal solid waste is a potentially valuable source of recycled raw materials (metals, glass, paper fibers). For such waste to be fed into a separation process, it must first be reduced to a manageable consistency by shredding. The cost of shredders is a very appreciable and often overlooked component in published projections of the economics of solid waste recycling.

Make a reasoned estimate of the generation rate of municipal solid wastes in a small city of 100,000 inhabitants. Use the article by Fang (1980) to estimate the current DFC of a shredding installation required to prepare the generated waste for materials separation.

5.21 Make a sketch (in the style of Fig. 5.3.4) of the schematic piping arrangement that you would recommend for steam-tracing lines. Show an initial connection to a steam header and a terminal connection to a condensate header (condensate is returned to the steam plant as boiler feed). The steam tracer line itself may be shown as an interrupted line, without the detail of tubing connections shown in Fig. 5.3.6.

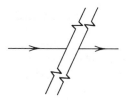

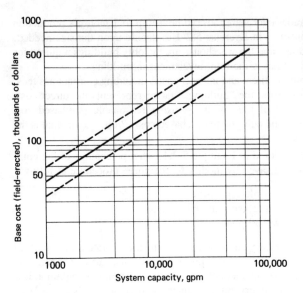

FIGURE P5.6
Cost of cooling towers. *(Guthrie, 1969; reproduced by permission, Chemical Engineering magazine.)*

Required:
Capacity, gpm
Temp. difference, °F

Basis of chart: Cooling range, 15° F

Time base: mid-1968

Size exponent: 0.60

Included:
Cooling tower
Concrete basin
Pumps and drives
Field erection
Subcontractor indirects

Cooling tower cost, $ = base cost × F_c

Adjustment factors: M&S Index = 273

Cooling Range, °F	F_c
15	1.00
20	1.55
25	1.95

Installation (Prime contractor work only):

Field installation (M & L)	1.16
Bare module factor	1.75
L/M ratio	0.85

5.22 A vertical tank, 6 ft inner diameter and 10 ft high, will be used as an emulsion settler in a mixer-settler extraction unit. The 10-ft dimension is between "tangent lines," i.e., the top and bottom termini of the cylindrical portion of the tank, at the point where the curvature of the heads commences.

Make a schematic sketch (elevation view) of a piping system that will automatically maintain the interface between the settled phases at the 4-ft level above the bottom tangent line, and the vapor-liquid interface at the 8-ft level. Both liquid phases are to overflow by gravity to tanks located well below the settler. The settler vapor space will be vented to the atmosphere. The two separated liquid phases have a specific gravity of 1.00 and 1.50, respectively. The emulsion will be fed as a continuous stream through a side nozzle on the tank located at the 4-ft level.

Show critical dimensions on your sketch.

5.23 A pipeline is to be designed to transport a continuous stream of 100 gpm of o-bromotoluene at 20°C a total linear distance of 300 ft. Use data and charts in Perry's handbook to pick the optimum pipe diameter (specify a standard steel pipe size).

Use the "pipe flow chart" in Perry to determine the likely frictional pressure drop in the pipe size you have chosen, accounting for normal fittings but neglecting the pressure drop across control valves. Check the accuracy of the chart by repeating the pressure drop calculation, using Fanning friction factor correlations.

5.24 A continuous nitration reaction takes place in a 4-ft-diameter kettle; the total holdup is 240 gal. There is some possibility that the reaction might "run away," and you are asked to size a quick-dumping pipeline that will allow the contents to be emptied in roughly 40 s. A temperature-controlled quick-opening valve should be located at the bottom of the reactor and is to be the same nominal size as the pipeline.

The bottom of the reactor is 15 ft above the level of a water-filled sump into which the reactor contents are to be dumped. Data include:

Viscosity of reactor contents: 500 cP
Density of reactor contents: 106 lb/ft^3
Maximum temperature of dumped material: 230°F
Reactor is vented and operates at essentially atmospheric pressure

5.25 The problem of the design of pipelines to handle the flow of two-phase fluids (liquid and vapor) was addressed in the classical publication of Lockhart and Martinelli (1949). Using this method (which has been reproduced in other publications such as Perry's handbook), generate a graph of the pressure drop (psi/1000 ft of pipe) versus the quality (that is, wt % vapor) for a stream of chlorine, 40 short tons/day, flowing through a standard Schedule 40 2-in commercial steel pipe at 60°F. The pressure drop will be much higher for the 40 ton/day stream if it is all vapor than if it is all liquid. You should find, however, that the pressure drop is the highest for some intermediate quality (i.e., when liquid and vapor coexist). For liquid chlorine at 60°F, use a specific gravity of 1.42 and a viscosity of 0.35 cP; other required data may be found in handbooks. Calculate vapor density at the vapor pressure of chlorine at 60°F; that is, even though a pressure *drop* is calculated, consider the flow to be isobaric.

Repeat the calculations and draw a graph for the following flows as well:

100 tons/day of chlorine in a 2-in pipe
100 tons/day of chlorine in a Sch. 40 3-in pipe
100 tons/day of benzene in a 3-in pipe

On the basis of your results, can you derive a rough rule of thumb that relates the maximum pressure drop to the pressure drop with all-vapor flow? How can such a rule of thumb be used to size pipes for the "worst pressure drop situation possible"?

The solution to this problem requires a great deal of "number crunching," and the use of a programmable calculator is recommended.

5.26 If a hot solution is pumped through a pipeline into a receiver vessel maintained below the saturation pressure, the solution may begin to flash well before reaching the vessel. Such flashing may result in supersaturation of the residual liquid and separation of solid crystals, and in many instances the solids deposit on the pipe walls and plug up the fluid passage, particularly if the flashing occurs in a restriction such as a flow control valve.

The solution to this very annoying problem has been examined by Gutierrez and Lynn (1969). The authors use theoretical arguments and experimental verification to show that if a solution flows through a pipeline at a velocity above a "minimum critical velocity," it will not flash even though the liquid may become significantly superheated by virtue of the pressure drop through the pipe. This apparently anomalous behavior occurs because above the critical velocity the energy available from the fluid expansion is insufficient to provide the required energy increase during flashing.

An approximate expression for the minimum critical velocity is

$$U_{min} = V_1 \sqrt{\frac{g_c \Delta H_v (dP/dT)_{sat}}{C_P V_g}} \tag{P5.1}$$

where U_{min} = critical velocity, ft/s
V_1 = specific volume of saturated liquid, ft^3/lb$_m$
V_g = specific volume of saturated vapor, ft^3/lb$_m$
g_c = gravitational constant, 32.2 ft·lb$_m$/(s^2·lb$_f$)
ΔH_v = specific enthalpy of vaporization, Btu/lb$_m$
C_P = heat capacity of liquid, Btu/(lb$_m$·°F)
$(dP/dT)_{sat}$ = slope of the bubble point pressure curve, lb$_f$/(ft^2·°F)

Suppose a stream of 150 gpm of an aqueous solution of 1-methionine is heated (under pressure) to 140°C and is then pumped into an evaporator vessel maintained at atmospheric pressure. What size of standard pipe would you recommend to avoid flashing and crystal plugging in the pipeline? For present purposes, use data for pure water found in handbooks.

5.27 The capacity of control valves is represented by the *capacity factor* (or *flow coefficient*) C_v, which is defined as the flow of water at 60°F, in gallons per minute, at a pressure drop of 1 lb/in^2 across the valve, at the valve's "normal" position. In Example 5.3.1 we saw that at the normal flow of 250 gpm of hot water, 30 ft of head is available for pressure drop through the control valve. What is the C_v required? Neglect the effect of temperature upon liquid water properties.

Of course, it is unlikely that a valve can be purchased with exactly the required C_v, although some manufacturers provide a choice of "trim" so that more than one value of C_v may be made available in a specific valve size (which is characterized by the pipe size of its connections). Suppose a 2-in control valve with $C_v = 38$ were to be used for the hot water application. What effect would the choice have upon the normal valve position?

5.28 Saturated pure process water at 55 psia and 287.1°F is stored in a horizontal cylindrical surge tank, 6 ft diameter and 20 ft long. The process calls for a stream of 15 gpm to be pumped from the tank with a spared centrifugal pump; the actual flow, however, is to be controlled by means of a motor valve on the pump discharge.

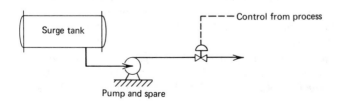

The pumps are to be located on a ground slab. Their maximum capacity is 30 gpm; they have a $1\frac{1}{4}$-in NPS suction and a 1-in discharge. The net positive suction head (NPSH) requirement may be approximated by

$$\text{NPSH (ft of fluid)} = 3.0 + 0.1 \text{ gpm}$$

where gpm is the pump throughput in gallons per minute.

State your recommendations on location and elevation of the tank as well as the piping scheme between the tank and pump.

Take the viscosity of the water as 0.2 cP.

5.29 In the sketch, 1000 gpm of water is fed to tank 1. An overflow pipeline maintains a constant level in the tank. The overflow passes into storage tank 2. Both tanks are open to the atmosphere. The downcomer AB discharges above the liquid level in tank 2.

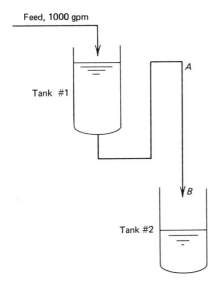

For proper tank 1 level control, the downcomer must remain air-filled, with liquid flowing downward along the walls. If the downcomer were to be filled entirely with liquid, a siphon would be established, and the contents of tank 1 would be rapidly dumped. In a review article, Simpson (1968) points out that a siphon will not form provided that the Froude number Fr in the downcomer is less than about 0.3. The expression for Fr is

$$\text{Fr} = \frac{U_L}{\sqrt{gD}} \sqrt{\frac{\rho_L}{\rho_L - \rho_G}} \tag{P5.2}$$

where U_L = liquid superficial velocity based upon the total flow cross section of the downcomer, ft/s
g = gravitational acceleration, 32.2 ft/s^2
D = inside diameter of downcomer, ft
ρ_L = liquid density, lb$_m$/ft^3
ρ_G = density of gas in downcomer, lb$_m$/ft^3

a What size of standard pipe do you recommend for the downcomer?

b If a siphon is established, what can you say about the internal pressure at A? What piping modification would prevent the formation of a siphon?

5.30 Dangerous overpressures in process vessels are avoided by providing the vessels with pressure relief devices such as pressure relief valves (PRVs). The design and specification of such devices is an important part of the preconstruction design package for full-scale plants and is often carried out by specialists. The design of individual PRVs is a chore that chemical engineers are often faced with during the preparation of pilot plant operations.

The design of PRVs is the subject of an important American Petroleum Institute publication (API, 1976). The procedure starts with evaluating a number of criteria to establish:

Can a situation arise whereby the internal pressure could exceed the maximum allowable working pressure (MAWP) of the vessel? PRVs must be preset to relieve at a pressure no higher than the MAWP.

If such a situation can arise, the PRV must then be sized to relieve vapors at a flow rate at least equal to the vapor generation rate in the vessel at the preset pressure conditions.

A frequent criterion which results in the worst internal pressure hazard is the possibility of an external fire in the structure housing the vessel. The recommended expression for the heat transfer Q (Btu/h) due to an external fire is

$$Q = 21{,}000 F A^{0.82} \qquad \text{(P5.3)}$$

where A is the wetted area (ft^2), that is, the total vessel surface area wetted by the internally stored liquid, and F is a factor which depends upon the vessel insulation; $F = 1$ for bare vessels.

The vapor generation rate can then be calculated by using the heat of vaporization of the contents.

The PRV itself is sized on the basis of the *orifice area* of the open valve. In the usual case, the vapor flow through the orifice is critical; that is, provided the valve relieves to a low enough pressure, the maximum flow through the orifice is that which occurs at the *critical pressure ratio* r_c. The expression for this ratio (see, for example, Perry's handbook) is

$$r_c = \left(\frac{2}{\gamma + 1}\right)^{\gamma/(\gamma-1)} \qquad \text{(P5.4)}$$

where γ is the ratio of specific heats for the vapor, C_P/C_V.

If this expression is substituted into the equation for flow through a nozzle, the standard design equation follows:

$$W = CKA_0 P \sqrt{\frac{M}{ZT}} \qquad \text{(P5.5)}$$

where W = maximum vapor generation rate, lb/h
$\quad K$ = coefficient of discharge (~ 0.98 for most commercial PRVs)
$\quad A_0$ = valve orifice discharge area, in^2 (note inconsistent units)
$\quad P$ = vessel relieving pressure, psia*

*A 10 percent "overpressure" may be allowed.

M = molecular weight of vapor
Z = compressibility factor at vessel conditions
T = absolute temperature in vessel, °R

$$C = 520 \sqrt{\gamma \left(\frac{2}{\gamma + 1}\right)^{(\gamma+1)/(\gamma-1)}} \tag{P5.6}$$

Use the equations given to design a PRV (i.e., orifice area) to protect a bare steel storage tank, 6 ft dia. × 25 ft, which will contain no more than 4500 gal of hexane. The vessel is rated for a maximum allowable working pressure of 150 psig, but the PRV will be set at 100 psig.
a What is the critical pressure ratio for hexane?
b What is your recommended orifice area?
c What do you think should be done with the PRV discharge?

5.31 Perry's handbook lists the following equation for the required pipeline velocity of a slurry to keep the slurry particles in suspension:

$$\frac{U^2}{gD_s} \frac{\rho_L}{\rho_s - \rho_L} = 0.0251 \left(\frac{UD_P \rho_m}{\mu_L}\right)^{0.775} \tag{P5.7}$$

where U = pipeline velocity, ft/s
g = gravitational acceleration, 32.2 ft/s²
D_s = particle diameter, ft, such that 85 wt % of particles are smaller than D_s
ρ_s = particle density, lb/ft³
ρ_L = liquid density, lb/ft³
D_P = pipe diameter, ft
ρ_m = slurry density, lb/ft³
μ_L = liquid viscosity, lb/(ft · s)

1000 short tons/day of anthracite coal (ρ_s = 97 lb/ft³) are to be transported in a cross-country pipeline as a 10 wt % aqueous slurry. The coal particles are −3 mesh, and D_s is about ¼ in (0.02083 ft). For normal flow conditions at 60°F, what nominal pipe diameter would you recommend?

5.32 Preliminary negotiations are under way between your company and a neighboring paper mill to supply them with available excess steam from your generator-boiler plant installation.

10,000 lb/h of steam is available; the pressure at the generating station is 165 psig (saturated steam). The customer wishes to obtain the steam at no less than 140 psig and dry.

A pipeline must be built between the boiler plant and the paper mill. It is estimated that the pipeline would be 3000 ft long. Make a rough estimate of the total cost of the installation, assuming there will be no expenses for rights of way. Use Bosworth's (1968) correlation reproduced in Fig. P5.7; add 15 percent for piping supports. Use M&S index = 750.

5.33 A new pipe connection is to be constructed in an existing plant; the layout is illustrated in Fig. P5.8. The connection will be used to transfer material from a tank with nozzle A to a tank with nozzle B, using nitrogen pressure above liquid in A. The line is to be 4-in Schedule 40 type 316 stainless steel pipe with welded construction. Because of the high melting point of the material transferred, the line will be traced and insulated. Two globe

332 CHAPTER 5: THE DIRECT FIXED CAPITAL INVESTMENT

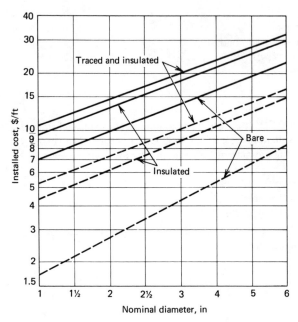

FIGURE P5.7
Cost of offsite pipelines. Longer runs of aboveground, external piping: dashed line shows installed cost in steel (standard weight); solid line in type 304L stainless steel (Schedule 10S). *(Bosworth, 1968; reproduced by permission, Chemical Engineering magazine.)*

M & S Index: 263

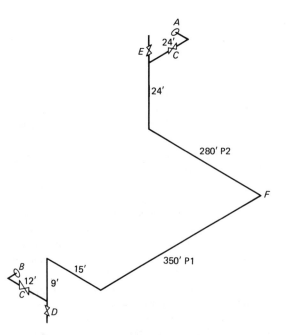

FIGURE P5.8
Isometric piping layout, Prob. 5.33.

valves C will be used;[1] the drain at D and vent at E (for draining line) are 2-in gate valves. A and B are 3-in flanged nozzles. Assume pipe is available in 20-ft lengths.

The two long pipeline runs will be accommodated by existing pipeways $P1$ and $P2$; other required pipe supports can be readily welded to the existing structures housing the tanks. At the intersection F, it is estimated that five welded elbows will be required to construct pipeline "kinks" to overcome the crossover spatial problems.

Use the article by Guthrie (Ref. 1 in Exhibit 5.3A) to estimate the total cost (including indirect costs) to build the system. Adjust costs in the article (M&S = 273) to current index value. Use Guthrie's recommended takeoff allowance, and provide for final pressure testing of the line for leaks.

5.34 Suppose the 4-in pipeline in Fig. P5.8 and Prob. 5.33 were to be constructed from flanged Saran-lined pipe (similar to pipe in Fig. 5.3.2). Estimate the man-hours of labor required to fabricate and erect such a pipeline, using the method of Clark (1957).

5.35 One thousand gallons of glycerol (sp gr = 1.26, μ = 300 cP) is to be transferred to an overhead atmospheric tank from a ground-level Pfaudler vessel by air-pressuring the Pfaudler to 100 psig. The transfer line is a 1-in NPS pipe (1.049-in ID) which enters the bottom of the overhead tank; the transfer line length, including equivalent length of fittings plus entrance losses, is 75 ft, and the average static head during transfer is 40 ft. How long will it take to transfer the glycerol?

---

[1] Use flanged valves (150-lb raised face) for easy replacement.

CHAPTER 6

DEPRECIATION

Our path toward the ultimate goal of defining and evaluating the criterion of economic performance of a project has taken us through various aspects of investment analysis. We now pause in this endeavor—for reasons which will become clear later—and turn our attention to the analysis of the operating costs associated with a project. Methods of estimating various categories of expenditures will be explored; these will then be compared with projected revenues to arrive at values of the net income. We will first focus upon defining and evaluating the expenses which constitute the *cost of manufacture* (COM) of a product—expenses directly associated with the manufacturing process and centered around the immediate production facility. The temporal aspect of the constituents of COM will be repeatedly emphasized. Such expenses occur continuously over the lifetime of the project; they may not be uniform, but they are time-dependent. In contrast, the investment is not a time-dependent entity; it occurs during one instant in time, or nearly so (provided the investment does not assume the form of an annuity).

In this chapter we start our examination of COM by focusing upon just one of its constituents, depreciation. Why should a whole chapter be devoted to a single aspect of COM? For one thing, the concept of depreciation has a number of nuances which constitute a source of puzzlement and confusion even to some experienced economic evaluators. One reason for such puzzlement is that, as we will see, all but one of the constituents of COM are easily understood "out-of-pocket" expenses representing the flow of "hard cash." The exception is depreciation, which is not a cash flow item; one does not reach into one's pocket to pay for depreciation, and yet it is considered to be an expense no different than, say, raw material costs.

Furthermore, depreciation serves to define certain concepts which turn out to be key ones in the computation of criteria of economic performance. Most of us have a certain amount of intuitive understanding of the concept of "profit," but we will find that profit has an exact definition in which depreciation plays a key role. Depreciation also helps to define the difference between the concepts of profit and cash flow. Most importantly, depreciation contributes to the establishment of the amount of income tax that must be paid. A discussion of depreciation therefore constitutes an introduction to the methodology of economic performance evaluation.

We will concentrate upon these important aspects of depreciation and perhaps devote less emphasis to the methods whereby depreciation charges are calculated.

6.1 THE ECONOMIC IMPACT OF DEPRECIATION

The Depreciation Concept

What is the meaning of the term "depreciation"? Strictly speaking, plant equipment ages; it wears out with use; some of it obsolesces; and as a result, it *decreases in "value"*—it *depreciates*. If a process plant is to be operated beyond the lifetime of a particular equipment item, that item must be replaced. On the other hand, if an operating plant is to be abandoned, the equipment in it, even that which happens to be in good shape, can be sold for very little, and it usually cannot be used as part of some new project. In either case, the initial plant DFC (or other fixed capital) is eventually lost; the initial investment amount cannot be recovered, and it thus becomes a very real charge against the process. In effect, this DFC loss is spread over the lifetime of the initially constituted plant, and it consequently may be considered a *time-dependent operating cost*.

The operating cost characteristic of depreciation is well-illustrated by an investment choice situation suggested by Woods (1975) and presented here (with permission) in somewhat modified form. Let us suppose that we have $10,000 to invest and that two investment opportunities are available to us:

Investment *A:* Taxicab business
Investment *B:* Banking investment

As part of alternative *A*, we invest our money in a new $10,000 cab, and we hire a cab driver to operate it around Manhattan's financial district. We anticipate the following schedule of revenues and operating costs (all costs in 1981 dollars):

Annual mileage: 20,000 miles
Revenues: Average of $1.50/mile
Operating costs:
  Oil and gas: 15 cents/mile
  License fees: $1500/year
  Maintenance: $1500/year
  Driver: $20,000/year

We calculate our annual cash "profit" as follows:

$$\text{Gross income} = 20{,}000 \times \$1.50 = \$30{,}000$$

Expenses:

$$\text{Oil and gas} = 20{,}000 \times 0.15 = \$\ 3{,}000$$
$$\text{License fees, maintenance} = 3{,}000$$
$$\text{Driver} = \underline{20{,}000}$$
$$\text{Total} = \$26{,}000$$

Net income = 30,000 − 26,000 = $4000/annum   (before taxes)

For the banking investment alternative B, let us deposit our $10,000 as a perpetuity (see Chap. 2) drawing 15 percent annually compounded interest.

From Eq. (2.3.11), annual net income (before taxes) is $1500.

Now, what do we have at the end of, say, 2 years? Well, investment A leaves $8000 in our pocket before taxes, investment B only $3000, and we might well be inclined to select investment A as the better one. Suppose, however, that we decide to terminate our project at the end of the 2 years. With alternative B, we simply go to the bank and withdraw our original principal of $10,000. However, had we chosen alternative A, we would have a rude awakening. We would quickly find out that 2 years of driving through the streets of New York rendered our cab something less than a collector's item, and we would be lucky indeed if we could sell our cab for a net of, say, $2500. Our relative *cash flow* situation would be the following:

Investment A		Investment B
$−10,000	Investment	$−10,000
+ 8,000	Revenues	+ 3,000
+ 2,500	Termination	+10,000
+ $ 500	Net cash flow	+$3,000

We see that investment A does not look so great after all. Why—what happened? It turns out that we made a serious error in calculating the "profit" from investment A. In cataloging the various expenses, we failed to include the *cost of depreciation*. This turns out to be a whopping $7500 for the 2-year period of our investment. Our cab does indeed lose a great deal of its value in 2 years—it depreciates. In fact, it probably depreciates to the point where it could not be driven as a cab much beyond the 2-year period; paying customers prefer not to ride around in junk. (New Yorkers will possibly find this conclusion amusing.)

The question is, "How is this depreciation cost to be charged against a process?" We seek a method, a "formula," to spread the initial cost of the fixed capital over the lifetime of the equipment to correspond to the equipment's loss of value, its depreciation. With such a formula, depreciation charges would be calculated as an *annual cost* charged against the process over its lifetime. Note that in this way the initial investment would be *recovered* by the time the process equipment became completely depreciated. That is, the sum total of all the depreciation "payments" would just equal the depreciable portion of the fixed capital.

It should be emphasized that depreciation charges cover the totality of the *original* (uninflated) fixed capital investment—not only the purchased cost of equipment and auxiliaries, but *all* the moneys invested in bringing a project to reality, right down to the contractor's fee. However, there are categories of invested capital that cannot be depreciated; these categories constitute *recoverable capital* that does not physically depreciate during the lifetime of the project. These categories include the following:

The *salvage value* (or scrap value) of equipment at the end of the project lifetime. The projected salvage value is the only portion of fixed capital that cannot be depreciated.

Land value. Land is not considered depreciable; it can be used indefinitely for succeeding projects on a specific site, or else it can be sold.

Working capital. The meaning of working capital was briefly introduced in Chap. 5. It is capital invested in various necessary inventoried items and which is presumably recoverable.

Two synonyms occasionally used for the term "depreciation" are *write-off* and *amortization*, although the latter is usually reserved for depreciation schedules which do not correspond to the equipment's useful life. Such accelerated schedules are occasionally promoted by the government as an incentive to invest in facilities such as pollution control equipment; why this is, in fact, an incentive will become clearer presently. Another associated concept is that of the *depletion* allowance, a depreciationlike charge to account for the exhaustion of natural resources (minerals, oil, timber). Depletion is not often involved in the evaluation of projects in the chemical process industry, and the complexities of establishing the depletion allowance will therefore not be further considered.

Depreciation and Its Impact upon Profit

Expenses directly associated with the manufacturing process of a specific product and centered around the immediate production facility are collectively termed the *cost of manufacture* (COM). It is a characteristic of these expenses that they are time-dependent—they occur more or less uniformly throughout the time continuum. Examples of components of COM are

Raw materials
Operating labor
Steam

Raw materials must be purchased at reasonably frequent intervals to replenish inventories which are so important to smooth and continuous production. The operating staff must obviously be paid regularly. Steam and other utilities in an integrated complex must be paid for by each user plant to the originating facility on a regular, usually monthly, basis; such "internal" payments, however, are normally bookkeeping transactions.

In contrast, the initial investment is effectively a *lump sum, one-time* expenditure. We use the concept of depreciation to convert this lump sum into a *time-dependent cost*. We will consider a number of methods of doing this, but regardless of the method, the depreciation charge thus arrived at is then considered as much an operating cost as raw materials or labor or steam. Depreciation, then, is just one component part of the COM. Furthermore, since COM constitutes expenses centered around the immediate production facility, it stands to reason that the depreciation charges in the COM are based upon the DFC only. This is not to say that other fixed capital (the project-oriented allocated fixed capital) does not depreciate; the way in which allocated capital depreciation is handled will be clarified in Chaps. 7 and 8.

Just as DFC is not the only capital associated with a project, so too is COM not the sum total of the expenditures needed to market the finished product. There are additional expenses, not centered around the immediate production facility, which are lumped together as *general expenses*. An obvious example of a general expense is the cost of selling a product (salespersons' salaries, sales offices overhead, etc.). Again, these general expenses are time-dependent. If the manufacturing (COM) and general expenses are subtracted from the product sales revenues (also time-dependent), the balance obtained is the *profit*, and the term "profit" is so defined:

Sales − COM − general expenses = gross profit   (profit before taxes)

where COM = labor, steam, . . . , depreciation.

Income taxes have not been subtracted in the scheme illustrated, and the resultant profit is therefore the *gross* (pretax) *profit*. Subtraction of income taxes yields the *net* (after-tax) *profit*. Since all the elements that go into the computation of profit are time-dependent, profit is also time-dependent (e.g., dollars per year). Note in particular that *depreciation defines the gross profit;* the import of this statement will be further explored presently.

The usual point of confusion in this definition of profit is that we pay, on a time basis, *out-of-pocket dollars* for raw materials, for labor; we even feel comfortable about looking at utility charges as out-of-pocket expenses, even though they may involve nothing more than the juggling of numbers in account books; but we feel distinctly uncomfortable about looking at depreciation as a cost, because we do not reach into our pocket to pay for it. Perhaps some degree of comfort may be imparted by the pretension that depreciation is, indeed, an out-of-pocket expense, paid out in hard dollars into a "kitty" which, at the end of the lifetime of the plant, just equals the depreciable amount of the investment. Such a kitty would eventually represent the initial investment, fully recovered. Actually, of course, this is never done; the "kitty" is continuously reinvested or used in other financial transactions by the corporation. In effect, the depreciation "payments" become part and parcel of the net *cash flow* generated by the project.

Depreciation cost is thus a *stratagem* to recover initial investment. As a practical matter, depreciation charges do not enter into the cash flow as a direct, identifiable component, a fact illustrated by the schematic cash flow diagram in Fig. 6.1.1. The cash flow network ties together the corporation and one of its constituent projects.

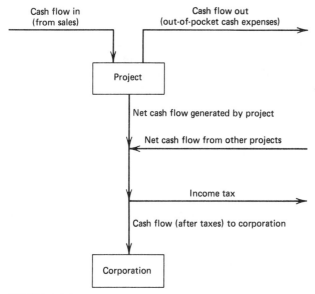

FIGURE 6.1.1
A diagram of continuous cash flow.

The cash flows shown are only those that are time-dependent, that is, cash flows that occur in a continuous manner rather than on a one-time basis. Cash flows into the project continuously from sales revenues. Cash flows out continuously as out-of-pocket cash expenses; since cash flow implies the transfer of hard cash only, this cash flow stream does not include any depreciation charges. The difference between the incoming and outgoing streams is the net continuous cash flow generated by the project. The combination of the net cash flows generated by all the corporation's projects is diminished by the income tax payment to yield the net continuous cash flow which accrues to the corporation. Income tax is, indeed, paid by the corporation, but for the purposes of project evaluation each project is considered as contributing its appropriate share of the tax burden.

Comparison of the definition of profit with the continuous cash flow diagram results in the relationships

Profit before taxes = continuous cash flow before taxes − depreciation
Profit after taxes = continuous cash flow after taxes − depreciation

all expressed on a common time basis, usually 1 year. In symbolic form,

$$\boxed{\begin{aligned} P_B &= C_B - D \\ P_A &= C_A - D \end{aligned}}$$

(6.1.1)

Depreciation is thus the only difference between the profit and continuous cash flow generated by a project.

There are also instantaneous, one-time cash flows associated with a project which are not shown in Fig. 6.1.1. Examples of these kinds of cash flows are the initial investment (negative) and the salvage value (positive). The reality of such one-time cash flow occurrences is another distinguishing characteristic between the concepts of profit and cash flow. It is useful to keep in mind the maxim

$$\boxed{\text{Cash flow} \neq \text{profit}}$$

particularly when we turn to the discussion of various criteria of economic performance, some of which are based upon profit considerations, others upon cash flow considerations.

Exercise 6.1.1

In the taxicab example (investment A), each year the cab depreciates 50 percent of its "beginning of the year" value. For each year of the 2-year life of the investment, draw up a schedule of before-tax profit, continuous cash flows, and total cash flows.

Equipment and Plant Lifetimes

The various components of a plant (equipment, buildings, improvements) are characterized by projected lifetimes which are based upon past experience. During its lifetime each item depreciates according to a predetermined depreciation schedule from its initial investment value V_0 to its salvage value S over the period of n years of its projected lifetime. At the end of any particular year k, the depreciated value, or *book value* V_k, is given by

$$\boxed{V_k = V_0 - \sum_{1}^{k} D_j} \tag{6.1.2}$$

where D_j is the annual depreciation charge for year j.

The useful service life of individual items of equipment is understandably extremely variable. Glass-lined reactors have an anticipated life of perhaps 5 years, whereas dryers may remain in service 25 years and concrete block buildings 30 years or more. Within each category considerable variability may be expected, depending upon:

Type of service; conditions of corrosion, vibration, exposure to the elements
Quality of maintenance
Forces of obsolescence

The Internal Revenue Service of the U.S. Treasury Department publishes estimated lifetimes of equipment in various areas of manufacturing activity to serve as a useful guide for drawing up depreciation schedules. Occasionally, individual items of equipment are identified and their estimated life is projected; for example, passenger automobiles are listed as having a useful life of 3 years (see also Prob. 6.1). In the usual case, however, the life of equipment typical of a given industry is given as a collective average; thus equipment in the chemical industry is given a projected lifetime of 11 years.

As a matter of fact, it would be just too much of an accounting challenge to carry each and every item of equipment in an integrated complex on the books under separate depreciation schedules. The usual practice is to depreciate equipment groupings and whole plants as depreciable units with a single projected lifetime. The projected life of plants is a composite of the projected lifetimes of their equipment components, as illustrated in Exercise 6.1.2 which follows.

Let us assume, for purposes of illustration, that each component of a plant (and the plant itself) will be depreciated by using the simplest method of depreciation, the *straight-line method,* over its particular projected lifetime. In this method of depreciation, the item in question depreciates the same amount during each and every year of its projected lifetime. Moreover, we neglect salvage value for the present, so that at the end of the projected life of n years the book value is zero. We wish to derive an expression for the "effective" lifetime of a plant, n_p, consisting of k groups of components each having a projected lifetime of n_k and each contributing a fraction f_k to the total DFC.

Exercise 6.1.2

Using the principle that the annual depreciation charge for the whole plant is the sum of the depreciation charges of the equipment components, derive the expression

$$n_p = \frac{1}{\sum_{}^{k} f_k/n_k}$$

Calculate the effective lifetime of a plant having the following component distribution:

k	f_k	n_k	Typical component items
1	0.25	5	Instruments
2	0.20	10	Reactors
3	0.45	15	Columns; electrical
4	0.10	30	Warehouses

The exercise demonstrates that there is actually no such thing as a "useful plant service life," that a plant represents equipment with a wide spectrum of lifetimes, that continuing operation of a plant is a matter of periodic regeneration of worn-out items. We may refer to an effective lifetime for depreciation purposes, but with proper equipment modification and replacement, a plant may be operated far beyond this projected lifetime limit, to conform to the product life cycle (Chap. 3). Nevertheless, since a plant may be thought of as being replaced at the end of its effective lifetime, it is customary and entirely proper, for economic evaluation purposes, to limit the productive life of a plant to some reasonable value. In the absence of compelling reasons (such as a product with limited sales potential beyond a certain date), it is customary to project *a plant life of 10 years*. The coincidence of equipment useful life and product life is implicit in this assumption.

The 10-year projection is one that the Internal Revenue Service will accept as a basis for the depreciation of chemical plants; shorter projected lifetimes (as low as 7 years) may be acceptable if properly documented from past experience. However, tax laws are always changing; in 1982 the permissible tax write-off time for equipment and some machinery was decreased to 5 years in the United States. This modification serves to emphasize the fact that depreciation time schedules do not really any longer reflect the useful life of equipment and plants, and that service lifetimes and lifetimes for tax purposes may be quite different. For purposes of preliminary project evaluation, the normal use of a projected plant lifetime of 10 years for both tax and service purposes is still recommended. This approximate number is not far from the actual *composite* tax lifetime for all DFC items, and it does serve to average out the effect of depreciation charges over the projected service lifetime.

Salvage Value

The projected salvage value of an item of equipment or plant is not depreciable. Equation (6.1.2) may therefore be rewritten as

$$V_n \equiv S = V_0 - \sum_{1}^{n} D_j \qquad (6.1.3)$$

where the salvage value S is the book value at the end of the projected lifetime of n years. For straight-line depreciation, since all D_j are equal to, say, D,

$$\sum_{1}^{n} D_j = nD$$

and

$$\boxed{D = \frac{V_0 - S}{n}} \qquad (6.1.4)$$

The projected salvage value of equipment at the end of its useful life is normally very small, particularly since any costs associated with the salvaging operation (such as removal from process tower and transportation to storage) are deductible. For purposes of preliminary project evaluation, *zero salvage value* is almost always assumed. At best, the salvage value of equipment turns out to be only a few percent of the purchased cost. An exception is afforded by equipment fabricated from valuable metals, for which metal scrap value may be obtained (for example, up to 90 cents/lb for copper in 1980).

An appreciable salvage value may be projected for equipment that is to be used in plants with short projected lifetimes. This is particularly true of equipment fabricated from scarce metals such as nickel and nickel alloys. Such equipment is sought out by firms which do business in the sale of reconditioned equipment, in many instances an economical substitute for long-delivery items required for a new plant or plant modification.

Exercise 6.1.3

At the end of the project life, is salvage a part of the profit for the last year? Is it part of the total cash flow?

Depreciation Methods and Tax Laws

During any one year of a project's lifetime, the net continuous cash flow generated by the project, diminished by that year's depreciation charges, equals the profit before taxes. This relationship is represented symbolically in Eq. (6.1.1). The relationship defines the profit before taxes, P_B, the quantity which establishes the amount of income tax to be paid. That is, if the tax rate is given by the fractional amount t, the income tax is

$$T = tP_B \qquad (6.1.5)$$

A very important characteristic of this expression is that the depreciation charges during any one year define the magnitude of the profit P_B; as a consequence, *depreciation establishes the magnitude of income tax payments*. This is why the Internal Revenue Service is so interested in depreciation scheduling and methodology; some of the depreciation methods allowed by the government will be outlined in Sec. 6.2.

The federal income tax rate for corporations has experienced historical fluctuations; in 1981 the rate was 46 percent. Actually, this rate applies to profits over $100,000, whereas the profits below this limit are taxed at a lower rate; for all practical purposes, however, the profits of the larger corporations in the chemical process industry are far enough above this limit to make the effective federal income tax rate close to 46 percent.

In addition to the federal income tax, corporations may have to pay state income taxes. State income taxes are paid to the states in which the manufacturing facilities are located. The state tax rates vary a great deal, and some states have decreased their rates as an incentive to attract new manufacture. The computation of the effect of state income taxes is further complicated by the fact that *state income taxes are deductible* from profits for federal income tax purposes. In California, for example, the corporate income tax rate is 9 percent (1981). Thus the state tax is

$$T_S = 0.09 P_B$$

and the federal tax is

$$T_F = 0.46(P_B - T_S)$$
$$= 0.46(0.91\, P_B) = 0.42\, P_B$$

The total income tax is

$$T = T_F + T_S = 0.51 P_B = t P_B$$

that is, in this particular case, $t = 0.51$.

For preliminary process evaluation purposes, the approximation

$$\boxed{t = 0.50}$$

is acceptable.

An additional important tax that must be paid by the corporation is the *property tax*. This tax is paid on the basis of the assessed valuation of the corporation's fixed assets. The taxes paid by the corporation are thus analogous to the taxes paid by the individual citizen—federal and state income taxes on income, property taxes on housing and possibly other fixed assets. The property tax will be considered in context of the discussion of COM (Chap. 7).

Let us return now to a simplified version of Fig. 6.1.1:

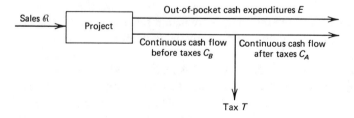

From the equation of continuity (no cash accumulation in project), during any one year,

$$C_A = \mathcal{R} - E - T \qquad (6.1.6)$$

We have previously defined P_B as

$$P_B = \Re - \text{COM} - G \tag{6.1.7}$$

where G represents the general expenses. Now, all components of COM and G, except depreciation, are out-of-pocket cash expenditures E; that is,

$$\text{COM} + G = E + D \tag{6.1.8}$$

so that

$$P_B = \Re - E - D \tag{6.1.9}$$

Substituting Eq. (6.1.5) into (6.1.9),

$$\boxed{T = t(\Re - E - D)} \tag{6.1.10}$$

This equation demonstrates the effect of the annual depreciation charge D upon the tax paid T.

If Eq. (6.1.10) is substituted into (6.1.6), we obtain a general equation for the computation of the after-tax continuous cash flow:

$$\boxed{C_A = (1 - t)(\Re - E) + tD} \tag{6.1.11}$$

For the approximation $t = 0.50$, Eq. (6.1.11) takes the form

$$C_A = \tfrac{1}{2}\Re - \tfrac{1}{2}E + \tfrac{1}{2}D \tag{6.1.12}$$

We also note that

$$P_A = P_B - T \tag{6.1.13}$$

Substituting Eq. (6.1.5),

$$P_A = (1 - t)P_B \tag{6.1.14}$$

and, with Eq. (6.1.9),

$$P_A = (1 - t)(\Re - E - D) \tag{6.1.15}$$

Subtracting Eq. (6.1.15) from (6.1.11),

$$C_A = P_A + D$$

which confirms Eq. (6.1.1) and our maxim

$$\text{Cash flow} \neq \text{profit}$$

The substance of the various mathematical relationships is summed up in Exhibit 6.1A, a convenient step-by-step procedure for computing the continuous after-tax cash flow for the case of $t = \frac{1}{2}$.

Example 6.1.1

A plant produces 100 million lb annually of a product that sells for 50 cents/lb. The plant DFC is $70 million, and the projected plant lifetime is 10 years, with zero salvage value. Straight-line depreciation is used. The COM is 22 cents/lb, and the general expenses are 8 cents/lb. Compute
    a Annual income tax payment
    b Profit after taxes (annual)
    c Cash flow after taxes (continuous, annual)
SOLUTION   Use method 1, Exhibit 6.1A.

EXHIBIT 6.1A

METHOD OF COMPUTING CONTINUOUS CASH FLOW FROM PROJECT

Method 1	Method 2
1 COM (includes depreciation)	1 COM (includes depreciation)
2 General expenses	2 Depreciation

3 Total operating expense:
    (3) = (1) + (2)
	3 COM less depreciation:
	    (3) = (1) − (2)

4 Total sales	4 General expenses

5 Profit before taxes:
    (5) = (4) − (3)
	5 Out-of-pocket expenditures:
	    (5) = (3) + (4)

6 Income tax: (6) ≈ ½(5)	6 Total sales

7 Profit after tax:
    (7) = (5) − (6) ≈ ½(5)
	7 Cash flow before taxes:
	    (7) = (6) − (5)

8 Depreciation	8 Profit before taxes: (8) = (7) − (2)

9 Continuous cash flow from project:
    (9) = (7) + (8)
	9 Income tax:
	    (9) ≈ ½(8)

	10 Profit after tax: (10) = (8) − (9) ≈ ½(8)
	11 Continuous cash flow from project: (11) = (10) + (2)

The results are identical.

1 $COM = 100 \times 10^6 \times 0.22 = \$22.0 \times 10^6$
2 $G = 100 \times 10^6 \times 0.08 = \$8.0 \times 10^6$
3 Total operating expense $= \$30.0 \times 10^6$
4 Total sales $\mathcal{R} = 100 \times 10^6 \times 0.50 = \$50.0 \times 10^6$
5 $P_B = (4) - (3) = \$20.0 \times 10^6$
6 Income tax $T \approx \frac{1}{2}P_B = \$10.0 \times 10^6$
7 $P_A \approx \frac{1}{2}P_B = \$10.0 \times 10^6$
8 $D = V_0/n = DFC/n = 70 \times 10^6/10 = \$7.0 \times 10^6$
9 $C_A = (7) + (8) = \$17.0 \times 10^6$

Exercise 6.1.4

Repeat the computations of Exercise 6.1.1 by using method 2 of Exhibit 6.1A.

The continuous cash flow after taxes C_A is the hard cash that is generated by the project and returned to the corporation (see Fig. 6.1.1). During the early stages of plant life, sales ($\mathcal{R}$) are often small, whereas out-of-pocket cash expenses (E) are inordinately large, perhaps because of unanticipated start-up problems. Reference to Eq. (6.1.11) shows that the quantity $\mathcal{R} - E$ becomes inordinately small, so that the cash flow C_A is reduced. To compensate for this sad state of affairs and to increase the generated cash flow at this stage of plant operations, the only recourse is to increase the depreciation charge D (in the term tD). By increasing the depreciation charge, the profit before taxes P_B is artificially reduced, and so is the required tax payment T, so that the net cash "left over" is increased. These considerations lead to a useful rule:

> Depreciate as rapidly as the law allows

If this is so, why not write off the investment right away, say within the first year of operations, even if this results in a giant loss, at least on paper? This may be an attractive concept, but as a matter of fact, tax laws do not allow this, and for a good reason. The government also looks at the present worth of money, and it prefers to receive its tax dollar as soon as possible. The allowable methods of depreciation scheduling (Sec. 6.2) represent a compromise between the interests of the corporation and the government.

Depreciation Accountancy

It is not our purpose here to explore the details of depreciation accounting methods. Suffice it to say that corporate depreciation regulations and income tax laws are complex and involve nuances that creative accountants may use to gain some financial advantage for the corporation. Two aspects of depreciation accountancy will be of some interest to project managers.

The first one involves the maintenance by some (but by no means all) corporations of two sets of books on equipment or plant depreciation. One set incorporates depreciation scheduling for income tax purposes and shows the "tax book value" of depreciable items. The other set incorporates depreciation scheduling for the corporation's financial statement purposes and shows the "book book value" of depreciable items. Why in the world complicate a complex enough situation still further in this particular fashion?

We saw that it is advantageous to a corporation to depreciate its assets for tax purposes as rapidly as possible. However, rapid depreciation does result in some disadvantages. For example, the book value of fixed assets on the corporation's annual balance sheet is diminished by rapid depreciation. This may be a discouraging factor to some investors who use the book value of fixed assets as one measure of the inherent value of the corporation's securities (see Chap. 12). It is argued that the book value of the fixed assets could be made to appear higher on the balance sheet by using a different, a less rapid, depreciation rate than is used for tax purposes. This can be accomplished by avoiding accelerated depreciation methods to establish the "book book value" and by using a longer projected lifetime than that used for tax purposes. The government permits two sets of books as long as reasonable consistency is maintained in the methods used to determine the two book values.

Suppose the annual depreciation on a plant for tax purposes is D, and the annual depreciation for internal book purposes is represented by the script $\mathscr{D}$. If the two are not identical, one of the quantities affected by the inequality is the magnitude of the net (after-tax) profit. The net profit for the corporation as a whole appears in the annual report and, naturally enough, is of no small interest to the investor. This key criterion of corporate performance can, then, be manipulated by means of the appropriate selection of depreciation scheduling for internal book purposes. For example, suppose that $\mathscr{D}$ is chosen so that

$$\mathscr{D} < D$$

In this case, a larger book value of the fixed assets would appear on the annual balance sheet. If we maintain the script convention to represent quantities based upon $\mathscr{D}$, it can be shown (see Prob. 6.5) that

$$\mathscr{P}_A > P_A$$

although, as might be expected,

$$\mathscr{C}_A = C_A$$

For the purpose of project evaluation, no distinction is made between the two depreciation schedules:

$$\mathscr{D} \equiv D$$

Exercise 6.1.5

Woods (1975) raises the question of ethics in maintaining duplicate books for depreciation scheduling. Do you consider the practice to be ethical?

The other aspect of depreciation accountancy which will be of interest to project managers is the question, "What happens if the projected lifetime of a plant for tax purposes turns out to be wrong?" This interesting question has been explored in some depth by Happel and Jordan (1975). Three situations may be distinguished:

1 The plant operates beyond its projected depreciation period. This is a common occurrence; because of proper maintenance and care, the plant continues to operate safely and efficiently beyond its projected lifetime. The equipment is fully depreciated, so that depreciation charges on the fixed capital no longer occur; the capital is *sunk*.

Exercise 6.1.6

In the case of sunk capital, what is the relationship between after-tax cash flow and profit?

2 The plant is shut down before the end of its projected life, and the equipment is disassembled and sold. Assuming that the assets have been held and used for more than 1 year, any gain (or loss) above (or below) the current tax book value by virtue of the sale is considered to be a *long-term capital gain* (or loss). Long-term capital gains are taxed at a rate lower than the federal corporate income tax; in 1980, the effective capital gains tax for large corporations was about 30 percent (including state obligations).

Example 6.1.2

A plant with DFC = 66×10^6 has a projected tax life of 10 years, a projected salvage value of 6×10^6, and a straight-line depreciation schedule. At the end of 8 years production is terminated because of unfavorable markets, and the equipment is sold for a net of 20×10^6 (after deducting disassembly costs). What tax burden does the sale involve?

SOLUTION   After 8 years,

$$\text{Book value} = 66 \times 10^6 - 8/10 \,(66 - 6) \times 10^6$$
$$= \$18 \times 10^6$$
$$\text{Long-term capital gain} = (20 - 18) \times 10^6 = \$2 \times 10^6$$
$$\text{Tax burden} = 0.30 \times 2 \times 10^6 = \$600{,}000$$

Exercise 6.1.7

In Example 6.1.2., suppose the plant had been shut down at the end of 7 years, and the equipment were sold for 20×10^6. What would be the tax situation?

Exercise 6.1.8

If a plant is shut down at the exact end of its scheduled tax lifetime and the equipment is sold for its projected salvage value, is the revenue from the sale taxable?

3 The plant is abandoned before the end of its projected life. In the extreme case, if the plant is abandoned (or sold below its book value) less than 1 year after acquisition, the loss may be taken as a *short-term capital loss*,[1] and tax credit may be taken at the normal corporate tax rate. This is not very likely to occur with a whole plant, but individual items of equipment may well turn out to be obsolete even at the time of delivery (see Prob. 6.8). If the plant operations are abandoned after several years and the equipment cannot be sold except for scrap value, the IRS will allow an accelerated depreciation for the last year so that the abandoned equipment will be fully depreciated. If, on the other hand, the equipment remains in usable condition, a joint decision by technical and financial management is required to determine the optimum disposal procedures for the equipment in question.

Example 6.1.3

A $50,000 on-stream atomic absorption analyzer is installed on the feed stream to a catalytic cracker. The analyzer has a projected life of 5 years with zero salvage value. Straight-line depreciation is used. At the end of 3 years the analyzer is considered to be obsolete and is taken out of service. What tax credit accrues to the corporation as a result of accelerated depreciation?

SOLUTION

Book value at end of 3 years = $20,000

= amount of accelerated depreciation

In Eq. (6.1.11), $\mathscr{R}$ and E are both zero, since transaction does not involve sales or expenditures. Thus

$$C_A = tD$$

and since $t \approx 0.50$,

$$\text{Tax credit} = 0.5 \times 20,000 = \$10,000$$

Example 6.1.3 raises some points worth keeping in mind. The transaction involving an isolated piece of equipment may result in a tax credit, and this is important in evaluating the economic justification of the isolated transaction. In reality, of course, the tax credit does not occur as a separate, identifiable cash flow; the credit occurs only in context of the tax obligations of the whole corporation.

An equation such as (6.1.11) is valid for an individual project (as in Fig. 6.1.1), but it is also valid for the corporate entity as well as an isolated equipment item.

---

[1] Corporate capital losses may be taken only up to the extent of compensating capital gains.

With isolated equipment items it is usually not possible to identify the sales revenue $\mathcal{R}$ ascribable to those items, but this is rarely a problem, since in the usual case the equations are used to *compare* two or more equipment items to accomplish the same job, and in that case $\mathcal{R}$ may be considered to be the same for each equipment alternative and cancels out in the course of analysis.

Expensed Capital

We have seen that the DFC incorporates all expenditures associated with the creation of an operational plant. The totality of these expenditures is depreciated, and the depreciation schedule must meet conditions imposed by the government; we cannot depreciate the plant in 1 year, for example. There are some capital expenditures, however, that may be effectively amortized in the year in which they occur—they become a one-time operating expense, and they are identified as *expensed capital*. To be eligible for IRS recognition as expensed capital, the DFC expenditure must be:

1 A nonrecurring expense
2 An investment with very limited life

Some examples of capital costs that may be expensed are

Training installations for plant operators (for example, dummy computer consoles)
Start-up labor from research
Consultants' fees
Equipment that obsolesces very rapidly

6.2 METHODS OF DETERMINING DEPRECIATION CHARGES

Straight-Line (SL) Depreciation

There are several commonly used methods of determining depreciation charges, although any consistent method is acceptable as long as it conforms to certain government-imposed limitations; these will be reviewed presently. We have already considered one of the methods, namely, straight-line depreciation. This is the simplest method to use; it is often used for tax purposes, it *must* be used if the projected lifetime of the item in question is less than 3 years, and it is almost invariably used for the early-stage economic analysis of projects. One reason for the method's popularity in preliminary economic analysis is that the depreciation charges are the same for each year of operations. In this way an operating cost analysis may be made for one "typical" year; other methods of depreciation calculations yield charges which are different for each year of the project's life.

We have seen (Eq. 6.1.4) that the annual depreciation charge with the straight-line method is

$$D = \frac{V_0 - S}{n} \qquad (6.1.4)$$

where, as before, V_0 is the initial fixed capital investment, and S is the projected salvage value at the end of n years of projected life. We may also consider the depreciation *rate* d_j for any particular year j; in general,

$$D_j = (V_0 - S)d_j \qquad (6.2.1)$$

and for the straight-line method, where all the d_j are the same, combining Eqs. (6.1.4) and (6.2.1),

$$\boxed{d = \frac{1}{n}} \qquad (6.2.2)$$

The book value at the end of any year k may be obtained from Eq. (6.1.2):

$$\begin{aligned}
V_k &= V_0 - \sum_{1}^{k} D_j \\
&= V_0 - \sum_{1}^{k} (V_0 - S)d_j \\
&= V_0 - (V_0 - S) \sum_{1}^{k} d_j
\end{aligned} \qquad (6.2.3)$$

For the straight-line method,

$$\sum_{1}^{k} d_j = \sum_{1}^{k} d = kd = \frac{k}{n}$$

$$V_k = V_0 - \frac{k}{n}(V_0 - S) \qquad (6.2.4)$$

Equation (6.2.4) is linear and gives rise to the term "straight-line" depreciation. If some particular depreciation method gives a book value V_k at the end of year k, and if the straight-line method is to be used for the balance of the item's projected life, the depreciation charges become, by extension of Eq. (6.1.4),

$$D = \frac{V_k - S}{n - k} \qquad (6.2.5)$$

Example 6.2.1

A new production facility has a DFC of $55 million, a projected salvage value of $5 million, and a projected life of 10 years. What is the depreciation schedule using the straight-line method?

TABLE 6.2.1
DEPRECIATION SCHEDULE EXAMPLE
$[V_0 = \$55 \times 10^6, S = \$5 \times 10^6, n = 10]$

Year	Depreciation charges, thousands of dollars			
	Straight-line	DDB	DDB + straight-line	SOYD
1	5000	11,000	11,000	9091
2	5000	8,800	8,800	8182
3	5000	7,040	7,040	7273
4	5000	5,632	5,632	6364
5	5000	4,506	4,506	5455
6	5000	3,604	3,604	4545
7	5000	2,884	2,884	3636
8	5000	2,307	2,307	2727
9	5000	1,845	2,114	1818
10	5000	2,382	2,114	909
Sum	50,000	50,000	50,000	50,000

From Eq. (6.1.4),

$$D = \frac{(55 - 5) \times 10^6}{10} = \$5 \times 10^6/\text{annum}$$

The depreciation schedule for this and the other methods is summed up in Table 6.2.1.

Double-Declining-Balance (DDB) Depreciation

In actual fact, equipment tends to depreciate, to lose value, more rapidly in the early stages of life. The IRS recognizes this by permitting depreciation scheduling based upon the declining book value balance,

$$D_j = d_j V_{j-1} \qquad (6.2.6)$$

In this method, as in the straight-line method, the d_j are the same for each year j; however, the depreciation charges decrease each year, since the book value decreases each year. We have seen that for the straight-line method, $d = 1/n$; for the declining-balance method, the IRS allows *the depreciation rate to be up to twice, but no more than twice, the straight-line rate:*

$$d_j = d = \frac{2}{n} \qquad (6.2.7)$$

The *double-declining-balance (DDB) method* is quite commonly used. The manner in which the method is used is illustrated by the example that follows; the initial rate of book value decline, of course, is higher than with the straight-line method in conformity with the rule to "depreciate as early and as quickly as possible."

Exercise 6.2.1

When the DDB method is used, the depreciation charge for the first year is often more than twice as large as the first-year depreciation charge with the straight-line method (compare, for example, Table 6.2.1). Why?

Example 6.2.2

Construct a DDB schedule for the facility described in Example 6.2.1.
SOLUTION   Refer to Table 6.2.1. For the first year, from Eq. (6.2.6),

$$D_1 = \frac{2}{n} V_0 = \frac{2 \times 55 \times 10^6}{10} = \$11 \times 10^6$$

$$D_2 = \frac{2}{n} V_1 = \frac{2}{n} (V_0 - D_1)$$

$$= \frac{2 \times 10^6 (55 - 11)}{10} = \$8.8 \times 10^6$$

$$D_3 = \frac{2}{n} V_2 = \frac{2}{n} (V_0 - D_1 - D_2)$$

$$= \frac{2 \times 10^6 (55 - 11 - 8.8)}{10} = \$7.04 \times 10^6$$

In this way, the depreciation charges for all years except the last one may be readily projected. During the last year, the balance of the depreciable book value, $V_{n-1} - S$, is written off (provided there is such a balance). Thus

$$V_{n-1} \equiv V_9 = V_0 - \sum_1^9 D_j$$
$$= 10^6 (55 - 47.618) = \$7.382 \times 10^6$$
$$D_{10} = V_9 - S$$
$$= 10^6 (7.382 - 5) = \$2.382 \times 10^6$$

The salvage value S does *not* enter into the calculation of the annual depreciation charges (except for the final year). Nevertheless, S does bear an important influence upon the DDB method, as we will see presently.

Both the straight-line and DDB methods result in the complete depreciation of the depreciable portion of the capital at the end of the projected lifetime. The DDB method reduces the book value more rapidly during the initial years of the project; this is illustrated in Fig. 6.2.1, which is based on the data in Table 6.2.1. (The graphs are drawn as continuous functions through the year-end book values.)

An analytical expression for D_j for the DDB method may be derived by induction. Thus

SECTION 6.2: METHODS OF DETERMINING DEPRECIATION CHARGES

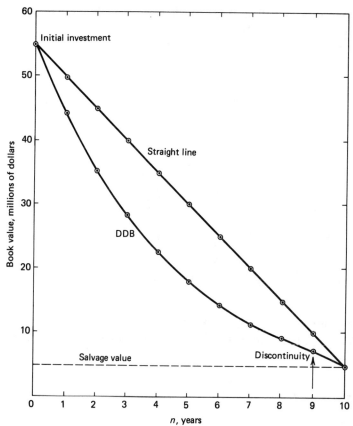

FIGURE 6.2.1
Comparison of annual book values (straight-line and DDB depreciation).

$$D_1 = \frac{2V_0}{n}$$

$$D_2 = \frac{2}{n}(V_0 - D_1) = \frac{2V_0}{n}\left(1 - \frac{2}{n}\right)$$

$$D_3 = \frac{2}{n}(V_0 - D_1 - D_2)$$

$$= \frac{2V_0}{n}\left[1 - \frac{2}{n} - \frac{2}{n} + \left(\frac{2}{n}\right)^2\right]$$

$$= \frac{2V_0}{n}\left(1 - \frac{2}{n}\right)^2$$

By induction,

$$D_j = \frac{2V_0}{n}\left(1 - \frac{2}{n}\right)^{j-1} \tag{6.2.8}$$

To complete the proof by induction, it remains to be shown that if the format of Eq. (6.2.8) holds for D_j, it also holds for D_{j+1}.

Now, $D_j = (2/n)V_{j-1}$ (by definition of DDB). Comparing with (6.2.8), we have

$$V_{j-1} = V_0 \left(1 - \frac{2}{n}\right)^{j-1} \tag{6.2.9}$$

But

$$\begin{aligned} V_j &= V_{j-1} - D_j \\ &= V_0 \left(1 - \frac{2}{n}\right)^{j-1} - \frac{2V_0}{n}\left(1 - \frac{2}{n}\right)^{j-1} \\ &= V_0 \left(1 - \frac{2}{n}\right)\left(1 - \frac{2}{n}\right)^{j-1} = V_0 \left(1 - \frac{2}{n}\right)^{j} \\ D_{j+1} &= \frac{2}{n} V_j = \frac{2V_0}{n}\left(1 - \frac{2}{n}\right)^{j} \end{aligned}$$

which has the same format as (6.2.8). Q.E.D.

Equation (6.2.9) is a general expression for the end-of-the-year book value; for year k, for example,

$$V_k = V_0 \left(1 - \frac{2}{n}\right)^{k} \tag{6.2.10}$$

Exercise 6.2.2

Derive an analytical expression for D_n, depreciation during the last year of the projected lifetime, for the DDB method. Assume that

$$V_0 - \sum_{1}^{n-1} D_j > S \tag{6.2.11}$$

Expression (6.2.11) is a manifestation of the effect that the salvage value S may have upon DDB depreciation. The effect is illustrated in Fig. 6.2.2. A graph of the book value of an installation with initial investment of V_0 and projected life n is shown. Three different salvage value situations may be recognized.

In case A (S_A), the book value decreases to S_A in N years, well before the projected life of n years. This means that the item is completely depreciated at time N, and no additional depreciation charges are incurred between N and n. This is a perfectly acceptable situation; the case is characteristic of high salvage values.

In case B, the DDB manages to depreciate the item to the salvage value S_B at the end of the projected life.

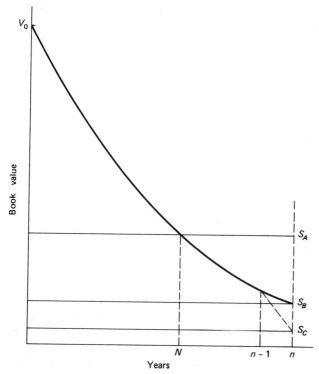

FIGURE 6.2.2
Effect of salvage value upon DDB depreciation.

Exercise 6.2.3

For a given V_0, what is the analytical relationship between n and S for case B? What value does the ratio S_B/V_0 assume for $n = 10$?

Case C is analogous to the situation illustrated in Fig. 6.2.1: DDB alone will not depreciate the item to its salvage value S_C at the end of n years, and the depreciable portion of the book value at the end of year $n - 1$, namely, $V_{n-1} - S_C$, is taken as the final year's depreciation.

Composite DDB plus Straight-Line (DDB + SL) Depreciation

Case C of the DDB method is actually a composite method; during the last year the remaining depreciable amount is taken as a lump sum—in effect, it is depreciated by the straight-line method over the final year. The disadvantage of this method is that the final depreciation charge may be higher than the one for the previous year, contrary to the maxim of depreciating as rapidly as possible at the earliest possible stage. For the situation in Table 6.2.1, for example, the DDB method yields a final depreciation charge which exceeds the charge for each of the previous two years.

To avoid this sort of delayed depreciation, it is permissible to switch over from DDB to straight-line depreciation at the end of some year m, where m is smaller than n, the projected life. This composite method is best illustrated by an example, such as the one in Table 6.2.1.

We start by noting that in the DDB column,

$$D_{10} > D_9 \quad (2382 > 1845)$$

We therefore try straight-line depreciation over the last 2 years by setting

$$D_9 = D_{10} = \frac{2382 + 1845}{2} = 2114$$

as shown in the DDB + SL column. We now ask, "Is the depreciation schedule now a monotonically declining one?" The answer is, "Yes"—$D_9 < D_8 < D_7$, etc. In this particular case, then,

$$m = 8$$

If it happened that D_9 (and D_{10}) were still larger than D_8, another trial would be required involving straight-line depreciation over the last 3 years (by averaging $D_{10} + D_9 + D_8$ in the DDB column). In this way, a trial-and-error method is used to arrive at a monotonically declining schedule and the appropriate value of m.

Note that the composite DDB + SL method is required only for case C of the DDB method. The trial-and-error procedure described is performed most easily in conjunction with tabulated values arranged as in Table 6.2.1. For instance, it is not necessary to use analytical expressions to determine which of the three cases of DDB depreciation actually obtains; the situation is evident in the process of developing the DDB tabulated values. The procedures are easily duplicated with computer programs (Prob. 6.11).

A plot of the book values resulting from the composite DDB + SL depreciation appears as in Fig. 6.2.3. The conversion occurs at point P, the tangency point for the straight line originating at (n, S). This suggests that m and the depreciation schedule may also be obtained graphically.

Exercise 6.2.4

Propose arguments that show that the trial-and-error procedure described for the determination of depreciation charges leads to the tangency principle illustrated in Fig. 6.2.3.

Sum-of-Years-Digits (SOYD) Depreciation

Still another permitted and frequently used depreciation method is the *sum-of-years-digits (SOYD)* method. It is an accelerated depreciation method not unlike the

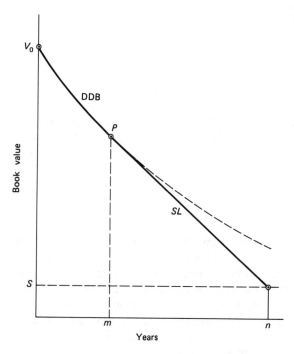

FIGURE 6.2.3
DDB + SL depreciation.

DDB method. It differs, however, in that the depreciation *rate* d_j declines throughout the lifetime of the depreciating item:

$$D_j = (V_0 - S)d_j \qquad (6.2.1)$$

d_j is computed as follows:

If the projected lifetime is n, form the sum-of-years digits:

$$1 + 2 + 3 + \cdots + n = \sum_{1}^{n} k$$

This is an arithmetic progression, and

$$\sum_{1}^{n} k = \tfrac{1}{2}n(n+1)$$

d_j is then given by

$$d_j = \frac{n+1-j}{\sum_{1}^{n} k} = \frac{2(n+1-j)}{n(n+1)} \qquad (6.2.12)$$

For example, if $n = 10$,

$$\sum_1^n k = 55$$
$$d_1 = {}^{10}/_{55},\ d_2 = {}^{9}/_{55},\ d_3 = {}^{8}/_{55}, \ldots, d_{10} = {}^{1}/_{55}$$

The SOYD column in Table 6.2.1 results from the use of these values of d_j in Eq. (6.2.1). A comparative SOYD and DDB + SL book value plot (based on Table 6.2.1) is shown in Fig. 6.2.4. The close resemblance of the results is apparent.

Preferred Depreciation Method

Several depreciation methods have been examined, and the question arises as to which one the corporation should choose for tax purposes. We have already seen that accelerated depreciation methods are preferred to maximize the after-tax cash flow that accrues to the corporation during the early stages of a project. A related argument is based upon Eq. (6.1.10). The term tD in it is negative—it is a *tax credit*. One should choose that depreciation method which results in the greatest present worth of all depreciation tax credits. Since the tax rate t is a common factor in all tax credits, the rule may be stated as follows:

> Choose the method of depreciation for which the present worth of all depreciation charges is a maximum

The rule does not lead to the choice of a unique method; the best one depends upon the project lifetime, the salvage value, and the time value of money. For instance, if salvage value is zero and $n > 5$, SOYD is preferred. Newnan (1976) has generated a useful chart to facilitate the proper choice. Practice in making the present worth comparison is given in the problem set.

For early-stage project evaluation, the accelerated depreciation methods are not used. For the sake of simplicity, the following rules are commonly followed:

> 1 Use straight-line depreciation
> 2 Use zero salvage value
> 3 Assume 10-year life

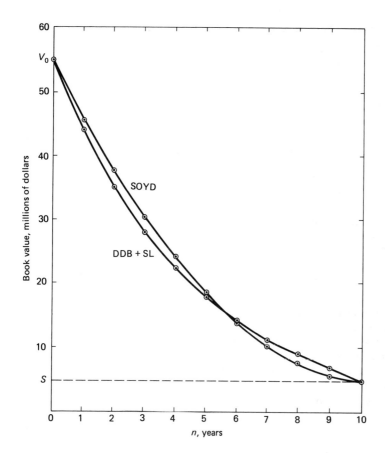

FIGURE 6.2.4
Comparison of two depreciation methods: SOYD and DDB + SL.

NOMENCLATURE

C	cash flow, \$/year
$\mathscr{C}$	cash flow on corporate books, \$/year
d	depreciation rate (fractional)
D	annual depreciation charge, \$/year
$\mathscr{D}$	annual depreciation charge on corporate books, \$/year
E	out-of-pocket cash expenses, \$/year
f	fraction of DFC represented by one class of equipment items
G	general expenses, \$/year
j, k	year number, corresponding to subscript index
m	year at the end of which DDB is switched to SL depreciation
n	project lifetime, years
P	profit, \$/year

$\mathcal{P}$ profit on corporate books, $/year
$\mathcal{R}$ sales revenue, $/year
S salvage value, $
t tax rate (fractional)
T annual income tax, $
V book value, $

Subscripts

A after taxes; also case A of DDB
B before taxes; also case B of DDB
C case C of DDB
F federal, as in T_F
j, k indices
0 original (zero time)
p plant
S state, as in T_S
$1, 2, 3, \ldots$ year numbers

Abbreviations

COM   cost of manufacture
DDB   double-declining balance
DDB + SL   double-declining balance plus straight line
DFC   direct fixed capital
IRS   Internal Revenue Service
SL   straight line
SOYD   sum-of-years digits

REFERENCES

Anonymous: "U. S. Productivity: What Can Be Done To Improve It?," Public Affairs Department, Caterpillar Tractor Co., Peoria, Ill., 1980.

Happel, John, and Donald G. Jordan: "Chemical Process Economics," 2d ed., Marcel Dekker, Inc., New York, 1975.

Naphtali, L. M., and R. Shinnar: Effect of Inflation on Energy Cost Analyses, *Chem. Eng. Prog.*, **77** (2):65 (February 1981).

Newnan, Donald G.: "Engineering Economic Analysis," Engineering Press, San Jose, Calif., 1976.

Peters, Max S., and Klaus D. Timmerhaus: "Plant Design and Economics for Chemical Engineers," 3d ed., McGraw-Hill, New York, 1980.

Woods, Donald R.: "Financial Decision Making in the Process Industry," Prentice-Hall, Englewood Cliffs, N. J., 1975.

PROBLEMS

6.1 A question of some importance to individuals and corporations both is, "What is the optimum length of time to hang on to newly purchased passenger vehicles?" Cars depreciate very rapidly, and their resale value declines correspondingly. At the same time, annual operating costs keep on increasing with age due to increasing maintenance costs and reduced fuel and lubricant efficiency.

One possible criterion for computing the optimum time is the concept of a *perpetuity* [Eq. (2.3.10)]. The principal P is compounded over the n years that the car is kept, and the interest thus earned is used to supply the required cash flows which occur throughout the period n, all compounded to the end of the period ($= R$). The problem then is to determine for what value of n is P a minimum; this value of n is the optimum.

R, the future worth of the periodic outgoing cash flows, has three components:

W, the purchased cost of the car, which occurs at the end of each period

S, the resale value of the old car, which also occurs at the end of each period (for present purposes, a negative quantity)

E, the annual operating costs, which, for present purposes, may be considered to occur as a discrete event at the end of each year, and which are all compounded to the end of the period n

Thus

$$R = W - S + \sum_{j=1}^{n} E_j(1 + i)^{n-j} \quad \text{(discrete compounding)} \quad (P6.1)$$

The car may be assumed to depreciate at the fixed rate d of its beginning-of-the-year book value, so that the sales value at the end of any year j is

$$S_j = W(1 - d)^j \quad \text{[see Eq. (6.2.10)]} \quad (P6.2)$$

Assume that annual operating expenses increase in proportion to the number of years,

$$E_j = jE_1 \quad (P6.3)$$

a Calculate and plot P for the case

$$W = \$10{,}000 \quad d = 0.5 \quad i = 0.20 \quad E_1 = \$1000$$

(i is assumed to be the same for drawing interest on P and compounding the E_j's.)

b Repeat the computation in part (a) for $i = 0.10$. You may wish to extend the computations to cases which, perhaps, you consider to be more realistic. What is your conclusion?

6.2 The expression

$$n_p = \frac{1}{\sum_k f_k/n_k}$$

was derived in Exercise 6.1.2 for the case of straight-line depreciation. It relates the effective plant lifetime n_p to the lifetime of the component equipment items in the several categories k. Assume that a large ensemble of items and categories exists and that a steady-state replacement regime has been reached. Show that the derived expression is independent of the method of depreciation (provided the method depreciates each item completely within its lifetime). The key step involves proof that the *average* annual depreciation of each item, D_k, is independent of the depreciation method.

6.3 The board of directors has suggested that for the contemplated plant in Example 6.1.1 the unit selling price be increased so that the annual cash flow after taxes will recover the DFC in just 2 years. Assuming no other costs are affected, what must be the selling price?

6.4 For the plant in Example 6.1.1, the IRS has granted permission to use a projected lifetime of 7 years instead of 10 years. If the plant is, indeed, operated 10 years with no equipment replacement, what is the present worth of the tax savings realized from the IRS dispensation? What percentage of the present worth of the total tax payments in the absence of such dispensation do the savings represent? Use 14 percent discrete annual discounting; neglect inflation.

6.5 If two sets of books are maintained by a corporation on equipment depreciation, show for the case

that $$\mathcal{D} < D$$ $$\mathcal{P}_A > P_A \quad \text{but} \quad \mathcal{C}_A = C_A$$

6.6 A heated discussion has arisen in the board room as to the justification for the proposal that the corporation adopt the practice of keeping two sets of books on depreciation. One faction argues that accelerated depreciation methods lower the average book value of equipment; this is fine for tax purposes, but the low value may not look attractive to investors. Therefore, the argument goes, why not use straight-line depreciation for internal book purposes; this method will increase the average book value of assets on hand, and this higher value will appear on the annual report balance sheet to bedazzle the investor.

The opposing faction calls this nonsense, pointing out that in a large ensemble of plants the average annual depreciation is independent of the depreciation method (see Prob. 6.2). Therefore the average book value cannot depend upon the depreciation method used, and the corporation might as well use the same depreciation method for both tax and internal purposes.

Can you help resolve this controversy?

6.7 We have seen that depreciation charges constitute a way for the investor to recover the initial investment. Discuss the impact that inflation has upon this concept. What are some ways that the impact of inflation upon investment recovery may be reduced? For some interesting views, refer to Naphtali and Shinnar (1981) and Anonymous (1980).

6.8 The construction of a new pesticide plant is approaching completion. The reaction is to take place in a large liquid-phase chlorinator (LPC) with an associated DFC of $4,000,000, which has already been expended. Annual operating charges associated with the chlorinator are expected to be $400,000, *excluding* depreciation. The economic life of the plant is projected to be 10 years.

At this time a research breakthrough indicates that the reaction may be carried out just as efficiently in a vapor-phase chlorinator (VPC) at much milder conditions. A careful estimate indicates that the liquid-phase chlorinator could be replaced for an additional DFC of $1,500,000, with no delay in the scheduled start-up. Operating charges (again excluding depreciation) associated with vapor phase chlorination are expected to be reduced to $100,000/annum. Unfortunately, no salvage value would be obtained for the obsolete, special-design liquid-phase chlorinator.

Which alternative should be chosen: LPC or VPC?

The decision process is an example of the *analysis of alternatives* (Chap. 10). Assume that if the VPC is chosen, the IRS will allow the liquid-phase chlorinator to be carried on the tax books as a one-time short-term capital loss ("obsolescence allowance"), with no depreciation charges. Make your choice on the basis of which alternative results in the higher present worth (at start-up time) of the sum of all the cash flows, both positive and negative, computed on an after-tax basis. Use 10 percent annually discounted interest for the time value of money.

6.9 Derive an expression for the book value at the end of any year k for the SOYD depreciation method. Using this result, formulate the expression for the time-average book value

$$\bar{V} = \frac{1}{n+1} \sum_{0}^{n} V_k \tag{P6.4}$$

for the SOYD method, and prove that it is smaller than $\bar{V}$ for the straight-line method for $n \geq 2$. Be sure that you understand the significance of Eq. (P6.4); if need be, demonstrate its validity graphically.

6.10 Consider a capital investment of \$100,000 for plant facilities, \$15,000 for working capital, \$10,000 for land, zero salvage value, and a projected plant life of 10 years. Determine the annual depreciation charges for the following types of depreciation:

1. Straight-line
2. Double-declining-balance
3. Double-declining-balance plus straight-line to get maximum earliest write-off
4. Sum-of-years digits

Consider each annual charge to be a single end-of-the-year payment; for each method of depreciation, compute the present worth of all the charges by using a 10 percent nominal *continuous* interest rate. Which method is the best?

6.11 Prepare a computer program flow chart for the computation of DDB depreciation charges. The program must recognize and identify for the user which of the three DDB cases obtains (see Fig. 6.2.2). For case C, the program is required to compute the optimum DDB + SL depreciation schedule.

6.12 Derive an analytical expression for m, the optimum year at the end of which to switch from DDB to SL depreciation, for the case of zero salvage value. Base the derivation on Fig. 6.2.3 and the observation that m occurs at P, the point of tangency (why?). Calculate m (rounded off to the closest integer) for the case of $S = 0$, $n = 20$. For $S \neq 0$, an explicit expression for m cannot be written, but m can be calculated from the derived equations by trial and error. What is m for the case $S = 0.1V_0$, $n = 20$?

The derivation based on Fig. 6.2.3 does not take into account the time value of money. Do you think the time value of money enters into this kind of derivation?

6.13 Another possible combined depreciation method involves both DDB and SOYD. The procedure for combining the two methods may be appreciated by reference to Fig. 6.2.4; at the intersection, the switch is made to the method giving the more rapid depreciation (DDB to SOYD in the particular case in Fig. 6.2.4). Write a general equation which will facilitate a trial-and-error solution of M, the year at the end of which the switch is to be made. Calculate M for the facility in Example 6.2.1. Does your result check with Fig. 6.2.4?

6.14 (Adapted from Woods, 1975, with permission.) Corporations generally peg stock dividends at a certain percentage of the after-tax profit. Why are the dividends not taken as a percentage of the after-tax cash flow?

6.15 A custom-made gold-lined reactor incurs an initial investment of $1,000,000. The reactor is expected to have a useful life of 10 years and a salvage value of $600,000. Make out a DDB annual depreciation schedule over the 10-year life of the equipment.

6.16 Make a literature search (for example, the textbooks in this chapter's list of references) to determine the meaning of the "sinking fund" method of depreciation. Based upon your understanding of this method, derive a general expression for the depreciation rate d_j, and compute the annual depreciation charges for the facility in Example 6.2.1. Plot the results as book value versus time, and compare with straight-line depreciation. Why is the sinking fund method not used in practice?

6.17 Make a schedule of the annual *continuous* after-tax cash flows generated by the following project:

Project life: 8 years
Rated production: 10 million lb/year
Projected product life cycle: Symmetrical triangular distribution, rated production at the end of year 4; 40 percent of rated production at start-up and end of year 8
Sales price: $5/lb
Out-of-pocket cash expenditures: $1/lb (assume these are independent of production level)
Direct fixed capital: $70,000,000
Salvage value: $4,000,000
Depreciation method: DDB (+ SL if required)

6.18 Derive an expression for the discretely discounted present worth of annual depreciation charges obtained with the DDB + SL method, with switchover at the end of year m.

6.19 We have seen that the best method of depreciation is the one for which the present worth of all depreciation charges is a maximum. There is, however, no simple analytical method available which will allow an a priori determination of the optimum depreciation method.

Construct a graph which will allow the choice to be made between SOYD and DDB + SL. The graph should have the depreciable life n as abscissa ($3 \leq n \leq 25$) and the salvage value, expressed as S/V_0, as ordinate ($0 \leq S/V_0 \leq 0.20$). Use the time value of money as the parameter for three separate curves ($i = 0, 0.1, 0.2$).

The graph can be constructed rather laboriously by using numerical computations. The labor demand may be reduced by deriving analytical expressions for the present worth of the depreciation charges (see also Probs. 6.12 and 6.18). The use of a programmable hand calculator is encouraged.

6.20 The initial installed cost for a new piece of equipment is $10,000, and its salvage value at the end of its projected life of 10 years is estimated to be $2000. After the equipment has been in use for 4 years, it is sold for $7000. If income tax rate is 50 percent, which depreciation method—straight-line or double-declining-balance—results in smaller total income tax payments, and what is the saving by using the better method?[1]

6.21 What is the *Matheson formula* and what relationship does it bear to the DDB method? See, for instance, Peters and Timmerhaus (1980).

---

[1] For long-term capital gains, if equipment had been depreciated by using accelerated methods, the capital gain *up to the book value that would be obtained with SL depreciation* is taxable as regular income.

CHAPTER 7

THE COST OF MANUFACTURE

The cost of manufacture (COM) has been defined as the expense directly associated with the manufacturing process to produce a specific commodity. As such, it is centered around the immediate production facility. Depreciation, one of the elements of COM, has received its share of emphasis, and we now turn our attention to some of the other constituent elements. All such elements have a temporal aspect; they represent expenditures which occur at widely differing rates or frequencies, but they all occur throughout the lifetime of a facility, and, as we have seen, they are important in determining the time-dependent continuous cash flow generated by a project.

Many elements of COM are reasonably obvious, and even the casual project evaluator would intuitively worry about items such as raw materials and labor costs. However, many elements are not quite so obvious, the many "hidden" costs. Insurance or payroll expenditures, for example, are just as real as expenditures for raw materials, and their omission from even a preliminary estimate of COM can lead to embarrassing errors and improper conclusions. It is true that during early stages of laboratory research the cost to produce a new chemical product is often projected in terms of the cost of the raw materials alone. This is a legitimate procedure for the early rejection of processes which just cannot cut it. In no way must such early approximations survive as lore purporting to reflect the anticipated cost of production.

There are those elements of COM, such as raw materials costs, which can be approximated with reasonable confidence once the process has been firmed up. Others are difficult to establish for a planned project even at advanced stages of planning, and recourse is therefore taken to the use of statistical factors which are not unlike the purchased cost factors used in DFC estimates. These statistical factors are based upon data from operating plants in an integrated complex. For such plants the costs incurred by each element of COM are known, or else prorated from the total integrated complex costs. Using such data collected over a period of many years, accountants and cost estimation specialists have developed factors which, in view of the statistical balancing effect (Chap. 4), yield estimates of acceptable reliability. As in the case of purchased cost factors, the factors used for estimating the elements of COM must be applied to some fundamental and reliably estimated entity. Two such entities are used for COM estimation:

Direct fixed capital
Operating labor

Sometimes it is not easy to fathom the relationship between an element of COM and the entity upon which it is based; for example, annual laboratory charges are sometimes taken as a percentage of the annual operating labor. In what follows an attempt will be made to elucidate these relationships and to indicate a range of factor choices so as to give the neophyte estimator some flexibility in fitting the estimate to a specific project.

Typical costs for COM elements such as labor and electric energy are scattered throughout the text. The intent is to give the project evaluator some feel for the general magnitude of such costs; it is recognized that in a period of rapid inflation the numbers are obsolete even as they are printed. As in previous chapters, the emphasis will be upon evaluation methodology and data sources which are most likely to contain reliable up-to-date costs.

7.1 THE ELEMENTS OF THE COST OF MANUFACTURE

Definition and Boundaries

The cost of manufacture encompasses expenses incurred in the course of production activities within the battery limits of a plant (Fig. 5.1.1). Thus the purchase of new gaskets for process turbine pumps is a direct expenditure (a "maintenance materials" expenditure) which falls within the scope of COM. The cost of shipping a tank car of product from Houston, Texas, to Troy, New York, is not related to a battery limits operation; it is therefore not a COM item. Activities within the scope of COM need not take place within a plant's geographical boundary, but they must be supporting facility activities within the integrated complex which contribute to the battery limits operations. Thus COM must reflect a proportional share of the expenses of operating a payroll department, for without such a department's help the operating crew would not receive its pay, with operating consequences not difficult to imagine. (Even grass roots facilities must expend part of their operating budget for payroll activities.) The COM must include a fair share for the operation of all integrated complex supporting facilities, down to the plant cafeteria and recreational centers.

Another category of expenses, the *general expenses*, has been briefly mentioned. These are expenses incurred in the course of operating the corporate entity, but they do not impact directly upon battery limits operations and are therefore not part of the COM. An example of a general expense is the cost associated with the selling of a particular product. This is an expense supporting an activity which is certainly important to the continuing successful production, but it is not immediately supportive of the day-to-day battery limits operations and is therefore considered to be a general expense rather than a cost of manufacture. The shipping cost that has been mentioned is part of the *cost of distribution,* a cost paid by the customer over and above the sales price at the distribution point site.

A distinction should perhaps be made between the COM for a process in the planning stages and the COM for an ongoing operation. The computational format is the same for both cases; in the first case, however, the computation involves the estimation of probable costs, whereas in the case of the ongoing operation, COM computation involves the proper accounting of known costs already expended. The methods of estimating the operating costs for planned processes are based upon the experience gained in computing the COM of existing processes. The techniques of computation may also be extended to limited-scope projects, such as the installation of equipment to improve an ongoing process. In computing the effect upon operating costs of a limited-scope improvement, the estimator may well ignore some of the elements of COM; whether this is justified is a matter of judgment and experience. For example, the additional operating labor demand incurred by the installation of a refrigerated vent condenser may be ignored, since the effect is likely to be entirely negligible.

There are three prerequisites for a reliable projection of the COM during the earlier stages of project development:

1 Proper process definition (flow sheet, mass and energy balances) as stipulated in Chap. 3.

2 A reliable DFC estimate (Chap. 5).

3 A definition of the level of production. The usual basis for the estimate of COM in a planned facility is *production at full plant capacity;* however, as we will see, it is useful to make other estimates at specified levels below capacity.

We have also referred to the temporal nature of the elements of COM. The frequency of expenditure occurrences ascribable to each element is quite variable. Depreciation is an annual book transaction; the payroll must be met at regular intervals (say 2 weeks); utility charges may involve more or less continuous intraplant transactions on books or regular payments to outside suppliers. For purposes of computing the COM, the complexities of timing and book accounting are bypassed; the expenses are assumed to be continuous and, except for depreciation, out-of-pocket. The COM is then calculated on an annual basis (dollars per year), or any shorter period basis desired. The COM (and all its elements) may also be computed as a *unit cost,* that is, cost per pound of product. The choice is a matter of individual preference or departmental tradition.

Computational Format

A typical format for the estimation of COM is shown in Fig. 7.1.1. The exact sequencing is unimportant; it is important, however, that all the elements listed in the figure be properly accounted for. Estimation methods for each category will be outlined in this chapter.

FIGURE 7.1.1
Manufacturing cost estimate form.

PRODUCT: _____ DATE: _____

PLANT: _____ BY: _____

PRODUCTION RATE: _____ (% OF CAPACITY)

RAW MATERIALS	QUANTITY—UNITS PER MONTH		PRICE PER UNIT	$ COST PER MONTH	
TOTAL RAW MATERIALS					
OPERATING LABOR					
FRINGE BENEFITS					
SUPERVISORY LABOR					
MAINTENTANCE LABOR					
MAINTENANCE MATERIALS					
OPERATING SUPPLIES					
UTILITIES:					
STEAM (M POUNDS) (150 PSIG)					
POWER (KWH), AC					
WATER (M GALLONS), COOLING					
COMPRESSED AIR (MCF)					
NATURAL GAS (MCF)					
WASTE DISPOSAL					
LABORATORY					
FACTORY EXPENSE					
INSURANCE AND TAXES					
DEPRECIATION	$	DIRECT INVESTMENT AT	%		
RESEARCH					
PATENTS AND LICENSES					
TOTAL BULK COST					
PACKAGING FOR TRANSFER					
TOTAL COST FOR TRANSFER					
PACKAGING FOR SALE					
TOTAL COST FOR SALE					

For raw materials and utilities, it is useful to jot down the quantity used (as pounds per month, thousands of gallons per pound of product, etc.) and the unit cost in corresponding units. For the other elements, it is also useful to indicate the estimating method employed to arrive at the cost. The costs of all the elements must, of course, be in consistent units (unit costs as cents per kilogram of product, periodic total costs as thousands of dollars per month, etc.). The total of all costs is then the COM, also called the *cost for sale,* as shown in Fig.7.1.1. Note that two subtotal costs are also indicated: a *bulk cost* and a *cost for transfer.* Chemical commodities are, of course, manufactured in bulk, and the bulk cost refers to the material issuing from the process equipment into bulk storage. Even bulk chemicals, however, must be distributed in some sort of container—tank cars, tank trucks, hopper cars, drums—and the transfer from plant to container generates a cost of *packaging* which escalates the total up to the cost for transfer. Additional costs may be incurred by *packaging for sale,* particularly if a consumer-oriented product is involved. As an example of the distinction between the various costs, consider the production of a dry-cleaning solvent (say, tetrachloroethylene). The solvent is stored in bulk at the production facility, and the COM up to this point is the bulk cost. The solvent may now be loaded into tank cars and transferred to a regional distribution point. The cost of loading, unloading, and (in this case) the freight is borne by the manufacturer; when it is added to the bulk cost, the cost for transfer is obtained. Finally, at the distribution point the solvent may be transferred into 55-gal drums and blended with a number of additives such as detergents, acid neutralizers, and garment conditioners. The cost of this final packaging for sale, added to the cost of transfer, yields the cost for sale.

The elements of COM listed in Fig. 7.1.1 are commonly subdivided into three categories:

1 *Variable charges.* These are charges which are directly proportional to production. An example is the cost of raw materials.

2 *Regulated charges.* Regulated charges vary somewhat with the production level, but not in direct proportion. An example is the cost of operating labor.

3 *Fixed charges.* Fixed charges are independent of production level; depreciation, for example, is not production-dependent.

As each element of COM is discussed in this chapter, the category to which it belongs will be specified (no attempt has been made to group the elements in Fig. 7.1.1). The category names are very commonly used in project evaluation parlance; the significance of the classification will be rendered clearer when profitability under variable conditions of production is discussed in the next chapter.

Not infrequently a process produces two or more *coproducts,* all of which are salable commodities in demand. An example is the chlorination of methane, a process which coproduces methyl chloride, methylene chloride, chloroform, and carbon tetrachloride in proportions which are to some degree at the discretion of the manufacturer. How is the COM prorated among the coproducts? In practice, one of the following prorating methods is used:

The COM is split up among the coproducts on the basis of their *unit values*. The unit values represent the total annual dollar sales of a product divided by the total weight. Unit value statistics may be located, for example, in the annual publication, *Synthetic Organic Chemicals* (U.S. International Trade Commission, Government Printing Office, Washington). Other statistical sources may be located in the "Chemical Statistics Directory," published by the U.S. Department of Commerce, and obtainable from the Government Printing Office, Washington.

A commonly used variant of the above is a split on the basis of the sales price of the products on corporate books.

The *value added by manufacture* is an alternate prorating basis. This is defined as the unit value of shipments diminished by the unit value of raw materials and energy. Statistics appear in the previously listed sources.

A common and easily used method involves a COM split on the basis of the total weight produced. For example, the chlorination of natural gas in a solvents plant (Fig. 3.3.10) coproduces CCl_4, C_2Cl_4, and HCl; the unit cost for sale is obtained by dividing the total annual cost of manufacture by the grand total annual weight of production—pounds of CCl_4 plus pounds of C_2Cl_4 plus pounds of HCl.

Exercise 7.1.1

The electrolysis of NaCl brine coproduces NaOH, Cl_2, and H_2. The data of Guthrie (1970), escalated to M&S = 750, suggest that the annual COM in a plant producing 100,000 annual short tons of Cl_2 is 10.6×10^6. If H_2 is credited at a rate of $6/$10^6$ Btu fuel value, what is the unit cost of manufacture per pound of NaOH?

An alternative approach to the cost accounting problem in the exercise would be to consider the unit costs of NaOH and Cl_2 to be the same coming out of the electrolytic cell, but to keep additional processing costs restricted to each product involved. Chlorine drying and compression costs, for example, are easily separable and can be added to the out-of-cell unit cost of Cl_2. In fact, any time that the processing costs of coproducts are readily separable, the COM of each product can and should be kept separated. An example of easily separable costs occurs in a *sequential process*. In the production of phenol via the air oxidation of toluene, benzoic acid is an intermediate product, purified and stored before being further oxidized to phenol; it is, in fact, an article of commerce in its own right. The two coproducts, benzoic acid and phenol, are each manufactured in their own production facility, and the COM of each is readily computable.

The accounting for *by-products* in a COM estimate is a matter of individual judgment. The implication of the term "by-product" is that the item it describes is not a primary production objective of the process. Nevertheless, if it is a desirable article of commerce, it may be handled as a coproduct, or else it may be shown in the COM computation as a *credit,* a negative quantity on the computation sheet. Credit is usually taken at the sales price for a readily marketable but minor product, or at an internally negotiated transfer price if the by-product is to be consumed within the integrated complex. Care must be taken that the by-product be obtained in marketable form if sales price credit is to be taken and that there be demand for the product; all too often by-products turn out to be a disposal burden to an otherwise acceptable process.

By-product utilities (process-generated steam, power, etc.) are credited at standard cost (see Sec. 7.2).

Exercise 7.1.2

A process produces a waste stream of 10 scf* of CO/lb of product. It is proposed that the stream be forwarded to the integrated complex steam plant for burning as fuel. If the standard cost of the steam plant fuel is $6/10^6$ Btu, what credit may be taken for the product's unit COM?

Stream Factors

The COM is computed for a specific installation with a specific rated capacity. The actual design plant throughput depends upon the installation's projected *stream factor* (S.F.). The S.F. is a design factor which accounts for anticipated downtime for maintenance or for plant shutdowns due to seasonal demand. The factor is defined as

$$\text{S.F.} = \frac{\text{number of operating days}}{365} \qquad (7.1.1)$$

Thus if a plant is designed for a rated capacity of r lb/year, the hourly production is $r/(365 \times 24 \times \text{S.F.})$.

For most chemical processing plants, the stream factor is in the range of 0.90 to 0.95. The actual value depends upon the complexities of the process—catalyst cleanup or replacement needs, time set aside for unfouling exchangers, need for careful inspection and repair of reactor linings. Seasonal demands may render the stream factor very low; tomato product plants, for example, operate at a stream factor of about 0.30 because of the relatively short growing season and the impossibility of long-range storage of the raw fruit.

A low stream factor is undesirable, for invested capital lies idle for extended time periods, and the labor force may face seasonal layoffs. In the chemical industry, production demand below full capacity is usually handled by a throughput slowdown, often with no change in the stream factor and only a modest reduction in the labor force. If several low-demand installations coexist in an integrated complex, and particularly if the demand is seasonal (antifreeze, planting season chemicals), management may resort to *block operation* of the facilities. With this approach, any one facility is operated at high rates of throughput for a limited time period; the labor force is then transferred to another facility for a limited time period, and so forth. Block operation does not eliminate the problem of idle capital, but it does increase the efficiency of labor utilization, provided the labor force is rapidly adaptable.

*Standard cubic feet.

7.2 RAW MATERIALS AND UTILITIES

Standard Cost

A projected plant may be scheduled to consume several of the commonly available utilities generated within an integrated complex (Exhibit 5.2A). Moreover, some of the scheduled raw materials may be, in turn, produced in some other plant within the complex; in the usual case, these raw materials are basic commodity chemicals such as chlorine, ammonia, purified salt, ethane, caustic soda, and hydrochloric acid. The question, then, arises as to how these particular utilities and raw materials should be charged to a contemplated or existing process. A common procedure is to charge in these entities at their *standard cost*.

The standard cost is nothing more than the COM incurred currently in a service facility supplying raw materials or utilities. In any one such facility the standard cost, often computed on a monthly basis, continuously changes, depending upon the service demand. For example, the standard cost of steam generated in an integrated complex steam plant will clearly depend upon the total steam demand, taken as a percentage of rated capacity. The steam cost is likely to fluctuate from month to month, but it is not difficult to compute a running average standard cost which may be used for new project evaluation (even though a new project, once operational, may well change the steam standard cost because of its effect upon demand—see Fig. 5.1.2).

The standard cost of a utility or raw material thus incorporates the associated out-of-pocket production costs plus depreciation, *but no return on the investment*, no profits. It therefore stands to reason that if a projected new process is to be charged standard costs, it is necessary to use the concept of allocated capital to determine the total investment that the project represents; the process must then show acceptable profit on *all* the invested capital.

Example 7.2.1

A proposed process will produce 100 million lb of product annually in a plant having a DFC of $20 million. The process will use 5 lb of steam/lb of product. Steam will be charged to the process at an integrated complex standard cost of $8/1000 lb; the steam plant allocated capital is $2/(1000 lb/year). Determine the unit cost for steam and the steam allocated capital to be added into the total investment.

SOLUTION

$$\text{Unit cost} = 5 \times {}^{80\cancel{c}}\!/_{1000} = 4\cancel{c}/\text{lb of product}$$
$$\text{Annual steam use} = 100 \times 10^6 \times 5$$
$$= 5 \times 10^8 \text{ lb of steam/year}$$
$$\text{Allocated steam plant capital} = \frac{5 \times 10^8 \times 2}{1000} = \$1 \times 10^6$$

This is 5 percent of the plant's DFC, an appreciable amount.

In the example, the profit from sales of the product must satisfy the demands of acceptable profitability not only on the $20 million DFC but on the $1 million steam plant allocated capital (and other allocated capital) as well. The allocated capital charge of $2/(1000 lb of steam/year) is determined for a particular integrated complex in pretty much the manner illustrated in Fig. 5.1.2. In the usual case the standard cost and the allocated capital charge are not corrected for the possible effect of the projected new plant, as suggested in Fig. 5.1.2, unless a very major impact is anticipated.

In some corporations the use of the standard cost, beefed up with an acceptable surcharge for return on investment, is preferred; allocated capital is then not added into the total investment. This approach has the advantage of simplicity, but it has two important disadvantages:

The return on investment surcharge is necessarily arbitrary—what is an acceptable return on the investment for a steam plant? With the allocated capital concept, each project is judged on its profitability performance with the allocated steam plant cost included; the profitability of the steam plant investment then becomes a blend of the profitabilities of processes making salable products.

An investment number excluding allocated capital does not reflect the true magnitude of the total investment that a proposed project represents; the project is not in proper perspective.

Example 7.2.2

The chlorination of ethylene to vinyl chloride in the presence of air may be represented by the overall reaction

$$C_2H_4 + \tfrac{1}{2}Cl_2 + \tfrac{1}{4}O_2 \longrightarrow C_2H_3Cl + \tfrac{1}{2}H_2O$$

Suppose an integrated complex has existing plants to produce ethylene from naphtha and chlorine via brine electrolysis (chloralkali complex), both with excess capacity. A 50,000 tons/year vinyl chloride plant is to be built which will use the excess chlorine and ethylene as complex-generated raw materials. Guthrie (1970) gives the following DFC values (in 1968 dollars) for the existing plants at their rated capacities:

Ethylene:

$$50{,}000 \text{ tons/year, DFC} = \$4.2 \times 10^6$$

Chlorine:

$$100{,}000 \text{ tons/year, DFC} = \$12 \times 10^6 \quad \text{(caustic coproduct)}$$

Guthrie estimates the DFC of the 50,000 tons/year vinyl chloride plant at 5×10^6. What would be the allocated capital investment for the complex-generated raw materials facilities (in 1968 dollars); what percentage of the proposed plant DFC does this represent?

SOLUTION   Assume stoichiometry as shown in the reaction. Molecular weights are 28 for ethylene, 71 for chlorine, and 62.5 for vinyl chloride. Therefore

$$C_2H_4 \text{ demand} = 50{,}000 \times 28/62.5 = 22{,}400 \text{ tons/year}$$

$$Cl_2 \text{ demand} = 50{,}000 \times \frac{71 \times 0.5}{62.5} = 28{,}400 \text{ tons/year}$$

$$C_2H_4 \text{ allocated capital charge} = 4.2 \times 10^6/50{,}000$$
$$= \$84.0/(\text{ton/year})$$

Thus

$$C_2H_4 \text{ allocated capital} = 22{,}400 \times 84 = \$1.9 \times 10^6$$

For Cl_2 allocated capital charge, let Cl_2 and NaOH bear the cost jointly on a total weight basis. Thus for brine electrolysis,

$$NaCl + H_2O \longrightarrow NaOH + \tfrac{1}{2}Cl_2 + \tfrac{1}{2}H_2$$

and the process produces 2 mol NaOH/mol Cl_2, or 1.127 lb NaOH/lb Cl_2. Thus the Cl_2 allocated capital charge is

$$\frac{12 \times 10^6}{(1 + 1.127)100{,}000} = \$56.4/(\text{ton/year})$$

$$Cl_2 \text{ allocated capital} = 28{,}400 \times 56.4 = \$1.6 \times 10^6$$

$$\text{Allocated capital for raw materials facilities} = \$3.5 \times 10^6$$

This is 70 percent of the vinyl chloride plant DFC.

In the above example, chlorine and ethylene could have been charged into the process at their sales price, and the allocated capital could have been neglected. The capital (DFC) requirement would then be listed as $5 million. Actually, however, the process will tie up an investment of $8.5 million, and this is a much more realistic appraisal of the project's economic impact. True, the $3.5 million balance has already been invested, but it is inseparable from the proposed plant capital; a portion of the service plants will remain inexorably dedicated.

We will return to the subject of allocated capital in the next chapter. For our present purposes we will use standard costs for nondedicated utilities, and for complex-generated raw materials in some situations. Given the standard cost and allocated capital charge, the informed estimator can readily convert these data into a cost which incorporates a *return on investment* (ROI), if this is the desired format. ROI, a widely used criterion of profitability, may be defined in a narrower sense for service facilities:

$$\text{ROI (\%)} = \frac{100(\text{actual unit cost charged} - \text{standard unit cost})}{\text{allocated capital charge per unit}} \quad (7.2.1)$$

Exercise 7.2.1

For the steam in Example 7.2.1, what should be the cost charged to the process for 1000 lb of steam to realize a 25 percent ROI on the prorated steam plant?

Raw Material Costs

The cost of raw materials is a major and often dominant element of the COM. The raw material *unit ratios* (i.e., weight of a particular reactant per unit weight of product) are determined from the appropriate process definition. Purchased raw material unit costs may be found in the weekly journal, *Chemical Marketing Reporter*, or other sources mentioned in Sec. 3.4. The estimator is again reminded to be on the careful lookout for pricing variations due to the effect of shipment size. USP anhydrous calcium chloride can be purchased for 90 cents/lb in 225-lb drums, but the cost skyrockets to $8.70/lb in 2½-kg fiber drums! (1981 costs.)

Exercise 7.2.2

In laboratory experiments, thiophene has been iodinated in the vapor phase over HgO catalyst to form 1-iodothiophene in 80 percent yield. Another 10 percent of the thiophene winds up as diiodothiophene, the rest as iodothiophene dimers and polymers having an average iodine-sulfur ratio of 1.0. Iodine vapor is fed in 25 percent stoichiometric excess, based upon desired monoiodo product. The excess iodine is recovered and recycled with a 10 percent loss of the recyclable material. What are the unit ratios of thiophene and iodine in a proposed plant, based on the laboratory experiments?

The estimator is also reminded that freight charges may add considerably to the listed prices of raw materials. Even bulk chemicals may incur shipping charges of 3 to 5 cents/lb, a factor in arriving at a decision whether to purchase aqueous or anhydrous chemicals. If 35 wt% hydrochloric acid may be shipped by tank car for 3 cents/lb, the actual cost per pound of contained HCl is 9 cents, not far from the sales price, FOB works! That is a lot of money to pay for shipping water, and the buyer may wish to consider the purchase of anhydrous HCl, in spite of its higher list price at the distribution point and the nuisance of handling liquefied gas.

On the other hand, many raw materials, particularly large bulk commodity chemicals, are sold at contract prices which may be well below the listed prices. During preliminary stages of project evaluation, the offsetting effects of freight charges and contract discounts are usually uncertain, and listed prices are best used for COM computations, unless there is reason to believe that freight charges will be a major cost factor. For example, 45% calcium chloride liquid is listed at about $30/ton in tank car lots at the works. If the liquid is required at a plant site in Seattle, since the nearest supply points are Michigan and southern California, it is quite likely that another 3 cents/lb, or $60/ton, will have to be added for freight, for a grand total of almost $100/ton for the raw material! The question of freight charges may often be settled rapidly by phoning listed distributors of specific raw materials (see, for example, the annual "Chemical Engineering Catalog," Van Nostrand Reinhold, New York).

For raw materials manufactured in another facility within the integrated complex, *in-plant transfer costs* may be calculated in various ways:

Large-volume basic commodity chemicals are best charged in at standard cost, with an allocated capital provision, as previously described.

Smaller-volume intermediates are often charged in at their sales price; the originating plant thus receives an acceptable return on investment.

The transfer cost is frequently internally negotiated. The negotiated price will be lower than the sales price, since some of the costs reflected in the sales price are saved—packaging, sales, storage, and other costs. If the material in question is in oversupply, or if it is a by-product difficult to market in the first place, it has to be disposed of anyway, and the internally negotiated transfer price can be quite low. The negotiation mechanism is a manifestation of corporate encouragement of internal competition; each plant superintendent has the responsibility of maximizing plant profits, and transfer cost negotiation is an opportunity for internal entrepreneurship.

In Chap. 5 we spoke of shortcut methods of estimating the DFC of integrated turnkey facilities to manufacture certain raw materials such as oxygen, and in Fig. 5.2.2 the DFC of oxygen plants was shown. The corresponding standard costs of the raw materials are of great utility for the purpose of early-stage estimates, and some have appeared in the cost literature (see, for example, Guthrie, 1970). Oxygen standard costs, corresponding to the plants in Fig. 5.2.2, are shown in Fig. 7.2.1.

Catalysts may be charged to a process in several ways. If a catalyst is expected to last for several years and represents an appreciable cash outlay (say, more than $10,000), it is often included as part of the plant DFC and depreciated on the plant's

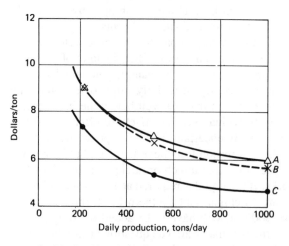

FIGURE 7.2.1
Standard costs of tonnage oxygen. Curve A: split-cycle plant; 450 psig discharge; 99.5% pure oxygen. Curve B: low-pressure plant; 0.5 to 5 psig discharge; compressed to 450 psig; 95% pure oxygen. Curve C: low-pressure plant; 0.5 to 5 psig discharge; 95% pure oxygen. (Source: Katell and Faber, 1959. Reproduced by permission from Chemical Engineering.)

Straight-line depreciation, 15-year life
Time basis: 1959 (M&S Index = 235)

schedule. If it must be substantially replaced within a time frame of, say, 2 years or less, it is then handled as a normal raw material. Catalysts, drying agents, and similar "stationary" chemical charges which are not in contact with primary process streams are usually included under "operating supplies" (see the discussion in Sec. 7.4).

Utilities Demand and Costs

The cost of raw materials is perhaps the most straightforward example of a variable charge—the cost is directly proportional to production level. The cost of utilities is also considered to be a variable charge, even though, strictly speaking, direct proportionality to production level does not prevail. The cost of illuminating a plant is not terribly production-dependent, provided the production level is controlled by varying throughput rate, not by periodic plant shutdowns.

The utilities demand in a projected plant can be read off directly from equipment lists such as the one in Table 4.1.1, and a grand total is generated for each utility category. Steam demand should be increased by about 10 percent to account for heating of buildings, and power demand should be similarly increased to provide for illumination and air conditioning. If detailed equipment lists have not yet been generated as part of the project evaluation effort, some idea of utility demand may be gained by analogy to similar processes within the integrated complex. Existing process utility demand information has been published (Guthrie, 1970). Some typical process utility demands are listed in Table 7.2.1.

The cost of utilities in the decade of the 1980s is a vexing subject. Rapid escalation of utility costs is an uncomfortable aspect of industrial as well as private existence. The escalation is the result of a number of interrelated factors, such as inflation, deregulation of domestic energy supplies, impact of OPEC pricing structure, and potential impact of nonconventional energy sources. The problem is illustrated by the recent and projected composite cost of petroleum to industrial consumers, shown in Fig. 7.2.2. The composite cost is an appropriate blend of domestic and import prices. Note the dramatic rise in the acquisition price of imported oil and the steady rise into the eighties as a result of domestic price escalation culminating with a projected cost of about $100*/bbl in 1990! The cost of low-sulfur fuel oil is about 15 percent higher than the costs in Fig. 7.2.2—$35/bbl in 1980, projected to $115/bbl in 1990.

For fuel oil delivered at a cost of $A/bbl, the equivalent energy cost is given

*All costs are in "cocurrent" dollars—inflated dollars for a particular year.

TABLE 7.2.1
TYPICAL PROCESS UTILITY DEMANDS

	Intensive	Normal
Power, kWh/lb product	1	0.1
Steam, lb/lb product	10	2.0
Cooling water, gal/lb product	40	5.0

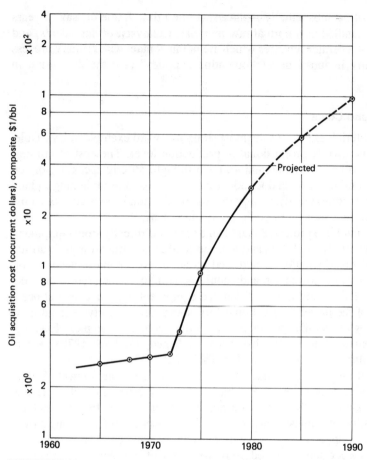

FIGURE 7.2.2
Oil acquisition cost to industrial consumers. *(Data from "U.S. Energy Data Book," 1980, by permission of Power Systems Sector, General Electric Company.)*

approximately by the expression,

$$\text{\$/million Btu} = \frac{A}{6} \tag{7.2.2}$$

Exercise 7.2.3

Derive your own version of Eq. (7.2.2) from available data in handbooks. Is the equation based on HHV or LHV?[*]

Natural gas prices (Gulf Coast location) are another example of rapid energy cost

---
[*]Higher or lower heating value.

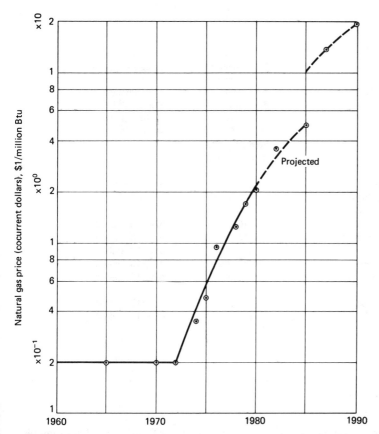

FIGURE 7.2.3
Natural gas prices, Gulf Coast. *(Data from Chemical Week, by permission. All rights reserved by McGraw-Hill Publishing Co., Inc.)*

escalation, compounded by anticipated deregulation after 1985 (Fig. 7.2.3). Projected prices beyond 1985 are predicated upon parity with low-sulfur fuel oil. The projected hundredfold escalation in two decades emphasizes the difficulty of obtaining reliable utility standard costs for use in COM computations; much the same kind of uncertainty exists in the costs of steam and power, both of which are strongly fuel cost–dependent. The estimator in the chemical process industry is advised to check frequently with the corporate accountants and planners responsible for monitoring and projecting utility costs. Good utility cost reviews do appear in the literature (Johnnie and Aggarwal, 1977), but projections should be constantly checked against current trends.

For present purposes, the utility costs in Table 7.2.2 may be taken as representative. It must be emphasized that the costs shown are *not* the prevailing ones in 1981. They are costs based upon projected energy costs beyond 1985, but discounted back to 1981 for an average inflation rate of 10 percent/year. The information in the table should therefore be useful for evaluating projects which are likely

TABLE 7.2.2
REPRESENTATIVE UTILITY COSTS
(In 1981 Dollars, Discounted from 1985)

Item	Standard cost	Allocated capital
Purchased power, distributed	4.8¢/kWh	1.7¢/(kWh/yr)
Steam, 150 psig	$8.90/$10^3$ lb	$1.50/($10^3$ lb/yr)
Steam, 600 psig	$11.60/$10^3$ lb	$2.90/($10^3$ lb/yr)
Natural gas, delivered	$6.80/$10^6$ Btu*	4¢/(10^6 Btu/yr)
Fuel oil, low-sulfur	$6.80/$10^6$ Btu	
Low-sulfur distillate	$8.60/$10^6$ Btu	
Cogenerated ac power	4.8¢/kWh	5.6¢/(kWh/yr)
Cogenerated steam, 150 psig	$7.80/$10^3$ lb	$2/($10^3$ lb/yr)
Clarified raw water	25¢/10^3 gal	57¢/(10^3 gal/yr)
Recycled cooling tower water	21¢/10^3 gal	56¢/(10^3 gal/yr)
Well water	59¢/10^3 gal	96¢/(10^3 gal/yr)
Process water (softened)	55¢/10^3 gal	86¢/(10^3 gal/yr)
Demineralized water (or condensate)	$5/$10^3$ gal	$20/($10^3$ gal/yr)
Compressed air, 100 psig	18¢/10^3 scf	10¢/(10^3 scf/yr)
Inert gas, low-pressure	$1/$10^3$ scf	60¢/(10^3 scf/yr)
Nitrogen, purchased	$2.20/$10^3$ scf	
Instrument air	$18/(month/scfm total cost) (includes ROI)	

*Approximately 1000 Btu/scf

to be realized in the middle eighties and beyond. The 1981 dollar numbers may be updated by assuming a 10 percent annual inflation rate.

The information in the table includes the standard costs and allocated capital *(backup capital)* associated with each utility. Both these numbers are, of course, dependent upon the size of the generating facilities; the figures shown are typical for the relatively large generating facilities in a diversified integrated complex. For example, the allocated capital for purchased power represents the investment in a 20,000-kVA substation and distribution feeders, a sizable industrial installation even for many electrochemically based operations. The tabulated information has been assembled from a number of sources and modified to conform to the projected energy costs in Figs. 7.2.2 and 7.2.3. The costs do not correspond to any one geographical area, although Gulf Coast costs are predominant in the data sources. Escalating energy costs tend to eliminate geographical differences in energy-intensive utility costs. Nevertheless, significant geographical differences may be anticipated for utilities such as clarified raw water; the numbers shown are a rough average value.

The indicated costs of power merit additional comment. The cost of purchased power has not kept up with escalating energy costs, partly because of the diversity of electric power sources (for example, hydroelectric power), partly because fixed capital charges in the power costs are often based on old facilities. The trend of purchased power costs to industry in the Gulf Coast is shown in Fig. 7.2.4. The 4.8 cents/kWh cost in Table 7.2.2 is discounted from the projected 1985 cost in the graph. It must be recognized that purchased power costs are likely to escalate more rapidly throughout the eighties (up to a projected 11 cents/kWh in 1990) as new power plants, many coal-fired, come on-stream; the power cost escalation rate is expected to conform more closely to the fossil fuel energy cost escalation rate. It must also be recognized that appreciable U.S. geographical differences did occur in purchased power costs at the start of the eighties, although these differences are likely to shrink with escalating energy costs.

The lag between purchased power and energy costs has not encouraged the chemical industry to invest in *cogeneration* facilities, in spite of the fact that such facilities lead to a very considerable conservation of energy. Cogeneration refers to the simultaneous onsite generation of electric power and steam. A variety of systems can be used for cogeneration—hot gas turbines followed by a boiler plant,

FIGURE 7.2.4
Cost of purchased power (Gulf Coast). *(Data from "U.S. Energy Data Book," 1980, by permission of Power Systems Sector, General Electric Company.)*

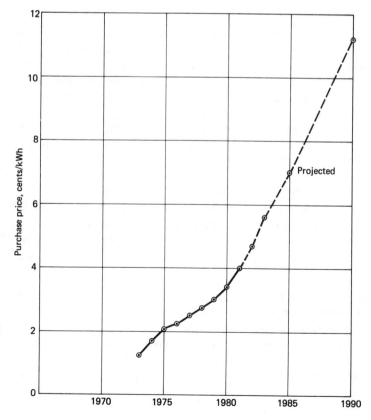

or a high-pressure steam plant followed by steam turbines. Many energy sources have been proposed for cogeneration plants, including coal, solid wastes, and nuclear fuels. For the most part, however, the chemical industry has found it more economical to buy the relatively cheap power generated by the giant electrical utility companies; this attitude may well change as purchased power costs escalate. Table 7.2.2 includes projected standard costs and allocated capitals for cogenerated power and steam.

Exercise 7.2.4

A process to produce 50×10^6 lb of a new product annually is expected to have normal utility demands (Table 7.2.1). If 25 percent ROI on allocated capital is required [see Eq. (7.2.1)], which of the following two choices will result in the lesser utility charge to the process?

    a Purchased power, generated 150 psig steam
    b Cogenerated power and steam

7.3 OPERATING LABOR

Estimation of Labor Requirements

With only minor exceptions, chemical process plants are operated 24 hours a day, 7 days a week. The workday is subdivided into three 8-h shifts, and normally the same number of operators works each shift, although additional "utility" operators may be employed on the day shift for product loading or routine maintenance.

The first step in estimating operating labor requirements is to determine the number of operators likely to be needed on each shift. The operating labor category includes individuals with "hands on" responsibilities of operating the plant, shift supervisors, and utility operators for semiskilled labor requirements. Several methods are commonly used for estimating operating labor.

Step-by-Step Analysis of Process Estimation of labor requirements by carefully considering all the jobs that need to be done to keep a plant smoothly operational is clearly the best approach, but one that is usually confined to advanced stages of a project. Nevertheless, a reasonable idea of the requirements may be gained by a commonsense review of the usual jobs that operators are called upon to perform:

    Process monitoring in control room (includes maintenance of logbooks, record keeping, instrument maintenance, control computer operation)

    Process monitoring in field and inspection tours

    Sample collection; simple analyses (such as titrations), control lab data monitoring

    Routine maintenance (lubrication, seal replacement, heat-exchanger cleaning, filter cartridge replacement)

Batch operations and other equipment manipulations not under fully automatic control

Loading, unloading, and transfer of raw materials, intermediates, and products

Packaging and warehousing operations

Troubleshooting, simple emergency repairs, unplugging lines and strainers

Equipment cleanup, housekeeping

Training and refresher courses, safety meetings

In dividing up the various chores, the planner must keep in mind that during routine operations operators are not "on the go" 100 percent of the time. Some time must be allowed, for example, for meal breaks, although these are best accounted for by proper scheduling and extension of the shift duty beyond the nominal 8 h; such extension also accommodates a shift "overlap" during which adjacent shift operators can exchange information to ensure smooth transition and operational continuity. Normally operators are asked to change clothes and shower on their own time. Nonroutine operational periods (major emergencies, equipment overhaul during plant shutdowns, plant start-ups) often demand considerable overtime work.

For further detail on step-by-step analysis, see Haines (1957).

Labor Requirement of Process Equipment Estimated labor requirements for operating specific types of process equipment (plus peripherals) have been published (Aries and Newton, 1955). These kinds of estimates are usually based upon labor-intensive situations and are of limited value for modern computer-controlled plant estimates, for they tend to predict an enormously excessive labor requirement. The equipment-based correlations are useful in projecting pilot plant and semiplant labor requirements.

Example 7.3.1

The key equipment components of a semiplant are listed, along with the estimated operator-hours per shift from Aries and Newton (1955):

Item	Operator-hours per shift
Reactor, batch	8.0
Crystallizer	1.0
Filter, rotary	2.0
Tray dryer	4.0
	15.0
Say, two operators per shift	

The semiplant in question is computer-controlled and is, in fact, operated by one operator plus an engineering supervisor each shift.

Estimate by Analogy with Other Similar Processes Estimation by analogy to other processes is perhaps the most common method of projecting labor requirements that is used in practice. In diversified chemical corporations, there is a considerable body of experience to draw upon, and errors in labor projection before plant start-up are adjusted as the need arises during subsequent operations. In projecting labor requirement by analogy, the estimator should keep a few points in mind:

The tendency in modern plants with automatic control is "to run labor-lean," to use as few operators as possible within the framework of demands for safety and reliable production.

Proper consideration must be given to the effect of plant size upon labor requirements (see below).

Automatic control reduces labor requirements.

Computer control (over and above automatic control) in general does *not* reduce labor requirements. The greatest benefit of computer control is in the improvement of product quality and in providing greater operational safety.

Guthrie (1970) lists operating labor demand for a number of common chemical processes.

Scale-up (or Scale-down) from Existing Processes with Known Operating Labor Requirements The plant size has some effect upon the total number of operators required to operate the plant. The relationship may be expressed approximately by using the format of the cost-capacity equation for the six-tenths factor rule for equipment [Eq. (4.2.3)]:

$$\frac{N_B}{N_A} = \left(\frac{r_B}{r_A}\right)^{0.25} \qquad (7.3.1)$$

Here N and r refer to the number of operators per shift and plant capacity, respectively, of two plants, A and B, different in capacity, but producing the same product. The capacity exponent is about 0.25.

Example 7.3.2

A new 2.5×10^5 ton/year propylene oxide plant will use a catalytic air oxidation of propylene process, not unlike existing ethylene oxide processes. Guthrie (1970) lists the operating labor requirements in a 1×10^5 ton/year ethylene oxide plant as about 0.25 operator-hours/ton. How many operators per shift are likely to be needed in the propylene oxide plant?

SOLUTION Using A as the subscript for ethylene oxide,

$$\text{Tons of ethylene oxide/shift} = \frac{1.0 \times 10^5}{365 \times 3}$$

$$= 91.32 \text{ tons/shift}$$

$$\text{Operator-hours/shift} = 91.32 \times 0.25 = 22.83$$

There are 8 operator-hours per operator; therefore

$$N_A = 22.83/8 \approx 3 \text{ operators/shift}$$

From Eq. (7.3.1),

$$N_B = 3\left(\frac{2.5}{1.0}\right)^{0.25} = 3.77 \approx 4 \text{ operators/shift}$$

(This is rather lean staffing for a plant of this magnitude.)

Wessel Correlation In the absence of any other information, the correlation of Wessell (1952), modified to conform to modern practice, may act as a useful guide (see Fig. 7.3.1). The operating labor demand is shown as operator-hours per ton per "operating block," as a function of plant capacity.

In the original publication, Wessel showed three lines, each one corresponding to a different degree of automation; only the line for *highly automated fluid processing plants* has been retained in Fig. 7.3.1, since this one appears to conform best to modern plant labor requirements. Furthermore, the original correlation was made in terms of operator-hours per ton per "step," where the step was defined in much

FIGURE 7.3.1
Operating labor demand. *(Adapted from Wessel, 1952; by permission of Chemical Engineering.)*

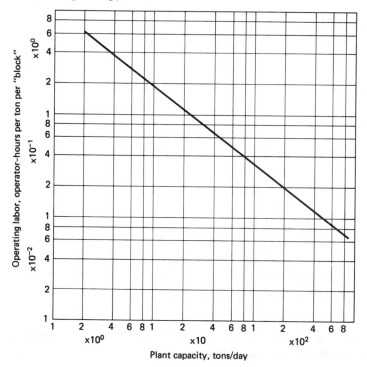

the same manner as the "functional step" in Sec. 5.1. This procedure again gives unrealistically high numbers for operator demand. A better procedure is one in which the labor-defining step is taken as a whole "operating block," analogous to the blocks of the prototypic chemical process of Fig. 3.5.1. As was the case with functional steps, the operating block concept is somewhat diffuse and uncertain, and its use represents a rough guess at best. A great many fluid processes may be represented by the four operating blocks in Fig. 3.5.1. For example, the ethylamine process in Fig. 3.5.3 may be subdivided as

Block 1: Raw material conditioning. Feed stream treatment, absorber.
Block 2: Reaction. Reactor, heat interchange, neutralizer-quencher.
Block 3: Product separation. NH_3 stripper, amines column, ethylamine distillation.
Block 4: Product conditioning. Cracking furnace, scrubber, brine treatment, product loading.

The idea is that the highly automated equipment in each operating block can be handled, on the average, by one shift operator. Solids handling, even when highly automated, requires more operator attention, and reasoned adjustments to the number of operators (or number of operating blocks) should be made.

Example 7.3.3

The urea process via ammonium carbamate intermediate involves four block operations:

Synthesis section
Carbamate decomposition and recycling
Evaporation and prilling of product
Pollution abatement

What is the estimated operating labor requirement in a ½ billion lb/year plant with a 0.95 stream factor?

SOLUTION

$$\text{Daily capacity} = \frac{0.5 \times 10^9}{365 \times 0.95 \times 2000} = 721 \text{ tons/day}$$

From Fig. 7.3.1,

$$\text{Operator-hours/ton/block} = 0.074$$
$$\text{Operator-hours/ton} = 0.074 \times 4 = 0.296$$
$$\text{Tons/shift} = 721/3 = 240$$
$$\text{Operator-hours/shift} = 0.296 \times 240 = 71.1$$
$$\text{Operators/shift} = 71.1/8 \approx 9 \text{ operators}$$

The figure in Example 7.3.3 of 0.296 operator-hours/ton is quite typical of large-scale organic synthesis processes (Guthrie, 1970).

Exercise 7.3.1

Using Fig. 7.3.1, make an estimate of the number of operating blocks in the propylene oxide process in Example 7.3.2.

Exercise 7.3.2

Figure 7.3.1 may also be thought of as representing the relationship between labor demand and size of plant for a particular process. How well does the graph conform to the capacity exponent rule for labor scale-up?

Once the number of operators per shift has been established by one of these methods, the total number of operators needed can be obtained from the approximate expression

$$\text{Total operators} = 4 \times \text{operators/shift} \qquad (7.3.2)$$

Now, even though there are three shifts per day, it is obvious that more than three times the number of operators per shift must be hired to accommodate a 40-h week, vacations, holidays, sick leave, and other emergencies. How well does Eq. (7.3.2) work out? If we assume that the average operator receives 4 weeks of vacation and holidays, the "normal" annual working hours are 48×40, or 1920. Actually, when overtime is included, the average operator works

$$2080 \text{ h/year}$$

Thus, if four operators are provided to satisfy the demand for each one operator per shift, the four work $4 \times 2080 = 8320$ h in 1 year. But there are 8760 h in a year; therefore, the formula (7.3.2) is valid for a stream factor of $8320/8760 = 0.95$, a common stream factor in chemical plants.

In point of fact, the simple relationship of Eq. (7.3.2) does not work out all that well in practice, even in a plant with the stream factor just equal to 0.95. For one thing, the operators do not just go home during scheduled plant shutdowns. Many, if not all, join maintenance workers in accomplishing needed repairs, overhauls, and improvements. Even if the operators' help were not needed during shutdown periods, the tight staffing based on Eq. (7.3.2) would result in almost insurmountable work period scheduling difficulties to accommodate vacation time requests, holidays, consecutive days off, and shift rotation (few of us cherish working the graveyard shift the whole year through). Scheduling flexibility is obtained by adding two or more operators to the total from Eq. (7.3.2); in fact, operators may be borrowed from and returned to an integrated complex "labor pool" as the need arises.

CHAPTER 7: THE COST OF MANUFACTURE

Labor costs are considered to be a *regulated* charge; that is, the costs decrease somewhat with a decrease in production rate, but not in proportion. It is true that if a plant is scheduled to be operated at reduced throughput, some operators may be returned to the labor pool, in rough conformity to the relationship in Eq. (7.3.1). Normally, however, the operating staff remains constant in numbers through upward and downward production swings.

Labor Rates

Operators are paid according to the "level" of their job category. For purposes of COM computation, an average rate typical of chemical plants is wanted. In 1981, this average at a Gulf Coast location was about

$$\boxed{\$13 \text{ per hour}}$$

This rate, however, is a base rate which does not account for the overtime which boosts the annual working hours from 1920 to 2080 h at an overtime rate of 1½ times the base rate, or more. The following rules account for this pay escalation for the purposes of COM computation:

The annual base rate is $13 \times 1920 \approx \$25,000$/year.

The base rate varies somewhat across the United States, with a range of about ±10 percent and an increasing gradient from east to west.

Even though the total operating staff equals four times the number of operators per shift, plus a few extras, by the time overtime and shift differentials are accounted for, the effective annual labor cost is equivalent to *five* times the base rate for the number of operators on each shift; that is,

$$\text{Annual labor cost} = \$125,000 \times \text{no. of operators/shift} \quad (7.3.3)$$

Equation (7.3.3) is valid for the year 1981. Short-range escalation of 10 percent/year is acceptable. Up-to-date labor rate information may be found in the *Monthly Labor Review* of the U.S. Bureau of Labor.

Example 7.3.4

What is the unit labor cost for urea production in Example 7.3.3?
SOLUTION With 9 operators per shift, from Eq. (7.3.3),

$$\text{Annual labor cost} = 9 \times 125,000 = \$1,125,000$$
$$\text{Production} = 0.5 \times 10^9 \text{ lb/year}$$

Therefore

$$\text{Unit labor cost} = \frac{1.125 \times 10^6 \times 100}{0.5 \times 10^9} = 0.225\text{¢/lb}$$

The results of the example demonstrate that labor costs tend to be small and not at all dominating in large, highly automated plants. In small plants, however, labor costs tend to be large and dominant, and small-scale projects may be killed by requirements of labor. If the semiplant in Example 7.3.1 produces 200,000 lb of a fine chemical/year, the unit labor cost is about $1.25/lb!

Exercise 7.3.3

Compare the average annual base rate for operators with the starting salary of B.S. chemical engineers from data in Chap. 1.

Fringe Benefits

A labor-related charge not included in the basic labor rate is that incurred for fringe benefits. These are expenditures paid out by the corporation for various benefits which accrue to its employees. The number and variety of such benefits are often quite impressive, but the actual spectrum depends upon the fundamental commitment of the corporation to the employees' welfare and, in some cases, upon a negotiated equilibrium between union demands and corporate concessions.

The following list incorporates some of the many fringe benefits:

Vacation pay	Social Security payments
Pension plans	Medical plans
Life insurance	Stock acquisition plans
Sick pay	Special courses, training programs
Social affairs	Civic activity support
Support for athletic programs	

Fringe benefit charges amount to an average of about 22 percent of the basic operating labor cost. If the corporation carries a dental plan, the factor increases to 26 percent.

The Cost of Supervision and Clerical Help

The efforts of operating labor are supplemented with appropriate help from an auxiliary clerical staff; both groups are supervised by the plant management. Minimum plant nonoperational staffing includes:

Plant superintendent	Plant supervisor
Assistant superintendent	Clerk
Plant engineer	Secretary

The number of nonoperational staff members may be considerably larger than this list would indicate, particularly in large plants or plants of unusual operational complexity. The cost of supervision and clerical help may be estimated as a percentage of the operating labor base rate (not including fringe benefits). The range is 10 to 25 percent, with an average of about 18 percent; this percentage *does* include the fringe benefits for the supervisors and clerks.

The indicated range and average both apply to the usual large production plant. For small units and limited-scope projects, separate estimation of the cost of supervision is preferred. In 1981, the "average" large-plant superintendent received $42,000/year (plus fringe benefits), junior supervisors and small-unit supervisors perhaps $10,000 less. Clerical help may be estimated at the pay scale of lower-level operators (about $20,000/year).

7.4 REGULATED AND FIXED CHARGES

Maintenance

The costs of plant maintenance are somewhat dependent upon production level, and as such are considered to be part of the regulated charges. Routine maintenance is performed by operating personnel and is therefore not separately costed. Some examples of routine maintenance are:

Heat-exchanger tube cleaning
Mechanical seal replacement
Lubrication

In properly managed production units routine maintenance operations are scheduled and assigned as an important part of the operators' duties. Major nonroutine maintenance and repairs are carried out by maintenance department employees (riggers, fitter-welders, electricians, painter-insulators, instrument technicians), by maintenance contractors, or by both working jointly.

About one-half of maintenance charges are expended for labor (including engineering department help), the other half for maintenance materials—items such as spare equipment, gaskets, piping supplies, and janitorial supplies. Maintenance supplies are often provisioned from a central storeroom, and the charges for maintenance materials should reflect prorated costs of maintaining the storeroom.

The total cost of maintenance (labor plus materials) is usually estimated as a percentage of the DFC per year. An average value is

$$6 \text{ percent of DFC per year}$$

although considerable variation is found in practice—up to 20 percent/year. Higher maintenance costs occur in plants with a high proportion of mechanical equipment (solids handling), those with unusually severe conditions of corrosion (which often dictate the use of exotic materials of construction), and, of course, those plants in which construction shortcuts and improvisations had been undertaken to reduce the DFC burden. The thoughtful estimator should escalate the recommended average judiciously to account for these unusual conditions.

The 6 percent of DFC maintenance cost is meant to be used for a plant operated at rated capacity. At less than rated capacity, maintenance costs are reduced to a level approximated by the following schedule:

Production level, as % of full capacity	Relative cost of maintenance %
100	100
75	85
50	75
25	60

Process Analysis and Quality Control

Chemical analysis and physical property characterization of process streams is a vital adjunct of chemical plant operations. Property monitoring spans the full spectrum of operations, from raw materials to final product.

Raw material shipments are analyzed to ensure compliance with required specifications. Spot checks may be used to avoid unloading and mixing of erroneous shipments which occasionally have led to catastrophic consequences. The analysis of intermediate streams is frequently an integral part of the operating sequence, and feedback information (often transmitted from on-stream analyzers) is used to control the progress of reactions or the performance of separation devices. Bulk products are frequently prestored in check tanks or bins and are not forwarded to final storage and packaging until analysis has shown that all specifications have been met. Failure to meet specifications may require blending with high-quality products, reprocessing, or, in the worst case, disposal as waste. Finished products must be analyzed at least by *lot,* although each *shipment* may have to be analyzed separately, particularly if different customers have differing sets of specification requirements.

Clearly, process analysis and quality control may impose a heavy cost burden upon a process. Intermediate stream analysis is often accomplished by on-stream instrumentation or by means of simple analytical techniques entrusted to the operators (titrations, pH measurements, refractive index measurements); such analyses need no separate costing. Samples submitted to a control laboratory are another matter, however. In 1981 the average wet analysis cost about $40, the average instrumental analysis perhaps $20. In view of these costs and the number of analyses demanded by process considerations, environmental monitoring, and abnormal conditions or emergencies, the annual control lab bill may grow to formidable proportions.

In the absence of more detailed estimates, the annual control laboratory charges may be estimated at *10 to 20 percent of the basic annual cost of labor*. The possibly puzzling relationship between laboratory charges and labor is explained by the common dependence upon the number of process operating blocks and the roughly common dependence upon plant production levels.

Waste Treatment

Waste treatment facilities should be designed and costed out as an integral part of the main process. Occasionally, however, there are generally available waste treatment facilities within the integrated complex for specific disposal tasks. If use of these kinds of facilities is contemplated, appropriate charges must be added to the other cost components of the COM.

Waste *gases* may, in certain instances, be forwarded as partial feed into integrated complex boiler furnaces. Clearly, such gases and their combustion products must be compatible with the furnace materials of construction and the furnace emission standards. If gross quantities of inerts are combined in the waste gases (N_2, CO_2, H_2O), an energy penalty may exceed the credit for the heat of combustion of the combustible component of the waste; the proper balance must be accounted for in the COM computations. Clearly, if waste gases are to be fed into a boiler furnace along with normal fuel and air, the waste stream must be small enough so as not to affect deleteriously the principal combustion process. Sulfur, nitrogen, or chlorine-containing molecules in the waste gas are generally cause for instant rejection of the idea of disposition in a boiler furnace. Gaseous wastes, such as CO, hydrocarbons, or organic oxygen-containing compounds, are well-suited for this method of disposal.

Very few other commonly shared facilities are available for waste gas disposal. Occasionally flare stacks may be available to accommodate the discharge of combustible vapors from emergency vents, but more usually such devices must be constructed as units dedicated to a particular plant. Similarly, scrubbers such as caustic soda scrubbers for acid gases (SO_2, H_2S, Cl_2, HCl) may be available at adjacent plants in the complex for handling modest volumes of waste gases.

Shared facilities for handling *liquid* wastes are more commonly available. Foremost among these are systems for handling aqueous process wastes—facilities for primary treatment (settling, filtration), secondary treatment (biooxidation), and even tertiary treatment (chemical treatment, adsorption). In dry weather areas, evaporation ponds (Chap. 5) offer an economical method for aqueous waste disposal. Combustible liquid wastes may, in certain instances, be burnt in existing boiler furnaces, and even aqueous wastes may be introduced as a spray into such furnaces to eliminate toxic contaminants. Phenolic water wastes may be handled in this manner; again, the volumes introduced into the furnaces must be compatible with the existing combustion regime, and a penalty charge equivalent to the additional energy requirement must be computed as part of the COM. *Thermal oxidizers* for burning liquid organic wastes and recovering the combustion products may occasionally be available as an auxiliary facility, but it is very difficult to design such units to handle a wide spectrum of feeds; usually they are designed as units dedicated to the disposal of a specific waste.

The costs of treating sanitary water are included in the *factory expense* category.

Disposal of *solid* wastes constitutes a particularly perplexing challenge. One of the few commonly available shared facilities for solids disposal is an incinerator, but its use is restricted to solids which burn without creating air pollution hazards. Land burial sites are often available within the confines of an integrated complex; again, however, this method of disposal is restricted to wastes which will not result in a potentially toxic leachate.

Reputable disposal companies exist which will accept both liquid and solid wastes, usually in 55-gal drums, for proper land burial. Only a relatively few sites have received federal and state agency approval for this method of disposal. It must be understood that the burial process does not guarantee permanent disposition; at properly managed sites, careful records are maintained on each drum, including the burial site coordinates, and the originator is identified as the "owner" with ultimate responsibility. The cost of this method of disposal is very high, up to $200 per drum, or about 40 cents/lb of contained waste (1981). Clearly, this is not the way to get rid of wastes from large-scale producing plants, and proper disposal methods remain within the realm of the process engineer's creative imagination. Nevertheless, external disposal of drummed (or trucked) wastes is practiced, especially in small-scale experimental units (pilot plants).

The costs of waste treatment in auxiliary facilities must be estimated for each situation (see Probs. 18, 20 to 22, 25 to 27), but frequently useful information does appear in the cost literature. Figure 7.4.1, for example, reproduces unit costs of primary, secondary, and tertiary water treatment processes.

Operating Supplies

Operating supplies include everyday items required to keep equipment in proper running condition, as well as tools, clothing, and protective devices for operators. Filter aids, instrument charts, lubricants, column packing, and respirators are some examples. The cost of operating supplies is estimated at *10 percent of the basic annual labor rate*.

Fixed Charges

Those components of the COM which remain constant regardless of the production level are called *fixed charges*. *Depreciation,* covered in some depth in Chap. 6, is one of the fixed charges; annual depreciation charges remain the same no matter what the production level is, even if the plant is shut down. For preliminary evaluation purposes, depreciation charges are computed as straight-line depreciation taken over a 10-year plant life, with zero salvage.

Insurance and taxes constitute another category of fixed charges. Insurance rates depend to a considerable extent upon the maintenance of a safe plant in good repair condition. The processing of flammable, explosive, or dangerously toxic materials

	Cost, cents/1000 gal For various plant sizes, mgd*					Cumulative percent removals				
Process	1	3	10	30	100	BOD	COD	S.S.	P	N
Primary treatment	4.4	3.2	2.4	2.0	1.7	35	35	50	5	5
Secondary treatment	5.5	4.0	3.2	2.9	2.7	80-90	50-70	80-90	25-45	30-40
Sludge handling	10.8	7.9	6.2	5.2	4.5					
Disinfection (Cl$_2$)	0.8	0.7	0.6	0.6	0.6					
Sand filtration	7.5	5.2	3.5	2.5	1.6	85-95		>95		
Activated carbon⁺	16.0	12.0	8.0	5.8	4.1	>95	>95			
Chemical precipitation	6.0	4.0	3.8	3.5	3.3				90-95	
Ammonia stripping⁺	3.3	2.0	1.6	1.5	1.4					90
Electrodialysis§	20.0	17.0	14.0	11.0	9.0				>95	>95

*mgd = millions of gallons per day
⁺Granular carbon
‡Not including pH adjustment
§Not including brine disposal

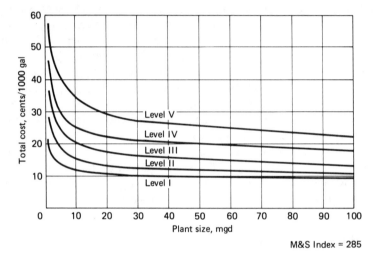

M&S Index = 285

FIGURE 7.4.1
Costs of wastewater treatment. Level I: primary and secondary treatment, sludge handling, and disinfection. Level II: level I plus sand filtration. Level III: level I plus activated carbon. Level IV: level I plus chemical coagulation, ammonia stripping, and activated carbon. Level V: level I plus activated carbon and electrodialysis. *(From Smith and DiGregorio, 1970. Reproduced by permission from Chemical Engineering.)*

may drive insurance rates far above the normal level of about 1 percent of DFC per year. The choice of a nonflammable solvent for vegetable oil extraction may be justified on the basis of lowered insurance rates alone.

The taxes within this cost category refer to local property taxes, not income taxes. Income tax is not considered to be part of the COM; it is introduced in the context of the analysis of the criteria of economic performance of the project and of the corporate entity. Property taxes are based upon the assessed value of the fixed plant (real property) and the movable fixtures. The DFC is the basis for the assessed value, but the latter may be appreciated or depreciated for tax roll purposes according to local rules of assessment. The property tax burden is quite dependent upon locale of the plant site; in some localities franchise taxes for the privilege of doing business are collected, whereas in others the local tax burden is minimized as a means of attracting new investments. An average value for the insurance and taxes category is *3 percent of DFC per annum.*

We have seen that plants cannot be operated without the support of process-oriented auxiliary facilities. There are many other facilities which are not directly involved in the manufacturing process but which, nevertheless, are important to the continuing smooth operation of a production unit. Some examples are:

Accounting department
Payroll department
Engineering
Fire protection
Security
First aid
Cafeteria
Administration
Planning

The costs incurred by the operation of these support organizations must be shared by all operating units (or by the one operating unit in a grass roots location). This shared "overhead" cost is often called *factory expense,* and it is still another fixed charge, charged against the plant even if it is shut down. A reasonable estimate of factory expense is

5 percent of DFC per annum

We continue to adopt the policy in this text of not including interest on invested capital as an operating expense in individual plants. This policy is based upon the uncertainty as to whether the capital for a projected plant will come from borrowed funds or from retained earnings. There are those who argue that retained earnings which are invested in new facilities should be considered to be equivalent to borrowed funds, borrowed from the corporate "bank," and that these too should draw interest. Therefore, the argument goes, interest based upon a prime borrowing rate should be incorporated in the COM as a fixed charge, regardless of the origin of the capital. There is validity to this argument, and each engineer must adapt to the particular corporate policy on this matter.

7.5 THE TOTAL COST FOR SALE

Cost of Process-Oriented Technology

Two items on the list in Fig. 7.1.1 have not so far been mentioned; both are concerned with technology acquisition. The first concerns *royalties.* If the process, any part of the process, or any of the equipment used in the process are covered by a patent not assigned to the corporation undertaking the new project, permission to use the teachings of the patent must be negotiated, and some form of royalties is usually required. The licensing agreement usually calls for a flat charge per unit of product or else a percentage on the sales dollar.

Royalties rarely exceed 5 percent of the COM: *2 percent of the COM* is the usual level of payments. Large corporations generally shy away from licensed processes, not only because of the expense involved but especially for fear of entering into a no-win competition with other licensees. The feeling is that "we are big enough to come up with our own way that is better," and attempts to bypass patents are commonplace (Chap. 1). Equipment or equipment components are more frequently licensed; examples of these are electrolytic cell dimensionally stable anodes and permanent diaphragms.

The other technology-acquisition-oriented item in Fig. 7.1.1 is marked down as *research*. It must be stressed that the research charges considered to be part of the COM are not prorated companywide R&D charges, not even the R&D charges to develop the process that is being evaluated. Such research charges are considered to be part of the general expenses which will be further treated in the next chapter. The research in question includes only those research charges which are directly related to the present plant operations, charges to cover research work to keep the plant in good operational condition. The production-plant-oriented research includes projects to improve the processing sequence or to solve production problems; it may also include extra costs of starting up the initial plant or subsequent improvements. Note that extra costs for a start-up which involves the use of research consultants equal $75,000/person-year (1980); for example, six researchers working 2 months will incur a charge of $75,000.

The research category may be estimated as *1 percent of sales*.

Bulk Cost: Summary and Shortcut Methods

Royalties and research charges complete the litany of items that add up to the "Total bulk cost" in Fig. 7.1.1. At this point we pause to arrange those items into three categories of fixed, regulated, and variable charges, a categorization which will prove particularly useful in carrying out sensitivity analyses outlined in succeeding chapters (see Table 7.5.1).

The various factors proposed in this chapter may be gathered to formulate a shortcut method of computing the bulk COM; the method is indicated in Exhibit 7.5A. As with any quick method, caution and judgment should be used. For example, maintenance charges for complex processes may exceed the 6 percent of DFC shown.

Cost of Packaging

Chemical products are obviously manufactured in bulk, and many are sold in bulk—in tank cars, tank trucks, or ships. The loading facilities for transferring the products from storage into carriers are normally included as part of the DFC; if the loading facilities are shared with other producing plants, a proportional share of allocated capital must be added to arrive at the effective investment. Failure to

TABLE 7.5.1
CATEGORIZATION OF COM ITEMS

Fixed charges	
Depreciation	Local taxes
Insurance	Factory expense

Regulated charges	
Operating labor	Laboratory
Fringe benefits	Maintenance labor and materials
Supervision, clerical	Royalties
Operating supplies	Waste treatment

Variable charges	
Raw materials	Research
Utilities	

account for the loading facility investment in preliminary states of evaluation may be a source of intense embarrassment; the overlooked need for a $5 million loading dock in conjunction with a proposal involving an overseas exchange of raw materials and products may doom an apparently attractive project—obviously.

EXHIBIT 7.5A

SHORTCUT METHOD, COST OF MANUFACTURE

1 *Labor-dependent items*

Operating labor:	$1.00 \times$ operating labor
Fringe benefits:	$0.22 \times$ operating labor
Supervision, clerical:	$0.18 \times$ operating labor
Operating supplies:	$0.10 \times$ operating labor
Laboratory:	$0.15 \times$ operating labor
	$\Sigma = 1.65 \times$ operating labor

2 *DFC-dependent items*

Maintenance:	$0.06 \times$ DFC/annum
Depreciation:	$0.10 \times$ DFC/annum
Insurance:	$0.01 \times$ DFC/annum
Local taxes:	$0.02 \times$ DFC/annum
Factory expense:	$0.05 \times$ DFC/annum
	$\Sigma = 0.24 \times$ DFC/annum

ADD

3 *Raw materials*
4 *Utilities*
5 *Royalties plus research*
6 *Waste treatment*

TABLE 7.5.2
SOME TYPICAL CONTAINER COSTS
(Spring 1981)

Container	Material	Cost
55-gal drum: Closed, bung holes	Steel Solvent-resistant lining 316 stainless	$22* $3 extra $250*
30-gal drum	Steel	$20*
5-gal pails, crimp lids	Steel	$373 per 100
50-lb bags	Plastic-lined	48¢*
Fiberpacs: ¾ gal 1 gal 2 ½ gal 4 gal 6 ½ gal 15 gal 30 gal 52 ½ gal	Plastic-lined	$163 per 100 $167 per 100 $200 per 100 $226 per 100 $279 per 100 $615 per 100 $760 per 100 $1656 per 100

*Cost per unit, delivered in carload lot quantities.

If loading facilities are an integral part of a production plant, all costs of loading are included in the various components of COM (loading labor by utility operator, for instance), and no additional cost need be added. For shared facilities, a reasonable operating cost of loading must be estimated (Prob. 7.15).

On the other hand, if the product is to be packaged in small containers, the cost of the container may turn out to be an appreciable fraction of the final cost for sale. Some typical container costs (spring 1981) are shown in Table 7.5.2; over a few years, an annual 10 percent cost escalation may be anticipated. Let us see just how important container costs may be.

Dry-cleaning solvent is distributed in 55-gal lined drums. The bulk selling price of the solvent (including all additives) is $2.50/gal. Thus the total cost of the drummed solvent is $137.50. The lined drum adds $25 to the final total of $162.50, or over 15 percent of the price of the contained solvent.

As the container gets smaller, it dominates the selling price of the product. In 1977, one of the cheapest shelf items in the supermarket was a small (8-oz) can of tomato sauce selling for 16 cents. The tomato canner received 5 cents for the tomato product; the can cost 6 cents.

7.6 THE RESPONSIBILITY FOR ENVIRONMENTAL PROTECTION

Concern for protection of the natural environment and for protection of human safety must constitute an integral part of the project evaluation sequence from its earliest stages. Chemical engineers, by virtue of their technological background and organizational skill, enjoy unique opportunities to contribute to the elimination of environmental problems, not only those which arise from the production, distribution, and use of chemical commodities but those caused by the full spectrum of human activity. Technological development dominates the world scene at an ever-increasing pace, and that dominance has opened up a Pandora's box of sociological and technological problems. The technological problems have technological solutions; the chemical engineering profession has assumed, and must continue to assume, a heavy share of the responsibility of finding solutions to those problems.

The individual engineer with project management responsibilities has an important role to play in the drama of the struggle to preserve the integrity of the environment. The initiative to participate in that struggle effectively must ultimately originate with each engineer's inner conviction, not with externally imposed regulations. A few maxims will serve as a guide to effective action in the context of project management and evaluation:

To meet the requirements of regulatory agencies, it is best to consult environmental control and safety specialists at all stages of project development.

Chemical engineers should make the effort to keep themselves informed on general trends in regulatory affairs, but only the specialist can keep up with the rapid changes in the law. Many corporations retain the services of such specialists in environmental and occupational health law, and these should be consulted for advice on specific problems:

Can I vent 5 kg/h of SO_2, or do I need a scrubber?
Is the working area noise level of a powder bin vibrator acceptable?

The early-stage assembly of reliable toxicological and chemical reactivity data on all process feeds, products, intermediates, and wastes is a primary requirement of effective project management.

Do not focus your process design upon the primary product alone.

Recognize that wastes are also products of the process and merit the same degree of processing attention as the primary product. Avoid those off-the-side-of-the-paper arrows on your flow sheets!

Design for total containment.

This maxim should perhaps be modified to say, "Design for total containment of all but the most innocuous materials." It is unrealistic to expect that all net feeds to the process "black box" will wind up exclusively in the product; any real process will have a multiplicity of exit streams, but none of those exit streams (including "fugitive" leaks) must pose a toxic hazard to the environment. Undesirable side streams must be converted into environmentally more acceptable products, a procedure that admittedly may lead to a sequence of product modifications which always ends with the question, "Is this final waste innocuous enough?"

Design for safety first.

Workplace safety concerns and environmental concerns are often inseparable; good practices in either area of concern go hand in hand. Safety considerations in the context of project evaluation may extend to the very substance of the process in question—process steps involving dangerously unstable or flammable materials are preferentially avoided. Safety has an impact upon all aspects of the design package—equipment design, emergency controls, fire protection, plant layout.

Avoid nuisances.

Considerations of *esthetics* should always be given some priority. An installation that constitutes a visual (or other sensual) nuisance in the landscape hardly contributes to the professional status of the originator, and it may well attract unwelcome displeasure or even harassment from the community. A waste steam plume may be toxicologically innocuous, but during certain weather patterns it may well become a prominent landscape feature which, to uninformed outsiders, may symbolize industrial pollution; a cooler-condenser may be a small price to pay to avoid community confrontations. "Lean" operating labor staffing may be *de rigueur* in highly automated plants, but if it will result in poor plant housekeeping and an eventual public (or internal) nuisance, a higher labor rate is the more acceptable alternative.

Recognize and accept the concept of ultimate responsibility.

Chemical engineers accept the responsibility for the proper and safe design and operation of chemical plants. This sense of responsibility must extend beyond the battery limits of the plant. Reputable corporations subscribe to the concept of *product stewardship,* which admits responsibility for the safety of corporate products in the human and natural environment (when properly used). The sense of ultimate responsibility must be similarly extended to all materials released beyond the boundary of the plant or complex. If a toxic material lies buried on the grounds of an approved outside disposal site, the ultimate responsibility for that material does not lie with the disposal site operator—it lies with the originator, with *you!*

Adherence to the enumerated maxims is nothing more than a manifestation of good individual and corporate citizenship. Government agency regulations which codify and promote the quantitative enforcement of some aspects of the maxims may often be burdensome and unreasonably excessive. It is the right and duty of chemical engineers and the corporations that employ them to oppose unreasonable regulations based upon false or unproven premises; the political action that such opposition entails must not be allowed to interfere with the tenets of good citizenship and respect for the integrity of the environment.

NOMENCLATURE

A cost of fuel oil, \$/bbl (1 bbl = 0.159 m^3)
N number of operators per shift
r plant rated production capacity, kg/year

"Good evening, sir. As you may know, the soaring costs of recent environmental-protection legislation have forced us to pass part of this burden along to the consumer. Your share comes to $171,947.65."

Drawing by Lorenz; © 1979
The New Yorker Magazine, Inc.

Abbreviations

COM   cost of manufacture
DFC   direct fixed capital
ROI   return on investment
scf   standard cubic feet (gas at 0°C and 1 atm)
scfm   standard cubic feet (gas at 0°C and 1 atm) per minute
S.F.   stream factor

REFERENCES

Aries, Robert S., and Robert D. Newton: "Chemical Engineering Cost Estimation," McGraw-Hill, New York, 1955.

Bevirt, Joseph L.: How Much Federal Regulations Cost Dow, *Chem. Eng. Prog.*, **75**(5):30 (May 1979).

Chauvel, Alain, et al.: "Manual of Economic Analysis of Chemical Processes," McGraw-Hill, New York, 1981.

Faith, W. L., Donald B. Keyes, and Ronald L. Clark: "Industrial Chemicals," 2d ed., Wiley, New York, 1957.

Guthrie, Kenneth M.: Capital and Operating Costs for 54 Chemical Processes, *Chem. Eng.*, **77**:140 (June 15, 1970).

Haines, Tom B.: Direct Operating Labor Requirement for Chemical Processes, *Chem. Eng. Prog.*, **53**(11):556 (November 1957).

Happel, John, and Donald G. Jordan: "Chemical Process Economics," 2d ed., Marcel Dekker, Inc., New York, 1975.

Johnnie, C. C., and D. K. Aggarwal: Calculating Plant Utility Costs, *Chem. Eng. Prog.*, **73**(11):84 (November 1977).

Katell, Sidney, and John H. Faber: What Does Tonnage Oxygen Cost?, *Chem. Eng.*, **66**:107 (June 29, 1959).

Smith, Clifford V., and David DiGregorio: Advanced Wastewater Treatment, *Chem. Eng.*, **77**:71 (Apr. 27, 1970).

Wessel, Henry E.: New Graph Correlates Operating Labor Data for Chemical Processes, *Chem. Eng.*, **59**(7):209 (July 1952).

PROBLEMS

7.1 Derive expressions for the unit cost of manufacture of a primary product which is produced at an annual rate of w_1 kg, along with by-products and coproducts $w_2, w_3, \ldots, w_n$. The expressions are to reflect three cases:

a All by-products and coproducts are to be credited at their unit sales values $s_2, s_3, \ldots, s_n$.

b All production costs are to be prorated between the primary product and by-products/coproducts in proportion to their annual production.

c All production costs are to be prorated in proportion to each product's annual sales value.

The total annual cost of manufacture may be designated as C. Calculate the unit COM for each of the three cases and the following data:

$$w_1 = 50 \times 10^6 \text{ kg/year}$$
$$w_2 = 20 \times 10^6 \text{ kg/year}$$
$$s_1 = \$2/\text{kg} \quad s_2 = \$1/\text{kg}$$
$$C = \$50 \times 10^6/\text{year}$$

7.2 In Prob. 7.1, suppose the annual general expenses are represented by G; what is the expression for the annual gross profit (before taxes) P_B for each of the three cases?

Suppose $G = 0.20C$. What is the profit before taxes for the numerical example?

7.3 It is proposed that pure oxygen requirements for a process be supplied with a 10,000-A water electrolyzer of the De Nora cell type. The cell is operated isothermally at 75°C at a voltage efficiency of 57 percent and a current efficiency of better than 98 percent. The overall electrolytic reaction may be written as

$$H_2O \longrightarrow H_2 + \tfrac{1}{2}O_2$$

The hydrogen product will be used as boiler plant fuel supplement. The electrolyte is 30 percent aq KOH, but the only feed to the cell is makeup water (at 30°C) to replace the water consumed by electrolysis plus water saturating the product gases.

The cell will be kept at 75°C by using recycled cooling tower water at 25°C. The gas products are at 1 atm.

Make a reasoned estimate of the following:

a Unit demand and cost (per kilogram of O_2) of ac power delivered to the dc rectifier (neglect rectifier losses).

b Unit demand and cost of cooling water.

c Maximum energy credit for hydrogen (at natural gas energy cost; use LHV* of H_2, and neglect effect of water vapor in H_2).

For liquid H_2O at 75°C,

$$\text{Heat of formation} = -284{,}090 \text{ J/(g}\cdot\text{mol)}$$
$$\text{Free energy of formation} = -229{,}280 \text{ J/(g}\cdot\text{mol)}$$
$$\text{Heat of vaporization} = 41{,}760 \text{ J/(g}\cdot\text{mol)}$$
$$\text{LHV of } H_2 = -241{,}910 \text{ J/(g}\cdot\text{mol)}$$
$$\text{Vapor pressure of water over 30\% KOH} = 33.33 \text{ kPa}$$

Use cost data in Table 7.2.2 scaled up at 10 percent/year.

In making an energy balance, note that the reversible electrical work (equivalent to the free energy change) is less than the enthalpy change of decomposing the liquid water; this energy difference comes from a sensible heat decrease of the electrolyte (the "entropy term").

7.4 Methylene chloride is to be brominated over a supported catalyst packed in a tubular reactor. The vapor-phase reaction to form the desired bromodichloromethane may be written as

$$CH_2Cl_2 + Br_2 \longrightarrow CHBrCl_2 + HBr \tag{P7.1}$$

Unfortunately, a second consecutive reaction also occurs,

$$CHBrCl_2 + Br_2 \longrightarrow CBr_2Cl_2 + HBr \tag{P7.2}$$

Laboratory data indicate that in an isothermal reactor at 200°C and 10 atm total pressure the reactions behave as ordinary homogeneous first-order reactions which are dependent upon the partial pressures of the halogenated hydrocarbons. The measured reaction rate constants are

$$k_1 = 0.20 \text{ s}^{-1} \quad k_2 = 0.05 \text{ s}^{-1}$$

In the proposed version of the process, twice the stoichiometric amount of Br_2 to satisfy reaction (P7.1) is to be fed; this is the same feed ratio as had been used in the laboratory experiments (at a total pressure of 10 atm). The process will be run to maximize the reactor exit concentration of the desired $CHBrCl_2$. Unreacted Br_2 and CH_2Cl_2 will be recycled, and the overbrominated by-product CBr_2Cl_2 will be forwarded to a thermal oxidizer for conversion to HCl and HBr:

$$CBr_2Cl_2 + 2CH_4 + 4O_2 \longrightarrow 2HCl + 2HBr + 3CO_2 + 2H_2O \tag{P7.3}$$

*Lower heating value, that is, H_2O of combustion as vapor.

HCl will be recovered as 18°Bé acid, and the HBr (including the HBr from the bromination reaction) as 48% acid.

Calculate:

a Raw material demand (CH_2Cl_2 and Br_2) per unit of primary product.
b Yield of 18°Bé HCl and 48% HBr per unit of primary product.
c Methane demand in thermal oxidizer per unit of primary product.
d Use current information in *Chemical Marketing Reporter* and adjusted natural gas (methane) cost in Table 7.2.2 to compute the net cost of chemicals to produce a kilogram of $CHBrCl_2$.

7.5 The liquid-phase oxidation of propylene to propylene oxide in the presence of air may be represented formally by the equation

$$CH_3CH = CH_2 + \tfrac{1}{2}O_2 \longrightarrow CH_3CH\underset{O}{-}CH_2$$

Miniplant experiments have demonstrated the coproduction in the oxidation reactor of three other carbon-containing species:

Acetaldehyde, $CH_3-\overset{O}{\overset{\|}{C}}H$
Carbon dioxide (and some light ends)
Oxygen-containing tar

To strike a balance between high PO (and acetaldehyde) and tar production, the *attack* upon the feed propylene to the continuous flow reactor must be limited to 10 percent. Under these conditions, the following *selectivities* have been measured in terms of the percentage of carbon atoms in the attacked propylene contained in each species:

PO	50%
CH_3CHO	30%
CO_2	10%
Tar	10%

In the proposed plant based on the miniplant experiments, the unreacted propylene will be separated from the products and recycled; calculations show that about 96 percent of the recyclable C_3H_6 will be recovered, the rest lost in the light-ends stream.

Calculate the kilograms of propylene required to produce 1 kg of propylene oxide.

7.6 Vanillin is an important flavoring agent that may be manufactured by means of nitrobenzene oxidation of eugenol, a component of oil of cloves. In an older edition of their book, Faith et al. (1957) list the following raw material unit ratios for this vanillin synthesis:

Eugenol	1.35
NaOH	0.55
Nitrobenzene	1.00
HCl*	1.60

Look up current prices of the raw materials in the *Chemical Marketing Reporter* and compare with the list price of vanillin. What are your conclusions?

*HCl bought as 20°Bé acid.

7.7 If you have access to historical price data (such as older issues of *Chemical Marketing Reporter*), compare the list price of aspirin (acetylsalicylic acid) over the past decade or two with the list price of the raw materials required for the synthesis starting with phenol. Information in the "Encyclopedia of Chemical Processing and Design"[1] suggests the following unit ratios:

Phenol	0.68
NaOH	0.29
H_2SO_4	0.38
Acetic anhydride	0.65

NaOH is supplied as 50% solution, H_2SO_4 as 98% acid. Another raw material, CO_2, is obtained by purification of flue gases.

Construct a time graph showing:

a The list price of aspirin in constant currency (use the Consumer Price Index for adjustment)

b The percentage of the list price ascribable to cost of raw materials

7.8 Estimate the operating labor requirement in a 1 ton/h adipic acid purification plant illustrated in Figs. 3.5.7 and 3.5.8. Use several approaches to reach a reasonable compromise; assume that the plant will be highly automated.

7.9 It has been implied in the text that computer control of chemical process plants does not reduce the operating labor requirement. This judgment is meant to apply to those plants which already possess a high degree of automatic control. Write a short (two-page) critique of this judgment. Base your conclusions upon your own reasoning as well as reports of experience you may be able to find in the process control literature. Is the judgment applicable to batch processes? See, for example, the AIChE Workshop booklet, "Computer Control of Batch Processes."

7.10 Use Wessel's correlation (Fig. 7.3.1) to estimate the labor requirement (operators per shift) to operate a modern sulfuric acid plant, illustrated in Fig. P7.1, producing 3 short tons of product/h (expressed as the H_2SO_4 species). If the stream factor is 0.92, what is the unit cost of labor in 1983? Include in the cost fringe benefits and supervision, and express the cost in cents per pound of the H_2SO_4 species.

7.11 Estimate the number of operators per shift (including the supervising engineer with operating responsibilities) required for the simple pilot plant operation illustrated in Fig. P7.2. Use the labor requirements for process equipment listed in Aries and Newton (1955) as a guide, but make your own reasoned adjustments based on the process analysis.

The operation will use automatic control (for example, distillation column operation), but operating conditions are expected to be changed frequently. Computer control will not be used. Sampling, some analysis, data acquisition, and maintenance will constitute important parts of operating responsibilities.

A reaction between a powder and a liquid reagent in a solvent carrier will take place in a steam-jacketed, stirred reactor. Three batch reactors will be operated on 4-h cycles—4 h for loading and reaction, 4 h for solvent evaporation, and 4 h for continuous feeding to the distillation columns. Each reactor has a nominal volume of 200 gal. The powder will be hand-dumped from 50-lb bags into a batch dissolver. Each batch will require one bag of powder and one 55-gal drum of liquid reagent (stored in an adjacent warehouse). Solvent makeup will be about one drum per week.

---

[1] Marcel Dekker, New York/Basel, 1977.

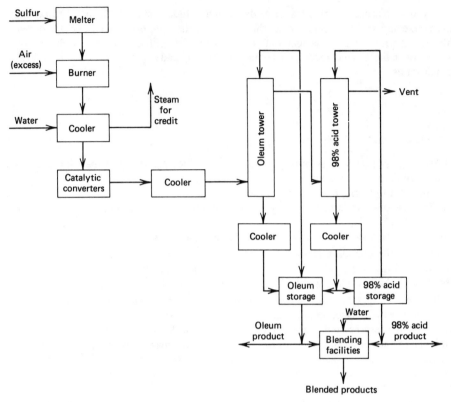

FIGURE P7.1
Contact process for sulfuric acid.

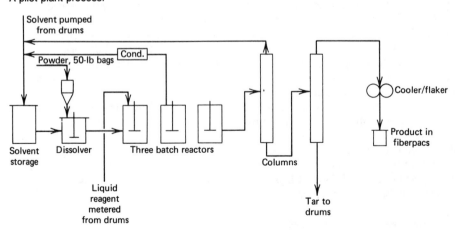

FIGURE P7.2
A pilot plant process.

408

Most of the solvent will be removed by simple evaporation in the reactors themselves; final removal will be done in a continuous distillation system with the usual auxiliaries. Another column will be required for tar removal; it is anticipated that one 55-gal drum of tar will be generated every other day. The drums will be picked up by incinerator operations every 10 days (not an operational responsibility in the pilot plant). The final product will be frozen and flaked directly on a water-cooled flaker, and the flakes will be collected in 50-lb Fiberpacs (10 drums per batch). The drums will be handled on pallets and moved to the adjacent warehouse with forklift trucks.

The principal anticipated problems will be at the bottom of the tar column and with the flaker operation.

7.12 What percentage of the standard cost of low-pressure oxygen (curve C in Fig. 7.2.1) is ascribable to the requirement for power shown in Fig. P7.3? Consider a 300 tons/day plant (low-temperature air distillation), and adjust costs to mid-1981 (M&S index = 710). Compare the power cost in a distillation plant to the power cost for the electrolytic generation of O_2 from water, which requires a cell voltage of about 2 V with an approximately 100 percent current efficiency. What do you conclude from this comparison?

7.13 Search the current literature and prepare a brief summary of the relative costs of electric energy generated from natural gas and from geothermal sources (steam or hot water).

7.14 In a typical power plant about 35 percent of the gross energy (HHV) of fossil fuels is converted into electric power. Use the current price of OPEC oil to compute the energy cost equivalent of electric power generated from oil (cents per kilowatthour).

FIGURE P7.3
Power requirements for O_2 generation (distillation). For identification of curves, see Fig. 7.2.1. *(Source: Katell and Faber, 1959. Reproduced by permission from Chemical Engineering.)*

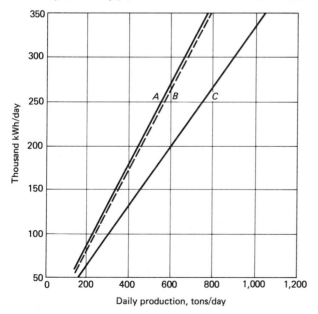

7.15 A plant produces 50 million lb/year of HCl which is marketed as 35% muriatic acid (specific gravity = 1.178). The product is shipped in 20,000-gal tank cars. Estimate the cost of loading the tank cars from storage (in cents per pound of contained HCl). Each tank car may be filled in 1 h, but 2 h is required for the loading turnaround, i.e., the period between bringing the tank car to the loading station and shuttling it back to a siding. Loading is accomplished by a single utility operator with a basic labor rate of $10/h. The loading pumps pump against an approximate head of 30 ft. The loading station DFC is $150,000.

How does the cost of loading compare with the current price of HCl in 35% acid?

7.16 Many fascinating problems arise from the consideration of optimum distribution points to serve a network of consuming localities. A simple example is illustrated in Fig. P7.4a. Suppose $A, B, \ldots, N$ represent warehouse localities along a contiguous transportation route (a railway or a highway), each serving to supply a packaged product for local

FIGURE P7.4
Transportation-distribution illustrations. (a) One-dimensional distribution problem. (b) The equilibrium analogy of the one-dimensional distribution problem. (c) Distribution point for four warehouses. (d) The equilibrium analogy of the four-warehouse problem.

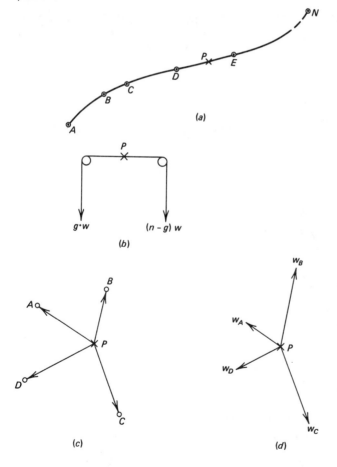

markets. The problem is to find a central distribution point P to which the bulk product is to be shipped from the factory for final packaging; the cost of bulk shipments is assumed, for the present, to be much lower per unit weight than the shipping costs of the packaged products. The optimum location of P will result in lowest shipping costs.

Problems of this nature can rapidly become quite complex, particularly if demand at each locality is different, and the techniques of linear programming, adapted to computer use, are often employed. For present purposes we will make some simplifying assumptions:

The unit cost of shipping the packaged product is directly proportional to the distance from P to the warehouse.

The *timing* of shipments will not be considered as an important variable (in other words, we assume that each warehouse will always have enough product on hand to meet local demand, and there is no penalty on excess inventory).

a Show that for the case of just two warehouse localities with equal demand (say, D and E), P may lie anywhere along the connecting route, including D or E.

b Show that for three localities, each with equal demand (say, C, D, and E), P will coincide with the middle locality, regardless of the distances involved!

c Show that for N localities with equal demand, the problem of locating P is equivalent to the trivial problem in *statics* illustrated in Fig. P7.4b. If we let w be the demand at each locality, n the total number of localities, then P is located so that $g = n - g$, where g is the number of localities to one side of P, regardless of the distances involved.

d As a corollary to part (c), show that for an even number of localities, P may be located anywhere between the central two localities, and for an odd number the central locality coincides with P.

e Extend the equilibrium analogy to the case of unequal demand among the warehouse localities.

f An agricultural product formulation and drumming facility is to be located on Highway 99 in California's central valley. Warehouse locations and local demand are listed below, in north-to-south sequence along the highway:

Location	Demand, drums/year
Redding	3,500
Chico	6,200
Sacramento	9,100
Stockton	10,500
Merced	4,500
Fresno	2,800
Bakersfield	8,900

Where should the distribution point be located? Check a map of California to identify a likely town.

g What happens if the demand at one locality exceeds the total demand at all other localities?

7.17 Another type of transportation-distribution problem involves the optimum location of a distribution point P in a two-dimensional network of warehouse localities. A relatively simple four-warehouse problem is illustrated in Fig. P7.4c. In addition to the assumptions stated in Prob. 7.16, the assumption is made that the highways (or railroads) from P to each warehouse lie along the connecting straight line without major detours. On the basis of the same arguments as were used in Prob. 7.16, it can be shown that the location of P coincides with the equilibrium point of the equivalent "demand forces," shown in Fig. P7.4d. Thus the location of P depends upon the relative demand at each warehouse, *but not upon the distances of the warehouses from P!* Does this mean that the cost of shipping is independent of the distances? Not at all. If P in Fig. P7.4c is, indeed, at the equilibrium point, its location would be exactly the same if B were moved anywhere along the vector PB from its origin at P; the cost of shipping, however, would increase with increasing magnitude of PB.

The establishment of P by analytical techniques is quite difficult. A trial-and-error solution involving a scale drawing is reasonably rapid and accurate. The four warehouses are located on a cartesian grid, and a trial point P is chosen. With a pair of dividers, the distances $r_1 \equiv PA$, $r_2 \equiv PB$, ... are determined, and the sum

$$\sum_1^4 r_i w_i$$

is computed. (This is proportional to the shipping cost—why? Note that direct shipping from A to B is not allowed in this simplified regime.)

Now choose another location for P along vector PA (on either side of the original P) and recompute the sum. Do this until a minimum is established. Now repeat the whole procedure for the *new* vector PB, followed by similar procedures for C and then D. Repeat another cycle; the optimum location of P will become closely bounded.

As an example of this technique, find the coordinates of a distribution point to serve warehouses in the following four cities (with approximate cartesian coordinates in miles):

Sioux Falls, S.D.	(0, 0)
Fargo, N.D.	(−15, 235)
Duluth, Minn.	(195, 245)
Minneapolis–St. Paul, Minn.	(160, 100)

Assume, as the simplest case, that demand will be the same at each of the warehouse locations. What town would you pick for the site of the distribution point? (Refer to a map of Minnesota.)

7.18 The biochemical oxygen demand (BOD) is the total amount of oxygen that a biological system would require to oxidize bionutrients in wastewater to CO_2 and water. The BOD is usually measured as parts per million by weight (or milligrams per liter) of O_2. The BOD may be substantially reduced before effluent discharge with treatment in aeration lagoons. One type of aerator that may be used for the air dispersion task is a floating aerator. This machine moves freely (within certain confines) over the surface of the lagoon, contacting a large throughput of pumped lagoon water with air.

A typical floating aerator has the following specifications:

Power consumption: 3 hp
Water pumping rate: 2750 gal/min
O_2 concentration in discharge: 8 ppm (parts per million by weight) at 1 ppm lagoon level
Lagoon area required per machine: 1000 ft^2

A petrochemical plant generates 5000 gal/h of an aqueous waste containing mixed glycols, with a BOD level of 1000 mg/liter. With the required O_2 added, 90 percent of the BOD is eliminated.

a How many aerators, and what lagoon area would you recommend?

b Calculate the total required investment from the following data:

DFC (M&S index = 750) of a 1-acre lagoon is $175,000, including pumps and piping. Scale up or down by using the 0.6 factor.

The purchased cost (M&S = 750) of a single 3-hp floating aerator is $4900. The DFC *modular factor* (see Chap. 5) is 3.2.

c Both feed and lagoon discharge pumps operate against about 50-ft heads. What is the estimated cost of treatment, in cents per 1000 gallons? Neglect labor.

d Can you reconcile this cost with that shown in the table of Fig. 7.4.1 for secondary treatment?

7.19 In Prob. 5.18, a decision between the use of air and oxygen in the synthesis of butylene oxide was asked for, based upon differences in DFC. Make a similar comparison on the basis of the differences in the BuO COM.

Assume that absolute pressure of O_2 in the feed must be 20 psia (about 5 psig) regardless of whether 95% O_2 or air is used; in other words, high-pressure air must be supplied, and its standard cost may be taken from Table 7.2.2. The standard cost of O_2 may be scaled up from Fig. 7.2.1 (use M&S = 700); this cost includes all fixed charges upon the oxygen plant. The only other difference in COM of the BuO is ascribable to the fixed charges associated with the DFC that are saved on the BuO plant by virtue of using 95% O_2. Labor demand is expected to be independent of the source of O_2.

7.20 A process generates 1 kg aqueous waste/kg primary product. The aqueous waste contains 5% ethylene glycol, 5% of glycerol, traces of phenol, and various "color bodies." It has been proposed that the waste be "burned" in an existing furnace for fuel credit. What energy credit (or penalty) does this method of waste disposal engender per kilogram of product?

Base the calculations upon the lower heating value of the combustibles, that is, the heat of combustion (at 25°C) with water *vapor* as product.

$$\text{LHV} = \begin{cases} -251 \text{ kcal/(g} \cdot \text{mol)} & \text{for ethylene glycol} \\ -361 \text{ kcal/(g} \cdot \text{mol)} & \text{for glycerol} \end{cases}$$

Heat of vaporization = 10.4 kcal/(g · mol)   for water

7.21 A cannery processes 500 tons of tomatoes/day; the delivered price of the fruit is around $80/ton (1981). About 10 gal of water is required to clean and process a pound of tomatoes. Because of the cost of this enormous demand for water, it has been suggested that facilities be built to treat the water for recycling into the fruit cleaning operation.

a It has been determined that primary treatment, secondary treatment, disinfection with Cl_2, and sand filtration are adequate to give an acceptable recycled water stream. Extrapolate the information in Fig. 7.4.1 and determine how much that proposed treatment would add to the delivered cost of $80/ton.

b How does the cost of treating the water compare with the cost of fresh clarified water (Table 7.2.2)?

7.22 Use data in Exercise 5.2.2 to calculate the current cost per thousand gallons of treating aqueous wastes by solar evaporation in the El Paso area. The main component of the cost is the fixed charges, but include a reasonable cost of power for pumping the wastes; neglect any labor charges.

7.23 Acetaldehyde may be prepared from ethylene by direct oxidation in the presence of a copper chloride catalyst containing palladium chloride. The reaction sequence may be represented by

$$C_2H_4 + 2CuCl_2 + H_2O \longrightarrow CH_3CHO + 2HCl + 2CuCl$$
$$2CuCl + 2HCl + \tfrac{1}{2}O_2 \longrightarrow 2CuCl_2 + H_2O$$

Two process alternatives are available. In the first, the two reactions are carried out in separate reactors, and air is used as oxidant. The second alternative combines the two reactions in a single vessel, but 99.5% O_2 is used as the oxidant.

The operating requirements are summed up in a brief note that appeared in the journal *Hydrocarbon Processing*:

OPERATING REQUIREMENTS PER SHORT TON OF ACETALDEHYDE

	Process	
	One-stage	Two-stage
Investment per ton/yr*	$53	$53
Raw materials:		
Ethylene	1340 lb	1340 lb
Oxygen (99.5%)	9460 scf	
Air		54,000 scf
Catalyst	~90¢	~90¢
HCl (supplied as 20° Bé acid)	30 lb	80 lb
Utilities†:		
Recycled tower water	48 × 10³ gal	48 × 10³ gal
Process water	1.7 × 10³ gal	7.2 × 10³ gal
Demineralized water	120 gal	
Steam (150 psig)	2400 lb	2400 lb
Power	45 kWh	270 kWh‡
Labor, operators/shift	3 to 4	3 to 4

  *The investment figure is based upon a 75,000 tons/year plant size. The quoted number *includes* the cost of the oxygen plant in the one-stage process, the cost of multistage air compressors in the two-stage process. M&S index basis: 263.
  †Oxygen plant utilities are *not* included.
  ‡Includes power for air compression.
  Source: *Hydrocarbon Process.*, **46**(11):135 (November 1967), reproduced with permission.

Calculate the COM of both processes (M&S = 710 basis), and recommend the more economical process, in a new 250 million lb/year plant (S.F. = 0.92).

Use scaled-up data on oxygen costs in Figs. 5.2.2 and 7.2.1, and note that the investment in the above table includes the cost of the oxygen plant.

Assume that an aqueous waste stream equivalent to the process water use will be generated and will be handled in an available 30 million gal/day level III treatment plant (Fig. 7.4.1).

Royalties equal 2 percent of COM.

7.24 (Modification of problem suggested by Happel and Jordan, 1975; adapted by permission of Marcel Dekker, Inc.). We saw (Chap. 2) that a project may be assessed on the basis of the present worth of all the cash flows, positive and negative, that the project generates.

The present worth concept may be used to assess a project at any stage of development. For example, research projects may be assessed by projecting future cash flows, both positive (generated by net income) and negative (capital expenditures, additional costs of development). Moreover, *past* expenditures may be accounted for by computing their present worth[1]; this is calculated simply as the future worth of the cash already expended.

Let us suppose that your company has completed a 5-year research project at a cost of $20,000/year and has just been granted a process patent. The decision has been made to license the process; at this time there are no licensees, but it is expected that income from royalties will increase linearly until the end of the seventeenth year, at which time the annual income is likely to reach $500,000. Process improvement and market development costs are expected to average $10,000/year. What is the present worth of the invention?

The tax rate may be taken as 50 percent, and the desired rate of return is 15 percent/year, *annually* compounded. In computing cash flow after taxes, consider past research costs as operating costs rather than capital expenditures. (Why is this important? What effect does this distinction have upon taxes paid by the corporation?)

Other companies are working on a competitive process, and it may well happen that at the end of year $k < 17$ your company's process may be rendered obsolete by patented improvements so that thereafter the royalties will drop to zero. How much lead time does your company need (i.e., what is k?) so that it will just recover all expenditures associated with process development (i.e., present worth = 0)?

7.25 A system for removing chlorine from a waste gas stream is illustrated in Fig. P7.5. The hot gases are forwarded through booster blower C-1, designed to provide for the pressure drop through the packed scrubbing column B-1. In the column, chlorine is removed by reaction with NaOH in the scrubbing liquid,

$$Cl_2 + 2\ NaOH \longrightarrow NaOCl + NaCl + H_2O$$

The pH at the bottom of the column is carefully controlled (pH = 8.0) by controlled addition of 50% NaOH from overhead tank T-2; in this way only a negligible amount of CO_2 in the feed gas is absorbed. The 50% NaOH is diluted at the point of addition with enough water to reduce the effective feed concentration to 5% NaOH; this step avoids salt precipitation in the packing. The 5% NaOH feed is introduced into a large circulating stream of recycled spent liquor, stored in T-1, and pumped with spared pump P-1. The large recycled liquid stream assures proper liquid irrigation and distribution in the packing of B-1. If necessary, the circulating stream is cooled in H-1 to maintain the desired operating temperature in B-1. The purge to waste treatment is controlled by the level in T-1.

The water-saturated gas from the top of B-1 passes through X-1, a demister and scrubber that removes any particulate matter passing through B-1 by means of a water spray directed against woven mats through which the gas must pass. An on-line Cl_2 analyzer X-2 monitors the Cl_2 in the gas discharged to the atmosphere (specified as 10 ppm or less) and actuates an emergency dump of excess 50% NaOH from the overhead tank T-2 if the allowable Cl_2 level is exceeded. Pump P-2 unloads tank trucks into T-2 as needed.

With the help of data listed below, calculate the cost of removing Cl_2 from the waste gases, as dollars per 1000 scf of waste gas feed and dollars per pound of Cl_2 removed. Use 1981 costs (M&S = 710).

---
[1] For a different viewpoint, see Sec. 9.3.

416 CHAPTER 7: THE COST OF MANUFACTURE

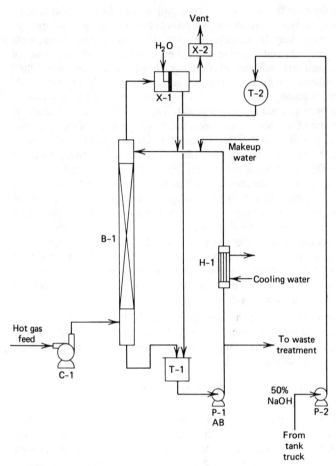

FIGURE P7.5
Caustic scrubber.

Data

Feed gas rate: 250 scfm
Feed gas at ~1 atm, 100°C, bone-dry
Composition (by volume):
  Cl_2 1%
  CO_2 2%
  O_2 18%
  N_2 79%
The gas contains colored particulates (iron, carbon).
Operating conditions in B-1:
  50°C at top
  Pressure drop through packing: 6 in H_2O
  H_2O vapor pressure at top: 80 mmHg
  Recycle liquor circulation rate: 50 gpm

B-1 temperature is controlled by cooling in H-1.
X-1 requires a flow of 2 gpm of process water.

Equipment

Purchased costs are given for M&S = 710

B-1: 16-in dia. × 20-ft FRP (fiber-reinforced plastic) column with 15 ft of 1-in Intalox saddles (Kynar, $180/ft^3), PVC support plates and distributors	$ 7,800
C-1: 1-hp steel turboblower	2,900
H-1: 20-ft^2 titanium plate exchanger	6,300
P-1(AB): 1-hp TFE (fluorocarbon) lined centrifugal pump plus spare, explosionproof motors	4,900
P-2: Rubber-lined steel centrifugal transfer pump (explosionproof motor)	1,200
T-1: 500-gal spent liquor storage tank, FRP	1,800
T-2: 2000-gal 50% NaOH storage tank, rubber-lined steel	$13,200
X-1: Demister-scrubber, 18 × 18 × 48 in, epoxy-lined, polyester mats	6,500

X-2, the on-line Cl_2 analyzer, costs $10,000, installed in place and ready to operate (i.e., the $10,000 may be appended to the DFC representing the rest of the equipment).

Other operating data

Labor requirements: 2 operator-hours/day
All makeup water is process water
Stream factor = 0.95
Cost of NaOH (in 50% NaOH): $290/contained short ton of Na_2O
Waste treatment charges are 2¢/gal

7.26 In the caustic scrubber of Prob. 7.25, the Cl_2 content is reduced from 1 percent to 10 ppm in 15 ft of packing. Representatives of the County Air Resources Board insist that the exit level be reduced to 1 ppm, the TLV of chlorine (see Prob. 3.2). By postulating that the same percent reduction in Cl_2 level is effected by a foot of packing regardless of Cl_2 level, how much additional packing depth would have to be added to column B-1 (Fig. P7.5)? Considering only the *differential* increase in COM due to the larger column size, what is the cost per pound of Cl_2 absorbed from the 10 to the 1 ppm level (the "marginal" cost of removing the last of the Cl_2)? Do you think that the demand for Cl_2 reduction to a 1 ppm level is justified? What is the fate of traces of Cl_2 in the atmosphere?

7.27 In Prob. 5.7 the calculation of the DFC of an installation to recover perchloroethylene from wastewater was asked for. Assume that the DFC is $390,000 (M&S index = 750). Estimate the cost of solvent recovery (cents per pound of solvent recovered) from previous and the following additional information.

The recovery unit will be operated around the clock for 350 days/year, and no more than ½ operator-hour of labor will be required per shift (not counting maintenance). The charcoal beds are 6 ft in diameter by 2.1 ft thick each, and the charcoal is estimated to have a life of 1 year. At the adsorption loading (0.4 lb C_2Cl_4/lb charcoal), the vapor pressure of C_2Cl_4 in the atmospheric vapor is 7 mmHg. The eluting atmospheric steam is passed through the beds at a superficial velocity of 0.5 ft/s in a direction opposite to airflow during adsorption. The wastewater pump consumes 10 hp and the blower 2½ hp. Cooling tower water is used in the condenser only during elution (automatic control); the latent heat of C_2Cl_4 is 100 Btu/lb.

7.28 Estimate the cost of manufacture of a product made in a complex plant having a DFC (including waste treatment facilities) of $6,000,000. The production rate (100 percent of capacity, stream factor = 0.95) is 20,000,000 lb/year. The plant is operated by 10 operators per shift. Raw materials cost 30 cents/lb of product, and utility needs are as follows:

150 psig steam, 20 lb/lb
Power, 0.5 kWh/lb
Clarified water, 10 gal/lb

The product is packaged in 50-lb plastic-lined bags. Express the COM as cents per pound of product.

7.29 Urea is synthesized from ammonia and carbon dioxide in two steps. The first step involves the synthesis of ammonium carbamate:

$$2NH_3 + CO_2 \longrightarrow NH_4OCONH_2$$

The carbamate is subsequently dehydrated to urea:

$$NH_4OCONH_2 \longrightarrow (NH_2)_2C=O + H_2O$$

The yield is reduced by the tendency of urea to degrade slowly to biuret at processing conditions,

$$2(NH_2)_2C=O \longrightarrow (NH_2C=O)_2NH + NH_3$$

Unconverted carbamate is removed from the urea solution out of the reaction train by stripping. Two methods of stripping are favored—one involves stripping with CO_2, the other with NH_3. The differences due to the two methods of stripping are summed up in the following table:

	Stripping agent	
	CO_2	NH_3
DFC, 250-million-lb plant, M&S index = 710 (mid-1981)	$11.0 × 10^6	$12.4 × 10^6
Raw materials unit ratios:		
Anhydrous NH_3 (at 9¢/lb)	0.57	0.57
CO_2 (at 2¢/lb)	0.75	0.75
Utilities unit ratios:		
Cooling tower water, gal/lb	9.3	7.5
Steam, 600-lb, lb/lb	0.93	1.29
Power, kWh/lb	0.010	0.017
Labor, operators/shift	4	4
Royalties, % of COM	2	2

What is the unit COM for each process in a 250 million lb/year plant with S.F. = 0.9?

7.30 Your plant normally purchases natural gas (400 psig pressure, 96% hydrocarbons—mostly CH_4, 4% inerts) at a rate of $6.80/1000 scf. The supplier offers, as an alternative, a low-grade natural gas (at 400 psig) containing 50 mol % CH_4 and 50 mol % N_2 for

$2.50/1000 scf. The low-grade material would have to be processed to reconstitute a more normal gas fuel. Should the offer be accepted?

The gas may be enriched by selective adsorption on activated charcoal. It is required that 1000 scfm of the low-grade gas be processed. Three charcoal beds will be used, each operating on a 1-h cycle; while one bed is on the adsorption cycle, the second will be desorbed (by lowering the pressure to 20 psig), and the third will be subjected to reactivation. The low pressure of the desorbed gas is acceptable.

During the 1-h adsorption cycle, 0.1 lb of gas is adsorbed per pound of charcoal. The adsorbed material is 96% CH_4, 4% N_2, and the same composition is obtained during desorption, although only 97% of the adsorbed material is recovered; the rest is lost during reactivation. The average composition of the gas leaving the charcoal during the adsorption cycle is 95% N_2, 5% CH_4; this gas will be sent to the boiler furnace for zero credit.

The required DFC estimate for handling 1000 scfm of feed is $2,000,000. Labor is projected at 1 operator-hour per shift, and the utilities will total $30,000/year. The charcoal sells for $3/lb; an annual replacement rate of 25 percent is projected. The DFC includes the original charge of charcoal. Assume S.F. = 0.95. Base your decision on the total cost of reconstituting the gas, with no surcharge to account for profit on the DFC.

7.31 Methanol is synthesized according to the reaction

$$CO + 2 H_2 \longrightarrow CH_3OH$$

over various catalysts. The feed mixture may be obtained by reforming of desulfurized natural gas:

$$3 CH_4 + CO_2 + 2 H_2O \longrightarrow 4 CO + 8 H_2$$

The CO_2 is recovered from flue gases in a Girbotol unit.

Guthrie (1970) gives the DFC for a 500 ton/day methanol plant, using the process given by the equations, as $8,000,000 (M&S = 285).

Make a back-of-the-envelope estimate of the current COM of methanol in a 500 ton/day plant, and compare the answer with quoted prices for methanol in the *Chemical Marketing Reporter*.

7.32 Chauvel et al. (1981) quote the following economic data for a plant to produce formaldehyde by the catalytic oxidation of methanol:

Capacity: 12,500 metric ton/year of pure formaldehyde
Battery limits investment: $590,000 in 1968 (M&S = 273)
Initial charge of catalyst: 2.2 metric tons valued in mid-1975 at $25/kg
Life of the catalyst: 1 year
Methanol: 1.15 ton/ton HCHO, value at $135/metric ton in mid-1975
Electricity: 256 kWh/metric ton HCHO
Cooling water: 75 m^3/metric ton
Demineralized water: 3 m^3/metric ton
Net steam *produced* (at 20 bars): 1.58 tons/ton HCHO
Labor: 2 operators/shift

Assume 10 percent/year increase in chemical prices. What is the 1981 (M&S = 710) COM?

7.33 A new antimildew agent for incorporation into paints shows a great deal of promise. It is currently little more than a laboratory curiosity, and a rough economic analysis is needed to ascertain its commercial prospects. Similar agents not possessing some of the desirable properties of the new product sell for about $5/lb.

Make an estimate of the COM of the product to meet all anticipated demand of 1 million lb/year. Some preliminary process synthesis work has been done, and the contemplated complex organic synthesis process is thought to consist of 15 functional steps, with a total of four operating blocks, each analogous to Fig. 3.5.1 (continuous processing).

The raw material unit ratios are thought to be:

Cl_2	0.50 lb/lb
NaOH	1.40 lb/lb (supplied as 50% NaOH)
SO_2	0.45 lb/lb
Thioglycolic acid	0.62 lb/lb
Epichlorohydrin	0.37 lb/lb
Octanol	0.19 lb/lb
Acetone	0.27 lb/lb
Demineralized water	4 gal/lb

Most of the input water will wind up as aqueous waste that will require filtration, oxidation, and activated charcoal treatment in large existing facilities. Five-gallon pails are used for packaging similar paint additives.

How does the COM compare with the likely sales price?

7.34 Bevirt (1979) has made a quantitative analysis of the costs of government regulation of the chemical business. Read the article, and write a two-page critical review of the methodology and conclusions.

CHAPTER 8

THE CRITERION OF ECONOMIC PERFORMANCE

The long litany is about done, the descriptive litany of items that constitute the investment analysis, as well as items that lead up to the computation of the net income, or profit, generated by a project. Indeed, we still have some polishing up to do, and we will explore some important components of investment and continuous cash flow that we have perhaps already mentioned in passing. But our main task in this chapter will be to combine the two elements of investment and net income, to analyze their interplay, and thus to answer the key question about a project, old or new: "Is the criterion of economic performance satisfied?"

The two basic concepts of direct fixed capital (investment analysis) and cost of manufacture (net income analysis) represent expenditures, one-time and continuous, respectively, which are confined to the operating environment of the producing plant. Other expenditures occur within the boundary or well beyond the boundary of the integrated complex, and yet they are equally important to the successful manufacture and marketing of the product. In fact, some of the expenditures may not seem to be directly supportive of the production effort in an individual plant; these are general corporate expenses, and yet each plant must be expected to share in those expenses, no matter how far removed from production they may appear to be. We have already seen an example of such shared expenses in the case of the *factory expense,* a prorated cost of operating some of the integrated complex support facilities.

We will first complete an outline of the investment analysis and focus upon some of the other forms of investment capital first mentioned in Chap. 5,

Working capital
Allocated fixed capital

and still another category,

Allocated market capital

The significance of working capital has been described in context of the retail store analogy; in what follows, the concept will be further explored, and methods of estimating this portion of the total invested capital will be outlined. The estimation of allocated fixed capital will be reviewed, and the concept of allocated market capital will be introduced. These three capitals, when added to the DFC, constitute the totality of the project-oriented invested capital—the capital for sale (CFS).

The general expenses are those continuous expenditures which are incurred outside the scope of the production-oriented facilities. In conjunction with the production-oriented COM they constitute the grand total of all continuing (recurring) expenditures. When they are subtracted from sales revenue (taken over the same time span), the resultant, defined in Chap. 6, is the profit (before taxes). We will see how the concept of CFS and profit are combined to define what is perhaps the most commonly used criterion of economic performance of projects in the chemical process industry, the *return on investment* (ROI). An attempt will be made to give the neophyte evaluator some feel for what acceptable levels of ROI are for various types of projects. Moreover, some of the shortcomings of ROI will be assessed and reasons will be given why other criteria of economic performance (to be taken up in the next chapter) are often preferred.

The evaluator's task does not end once the project's ROI has been determined. The initial evaluation attempt is based upon CFS and profit values for the projected plant at full capacity. But what happens to the ROI at reduced plant throughputs dictated by demands of the marketplace? And what of considerations of *uncertainty*—the estimates leading to the ROI computation abound with uncertainties, and how do we know what effect those uncertainties might have upon an apparently attractive project? We will learn how to estimate profitability under variable conditions and how to apprise the decision makers of the effects of uncertainty upon economic results. We will, however, postpone the quantitative analysis of *risk* to a later chapter. We will specifically characterize risk in terms of quantified estimates of probability; uncertainty has no associated levels of probability.

8.1 THE CAPITAL FOR TRANSFER

Allocated Fixed Capital

The capital for transfer (sometimes called *plant gate* capital) is the total investment backing up the manufacture of a product within the geographic confines of the integrated complex. It consists of three categories:

Direct fixed capital
Allocated fixed capital
Working capital

The significance of each category has been explained in context of the retail store analogy (Sec. 5.2). The format for computing the capital for transfer is shown in Fig. 8.1.1.

Once the DFC has been established, the next step is to generate an estimated total for the allocated fixed capital. We again remind ourselves that the allocated fixed capital represents a prorated share of the investment in auxiliary facilities shared with other production plants. The concept is illustrated in Fig. 5.1.2, and again in Fig. 8.1.2 in what is hoped to be a self-explanatory manner. The allocated fixed capital (*backup* capital) for the principal utilities is shown in Table 7.2.2. The format of the backup capital factor is

FIGURE 8.1.1
Capital summary form.

PRODUCT _____ DATE _____
PLANT _____ BY _____
PRODUCTION RATE _____ % CAPACITY

MANUFACTURING CAPITAL					
ITEM	UNITS PER POUND	UNITS PER YR	CAPITAL PER UNIT/YR		TOTAL CAPITAL $
DIRECT FIXED					
ALLOCATED FIXED			TOTAL DIRECT FIXED		
RAW MATERIALS					
UTILITIES					
POWER – AC, DC					
WATER – RAW, CLAR.					
STEAM					
GAS					
PACKAGING					
FACTORY EXPENSE:		FACTOR: 5.2% OF DFC			
WORKING CAPITAL:			TOTAL ALLOCATED		
M & S – SPARES 2% OF DFC					
PURCH. RAW MATERIALS					
IN – PROCESS INVENTORY					
FINISHED INVENTORY					
			TOTAL WORKING		
TOTAL CAPITAL FOR TRANSFER					

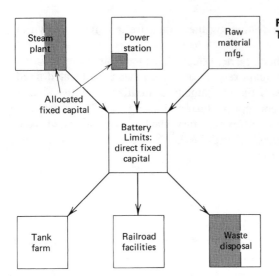

FIGURE 8.1.2
The allocated fixed capital concept.

Dollars per unit per year

so that once the utilization in units per year has been established (for instance, 10^3 gal/year of demineralized water), the allocated capital can be readily computed as the product of the utilization and the appropriate factor. An instance of this technique is given in Example 7.2.1. When the allocated capital factors listed in Table 7.2.2 are used, it is important to remind oneself that each factor is a typical one for large integrated complex installations; the factors do depend upon the size of the auxiliary facilities.

Four categories of allocated fixed capital are indicated in Fig. 8.1.1:

Raw materials. We have seen (Sec. 7.2) that if the standard cost is used for the transfer price for complex-generated raw materials, a prorated part of the DFC of the generating plants must be assigned as allocated capital. The method of computing the allocated capital factors for raw materials is illustrated in Example 7.2.2. Shared raw material storage facilities (such as warehouses) must also be accounted for with allocated capital.

Utilities. As is the case with raw materials, if the standard cost of any utility is used in generating the COM, the utility allocated capital must be accounted for.

Packaging. Any loading, formulating, and packaging facilities shared with other producing plants must be reflected as appropriate allocated capital. Dedicated packaging facilities should be included in the DFC and need not be shown in the allocated capital category.

Factory expense. This is factory expense *capital*, related to, but not to be confused with, the *recurring* factory expense category in the COM. The recurring factory expense (amounting to about 5 percent of DFC *per year*) does *not* include any ROI upon the physical assets housing fire protection, engineering, first aid, and the many other support organizations. Therefore, allocated capital to account for these physical assets must be included. This may be estimated as a lump sum equal to 5.2 percent of the DFC.

Exercise 8.1.1

In Prob. 7.29 economic parameters are listed for a plant to produce 250 million lb/year of urea by using CO_2 as the stripping agent. Assume for the moment that both raw materials are generated in large plants within the complex and are transferred at standard cost. The raw materials plants are (M&S = 710), for a 300,000 short tons/yr total production,

$$\text{DFC} = \begin{cases} \$37 \text{ million} & NH_3 \\ \$19 \text{ million} & CO_2 \end{cases}$$

The urea loading facilities are dedicated. What is the estimated total allocated fixed capital?

Note that the *allocated* fixed capital is never to be added to the *direct* fixed capital as a basis for COM computations.

Working Capital

In Chap. 5, the working capital was defined as the investment in portable, temporary, or consumable materials. It represents tied-up funds required to operate the business. The working capital *cannot* be depreciated since, at least in principle, it is recoverable capital. The working capital, as we will see, changes somewhat with the level of production; for the present, we will look at methods of estimating the working capital for a plant at full capacity (the normal production criterion for the initial evaluation study).

The component parts of working capital are tabulated in Fig. 8.1.1.

Maintenance and Stores Inventory The prorated working capital charge for spare parts on hand averages about 2 percent of the DFC. Note that this addition to the investment has nothing to do with the operating and maintenance supplies costs in the COM; these costs are for those supplies that are actually consumed on a recurring basis, whereas the M&S inventory working capital represents funds tied up in parts which are sitting around, ready to be used.

Purchased Raw Materials The same distinction as has been emphasized for spare parts holds for raw materials. Raw materials consumed in the process represent a continuous operating cost within the COM, not a capital expenditure. However, to render a new plant operational, the initially required quantities of raw materials must be purchased, and a certain amount must be kept on hand at all times. Stored raw materials represent a capital investment, and a return on that investment is expected just as much as a return on the DFC.

The amount of raw materials that is stored depends upon a number of parameters:

Size of shipments. Unit cost of raw materials decreases with the size of the shipments, and it is usually better to purchase the required materials in tank car or some other large lots. If delivery is, indeed, by railroad, additional economy is obtained by multiple-car shipments. On the other hand, shipments that are too large require larger storage facilities (and more DFC) and tie up excessive working capital. The optimum choice is a matter of judicious economic balancing and compromise.

Reliability of supply. The interruption of the raw materials supply stream can have catastrophic consequences to the profitability of a project. If there is a question about the reliability of the supply, large amounts of raw materials may be kept on hand to protect the production plant from the consequences of strikes, transportation breakdowns, or supplier production problems.

Seasonal variations. In some instances, particularly in the case of food and natural product processing, the raw material supply is subject to seasonal variations. Since it is preferable to operate the processing plant on a year-round basis if at all possible, large quantities of the raw material are accumulated during the growing season to last throughout the year. For example, cottonseed oil extraction plants are characterized by giant stockpiles of cottonseeds accumulated during the cotton harvest.

Available storage facilities. Dedicated storage facilities are designed to accommodate the parameters already mentioned. The size or frequency of shipments may be limited by the size and availability of *shared* storage facilities. Oil refineries are characterized by giant storage tank farms, and the scheduling of incoming crude shipments and proper distribution to tankage for subsequent blending is a major technological enterprise. Integrated complex–generated raw materials require little additional storage over that provided in the generating plant; such raw materials represent a smaller working capital investment, a distinct advantage.

The working capital associated with purchased raw materials is usually estimated as *1 month's supply at the purchased price* (FOB production plant). Complex-generated raw materials are not included in the total unless, for some reason, additional project-dedicated storage facilities are to be used.

Product Inventory The criteria which establish the quantity of product to be kept on hand in storage are much the same ones as those which dictate the raw materials storage policy. Thus the quantity may be limited by availability of shared storage in tank farms; it may be dictated by the size of shipments, or by the desire to improve the reliability of supply. Large quantities of product may have to be stored to meet seasonal demand; an example of this is caustic soda, stored to meet demand as a peeling agent for fruit processed in canneries during the harvest season.

The product inventory working capital may be estimated as *1 month's supply at unit value equal to COM*. (Why COM? Why not sales price?)

In-Process Inventory The working capital tied up with the inventory of chemicals present at any one time within the process train may be estimated from the residence time θ in the train. θ is not really the residence time of the product—after all, the product does not exist as such in some portions of the plant. θ may be thought of as the sum of the individual residence times of the various *primary* process streams in the various items of equipment; the primary process streams may incorporate raw materials (diluted in carriers or not), products, and intermediates. The in-process inventory is then the product of θ and the production rate. The unit value of the inventory is the average of the cost of raw materials and the COM, both on a product unit weight basis. A symbolic expression for the working capital generated by in-process inventory $(WC)_i$ is

$$(WC)_i = \tfrac{1}{2}\theta r[(\Sigma\, w_k c_k) + \overline{COM}] \tag{8.1.1}$$

where θ = residence time, years
r = production rate, kg/year
w_k = unit ratio of raw material k
c_k = cost (dollars) per kg of raw material k
$\overline{COM}$ = unit product cost of manufacture, \$/kg

Exercise 8.1.2

Does Eq. (8.1.1), as written, hold for throughputs other than that at rated capacity? Will $(WC)_i$ increase or decrease at reduced (but continuous) throughput?

$(WC)_i$ is usually quite negligible compared with the other working capital categories, but there are exceptions:

Long residence time processes (such as fermentation)
Large intermediate storage processes (for instance, the toluene $\longrightarrow$ benzoic acid $\longrightarrow$ phenol process in which about 1 week's storage of benzoic acid is provided to lessen interdependence between the two halves of the plant)

Exercise 8.1.3

A plant to produce ethanol by means of mixed biomass fermentation is expected to be operated with S.F. = 0.9. The annual production will be 10 million gal of anhydrous ethanol, and 4 lb of mixed biomass (waste agricultural products generated throughout the year) are required per pound of ethanol. The estimated DFC is \$30 million, and the estimated annual COM is \$20 million. The delivered cost of the biomass is \$60/short ton. The residence time in the fermenters is 4 days; the residence time in other process equipment totals 6 h. About 2 weeks' supply of biomass will be stored; longer storage time creates the hazard of spoilage and spontaneous combustion.
Estimate the total working capital.

Many estimators consider the category of *cash and accounts receivable* to be part of working capital. A certain amount of cash must be kept on hand (in rapidly accessible deposits) to meet short-term obligations (payroll, accounts payable). The "accounts receivable" category represents capital in the form of short-term credit, capital that would otherwise be usable for other investments but is not currently available for that purpose. It is capital that does not draw interest and is therefore often incorporated as part of the working capital (which must draw a return from the project along with the rest of the invested capital). We will consider the cash and accounts receivable category to be part of the allocated market capital.

Similarly, *land* values are recoverable, and the land site of a plant, reusable at the end of the plant's lifetime, could logically be considered part of the working capital. The usual procedure is to consider land as part of the corporation's overall assets; these assets are prorated to each project as part of the allocated market capital. Land values are usually small in comparison with the DFC of the plant on the land it occupies. Even prime industrial land is valued at perhaps $25,000/acre (1981).

The magnitude of the working capital is not difficult to estimate in terms of the four categories in Fig. 8.1.1. For very rough preliminary estimates, the working capital may be taken as *12 percent of the DFC* (excluding cash and accounts receivable).

8.2 RETURN ON INVESTMENT

General Expenses and Profit

We are now ready to complete the economic evaluation of a project by computing one of the most commonly used criteria of economic performance, the *return on investment before taxes* (ROIBT). ROIBT is defined very simply as

$$\text{ROIBT} = \frac{100 P_B}{\text{CFS}} \qquad (8.2.1)$$

where P_B is the profit before taxes, and CFS is the *capital for sale*. P_B is the representation of the analysis of income, and CFS is the representation of the analysis of investment; the ratio of the two is the blending of the two analyses which yields a criterion of economic performance (ROIBT). The ratio is a simple measure of the project's estimated profitability: what percentage of the total capital investment will be realized as profit each year?

The computational format for ROIBT is shown in Fig. 8.2.1. The top half of the format is devoted to the computation of the general expenses and profit before taxes. The general expenses are those expenditures which are not centered around the immediate production facility; they are recurring (continuous) expenses incurred by the corporate entity and shared by each production unit as part of its profitability assessment. The format conforms to Eq. (6.1.7), the definition of P_B:

$$P_B = \mathcal{R} - \text{COM} - G \qquad (6.1.7)$$

PRODUCT:				
ACCOUNT NO.: _____ DATE: _____				
PLANT: _____ BY: _____				
PRODUCTION RATE: _____ % OF CAPACITY				

SALES PRICE, $/LB				
ANNUAL SALES, M$/YR				
MANUF. COST, M$/YR (COM)				
SELLING COST, M$/YR (7% OF SALES)				
G and A, M$/YR [1.6% (DIR. & ALLOC. F.C. PER YEAR)]				
RESEARCH, M$/YR (4.6% F.C. PER YEAR)				
PROFIT BEFORE TAX				
CAPITAL FOR TRANSFER				
SELLING CAPITAL ($0.367)/$ SELL. COST				
G AND A CAPITAL (0.7%) (DIR. & ALLOC. F.C.)				
RESEARCH CAPITAL (5.5%) (DIR. & ALLOC. F.C.)				
CASH AND ACCTS. RECEIVABLE (10%) SALES				
TOTAL CAPITAL FOR SALE				

ROIBT = 100 (PROFIT BEFORE TAX/TOTAL CAPITAL)

FIGURE 8.2.1
ROIBT calculation form.

The annual sales revenue $\mathcal{R}$ is the product of the plant's rated capacity r and the unit sales price s established as the most likely one by marketing research efforts (Sec. 3.2). Note that the first ROIBT calculation serves as a keystone; it is based upon rated capacity and the most likely sales price, but other ROIBT calculations are made subsequently as part of a sensitivity analysis (Sec. 8.3).

The manufacturing cost is the plant-oriented COM. The general expenses incorporate three categories, each one of which may be approximated as follows:

Selling cost, 7 percent of annual sales

General and administrative (G&A) costs, 1.6 percent of the sum of direct and allocated fixed capitals per year

Research costs, 4.6 percent of the sum of direct and allocated fixed capitals per year

The recommended calculation format provides for all components of Eq. (6.1.7) to be expressed as thousand-dollar units per year (M$ stands for 10^3 dollars).

The *selling cost* incorporates the salaries of sales and marketing personnel, operating costs (including depreciation) of sales facilities, field expenditures (travel expenses), and TS&D expenditures. An average value of 7 percent of sales is given for estimating purposes, but it must be understood that this value varies a great deal among industries and products. For bulk commodity chemicals the selling cost ranges from about 2 percent to 10 percent of annual sales. The actual cost is a matter of the complexity of the selling job, and the estimated selling cost for a new product should be tailored to conform to the estimate of the complexity. The selling of caustic soda during periods of high demand requires only a modest marketing effort, much more modest than the one required, for example, to penetrate the markets for pulp drainage aids (on Fourdrinier machines) with a new product. The relative difficulty of marketing gasoline on the one hand, and high-sulfur bunker fuel on the other, may be readily appreciated. Consumer products often entail high and sustained sales costs, as high as 35 percent of annual sales.

Both the *G&A* and *research costs* are estimated as a certain percentage of the direct plus allocated fixed capital per year. The G&A costs incorporate such diverse corporate expenditures as salaries of the board of directors and staff administrators, depreciation on buildings and furniture at corporate headquarters, and expenses incurred by public relations, government interactions such as lobbying, and employee recruiting. R&D costs are finally accounted for; the successful project must share the cost of *all* research, including the costs incurred by the 91 out of 92 projects that fail (Table 1.1.5)!

The profit before tax may now be computed according to the format in Fig. 8.2.1 and in conformity with Eq. (6.1.7).

Capital for Sale (CFS)

In Sec. 8.1 the *capital for transfer* was described as the total investment backing up the manufacture of a product within the geographic confines of the integrated complex. Each production unit must also share the burden of corporate investments in facilities beyond the integrated complex boundary. These extraterritorial investments are collectively termed the *allocated market capital*. The allocated market capital, when added to the capital for transfer, yields the *capital for sale* (CFS), truly the total investment associated with a project. It is the total capital required to bring a product up to and including the final sale—the market.

The format for calculating CFS is given in the lower portion of Fig. 8.2.1. Three of the four categories of the allocated market capital correspond closely to the three categories of general expenses:

Selling capital (36.7 percent of annual selling cost)
G&A capital (0.7 percent of direct plus allocated fixed capitals)
Research capital (5.5 percent of direct plus allocated fixed capitals)
Plus cash and accounts receivable (10 percent of annual sales)

The suggested estimating factors are actually quite variable among chemical corporations; their magnitude reflects corporate attitudes and policies toward research, for example. The selling capital refers to sales offices and TS&D labs; research capital incorporates *all* research facilities (some of which may, in fact, be located adjacent to production units within the physical boundary of an integrated complex—the implied extraterritoriality is an abstraction). G&A capital includes corporate headquarters and administrative offices, engineering and production planning offices, computer centers—any physical assets that creative management may conjure up beyond the divisional jurisdiction.

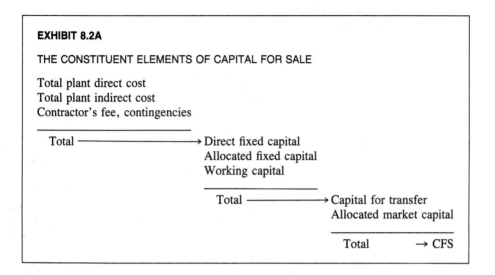

EXHIBIT 8.2A

THE CONSTITUENT ELEMENTS OF CAPITAL FOR SALE

The sundry categories of capital which constitute CFS are reviewed for the reader's convenience in Exhibit 8.2A. Occasionally estimates of CFS appear in the cost literature which include *start-up costs* as a separate category, treated as a one-time investment cost. The start-up costs have been included as part of the plant indirect costs in computing DFC (Sec. 5.1).

The ROIBT is now calculated in Fig. 8.2.1, in conformity to Eq. (8.2.1).

Example 8.2.1

A projected plant is rated at 100 million lb/year of a product to sell at 40 cents/lb. The 1981 DFC is estimated as $30 million, the $\overline{\text{COM}}$ 20 cents/lb. Normal utility consumption (Table 7.2.1) is anticipated. What is the ROIBT? (All raw materials will be purchased.)

SOLUTION

$$\text{Annual sales} = 100 \times 10^6 \times 0.40 = \$40 \times 10^6$$

Utilities:
$$\text{Power} = 100 \times 10^6 \times 0.1 = 10 \times 10^6 \text{ kWh/yr}$$
$$\text{Steam} = 100 \times 10^6 \times 2 = 200 \times 10^6 \text{ lb/yr}$$
$$\text{Water} = 100 \times 10^6 \times 5 = 500 \times 10^6 \text{ gal/yr}$$

Allocated fixed capital (data in Table 7.2.2):

Power:
$$10 \times 10^6 \times 0.017 = \$170{,}000$$

Steam:
$$\frac{200 \times 10^6 \times 2.90}{1000} = \$580{,}000$$

Water:
$$\frac{500 \times 10^6 \times 0.56}{1000} = \$280{,}000$$

Raw materials: none
Packaging: unknown, assume included in DFC
Factory expense (5.2% of DFC): 1.560×10^6
Total allocated fixed capital: 2.590×10^6

Total fixed capital: 32.590×10^6
General expenses:

Selling cost (7% of sales):	2.800×10^6
G&A cost (1.6% fixed capital):	0.521×10^6
Research cost (4.6% fixed capital):	1.499×10^6
Total:	4.820×10^6/yr

$$\begin{aligned} \text{Profit (before taxes)} &= \mathcal{R} - \text{COM} - G \\ &= (40 - 20 - 4.820)10^6 \\ &= \$15.180 \times 10^6/\text{yr} \end{aligned}$$

CFS:

DFC:	30.000×10^6
Allocated fixed capital:	2.590×10^6
Working capital (12% DFC):	3.600×10^6
Selling capital, $0.367 \times 2.800 \times 10^6$	1.028×10^6
G&A capital, $0.007 \times 32.590 \times 10^6$	0.228×10^6
Research capital, $0.055 \times 32.590 \times 10^6$	1.792×10^6
Accounts receivable, $0.10 \times 40 \times 10^6$	4.000×10^6
CFS:	43.238×10^6

$$\text{ROIBT} = \frac{100 \times 15.180 \times 10^6}{43.238 \times 10^6} = 35.1\%$$

Acceptable Return on Investment

In Example 8.2.1, the ROIBT turns out to be about 35 percent. What does this mean—does the result signify an economically attractive venture? How good, or how bad, is 35 percent? Regrettably, as seems to be the case so often in matters economic, there is no straightforward answer. To most individuals who consider any investment yielding more than, say, 15 percent ROIBT exceptionally good, 35 percent sounds downright fabulous. But then again, how fabulous would the 35 percent sound if the probability of obtaining such a return were only 50 percent? The point is that the attractiveness of a particular ROIBT must be evaluated in conjunction with considerations of *risk,* the *availability of capital,* and the *inherent value of money.*

A number of years ago, Aries and Newton (1955) proposed minimum acceptable values of ROIBT as a function of both risk (qualitatively described as low, average, or high) and the category of process industry. A tabulated summary is shown in Table 8.2.1. The suggested values are still reasonably good guidelines, but they

TABLE 8.2.1
MINIMUM ACCEPTABLE ROIBT

	% ROIBT for		
Industry	Low-risk projects	Average projects	High-risk projects
Industrial chemicals	11	25	44
Petroleum	16	25	39
Pulp and paper	18	28	40
Pharmaceuticals	24	40	56
Metals	8	15	24
Paints	21	30	44
Fermentation products	10	30	49

Source: Adapted from Aries and Newton (1955), by permission, McGraw-Hill Book Co.

must be accepted as nothing more than that; in the final analysis, the decision as to what is an acceptable ROIBT is up to the corporate management, guided by a combination of current economic facts and intuition.

The decision as to whether a new project is "high risk" or "low risk" is subjective, but in most cases the correct choice may be based upon the various aspects of the project, and past experience. A new caustic soda plant in a caustic-short region is an example of a low-risk project. Construction of an ethylene oxide plant during a period of steady ethylene oxide growth involves an average risk. A plant to produce a new anticoccidiostat (to combat coccidiosis in fowl) represents a high risk. The risk is dependent upon considerations such as:

Product commercial stability (commodity versus experimental products)
Product obsolescence (due to intense developmental activity in areas of use)
Competition from existing or prospective plants
Market penetration difficulties

High risk implies a higher likelihood that a proposed plant will not attain production goals upon which the ROIBT is based. Note that Table 8.2.1 implies that some of the process industries are inherently riskier than others. For example, intense developmental activity and competition in the pharmaceutical industry make it a sufficiently risky one to justify the extraordinarily high minimum acceptable ROIBT.

How does the performance of the chemical industry compare with the guidelines in Table 8.2.1? The following numbers[1] are typical:

Chemical corporations with annual sales exceeding $1 billion had a median ROIBT of 8.8 percent (1978), 10.2 percent (1979).

[1]Adapted from *Chem. Eng. News* (June 9, 1980).

Smaller corporations with a sales volume below $1 billion but above $160 million had corresponding median ROIBTs of 12.2 and 14.4 percent, respectively.

These numbers are well below the 25 percent average ROIBT for industrial chemicals in Table 8.2.1. Yet many decision makers in the industry would consider the guidelines too low for the proper evaluation of new projects.

Returning now to the result of Example 8.2.1, note that the 35 percent ROIBT would be acceptable if the product represented an average-risk project in most process industries. The project would *not* be acceptable, however, if the risk were judged to be high!

The availability of capital for new investments has a marked effect upon guidelines such as those in Table 8.2.1. During periods of high interest rates on borrowed capital or instances of economic factors limiting retained earnings, capital for new investments may become "tight," and new projects must be economically more competitive to survive. Corporate management may, for instance, decide that no project will be considered for approval unless it exhibits ROIBT over, say, 50 percent. The prime interest rate on borrowed capital also influences the minimum levels of acceptable ROIBT.

Exercise 8.2.1

Explain how a high prime rate of interest (say, 15 percent) affects the level of minimum acceptable ROIBTs.

The ROIBT concept is just one of several criteria of economic performance commonly used in the chemical process industry; we will take a look at some of them in the next chapter. The rest of this chapter will be devoted to a further exploration of ROIBT, for it is still considered to be the *primary engineering profitability criterion,* even for projects in advanced stages of development. ROIBT has some serious disadvantages:

The time value of money is not properly accounted for; the format of ROIBT is that of simple interest (Sec. 2.1).

The assumption of an invariant annual production rate is implicit. In Sec. 8.4 some modified versions of ROI that attempt to account for nonuniform annual cash flow will be introduced.

As a matter of fact, none of the commonly accepted criteria of economic performance is perfect, and ROIBT does have the virtue of simplicity. It is useful for assessing the venture worth and for establishing acceptable pricing. Most engineers and managers are used to the concept; they understand its limitations, and they find that it serves as a useful basis of comparison.

Exercise 8.2.2

Derive a simple relationship between ROIBT and ROIAT, the return on investment *after* taxes, defined as the ratio (in percent) of profit after taxes to CFS.

8.3 PROFITABILITY UNDER VARIABLE CONDITIONS

Inflation and Profitability

So far our consideration of ROIBT has focused upon the value that is computed for one set of "design" conditions. The initial ROIBT of a new project reflects one product selling price (perhaps one established by marketing research), one production throughput (usually rated capacity), one estimate of COM (again normally based upon rated capacity). We now wish to investigate how the profitability, measured as ROIBT, changes as certain input variables are changed systematically. As we will see, such procedures allow us to gain a great deal of additional insight into the nature of the project.

One potential effect upon profitability that may not have escaped the observant is the effect of inflation. It is a matter of common experience that production costs increase in conformity to the pressures of inflation; they do not remain constant as implied by the ROIBT computation. Does this mean that we have to predict what the inflation rate will be in future years in order to assess the future economic performance of a venture? Fortunately, the answer is no (well, a qualified no). The economic evaluation of projects is best expressed in terms of *current, uninflated (constant-value) dollars*. The basis for this manner of assessing profitability is the reasonable assumption that both revenues and costs will increase in proportion to the same inflation rate factor **f**. If, then, all the ROIBT input parameters at some future time are jointly adjusted for inflation, the resulting computation is effectively performed in constant-value, uninflated dollars, no matter how far into the future we choose to gaze; that is, the adjusted numbers are always the same, and the computed ROIBT is the same. We therefore need not be concerned about what the value of the factor **f** is likely to be. This is precisely what we would wish a criterion of economic performance to do—to remain invariant, independent of factors which really cannot be predicted.

There is, regrettably, a small fly in the ointment, and that is depreciation. Depreciation charges do not change with inflation; but in *uninflated* dollars even straight-line depreciation charges do effectively decrease each succeeding year. As a matter of fact, during periods of inflation, it may be argued that none of the depreciation methods allows the investor to recover the full value of the investment!

What effect does this depreciation phenomenon have upon ROIBT? Since depreciation is one item in the COM, the future COM in current dollars decreases, and P_B therefore increases! In other words, uninflated depreciation charges tend to *improve* the ROIBT of a project with time. This small inconsistency is particularly troublesome in comparing projects involving large investments, as pointed out by a number of writers, including Naphtali and Shinnar (1981).

Exercise 8.3.1

Uninflated depreciation charges result in a slowly increasing profit before taxes, expressed in current, constant-value dollars. What is the effect upon income tax paid and cash flow after taxes, both converted to current dollars?

Naphtali and Shinnar (1981) point out still another inflation-caused inconsistency. We have seen (for example, Exercise 8.2.1) that a high interest rate on borrowed capital tends to increase the minimum level of ROIBT that is found to be acceptable. After all, if the interest that will have to be paid on borrowed capital for a project is 15 percent, it would be foolish to proceed with that project if its ROIBT will be only 10 percent. The same kind of argument would hold even if the project were financed from retained earnings; why invest the earnings into a project at 10 percent when the money can be loaned at 15 percent?

The fallacy in this argument is that a high interest rate usually goes hand in hand with high inflation (Prob. 1.15); banks increase the prime interest rate to stay ahead of anticipated inflation. This means that, in current dollars, the annual payments R on the borrowed principal P in effect decrease year by year according to the relationship

$$R'_n = \frac{R}{(1 + \mathbf{f})^n} \qquad (8.3.1)$$

where R'_n is the uninflated dollar equivalent (indicated by the prime) of R paid in inflated dollars n years from the present; $\mathbf{f}$ is the average annual inflation rate over n years, as defined by Eq. (1.3.5).

Inflation thus reduces the equivalent constant-dollar interest income that the lending institution receives. The equivalent interest rate received becomes

$$\mathbf{i}' = \frac{1 + \mathbf{i}}{1 + \mathbf{f}} - 1 \qquad (8.3.2)$$

Exercise 8.3.2

Derive Eq. (8.3.2).

On the other hand, as we have seen, a project continues to realize roughly the same ROIBT regardless of inflation. The 10 percent ROIBT project thus may not look quite so bad in conjunction with a 15 percent prime rate; if the inflation rate happens to be 11 percent, the effective interest rate [from Eq. (8.3.2)] is 3.6 percent, well below the ROIBT! Therefore, minimum acceptable levels of ROIBT should not be raised because of anticipated high inflation rates.

The imbalance between ROIBT and interest payments on borrowed capital that is caused by inflation is the subject of an amusing "Economic Fable" that appeared in the journal *CHEMTECH* (April 1980, p. 201); it is recommended reading.

Pricing for Acceptable Profitability

The criterion of acceptable economic performance is the dominant one in setting prices in the chemical process industry (Sec. 3.4). The final selling price of a

SECTION 8.3: PROFITABILITY UNDER VARIABLE CONDITIONS

product must therefore be chosen to match (or surpass) the ROIBT (or other criterion of economic performance) prescribed by the corporate management as the minimum. If the price thus determined exceeds the current list price of an existing product or the customer's economic benefit price for a new product, then some form of compromise must be reached (perhaps a different plant size), or else the project must be rejected as unattractive.

The initial sales price used for the ROIBT computation is one that is judged to be, for one reason or another, the most likely one. The ROIBT thus computed may exceed the estimator's fondest expectations, or it may turn out to be shockingly low, perhaps even negative! Regardless of the initial outcome, the computation should be repeated for a range of sales prices to establish the one to match minimum acceptable ROIBT, and to determine the *sensitivity* of the profitability criterion to price changes. The format of Fig. 8.2.1 provides for several pricing trials.

Once a computation has been made for one price, the succeeding computations may be done easily and rapidly, since only a small proportion of the input parameters is price-dependent. Table 8.3.1 is intended to facilitate such computations; problem type A is the one under consideration. Changes in price have a proportional effect upon:

Selling capital
Accounts receivable
Annual sales
Costs of selling

We agree to make the costs of selling proportional to the sales price on the premise that one must sell harder to sell a more expensive product.

A series of ROIBTs may now be recalculated in Fig. 8.2.1, although for problem type A an analytical method may be used. Thus, for unit sales price s, the annual revenue $\mathcal{R}$ is the product of s and r, the annual production. Let K_0 represent the invariant capital items (such as DFC), G_0 the invariant general expenses (such as research). From Eq. (6.1.7),

$$P_B = sr - \text{COM} - G_0 - k_G s \tag{8.3.3}$$

where k_G is the proportionality constant for the sales-dependent general expenses (cost of selling). Similarly,

$$\text{CFS} = K_0 + k_c s \tag{8.3.4}$$

where k_c is the proportionality constant for the sales-dependent capital items. From Eq. (8.2.1),

$$\text{ROIBT} = 100 \, \frac{(r - k_G)s - (\text{COM} + G_0)}{K_0 + k_c s} \tag{8.3.5}$$

TABLE 8.3.1
PROFITABILITY UNDER VARIABLE CONDITIONS

	Problem type		
	A ROI vs. sales price, 100% production, given plant size	**B** Annual expenditure or ROI vs. production rate, given plant size, given sales price	**C** Annual expenditure or ROI vs. plant size, 100% production; selling price is a function of total production
Capital:			
DFC	Constant	Constant	$\propto r^{0.7}$
Working	Constant	Almost $\propto r$	Approx. $\propto r$
Allocated fixed	Constant	Constant	$\propto r$
Allocated market:			
G&A, research	Constant	Constant	$\propto (FC)^*$
Selling	$\propto$ Sales price	$\propto r$	$\propto r \times$ sales price
Accounts receivable	$\propto$ Sales price	$\propto r$	$\propto r \times$ sales price
COM (annual):			
Fixed charges (depreciation, insurance, taxes, factory expense)	Constant	Constant	$\propto r^{0.7}$
Regulated charges:			
Labor-dependent	Constant	$\propto r^{0.25}$	$\propto r^{0.25}$
Maintenance	Constant	85% at 0.75r 75% at 0.50r 60% at 0.25r	$\propto r^{0.7}$
Variable charges (raw materials, utilities, royalties, packaging)	Constant	$\propto r$	$\propto r$
Annual sales	$\propto$ Sales price	$\propto r$	$\propto r \times$ sales price
General expenses (annual):			
Selling cost	$\propto$ Sales price	$\propto r$	$\propto r \times$ sales price
G&A	Constant	Constant	$\propto (FC)^*$
Research	Constant	Constant	$\propto (FC)^*$

*FC is direct plus allocated fixed capital.

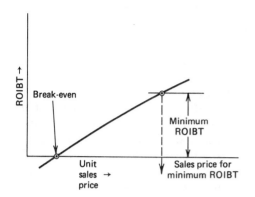

FIGURE 8.3.1
Profitability and pricing.

This relationship may be plotted as in Fig. 8.3.1 to find the price for the minimum acceptable ROIBT and the price at the *break-even* point, where revenues just match expenses.

Break-Even Charts

Suppose the optimum selling price at rated capacity has been established. Another study of considerable importance involves the change in the ROIBT as production level in a given plant decreases below the rated production. After all, it is not very realistic to anticipate that a plant will produce at full capacity from the date it is started up. We have seen that the annual production level may follow a life cycle pattern, perhaps as shown in Fig. 3.3.9, or similar to the distribution patterns in Fig. 2.2.1. We need a method of assessing the effect upon profitability of reduced production; such a method is afforded by the commonly constructed *break-even charts*.

The construction is based upon the relationships that are summed up as problem type *B* in Table 8.3.1. In particular, the dependence upon production level of the three principal categories of charges within the annual COM should be noted:

The fixed charges remain constant.
The variable charges are directly proportional to production.
The regulated charges are somewhat dependent upon production.

The components of the regulated charges are listed in Table 7.5.1. Labor and maintenance are the costliest components, and their variation with production level is indicated in Table 8.3.1. The use of a 0.25 exponent for the labor load in a given plant is particularly debatable; usually at fairly high levels of production the full complement of operating labor is kept, but drastic layoffs may occur at low levels. As a matter of fact, a frequently used rule of thumb in the construction of break-even charts is that

> Regulated charges at zero production equal 30% of those at full production

A linear variation between zero and full production is then assumed.

With this background, we are ready to construct the chart, plotted as annual charges *versus* the production level as percent of full capacity. The features of a typical chart are illustrated in Fig. 8.3.2. The fixed charges are plotted first as a horizontal line. The variable charges are *added on* next; the variable charges at 100 percent capacity are stepped off, and a straight line is drawn to meet the fixed charges line at zero capacity. Again, the regulated charges are calculated and stepped off at 100 percent capacity; 30 percent of this amount is stepped off at zero capacity, and the two points are connected with a straight line. Finally, general expenses are calculated and stepped off at 100 percent and zero capacity; the calculations are based upon the information in Table 8.3.1. This final line represents the total cost as a function of production throughput.

The annual sales at capacity are calculated and plotted, and an independent "sales" line is drawn from this point to the origin (dashed line). The resulting diagram has a number of features worth noting.

The area above the total costs line but below the sales line represents gross

FIGURE 8.3.2
Break-even chart.

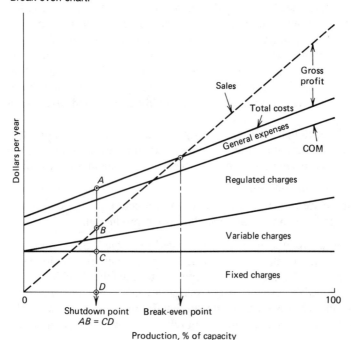

(before-tax) profit; it is the area of positive ROIBTs. The profit (and ROIBT) is reduced to zero at the *break-even point*. At this point the sales revenue just matches total costs; the plant "breaks even." The break-even point is characterized by the percent of plant capacity at which it occurs ("break-even is 50 percent"). It is desirable that break-even occur at a low percentage of capacity, for this condition gives the plant the flexibility to be operated at widely varying throughputs and still to make money. If break-even occurs at, say, 80 percent, the project is in trouble, for even a small reduction in demand will reduce profitability to near zero. A 40 percent break-even represents a comfortable margin for throughput flexibility. Very low break-evens suggest an overpriced product.

Another key item of information that can be obtained from the break-even chart is the *shutdown point*, the production level at which it is more economical to shut down the plant entirely. The conditions leading to shutdown are illustrated in Fig. 8.3.2. In the region of production below break-even, the segment AB represents the annual *loss*; total costs (A) are larger than annual revenues (B). The segment CD represents the annual fixed charges; these accrue no matter what the production level is, even if the plant shuts down. At the shutdown point, $AB = CD$; that is, the fixed charges are equal to the annual loss. Why is this the shutdown criterion? Below the shutdown point the annual loss, if the plant were to be operated, would be larger than the inevitable fixed charges. Therefore it would be best to "bite the bullet" and shut the plant down, for then the annual loss would be limited just to the fixed charges.

The break-even chart demonstrates the point at which ROIBT becomes zero, but it does not give information on how ROIBT varies with production above break-even, or specifically at what production level the minimum acceptable ROIBT is achieved. The graphical construction shown in Fig. 8.3.3 may be used for that

FIGURE 8.3.3
Production level at minimum acceptable ROIBT.

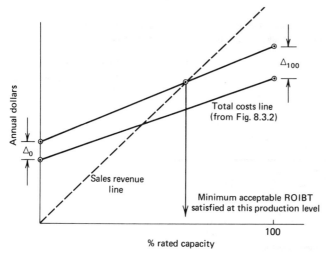

specific purpose, in conjunction with the problem type B relationships indicated in Table 8.3.1.

The total costs and sales revenue lines are reproduced from Fig. 8.3.2. A second line is constructed above the total costs line, offset by the amount Δ, where

$$\Delta = \text{CFS} \times (\text{minimum acceptable ROIBT, fractional})$$

In general, it is acceptable to determine Δ at 100 percent and zero production and to connect the two points with a straight line. The intersection of this line and the sales revenue line pinpoints the production level at minimum acceptable ROIBT (why?).

The implied linear relationship between Δ and the production level r is almost, but not quite, correct. The allocated market capital components which vary with production level are, indeed, proportional to r (Table 8.3.1), but the working capital is generally not. Nevertheless, it is acceptably accurate to base Δ_0 upon zero inventory capital costs.

The results of break-even analysis are sometimes shown on a unit cost basis, as illustrated in Fig. 8.3.4.

It may be noted that considerations of ROIBT have been limited to rated capacity or lower. But what of production levels above rated capacity—how far can a plant be pushed above the design throughput, and what effect upon profitability does this have? Most plants exhibit a "superproductivity" of perhaps 10 to 20 percent with only minor "debottlenecking" demands; such superproductivity may be ascribed to the safety factors used as part of the design. Resnick (1981) discusses "learning curves" which correlate the time growth of production beyond design that results from debottlenecking and learned efficient operation. The relationships in Table 8.3.1 must be used with caution in the region of superproductivity, for forced excess production may reduce operational efficiency and increase relative costs.

Exercise 8.3.3

A reactant dissolved to saturation in a solvent carrier is converted to a product via a first-order reaction. The reaction takes place in a continuous stirred-tank reactor (CSTR)

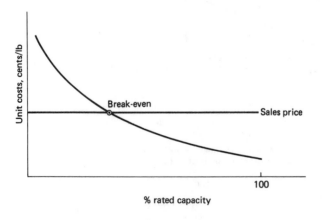

FIGURE 8.3.4
Break-even chart with unit costs.

SECTION 8.3: PROFITABILITY UNDER VARIABLE CONDITIONS

FIGURE 8.3.5
An idealized demand-price curve.

which is completely liquid-filled. At rated capacity, the conversion is 70 percent; the unreacted feed is then recycled. What percent increase in productivity may be expected if the flow through the reactor can be increased up to 120 percent of design?

Economic Optimization of Plant Size

We saw in Chap. 3 that marketing research techniques may be used to project the size of a plant. The size selected, however, may not represent the highest attainable ROIBT, or even the minimum acceptable ROIBT. The analysis required to establish the optimum plant size involves a compromise between demand and economic factors.

Unit production costs almost invariably decrease with increasing plant size, but the attainable sales price may also go down with increasing production. Under such circumstances, there may be a plant size optimum. If the attainable price is not production dependent, then there is no plant optimum—the bigger, the better. On the other hand, if a demand-price curve (Fig. 8.3.5) may be established, then a ROIBT versus plant size curve (Fig. 8.3.6) may be computed by using the relationships listed as problem type C in Table 8.3.1. The relationships are based upon material which should be familiar at this point; for example, DFC is proportional to plant size taken to the 0.7 power [Eq. (5.1.3)]. The working capital may be taken as directly proportional to plant size, a good approximation.

The maximum ROIBT in Fig. 8.3.6 may not turn out to be the best criterion for choosing plant size. For example, the demand-price curve used for the analysis may

FIGURE 8.3.6
ROIBT versus plant size curve with a maximum.

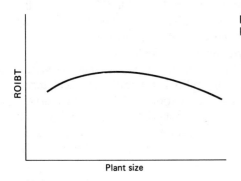

be an overall consumer demand curve; perhaps the projection of the market penetration does not justify the large size at the maximum. Management may not wish to commit itself to the capital investment represented by the optimum plant. Finally, the *maximum* may not be the *optimum,* as we will see in Chap. 10 when we explore the concept of *differential* ROI.

Input Parameter Uncertainties: Sensitivity Analysis

What is meant by the term "sensitivity analysis"? We have seen that all input components of a profitability estimate encompass estimating uncertainties. Sometimes the effects of such uncertainties tend to balance each other out, but there is always the suspicion that one or more of the input parameters may be dominant in establishing the level of profitability. If these dominating parameters incorporate large potential errors that have escaped the attention of the project evaluator, an apparently attractive project may turn out to be a disaster. How can these input parameter uncertainties be accounted for?

It is the purpose of sensitivity analysis to afford some insight into the quantitative effect of major uncertainties upon profitability. Since uncertainty cannot be quantified, how can a quantitative effect be computed? The answer is that a series of quantitative errors in the input parameters is assumed, and the effect of each upon the profitability is computed. This says nothing about the *likelihood* of each error, but the procedure does reveal just how serious such an error would be should it occur.

Specifically, as part of the sensitivity analysis procedure, the profitability criterion (such as ROIBT and others that will be discussed) is recomputed for a range of values of a *single* input item, with that one item and one item only taken at a time, to cover a reasonable span of uncertainty (say, from $+30$ to -30 percent of the estimated value of each item). In a way, this procedure has already been described for two input parameters: the sales price and the production level.

Suppose, for example, that the effect upon ROIBT of uncertainties associated with the estimation of DFC is to be explored. The basic ROIBT value has presumably been computed from the best estimate of DFC already. The ROIBT is now recomputed for, say, six other values of DFC which cover the range between the worst anticipated errors in DFC estimation, perhaps 0.7, 0.8, 0.9, 1.1, 1.2, and 1.3 DFC. Each computation must account for the effect of variable DFC upon all the parameters which enter into the ROIBT computation. For instance, a changing DFC affects the fixed charges in the COM and some of the general expenses. Table 8.3.1 can be used as a guide for sensitivity analysis calculations.

Sales price, production level, and DFC have been mentioned as input parameters which may have an important effect upon profitability because of associated uncertainties. Other parameters which often merit investigation are labor, particularly in small, only modestly automated plants, and raw materials costs. Sensitivity of ROIBT to energy (as utilities) should be explored for energy-intensive processes.

The format for reporting the results of a sensitivity analysis is shown in Fig.

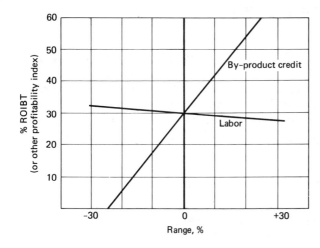

FIGURE 8.3.7
Sensitivity diagram.

8.3.7, a sensitivity diagram. The profitability index is plotted as a function of the values of each one of several input parameters; the values are expressed as a percentage above or below the best estimate. In the example shown, the base-case ROIBT is about 30 percent. The effect of two input parameters is shown. The effect of uncertainties in estimating the labor is evidently minor in this case, but the project is seen to be very sensitive to by-product credit. The sensitivity diagram thus serves the function of alerting management to problem areas which are likely to have an important impact upon project profitability. The decision maker, given Fig. 8.3.7, would look at the near-zero slope of the labor sensitivity line and immediately recognize that labor is not a problem area. Even gross errors due to the uncertainties of properly estimating labor costs will have little impact upon the profitability of this project. On the other hand, even a small overestimate of by-product credit could have a disastrous effect upon profitability, and the decision maker is forewarned to pin down the credit to be received very firmly before the decision to proceed with the project is made. Chemical engineers responsible for project evaluation owe it to the decision makers to warn them of the effect of uncertainties by using sensitivity diagrams.

A great deal of computation goes into the construction of a sensitivity diagram. The procedure is best carried out with the help of computers, and computer practice is afforded by the special project at the end of Chap. 9.

The sensitivity analysis procedure says nothing about the probability of an item assuming a particular value within the range explored, except that the 0 percent value is the most likely one, by definition. If probabilities can be rationally assigned to various value ranges of an input parameter, a quantitative *risk analysis* may be undertaken. The result of such an analysis, described in the next chapter, is intended to apprise the decision maker of the probability that a profitability index will assume a certain value.

8.4 PROFITABILITY CRITERIA RELATED TO ROI

ROI and Variable Cash Flow

One of the limitations of the ROIBT concept is the implicit assumption of invariant annual revenues and expenditures. This limitation is not serious during early stages of project development; an idealized projection of constant annual production at rated capacity is commonplace. Realistically, the annual profit will vary from year to year; it certainly will do so if accelerated depreciation methods are used (why?), and more complex variations may be projected in light of anticipated product or plant life cycles (Fig. 3.3.9).

The definition of ROIBT in Eq. (8.2.1) may be expanded to the case of *variable annual continuous cash flow* by using time averages:

$$\text{ROIBT} = \frac{100\tilde{P}_B}{\widetilde{\text{CFS}}} = \frac{100(\tilde{C}_B - \tilde{D})}{K_F + \tilde{K}_R} \qquad (8.4.1)$$

$\tilde{C}_B$ is the time-averaged continuous cash flow before taxes. It is important to include here only the continuous cash flow components, not the one-time cash flows which represent investments. In particular, it is important *not* to include *recoverable capital* as a continuous cash flow, even though such capital may be recovered as cash revenue in several installments throughout the project's life. After all, if cash revenue from recoverable capital were included as a continuous cash flow (or profit) item, a project could have a positive ROIBT without ever generating a dime of product sales revenue!

The time-averaged depreciation charge $\tilde{D}$ is given by

$$\tilde{D} = \frac{\text{DFC} - S}{n} \qquad (8.4.2)$$

where S is the salvage value and n is the project lifetime.

Exercise 8.4.1

Show that $\tilde{D}$ is independent of the method of depreciation.

$\widetilde{\text{CFS}}$ may be decomposed into two components. K_F is the sum total of the fixed capitals, both direct and allocated.[1] K_F is normally invariant (initial investments), although additional fixed capital investments may well be planned during the lifetime of a project (say, a plant expansion). $\tilde{K}_R$ represents the time-averaged recoverable capital (working capital, accounts receivable).

Exercise 8.4.2

Why may the recoverable capital investment change under conditions of variable cash flow?

---
[1] Nonrecoverable portions of the market capital are included.

By analogy to Eq. (8.4.1) the return on investment after taxes (ROIAT) may be written as

$$\text{ROIAT} = \frac{100(\tilde{C}_A - \tilde{D})}{K_F + \tilde{K}_R} \qquad (8.4.3)$$

Here $\tilde{C}_A$ is the time-averaged continuous cash flow after taxes.

Example 8.4.1

For the project outlined in Table 8.4.1, determine the value of ROIAT.
SOLUTION

$$\tilde{C}_A = \frac{(\Sigma^n C_A)}{n} = \frac{50,000}{5} = \$10,000$$

$$\tilde{D} = {}^{20,000}\!/_5 = \$4000$$

$$K_F = \$20,000 \qquad \tilde{K}_R = \$5000$$

$$\text{ROIAT} = \frac{100(10,000 - 4,000)}{20,000 + 5,000}$$

$$= 24\%$$

Exercise 8.4.3

What is ROIBT for the above example if the tax rate is 50 percent?

ROIBT (or ROIAT) is often computed for ongoing projects as a means of monitoring financial performance. The profitability criterion may be derived for 1 year's operation, or for several years (to answer the question, "How well has the project done?"). It is, in fact, used to monitor the performance of the whole corporation, or even a whole industrial grouping, as we will discover in Chap. 12.

TABLE 8.4.1
FINANCIAL DATA FOR A SIMPLE PROJECT

Five-year life, zero salvage value, no allocated capital investment:
  Direct fixed capital $20,000
  Recoverable capital (invariant)   5,000
Cash flow (continuous) *after* taxes using *straight-line* depreciation:

Year	C_A
1	$ 6,000
2	8,000
3	10,000
4	12,000
5	14,000

The Return on Average Investment

One of the perceived shortcomings of the ROI criterion of economic performance is the fact that the total investment (represented by $\widetilde{CFS}$) is actually decreasing with time, since the DFC is gradually recovered with depreciation payments. Some analysts recommend the use of an *average* direct fixed capital,

$$\tilde{F} = \frac{\sum_1^n \tilde{U}_j}{n} \qquad (8.4.4)$$

where $\tilde{U}_j$ is the *average undepreciated capital during year j*. For instance, if the initial DFC is \$20,000 and depreciation during year 1 is \$4000,

$$\tilde{U}_1 = \frac{20{,}000 + 16{,}000}{2} = \$18{,}000$$

that is,

$$\tilde{U}_j = V_{j-1} - \tfrac{1}{2} D_j \qquad (8.4.5)$$

Allocated fixed capital is considered to be invariant for present purposes; actually, of course, it also depreciates, and the allocated capital depreciation charges are included as part of the COM and general expenses. With K_{AF} representing the invariant allocated fixed capital,[1] Eq. (8.4.3) becomes

$$\text{ROAIAT} = \frac{100(\tilde{C}_A - \tilde{D})}{\tilde{F} + K_{AF} + \tilde{K}_R} \qquad (8.4.6)$$

where ROAIAT is the return on *average* investment after taxes. ROAIBT may be similarly defined. The ROAI profitability criterion is not often used.

Example 8.4.2

Calculate ROAIAT for the project of Table 8.4.1.
SOLUTION

Year no.	Book value of capital, start of year	Depreciation	Book value of capital, end of year	Av. undepreciated capital during year
1	20,000	4000	16,000	18,000
2	16,000	4000	12,000	14,000
3	12,000	4000	8,000	10,000
4	8,000	4000	4,000	6,000
5	4,000	4000	0	2,000
				$\sum_{}^{n} \tilde{U}_j = 50{,}000$

[1] Including nonrecoverable portions of the market capital.

$$\sum_{}^{n}(C_j - D_j) = \sum_{}^{n} C_j - \sum_{}^{n} D_j \equiv \sum_{}^{n} P_j = 50{,}000 - 20{,}000 = 30{,}000$$

$$\text{ROAIAT} = 100 \times \frac{1/5(30{,}000)}{1/5(50{,}000) + 5000} = 40\%$$

Exercise 8.4.4

Show that for straight-line depreciation,

$$\tilde{F} = 1/2(V_0 + S) \qquad (8.4.7)$$

where V_0 is the initial DFC and S is the salvage value.

The numerator in Eq. (8.4.6) is independent of the method of depreciation (Exercise 8.4.1), but $\tilde{F}$ is not. This means that the ROAIAT depends upon the method of depreciation, as shown in the next example.

Example 8.4.3

Suppose the cash flows in Table 8.4.1 were based upon DDB depreciation. What would be the ROAIAT?

Year no.	Book value of capital, start of year	Depreciation	Book value of capital, end of year	Av. undepreciated capital during year
1	20,000	8,000	12,000	16,000
2	12,000	4,800	7,200	9,600
3	7,200	2,800	4,320	5,760
4	4,320	1,728	2,592	3,456
5	2,592	2,592	0	1,296
				$\sum_{}^{n} \tilde{U}_j = \$36{,}112$

$$\sum_{}^{n}(C_j - D_j) = 30{,}000 \quad \text{as before}$$

$$\text{ROAIAT} = 100 \times \frac{1/5(30{,}000)}{1/5(36{,}112) + 5000} = 49.2\%$$

As might have been expected, more rapid depreciation results in a higher profitability criterion.

Payout Time (Payback Period)

A popular criterion of profitability, simple to compute and readily understandable, is the *payout time* (payback period). It is defined as the number of years required to recover the DFC *only* with continuous *after-tax* cash flow. In a sense it is the reciprocal of the ROIAT, although it encompasses fewer input parameters and thereby suffers from lack of completeness; return on working and allocated capitals is essentially ignored. Economic performance is assessed on the basis of whether the payout time is much less than the project lifetime; of course, a short payout time is desirable, and values of 3 to 4 years for long-range projects are generally acceptable.

An example will serve to illustrate the method.

Example 8.4.4

What is the payout time for the project in Table 8.4.1?
SOLUTION
Year 1:

$$\text{Investment} - \text{cash flow (year 1)} = 20{,}000 - 6{,}000$$
$$= \$14{,}000$$

Year 2:

$$\text{Unrecovered investment (end of year 1)} - \text{cash flow (year 2)} = 14{,}000 - 8{,}000$$
$$= \$6{,}000$$

Cash flow (year 3) is larger than unrecovered capital at end of year 2; therefore only a fraction of a year, or

$$\frac{6{,}000 \text{ (unrecovered investment)}}{10{,}000 \text{ (cash flow)}} = 0.6$$

is required to recover all depreciable capital.

$$\text{Payout time} = 2.6 \text{ years}$$

This is less than the lifetime of 5 years, and the project is therefore acceptable in that sense.

Another popular profitability criterion related to ROI is the *percent return on sales*,[1] which may be computed on either a before-tax or after-tax basis. The criterion is often used as a project or corporate evaluation yardstick by the sales and marketing function. It is, very simply, the ratio of profit (say, P_B) to the annual sales revenue $\mathcal{R}$ (expressed as a percentage). For projects that have K_{AF} and $K_R \ll \text{DFC}$,

$$\frac{P_B}{\mathcal{R}} \approx \frac{\text{ROIBT}}{T} \tag{8.4.8}$$

where T is the turnover ratio.

[1] Also called *profit margin*.

Exercise 8.4.5

Derive Eq. (8.4.8).

8.5 THE COSTS OF PRODUCT TRANSPORTATION

Rail Freight

Once a product has left the confines of the manufacturer's distribution warehouse or tank farm, the additional costs of transportation are normally borne by the buyer. That is, the product is sold FOB (free on board) at the final distribution point, and the final cost borne by the manufacturer (and reflected in the profitability computation) is that of loading the product into the carrier.

This is not to say, however, that transportation costs do not influence the pricing and profitability of a product. On the contrary, we have already seen that transportation costs may play a key role in the decision to site a plant in a particular geographic area because of a competitive advantage in pricing to regional consumers. On the other hand, distant markets may force the producer to "swallow" at least part of the shipping cost to meet competition on a freight-equalized basis. Transportation costs enter profitability considerations most directly in the way they influence the producer's raw materials costs.

Corporations manufacturing chemical products are well aware of the role that transportation costs play in meeting competition, and many of the larger corporations possess their own transportation fleets—tank trucks, railroad tank cars, ships and barges, cross-country pipelines. The large integrated oil companies control transportation and distribution from the raw material source to the retail outlets, often through the agency of subsidiary or associated companies.

Transportation costs are reflected in the profitability of a product in still another subtle way, as costs associated with transportation accidents. The accidental spillage of toxic and otherwise hazardous chemicals continues to plague railroad, highway, and waterway transport. The *product stewardship* concept has led responsible corporations to organize distribution emergency response teams that are sent out to cope with and clean up accidental spills and releases of the corporation's products. These teams often act in an advisory capacity during community emergencies involving *any* chemical release hazard. Corporate transportation specialists are involved in the never-ending search for safer methods of packaging and transportation to minimize the risks associated with the shipping of chemical products. The industry (through the auspices of the Chemical Manufacturers' Association) supports the services of a Chemical Transportation Emergency Center (CHEMTREC).

Railroads represent the primary mode of overland transportation of chemical products, especially over long distances. The railroad cars commonly used are:

Tank cars (for bulk liquid products)
Hopper cars (for bulk solids)
Boxcars (for drummed and packaged products)

Tank cars are available in a sequence of sizes to fit a variety of transportation needs, starting with diminutive 4000- and 6000-gal units and ranging through standard 10,000- and 20,000-gal sizes to "jumbos" of 30,000 gal or more. A variety of materials of construction is available, including rubber lining, and units may be provided with steam coils for the transportation of high-melting products. The loading and unloading of liquefied products is particularly convenient. They may be pumped or displaced with compressed gas; often the tank cars may be vented to the storage tanks. Properly designed loading stations incorporate ingenious piping systems replete with swivel joints, flexible couplings, and quick connectors to accelerate the loading task and to reduce the inherent hazards of temporary connections handling high liquid rates of transfer. Ingenious solutions have been found to reduce the cost of the normally more complex and labor-intensive operations of loading solids and packaged products. For example, salt for the operation of chloralkali units may be brought in hopper cars and dumped through bottom car gates directly into sumps below the rails, where water is circulated to dissolve the salt crystals and carry the raw material into subsequent processing steps as a saturated brine.

The cost of rail freight is extremely variable and depends upon a number of factors, including the distance shipped, the inherent shipping hazards, the type of cars required for best handling and corrosion resistance, the quantity shipped and the frequency or regularity of shipment, competition with other modes of transpor-

FIGURE 8.5.1
A jumbo railroad tank car. *(General American Transportation Corporation.)*

tation to the destination, the negotiating skills of the traffic department representatives, and the ownership of the rolling stock. Tank cars may be leased from companies specializing in this type of rolling stock; some chemical corporations own their own tank cars. A typical 20,000-gal steel tank car represents an investment of about $60,000 (1981) and may be expected to be used for 10 to 12 annual consignment trips.

Example 8.5.1

Estimate the cost of shipping 1 lb of a product (specific gravity = 1) in a corporation-owned tank car. Assume that operating costs are 25 percent of the investment per year and that a 30 percent ROIBT is required. Add on a charge of 1 cent/lb for the services of the railroad company.

$$\text{Total shipped per year} \approx 20{,}000 \times 8.34 \times 10$$
$$= 1.67 \times 10^6 \text{ lb/yr}$$
$$\text{Annual costs and return} = (0.25 + 0.30)60{,}000$$
$$= \$33{,}000$$
$$\text{Shipping cost} = 1.0 + \frac{33{,}000 \times 100}{1.67 \times 10^6} \approx 3\text{¢/lb}$$

Shipments handled entirely by the railroad company are higher, and unit rates of 5 to 6 cents/lb are more usual (1981). Special circumstances will reduce these rates; for instance, "water-compelled rates" (due to competition of marine transport through the Panama Canal) from the West Coast to the Gulf Coast are 4 cents/lb or even less.

Tank Trucks

Tank trucks represent an economical mode of transport of bulk liquids over relatively short distances and to destinations not well served by railroads. They share with railroad tank cars the advantages of easy loading and availability in a variety of materials of construction. Local supply from central distribution points is usually handled in trucks (including transport of bulk solids and packaged products in applicable truck types).

Federal and state regulations limit the capacity of tank trailers to about 7500 gal; some are compartmented. Trucking rates depend upon some of the same factors that establish railroad rates. In addition, the rates are determined by requirements of the driver; overnight trips tend to be more costly. Relative trucking costs are illustrated by the following (1979) figures for shipping 5000 gal of solvent from the San Francisco Bay Area to various destinations:

Modesto (central California)	$ 150
Los Angeles	$ 350
Seattle	$1300

A considerable cost reduction may be effected by making arrangements for trailer occupancy during the return trip.

Marine Transport

Marine transport represents the cheapest mode of long-distance shipment of chemical products. Crude oil tankers, petroleum product tankers, chemical tankers, and dry cargo carriers are available for coastal and overseas service. The size of bulk product shipments for this kind of service is quite large, on the order of 1 million gal or more. Crude oil and petroleum products are shipped in much larger quantities than that, and economic pressures have resulted in the construction and operation of very large crude carriers (VLCCs, or supertankers) during the decade of the seventies. These giant ships boast displacements of between 200,000 and 350,000 tons deadweight and higher. A 200,000-ton ship has a maximum carrying capacity of about 1,300,000 bbl or 55 million gal. The construction rate of the VLCCs has exceeded the demand for crude transport, and at the beginning of the decade of the eighties, many of those ships were relegated to the ignominious duty of floating storage tanks (Aitken, 1981).

Smaller volumes may be transported in barges, which afford the additional transportation dimension of the inland waterways. Federal and state regulations may limit the nature of the cargo thus transported in certain areas; for example, the

FIGURE 8.5.2
A supertanker. *(Standard Oil Company of California.)*

barge transport of ethylene oxide is restricted because of potential safety hazards to busy ports and waterways.

Marine transport rates are not strongly distance-dependent; the cost of shipping million-gallon lots of organic liquids from the Gulf Coast to the West Coast was 1.5 cents/lb in 1981, about the same cost as that for shipments from the Gulf Coast to Japan. Some other representative costs of long-distance marine transport are given by Aitken (1981):

Coal, U.S. East Coast to Japan: $29/metric ton
Liquid ammonia, Black Sea to U.S. Gulf Coast: $29/metric ton
Phosphate rock, Aqaba to India: $22/metric ton.

NOMENCLATURE

c	unit cost of a raw material, $/kg; specifically, c_k, unit cost of raw material k
C	cash flow, $/year; C_A is cash flow after taxes, C_B is cash flow before taxes
D	depreciation charges, $/year
f	annual inflation rate (fractional)
$\tilde{F}$	time-averaged direct fixed capital investment
G	general expenses, $/year; G_0 are general expenses which are invariant under conditions of changing product sales price
i	interest rate, compound (fractional)
k	proportionality constants (k_C and k_G) defined by Eqs. (8.3.3) and (8.3.4)
K	capital; K_{AF} is allocated fixed capital, K_F is fixed capital, K_0 is capital that is invariant under conditions of changing product sales price, K_R is recoverable capital
n	number of years (for example, project lifetime)
P	profit; P_B is profit before taxes, P_A is profit after taxes
r	production rate, kg/year
R	annuity payment, $/year
$\mathscr{R}$	annual revenue, $/year
s	unit sales price of product, $/kg
S	salvage value, $
$\tilde{U}$	average undepreciated capital during a particular year
V	book value
w	unit ratio of a raw material, kg/kg of product
$(WC)_i$	working capital tied up in inventory
Δ	product of CFS and minimum ROIBT
θ	residence time in total plant, years
T	turnover ratio

Auxiliary symbols

Bars above a symbol indicate unit quantities (per kg of product); for instance, $\overline{G}$ is the general expense per kg of product.

Tildes above a symbol indicate a time-averaged quantity; for instance, $\tilde{C}_B$ is the annual cash flow (before taxes) averaged over the project lifetime.

Primes used as superscripts indicate value in uninflated, constant-value currency (as in R').

Subscripts refer to a particular year (j, n) or an invariant, base-case condition (0).

Abbreviations

COM	cost of manufacture, $/year; $\overline{\text{COM}}$ (with bar) is unit cost of manufacture, $/kg of product
CFS	capital for sale
CSTR	continuous stirred-tank reactor
DDB + SL	double-declining-balance plus straight-line depreciation
DFC	direct fixed capital
FC	fixed capital (direct plus allocated); also, fixed charges (can be distinguished in context)
FOB	free on board
G&A	general and administrative
M&S	Marshall & Swift (cost index)
RC	regulated charges
R&D	research and development
ROI	return on investment; includes ROIBT (return on investment before taxes), ROIAT (return on investment after taxes), ROAIBT (return on average investment before taxes), and ROAIAT (return on average investment after taxes)
S.F.	stream factor
TS&D	technical service and development
VC	variable charges
VLCC	very large crude carrier
WC	working capital

REFERENCES

Aitken, D.: Ocean Freights, *Process Econ. Int.*, **II** (2):38 (Winter 1980/81).

Aries, Robert S., and Robert D. Newton: "Chemical Engineering Cost Estimation," McGraw-Hill, New York, 1955.

Malina, Marshall A.: Storage Capacity: How Big Should It Be?, *Chem. Eng.*, **87** (2):121 (Jan. 28, 1980).

Naphtali, L. M., and R. Shinnar: Effect of Inflation on Energy Cost Analyses, *Chem. Eng. Prog.*, **77** (2):65 (1981).

Resnick, William: "Process Analysis and Design for Chemical Engineers," McGraw-Hill, New York, 1981.

PROBLEMS

8.1 Market studies indicate that 8 million lb/year of a new chemical may be sold at 40 cents/lb. The capital for sale for such a plant has been estimated at $2,000,000. The manufacturing cost has been ascertained at 22 cents/lb, general expenses at 8 cents/lb. What is the ROIBT?

8.2 A new project involves a DFC of 2×10^6 and CFS of 2.6×10^6. The investment may be written off in 5 years by using straight-line depreciation, zero salvage. Five years hence the anticipated annual continuous cash flow is 1×10^6 in inflated dollars (based on a projected $f = 0.10$). What is the ROIBT?

8.3 Compound XC-5732.3 is an extremely promising anti-inflammatory agent which has been synthesized in your company's laboratories in batches up to 1000 g with 98% purity.

Preliminary toxicological data indicate no observable side effects of the drug in doses up to 10 times the recommended level. The drug is easily competitive with other antiarthritic agents sold to formulator-distributors in the price range of $4 to $6/kg of active ingredient. Because of the drug's effectiveness, a quick market penetration of up to 3 million kg/year is anticipated.

An early decision is needed whether to proceed with the project and the costly testing to obtain FDA approval. Rough preliminary estimates based upon laboratory procedures have produced the following figures:

Total out-of-pocket expenditures: $1.50/kg produced
DFC, 3-million-kg plant: $5 million
CFS: $6 million

Because of intense competition and high risk, the plant is expected to remain in production for only 5 years.

Do you recommend that the project be pursued?

8.4 (Adapted from 1972 Chemical Engineering Registration Examination, State of California. By permission.) A new silver catalyst is available which, if installed in an existing ethylene oxide plant, will raise production 80 percent and lower the amount of ethylene consumed per pound of product by 10 percent.

The present plant produces 10^8 lb of ethylene oxide/year, which sells for 12 cents/lb at the plant. Fixed charges are 2×10^6/year. Variable charges are 8 cents/lb, of which 5 cents/lb goes for ethylene. The new catalyst and modification of the feed and recovery system will cost 2.6×10^6 in capital, and the catalyst will have a scrap value of 0.2×10^6 after 3 years (10 percent of its original value).

On the basis of ROIBT, do you recommend use of the new catalyst? Amortize the change in 3 years by the straight-line method. Neglect any effect upon general expenses.

For this type of problem, the ROIBT is the ratio of extra profit to new capital.

8.5 Analyze the statement, made in the text, that, during periods of inflation, none of the methods of depreciation allows the investor to recover the full value of the investment. What considerations would reduce the impact of this problem on the investor? Do current government policies promote such considerations?

8.6 Inflation rates do not enter into ROIBT computations because both expenses and revenues (as well as some components of CFS) are assumed to increase in direct proportion to the inflation rate—although it has been pointed out that COM is not exactly proportional to the inflation rate because depreciation payments remain uninflated.

It often happens, however, that *sales price increases lag inflation*. Assume that the sales price increases lag inflation by j years. For instance, if $\mathbf{f} = 0.10$ and $j = 2$, all expenditures following start-up will increase by 10 percent/year; during the third year, the sales price is increased by 10 percent over the amount at start-up, but by that time expenditures are $100[(1.10)^3 - 1]$ percent over the initial values. During year 4 the price is increased another 10 percent, to $100[(1.10)^2 - 1]$ percent of the start-up price, but now expenditures are $100[(1.10)^4 - 1]$ percent of their initial values, and so forth.

a Using the format of Eqs. (8.3.3) and (8.3.4), derive an expression for ROIBT for the case of the lagging price increase. Use primes [as in Eq. (8.3.1)] to indicate uninflated dollar equivalents *at start-up time*. Assume that COM is proportional to inflation; i.e., neglect the fact that depreciation charges are not inflated.

b What is the relationship between ROIBT and ROIAT (see Exercise 8.2.2) for this case?

c Calculate ROIBT for $f = 0.15$ and the two cases, $j = 0$ and $j = 2$, for a project with the following parameters:

$$r = 100 \times 10^6 \text{ kg/yr} \qquad \overline{G_0'} = 4¢/\text{kg}$$
$$s = 100¢/\text{kg} \qquad K_0 = \$80 \times 10^6$$
$$\overline{(\text{COM})'} = 40¢/\text{kg} \qquad k_c = 26.4 \times 10^6$$
$$k_G = 14.0 \times 10^6$$

8.7 There are several approaches to the selection of the optimum size of raw material storage; the following procedure is one such approach.

Let us suppose that a 100×10^6 lb/year sodium dodecylbenzene sulfonate detergent plant (see Prob. 3.35) is to be built. One of the raw materials is NaOH, delivered as a 50% aqueous solution (density = 12.7 lb/gal), which will be stored in steel tanks; 70.8 gal of the 50% solution is required per ton of detergent product.

The *strategy* adopted for the storage facilities is the following:

1 Provide two large storage tanks.
2 Size tanks so that they will jointly hold twice the volume of each caustic soda shipment. In this way the tanks will be half full (or one will be completely full, the other empty) by the time the next regularly scheduled shipment arrives.
3 If shipment delays do occur, some raw material is on hand as a safeguard so that operations may continue.
4 Normally the average inventory volume is therefore three-quarters of the storage volume.

The caustic soda will be delivered in 20,000-gal tank cars. The delivered price of the caustic (quoted as dollars per ton of contained Na_2O species) decreases with increasing size of shipment according to the schedule tabulated below. The table also shows the estimated annual loss of production revenue as a result of undersized storage facilities and resulting estimated production interruptions.

Number of cars per shipment	Delivered price, $/ton Na$_2$O	Estimated annual production loss, $
1	310	2,000,000
2	306	1,000,000
5	300	500,000
10	297	300,000
20	292	100,000

The various charges incurred by the storage facilities may be taken as follows:

1 Annual cost of delivered caustic.
2 Annual loss of production.
3 Annual fixed charges (assume 20 percent of DFC, including depreciation).
4 Annual ROIBT on DFC and working capital; i.e., in making the size selection, one must account for the expected ROI on more expensive equipment. Take ROI = 25 percent.

As the number of cars per shipment and the corresponding size of storage facilities decrease, the first two items increase, the last two decrease.

Labor may be considered to be independent of size of shipment (constant labor per car, constant annual labor charges). The DFC may be scaled up and down from the known battery limits DFC contribution of a *single* 100,000-gal steel tank storage facility — $325,000. Allocated capital may be considered to be independent of the size of the facilities.

Which of the five options in the table is best, and what size of storage tank do you recommend?

8.8 a For the plant in Example 8.2.1, use Eq. (8.3.5) to generate a plot similar to Fig. 8.3.1. What is the unit sales price at break-even?

b Assume that the product is an average-risk petroleum commodity. On the basis of Table 8.2.1, what minimum selling price is indicated?

8.9 The COM of the product in Example 8.2.1 has the following components at rated production:

Raw materials	4.7 ¢/lb
Utilities	5.0 ¢/lb
Fixed charges	7.2 ¢/lb
Regulated charges	2.1 ¢/lb
Royalties	1.0 ¢/lb
COM	= 20.0 ¢/lb

Draw a break-even chart.

8.10 The working capital at rated capacity in the plant of Example 8.2.1 and Prob. 8.9 is 12 percent of DFC. The product is an average-risk petroleum commodity. Generate a graph analogous to Fig. 8.3.3, and determine at what level of production the minimum acceptable ROIBT is attained.

8.11 Preliminary cost estimate for a grass roots plant producing 10,000,000 lb of a product/year yields the following cost figures:

Direct fixed capital:	$2,500,000
Working capital (including accounts receivable) (taken as 22 percent of fixed capital):	$550,000
Total production cost:	
A. Direct production costs (raw materials, labor, utilities, overhead)	15.7¢/lb
B. Fixed charges (18% of fixed capital investment per year)	4.5¢/lb
C. General expenses (10% of A & B)	2.0¢/lb
Total production cost	22.2¢/lb

Market research people suggest that up to 20,000,000 lb/year of product could be sold at a price of 32 cents/lb.

The decision is made to construct a plant to produce 20,000,000 lb/yr. What is the break-even point at a selling price of 32 cents/lb, that is, the annual production rate at which gross sales would just equal total production cost?

8.12 Preliminary cost analysis may be performed with the aid of computers, but for that purpose the analytical criteria must be expressed in algebraic form.

Derive algebraic expressions for the relative production rate (as a fraction of rated capacity) at the break-even point ($\bar{r}_b$) and at the shutdown point ($\bar{r}_s$) in terms of the charges incurred at rated capacity:

$(FC)_r$ = fixed charges (including depreciation)
$(VC)_r$ = variable charges
$(RC)_r$ = regulated charges
$\mathcal{R}_r$ = annual sales revenue at rated capacity
K_F = total fixed capital investment (direct and allocated)

Take regulated charges at zero production rate = $0.30(RC)_r$.

8.13 A small new plant has been constructed to extract lycopene from carrots by using supercritical carbon dioxide. The plant represents a DFC of $2 million. Operating problems and poor markets have forced production down to 27 percent of rated capacity, and total production costs have been exceeding sales by about $¼ million/year (negative operating margin). Should the plant be shut down? Give your reasoning.

8.14 (Adapted from the August 1970 Registration Examination for Chemical Engineers, State of California. By permission.) It is proposed that a plant to manufacture 10 million lb/year of highly purified m-cresol from m-xylene be constructed. The proposed process has a number of attractive features:

1 The mechanism involves a catalytic air oxidation,

$$CH_3\text{-}C_6H_4\text{-}CH_3 \xrightarrow{O_2} CH_3\text{-}C_6H_4\text{-}OH + CO_2 + H_2O$$

The yield is 90 percent, and catalyst costs are negligible.

2 m-xylene at 31 cents/lb is upgraded to a product selling for $1.50/lb.

3 There is considerable demand for pure m-cresol.

4 A similar process to oxidize toluene to phenol has been operated successfully for a number of years.

From your general knowledge of economic factors and the few data supplied below, make a rough estimate of the return on investment (before taxes) for the proposed plant. In your opinion, is the proposal attractive?

The existing toluene oxidation plant manufactures 5 million lb/year of very pure phenol. The direct fixed capital in the existing plant is $4,000,000, with another $1,000,000 of allocated capital (primarily utilities). The plant is operated with four operators per shift.

By analogy, it is estimated that the process will require 2 kWh and 10 lb of 150 psig steam per pound of cresol. Other utilities are negligible.

8.15 a The *marginal cost* of a product is the cost of producing an additional unit weight without additional investment. For instance, if a plant is operating at rated capacity r, the marginal cost is the difference in total annual costs when the plant is operated at a level of $r + 1$ units per year and when it is operated at the level r. Marginal costs are generally lower than the unit costs of production; they are important in the course of justifying an expansion of production, perhaps into the region of superproductivity.

Suppose the total annual costs of producing r units of a product are given by

$$\sigma = (FC)_r + (VC)_r + (RC)_r + G_0 + [G(r)]_r$$

where FC, VC, and RC are the fixed, variable, and regulated charges, respectively; G_0 is the production-independent general expense; and $G(r)$ is the production-dependent general expense.

How would you use an equation of this format to derive a marginal cost?

b A plant is operated at rated capacity of 10 million kg/year with the following unit costs:

Raw materials	20¢/kg
Utilities	10¢/kg
Fixed charges	15¢/kg
Labor-dependent items	6¢/kg
Waste treatment	2¢/kg
Royalties	1¢/kg
Selling cost	4¢/kg
Other general expenses	6¢/kg
$\sigma =$	64¢/kg

What is your best estimate of the marginal cost?

8.16 The demand-price curve for a modified polyacrylonitrile fiber for carpeting applications has been developed; see Fig. P8.1. A large plant is indicated, and because of the risks

FIGURE P8.1
Demand-price curve, modified polyacrylonitrile fibers.

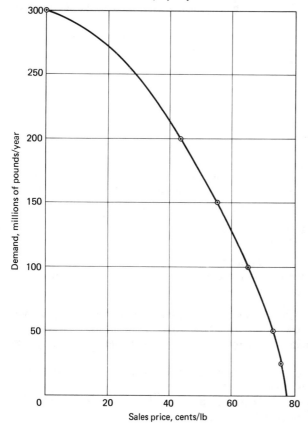

and huge investments involved, corporate management insists upon a minimum ROIBT of 35 percent. What size of plant will result in the highest ROIBT, and will the minimum ROIBT criterion be satisfied?

The following data are available for a 100-million-lb plant:

$$DFC = \$50 \times 10^6$$
$$\text{Allocated fixed capital} = \$5 \times 10^6$$
$$\text{Working capital} = \$10 \times 10^6 \text{ (excludes accounts receivable)}$$

Annual COM:

$$\text{Fixed charges} = \$12 \times 10^6$$
$$\text{Variable charges} = \$15 \times 10^6$$
$$\text{Labor-related charges} = \$ 1 \times 10$$
$$\text{Maintenance} = \$ 2 \times 10^6$$

8.17 [This problem is adapted from the book of Aries and Newton (1955). The investment and revenue dollar numbers are nostalgically low, but the general relationships are still valid.]

A market survey has indicated the existence of a potential sales volume on a new product at the noted prices:

Market, lb/yr	Sale price per lb
200,000	$0.85
500,000	0.85
1,000,000	0.75
2,000,000	0.60

For a plant of 200,000 lb/year capacity, a fixed capital cost estimate has been prepared. This has been analyzed to ascertain the cost of those components which would be increased in number and those which would be increased in size in a plant of greater capacity.

Units varying by number: $100,000
Units varying by size: $\underline{152,000}$
Fixed capital $252,000 (direct)
Working capital $ 50,000 (including cash and accounts receivable)
Allocated fixed capital   None

Similarly, manufacturing cost and general expense estimates have been determined for the plant of 200,000-lb annual capacity. The manufacturing cost estimate has been analyzed to determine the expenses on units varying by number and those on units varying by size in larger plants.

Management has declared that a 20 percent return on investment before taxes must be earned in view of the relative risk involved.

a What will be the direct fixed capital costs of plants operating at rated annual capacities of 500,000, 1,000,000, and 2,000,000 lb?

Cost Item	Cost per lb
Raw materials, utilities, packaging	$0.200
Fixed charges and maintenance on units varying by number	0.080
Fixed charges and maintenance on units varying by size	0.075
Labor-related expenses on units varying by number	0.160
Labor-related expenses on units varying by size	0.071
Manufacturing cost	$0.586
General expenses	0.059
Total cost	$0.645

b What is the unit manufacturing cost for each of the plant sizes considered?
c What is the unit general expense at each of the plant sizes considered?
d At which of the plant sizes considered will the maximum unit profit be secured?
e At which of the plant sizes considered will the maximum annual profit be secured?
f At which of the plant sizes considered will the maximum percent return on the investment before taxes be secured?

Assume that annual general expenses are a constant percent of annual COM. Assume that working capital is a constant percent of total fixed capital.

8.18 Draw a sensitivity diagram for the project represented by the data in Example 8.2.1 and Prob. 8.9. Incorporate the following parameters into the diagram:

Raw materials
Labor (represented by regulated charges)
DFC
Working capital
Energy (represented by total utilities)

8.19 Devise your own symbols to derive an equation showing the relationship between ROIBT and DFC, with all other input parameters held constant. Obtain the proper numerical coefficients for the DFC variable from Figs. 8.1.1 and 8.2.1. Form the derivative of your expression with respect to the DFC percent error range, and write out its format at the origin [zero error, or most likely value = $(DFC)_0$]. The resulting expression is the slope of the sensitivity line at the origin.

What is the numerical value of the slope for the case represented by the data in Example 8.2.1 and Prob. 8.9? Do you think the slope at the origin is a valid quantitative indication of sensitivity?

8.20 A plant to produce 10 million lb/year of a chemical product has a projected 10-year life and a capital for sale of $6 million (DFC = 5×10^6). The cost of manufacture is 25 cents/lb, general expenses are 8 cents/lb. If the sales price is such as to give a payout time of 4 years, what is the ROIBT?

8.21 a A new plant project is described by the following parameters:

DFC	$3,400,000
Working capital (including accounts receivable)	510,000
Allocated capital	150,000
CFS	$4,060,000
Annual raw materials:	$6,240,000

Other annual *out-of-pocket* expenses: $1,247,000
Annual sales revenues (invariant): $9,770,000
Plant life: 10 years
Depreciation: DDB + SL

What is the continuous cash flow after taxes (C_A) for each of the plant's 10 years?
b Calculate the following after-tax profitability yardsticks:

Return on original investment
Return on average investment
Payout time

8.22 The production and marketing of a new compound are under study by a chemical manufacturing concern. Preliminary estimates based on a production capacity of 36 million lb/year indicate a direct fixed capital of $6 million, allocated capital of $500,000, and working capital of $300,000.

Annual manufacturing costs and charges have been approximated:

Raw materials, catalyst, and utilities: $650,000
Maintenance: $360,000
Insurance and taxes: $160,000
Factory expense: $130,000
Labor-related items: $100,000
General expenses: $100,000

A 10-year plant life has been estimated, and depreciation is to be calculated by using the straight-line method.

a Estimate the unit price that must be obtained to realize a return on *average* investment of 10 percent after taxes.
b What would be the actual ROAIAT if a slow market forced a decrease in production to 80 percent of capacity, and the price had to be held to that estimated above?

8.23 For average-risk projects in the chemical industry, what is the expected percent return on sales? Assume CFS is predominantly DFC.

8.24 Show that the average DFC of a particular project, given by Eqs. (8.4.4) and (8.4.5), is the same as the average book value, given by Eq. (P6.4), for *straight-line depreciation only,* and that the two concepts are *not* the same for accelerated depreciation.

8.25 Derive an analytical expression for the average DFC [$\bar{F}$ in Eq. (8.4.4)] of a plant depreciated with the SOYD method.

8.26 Several criteria may be used for sizing product storage facilities. One such criterion is the penalty in lost sales if the facilities are not large enough to fill accumulating orders. Malina (1980) used Monte Carlo techniques to generate probabilities of order delays of various lengths as a function of relative storage capacity.

Refer to Malina's article to establish the optimum capacity of liquid product storage facilities for the following case:

Annual sales: 51,000 tons
Average order size: 50 tons
Average frequency of orders: 10 orders in 3 days
COM per order: $4385
Number of orders per year: 1020 (Sundays are not counted in frequency of orders)

Fixed charges per order: $1027
Gross margin per order: $1600

The *gross margin* is the difference between sales price and COM. Malina considers the cost of lost sales to be proportional to the sum of fixed charges and gross margin (why?).
Other required costs are given in the reference.

8.27 The continuous before-tax cash flow projection for a plant may be represented by a truncated normal distribution curve (see, for instance, Fig. 2.2.1d). The maximum cash flow of $10 million is expected to occur during year 4, with the first-year cash flow projected at 70 percent of maximum, and the 10-year cash flow (last year before shutdown) pegged at 23.2 percent of maximum. The working capital (including accounts receivable) is expected to be roughly proportional to the cash flow, reaching a maximum of $2 million in year 4. The estimated DFC is $10 million (no salvage value), and straight-line depreciation will be used. Allocated capitals will add another $1 million.

Use your favorite normal distribution function table to generate the required input data format to calculate the ROIBT.

8.28 The following production strategy has been devised for a new hormone preparation for supplementing turkey feeds (initial DFC = 3.5×10^6):

Year of operation	DFC investment,* millions of $	Production, 10^6 kg/yr	Sales price,* $/kg	Out-of-pocket expenses,* $/kg
1	3.50	0.50	25.00	13.20
2		0.75	24.00	10.45
3		1.00	22.00	10.45
4	1.50	2.00	19.00	8.25
5		2.50	19.00	7.70
6		2.75	18.00	7.70
7		2.50	18.00	7.70
8		2.30	18.00	8.25
9	1.00	2.00	20.00	7.70
10		1.50	20.00	7.70
11		1.00	20.00	7.70
12†		0.80	20.00	7.70

*Constant-value money.
†Shutdown at end.

The investments are made at the beginning of the year shown. The first one is the initial DFC, which will be depreciated over 10 years. The second investment (year 4) is for a planned plant expansion; this will be depreciated over 5 years. The last investment (year 9) is expected to be required to replace worn equipment; it will be depreciated over the remaining 4 years. Straight-line depreciation will be used, and no salvage value is anticipated.

The working capital for each year may be estimated as 20 percent of annual sales plus 10 percent of the *total* current DFC investment. Allocated capital is 10 percent of the DFC investment.

What is the return on average investment after taxes (ROAIAT)?

CHAPTER 9

CASH FLOW ANALYSIS

In Chap. 1 the formal approach to the economic evaluation of a project was sketched out in terms of two parallel paths—one the investment analysis, the other the net income analysis, the two eventually to meet within the context of a defined criterion of economic performance to complete the economic evaluation. The two tortuous paths have now been outlined, and in Chap. 8 they were joined to define a widely used profitability criterion, the return on investment, and its variants. The ROI class of profitability yardsticks is quite commonly used, but it does have its drawbacks, not the least of which is that it does not evaluate a project in terms of any investment methodology that recognizes the time value of money.

We now turn our attention to the definition of profitability yardsticks that conform more closely to such investment methodologies. Projects will be evaluated in terms of an analogous cash investment drawing a *rate of interest,* and this equivalent interest rate will then be compared with prescribed standards—the criteria of economic performance. In this aspect of definition plus comparison, the approach is not greatly different from that used with the ROI concept, which may even be thought of as an investment evaluation using simple interest. In the final analysis, however, the use of an equivalent compound interest rate is more realistic, and the format of the methods to be presented has some additional advantages; continuous cash flow, for instance, may be handled better in the mathematical sense. Note that when we talked about "continuous cash flow" in context of ROI, the cash flow was still handled as an annual event, an integral; this need not be the case if *continuous discounting* is used. In any case, the key distinguishing word between ROI methods and methods employing compound interest is "rate," referring to the rate of interest; note that this word does not appear in the ROI terminology.

The proper evaluation of projects using compound interest methodology requires a rather detailed definition of the various kinds of cash flows, and these will be presently examined—particularly the "one-time" cash flows, and the influence upon them of various tax credits. The thrust of the chapter will be devoted to two methods of assessing profitability: the *net present worth* (NPW) method and the *discounted cash rate of return* (DCRR) method. It turns out that both methods have some associated problems, just like the ROI concept, and a number of conventions must be followed, some of which, perhaps, are debatable. Both methods, however, have their place; the utility of each in particular situations will become clearer when we take up the subject of the analysis of alternatives in Chap. 10.

A number of special subjects will be introduced in context of the compound interest methods. For instance, the quantitative analysis of risk will be explored in an attempt to improve the credibility of the results of project evaluation and to facilitate the decision-making process. We will again look at the problem of forecasting the productive life cycle of a plant and the effect the projection will have upon the DCRR profitability index. Some of the manipulative problems of the DCRR method will be emphasized, particularly those associated with projects involving time-distributed investments. Finally, the relationship between ROI and the DCRR method will be examined to compare the two criteria of economic performance side by side.

9.1 CASH FLOW CONCEPTS

The Cash Flow "Black Box"

The term "cash flow" has now been bandied about on a number of occasions, and some aspects of the concept should perhaps be reviewed. The temporal aspects of cash flow have been emphasized; simplistically, cash flow is either *continuous* (recurring) or *discrete* (one-time). The continuous versus discrete classification has mathematical overtones explored in Chap. 2; continuous cash flow is represented by a continuous analytical function, whereas discrete cash flow is represented by a discontinuous function which may assume discrete values at regular intervals of time. In common usage, the classification has less exact implications. Continuous (or recurring) cash flows are associated with incomes and out-of-pocket expenditures which do take place "all the time," but are not truly continuous. In fact, often (as is the case in ROI computations), continuous cash flows are handled as an integral, once-a-year event; the reasons are related to the question raised in Exercise 2.2.1. Common usage also tends to lump all the continuous cash flow components into the single term "cash flow," actually the net continuous cash flow—the algebraic sum of the components which, by convention, are positive if they flow into the project, negative if they flow out.

The discrete cash flow events, the "one-time" occurrences, may, on the other hand, have a recurring or continuous characteristic of their own. The initial DFC investment would certainly be classified as a discrete event, and yet, as we will see, the actual disbursement process may span 2 or more years. Nevertheless, the distinction between discrete and continuous cash flows is usually intuitively obvious; the discrete events are associated with the project investment, and the continuous events are associated with the net income. The two cash flow categories are rigorously separated in the ROI computation, but they are blended as part of the mainstream cash movement in the analysis that follows.

The project may be thought of as a "black box" with incoming and outgoing cash flow streams, as shown in Fig. 9.1.1. Some of the streams are one-time or occasional discrete events, others are recurring or continuous. The box is a cash reservoir that obeys the equation of continuity,

$$\text{Input} - \text{output} = \text{accumulation}$$

Hopefully, at the end of the project's lifetime, there is, indeed, a cash accumulation in the box, signifying that the positive cash flows have exceeded the negative ones over that lifetime. Of course, since the first cash flow events in the project's life are negative (research expenses, capital investment), the box contents must be "primed" by infusion of corporate funds.

The continuous cash flows are shown in Fig. 9.1.1 as two streams, the incoming sales revenue and the outgoing out-of-pocket expenditures. The difference between the two, the net continuous cash flow, is not shown on this diagram as it was in some other diagrams in Chap. 6; here the net continuous cash flow accumulates in the box. The sales income is shown as being *after taxes;* by convention, cash flow analysis is performed usually on an after-tax basis, in contrast to ROI computations which are more popularly kept on a before-tax basis. (If desired, before-tax income and taxes could be shown as separate streams.)

Three kinds of discrete cash flows are shown in Fig. 9.1.1:

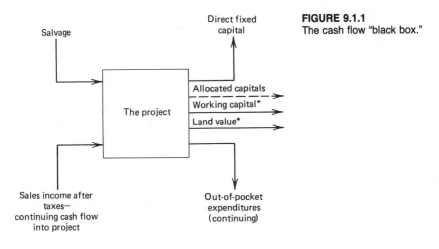

FIGURE 9.1.1
The cash flow "black box."

1 *Recoverable discrete cash flows.* Land value and the various components of working capital (including accounts receivable) represent an initial investment, but that investment is recovered (at least in theory) when the project is terminated. Moreover, working capital may change in magnitude throughout the course of a project's lifetime due to the variable size of inventories; the stream arrow that represents it may therefore sometimes point in, sometimes out. At any rate, the initial investment (outgoing stream) is eventually recovered as an incoming stream; this dual sense of the recoverable cash flow streams is indicated on the diagram by the asterisks.

2 *One-time cash flows.* The DFC may be considered a one-time investment event at the start of the project (time zero). In point of fact, the initial investment is usually disbursed over the time period of construction, and many projects involve additional investments during their lifetimes. Salvage value is a one-time positive cash flow at the termination point of the project.

3 *Allocated capitals.* The subject of allocated capital in cash flow analysis is a sticky one. It may be quite legitimately pointed out that the allocated capital, in the usual case, has already been expended for facilities which will be operational at the time of the new project's start-up. Why, then, burden the project black box with a cash stream not expended as part of the project? The answer is that cash flow analysis, just like the ROI analysis, is done in such a way as to assess the profitability on *all* the capital associated with a project. Therefore the allocated capital is shown as a one-time negative cash flow as if it were, indeed, expended as part of the project. The artificiality of the concept is indicated by the dashed stream line in Fig. 9.1.1.

Some allocated capital items may have a projected lifetime well beyond the lifetime of the project DFC. Utilities, for example, are often characterized by a 20-year lifetime projection. In that case, the undepreciated portion of the allocated capital should be shown as recovered capital at the time of the project termination, but this is usually not done. In most cases this is not a serious omission, since its impact upon the finally computed criterion of economic performance is negligible.

Cash Flow Diagrams

The temporal sequencing of the cash flow streams in and out of the project black box may be represented with *cash flow diagrams,* such as the one illustrated in Fig. 9.1.2. Cash flow diagrams give a quick overview of the projected economic performance of a project and the anticipated *cumulative cash position* during any one year.

The cumulative cash position is the arithmetic sum of all the after-tax cash flows preceding any chosen point in time. The cash flow diagram is a plot of the cumulative cash position as a function of time, with plant start-up arbitrarily taken as the origin. The various sequential cash flows are readily distinguished in Fig. 9.1.2. Quite typically, any project begins with a period of negative cash positions, which reflect the investment demand that precedes the generation of revenues.

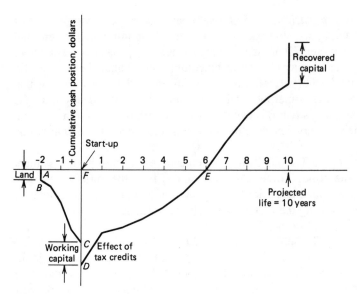

FIGURE 9.1.2
Typical projected cash flow diagram.

Once revenues are generated, the positive cash flows gradually reduce and eliminate the negative cash position, and hopefully during later stages of the project a healthy positive cumulative cash position is attained (the accumulation in the black box).

In the illustration, it has been assumed that the construction period required to assemble the project hardware begins 2 years before start-up. The first cash flow shown is the investment for land, which will eventually be recovered. The cash outflow curve during the construction period is shown to be nonlinear but uniform over the 2-year period of design and construction. Past experience may be used to project the cash disbursement sequence during this period, and each corporation and construction company seems to have its favorite cash outflow curve. All projected sequences are characterized by some form of an S-shaped curve, such as the one in Fig. 9.1.3. A 5 percent "priming" cash supply is indicated at time zero; the cash outflow is relatively slow during the design and equipment ordering period, increases to a maximum about halfway through the design-construction period because of payments for delivered equipment and basic construction, and again decreases at the end of the period in the course of putting on the plant's "finishing touches."

Actually, at the time that the first cash flow diagrams are constructed, the shape of the construction period cash outflow curve is not very important—it is what happens after start-up that is important. The shape of the curves becomes more of a concern during advanced stages of a project, when the time worth of the disbursement sequence at zero time is computed as part of the more detailed financial

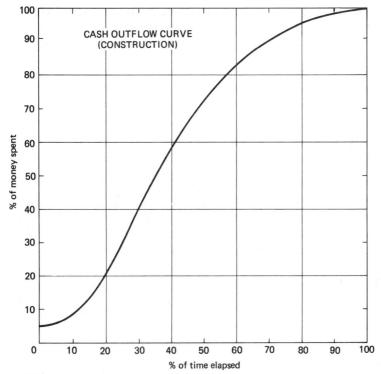

FIGURE 9.1.3
Cash outflow curve.

analysis of the project, or when the curve is used as a guide during the actual construction as a cost control tool (Chap. 11).

Returning to Fig. 9.1.2, we see that a further decline in cash position occurs at start-up due to the working capital and allocated capital investment (these are usually indicated as a one-time cash flow). Thereafter, the cumulative cash position curve hopefully reverses direction toward more positive values, although this does not necessarily occur during the first year or two, particularly if slow market penetration or start-up problems are anticipated. The curve during the productive years may be straight; however, the actual plant life cycle is more likely to contribute to an S-shaped curve as shown, with the steepest slope occurring during the middle years of maximum demand. Further irregularity or nonlinearity is caused by accelerated depreciation methods and by tax credits during the early years of project lifetime. The final cumulative cash position at the project's termination is enhanced with the value of the capital recovered (land, working capital, salvage).

Exercise 9.1.1

In Fig. 9.1.2 the cumulative cash position changes to a positive value at the end of year 6. Is this equivalent to the payout time (payback period)?

During early and middle stages of a project, cash flow diagrams are often drawn in simpler form than in Fig. 9.1.2. Capital investment is taken as a one-time cash flow at zero time, and the effects of tax credits are ignored. In the absence of time-dependent sales projections and with straight-line depreciation, the cumulative cash position is itself a straight line.

Tax Credits

From time to time the federal government grants a variety of tax credits in an effort to stimulate business investments. The laws governing these credits are frequently changed, and the substance of those laws is complex enough so that the monitoring of such benefits is best left to the corporate accounting experts. Two forms of tax credit will be briefly mentioned, since their effect is incorporated into advanced-stage cash flow analyses, and their impact can be, indeed, significant.

The *investment tax credit* is an income tax credit for new investments. The detailed provisions are rather involved, but basically a 1-year tax credit of 10 percent (1981) of the depreciable plant investment (except buildings) is allowed for the first year of operations. The credit is reflected in the corporation's overall tax payment, but it is internally credited to the plant's operations. This one-time cash flow, however, does *not* contribute to taxable profit and is not itself taxable; it does not have any effect upon the basis used for depreciation calculations.

For preliminary estimates, no large error is incurred by taking the investment tax credit as 10 percent of DFC.

Certain eligible investments qualify for an *energy tax credit*. Such eligible investments are limited to specific equipment devoted to pollution control, energy conservation, and utilization of "alternative" energy sources. Much of that kind of equipment is part and parcel of new chemical processing plants as well as plant improvements, and the 10 percent tax credit (1981) may be a strong incentive to use such equipment. Note that with the investment tax credit added, the total tax credit amounts to 20 percent. Pavone and Patrick (1981) describe eligible investments in some detail and show how the energy tax credit increases the economic attractiveness of certain projects.

Depreciation on Allocated Capital

Without wishing to belabor the point, we again point out that depreciation is not part of the cash flow, and cash flow computations based on COM must have the depreciation on the DFC properly added back in. Specifically, if Eqs. (6.1.8) and (6.1.11) are combined, the after-tax cash flow becomes

$$C_A = (1 - t)(\mathcal{R} - \text{COM} - G) + D \qquad (9.1.1)$$

So far the term D has been associated with the DFC only, and nothing has been said about depreciation on allocated capital. We have seen that if standard costs were to be used in the COM for raw materials or utilities, then an allocated capital accounting for the facilities to produce those raw materials or utilities had to be added to the various investments constituting the CFS. The standard costs, however, *include* the depreciation on the associated allocated capitals. We must therefore find a way to identify and extricate the allocated capital depreciation costs from our computed cash flows. Similarly, the general expenses incorporate depreciation charges for the allocated market capital, and these must be identified and extricated.

This is usually done by assuming that the total allocated capital, both fixed and market, has a 20-year life and that it is depreciated by the straight-line method. The term D in Eq. (9.1.1) therefore is the sum of the annual DFC depreciation (which may be accelerated) and 0.05(AC), where AC is the total allocated, nonrecoverable capital.

In most cases, the correction for allocated capital depreciation is small relative to DFC depreciation, and it is often ignored in early-stage project evaluations.

Equivalent Maximum Investment Period

Allen (1967) has defined a criterion of economic performance based upon the specific shape of the cash flow diagram. It is the *equivalent maximum investment period* (EMIP), and its definition is best appreciated by reference to Fig. 9.1.2:

EMIP = area enclosed by cumulative cash position curve up to break-even point E, divided by the maximum cumulative investment

or

$$\text{EMIP} = \frac{\text{area }(ABCDEFA)}{\overline{FD}}$$

EMIP is analogous to the payout time, but it accounts for the sequence of cash flow events in a way that the payout concept cannot. An EMIP of 3 years is considered acceptable for average-risk projects.

Example 9.1.1

The following parameters characterize a project (all in millions of dollars):

Initial land investment:	0.1
DFC investment, disbursed uniformly over 3 years:	25.0
Working capital:	4.0
Allocated capital:	5.0
Annual net positive cash flows after start-up, uniform (after taxes):	12.5

What is the EMIP?

SOLUTION

$$\text{Area to left of cash position axis} = 0.1 \times 3 + \tfrac{1}{2} \times 3 \times 25$$
$$= 37.8$$
$$\overline{FD} = 0.1 + 25.0 + 4.0 + 5.0 = 34.1$$
$$\overline{FE} = \frac{\overline{FD}}{C_A} = \frac{34.1}{12.5} = 2.73$$
$$\text{Area to right of cash position axis} = \tfrac{1}{2} \times 2.73 \times 34.1$$
$$= 46.6$$
$$\text{EMIP} = \frac{37.8 + 46.6}{34.1} = 2.48 \text{ years}$$

Note that the EMIP criterion, just like payout time, ignores events beyond break-even.

9.2 NET PRESENT WORTH

Payout Time with Interest

In Chap. 2 the net present worth (NPW) of a project was defined as the arithmetic sum of all project cash flows discounted to zero time. A few simple examples served to demonstrate that the NPW concept could be used as a criterion of economic performance. The idea behind the method is to see whether the revenues generated by a project investment will match revenues generated by the same amount of capital invested at some prescribed interest rate. The interest rate to be used in the comparison is established by the corporate management; it is usually significantly higher than the interest rate currently realized by the sum total of the corporate investments.

The ROI concept and its variants do not account for this time value of money. The payout time, for example, is the simple ratio of DFC to undiscounted continuous after-tax cash flow. A more reasonable approach might be to have the cash flow recover not only the DFC but the interest on the undepreciated capital for each year as well. This modified method is called *payout time with interest;* it is not used a great deal, but it does serve as an illustrative bridge between profitability indices ignoring the time value of money and the NPW method.

Payout time with interest calculations is best demonstrated with an example.

Example 9.2.1

For the simple project with parameters listed in Table 8.4.1, determine the payout time with interest for a 10 percent interest rate.
SOLUTION
  Year 1: The initial undepreciated DFC is $20,000. The simple interest for the year is

$$0.10 \times 20{,}000 = \$2000$$
$$\text{Total to be recovered} = 20{,}000 + 2{,}000 = \$22{,}000$$

But C_A for year 1 is $6,000. Therefore

$$\text{Unrecovered balance at end of year 1} = 22{,}000 - 6{,}000$$
$$= \$16{,}000$$

Year 2: Depreciation during year 1 = $20{,}000/5$ = $4,000. Therefore

$$\text{Initial undepreciated DFC, year 2} = 20{,}000 - 4{,}000$$
$$= \$16{,}000$$
$$\text{Simple interest} = 16{,}000 \times 0.10 = \$1600$$
$$\text{Total to be recovered} = \text{unrecovered balance at end of year 1 plus interest}$$
$$= 16{,}000 + 1{,}600$$
$$= \$17{,}600$$

C_A for year 2 is $8,000. Therefore

$$\text{Unrecovered balance at end of year 2} = 17{,}600 - 8{,}000$$
$$= \$9{,}600$$

Year 3:

$$\text{Initial undepreciated DFC, year 3} = 16{,}000 - 4{,}000$$
$$= \$12{,}000$$
$$\text{Simple interest} = 12{,}000 \times 0.10 = \$1{,}200$$
$$\text{Total to be recovered} = 9600 + 1200 = \$10{,}800$$

C_A is $10,000. Therefore

$$\text{Unrecovered balance at end of year 3} = 10{,}800 - 10{,}000$$
$$= \$800$$

Year 4:

$$\text{Initial undepreciated DFC, year 4} = 12{,}000 - 4{,}000$$
$$= \$8000$$
$$\text{Simple interest} = 8000 \times 0.10 = \$800$$
$$\text{Total to be recovered} = 800 + 800 = \$1600$$

The C_A of $12,000 will recover this in less than a year:

$$\frac{1{,}600}{12{,}000} = 0.13$$

Therefore the payout time with interest is 3.13 years. This is below the project life of 5 years, and the project is acceptable.

The payout time without interest was 2.6 years (Example 8.4.4); the payout time obviously increases with the interest rate.

The interest rate to be used with this or the NPW method is one which, in the judgment of management, constitutes the acceptable minimum for a project with the same degree of risk.

The NPW Criterion of Profitability

The net present worth method of evaluating the economic performance of projects is one that is quite commonly used during the more advanced stages of project development. To find the NPW of a project at a prescribed interest rate, *all* the associated cash flows, both discrete and continuous, are discounted to zero time ("the present") and summed up arithmetically. If the resultant is zero or a positive sum, then the project meets the criterion of economic performance *for the prescribed interest rate*. If the interest rate is increased, the NPW decreases; that is, the criterion of performance is more difficult to reach at higher prescribed interest rates, as was the case with the "payout time with interest" criterion.

By convention, the continuous cash flows are expressed either as integral annual events which occur at the end of each year or as cash flows occurring uniformly over the full year. Either discrete or continuous discounting may be used. The simple project characterized in Table 8.4.1 will serve to illustrate the method.

Example 9.2.2

Corporate management prescribes a 10 percent discretely discounted rate of interest for the NPW evaluation of projects such as that outlined in Table 8.4.1. Does the project meet the criterion of economic performance? Consider cash flows to be discrete events.
SOLUTION   The required computation is summed up in Table 9.2.1. Since the NPW is positive (+$14,570), the project does, indeed, satisfy the criterion of economic performance.

Exercise 9.2.1

Repeat the calculation of Example 9.2.2 for a nominal continuous interest rate of 10 percent, considering the continuous cash flows to occur uniformly throughout each year.

The result of a net present worth calculation is a dollar sum which, on an absolute basis, may not have much meaning—except that it is positive, or negative, or zero. There is a commonly used profitability index used in conjunction with the NPW method which gives more of a feel for how closely a project has met the criterion of economic performance. This index is called the *present worth ratio,* and

TABLE 9.2.1
ILLUSTRATION OF PRESENT WORTH METHOD
(Basis: Project, Table 8.4.1. Discrete Compound Interest, 10%)

Year	Cash flow	Present worth factor* at 10% interest	Present worth
0	−(25,000)	1.000	−(25,000)
1	6,000	0.909	5,450
2	8,000	0.826	6,610
3	10,000	0.751	7,510
4	12,000	0.683	8,200
5	14,000	0.621	8,690
Recoverable capital	5,000	0.621	3,110
Net present worth (NPW)			$ 14,570

*For year k, the present worth factor is $(1 + i)^{-k}$.

it is defined as

$$\text{Present worth ratio} = \frac{\text{present worth of all positive cash flows}}{\text{present worth of all negative cash flows}}$$

The positive cash flows are the net continuous cash flows and recovered capital cash flows; the negative cash flows are the investments. For the case in Table 9.2.1, for example, the present worth ratio is

$$\frac{5450 + 6610 + 7510 + 8200 + 8690 + 3110}{25,000} = 1.581$$

Essentially, the present worth ratio gives an indication of how much the project makes relative to the investment. A ratio of 1.0 means that the income just matches the expected income from capital invested at the prescribed interest rate. A ratio of less than 1.0 indicates that the income does not come up to minimum expectations. A ratio of more than 1.0 shows that the project exceeds minimum expectations; the ratio of 1.581 is clearly very good, some 60 percent above the minimum expected income.

Exercise 9.2.2

Suppose one or more of the annual cash flows in Table 9.2.1 were negative (indicating an operating loss for the year). Should these appear in the numerator or the denominator of the present worth ratio?

NPW calculations may, of course, turn out to be more involved than the simple case in Example 9.2.2. Past expenditures such as nonuniform DFC disbursements (Fig. 9.1.3) or project-specific research expenditures may be compounded to zero time as part of the overall NPW balance. Multiple investments may be projected in the years to come, and fluctuations in the working capital may be anticipated. Various components of the continuous cash flow may well be written in analytical form, and in that case the NPW integrations may be performed analytically; some instances were explored in Chap. 2. We may, if so desired, decompose the after-tax continuous cash flow into its components according to Eq. (6.1.11),

$$C_A = (1 - t)(\mathcal{R} - E) + tD \qquad (6.1.11)$$

and sum up the present worths of each component, for example,

$$t \sum_{k=1}^{n} \frac{D_k}{(1 + \mathbf{i})^k}$$

to arrive at the present worth of all continuous cash flows. Some authors thus write out algebraic expressions for the NPW computational format which appear most formidable, indeed, in terms of apparent mathematical complexity. The simplest procedure, however, is to calculate C_A for each year and to perform the numerical integration as in Table 9.2.1.

Happel and Jordan (1975) describe a variant upon the NPW method which they call *venture worth*. The variant involves a NPW calculation using the corporation's "internal" interest rate actually realized on the totality of its investments, rather than the prescribed minimum acceptable interest rate, which is almost always higher. However, a term is subtracted from the venture worth thus calculated which represents the present worth of the incremental return on the total capital for sale, as determined by the difference in the two interest rates. The venture worth thus calculated will be different than the NPW; the preferred method is a matter of corporate convention.

What value of i (continuous discounting) is considered acceptable for average risk projects in the chemical process industry? The actual value is a matter of individual corporate lore and policy, but a value of 15 percent is commonly accepted. As was the case with acceptable ROIBT, the acceptable interest rate depends upon current availability of capital.

9.3 DISCOUNTED CASH FLOW

The DCRR Criterion of Profitability

The *discounted cash rate of return* (DCRR) criterion of economic performance is almost universally used for advanced-stage evaluation of projects. The method is analogous to the NPW method, but instead of asking what the NPW is for a prescribed interest rate, we seek a value of the interest rate which will make the

NPW of all the cash flows just equal to zero. In effect, it is the interest rate at which money would have to be invested, over the same period of years, to generate the same NPW of annuities.

As is the case with the NPW procedure, any combination of discrete or continuous discounting methods and cash flows may be used to determine a discrete interest rate $i = r$ or a nominal continuous interest rate $i = r$. In the usual case, continuous discounting of cash flows is chosen, and the cash flows are assumed to be uniform over each year. No matter what combination is used, a trial-and-error solution is required, except for the very simplest situations.

The project in Table 8.4.1 will again be used to illustrate the DCRR method, and discrete compounding of discrete annual cash flows will be chosen for the solution format.

Example 9.3.1

Calculate the discounted cash rate of return for the project in Table 8.4.1.

SOLUTION The computations are given in Table 9.3.1. The result ($r = 0.268$) represents a very attractive after-tax rate of return.

In the above example, linear interpolation was used to obtain the desired solution. It is better to determine the NPW for three or more trial values of the DCRR and to plot the results, since the relationship between NPW and the trial DCRR may be quite nonlinear.

A typical format for calculating the discounted cash rate of return is shown in Fig. 9.3.1. The format indicates how the various cash flows may be calculated,

TABLE 9.3.1
ILLUSTRATION OF DISCOUNTED CASH FLOW METHOD
(Basis: Discrete Discounting, Discrete Cash Flows; Project, Table 8.4.1)

Year	Cash flow	At 25% DCRR		At 30% DCRR	
		Discount* factor	Present worth	Discount* factor	Present worth
0	-(25,000)	1.000	-(25,000)	1.000	-(25,000)
1	6,000	0.800	4,800	0.769	4,610
2	8,000	0.641	5,130	0.592	4,740
3	10,000	0.513	5,130	0.455	4,550
4	12,000	0.410	4,910	0.350	4,200
5	14,000	0.318	4,450	0.269	3,770
Recovered capital	5,000	0.318	1,590	0.269	1,350
NPW			+1,010		-1,780

By linear interpolation,

$$r = 25 + \frac{1010}{2790} \times 5 = 26.8\%$$

*Also called present worth factor.

PROJECT: ______________________

CAPITAL		
DIRECT FIXED		
ALLOCATED FIXED		BY
WORKING, ACCTS. REC.		DATE
TOTAL		NO.

YEAR:	CASH FLOW BEFORE DEPRECIATION				CASH FLOW FROM DEPRECIATION			CAPITAL	NET CASH FLOW (A + B + C)		DISCOUNTING FACTORS							
	(1) SALES	(2) COSTS EXCL. DEPR.	(3) (1-2)	CASH A	(4) DIRECT —YRS.	(5) ALLOCATED —YRS.	CASH B	CASH C	Δ	Σ	8% × f	=	12% × f	=	15% × f	=	—% × f	=
-1½-0											1.0625		1.0956		1.1214			
-0																		
TIME 0											1.0000		1.0000		1.0000		1.0000	
0 TO 1											.9610		.9423		.9286			
1 TO 2											.8872		.8358		.7993			
2 TO 3											.8189		.7413		.6879			
3 TO 4											.7560		.6574		.5921			
4 TO 5											.6979		.5831		.5096			
5 TO 6											.6442		.5172		.4386			
6 TO 7											.5947		.4588		.3775			
7 TO 8											.5490		.4069		.3250			
8 TO 9											.5068		.3609		.2797			
9 TO 10											.4678		.3201		.2407			
10 TO 11											.4318		.2839		.2072			
11 TO 12											.3986		.2518		.1783			
12 TO 13											.3680		.2233		.1535			
13 TO 14											.3397		.1981		.1321			
14 TO 15											.3136		.1757		.1137			
								SUMMATION OF CASH FLOWS										
															NPW = 0 @			

CASH A = (SALES MINUS COSTS EXCL. DEPR) X (1 − TAX RATE)
CASH B = TOTAL DEPRECIATION X TAX RATE
CASH C = CAPITAL EXPENDITURE, HAS A MINUS SIGN

WORKING CAPITAL DEBITED AS—CASH AT TIME 0
U.S. INVESTMENT TAX CREDIT ENTERED AS + UNIFORM CASH FLOW DURING 1ST, YEAR

FIGURE 9.3.1
Cash flow calculation sheet (in \$1000 units).

including some (such as the investment tax credit) which were ignored in Example 9.3.1. The procedure parallels the development of cash flow expressions that has been previously outlined. As a first step, the various capitals are listed at the top right: direct fixed (DFC), allocated fixed (including nonrecoverable portions of allocated market capital), and working (including accounts receivable). Land value may also be listed separately.

The columns are then filled for each year indicated:

(1) Sales revenue $\mathcal{R}$
(2) The out-of-pocket expenditures (COM + G − D)
(3) The indicated difference

$$\text{Cash A} = \text{col. } 3 \times (1 - t)$$

(4) Depreciation on DFC (normally based on 10 years)
(5) Depreciation on allocated capitals (normally based on 20 years)

Cash B = (sum of cols. 4 and 5) × t

Cash A + cash B = C_A, the after-tax continuous cash flow, as defined in Eq. (6.1.11).

Cash C represents the various investments written as negative quantities. Footnotes indicate some of the conventions regarding investments.

The "Net Cash Flow" is the sum of the three computed cash flows A, B, and C. The net cash flow is computed for each year (Δ) for subsequent discount calculations, and a cumulative cash flow (Σ) is computed as an aid in constructing a cash flow diagram.

The discounting factors shown are those for continuous discounting of uniformly distributed cash flows for each full year. Even the one-time (investment) cash flows are taken as uniformly disbursed (or recovered) throughout one year. Three DCRR trial values are shown to reflect perceived (but by no means generally accepted) levels corresponding to an unfavorable, average, and good rate of return. An optional calculation is suggested (and perhaps additional ones should be urged) so as to accumulate enough points to plot a curve from which the DCRR (at NPW = 0) may be picked.

DCRR calculations are easily programmable for computers or hand-held calculators,[1] and practice is afforded by the special project at the end of this chapter. Quite often the DCRR computation is programmed as part of a computer-based complete economic analysis of a project; such programs are commonly used by corporate economic evaluation departments on a routine basis. The more complex programs are embellished with many attractive output features such as break-even charts and sensitivity diagrams.

Project economic evaluation is certainly systematic enough to be handled by computer programs. The problem with such programming is that there is a relatively large demand for input data, and that the computations involving the data are not particularly complex and sophisticated. Why are computers used so often, then? The answer to this is the desire to achieve consistency—an internal consistency of methodology, throughout all levels and departments of a corporation, that promotes confidence in the results of economic analysis. An economic evaluation program that is adopted on a corporatewide basis assures that all items necessary for an adequate evaluation are properly included and accounted for as part and parcel of every definitive effort. Differences in approach and opinion of different estimators are eliminated; acceptable estimating factors are incorporated into the unique program. In this way the corporate decision makers can compare a variety of economic analyses with some confidence that departmental, divisional, or regional prejudices have been moderated.

The Projection of Production Life Cycles

We saw in Chap. 3 that the projection of the production life cycle of a plant was an uncertain and difficult task. In the absence of detailed potential market surveys,

[1] See, for example, Wild (1977).

the assumption is often made that the plant will be operated at the rated capacity over its life span, and the DCRR is then calculated on this basis. With the added stipulations that straight-line depreciation will be used and that the sales price (in uninflated dollars) will remain constant, the annual continuous cash flows are then constant, and the present worth expression is that for an annuity. For discrete discounting of discrete annuities, for example, the present worth of the annuities is given by Eq. (2.2.21),

$$P = R \frac{(1 + i)^n - 1}{i(1 + i)^n} \qquad (2.2.21)$$

The DCRR computation involves finding the value of $i = r$ so as to make the NPW of all cash flows zero. Even in the simplest case of invariable annual cash flows a trial-and-error calculation is involved, since the roots of the analytical expression for the sum of all cash flows cannot be obtained explicitly. Nevertheless, the trial-and-error solution is facilitated by using integrated expressions such as Eq. (2.2.21). Other integrated expressions for simple cash flow distributions which can be represented analytically were derived in Chap. 2.

The question naturally arises as to how valid a project evaluation is afforded by the DCRR criterion based upon the assumption of invariant annual production at rated capacity. It has been pointed out (Chap. 3) that production is likely to be below rated capacity at the early and late stages of the plant life (see, for example, Fig. 3.3.9). Any such reduction in productivity below previously anticipated levels would render the actual DCRR criterion of profitability smaller than that predicted on the basis of the simple projection of constant rated production—clearly, a potentially serious error of overestimating the project profitability. On the other hand, low production at early and late stages of plant life could well be counterbalanced by production levels above rated capacity (superproductivity) during the middle stages of plant life, the period of maximum demand. Is it possible, in fact, to project a likely production life cycle of a plant, based upon past experience with similar plants, which would result in a more realistic estimate of the DCRR criterion than can be afforded by the assumption of invariant production? This question has been addressed by Valle-Riestra (1979).

The often observed lifetime distribution of plant productivity suggests that the distribution could be approximated with a bell-shaped curve such as the normal distribution curve. A section of the normal distribution curve and some production-oriented parameters are illustrated in Fig. 9.3.2. The parameters shown are the following:

p, the production rate. This is shown as a continuous function in Fig. 9.3.2. It is also possible to define p_k as the (discrete) annual production during year k; as a close approximation, p_k equals the instantaneous production rate at $k - \frac{1}{2}$.

p_R, the rated capacity of the plant.

$p_{\max}$, the maximum sustained plant capacity. This almost always is above the rated capacity due to equipment overdesign, or else it can be increased with only

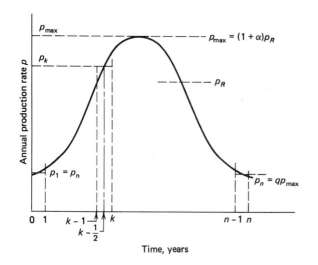

FIGURE 9.3.2
Parameters for normally distributed production. (Valle-Riestra, 1979. Reproduced by permission.)

minor expenditures for debottlenecking. The relationship between p_R and p_{max} is given by

$$p_{max} = (1 + \alpha)p_R \qquad (9.3.1)$$

where α, the superproductivity, typically has a value of about 0.2.
n = the productive life span of the plant.
q = the ratio of the initial production rate, or production during the first year, to p_{max}. A typical value of q is 0.3 to 0.5.

If $q = 1.0$ and $\alpha = 0$, the case of constant annual rated production obtains. A symmetrical distribution is shown in Fig. 9.3.2 but, of course, it need not be so, and a truncated normal distribution curve (Fig. 2.2.1d) may be judged to be a more likely projection of lifetime production distribution.

For the discrete cash flow case, the annual sales revenue during year k is $p_k s$. If $p_1 = p_n$ (symmetrical distribution), it may be shown that $\mathcal{R}_0$, the present value of lifetime sales revenues, is given by

$$\mathcal{R}_0 = p_{max} s \sum_{k=1}^{n} \frac{q^A}{(1 + r)^k} \qquad (9.3.2)$$

where

$$A = \left(\frac{2k - 1 - n}{n - 1}\right)^2 \qquad (9.3.3)$$

The summation term cannot be evaluated analytically, but Valle-Riestra (1979) gives a graphical summary of the numerical values of the function

$$\phi(n, \mathbf{r}, q) = \sum_{k=1}^{n} \frac{q^A}{(1 + \mathbf{r})^k} \qquad (9.3.4)$$

The function ϕ can be used to obtain the present worth of all other cash flow components which depend upon the postulated normal distribution. For the simpler case of discrete discounting, invariable costs, and straight-line depreciation, the DCRR may be obtained (by trial and error) by solving the equation

$$0 = -(K_F + K_w) + K_w(1 + \mathbf{r})^{-n} - \frac{1}{2}\left(F - \frac{K_F}{n}\right)\frac{(1 + \mathbf{r})^n - 1}{\mathbf{r}(1 + \mathbf{r})^n}$$
$$+ \tfrac{1}{2}p_{\max}(s - c_v - g)\phi \qquad (9.3.5)$$

The various terms in Eq. (9.3.5) are explained in the Nomenclature. The unit variable charges c_v and general expenses g are assumed to be independent of the production level. The assumption is also made that regulated charges do not change appreciably over the production levels involved and that they may be considered a part of the *annual* fixed charges F. Note that F does *not* include depreciation.

Exercise 9.3.1

We cannot really speak of "deriving" Eq. (9.3.5); it is simply a statement, in mathematical terminology, of a procedural sequence. Identify and describe the significance of each term in Eq. (9.3.5), and list some of the simplifying assumptions, not so far mentioned, inherent in Eq. (9.3.5) as written.

Example 9.3.2

Initial market evaluations call for a plant to produce 100,000 kg/day at rated capacity. The product will sell for 50 cents/kg. What is the anticipated present worth of sales revenues for the following parameters:

    Time value of money: 15% (= $\mathbf{r}$)
    Plant life: 10 years (= n)
    Symmetrical normal distribution of the production is projected, with $q = 0.30$ and $\alpha = 0.20$
    Stream factor = 0.95

Use discrete discounting.
SOLUTION   The stated reference gives $\phi(10, 0.15, 0.30) = 3.26$.

$$p_{\max} = (1.20 \times 100{,}000) \times 365 \times 0.95 = 4.16 \times 10^7 \text{ kg/yr}$$
$$\mathcal{R}_0 = 4.16 \times 10^7 \times 0.50 \times 3.26 = \$67.8 \times 10^6$$

It should be stressed that the data to support the perception of a normally distributed lifetime productivity of chemical plants is not readily available. Corporations have the data on the lifetime production history of their own plants, but much of that information is considered proprietary. There is scattered information in the published literature (Cox, 1967; Frederixon, 1969) to indicate that normal distribution does characterize the life cycle of *products*. If the product is produced in a unique plant, then the two life cycles coincide.

A projection of a normally distributed plant production life cycle is therefore only an informed guess, and a more legitimate question might be this: Granted that a normally distributed production is likely, at what fraction of the *rated* productivity would the plant have to be operated over its full lifetime to realize the same DCRR? Given the answer, the effect of the uncertain distribution of productivity could be explored as part of the project sensitivity analysis, with diagrams such as Fig. 8.3.7.

Suppose a plant were to be operated at the invariable fraction ω of its rated capacity. The DCRR equation corresponding to Eq. (9.3.5) (and based upon the same set of simplifying assumptions) is

$$0 = -(K_F + K_w) + K_w(1 + r)^{-n} - \frac{1}{2}\left(F - \frac{K_F}{n}\right)\frac{(1 + r)^n - 1}{r(1 + r)^n}$$
$$+ \frac{1}{2}\omega p_R(s - c_v - g)\frac{(1 + r)^n - 1}{r(1 + r)^n} \quad (9.3.6)$$

If Eqs. (9.3.5) and (9.3.6) are combined, the result is

$$\frac{\omega}{1 + \alpha} = \frac{\phi r(1 + r)^n}{(1 + r)^n - 1} \quad (9.3.7)$$

Ostensibly, the ratio $\omega/(1 + \alpha)$ is a function of n, r, and q, but, interestingly, it turns out to be a very weak function of n and r within the normally anticipated ranges of those variables. Thus ω depends essentially upon α and q only, and for the anticipated values of those variables,

$$0 \leq \alpha \leq 0.2 \quad \text{and} \quad 0.3 \leq q \leq 0.5$$

it can be shown that

$$\boxed{\omega \approx 0.7}$$

In other words, the more likely project profitability is the one corresponding to 70 percent of rated capacity, rather than full rated capacity, the usual earlier-stage project basis. In view of the reality of periods of low production in most chemical plants, it would be foolhardy, indeed, to bet on obtaining profitability levels corresponding to steady full production.

Time-Distributed Investments

It may happen that a project involves one or more expansions of facilities in the future to match anticipated increases in demand. In such a case the investment is no longer a single event at time zero, or a series of disbursements before time zero; negative cash flows are scheduled for some future year or years. DCRR computations for this sort of situation raise a peculiar problem, namely, that more than one value of **r** may satisfy the criterion of zero NPW.

The reasons for this unwelcome multiplicity have been explored, among others, by Newnan (1976). The basis is a matter of the multiplicity of roots of polynomial equations. Consider, for instance, the trivial case of a zero-time investment, $(-K_F)$, followed by the two positive cash flows C_1 and C_2 during years 1 and 2. The DCRR computation would assume the format

$$0 = -K_F + C_1(1 + \mathbf{r})^{-1} + C_2(1 + \mathbf{r})^{-2}$$

Let $(1 + \mathbf{r})^{-1} = x$:

$$C_2 x^2 + C_1 x - K_F = 0$$

This is a quadratic equation in x with two roots.

In the same way, any DCRR equation, no matter how complex, can be resolved into a polynomial of order n, where n is the number of years which the project spans:

$$C_n x^n + C_{n-1} x^{n-1} + \cdots + C_2 x^2 + C_1 x - K_F = 0$$

Such a polynomial equation has, in general, n roots, although only those roots which yield a real, positive value of $\mathbf{r}$ $[=(1/x)-1]$ are acceptable as a solution. It turns out that the number of such acceptable roots is a function of *the number of sign changes* in the nonzero terms, written in sequential descending powers as shown. In the polynomial equation, as written, if all the coefficients are positive (i.e., all cash flows are positive), there is only one sign change, namely, $-K_F$, the negative initial investment. However, if another investment is made during year k, the result may very well be a net negative cash flow for that year:

$$C_n x^n + C_{n-1} x^{n-1} + \cdots + C_{k+1} x^{k+1} - C_k x^k$$
$$+ C_{k-1} x^{k-1} + \cdots + C_2 x^2 + C_1 x - K_F = 0$$

Here there are three sign changes:

1. The + sequence ending with $+ C_{k+1} x^{k+1}$ changing to $- C_k x^k$
2. $- C_k x^k$ changing to the + sequence ending with $+ C_1 x$
3. $+ C_1 x$ changing to $- K_F$

Newnan formulates a *rule of signs*, which states that

> There may be as many positive values of **r**
> as there are sign changes in the cash flow

The rule holds for both discrete and continuous discounting. Trivial exceptions may occur (Rapp, 1980).

A simple example will serve to demonstrate the rule.

Example 9.3.3

The following relative cash flows at the end of each year characterize a project:

Year	Cash flow
0	+ 40
1	+ 20
2	−100
3	−100
4	+ 40
5	+120

This kind of cash flow distribution might occur, for example, if customers were willing *to prepay* the future delivery of a high-demand product to guarantee such delivery, and if the prepayments preceded the commitment of construction funds for the production facility. In the table, prepayments occur during the first two years; investment funds are committed during years 2 and 3, and additional positive cash flows occur during years 4 and 5.

What is the DCRR?

SOLUTION Two sign changes occur in the cash flow sequence, and the rule of signs implies that two values of **r** may be computed. A plot of NPW versus discrete interest rate is given in Fig. 9.3.3. Indeed, two values of DCRR are indicated:

$$\mathbf{r} = 13.0\% \quad \text{or} \quad \mathbf{r} = 41.4\%$$

One may be willing to accept the mathematical reality of the two roots in the example given, but a question arises as to the significance of the multiple results in the context of a realistic investment analysis. What do the two different results *mean?* A DCRR of 41.4 percent is very good, indeed, but perhaps the result should really be 13.0 percent, a much more modest performance.

A review of the cash flow sequence of Example 9.3.3 reveals the cause of the dilemma. We see that the negative cash flows, which represent the investment, occur during years 2 and 3, well after the "start" of the project. If these investment cash flows are discounted back to zero time by using some specific interest rate, this is really the same thing as saying that these investments originated as *a principal at time zero drawing the same specific interest rate*. For example, the two negative

488 CHAPTER 9: CASH FLOW ANALYSIS

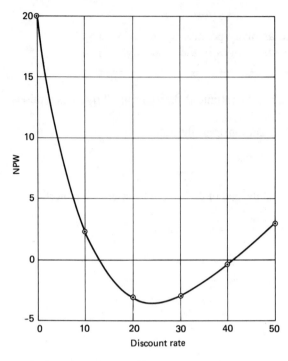

FIGURE 9.3.3
DCRR computation for Example 9.3.3.

cash flows may be discounted to time zero using a 41.4 percent interest rate to obtain an equivalent zero-time investment of −85.4. The positive cash flows for years 0, 1, 4, and 5 do, indeed, then represent a 41.4 percent rate of return on the zero-time investment!

Similarly, the negative cash flows may be discounted to zero time at a 13.0 percent interest rate to give a zero-time investment of −147.6, and the positive cash flows then represent a 13.0 percent return on *that* investment.

Exercise 9.3.2

Verify the two values of zero-time equivalent investment and the rates of return represented by the positive cash flows.

We must ask ourselves whether it is reasonable to expect the initial investment to sit around for the 2 or 3 years, drawing 41.4 percent interest (or, for that matter, 13.0 percent interest). In general, the answer is no, and there is certainly no reason why such "externally" held investment should draw the same rate of interest *before* becoming a part of the project cash flow as it does *after*!

Clearly, such time-distributed investments must be held at some rational, acceptable interest rate until such time as they are invested. The most acceptable rate would seem to be the "internal" interest rate earned by the totality of the corporation's investment after taxes. The rule is to discount all negative cash flows at nonzero times to a single zero-time investment by using the internal interest rate i_0 (or i_0). In this way, a single DCRR solution is obtained.

Example 9.3.4

Rework Example 9.3.3 for the case of $i_0 = 0.10$.

SOLUTION The investments are discounted to zero time:

$$(K_F)_0 = 100(1 + 0.10)^{-2} + 100(1 + 0.10)^{-3} = 157.7$$

The DCRR is the root of the equation

$$157.7 = 40(1 + r)^0 + 20(1 + r)^{-1} + 40(1 + r)^{-4} + 120(1 + r)^{-5}$$

A plot of this equation is shown in Fig. 9.3.4; for zero NPW,

$$r = 10.5\%$$

This is clearly the only root, since the right-hand side of the equation decreases monotonically as r increases.

The rule should be used even if the investment during some nonzero year is small enough so that the total net cash flow for that year remains positive—*all* one-time

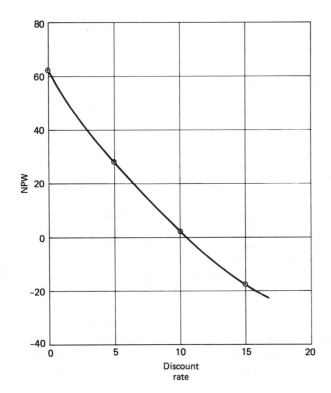

FIGURE 9.3.4
DCRR computation for Example 9.3.4.

fixed capital investments should be discounted to zero time by using the internal interest rate. If the fixed capital investment is disbursed over a period of time before zero, for example, as in Fig. 9.1.3, it should be compounded to zero time at the rate $\mathbf{i}_0$ (or i_0).

Exercise 9.3.3

Suppose prepayments occur as part of a project during some years preceding time zero, followed by a single investment at time zero and additional positive cash flows in subsequent years. Will the mathematical solution for the DCRR result in multiple roots, and, if so, how can a unique solution be obtained?

Exercise 9.3.3 demonstrates that judgment must be exercised in solving for the DCRR for unusual cash flow sequences. Another example is the case of variable working capital. If production levels are increasing, the required working capital investment may increase each year. Should each year's differential be discounted back to zero time at the rate $\mathbf{i}_0$? Not in the normal case, because the increasing working capital requirement can be readily satisfied by current revenues. Newnan (1976) gives examples of time-distributed investments which may be at least partly met by compounded net positive cash flows from previous years.

Exercise 9.3.4

What is the significance of a negative DCRR?

The Effect of Debt upon Discounted Cash Flow

The policy adopted in this book has been not to consider the effect upon individual projects of debts which stem from borrowed capital. Interest upon borrowed capital will be dealt with on the overall corporate level in the final chapter.

Cases may arise, however, when it is better to account for the effect of debt upon discounted cash flow evaluations of a project. In smaller business enterprises in particular, a proposed project may represent an appreciable fractional increase in the corporate assets, and, if the project is to be financed to a substantial degree from borrowed capital, the proper accounting of debt in discounted cash flow calculations is indicated. It may be of some interest that there are engineering economists who feel that *any* project should be evaluated in this fashion, that all invested moneys associated with a specific project should be considered as "borrowed" from the corporation and "paid off" by a part of the revenues.

The consideration of debt in discounted cash flow calculations does not result in any undue complications. An analysis of the problem has been published by Holland (1976).

Consider a project characterized by positive cash flow C_k during year k; the cash flow may include investment tax credits, salvage, and other unusual or infrequent components. For the sake of simplicity, take the investment to be a one-time cash flow at time zero (no time-distributed investments). The cash flows C_k may be considered to be discrete annual events, and discrete discounting will be used in the analysis that follows. For the case of no borrowed capital (investment of retained earnings), the NPW balance may be written as

$$\text{NPW} = -(K_F + K_w) + K_w(1 + i)^{-n} + \sum_1^n [C_k(1 + i)^{-k}] \qquad (9.3.8)$$

or, for DCRR computation,

$$0 = -(K_F + K_w) + K_w(1 + r)^{-n} + \sum_1^n [C_k(1 + r)^{-k}] \qquad (9.3.9)$$

Suppose, now, that some portion of $K_F + K_w$, say K_B, will be borrowed, and will be repaid at the interest rate i_B with annual payments R_k (not necessarily equal) over a span of m years. K_B may equal $K_F + K_w$ and, if desired, may be even higher. Three additional cash flows will now appear in Eq. (9.3.8):

1 $+K_B$. This is the amount borrowed; it appears as a positive cash flow at time zero. Note that if $K_B = K_F + K_w$, the net investment at time zero is *zero;* the actual investment is made in the form of the payments R_k.

2 $-\sum_1^m [R_k(1 + i)^{-k}]$. This is the NPW of the annual repayments on the loan (negative cash flows). For the special case of equal annual repayments R, we note that R is given by

$$R = K_B \frac{i_B(1 + i_B)^m}{(1 + i_B)^m - 1} \qquad (9.3.10)$$

so that

$$-R \sum_1^m (1 + i)^{-k} = -K_B \frac{(1 + i)^m - 1}{i(1 + i)^m} \frac{i_B(1 + i_B)^m}{(1 + i_B)^m - 1} \qquad (9.3.11)$$

3 $+t\sum_1^m [I_k(1 + i)^{-k}]$. This is the NPW of the tax credit on the interest portions I_k of the annual loan repayments R_k; t is the tax rate. For the special case of equal R's, the NPW sum,

$$\sum_1^m [I_k(1 + i)^{-k}]$$

is given by Eq. (2.3.23).

Exercise 9.3.5

Explain the meaning of the third additional cash flow.

Equation (9.3.8) may now be modified to read

$$\text{NPW} = (K_B - K_F - K_w) + K_w(1 + i)^{-n} + \sum_{1}^{n} [C_k(1 + i)^{-k}]$$
$$- \sum_{1}^{m} [(R_k - tI_k)(1 + i)^{-k}] \quad (9.3.12)$$

A similar equation can be written for the DCRR computation, although it is debatable whether such a computation has any meaning when loan repayments are discounted at the DCRR rate. Often the results from such a computation verge on the absurd (see, for example, Prob. 9.23), and in some cases no real solution exists. It is best to perform the DCRR calculation by discounting the debt repayments using the *loan rate*; i.e., the discounted value of the repayments is just equal to the loan amount. However, the interest tax credits, $t \cdot I_k$, may be discounted at the DCRR.

Exercise 9.3.6

Can borrowing of capital ever increase the NPW of a project? (*Hint:* Subtract Eq. (9.3.8) from Eq. (9.3.12) to obtain a ΔNPW.)

An example will serve to demonstrate the application of Eq. (9.3.12).

Example 9.3.5

Calculate both the no-loan and the loan case NPW for a project with the following parameters:

$$K_F = £50 \times 10^6 \qquad K_w = £2 \times 10^6$$
$$C_k(\text{all equal}) = £15 \times 10^6$$
$$i = 0.10 \qquad i_B = 0.15 \qquad n = m = 10 \qquad t = 0.50$$

For the loan case, $K_B = K_F$.
SOLUTION   For the no-loan case, from Eq. (9.3.8),

$$\text{NPW} = -52 \times 10^6 + 2 \times 10^6(1.1)^{-10} + 15 \times 10^6 \frac{(1.1)^{10} - 1}{0.1(1.1)^{10}}$$
$$= £40.9 \times 10^6$$

For the loan case,

$$\text{NPW} = -2 \times 10^6 + 2 \times 10^6(1.1)^{-10} + 15 \times 10^6 \frac{(1.1)^{10} - 1}{0.1(1.1)^{10}}$$
$$- 50 \times 10^6 \frac{(1.1)^{10} - 1}{0.1(1.1)^{10}} \frac{0.15(1.15)^{10}}{(1.15)^{10} - 1} \quad \text{[from Eq. (9.3.11)]}$$
$$+ \frac{50 \times 10^6 \times 0.15}{(1.15)^{10} - 1} \left[\frac{1 - (1.15/1.10)^{10}}{0.15 - 0.10} + \frac{(1.15)^{10}[(1.1)^{10} - 1]}{0.10(1.1)^{10}} \right] 0.5$$
$$\text{[Eq. (2.3.23)]}$$
$$= £46.6 \times 10^6$$

In the example, the loan does, indeed, increase the NPW of the project, even though $i_B > i$. Holland (1976) solves the same problem for the case of loan payments which involve equal annual reductions of the principal (see Prob. 9.18).

The Evaluation of Research Projects

The importance of early-stage evaluation of projects has been repeatedly emphasized throughout the preceding text. In early project stages, by definition, a great deal of research and development remains to be done, and yet, in what has been presented up to now, the remaining R&D costs have not been considered in the project evaluation. Indeed, prorated corporate R&D costs have been included as part of the project evaluation in the general expense category, but this is not really the same thing as asking whether the project will be able to survive economically *the anticipated project-specific R&D costs*. It may happen, for example, that a project is found to be marginally acceptable when such R&D costs are neglected, but that inclusion of the R&D costs in the cash flow balance forces the criterion of economic performance below the acceptable level. Anticipated R&D costs rarely have a serious impact upon projects with outstanding profitability, but the question of their effect upon marginal projects justifies their inclusion in early-stage cash flow analyses.

Inclusion of R&D costs in NPW or DCRR computations results in no undue mathematical complication. The largest obstacle is the estimation of the R&D costs themselves, still another component in the matrix of estimation uncertainties characteristic of early-stage projects. Many corporations maintain statistical documentation to facilitate the prediction of such costs for specific types of projects.

The additional computation required in the NPW balance involves the future worth of the after-tax R&D costs,[1] compounded to time zero (plant start-up). The question here is whether the compounding should be done at the "internal" interest rate i_0 or at the same rate as the subsequent cash flow discounting (i.e., at the interest rate i or r). Anderson et al. (1965) and Cevidalli and Zaidman (1980) favor the latter, since the R&D costs are an expenditure just like the COM; this is probably the most reasonable approach.

It should be noted that the inclusion in cash flow analysis of both anticipated R&D costs as well as R&D related general expenses will result in a certain amount of "overlap." Nevertheless, the general R&D expenses should be kept in the analysis even at the risk of a conservative error, since any project which eventually proves successful must share the burden of costs of the unsuccessful projects.

The economic analysis of developing projects is a continuing task; and the results of such analysis change with the technological and business climates. As the project advances, inclusion of anticipated additional R&D expenditures in the cash flow analysis may still be justified, but, in general, *the anticipated total becomes gradually smaller*. What of the R&D expenses already incurred? These may be considered to be *sunk costs*, past expenditures which no longer have any effect upon the project. Past expenditures, unless they in some fashion affect future cash flows,

---

[1] R&D costs are tax-deductible corporate expenses; some are eligible for additional tax credits.

no longer affect the current nature of the project, and they should be excluded from the cash flow analysis. A corollary to this rule is that, all other things being equal, a project in R&D stages improves its profitability with time! We will return to the concept of sunk costs in discussing replacement analysis in the next chapter.

9.4 RELATIVE MERIT OF PROFITABILITY CRITERIA

The Low Equation*

Only the best known of the criteria of economic performance have been considered in the preceding pages. There are many more—Phung (1980) identifies 32—but most are variants or nuances of the two basic concepts of project evaluation, return on investment and net present worth. Phung explores a number of mathematical interrelationships to conclude that the various cost analysis methods are, in fact, equivalent. We have emphasized the relationships between methods in each one of the two principal categories; for example, the payout time has been identified with the reciprocal of ROI, and the similarity between NPW and DCRR methods is quite evident. No attempt has been made, however, to tie the two basic concepts together, to explore their interrelationships and to determine whether the corresponding criteria of risk and acceptability (Table 8.2.1 and Fig. 9.3.1) are consistent. The relationship between the two concepts is expressed by the Low equation (see also Barreau, 1978).

To derive an equation that may be readily interpreted, it is necessary to assume a certain number of project simplifications which, however, do not change the sense of the conclusions. We start with previously formulated basic expressions:

$$\text{ROIBT} = 100 \cdot \frac{P_B}{\text{CFS}} \quad (8.2.1)$$

$$P_A = P_B - T \quad (6.1.13)$$

$$P_A = C_A - D \quad (6.1.1)$$

$$0 = -(K_F + K_w) + K_w(1 + r)^{-n} + \sum_{1}^{n} C_k(1 + r)^{-k} \quad (9.3.9)$$

We make the following simplifying assumptions:

1 Annual cash flows C_A are constant.
2 Allocated fixed capital is ignored in the analysis. (The analysis may be readily modified to incorporate allocated fixed capital as a fixed percentage of DFC $\equiv K_F$.)
3 The working capital K_w is invariant and is equal to 22 percent of K_F (including accounts receivable).
4 Straight-line depreciation.

---

*Named after Gordon Low, a graduate student of chemical engineering at the University of California, Berkeley, who first derived it.

Now,
$$T = tP_B$$
and, for $t = 0.5$,
$$P_A = \tfrac{1}{2}P_B$$

For straight-line depreciation, $D = K_F/n$. Therefore, from Eq. (6.1.1),

$$C_A = P_A + D = \tfrac{1}{2}P_B + \frac{K_F}{n} \qquad (9.4.1)$$

With assumed restrictions,

$$\text{CFS} = K_F + K_w$$

and

$$\text{ROIBT} = \frac{100\,P_B}{K_F + K_w} \qquad (9.4.2)$$

For constant C_A, from Eq. (9.3.9),

$$0 = -(K_F + K_w) + K_w(1 + r)^{-n} + C_A\frac{(1 + r)^n - 1}{r(1 + r)^n} \qquad (9.4.3)$$

Substitute (9.4.1) into (9.4.3); with $K_w = 0.22 K_F$,

$$0 = -1.22 K_F + 0.22 K_F(1 + r)^{-n} + \frac{(1 + r)^n - 1}{r(1 + r)^n}\left(\tfrac{1}{2}P_B + \frac{K_F}{n}\right)$$

Substitute (9.4.2):

$$0 = -1.22 K_F + 0.22 K_F(1 + r)^{-n}$$
$$+ \frac{(1 + r)^n - 1}{r(1 + r)^n}\left(\frac{1}{2}\frac{\text{ROIBT}}{100} \times 1.22 K_F + \frac{K_F}{n}\right)$$

Rearranging,

$$\boxed{\frac{\text{ROIBT}}{100} = \frac{2r}{(1 + r)^n - 1}[(1 + r)^n - 0.180] - \frac{1.64}{n}} \qquad (9.4.4)$$

The table shows corresponding values of ROIBT and **r** (for $n = 10$):

ROIBT, %	r
17.0	0.12
24.9	0.17
29.9	0.20
47.4	0.30

Note that if 25 percent ROIBT is taken as the acceptable level for average-risk projects in the chemical industry (Table 8.2.1), the corresponding DCRR of 17 percent is higher than the perceived "good" rate of return in Fig. 9.3.1.

Exercise 9.4.1

Suppose the allocated fixed and market capital of a project is a very substantial percentage of the DFC. What effect will this have upon the relative spread between DCRR and ROIBT?

Critique of the Criteria

Which one of the criteria of economic performance should one use to evaluate a project? Since a predictable relationship exists between ROIBT and DCRR, one is tempted to conclude that it really does not make much difference which one of the two basic concepts is used. Nevertheless, particularly for projects characterized by a more complex temporal distribution of both positive and negative cash flows, the DCRR (or NPW) method is better adapted to cope with such complexities, and it possesses the additional advantage of being rationally identifiable with and comparable to cash investments in financial institutions. The majority of corporate decision makers, as a matter of fact, use the DCRR method for ultimate project evaluation. Rapp (1980) finds that about 70 percent of the corporations in the United States and Great Britain use the method as the ultimate evaluation tool; the percentage is somewhat lower in the Scandinavian countries.

Methods based upon the ROI concept, and specifically the ROIBT method, are nevertheless commonly used for the evaluation of projects, particularly in the early project stages when only approximate estimates of costs and revenues are available. Some of the more limited variants of ROI are used in more narrowly defined situations in spite of their readily identifiable shortcomings. The percent return on sales, for example, is a profitability criterion that certainly falls short of accounting for all the cash flow aspects of a project, but it is often used for the monitoring of *ongoing* projects because it is easy to understand and to compute from readily available data. In the final analysis, the specific method used for project evaluation is a matter of corporate tradition and lore.

Criticism of one method or another is occasionally seen in the engineering economics literature. [See Rapp (1980) for an interesting analysis of often-voiced criticisms of the DCRR method.] The usual gist of such criticism is that a method does not yield a reliable, unequivocal measure of the degree of attractiveness of any project that is submitted for analysis. Many such criticisms are misguided, because the fault does not lie with the method but rather with the complex and uncertain nature of the real economic world. For example, the DCRR method is occasionally criticized because it may yield multiple values of the discount rate. We have seen that the problem is not with the method but with the proper interpretation of the compounding rates of "externally held" capital. The DCRR may sometimes be particularly sensitive to small errors in one of the input parameters. Again, this observation cannot be used as a criticism of the *method;* the answer to the problem lies in a proper sensitivity analysis, a supplementary requirement in any evaluation methodology. All methods of defining project profitability have been criticized for not accounting properly for the elements of risk. This perceived shortcoming, however, is not inherent to the methods of calculating the profitability criteria; there is nothing in the mathematical formulation of ROIBT, for instance, that is related to the quantification of risk. The elements of risk must be accounted for subsequently by making the comparison between the calculated values and some pre-established standards, the true criteria of acceptable economic performance. Indeed, this is probably where much of the unhappiness with the methodology of project evaluation lies—there are no unequivocal guidelines that account for the degree of risk associated with a given project; they are all a matter of subjective judgment. In Sec. 9.5 we will see how the separate elements of risk may be roughly quantified and incorporated into calculations of economic profitability.

As a matter of fact, the profitability methods that we have described do raise some problems when *mutually exclusive* projects are compared. These difficulties will be considered as part of the discussion of the analysis of alternatives in Chap. 10.

9.5 THE ANALYSIS OF RISK

The Quantification of Risk

In context of an industrial project evaluation, what is the meaning of risk? We can shed some light upon this question by considering a mythical project which is blessed with a most unusual array of input parameters—the value of each is known with absolute certainty. If the parameters are used in a DCRR computation, the resulting value of **r** will be established with equally absolute certainty. Acceptance of the project entails *zero risk*, risk that the profitability will, in fact, turn out to be less than expected. Of course, the decision maker still must decide whether the absolutely certain **r** is good enough, whether some other investments might not be better. A decision to proceed entails another kind of risk, the risk that the best possible investment has not been made. This, however, is a risk associated with *the choice among investment alternatives*. We will address this kind of risk in the next chapter; for the present, we will concern ourselves with the risk engendered by input parameter uncertainties associated with a specific single project.

Let us suppose that corporate management in its wisdom has prescribed some minimum DCRR for the acceptability of new projects, r_{min}, and that a particular project is characterized by $r > r_{min}$ as a result of a project evaluation team effort. Suppose, however, that one of the input parameters used in the evaluation, say, the DFC, has an uncertain value, one that is subject to error; we have seen that methods of DFC estimation offer ample opportunities for such error. The risk we speak of is this: suppose the error in estimating DFC has been so large that, after project approval, plant construction, and start-up, it turns out that $r < r_{min}$, and the project becomes a burdensome failure. Of course, we would like to have some sort of guarantee that this will not happen, but we will never get it from any rational evaluator. As a second best choice, we would like to receive some kind of assessment of the "chances," the *probability*, that such a terrible situation will not occur. If we are told that the worst possible error in estimating the DFC will result in a probability of 1 percent ($\mathbf{p} = 0.01$) that r will be less than r_{min}, we at least have the option of deciding (subjectively) whether this is a risk worth taking. The risk has been *quantified;* it is now associated with the probability of occurrence of some desirable or undesirable event.

How can this risk quantification be carried out? There is, first of all, the subjective method which classifies projects as *low, medium,* or *high* risks by virtue of the degree of perceived uncertainty in input parameters, including projected sales. The proper placement of projects into subjective risk categories is a matter of intuition, common sense, and a great deal of experience. The same subjective qualities are required to generate a set of reasonable standards (such as the ROIBT standards in Table 8.2.1) which define acceptable profitability in each risk category. High-risk projects are expected to meet high values of minimum ROIBT or r_{min} because their probability of success is low. The attempt is made, at least conceptually, to have all projects satisfy a minimum *expectation* ε, where

$$\varepsilon = \mathbf{p} r_{min} \qquad (9.5.1)$$

A very low risk project has a high degree of probable success ($\mathbf{p} \to 1$), and the prescribed r_{min} may therefore be set low—almost as low as ε, which in this case is the bottom value of the criterion of profitability below which no project money will be invested. High-risk projects ($\mathbf{p} \to 0$) correspondingly require high values of prescribed r_{min}. The high standard may also be thought of as a "cushion," so that even in cases of gross error in input parameters, ε may still be exceeded.

The subjective method, although commonly used, clearly has its limitations. An alternative is to define the probability distribution of the computed criterion of economic performance. This method is a variant upon sensitivity analysis, except that the result gives a quantitative assessment of the risk of not attaining the expected profitability. In fact, as we will see, it is possible to use the so-called *Monte Carlo techniques* to generate a probability distribution of profitability by varying all input parameters simultaneously, each with its own probability distribution.

SECTION 9.5: THE ANALYSIS OF RISK

We must first ascertain, however, how to obtain the probability distribution of an input parameter. If past performance data is available, the task is a matter of a statistical analysis typified by the following example.

Example 9.5.1

The corporate marketing research department has accumulated the following record of predicting first-year sales of new products in the last 2 years:

Predicted first-year sales, $ × 10⁶	Actual first-year sales, $ × 10⁶
27.0	21.0
1.4	1.6
11.7	9.7
0.65	0.51
21.0	23.3
4.7	3.0
15.1	10.9
1.5	2.1

The department now predicts first-year sales of $8 × 10⁶ for a new line of PVC pipe. Assuming a normal distribution hypothesis, compute the 95 percent confidence limits for projected first-year sales and the probability that the sales will be between $7 × 10⁶ and $9 × 10⁶.

SOLUTION Any possible trends in the data will be neglected. The information is recalculated as a percentage by which actual sales exceed or lag behind predicted sales:

N	x, %
1	−22.2
2	+14.3
3	−17.1
4	−21.5
5	+11.0
6	−36.2
7	−27.8
8	+40.0

For this set of data, the mean $\bar{x} = -7.4$ percent.

$$\text{Standard deviation } S(x) = \sqrt{\frac{\Sigma (x_i - \bar{x})^2}{n - 1}} = 26.2\%$$

For 95 percent confidence limits, from any table of normal distribution curve areas,

$$x_{95} = \bar{x} \pm 1.96S(x)$$
$$= -7.4 \pm 51.4 = -58.8\% \text{ and } +44.0\% \text{ of the predicted value}$$

Therefore for the forecast of $\$8 \times 10^6$, the 95 percent confidence limits are $\$3.3 \times 10^6$ and $\$11.5 \times 10^6$—a very uncomfortable range!

For $\$7 \times 10^6$, the percent lag from predicted value = -12.5 percent.

$$x_7 - \bar{x} = -12.5 + 7.4 = -5.1\% \equiv 0.195S(x)$$

From tables, the area under normal distribution curve is 0.0773. For $\$9 \times 10^6$, excess = $+12.5$ percent;

$$x_9 - \bar{x} = 12.5 + 7.4 = 19.9 \text{ percent} \equiv 0.760S(x)$$
$$\text{Area} = 0.2764$$
$$\text{Probability} = \sum \text{areas} = 0.35$$

A more thorough analysis might check the validity of the normal distribution hypothesis (using the chi-square test). Since x cannot assume values of less than -100 percent but can assume positive values without limit, it might be worthwhile to try and fit the data to skewed distributions such as the Poisson or gamma distributions.

If past performance data are not available, as is frequently the case, the prospects for risk quantification are not hopeless. In his fascinating treatise, Rose (1976) describes some ways in which subjective judgments of knowledgeable individuals may be converted into quantitative probability distributions. Project evaluators with many years of experience acquire an intuitive feel for the likely error distribution of certain input parameters; the task is to draw out the opinions of these "experts" with a line of questioning that will result in a consistent and credible quantified distribution. Rose concedes that many mathematicians cannot accept the concept of applying the theory of probability to opinions, and, indeed, the results of such a procedure may stir up a great deal of controversy (witness the Rasmussen Report controversy of the seventies).

Rose favors the trapezoidal distribution for quantifying subjective judgments. The distribution is illustrated in Fig. 9.5.1. The format reflects a set of judgments

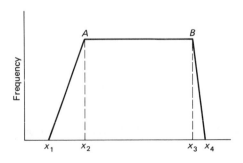

FIGURE 9.5.1
The trapezoidal probability distribution.

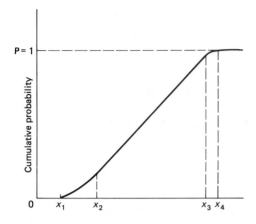

FIGURE 9.5.2
Trapezoidal cumulative probability distribution.

which is not too difficult to evoke. The range of parameter values between x_2 and x_3 is judged to be "equally likely"; also incorporated is the opinion that there is "no chance" that the parameter will assume values of less than x_1 or more than x_4. Most reasonably knowledgeable evaluators, if pressed hard enough, would be able to formulate such a distribution for input parameters such as DFC, or labor, or sales price. As shown in the figure, the distribution need not be symmetrical.

The distribution format in Fig. 9.5.1 is not immediately useful, since the required probability values are proportional to the area of the trapezoid (the total area represents $\mathbf{p} = 1$). A cumulative distribution curve (Fig. 9.5.2) is easily constructed; for instance, the cumulative probability $\mathbf{p}(x_2)$ that the parameter will assume a value of x_2 or less is the area of the triangle $x_1 A x_2$ divided by the area of the trapezoid $x_1 A B x_4$. A plot such as Fig. 9.5.2 may now be used to generate probabilities that the parameter will assume certain *discrete* values; this is the preferred probability distribution format for Monte Carlo computations. In Fig. 9.5.3, the cumulative probability curve is reproduced, and the full range of parameter values is divided up into several intervals 0–1, 1–2, . . . , not necessarily equally spaced. Each interval is characterized by the midpoint value of the parameter; for example, x_{III} is

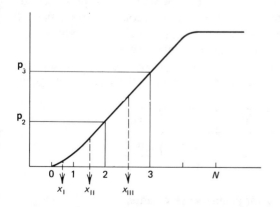

FIGURE 9.5.3
Conversion of cumulative to discrete distribution.

the midpoint of the interval 2–3. Then, as a reasonable approximation, the probability $p(x_{III})$ is given by

$$p(x_{III}) = p_3 - p_2$$

Of course, the sum of the discrete value probabilities must equal 1.0.

Other simple distributions sometimes used to quantify subjective judgments are the rectangular and the triangular distributions; these can be seen to be special cases of the trapezoidal distribution.

Exercise 9.5.1

In calculating the cumulative distribution of Fig. 9.5.2 from the trapezoidal distribution in Fig. 9.5.1, is it necessary to know the value of the frequency at points A and B?

Exercise 9.5.2

Project a five-value discrete probability distribution of your likely salary 10 years from now.

Risk Analysis Methodology

Once the probability distribution of input parameters has been established, the quantification of project risk may proceed. The objective is to establish the probability distribution of the profitability index such as the DCRR. The risk may then be assessed by answering quantitatively questions such as:

"What is the profitability index at the 50 percent cumulative probability level?"
"What is the probability that the profitability index will be below the minimum standard?"

The calculational procedure requires the use of computers, but we will not concern ourselves here with the programming; the methodology is the subject of interest. Rose (1976) has published a program for the economic evaluation of projects which includes subroutines for the Monte Carlo simulation of profitability distribution based upon trapezoidal distributions of input parameters.

Suppose the probabilities of discrete values of the input parameters are available. We may have, for instance, an array of values of the direct fixed capital estimate:

DFC, $ \times 10^6$	8	9	10	11	11.5
p	0.15	0.35	0.25	0.15	0.10

Note that $\Sigma p = 1.0$.

Similarly, we may have an array for annual sales volume,

Sales, lb × 10^6	80	90	100	110
p	0.10	0.20	0.50	0.20

and other input parameter arrays: Selling price, COM, etc.

To employ the Monte Carlo simulation, it is necessary to generate a sequence of random numbers. Such sequences are tabulated in many handbooks of mathematics, but it is easy enough to write a subroutine to generate random numbers as part of the computer simulation. Several methods to do this are available; none results in "perfect randomness," but the degree of randomness is more than adequate for present purposes. Perhaps the most amusing method is the *midsquare method* (Meyer, 1956). For instance, to generate four-digit random numbers, start with any such four-digit number, say,

$$x_0 = 2061$$

Now square it:

$$x_0^2 = 04247721$$

expressed as an eight-digit number. Take the middle four digits for the next random number,

$$x_1 = 2477$$

Continue the procedure as long as needed:

$$x_1^2 = 06135529$$
$$x_2 = 1355$$
$$\ldots\ldots\ldots\ldots$$

Let us for the moment assume, however, that we will use two-digit random numbers for our simulation, ranging from 00 to 99. Now, assign to each value of an input variable a number of sequential digits in proportion to the associated probability. For example, for the DFC array, the $8 million value has an associated probability of 0.15; therefore, assign the first 15 of the 100 two-digit numbers (00 to 14) to this value. In this fashion, the following assignment is generated for the first array:

DFC, $ × 10^6	Assigned random numbers
8	00–14
9	15–49
10	50–74
11	75–89
11.5	90–99

For the second array of sales volume, the assignments turn out to be:

Sales, lb × 10⁶	Assigned random numbers
80	00–09
90	10–29
100	30–79
110	80–99

Now generate a sequence of random numbers equal to the number of input variables, say,

$$26, 34, 65, 22, \ldots$$

26 belongs to the first array (DFC), and from the assignment table, 26 is associated with DFC = $9 × 10⁶.

The second number in the sequence, 34, is associated with 100 × 10⁶ lb annual sales volume.

When all the input variables are supplied with values, the DCRR is calculated. The procedure is then repeated for another DCRR calculation, and so on, for perhaps several thousand trials. From the results, the percentage of the total in each selected ΔDCRR range is calculated; this then is the probability of the mean value of DCRR for that range. In this manner, a DCRR cumulative probability distribution curve may be generated, similar to the one in Fig. 9.5.4. Such a graph represents, at a glance, the risk associated with the acceptance of a project; we are now in position to answer some of the questions posed at the start of the risk analysis methodology discussion.

If the probability distributions of all the input parameters happen to be normal distributions, the computation of the project profitability distribution may be mathematically systematized, and a computer simulation is not required. A detailed outline of the procedure is given in Holland et al (1974).* The basis is the

*Copyright 1974, John Wiley & Sons, Ltd. Used by permission.

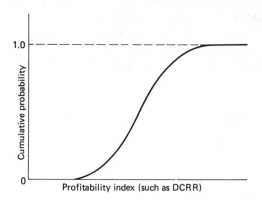

FIGURE 9.5.4
Probability distribution curve for project profitability.

relationship expressed by Eq. (4.5.12),

$$S^2(Y) = \sum_1^n S^2(y_i) \tag{4.5.12}$$

which is valid for the linear relationship

$$Y = \sum_1^n y_i$$

Equation (4.5.12) may be generalized for any functional relationship with normally distributed y_i,

$$Y = \phi(y_1, y_2, \ldots, y_n)$$

to read

$$S^2(Y) = \left(\frac{\partial Y}{\partial y_1}\right)^2 S^2(y_1) + \left(\frac{\partial Y}{\partial y_2}\right)^2 S^2(y_2) + \cdots + \left(\frac{\partial Y}{\partial y_n}\right)^2 S^2(y_n) \tag{9.5.2}$$

Suppose, as an example, we have an NPW relationship

$$\text{NPW} = k_1 \mathcal{R} - k_2(\text{DFC}) + k_3 \tag{9.5.3}$$

where the annual revenue $\mathcal{R}$ and the direct fixed capital investment DFC are both characterized by normal probability distributions with variances $S^2(\mathcal{R})$ and $S^2(\text{DFC})$, respectively. Since

$$\frac{\partial(\text{NPW})}{\partial \mathcal{R}} = k_1 \quad \text{and} \quad \frac{\partial(\text{NPW})}{\partial(\text{DFC})} = -k_2$$

we have, from (9.5.2),

$$S^2(\text{NPW}) = k_1^2 S^2(\mathcal{R}) + k_2^2 S^2(\text{DFC}) \tag{9.5.4}$$

The annual revenue $\mathcal{R}$ may be the product of the annual production p and the sales price s. For the product,

$$\mathcal{R} = ps \tag{9.5.5}$$

if p and s are both characterized by normal probability distributions,

$$\frac{\partial \mathcal{R}}{\partial p} = \bar{s} \quad \text{and} \quad \frac{\partial \mathcal{R}}{\partial s} = \bar{p}$$

where the bars represent the mean values.[1] Equation (9.5.2) then becomes

$$S^2(\mathcal{R}) = \bar{s}^2 S^2(p) + \bar{p}^2 S^2(s) \tag{9.5.6}$$

which may be substituted into Eq. (9.5.4).

The mean of the NPW, $\overline{\text{NPW}}$, is given by

$$\overline{\text{NPW}} = k_1 \overline{\mathcal{R}} - k_2(\overline{\text{DFC}}) + k_3 \tag{9.5.7}$$

which is obtained from Eq. (9.5.3) and the relationship in Eq. (4.5.7).

We thus have both $\overline{\text{NPW}}$ and $S^2(\text{NPW})$; the probability distribution of NPW can now be computed from standard normal distribution tables.

Example 9.5.2

For a particular project, calculations show that

$$\text{NPW} = 2.085 p(s - 135) - 0.892(\text{DFC}) - 1.231 \times 10^6$$

The input parameters p, $s - 135$, and DFC are characterized by normal distributions:

	Mean, $\bar{y}$	Variance, $S^2(y)$
p	15,000	250,000
$s - 135$	165	1.563
DFC	4.1×10^6	0.104×10^{12}

What is the probability that NPW will be less than zero?

SOLUTION

$$\overline{\text{NPW}} = 0.272 \times 10^6$$
$$S^2[p(s-135)] = (165)^2 \times 250{,}000 + (15{,}000)^2 \times 1.563$$
$$= 0.7158 \times 10^{10}$$
$$S^2(\text{NPW}) = (2.085)^2 \times 0.7158 \times 10^{10} + (0.892)^2 \times 0.104 \times 10^{12}$$
$$= 11.387 \times 10^{10}$$
$$S(\text{NPW}) = 0.337 \times 10^6$$

For NPW $= 0$,

$$\overline{\text{NPW}} - \text{NPW} = 0.272 \times 10^6$$
$$\equiv 0.807 S(\text{NPW})$$

From tables of normal distribution curve areas,

---
[1] See also discussion of the transmission of error in nonlinear functions in Box et al. (1978).

$$\mathbf{p} = 0.500 - 0.290 = 0.210 \quad \text{or} \quad 21.0\%$$

If DCRR computations are based on input parameters with normally distributed probabilities, a similar procedure may be used to generate a DCRR probability distribution curve. The reader is referred to Holland et al. (1974) for detail.

If the effect of only one distributed input parameter upon NPW is to be calculated, the NPW is given by

$$\text{NPW} = \sum_1^n \varepsilon_i(\text{NPW}) \tag{9.5.8}$$

where the expectation $\varepsilon_i(\text{NPW})$ is

$$\varepsilon_i(\text{NPW}) = (\text{NPW})_i \mathbf{p}_i \tag{9.5.9}$$

and $(\text{NPW})_i$ is calculated using the ith value of the input parameter with probability $\mathbf{p}_i$.

The controversial nature of probability distribution projections for economic evaluation input parameters limits the popularity of the quantification of project risk. There are those corporations in which quantitative risk analysis is a prerequisite of advanced-stage project evaluations; in other corporations, currently available risk quantification methods are not viewed seriously.

NOMENCLATURE

c_v	unit variable charges, $/kg
C	cash flow, $/year: a continuously generated cash flow, but expressed as a single discrete annual event (i.e., an integral). C_A is cash flow after taxes; C_k is cash flow during year k
$\bar{C}$	cash flow *rate*, $/year; a continuous function of time
D	annual depreciation charges, $/year
E	annual out-of-pocket expenditures, $/year
$f(x)$	frequency function
F	annual fixed charges excluding depreciation, $/year
$F(x)$	a mathematical function
g	unit general expenses, $/kg (also used as time span designation in Fig. P9.1)
$g(x)$	a mathematical function
G	annual general expenses, $/year
h	an index year
i	nominal interest rate, fractional
$\mathbf{i}$	effective discrete interest rate, fractional; $\mathbf{i}_0$ is the corporate "internal" interest rate, $\mathbf{i}_B$ is the interest rate on a loan
I	interest portion of annual loan repayment R, $/year
k	index year
K	capital investment, $; K_F is fixed capital, K_w is working capital, K_B is borrowed capital

m time span of loan, years
n lifetime of project, years
N time span between start of research and project termination, years
p production, kg/year; p_R is rated production capacity of plant, p_{max} is the maximum sustained production rate
p   probability
P principal, or present worth of annuities (no subscript). With subscript, annual profit, \$/year; P_B is profit before taxes, P_A is profit after taxes
q fraction of maximum sustained production rate
r discounted cash rate of return for continuous discounting (nominal)
r   discounted cash rate of return for discrete discounting (effective)
R total annual repayment on loan, \$/year
$\mathcal{R}$ annual revenue from sales, \$/year; $\mathcal{R}_0$ is the revenue discounted to zero time
s unit sales price, \$/kg
$S(x)$ standard deviation of x; $S^2(x)$ is the variance
t tax rate
T annual taxes, \$/year
x a variable
y an independent variable
Y a dependent variable
z characteristic constant for exponential decline
α superproductivity
ε expectation
θ construction period time, years
ϕ discrete present worth function
Ψ continuous present worth function
ω fraction of rated capacity

Subscripts

A after taxes
B borrowed; or before taxes
F fixed
i an index
k year index
max   maximum
min   minimum
n for year n
0   at zero time
R rated
t at time t
w working

Auxiliary symbols

A bar over a symbol indicates the mean value ($\bar{x}$). $\overline{C}$, however, indicates a continuous cash flow function, distinct from the discrete function C.

Abbreviations

AC   allocated capital
CFS   capital for sale

COM   cost of manufacture
DCRR   discounted cash rate of return
DFC   direct fixed capital
EMIP   equivalent maximum investment period
NPW   net present worth
PVC   polyvinyl chloride
R&D   research and development
ROI   return on investment
ROIBT   return on investment before taxes

REFERENCES

Allen, D. H.: Two New Tools for Project Evaluation, *Chem. Eng.*, **74**:75 (July 3, 1967).

Anderson, M. L., J. Eschrich, and R. C. Goodman: Economic Analysis of R & D Projects, *Chem. Eng. Prog.*, **61** (7):106 (1965).

Barreau, Jean: Relation entre le Taux de Rentabilité Interne des Investissements et le Taux de Rendement Comptable, *RAIRO Rech. Oper.*, **12** (1):15 (1978).

Box, George E. P., William G. Hunter, and J. Stuart Hunter: "Statistics for Experimenters," Wiley, New York, 1978.

Cevidalli, Guido, and Beno Zaidman: Evaluate Research Projects Rapidly, *Chem. Eng.*, **87**:145 (July 14, 1980).

Cox, W. E., Jr.: Product Life Cycles as Marketing Models, *J. Bus.*, **40** (4):375 (1967).

Frederixon, M. S.: "An Investigation of the Product Life Cycle Concept and Its Application to New Product Proposal Evaluation within the Chemical Industry," Ph.D. thesis, Michigan State University, East Lansing, 1969 (University Microfilms, Inc., Ann Arbor, Mich.).

Happel, John, and Donald G. Jordan: "Chemical Process Economics," Marcel Dekker, Inc., New York, 1975.

Holland, F. A.: The Effect of Debt on Discounted Cash Flow Calculations, *Eng. Process Econ.*, **1**:223 (1976).

———, F. A. Watson, and J. K. Wilkinson: "Introduction to Process Economics," Wiley, London, 1974.

Meyer, Herbert A. (ed.): "Symposium on Monte Carlo Methods," Wiley, New York, 1956.

Newnan, Donald G.: "Engineering Economic Analysis," Engineering Press, San Jose, Calif., 1976.

Pavone, Tony, and Greg Patrick: Energy Tax Credit Aids Investment Projects, *Chem. Eng.*, **88**:99 (Feb. 23, 1981).

Phung, Doan L.: Cost Analysis Methodologies: A Unified View, *Cost Eng.*, **22**(3):139 (1980).

Rapp, Birger: The Internal Rate of Return Method—A Critical Study, *Eng. Costs Prod. Econ.*, **5**:43 (1980).

Rose, L. M.: "Engineering Investment Decisions: Planning under Uncertainty," Elsevier, Amsterdam, 1976.

Valle-Riestra, J. Frank: The Evaluation of the Present Worth of Normally Distributed Cash Flows, *Eng. Process Econ.*, **4**:37 (1979).

Wild, N. H.: Program for Discounted-Cash-Flow Return on Investment, *Chem. Eng.*, **84**:137 (May 9, 1977).

SPECIAL PROJECT: A COMPUTER STUDY OF RESEARCH ECONOMIC INCENTIVES

The purpose of the project is to develop and test a computer program for the prediction of the long-range profitability of a venture in early stages of research and development.

The profitability is to be measured in terms of a discounted cash rate of return (DCRR) profitability index, discounted to zero time. A sensitivity analysis is to be performed by estimating the variation in profitability index as input parameters are varied over a ±30 percent range.

In early stages of R&D all input parameters are simply intelligent estimates. They should include the following:

1 Number of years N from present to estimated end of project life (i.e., production plant shutdown). This span includes estimated years of research, development, and construction up to plant start-up.

2 Year-by-year research expenditures, entirely project-oriented before plant start-up, but including general research and administrative expenses after start-up.

3 Total fixed capital investment estimates and years in which expended. (As an approximation, allocated fixed capital is handled as part of total fixed capital, with same depreciation schedule.)

4 Year-by-year estimates of:
Marketing expenses
Production rate
Manufacturing costs (COM)
Selling price

In addition, input information must include:

1 Royalty rate, if any (in $/$ sales revenue, not part of COM).

2 Working capital expenditures. These may be approximated for any given year as 25 percent of the increase in sales revenue over the year before. For example, if sales are projected at $1,000,000 for 1995 and $2,000,000 for 1996, the working capital expenditure for 1996 is $0.25(2,000,000 - 1,000,000) = \$250,000$. If annual sales are constant, there is no additional working capital expense; if sales decrease, there is additional *income* from corresponding working capital decrease.

3 Income tax rate. The rate applies equally to negative profit (expenditures larger than income), since losses are accounted for in overall corporate income; e.g., pre-start-up research expenses receive a tax credit. (In numerical example, use current tax rate.)

4 Investment and other tax credits: take as current percent of fixed capital investment during the year investment is made.

5 Salvage value of equipment: take as 3 percent of original fixed capital investment at end of project life.

6 Internal interest rate for discounting future investments to present (use 12 percent in example).

The computer program must handle the following:

1 Input data (see Table SP9.1).
2 Calculate discount factors for both year-end and continuous discounting (years

TABLE SP9.1
INPUT DATA FOR RESEARCH ECONOMIC INCENTIVES EXAMPLE

Year	Sales, lb × 10³	Selling price, ¢/lb	Investment, $ × 10³	Research expenses, $ × 10³	Marketing expenses, $ × 10³	COM (without depreciation), ¢/lb
1				1000		
2				4000		
3				5000	300	
4			9,700	6000	300	
5	1,000	71.0	14,100	3000	2500	300
6	5,000	71.0		1200	2000	70
7	10,000	71.0		1200	2000	40
8	25,000	70.0		1200	3000	30
9	40,000	68.0	7,200	1200	4000	25
10	60,000	65.0	9,500	1200	5000	20
11	80,000	62.0		1200	5000	18
12	110,000	62.0		1200	5000	18
13	130,000	62.0		1200	7000	17
14	140,000	60.0		1200	7000	17
15	150,000	60.0		1000	9000	17
16	150,000	60.0		1000	9000	17
17	150,000	60.0		1000	9000	16
18	170,000	60.0		1000	9000	16
19	170,000	58.0		1000	9000	16
20	170,000	58.0		1000	9000	16

from 0 to N, any interest rate); choice to be part of input information. For continuous discounting, consider all cash flows to be uniform (continuous) throughout each year.

3 Compute depreciation charges by sum-of-years-digits method over tax life of equipment. In example, assume tax life of 10 years beyond year of investment.

4 Compute year-by-year cash flow.

5 Compute DCRR profitability index. Repeat for ±30 percent variation in input parameters, each considered separately in 10 percent increments. The input parameters to be varied include:
Research expenses
Investment (fixed)
Sales in pounds per year of product
Selling price
Manufacturing cost

In varying a particular input parameter, apply the same percentage change equally to each annual value. For example, if the base case prediction of sales is 1000 tons for year 1 and 2000 tons for year 2, for a −10 percent increment use 900 tons for year 1 and 1800 tons for year 2.

The output format should include:

1 Input data (similar to Table SP9.1).

2 Significant intermediate results on a year-by-year basis; for example, sales revenue, depreciation charges, gross profit, cash flow, etc.

3 DCRR profitability index versus varied input parameters in tabular form:

	−30%	−20%	−10%	0	+10%	etc.
1. Research expense 2. Investment 3. Etc.		Computed values of index				

4 Any other identifying printout judged important (code numbers, comments such as "continuous discounting," etc.).

Use the data in Table SP9.1 to run your program by using both discounting methods. Ignore royalty charges.

The project report should incorporate:

1 Program (identify user language)
2 Printout based on data in Table SP9.1
3 A short written report outlining methods, difficulties encountered, suggestions, etc.

PROBLEMS

9.1 Draw a cash flow diagram for the project described in Prob. 8.28.

9.2 a See if the format of the logistic curve, Eq. (2.2.14), can accommodate the shape of the *cumulative* cash outflow curve in Fig. 9.1.3. Plot the derived curve and compare with the original.

b The instantaneous cash flow rate $\overline{C}_t$ is actually the *slope* of the curve in Fig. 9.1.3. Use graphical methods to generate a plot of $\overline{C}_t$ versus t, the percent of time elapsed. What are the units of $\overline{C}_t$?

c Generate a second plot of $\overline{C}_t$ compounded to time zero (end of construction period) versus t for a construction period θ equal to 1 year and a *continuous* (nominal) interest rate $r = 0.10$. Integrate the curve graphically, and express the "present worth" of the disbursements as a multiple of K_F, the total of the disbursements. Be sure to account for the 5 percent of K_F disbursed as a lump sum at the start of the construction period.

d Prepare a plot which shows the multiplier that must be used as a function of the product $r \cdot \theta$ (from zero to 1).

9.3 Determine the EMIP for the project outlined in Table 8.4.1. The cash flow for year 1 includes all tax credits.

9.4 In Example 9.2.1, find the value of the interest rate which will make the payout time just equal to the project lifetime.

9.5 If you have not done so before, solve part (a) of Prob. 8.21 and calculate
   a NPW at 10 percent discrete annual discounting. What is the present worth ratio?
   b The discrete DCRR.

9.6 A large proposed project has the following characteristics:

DFC investment: 50×10^6
Allocated fixed capital: 10×10^6
Recoverable capital: 10×10^6

Constant annual production: 100×10^6 kg
Out-of-pocket expenditures: 30¢/kg
Project life (= tax life): 8 years
Tax rate: 45%

Management insists that for this type of a project a present worth ratio of at least 1.0 must be obtained at a discrete discounting rate of 18 percent. Straight-line depreciation on the DFC is to be used. What must be the sales price?

For present purposes, neglect investment tax credits and corrections for allocated capital depreciation.

9.7 The board of directors has stipulated that no project is to be presented for approval unless it exhibits a positive NPW at 15 percent discrete annual discounting. The project you have been trying to ramrod through for board approval unfortunately exhibits a present worth ratio of only 0.95, based on a total zero-time negative cash flow of $70 million. You have discovered, however, that some of the investment qualifies for various tax credits which were not properly considered in the unfavorable present worth evaluation:

10 percent investment tax credit on a total of $40 million of invested capital
10 percent energy tax credit on a total of $20 million of invested capital

What is the project present worth ratio when tax credits are accounted for? Assume tax credits are discounted from the end of the first year.

9.8 The tax life of a plant for depreciation purposes and the service life need not be the same thing; in fact, the service life is often longer. Repeat Prob. 9.6 for the case of tax life of 8 years and service life of 14 years.

9.9 Derive an expression for the present worth of revenues $\Re_k$ (discrete annual discounting at rate **i**) with the following distribution pattern:

Constant annual revenues $\Re_0$ from the end of year 1 to the end of year h
Declining annual revenues $\Re_k$ from the end of year $h + 1$ to the end of year n, characterized by the exponential decline

$$\Re_k = \Re_0 e^{-z(k-h)}$$

9.10 We have seen that time-distributed investments can result in multiple-root solutions of DCRR equations unless the future investments are discounted to zero time at the predetermined interest rate i_0, the corporate "internal" interest rate. Should future investments be discounted in the same way with the NPW method?

9.11 Repeat Example 9.3.1 for the two cases of continuous discounting of the cash flows, considering them to be discrete in the one case and uniformly spread over each year in the other (DFC expended over 1 year before plant start-up).

9.12 Use the graphical correlations in the article by Valle-Riestra (1979) to compute the likely DCRR for a projected plant characterized by the following parameters:

Rated production capacity: 25×10^6 kg/yr
Estimated superproductivity: 0.10
First year's production: 50% of *rated* capacity
Last year's production: 50% of rated capacity
Service life = tax life = 10 years
Straight-line depreciation

DFC = $17.5 × 10^6$
Allocated capital (nonrecoverable) = $1.3 × 10^6$
Recoverable capital = $3.8 × 10^6$
Sales price = 70¢/kg
Variable charges = 25¢/kg
General expenses = 10¢/kg
Fixed and regulated charges (excluding depreciation): 14¢/kg
Tax rate = 45%

Equation (9.3.5) may be used as a guide to the computations, which should be based on discrete discounting.

9.13 In Example 9.3.2, what would be the present worth of sales revenues for the case of constant annual production at rated capacity? What percentage of this value do the more realistically projected revenues represent?

9.14 Suppose a continuous cash flow $\overline{C}_t$ follows the normal distribution pattern of Fig. 9.3.2. For the continuous cash flow case, let

$$\overline{C}_0 = \overline{C}_n = q\overline{C}_{max}$$

a Derive an expression for the continuously discounted present worth of the normally distributed cash flow. The desired format of the derived expression is

$$\overline{C}_0 = \overline{C}_{max} \Psi$$

where Ψ, the present worth function for the continuous discounting case, is not formally integrable.

b Evaluate Ψ by graphical integration for the case of $n = 10$, $q = 0.30$, and **r**, the *effective* interest rate, equal to 0.20. (*Note:* For the discrete discounting case, $\phi = 2.68$.)

9.15 A projected normal distribution of cash flows need not necessarily be symmetrical. Figure P9.1 illustrates the case of a truncated normal distribution, with $q_0 > q_n$. To compute the present worth of cash flows distributed as shown, proceed as follows:

a The equation of the normal distribution curve may be written as

$$f(x) = \frac{1}{\sqrt{2\pi}} e^{(-1/2)x^2} \qquad (P9.1)$$

The first step is to convert this equation into the form

$$\overline{C}_t = F(t) \qquad (P9.2)$$

where $F(t)$ is the appropriate functional analog of the right side of Eq. (P9.1) in terms of time (years). Referring to Fig. P9.1, note that the truncated curve may be extended backward by $-g$ years to form a symmetrical curve. Let

$$m = n + g \qquad (P9.3)$$

and

$$(\overline{C})_{-g} = (\overline{C})_n = q_n \overline{C}_{max} \qquad (P9.4)$$

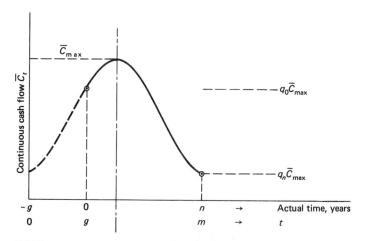

FIGURE P9.1
Truncated normal distribution of cash flow.

Show that the congruence of Fig. P9.1 and Eq. (P9.1) leads to the result[1]

$$x_m = \sqrt{-2 \log q_n} \tag{P9.5}$$

b Using Eq. (P9.5) and the principle of congruence, derive the general expression for the continuous cash flow at t (measured from $-g$),

$$\overline{C}_t = \overline{C}_{max}(q_n)^{[(2t-m)/m]^2} \tag{P9.6}$$

c With the help of Eq. (P9.6), show that the following relationship holds:

$$q_0 = q_n^{[(2g-m)/m]^2} \tag{P9.7}$$

The value of g may now be obtained from Eq. (P9.7) by substituting Eq. (P9.3).

d The present worth (at $t = g$) of the cash flows, C_0, is given by

$$C_0 = \overline{C}_{max} \int_g^m (q_n)^{[(2t-m)/m]^2} e^{-r(t-g)} dt \tag{P9.8}$$

Explain the derivation of Eq. (P9.8).

e Make a plot of the integrand in Eq. (P9.8) versus t for the following parameters:

$$n = 10 \quad q_0 = 0.50 \quad q_n = 0.30 \quad r = 0.15$$

f Integrate the plot graphically and determine C_0 for $\overline{C}_{max} = \$1 \times 10^6/\text{year}$.

9.16 The net continuous cash flows generated by a project are shown in Fig. P9.2. The distributions are ramp distributions (see also Example 2.2.6, Exercises 2.2.6 and 2.2.7).

---
[1] The logs are napierian.

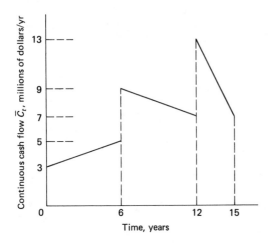

FIGURE P9.2
Ramp cash flows, Prob. 9.16.

Compute the continuously discounted DCRR for the project with the following time-distributed investments:

Fixed capital: Initial, 10×10^6, expended uniformly over 2 years before zero time; 6×10^6, expended uniformly during year 6; and 4×10^6, expended uniformly during year 12.

Working capital: Approximate as 30 percent of net continuous cash flow at any one time. (For example, at time zero, working capital is $0.30 \times 3 \times 10^6 = \$0.9 \times 10^6$; at the end of 12 years, working capital jumps to $0.3 \times 13 \times 10^6 = \$3.9 \times 10^6$, etc. Assume working capital is obtained from current income.)

Discount fixed capital investments continuously at a nominal rate $i = 12$ percent. The salvage value at the end of 15 years is 10 percent of the fixed capital investment.

9.17 Holland (1976) considers the case of loan repayments which involve equal annual reductions of the principal. Derive an expression for R_k, the annual payment during any year k, for this method of repayment. Does your derived equation possess the required property that

$$\sum_1^n R_k(1 + i_B)^{-k} = P$$

Derive an expression for the present worth of interest payments for this repayment method, that is,

$$\sum_1^n I_k(1 + i)^{-k}$$

In general, how will the present worth of interest payments for this method compare with the present worth of interest payments for the case of equal annual payments [Eq. (2.3.23)]? Confirm your judgment by plotting the present worth (as a multiple of P) versus i_B, with $0 > i_B > 0.20$, $i = 0.10$, $n = 10$, for the two cases.

9.18 Rework the loan case of Example 9.3.5 for the case of loan payments which involve equal annual reductions of the principal.

9.19 In Example 9.3.5, suppose the equal £15 × 10^6 annual cash flows are uniformly distributed throughout each year; i.e., they are, indeed, continuous. What would be the project NPW if K_F, K_w, and the loan payments are discrete cash flows, but if they, as well as the continuous cash flows, are discounted continuously at the *nominal* interest rate of $i = 0.10$?

9.20 Repeat Example 9.3.5 for the case of $i = i_B = 0.10$.

9.21 What must be the maximum value of i_B in Example 9.3.5 to keep the NPW the same as for the no-loan case?

9.22 Consider the project outlined in the statement of Prob. 9.6. Suppose the product is to be sold for 58 cents/kg. Calculate the percent error in computed DCRR by neglecting

a Depreciation on allocated capital (20-year tax and service life).

b Ten percent investment tax credit on DFC during first year of operations. [Determine effect of parts (a) and (b) separately.]

9.23 A tomato canner wishes to borrow $1,000,000 to install a new concentrator for tomato puree. The new installation will increase the net cash flows (after taxes) by $270,000/year for 20 years (the life of the concentrator). However, this increase in cash flow from the tomato processing facility does not account for the time payments on the borrowed cash; these time payments will decrease the $270,000 cash flows during the years the loan payments are made.

Suppose the canner takes out a 10-year loan, to be paid off in equal annual payments. What is the maximum interest rate on the loan (discrete compounding) which will allow the canner to realize a 15 percent DCRR? Consider all cash flows to be discrete annual events, and assume a 50 percent tax rate. Use the DCRR to discount the loan repayments.

9.24 In Prob. 9.23, what is the NPW of the project if a loan can be secured at 10 percent? In that case, what is the dollar amount of the total of the interest portions of the payments? Use a 15 percent discount rate on all cash flows, including the loan repayments.

9.25 What is the DCRR for the loan case in Example 9.3.5? Discount the loan repayments at a 15 percent rate.

9.26 The following problem is similar to one suggested by Anderson et al. (1965). Optimistic estimates of a proposed R&D process development project predict the following:

R&D cost: $3 million over a 3-year period, with annual R&D growth rate of 50 percent/year

Investment: $4 million depreciated (straight-line) over a 10-year period (considered as single zero-time cash flow)

Total sales: $100 million with annual sales growth rate of 10 percent/year over a 10-year process life

Out-of-pocket expenditures: 80 percent of sales

Working capital: 10 percent of annual sales (considered as investment level at *start* of each year)

Company policy requires a 20 percent DCRR (discrete) for this type of project. Should the project be pursued?

Be sure to remember that R&D costs are tax-deductible expenses; take $t = 0.50$.

9.27 A project is expected to generate a uniform cash flow of $15 million over 5 years. The estimated fixed capital investment is $5.0 million, the recoverable capital another $1 million. A 1-year crash R&D effort will be required to bring the project to reality. If corporate policy requires an 18 percent discrete DCRR on R&D projects, how many R&D personnel (1983 R&D costs) can be assigned to the project?

9.28 A project is characterized by allocated fixed capital equal to 15 percent of DFC, recoverable capital equal to 15 percent of DFC, and a service life of 15 years. At a tax rate of 45 percent, the discrete DCRR is 20 percent. For straight-line depreciation, what is the ROIBT?

9.29 A major plant modification is needed to modernize the facilities and expand production. An investment of $8 million (none of it recoverable) will result in a net annual increase in cash flow of $1.5 million after taxes. How many years beyond the time of investment must the plant be operable to realize a continuous DCRR of 15 percent?

9.30 Rapp (1980) suggests that projects with a service life of less than 5 years should not be evaluated with the DCRR method. Review the arguments in the publication and write a brief summary and analysis of the author's proposals.

9.31 The DFC estimate of a plant has a trapezoidal probability distribution with the following parameters:

$$x_1 = \$13.5 \times 10^6 \qquad x_3 = \$18.0 \times 10^6$$
$$x_2 = \$14.5 \times 10^6 \qquad x_4 = \$20.0 \times 10^6$$

Convert the data to a five-value discrete distribution of your choice.

9.32 A triangular distribution of a sales forecast may be described by equations of the two lines forming the apex of the triangle,

$$f(x) = m_1 x + b_1$$
$$g(x) = m_2 x + b_2$$

where $f(x)$ and $g(x)$ are frequencies, and x is the value of the forecast.

Derive an analytical expression for the cumulative probability **p** as a function of the value of x.

9.33 Use Fig. 9.5.1 to derive an analytical expression for the mean value $\bar{x}$ of a trapezoidal distribution.

9.34 Your company has just invested $450,000 of fixed capital in a manufacturing process which is estimated to give positive annual cash flow after tax A of $100,000 in each of the next 10 years. At the end of year 10 there will be no further market for the product. If marketing problems delay plant start-up for 1 year, what value of A will be needed to maintain the same DCRR as would be experienced if no delay occurred?

9.35 Equal annual cash flows over a period of 12 years are generated by a fixed capital investment of K_F (no recoverable capital). At a discrete discounting rate of $i = 10$ percent, the NPW of the project is just equal to K_F. What is the DCRR?

9.36 Holland et al. (1974) suggest that the maximum estimated range of a parameter, such as the DFC estimate, may be properly considered to encompass the 99.99 percent confidence limits. If the distribution of estimates is hypothesized to be normal, 99.99 percent confidence limits incorporate a range of $\pm 4S(x)$, or a range of 8 standard deviations. Thus for a maximum range from $x = A$ to $x = B$,

$$S(x) = \frac{|A - B|}{8}$$

Let us suppose that an estimator characterizes a project as generating an annual sales volume of no less than 50 million and no more than 70 million lb, and the selling price will almost certainly be in the range of 40 to 60 cents/lb.
  a What is the most probable annual sales revenue?
  b What are the 95 percent confidence limits of the annual sales revenue?

9.37 A symmetrical trapezoidal distribution may be interpreted as representing a "likely" range (x_2–x_3 in Fig. 9.5.1) centered within an "extreme" range (x_1–x_4). Rose (1976) suggests that a reasonable normal distribution fit to the symmetrical trapezoidal distribution may be accomplished by setting

$$x_1 = \bar{x} - 2.6\, S(x)$$
$$x_2 = \bar{x} - 0.4\, S(x)$$

The justification for this rather arbitrary looking relationship is the close coincidence of the cumulative probability curves generated from either distribution, using the written relationship.

Suppose the NPW equation in Example 9.5.2 holds for the following input parameter *extreme* ranges:

Production p 8,000–22,000
Sales price s 250–350
DFC   3.5–4.7 × 10^6

  a What are the parameter likely ranges?
  b What is the probability that NPW will be more than zero?

9.38 In Example 9.5.2, the NPW is written as

$$\text{NPW} = 2.085\,p(s - 135) - 0.892(\text{DFC}) - 1.231 \times 10^6$$

The temptation may be to write this out as

$$\text{NPW} = 2.085\,ps - 281.5\,p - 0.892(\text{DFC}) - 1.231 \times 10^6$$

and to proceed by setting

$$S^2(\text{NPW}) = (2.085)^2[(300)^2 \times 250{,}000 + (15{,}000)^2 + 1.563]$$
$$+ (281.5)^2 \times 250{,}000 + (0.892)^2 \times 0.104 \times 10^{12}$$

This would be quite wrong. Why?
  Hint: If $Y = 2y$, show that

$$S^2(Y) = S^2(2y) \neq S^2(y + y)$$

9.39 Let the NPW of a project be represented by

$$\text{NPW} = 10 \times 10^6 - 0.5(\text{DFC})$$

where the DFC has the discrete probability distribution:

DFC	p
8×10^6	0.10
9×10^6	0.20
10×10^6	0.40
12×10^6	0.25
14×10^6	0.05

What is the expected NPW?

9.40 An investment K_F to improve portions of an existing production facility is expected to generate increased net cash flows. Preliminary analysis shows a very attractive payout time of about 1 year but a dismal DCRR of 5 percent. What can you say about the time distribution of the generated cash flow increments?

9.41 The text of Prob. 7.30 describes a proposal to enrich a low-grade natural gas. A fabricator of turnkey adsorption units has submitted a bid to provide a "ready to go" facility for just $1 million, with a guaranteed service life of 10 years. What would be the DCRR on this investment? Would you recommend to your management that the investment be made?

CHAPTER 10

THE ANALYSIS OF ALTERNATIVES

(1) It is difficult to forecast, especially about the future.
(2) He who lives by the crystal ball soon learns to eat ground glass.
(3) The moment you forecast, you know you're going to be wrong—you just don't know when and in which direction.
(4) If you're ever right, never let them forget it.

Edgard R. Fiedler

The focus of discussion in the middle chapters of this book has been the individual project, its economic definition and evaluation. It turns out, however, that many of the decisions which the chemical engineer must make as part of the everyday industrial work duties involve a choice among alternative projects. This choice can assume many different aspects, such as the choice among alternative items of equipment to accomplish a specific task, the choice among alternative processes to manufacture a specific product, even the choice among alternative ways to invest a specific amount of money. The economic optimization of technological systems can be considered an analysis of a large number of sequentially ordered alternatives. Many decisions demanded of the engineer do not appear to involve a choice among definable alternatives upon first consideration, but it is surprising how facilitated the decision-making process becomes when the alternatives are properly recognized in the first place, and then economically defined. Quite often, the decision as to whether to proceed with a certain course of action depends upon the proper consideration of the alternative, namely, to do nothing. It is, of course, necessary then to define the economic consequences of doing nothing.

In many cases of the analysis of alternatives, the key to the proper decision is to compute the appropriate set of *differential* quantities, for example, the differential investments and costs between two choices or the differential benefit when one alternative is chosen over the other. The economic decision then involves the appropriate combination of the differential quantities by means of essentially the same methods that have been described for the evaluation of individual projects. In the case of the analysis of multiple alternatives, the procedure is to compare the choices two at a time, so that the best of the alternatives survives each paired comparison to come out the "winner" in the end. Another key consideration in such a comparison is the requirement that the choices be *mutually exclusive;* when two alternatives are compared, we choose one or the other, but not both. We either decide to take a particular course of action, or we do nothing; we cannot have it both ways. There are, of course, occasions when the alternatives are not mutually exclusive, in which case the economic analysis is performed on each alternative with no reference to any of the others. We may have, for example, a series of projects, and we may ask if they meet some desired economic criterion. We may very well determine that all the projects are acceptable. But the moment that we ask which one of the projects is *best,* we are faced with the analysis of mutually exclusive alternatives; only one project can be best.

As we delve into the various kinds of alternatives which are subject to economic comparison, we will discover that the criteria of economic performance which we have applied to individual projects raise some peculiar difficulties. To obviate these difficulties, we will categorize the substance of the alternatives, and variants of the familiar profitability criteria will be constructed to fit each category best. We will find that it matters a great deal whether our analysis is addressed toward the choice of equipment alternatives on the one hand, or the choice of alternative ways to invest capital on the other; each category of alternatives demands its own analytical methodology. Initially the neophyte estimator may be taken aback by the uncertainty as to which method should be used under a given set of circumstances, but, in the long run, most engineers are fascinated and challenged by the many variants and nuances of economic choice subject to rational analysis. We will briefly examine subjects such as replacement analysis (dealing with the decision to replace worn or obsolete equipment and processes), the advisability of leasing specialized equipment, plant expansion and shutdown decisions, the optimum allocation of resources, and even the justification for investing money into research. It is clearly not possible to cover all the related subjects, but hopefully there is enough to establish a basic understanding of the techniques of the analysis of alternatives and to stimulate the interested chemical engineer to explore further.

10.1 THE ANALYSIS OF EQUIPMENT ALTERNATIVES

The Principle of Minimum Investment

Within the context of the chemical process industry, economic alternatives may be classified into two broad categories:

1 Two or more alternative investments *to do the same job*. The investments may involve different items of equipment, different equipment arrays, or even whole processes. The responsibility for the proper decision lies with the technologists and line managers.

2 Different ways to invest money (not necessarily to do the same job). The responsibility for proper decisions lies with the corporate staff management, but input from technologists may be solicited.

We will concern ourselves in the first instance with alternatives to do the same job, and we will concentrate specifically upon equipment alternatives. We wish to determine, for example, whether a specific filtration task is better accomplished with a vacuum filter or a centrifuge (assuming, of course, that either choice is not largely excluded by technological considerations). Or, we may wish to decide whether, for a particular pumping demand, it is better to use a short-life steel pump or an extended-life alloy pump.

Decisions of this sort are sometimes facilitated by calling upon a principle that verges upon the trivial, the principle of minimum investment:

> If alternatives are available for a specific job, all other things being equal, pick the alternative with the lowest investment

The key to the application of this principle is the words, "all other things being equal." If, in the case of the filtration alternatives, both choices meet the throughput demand, produce a cake of equal quality, incur equal costs of maintenance and labor, and are expected to last through the lifetime of the plant, then the thing to do is to buy the cheaper piece of equipment. Suppose, however, that in the case of the pump alternatives both pumps deliver the same performance at the same energy efficiency, but the alloy pump lasts longer than the steel pump. The alloy pump is likely to be the more expensive one, but we can no longer apply the principle of minimum investment and buy the steel pump; "all other things" are no longer equal, and the alternatives need further analysis.

Clearly, the choice of the more expensive equipment must be justified by some other economic benefit. If there is no such benefit, or if there is actually an economic penalty in addition to the higher cost, then the principle of minimum investment applies.

The Differential Return on Investment

The return on investment before taxes (ROIBT) has been defined as the percentage ratio of profit (before taxes) to total investment (CFS). The ROIBT criterion may be extended to make the proper choice between two or more investment opportunities. Let K_1 and K_2 represent the required capital investments associated with two such alternatives, and P_1 and P_2 the corresponding annual profits. If $K_2 > K_1$, clearly P_2 must be greater than P_1 for alternative 2 to be considered at all. How much greater must P_2 be? By computing $\Delta K = K_2 - K_1$ as well as ΔP, we may define a *differential return on investment*, ΔROIBT, the ratio of ΔP to ΔK. If alternative 2 is to be the preferred choice, ΔROIBT must exceed a minimum usually prescribed by corporate management.

The ΔROIBT criterion is a popular one for evaluating equipment alternatives for a specific job. For individual items of equipment, however, we must use somewhat restricted definitions of profit and investment. In fact, it is rather difficult to think in terms of a "profit" associated with an individual piece of equipment that is part of a large array that defines the real plant. Profit before taxes may be expressed as in Eq. (6.1.7),

$$P_B = \mathcal{R} - \text{COM} - G \tag{6.1.7}$$

Two equipment alternatives in general will not have any effect upon the revenues $\mathcal{R}$ from sales of the full plant's production (there may, of course, be exceptions), and they will hardly have much influence upon the general expenses G. Therefore

$$\Delta P = (P_B)_2 - (P_B)_1 = (\text{COM})_1 - (\text{COM})_2 \tag{10.1.1}$$

that is, the "differential profit" between two equipment items is the *saving* in the cost of manufacture assignable to the items. When that saving is computed, COM items common to both units that are compared need not be considered; for example, if the choice is between two pumps with different materials of construction, the required operating labor to keep either pump operational is quite likely to be the same, and operating labor cost components in ΔP cancel out. The most likely COM components that will be different are depreciation and other fixed charges which are directly proportional to the equipment cost, maintenance costs, and energy costs.

In generating ΔK, the differential investment between two equipment alternatives, consider only those investment items which are truly different. For example, two pump alternatives are hardly likely to have an effect upon the size and cost of the building that will house them. The usual approach is to compare the *installed* cost plus any other investment costs which do depend upon the choice (such as the cost of electric auxiliaries).

A commonly used criterion for accepting a more costly equipment alternative is

$$\boxed{\Delta ROIBT \geq 15\%}$$

This may seem to be a rather forgiving criterion for choosing the higher investment, but remember, a differential investment of ΔK immediately incurs an *increase* in fixed charges of some $0.2\Delta K$; to realize a sufficient net saving, the more expensive choice must result in a considerable saving in some other component of COM.

Example 10.1.1

Two equipment items, A or B, may be used in a plant under design to accomplish a particular unit operation. Using data in the table and the ΔROIBT criterion, recommend which alternative should be used.

Solution

	Equipment	
	A	B
Investment, $	10,000	15,000
Depreciation (10 yr), $	1,000	1,500
Other annual operating costs, $	3,000	1,500
Total costs, $	4,000	3,000
Δ investment, $		5000
Δ savings (=Δ profit), $		1000

$$\Delta \text{ROIBT} = \frac{100 \times 1000}{5000} = 20\%$$

This is more than the minimum criterion of 15 percent; therefore, recommend alternative B.

If several alternatives are available for a given job, the procedure calls for lining them up in order of increasing investment,

$$A, B, C, \ldots, N$$

and starting with the pair (A, B). If A is chosen (ΔROIBT < 15 percent), (A, C) is the next comparison, and so forth, until the last "survivor" is compared with N.

The most serious disadvantage of the ΔROIBT concept is that it does not properly account for possible differences in the service lives of the alternatives. This may be appreciated, for example, by considering the case of two pumps, one steel and the other an expensive alloy. Suppose the alloy pump is twice as expensive, and suppose, for the sake of argument, the total annual operating charges are the same for both. The ΔROIBT concept would reject the more expensive pump (Δ savings = 0). But suppose the alloy pump lasts five times as long. It then turns out that over a sufficiently long period of time it is the steel pump (and its replacements) that is the more expensive investment! A more rational ΔROIBT comparison might be performed on the basis of the sum total of investments for one or the other alternative over a fixed "study" period of time, but this is rarely done. The ΔROIBT method is almost always restricted to alternatives having the same service life. Comparison methods which account for the time value of money are usually applied to alternatives with different service lives.

Exercise 10.1.1

Perform a ΔROIBT analysis on the 10-year study period investments associated with the two types of pumps described in the text. Assume the stated relative costs include installation. Depreciate each pump over its service life by the straight-line method. Take other fixed charges as 15 percent of the installed cost of each pump per year; other operating costs (for instance, power) are comparable for the two pump types. Which pump do you recommend?

Capitalized Cost

In Chap. 2 the capitalized cost of an item of equipment was defined as the sum of money S which, when deposited at some prescribed interest rate i, would guarantee that enough could be withdrawn at required times to purchase replacements for worn-out items indefinitely. The capitalized cost is thus a form of perpetuity, illustrated schematically in Fig. 10.1.1. The magnitude of S in terms of the purchased cost W was given by Eq. (2.3.13),

$$S = W \frac{(1 + i)^m}{(1 + i)^m - 1} = W \mathcal{F}_K \qquad (2.3.13)$$

where

$$\mathcal{F}_K = \frac{(1 + i)^m}{(1 + i)^m - 1} \qquad (2.3.14)$$

is the *capitalized-cost factor*.

We also saw that the purchased cost W could be supplanted with the replacement cost C_R given by

$$C_R = W(1 + b - s_v) \qquad (2.3.15)$$

where b is an installation-cost factor (i.e., installation cost = bW), and s_v is the salvage factor. Equation (2.3.13) is then written as

$$S = W[\mathcal{F}_K(1 + b - s_v) + s_v] \qquad (2.3.16)$$

FIGURE 10.1.1
Schematic interpretation of capitalized cost.

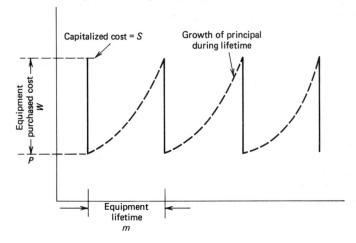

where the last s_v term is inserted to account for the fact that no salvage value is received the first time that the investment is made.

Equations (2.3.13) to (2.3.16) are written for the case of a true perpetuity; in Chap. 2, somewhat more complex equations were derived for the capitalized cost of items to be used in a plant with a finite life.

Now, the capitalized cost may be thought of as a measure of the cost of an item of equipment, which does account for the equipment lifetime as well as the time value of money. The concept therefore appears to be applicable to the evaluation of equipment alternatives with different replacement costs and different lifetimes; all else being equal, the alternative with the lowest capitalized cost should be the preferred one. Such comparison, however, does not account for any differences in the operating costs associated with the alternatives.

There are situations where differences in the associated operating costs may, indeed, be neglected, in which case a capitalized-cost comparison is valid. Admittedly, we saw in Chap. 7 that certain fixed charges are *projected* as a particular percentage of the plant DFC per annum. By extension, each item of equipment would be expected to share in those fixed charges in proportion to its contribution to the DFC. If alternatives are available for a given job, the costlier alternative will incur higher fixed charges. In practice, accounting procedures that are used to prorate complexwide fixed charges among the various operating plants are more subtle than a simple proportioning on the basis of DFC, and minor variations within a particular plant's hardware will not affect the assigned burden of fixed charges. It is hardly likely that the choice of rubber lining to extend the life of a storage tank will have much effect upon the plant's share in the operations of the employee relations department.

Therefore, if alternatives are available to accomplish a particular job, and if those alternatives have no obvious operating costs associated with them, a capitalized-cost comparison is often used. The method, however, is restricted to the case of alternatives with different service lives and a relatively minor impact upon the total plant DFC—for instance, alternative choices of materials of construction for a storage tank.

To use Eq. (2.3.13) for the computation of the capitalized cost S, the interest rate **i** must be properly established. Of the three interest rate classifications described in Chap. 1, which is to be used? Perhaps the prime interest rate? Indeed, if the amount P in Fig. 10.1.1 were deposited in a bank, a system of perpetuating the item of equipment would be established.[1] However, as has already been pointed out, nobody sets an amount of money aside in a bank just to perpetuate equipment; there is always the expectation that the principal can be invested at a higher rate of return than from a bank deposit. Should **i**, then, be the minimum acceptable rate of return that management has prescribed for the project? Again, this would imply that somehow only revenues generated by the project will be used to realize the capitalized cost. The best choice appears to be the after-tax "internal" rate of return i_0 on the totality of the corporation's current investments.

---

[1] Remember, however, that income tax must be paid on interest received from bank deposits.

Example 10.1.2

A high-temperature vapor-phase reactor must be lined with a heat- and chemical-resistant lining to protect the metallic reactor walls. Two kinds of linings with different service lives have been proposed; the costs include cost of installation:

Lining	Service life, years	Installed, cost, $
Graphite brick	4	23,000
Rigid glass foam	7	39,000

If $i_0 = 0.08$, which lining should be used?

SOLUTION

$$S_1(\text{graphite}) = 23{,}000 \,\frac{(1.08)^4}{(1.08)^4 - 1} = \$86{,}800$$

$$S_2(\text{glass}) = 39{,}000 \,\frac{(1.08)^7}{(1.08)^7 - 1} = \$93{,}600$$

The graphite brick is recommended.

Exercise 10.1.2

Is the *choice* based upon a capitalized-cost comparison dependent upon the value of **i** used for the comparison?

Note that differences in *depreciation* charges need not be of concern in making a capitalized-cost comparison. We have seen that depreciation charges constitute a way of recovering the capital investment; the capitalized-cost method is really doing the same thing. The method does have one disadvantage: the choice among alternatives does depend upon *inflation*. The expressions for the capitalized cost may be corrected for inflation [see Prob. 10.2; also Eq. (2.3.19)], but the choice must now be based upon the projection of a most uncertain quantity, indeed. The best way out of the inflation dilemma is to use a value of the internal rate of return i_0 that has already been corrected for the effects of inflation. That is, when i_0 is calculated from corporate after-tax earnings, those earnings must first be converted into constant-value (uninflated) dollars for some common time base. If such value of i_0 is now used to calculate the capitalized cost of an item, it is understood that corporate assets invested at that rate will generate the appropriate revenues expressed in constant-value, uninflated currency. In actuality, those revenues will be received in inflated currency, inflated to match the inflation of an item's purchase price, *no matter what the inflation rate may be*. The internal rate of return i_0 is used in this inflation-adjusted sense throughout this chapter.

The capitalized-cost method may, in fact, be extended to account for the associated operating expenses by including a capitalized cost of such annual expenses (Prob. 10.24). The procedure then becomes essentially analogous to the break-even analysis, a description of which follows.

Break-Even Analysis

The most generally applicable method for the analysis of equipment alternatives is break-even analysis. In this method, the *equivalent annual cost* of operating each alternative is computed. The alternative with the lowest equivalent annual cost is the preferred one. The method may also be used to determine the required relative service lives of two alternatives so that the annual cost will be the same—the two alternatives then "break even," hence the name. The method is applicable when alternatives have different projected lives and different associated operating costs.

The equivalent annual cost is the sum of the cost of recovering the capital and all other operating costs *except depreciation charges*. The cost of recovering the capital is taken as the annuity required to repay the capital over the service lifetime:

$$R = W \frac{i_0(1 + i_0)^m}{(1 + i_0)^m - 1} = W \mathcal{F}_R \qquad (10.1.2)$$

where the *capital-recovery factor* $\mathcal{F}_R$ is given by

$$\mathcal{F}_R = \frac{i_0(1 + i_0)^m}{(1 + i_0)^m - 1} \qquad (10.1.3)$$

The capital-recovery and capitalized-cost factors differ in the single term i_0. Again, the internal rate of return i_0 is used for computing the cost of recovering the capital, for much the same reason it was used in computing capitalized costs. The purchased cost W may be expanded to include installation and any other components of the DFC peculiar to the particular item of equipment. The salvage value S_v, however, should be accounted for by subtracting from the equivalent annual cost an equivalent annuity credit on S_v:

$$R' = S_v \frac{i_0}{(1 + i_0)^m - 1} \qquad (10.1.4)$$

Only those operating costs which are different for the two alternatives need be considered in computing the equivalent annual cost, as was the case in differential ROI calculations. Depreciation charges must not be included as part of the operating costs, since capital recovery costs are already accounted for.

A more subtle distinction, not often emphasized in the engineering economics literature, has to do with the relationship of the capital-recovery cost and other operating costs to income taxes. Normally, the internal rate of return i_0 on the totality of the corporation's investments is computed on the basis of the after-tax income. If $(i_0)_A$ (the A subscript emphasizes an after-tax situation) is to be used for the capital-recovery cost, it stands to reason that the operating costs should be corrected for the effect that income tax has upon them. From Eq. (6.1.11), the after-tax cash flow C_A is

$$C_A = (1 - t)(\mathcal{R} - E) + tD \qquad (6.1.11)$$

For items of equipment destined to do the same job and subjected to break-even analysis, the revenue $\mathcal{R}$ refers to any *savings* (such as value of heat saved) ascribable to a particular item, E are the operating expenses (excluding depreciation), and tD is the tax credit generated by depreciation charges. The equivalent *operating* cost after taxes is just the negative of the equipment-generated cash flow after taxes. Therefore, the total equivalent annual cost after taxes c_A is

$$c_A = (1 - t)(E - \mathcal{R}) - tD + R_A - R'_A \qquad (10.1.5)$$

where R_A and R'_A are computed from Eqs. (10.1.2) and (10.1.4) by using $(i_0)_A$.

A before-tax break-even analysis may also be performed by using a before-tax internal rate of return, $(i_0)_B$. The computation mechanics for this case are simpler:

$$c_B = E - \mathcal{R} + R_B - R'_B \qquad (10.1.6)$$

This is the format often seen in the engineering economics literature.

Example 10.1.3

A large pump and motor cost $12,000 installed and are expected to have a service life of 10 years with no salvage. It has been proposed that a variable-speed drive be used to control the pump discharge flow rather than a control valve; it is anticipated that an annual energy saving of $3200 can be realized. However, the inclusion of the variable-speed drive is expected to boost the installed cost to $17,000, although a $1000 control valve cost will be eliminated. The useful life of the combination is expected to be again 10 years; annual maintenance costs will increase by $1500, and no net salvage after disassembly costs is anticipated.

Neglecting minor fixed charge differences such as insurance costs, which of the alternatives should be selected? Use $(i_0)_A = 0.10$, $t = 0.5$.

SOLUTION For pump and motor alone plus control valve (10-year life),

$$R_1 = \frac{(12{,}000 + 1000)(0.10)(1.10)^{10}}{(1.10)^{10} - 1}$$
$$= \$2116$$

With variable-speed drive added,

$$R_2 = \frac{(17{,}000)(0.10)(1.10)^{10}}{(1.10)^{10} - 1}$$
$$= \$2767$$

Thus

$$\text{Annual cost for (1)} = 2116 + 0.5(3200) - 0.5(^{13{,}000}\!/_{10})$$
$$= \$3066$$
$$\text{Annual cost for (2)} = 2767 + 0.5(1500) - 0.5(^{17{,}000}\!/_{10})$$
$$= \$2667$$

The variable-speed drive should be installed.

10.2 THE ANALYSIS OF PROCESS AND INVESTMENT ALTERNATIVES

The Ranking of Mutually Exclusive Projects

We now turn our attention from the analysis of equipment alternatives to the analysis of major investment alternatives. These may take the form of different systems or processes to do the same job, or different ways to invest money. Again, the emphasis will be upon mutually exclusive choices; that is, we will seek the one best choice among alternatives.

In Chaps. 8 and 9 a variety of criteria of economic performances was introduced, and one possible method of evaluating alternative projects is to determine the value of one or another of such profitability criteria for all the alternatives and then to rank the choices. For instance, the payout time criterion might be chosen and calculated for each of several alternatives; the alternatives might then be ranked in order of increasing payout time, and the one with shortest payout time would be chosen as the best one. In Table 10.2.1, six hypothetical projects have been ranked in such a manner by using several different profitability criteria.

The resultant rankings emphasize an unfortunate problem: the choice of the best alternative depends upon the criterion used for the evaluation. Even if we are ready to eliminate methods with obvious shortcomings such as the payout time method, we are still faced with uncertainties as to the meaning and interpretation of the rankings. Suppose we decide that the DCRR method is the best for evaluating investment alternatives. $r = 0.56$ for project E, and $r = 0.46$ for project D; with equal investments, project E is clearly the better choice. Suppose project E, however, involved just twice the investment (\$200,000) and twice the cash flows; r would still be 0.56, but the question would be whether the additional \$100,000 investment over project D would be justified by a Δr of 0.10. Clearly, what is needed is a closer examination of the nature of each profitability method as applied to the analysis of alternatives, and the identification of the methods which do the best job of distinguishing alternatives under a variety of limiting conditions.

TABLE 10.2.1
SIX HYPOTHETICAL INVESTMENTS

		After-tax cash flow, $					
Project	Capital expenditure, $	1st year	2d year	3d year	4th year	Total	Average per year of life
A	100,000	100,000				100,000	100,000
B	100,000	50,000	50,000	50,000	50,000	200,000	50,000
C	100,000	20,000	40,000	60,000	80,000	200,000	50,000
D	100,000	80,000	60,000	40,000	20,000	200,000	50,000
E	100,000	80,000	70,000	50,000	40,000	240,000	60,000
F	100,000	40,000	50,000	80,000	90,000	260,000	65,000

	Comparison of rankings							
Rank	Return on original investment	Return on average investment	Payout	Payout with interest	Present worth			Discounted cash rate of return
					6%	30%	40%	
1	F	F	A	E	F	E	E	E
2	E	E	E	D	E	F	D	D
3	BCD	BCD	D	F	D	D	F	F
4			B	B	B	B	B	B
5			F	C	C	C	C	C
6	A	A	C	A	A	A	A	A

Exercise 10.2.1

Rank the projects in Table 10.2.1 by using the DCRR criterion without actually calculating the DCRR; that is, see if you can justify the displayed ranking "by inspection."

Differential Profitability Criteria

By analogy to the ΔROIBT concept used in the analysis of equipment alternatives, differential profitability criteria may be used for making the best choice among investment alternatives. In conformity to the two principal methods of judging profitability, two differential methods are commonly used:

Differential ROIBT
Differential DCRR

Projects are more thoroughly defined and limited by economic parameters than are individual items of equipment, and the differential evaluation must first be preceded by the economic evaluation of the individual projects. That is, if the ΔROIBT concept is to be used for the evaluation of two or more alternatives, each alternative must first be evaluated separately to establish its ROIBT, and that ROIBT must exceed the minimum prescribed by management for that type of project (as shown, for example, in Table 8.2.1). Only then can each project enter the ΔROIBT competition, to be compared with other projects which have made the grade on the basis of still another numerical criterion, a minimum ΔROIBT criterion, also prescribed by management, and not necessarily the same as the minimum criterion for individual projects.

For two investment alternatives, the differential ROIBT may be written as

$$\Delta \text{ROIBT} = \frac{(P_B)_2 - (P_B)_1}{(\text{CFS})_2 - (\text{CFS})_1} \times 100 \qquad (10.2.1)$$

It is, of course, implied that if $(\text{CFS})_2 > (\text{CFS})_1$, then $(P_B)_2 > (P_B)_1$; otherwise, the principle of minimum investment applies. For process alternatives, the minimum acceptable ΔROIBT criterion is usually considerably above the 15 percent which is acceptable for equipment alternatives, since the risk associated with choosing the more expensive alternative applies to a much higher investment level.

Example 10.2.1

Two processes A and B are available to produce a desired compound. The parameters of the two alternatives are given in the table. Either process may be considered to be a low-risk industrial chemical project (see Table 8.2.1). Use the ΔROIBT concept to decide which process should be used.

	Process A	Process B
Capital for sale, $	1.2×10^6	2.0×10^6
Annual profit before taxes, $	0.24×10^6	0.36×10^6

SOLUTION

$$(\text{ROIBT})_A = \frac{100 \times 0.24 \times 10^6}{1.2 \times 10^6} = 20\%$$

$$(\text{ROIBT})_B = 18\%$$

That is, both processes meet the minimum ROIBT criterion for low-risk industrial chemical processes, and they may therefore be compared. From Eq. (10.2.1),

$$\Delta \text{ROIBT} = 100 \times \frac{(0.36 - 0.24)10^6}{(2.0 - 1.2)10^6} = 15\%$$

For process alternatives, this is marginal; it is best to choose process A.

For projects having varying annual profit, a time-averaged $\tilde{P}_B$ may be used in Eq. (10.2.1).

A differential DCRR may be defined for two alternatives, each characterized by a constant annual cash flow (*after* taxes). The differential cash flow ΔC_A is then the same for each year of the life of the alternatives (the same life is assumed for both choices); the ΔC_A may then be thought of as an annuity on the differential total investment ΔK. If the recoverable capital cash flow is neglected for the moment, by analogy to Eq. (2.2.21),

$$\Delta K = \Delta C_A \frac{(1 + r_D)^n - 1}{r_D (1 + r_D)^n} \qquad (10.2.2)$$

where r_D is the differential DCRR.

Example 10.2.2

Corporate management insists that acceptable projects must meet a minimum DCRR criterion of 14 percent. Evaluate the two tabulated projects by using the differential DCRR criterion and pick the better one.

	Project A	Project B
Total investment, $	180,000	110,000
Continuous annual cash flow after taxes, $	48,888	26,240
Life, years	10	10

SOLUTION   For the two projects, $(r)_A = 24$ percent, $(r)_B = 20$ percent; i.e., both projects satisfy the minimum DCRR criterion.

$$\Delta K = K_A - K_B = \$70,000$$
$$\Delta C_A = (C_A)_A - (C_A)_B = \$22,648$$

Substituting into (10.2.2) and solving for r_D,

$$r_D = 30.0\%$$

A minimum acceptable differential DCRR has not been stipulated, but a value of 30 percent is excellent. Choose the more expensive project A.

The method illustrated by Example 10.2.2 is rather restricted in scope, since it does not account for the more general case of varying annual cash flows, recoverable capital, time-distributed investments, and other complications. Nevertheless, the differential DCRR concept may be adapted to account for such complications. The first step is to find r by trial and error for each alternative, using equations analogous to Eq. (9.3.9). Let us say that these are r_1 and r_2 for alternatives 1 and 2, respectively. Next, a "phantom" equal annual cash flow, C_{AE}, is calculated for each alternative on the basis of the total investment K for that alternative and the annuity equation format, Eq. (2.2.21). For example, for alternative 1,

$$K_1 = (C_{AE})_1 \frac{(1 + r_1)^n - 1}{r_1 (1 + r_1)^n} \tag{10.2.3}$$

K_1 is the present worth of all investments, including time-distributed investments, discounted at the rate i_0; r_1 is the DCRR determined for investment 1 alone.

Finally, r_D is calculated from Eq. (10.2.2) by setting $\Delta C_A = |(C_{AE})_1 - (C_{AE})_2|$.

Both differential methods are relatively independent of the effects of inflation (except for the usual small inconsistencies, such as the fact that depreciation charges are not inflated). However, both methods suffer from a serious disadvantage; they cannot be used to compare alternatives with different lifetimes. This may be illustrated by considering the trivial ΔROIBT case of two alternatives having the same total investment K and the same annual profit before taxes P_B, but different lifetimes. The ΔROIBT concept cannot distinguish between the two alternatives, and yet the one with the longer lifetime is clearly preferable. The differential DCRR method also breaks down for the case of equal K and equal C_A. For unequal lifetimes, comparison of present worth ratios is the best approach to the proper analysis of investment alternatives.

Exercise 10.2.2

If two alternatives have equal K and equal C_A but unequal lifetimes, which one will have the higher DCRR? Can it be concluded that the alternative with the higher DCRR is the better one?

Present Worth Ratio Analysis*

Possibly the best approach to the analysis of investment alternatives is the formulation of an incremental present worth ratio (ΔPWR) for pairs of alternatives by using a rate of return stipulated by corporate management as the desirable one for the kinds of projects under consideration. The alternatives are ranked in increasing order of investment and evaluated in pairs; the "winner" of each paired evaluation is then compared with the next ranked alternative. For two projects A and B and investment $K_B > K_A$,

$$\Delta\text{PWR} = \frac{(\text{PW})_B - (\text{PW})_A}{K_B - K_A} \tag{10.2.4}$$

where PW is the present worth of all positive cash flows for each project, computed at the stipulated discount rate.

For any paired comparison, if

$$\Delta\text{PWR} > 1.0$$

*Sometimes called *benefit-cost ratio analysis*.

then the project with the larger investment is the better one; a ratio in excess of 1.0 implies that the benefit (incremental PW of cash flows) exceeds the cost (incremental investment).

Example 10.2.3

Consider the two projects A and B with the following parameters;

	Project A	Project B
Investment, $ × 10^6	1.0	2.0
PW of all positive cash flows, $ × 10^6	1.3	2.4

The values of NPW and the present worth ratios are readily determined:

NPW, $ × 10^6	0.3	0.4
Present worth ratio (PWR)	1.3	1.2

Both projects are acceptable—the PWs match and exceed the minimum at the stipulated rate (whatever it may be in this case).

$$\Delta \text{PWR} = \frac{2.4 - 1.3}{2.0 - 1.0} = 1.1 > 1.0$$

Therefore project B is the better one.

Note that the parameters defining the projects in Example 10.2.3 say nothing about the lifetime of the projects; they may be the same, or they may be different. It is a particular advantage of present worth ratio analysis that it can accommodate alternatives with different lifetimes, as illustrated by the following rather extreme example.

Example 10.2.4

Evaluate the relative merit of projects I and II by computing the present worth ratios at 10 percent discrete rate of return.

	Project I	Project II
Investment, $ × 10^6	10.0	10.0
Annual positive cash flow, $ × 10^6	2.0	2.0
Lifetime of project, years	7	10
PW of positive cash flows, $ × 10^6	9.70	12.28
NPW, $ × 10^6	−0.30	2.28
Present worth ratio	0.970	1.228

SECTION 10.2: THE ANALYSIS OF PROCESS AND INVESTMENT ALTERNATIVES

$$\Delta \text{PWR} = \frac{12.28 - 9.70}{2.0 - 2.0} = \infty > 1.0$$

In the example, project II is clearly the better one; after all, the cash flows keep coming for an extra 3 years with no extra investment. The incremental present worth ratio reflects this conclusion, yet differential ROIBT analysis in particular could not handle this situation readily.

One thing must be considered if, as a result of PWR analysis of mutually exclusive alternatives, a higher investment is chosen, and that is whether the corporation has the additional money to invest. If the money is not available, then the lower cost alternative must, of course, be chosen provided its own PWR > 1.0; it is not the best choice, but it is "good enough." Even if the money is available, the question always arises as to whether there is some other investment opportunity which would result in an even better ΔPWR on the incremental capital. The choice is a matter of appropriate capital budgeting.

Table 10.2.2 is a selection chart for the various methods that may be used for the analysis of equipment and process (or investment) alternatives that are mutually exclusive.

Capital Budgeting

The exploration of the subject of the analysis of alternatives has so far focused upon mutually exclusive alternatives. Situations do arise, however, when one is asked to make the best selection among *nonexclusive* alternatives. The corporation, or a corporate department (such as R&D), may have a certain amount of investment capital available, and the question is how to ration that capital among an array of possible investment opportunities in the best way. Capital budgeting problems of this kind are rarely within the province of the project evaluator, but some familiarity with the problems and their solutions is beneficial.

Suppose the problem is formulated as one involving a fixed available capital K_F that is to be rationed out among n projects, each involving a capital investment K_j. We seek m projects among the n available so that

$$\sum_{1}^{m} K_j = K_F$$

The m projects are to be selected to maximize the cash return on K_F. The first step is to rank the projects in order of economic attractiveness. How is this to be done? We saw in Table 10.2.1 that the ranking depends upon the method chosen for the evaluation of profitability. The most reasonable approach would appear to be a ranking according to each project's PWR, the present worth ratio. This ratio tells us essentially how many dollars each invested dollar generates on a present worth basis; the more dollars generated per dollar invested, the better. The dimensionless PWR accommodates variations in investment size and project lifetime. Moreover,

TABLE 10.2.2
SELECTION CHART FOR THE ANALYSIS OF ALTERNATIVES

Equipment alternatives		Process/Investment alternatives	
Boundary conditions	Method	Boundary conditions	Method
Equal lifetimes only Operating cost savings No time value of money	Differential ROIBT	Equal lifetimes only No time value of money	Differential ROIBT
Different (or equal) lives No associated operating costs Time value of money	Capitalized cost	Equal lifetimes only Time value of money	Differential DCRR
Different (or equal) lives Operating cost savings (or none) Time value of money	Break-even analysis* (equivalent annual cost)	Different (or equal) lives Time value of money	Present worth ratio* analysis (benefit-cost ratio)

*Preferred methods.

the rate of return **i** (or *i*) used to compute the PWR may vary from project to project, depending upon management's perception of the risks involved in each.

The projects are now arranged into a column array of diminishing PWR:

	Project	PWR	K_j	Cumulative K_j
	1	$(PWR)_1$	K_1	K_1
	2	$(PWR)_2 < (PWR)_1$	K_2	$K_1 + K_2$
	3	$(PWR)_3 < (PWR)_2$	K_3	$K_1 + K_2 + K_3$
	⋮	⋮	⋮	⋮
Cutoff →	m	$(PWR)_m < (PWR)_{m-1}$	K_m	K_F
	⋮	⋮	⋮	
	n	$(PWR)_n < (PWR)_{n-1}$	K_n	

At the cutoff point, the cumulative K_j just match K_F. The cutoff defines the optimum project selection, provided, of course, that $(PWR)_m \geq 1.0$.

In practice, it may turn out that this simple ranking procedure precludes a precise match between ΣK_j and K_F. Adjustments may be made in several ways. For instance, some projects may consist of two or more mutually exclusive alternatives; judicious juggling of the alternatives may lead to a close match (see, for instance, Newnan, 1976). The sum total of attractive investment opportunities may not match the total available capital; in such a case, the financial officers of the corporation must be on the lookout for external opportunities to "plug the gap," perhaps even to displace some of the lower-ranked internal projects. Corporations often have a substantial portion of their capital invested in securities, for example.

Equipment Leasing

An occasionally encountered exercise in the analysis of alternatives involves the decision whether to buy or lease a particular item of equipment. In the chemical process industry this kind of choice is associated with highly specialized equipment faced with early obsolescence because of unusually rapid technological developments. Examples are control computers, advanced analytical instruments, and copying machines. The comparison here is not between equipment alternatives but between *alternative methods of financing the investment* associated with a particular item of equipment.

The criterion for making the decision is based on the computed value of the after-tax cost of leasing, expressed as an effective interest rate. This is then compared with the after-tax prime borrowing rate; if the leasing interest rate is higher, then the equipment is purchased. True, we do not know a priori whether the equipment will be purchased with borrowed funds or from retained earnings, but this does not matter. We make the comparison as if the purchase funds *were* borrowed; i.e., we compare alternative methods of financing. Leasing has the advantage that the rental payments are tax-deductible operating expenses; on the other hand, purchasing results in a tax advantage on depreciation charges. The following simple example illustrates how these effects are balanced out.

Example 10.2.5

A Van de Graaff generator may be purchased for $100,000. It is expected to have a useful life of 5 years and no salvage value because of obsolescence. The machine may also be leased for $40,000/year for 5 years. Should the machine be purchased or leased? Use straight-line depreciation and a (before-tax) prime rate of 20 percent; the tax rate is 50 percent.

SOLUTION During each year the after-tax cost of leasing is $20,000. If the equipment is purchased, the annual depreciation tax credit is $10,000. Leasing therefore costs annually $30,000. The after-tax cost of leasing may be calculated from the expression

$$P = R\frac{(1 + i)^n - 1}{i(1 + i)^n}$$

or

$$100{,}000 = 30{,}000\frac{(1 + i)^5 - 1}{i(1 + i)^5}$$

whence, by trial and error, $i = 15.2$ percent.

After-tax borrowing rate = $0.5 \times 20 = 10.0$ percent

In this case, then, the equipment should be purchased.

Exercise 10.2.3

Suppose in Example 10.2.5 the item under scrutiny were a plot of land. What would be your recommended strategy in that case?

Leasing may have other benefits which must be properly accounted for in analyzing the alternatives:

1 Leasing agreements may include maintenance contracts which guarantee repair and routine maintenance by the leasing institution's technical staff, up to some maximum number of annual hours. The annual savings engendered by such an agreement must be evaluated.

2 The leasing institution is often the manufacturer who may be using leasing agreements as a promotional device. Often the rent may be based upon an equipment cost that is well below list price; in effect, the cost of leasing is forced down closer to the cost of borrowing, which is based upon the list price. At other times, the manufacturer may be willing to give liberal "trade-ins" for existing obsolete equipment that is owned.

3 The rental agreement may have a clause which allows rental payments to be applied toward the eventual purchase of the equipment.

4 Leasing has the advantage that obsolete or unsatisfactory equipment can be returned. However, even if equipment is owned, there are ways of obtaining tax credits if the item is scrapped due to obsolescence; this point will be further explored as part of the discussion of replacement analysis (Sec. 10.4).

One benefit of ownership is that the equipment may have an appreciable salvage value. If equipment is leased, its salvage value accrues to the leasing institution.

Further detail on leasing decisions is given by Kroeger (1965).

Economic Justification of Research Investments

The methodology of the analysis of alternatives is a powerful one, and mathematically it is not at all difficult. The difficulty usually arises in defining the exact nature of the alternatives—in fact, in perceiving that a particular question or problem can be resolved by recourse to the analysis of alternatives. There is no systematic way of classifying, setting up, and solving such problems; the keys to success are intuition and experience. The economic justification of research investments will be used as an example of the general approach.

In the chemical industry, the only justification for research investments is the expectation of increasing corporate profits. In the case of offensive research, the end goal is usually a plant that will produce the new product or incorporate the new process. The question then arises as to how much new *capital investment opportunity* each dollar invested into research should generate. By capital investment opportunity we mean the number of dollars that actually *will* be invested in new plants that then, in turn, will generate enough profit to more than justify the research dollar investment.

Let us assume that K_R is an annual research investment and that ωK_R is the amount of new capital investment opportunity which K_R must generate. The factor ω is thus the value we seek—dollars of new capital per dollar of research. The evaluation of ω is a matter of analysis of two alternatives:

1 Do not invest in research. Invest $K_R + \omega K_R$ in established corporate facilities and investment opportunities at the corporation's internal rate of return i_0 over n years.

2 Invest in research. Invest K_R in research (which results in a tax credit for the year in which the deductible research investment was made). Invest ωK_R in a plant which will yield cash flows (over n years) that represent a rate of return of i on ωK_R.

Refer now to Fig. 10.2.1, in which the two alternatives are represented symbolically by the bar charts. For alternative 1, the total investment $(1 + \omega)K_R$ generates the total cash flow $(\Sigma C_A)_1$. For the sake of simplicity, we assume that the cash flows are the same for each of the n years. The investment generates cash flows at a rate of return of i_0; therefore, the PW of the cash flows, discounted at the rate i_0, is precisely $(1 + \omega)K_R$, and the NPW of alternative 1 is zero.

For alternative 2 to be at least as good as alternative 1 (break-even), its NPW must also be zero (or higher). Note, however, that the research investment K_R generates no cash flow, except the tax credit $0.5K_R$ (at a 50 percent tax rate). Therefore, the balance of the investment, ωK_R, must generate cash flows at a rate

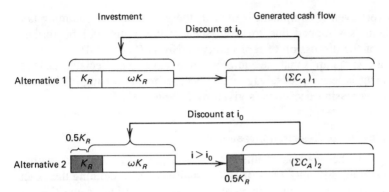

FIGURE 10.2.1
Research investment alternatives.

of return of $i > i_0$ so that the NPW (discounted again at i_0) will still be zero. One feels intuitively that the required value of ω will depend upon the value of i, and the mathematical treatment confirms this.

For alternative 1, then, NPW = 0. For alternative 2, assume again that ωK_R generates n equal annual cash flows $(C_A)_2$. Then, from Eq. (2.3.4),

$$(C_A)_2 = \omega K_R \mathscr{F}_R \qquad (10.2.5)$$

where the capital-recovery factor $\mathscr{F}_R$ [Eq. (10.1.3)] is

$$\mathscr{F}_R = \frac{i(1+i)^n}{(1+i)^n - 1} \qquad (10.2.6)$$

The PW of $(\Sigma C_A)_2$, discounted at i_0, is

$$\omega K_R \frac{\mathscr{F}_R}{(\mathscr{F}_R)_0}$$

where $(\mathscr{F}_R)_0$ is the capital-recovery factor computed with i_0 and n.

The total PW of cash flows is therefore

$$0.5 K_R + \omega K_R \frac{\mathscr{F}_R}{(\mathscr{F}_R)_0}$$

and this must equal the investment $(1 + \omega)K_R$ for NPW to be zero:

$$0.5 K_R + \omega K_R \frac{\mathscr{F}_R}{(\mathscr{F}_R)_0} = (1 + \omega)K_R$$

whence

$$\omega = \frac{(\mathscr{F}_R)_0}{2[\mathscr{F}_R - (\mathscr{F}_R)_0]} \qquad (10.2.7)$$

For the chemical industry, typical values of the parameters are

$$n = 10 \text{ years} \qquad i_0 = 0.10$$

With these values, Eq. (10.2.7) may be used to develop a relationship between ω and i; this is plotted in Fig. 10.2.2. The graph tells us, for example, that each dollar of R&D should generate approximately $1 of new plant construction that will generate cash flows at a 20 percent rate.

The discussion so far has incorporated the assumption that the research effort generates the new production facility instantaneously, with zero "lead time." This, of course, is not true; in Chap. 1 we learned that successful projects may take 6 years or more to reach fruition at the plant start-up stage. Let us suppose that the lead time between research and the first generation of cash flows is m years. Then the PW of the investment at the time of plant start-up is, for either alternative,

FIGURE 10.2.2
Capital investment opportunity generated by research (zero lead time).

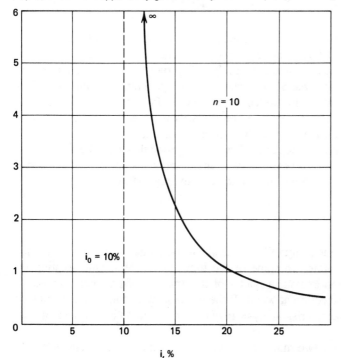

$$K_R (1 + i_0)^m + \omega K_R$$

For alternative 1, the portion K_R generates cash flows at the rate i_0 from year $-m$ to year $n - m$; ωK_R generates cash flows from time zero to year n. A little reflection will convince the reader that the PW of the generated cash flow, discounted (and compounded) at i_0, is just equal to the PW of the investment, and, as before, the NPW of alternative 1 is zero.

Again, therefore, the NPW of alternative 2 must be zero or better. $(C_A)_2$ again equals $\omega K_R \mathscr{F}_R$, and the PW of all cash flows is therefore

$$0.5 K_R (1 + i_0)^m + \omega K_R \frac{\mathscr{F}_R}{(\mathscr{F}_R)_0}$$

For zero NPW, the PW of the cash flow equals the PW of the investment:

$$0.5 K_R (1 + i_0)^m + \omega K_R \frac{\mathscr{F}_R}{(\mathscr{F}_R)_0} = K_R (1 + i_0)^m + \omega K_R$$

whence

$$\omega = \frac{(\mathscr{F}_R)_0 (1 + i_0)^m}{2[\mathscr{F}_R - (\mathscr{F}_R)_0]} \tag{10.2.8}$$

The graph in Fig. 10.2.3 was developed with the aid of Eq. (10.2.8). If the average lead time of relatively modest projects is taken as 3 years, and 15 to 17 percent interest is taken as characteristic of "good" projects, then the rule of thumb expounded in Chap. 1 follows: $1 of research should generate a minimum of $2½ of production investment, and preferably much more. For projects with very long lead times, the research effort must be extraordinarily productive, or the anticipated return on investment must be extraordinarily high, or both.

Another rule of thumb mentioned in Chap. 1 suggested that TS&D efforts should generate, over a period of 10 years, $1/year increased sales per dollar invested in the TS&D effort. Proof of this rule is left as an exercise (Prob. 10.20).

Decision Trees

In Chap. 9 we saw that the profitability analysis of individual projects was clouded by uncertainties incorporated in some of the input parameters. One solution involved the quantification of risk, a matter of the analysis of past performance at best, or a subjective projection at worst. If the probability of future events may be quantified in such manner, a risk analysis may be performed on investment alternatives.

Suppose there exist the two mutually exclusive investment alternatives A and B, and suppose, in one way or another, the probabilities of success of either alterna-

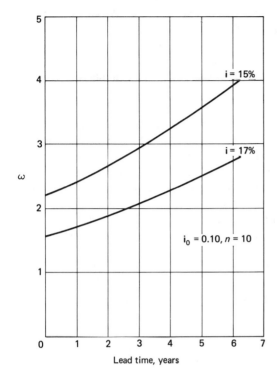

FIGURE 10.2.3
Effect of lead time upon required research performance.

tive, p_A and p_B, can be obtained. The probabilities of failure are then given by

$$q_A = 1 - p_A \qquad q_B = 1 - p_B$$

It is now necessary to compute the economic consequence of each success and failure by using the various methods we have already looked at. Of course, we have to define what we mean by "success" and "failure," but once these are defined, then we can calculate the economic consequence in terms of, say, the NPW. Then, for each project, it is possible to calculate the *expected value* of the project's NPW, just as we did in arriving at Eq. (9.5.8): thus

$$(\text{NPW})_A = (\text{NPW})_{A,S} p_A + (\text{NPW})_{A,F} q_A \qquad (10.2.9)$$

where $(\text{NPW})_{A,S}$, for example, refers to the net present worth of project A, if successful, $(\text{NPW})_{A,F}$ if not. A similar expression holds for $(\text{NPW})_B$. Now, we have no way of controlling the destiny of either choice—we cannot be sure whether either project will, in fact, be successful. We do, however, have full control over the initial choice of project, and the criterion of choice is the expected value: if $(\text{NPW})_A > (\text{NPW})_B$, then project A is the choice. Note that the expected value is not the most likely value; the latter is the one with the highest probability of occurring.

Even a specific project may actually involve two or more investment alternatives; any project which is at the stage where a decision to build must be taken involves the two alternatives, to build or not to build. A decision among alternatives of this sort may be facilitated by constructing a *decision tree*. To do so, however, we must be ready to project the probabilities of each defined event that results from each controlled decision.

A simple example of a decision tree is shown in Fig. 10.2.4. Two alternatives are involved—to build a plant at a cost of $10 million, or not to build a plant, at a cost of nothing. In the diagram, the *decision point* is symbolized by the triangle; the time frame of this first decision point is *now*, and we have full control over the decision. The two options (and the associated investments) are shown as two branches pointing to *chance event points*. For the sake of simplicity, we show only two consequences emanating from each chance event point—a good market or a poor market for the product. Probabilities have been assigned to each consequence, and the present worth of the net cash flows generated by each consequence have been calculated. If no plant is built, the PW is, of course, zero.

The expected value of each option is now calculated as in Eq. (10.2.9). Thus

$$(NPW)_{build} = (15.0 - 10.0)0.67 + (3.0 - 10.0)0.33$$
$$= 1.0*$$

Clearly,

$$(NPW)_{don't\ build} = 0$$

and the proper decision between the two alternatives is to build.

*If desired, this can be computed as $15.0 \times 0.67 + 3.0 \times 0.33 - 10.0$.

FIGURE 10.2.4
A simple decision tree.

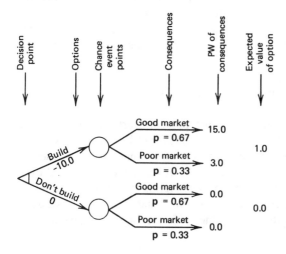

Exercise 10.2.4

Suppose in Fig. 10.2.4 the chances of a good market were no better than even. Would this affect the final decision?

A more complex but particularly important example is taken from the booklet of Allen (1972) (Fig. 10.2.5). The decision tree network shown involves the construction of a semiplant and, 2 years later, a full-scale plant. The semiplant test market will be used as a guide for making the main plant decision, a common practice in product or process development.

We are at decision point 1, and we must decide whether to build a large (L), medium (M), or small (S) semiplant, with the indicated investments. The first set of chance event points generates test market consequences, each marked with the PW of the net cash flow and an estimated probability of occurrence. Thus, as a consequence of building a large semiplant, three different test markets are possible: large, medium, and small. Note, however, that a medium semiplant cannot result in a large test market, and therefore only two consequences are possible. The small semiplant can satisfy a small test market only. Two years after the semiplant has been built, the decision must be made as to what size of main plant to build: again, small, medium, or large. There are six such decision points (2 to 7); each decision point thus leads to three chance event points which, in turn, lead to several consequences, namely, the size of the main market (small, medium, or large). Of course, a small plant, as before, can again result in only one consequence, a small main market.

The PW (at $t = 2$ years) of the net cash flows generated by each size of main market is shown. For example, for the M Market, PW = 600 regardless of previous decisions (a medium market is a medium market, and that is how much it is worth). However, the estimated probability of the M market is very much dependent upon previous decisions, a consequence of the *correlation between events* (Rose, 1976). For example,

For L semiplant, L test market, M plant, $p_M = 0.9$

For L semiplant, S test market, M plant, $p_M = 0.35$

In the event of an S test market, the indication would be that demand for the product was not large, and therefore the probability of even an M plant selling its full output would be drastically reduced. Just how much the probability should be reduced is still, for the most part, a matter of subjective judgment; clearly, some care and logic must be used in estimating the probabilities. Note that if an S plant is built, it can satisfy an S market only, and p_S must perforce be 1.0.

To establish the optimum choice at decision point 1, we use the so-called *rollback technique*. First, the expected value of NPW of *each controlled decision* at the second set of decision points is established. Consider decision point 2. For the large plant option,

$$(\text{NPW})_{2,L} = (1000 \times 0.6) + (600 \times 0.3) + (400 \times 0.1) - 560 = 260$$

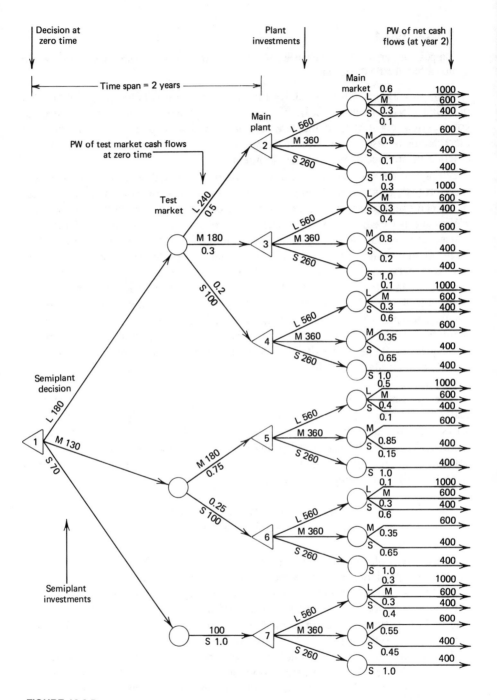

FIGURE 10.2.5
Decision tree network, semiplant operation. S-small, M-medium, L-large. *(Taken from Allen 1972, by permission of the Institution of Chemical Engineers, London.)*

Similarly,

$$(NPW)_{2,M} = 220 \quad \text{and} \quad (NPW)_{2,S} = 140$$

The "position value" of decision point 2 is the largest of these values—260, equivalent to the choice of an L plant.

Exercise 10.2.5

What is the position value of decision point 5?

Once all intermediate decision point position values have been established, we can roll back these values to the initial decision point, 2 years back (10 percent discrete discounting is used in this example). With $(NPW)_2 = 260$, $(NPW)_3 = 200$, and $(NPW)_4 = 140$, we have

$$\begin{aligned}(NPW)_{1,L} &= [260\,(1 + 0.10)^{-2} + 240]0.5 \\ &\quad + [200\,(1 + 0.10)^{-2} + 180]0.3 \\ &\quad + [140\,(1 + 0.10)^{-2} + 100]0.2 - 180 \\ &= 194\end{aligned}$$

Similarly,

$$(NPW)_{1,M} = 195 \quad \text{and} \quad (NPW))_{1,S} = 154$$

Therefore, since $(NWP)_{1,M}$ is the largest, the conclusion is to *build the medium semiplant*. Admittedly, the L semiplant case is, in this instance, so close that other factors might influence the choice of an L semiplant.

The method of decision trees not only tells us what the best decision is now; for the case of two or more sets of decision points in time, it points out an *optimum strategy*. For the system of Fig. 10.2.5, the optimum strategy is, to start with, to build a medium semiplant. *If* the test market turns out to be M (decision point 5), choose the option with the largest expected value of NPW; in this case, this turns out to be the L plant. *If* the test market turns out to be S (decision point 6), the best option turns out to be an S plant.

It is important to remember, however, that no strategy is frozen in time. Decision networks such as decision trees (and others to be outlined in Chap. 11) must be constantly revised to accommodate new information. For instance, with the M semiplant, an M test market might be prevalent, but perhaps weaknesses might be detected which would then lead to a modification of the main market probabilities and, consequently, a change in the numerical order of the present worth of the options. Rose (1976) shows how a sales forecast may be used to modify the strategy obtained from a decision tree.

Suppose in the scheme of Fig. 10.2.5 we decide to analyze the test market and financially support a sales forecast before making the final main plant decision. The network will then be further expanded in a manner indicated in Fig. 10.2.6. Only a small portion of the expanded network is shown. The first chance event point represents the terminus of the L decision emanating from decision point 1; we now consider the case of the L test market. Instead of proceeding to the main plant decision point 2, we interject a decision point X, with the two decisions of making a forecast or not. If no forecast is made, the network branch continues on to decision point 2, with options as in Fig. 10.2.5. If the decision is made to make a forecast (with an additional investment of -100), then three possible consequences result: forecasts of an L market, an M market, or an S market. With forecast in hand, we now make the plant decision. For example, if the forecast is an L market, we reach a decision point A with three options; the L plant option leads to three consequences with the probabilities p_L, p_M, and p_S. How can these probabilities be estimated?

Let us say that past experience has taught us that, with three possibilities (S, M, and L), the sales forecasts predict the correct market 80 percent of the time; the remaining 20 percent is split evenly between the wrong predictions. Now, at decision point 2, the probability of an L market was 0.6 (see Fig. 10.2.5). Therefore, the probability of an L *forecast and* an L market is

$$0.6 \times 0.8 = 0.48$$

The probability of an M market was 0.3, and the probability of an L forecast and

FIGURE 10.2.6
Expansion of decision tree network to accommodate sales forecast.

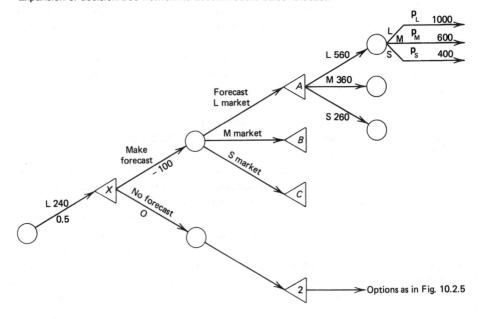

an M market is

$$0.3 \times 0.1 = 0.03$$

The probability of an L forecast and an S market is

$$0.1 \times 0.1 = 0.01$$

The *total* probability of an L forecast is therefore

$$0.48 + 0.03 + 0.01 = 0.52$$

The probability that, *given* an L forecast, the market will be large is then

$$\mathbf{p}_L = 0.48/0.52 = 0.92$$

Similarly,

$$\mathbf{p}_M = 0.03/0.52 = 0.06$$

Exercise 10.2.6

What are the probabilities of the M option consequences at decision point A? What is the position value for decision point A?

It is obvious that decision trees can rapidly involve huge amounts of calculation, and computer-aided solutions are required. The preponderance of computation involves the present worth of cash flows generated by each of the many paths; an appropriate algorithm is needed for computer-based solutions.

10.3 ECONOMIC OPTIMIZATION

Economic Optimization in Plant Design

Economic optimization may be thought of as the analysis of a large number of sequentially ordered alternatives. If the economic *objective function* (for instance, NPW) is continuous, then, of course, the number of alternatives to choose from is infinite. Many chemical engineering operations consist of discrete stages; for such systems the number of alternatives is finite, provided that the discrete objective function is bounded.

In most cases economic optimization is used for the purposes of designing an item of equipment or a system of interacting items of equipment to satisfy some most desirable criterion for the objective function—perhaps the lowest investment, or the highest NPW. Economic optimization in this sense lies within the province of plant design. In Fig. 1.3.1, for instance, the criterion of maximum *net income* was used to specify the optimum thickness of heat insulation. Economic analyses of this kind are part and parcel of routinely used design procedures; economic considerations are inseparable from applied chemical engineering methodology. A limited number of such economic optimization problems is subject to an analytic solution, and several of these solutions are found in standard chemical engineering texts and Perry's handbook.

Perhaps the best known of the lot is the problem of specifying the optimum steel pipe diameter to accommodate a given flow rate of a specified fluid. The problem has been solved and updated a number of times, and convenient nomographs have been published (Perry, for example) that may be used for quick preliminary piping design (Chap. 5). The graphs must be used with a certain degree of caution, however, and the user should understand their basis. Some, for example, do not account for return on investment (or the capital recovery cost) represented by the installed pipeline; the cost of pumps may be ignored; the cost basis for the pipe, and electric energy in particular, may be out of date. Careful distinction must be made between laminar and turbulent flow (separate nomographs apply to each), and newtonian fluids only are usually accommodated.

Another common problem often subject to an analytic solution is that of specifying the optimum thermal insulation thickness for hot surfaces. No general correlation or nomographs have been developed analogous to the optimum pipe diameter solution. However, the insulation optimization has been performed so many times that a pattern has emerged, and recommended insulation thickness has been tabulated as a function of parameters such as surface temperature and the time value of money (Thermal Insulation Manufacturers' Association, 1973). Other commonly treated design optimization problems include:

Optimum reflux ratio in distillation columns
Optimum coolant flow in condensers
Optimum design of heat exchangers
Optimum design of gas absorbers
Determination of the optimum number of stages in staged operations

Much has been written on the economic optimization of reactors. In the area of production operating discipline, economically defined optimum maintenance schedules and cyclic production schedules may be computed analytically. The results of such analysis, combined with accumulated input data on subjects such as the typical time decay of exchanger overall coefficients of heat transfer, are used in well-organized operations to schedule the appropriate shutdown and cleaning of heat exchangers, for example.

Many of the published optimization studies are of considerable value in accelerating plant design, including the preliminary plant design, which is such an important prerequisite of project evaluation. More often, however, such studies serve best as a guide, as an example of how economic optimization works, rather than as a direct aid to design. Chemical engineering design is simply much too variable to be classified within the context of one published method or another; each design problem is, in a sense, unique. The successful plant designer learns how to apply the various optimization *techniques* that are available.

If the economic objective function can be expressed analytically in terms of one or more independent variables, then often the optimum (which need not be the maximum or minimum) may be obtained by differentiation—provided that an optimum does, indeed, exist within the confines of the boundary conditions. In the more usual case, however, the chemical engineering economic optimization problem cannot be formulated by a single analytic function; it is characterized, typically, by several related ensembles of numerical data. The solution to such problems involves the application of a number of *search* techniques, many of which realistically require the use of computers for even relatively simple problems. An overview of various economic optimization techniques useful in chemical engineering design may be found in the following texts: Happel and Jordan (1975), Peters and Timmerhaus (1980), and Baasel (1980).

Our purpose here is not to dwell so much upon the techniques, but rather to see how the principles of the analysis of alternatives impact upon the definition of the desirable criterion of optimization.

Optimization of Continuous Objective Functions

Figure 10.3.1 represents the familiar relationship of the effect of reflux ratio upon the total annual operating costs in a distillation operation. The objective function here is the total cost; in this particular case, it is a continuous function of the single

FIGURE 10.3.1
Effect of reflux ratio upon annual operating costs.

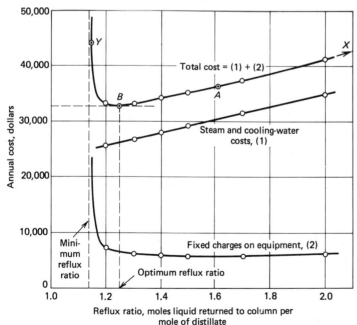

variable, the reflux ratio. An analytic relationship between total annual cost and reflux ratio may be derived, but the usual procedure is to design the system for each of a set of reflux ratios, to calculate the appropriate costs, and then to plot the results. The optimization *technique* is therefore graphical; the traditional criterion is *minimum* cost, at the point where the tangent line has a zero slope. The minimum cost, in turn, defines the optimum reflux ratio. This traditional criterion, however, does not quite conform to the requirement that any investment must generate some minimum prescribed return.

Suppose the information in Fig. 10.3.1 were to be replotted as annual cost versus the investment associated with each reflux ratio. The resulting curve, drawn in Fig. 10.3.2, is a bit strange, but we need not be concerned about its shape for the moment. The four points, X, A, B, and Y, are shown on both graphs. The minimum *investment* occurs at point A; as we proceed from X to A, the investment and annual cost both decrease, and the optimum then certainly lies at A or beyond (why?). The minimum annual cost occurs at point B; if we pass from Y to B, again the investment and annual cost both decrease, and the optimum lies at B or beyond. In other words, the optimum lies somewhere along the segment AB. Let us look at an enlarged graph of this segment (Fig. 10.3.3).

As we move from A to B, the investment goes up as the annual cost goes down. At some point P, the slope is

$$\left(\frac{dc}{dK}\right)_P$$

But $dc = -dP_B$, the differential profit before taxes [see (10.1.1)], and

FIGURE 10.3.2
Replot of reflux ratio data.

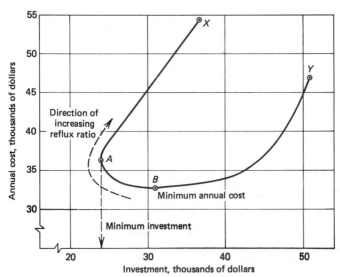

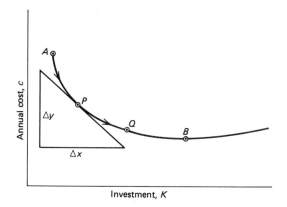

FIGURE 10.3.3
Differential return on investment.

$$\left(\frac{dP_B}{dK}\right)_P = -\left(\frac{dc}{dK}\right)_P = \left|\frac{\Delta y}{\Delta x}\right|$$

where Δy and Δx can be measured. But $(dP_B/dK)_P$ is the *differential* ROIBT which, we agreed, should exceed 15 percent (or some other prescribed value) for equipment alternatives. Therefore, starting at point A (the low investment side), we measure $|\Delta y/\Delta x|$, and we keep moving toward B as long as the ratio exceeds 15 percent, until we reach some point Q where the ratio just equals 15 percent. This is the optimum point, not B. At Q, a small investment increment will just generate a 15 percent annual profit on the increment; a little beyond Q, that criterion will no longer be met.

The differential ROIBT criterion is not often used for economic optimization, for a number of reasons. The accuracy of computations does not often justify such "fine tuning," and, indeed, the economic minimum curves are usually very flat in the region of interest.

The equivalent annual cost (*break-even analysis*), the other principal (and preferred) method of choosing equipment alternatives, may also be applied to the optimization of continuous objective functions, as illustrated by the example which follows. The optimum criterion for break-even analysis is

$$-\frac{dc_A}{dK} = 0$$

since at this point the differential investment is, indeed, earning the rate of return $(i_0)_A$.

Example 10.3.1

A cryogenic fluid is pumped through a long 12-in-diameter pipeline. With 1.3 in of insulation, the heat gain from the environment results in an additional operating cost of $100/day. The installed cost of 1.3 in of insulation is $26,000.

556 CHAPTER 10: THE ANALYSIS OF ALTERNATIVES

Calculate the optimum insulation thickness for the following economic criteria:

Insulation cost varies with 0.6 power of thickness.
Heat gain is inversely proportional to insulation thickness.
Maintenance costs are independent of insulation thickness.
The insulation has a service life of 2 years.
Money is valued at 10 percent/annum after taxes, discrete compounding.
Tax rate = 48 percent.

SOLUTION

1 Differential ROIBT = 15 percent method

Annual cost of operation = depreciation + cost penalty of heat leak
$$\qquad\qquad\qquad\qquad\qquad + \text{other fixed charges (maintenance, etc.)}$$
$$\text{Investment} = 26{,}000 \left(\frac{\theta}{1.3}\right)^{0.6} = \$22{,}200\theta^{0.6} = K$$

For 2-year life,

$$\text{Depreciation} = \$11{,}100\theta^{0.6}/\text{yr}$$
$$\text{Annual heat leak penalty} = \$100 \times 365 \times \frac{1.3}{\theta}$$
$$= \$47{,}500\theta^{-1}/\text{yr}$$

Fixed charges F are independent of θ:

$$c = 11{,}100\theta^{0.6} + 47{,}500\theta^{-1} + F = \text{annual cost of operation}$$
$$\text{Differential ROIBT} = 0.15 = -\frac{dc}{dK} = -\frac{dc/d\theta}{dK/d\theta}$$
$$\frac{dc}{d\theta} = 0.6 \times 11{,}100\theta^{-0.4} - 47{,}500\theta^{-2}$$
$$\frac{dK}{d\theta} = 0.6 \times 22{,}200\theta^{-0.4}$$

Thus

$$0.15 = \frac{-0.6 \times 11{,}100\theta^{-0.4} + 47{,}500\theta^{-2}}{0.6 \times 22{,}200\theta^{-0.4}} = 3.566\theta^{-1.6} - 0.50$$
$$\theta = 2.9 \text{ in}$$

2 Break-even analysis

$$\text{Capital recovery cost} = K \frac{i_0(1 + i_0)^n}{(1 + i_0)^n - 1}$$
$$= 22{,}200\theta^{0.6} \frac{0.10(1.10)^2}{(1.10)^2 - 1} = 12{,}790\theta^{0.6}$$

Cost penalty of heat leak (after tax deduction) = $47,500\theta^{-1}(1 - 0.48)$
$$= 24,700\theta^{-1}$$

From Eq. (10.1.5), therefore,

$$c = 12,790\theta^{0.6} + 24,700\theta^{-1} - 0.48 \times \tfrac{1}{2} \times 22,200\theta^{0.6}$$

$$-\frac{dc}{dK} = -\frac{dc/d\theta}{dK/d\theta} = 0$$

Thus

$$0 = \frac{24,700\theta^{-2} - 0.6 \times 7,462\theta^{-0.4}}{0.6 \times 22,200\theta^{-0.4}}$$

$$\theta = 2.9 \text{ in}$$

The two methods give the same answer; i.e., the two criteria, ROIBT = 15 percent and $(i_0)_A = 0.10$, are compatible.

Exercise 10.3.1

What insulation thickness would be obtained if the traditional criterion of minimum annual cost were used?

Optimization of Discrete Objective Functions

We have seen that the relative attractiveness of two alternatives may be assesed with the ΔROIBT concept. The concept may be extended to the comparison of a whole sequence of discrete alternatives, arranged in increasing order of capital investment, to establish the economic optimum. For example, in multistage operations, greater efficiency of operation is generally attained by increasing the number of stages, but, with increased capital costs, a "point of diminishing returns" is eventually reached, i.e., that stage for which ΔROIBT = Δ profit/Δ investment is less than some acceptable minimum.

There are three approaches that may be used for the optimization of discrete objective functions with the ΔROIBT method:

1 Each discrete member of the sequence is evaluated separately to establish a trend. With countercurrent operations, for instance, the evaluation sequence might start with a single stage, then two stages, etc. The optimum is obtained by trial and error, aided by graphical interpolation. This is the most time-consuming method, but it may have to be used if an analytic expression for the objective function cannot be derived.

2 A finite difference expression for ΔROIBT is derived and is optimized by using the calculus of finite differences.

3 A continuous analytic objective function is derived and optimized, but the result is rounded off to the closest integral value. This is the approach used in the example that follows.

When comparing equipment alternatives in a given processing plant, a minimum acceptable ΔROIBT of 15 percent is frequently stipulated, although a higher value may be prescribed for the comparison of expensive, complex equipment alternatives.

Example 10.3.2

Using the ΔROIBT concept, estimate the optimum number of effects in a forward-feed, multiple-effect evaporator system to evaporate 100,000 lb/day of water from an aqueous salt solution. The system will be operated 300 days/year, and each effect will evaporate 0.8 lb of water/lb of steam fed to that effect.

The direct fixed capital associated with each effect is $35,000. Maintenance and fixed charges (not including depreciation) are 10 percent of DFC per year; steam supplied from plant site boilers costs $2/1000 lb; labor and other regulated and variable charges are independent of the number of effects. Use 10 percent straight-line annual depreciation; stipulate minimum acceptable ΔROIBT = 25 percent. You may neglect steam plant allocated capital variation.

SOLUTION First, we must determine the net amount of steam used as a function of number of stages. This may be derived by induction.
$n = 1$.

$$\text{Net steam requirement} = \frac{1 \times 10^5}{0.8} \text{ lb/day}$$

$n = 2$. Let x lb be evaporated in no. 2 effect; then $(1 \times 10^5 - x)$ lb is evaporated in no. 1 effect. But vapor from no. 1 is used as steam in no. 2;

$$1 \times 10^5 - x = \frac{x}{0.8}$$

$$x = 1 \times 10^5 \frac{1}{1 + 1/0.8}$$

$$\text{Net steam is fed to no. 1} = \frac{1 \times 10^5 - x}{0.8}$$

$$= \frac{1 \times 10^5}{(0.8)^2} \frac{1}{1 + 1/0.8} \text{ lb/day}$$

$n = 3$. Let x lb be evaporated in no. 3 effect and y lb be evaporated in no. 2. Then $1 \times 10^5 - x - y$ lb is evaporated in no. 1 effect. Then, as before,

$$y = \frac{x}{0.8}$$

$$1 \times 10^5 - x - y = \frac{y}{0.8} = \frac{x}{(0.8)^2} = 1 \times 10^5 - x - \frac{x}{0.8}$$

or

$$x\left[1 + \frac{1}{0.8} + \frac{1}{(0.8)^2}\right] = 1 \times 10^5$$

$$\text{Net steam to no. 1} = \frac{1 \times 10^5 - x - y}{0.8}$$

$$= \frac{1 \times 10^5}{(0.8)^3} \frac{1}{1 + 1/0.8 + 1/(0.8)^2} \text{ lb/day}$$

$n = k$. By analogy and induction,

$$\text{Steam use} = \frac{1 \times 10^5}{(0.8)^k} \frac{1}{1 + 1/0.8 + 1/(0.8)^2 + \cdots + 1/(0.8)^{k-1}}$$

$$= \frac{1 \times 10^5}{(0.8)^k} \frac{1}{\sigma}$$

where σ is the sum of a geometric progression. From standard summation formula,

$$\sigma = \frac{(1/0.8)^{k-1}(1/0.8) - 1}{1/0.8 - 1} = \frac{(1/0.8)^k - 1}{1/0.8 - 1}$$

Thus

$$\text{Steam use} = \frac{1 \times 10^5}{(0.8)^k} \frac{1/0.8 - 1}{(1/0.8)^k - 1}$$

$$= 0.25 \times 10^5 \frac{1}{1 - (0.8)^k} \text{ lb/day}$$

$$\text{Annual steam cost} = \frac{300 \times 2.00}{1000} \times 0.25 \times 10^5 \frac{1}{1 - (0.8)^k}$$

$$= \frac{\$0.15 \times 10^5}{1 - (0.8)^k}$$

$$\text{Capital investment} = I = \$35,000k$$

Fixed charges, maintenance, depreciation = $0.20I$. Other operating costs are constant. Therefore operating costs varying with k are

$$B_k = 7000k + \frac{0.15 \times 10^5}{1 - (0.8)^k}$$

$$\Delta\text{ROIBT} = 0.25 = \frac{dP}{dI} = \frac{-dB_k}{dI} = \frac{-dB_k/dk}{dI/dk}$$

$$\frac{dB_k}{dk} = 7000 + 0.15 \times 10^5 \frac{(0.8)^k \ln 0.8}{[1 - (0.8)^k]^2}$$

$$= 7000 - 3.346 \times 10^3 \frac{(0.8)^k}{[1 - (0.8)^k]^2}$$

$$\frac{dI}{dk} = 35,000$$

Thus

$$0.25 = \frac{3.346 \times 10^3 \frac{(0.8)^k}{[1-(0.8)^k]^2} - 7000}{35,000}$$

Let $(0.8)^k = x$. Then

$$0.25 = \frac{0.0956x}{(1-x)^2} - 0.20$$

$$\frac{x}{(1-x)^2} = 4.705$$

$$x = 1.577 \quad \text{or} \quad 0.633 = (0.8)^k$$

For integral k, $x < 1$. Therefore

$$(0.8)^k = 0.633 \qquad k = 2.06$$

Optimum number is next lowest integer. That is, the optimum number = 2

A more detailed consideration of the economic optimization of evaporator trains is given by King (1980).

As was the case in continuous function optimization, break-even analysis may be used for the optimization of discrete sequences.

The Optimum Allocation of Resources

A problem which frequently arises in the chemical process industry is this:

A number of raw materials A, B, C, ... may be processed to produce a spectrum of products W, X, Y, ... in facilities having a specified maximum throughput. It may happen that the supply of the raw materials is in some fashion limited. Also, it happens as a matter of course that the products have different market values. The question is, "What is the optimum way of allocating the scarce raw material resources to maximize the profitability of the venture?"

Problems of this nature are characterized mathematically by linear equations of the form

$$\phi = a_0 + a_1 x_1 + a_2 x_2 + \cdots + a_k x_k \qquad (10.3.1)$$

where the a's are constants, the x's are independent variables, and ϕ is an *objective function* that is to be maximized (or minimized). In economic investigations, ϕ is usually the profit or total cost, and the independent variables represent items which contribute to the cost or profit. (ϕ may be, for example, the total value of inventory in a warehouse; the x's are the numbers of specific inventoried items, and the a's are their unit costs.)

Such problems are solved by techniques of *linear programming*. The term "programming" is used in the sense of planning rather than computer programming, although all except the very simple problems do, indeed, require computer help.

The linear equation (10.3.1) is the equation of a hyperplane (a plane in $k + 1$ dimensions) with no maxima or minima. If the solution to the problem is to exist—i.e., if there is to be a Max (ϕ) or Min (ϕ)—then the hyperplane must be bounded by *constraints*. One very common constraint is that the x_i's must all be zero or positive. Additional constraints are represented by linear inequalities,

$$b_1 x_1 + b_2 x_2 + \cdots + b_j x_j \leq b_0 \tag{10.3.2}$$

Max (ϕ) or Min(ϕ) then occur at one of the vertices formed by the intersection of the objective function hyperplane with planes representing the constraints. The vertices are *points,* and the maximum/minimum must occur at some such point, since it cannot occur in the middle of a bound line, in the middle of a bound plane, etc. (Why?)

Exercise 10.3.2

Can you think of a condition which would result in the absence of a maximum or minimum?

The restriction of the required solution to a few discrete possibilities, of course, greatly simplifies the process of optimization. The reason why the solution lies on one of the vertices may perhaps be better appreciated by considering the simple three-dimensional case illustrated in Fig. 10.3.4.

For this simple case, the objective function ϕ is determined by just two independent variables, x_1 and x_2. We wish to optimize

$$\phi = f(x_1, x_2)$$

where the functional relationship is given by the linear equation

$$\frac{x_1}{a} + \frac{x_2}{b} + \frac{\phi}{c} = 1 \tag{10.3.3}$$

This is an equation of a plane. It is shown in Fig. 10.3.4 as the plane 123. The constraints

$$x_1 \geq 0 \quad x_2 \geq 0 \quad \phi \geq 0 \tag{10.3.4}$$

bound the plane as illustrated. Another constraint may be written

$$A x_1 + B x_2 \leq 1 \tag{10.3.5}$$

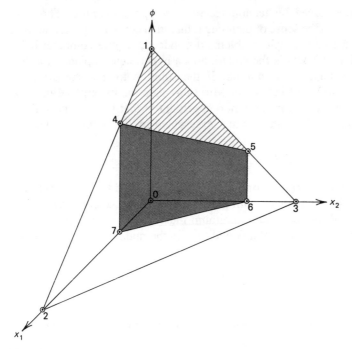

FIGURE 10.3.4
A three-dimensional resource allocation problem.

$Ax_1 + Bx_2 = 1$ is the equation of a plane parallel to the ϕ axis; in the illustration, it is the plane 4567 which intersects the plane 123 along the bounded line 45. The inequality expressed by Eq. (10.3.5) restricts the solution to the surface 145. Now, *any* line drawn on surface 145 has no maximum or minimum of ϕ except at the points where the line intersects the edges of the plane 145. Therefore the solution must lie somewhere on the line segments 15, 54, or 41. But these lines also have no maximum or minimum except as the vertices 1, 4, or 5. Thus the optimum value of ϕ occurs at one of these three vertices; an infinite number of possibilities has been reduced to just three!

For the simple, three-dimensional case of Fig. 10.3.4, the solution is straightforward. The vertices 1, 4, and 5, projected onto the x_1x_2 plane, are 0, 7, and 6, respectively; the 076 region is reproduced in Fig. 10.3.5a. The values of x_1 and x_2 at the vertices are readily determined; each set (x_1, x_2) is substituted into Eq. (10.3.3) to obtain three values of the objective function, and the optimum value (maximum or minimum, depending upon nature of problem) is selected.

The situation to be optimized may involve more than one linear constraint, as shown in Fig. 10.3.5b. Here, in addition to

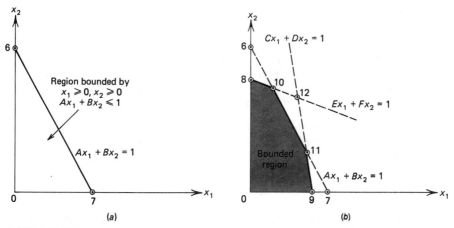

FIGURE 10.3.5
(a) A two-variable region with single linear constraint. (b) Multiple linear constraints.

$$x_1 \geq 0 \quad \text{and} \quad x_2 \geq 0$$

the linear constraints are Eq. (10.3.5) plus

$$Cx_1 + Dx_2 \leq 1 \tag{10.3.6}$$
$$Ex_1 + Fx_2 \leq 1 \tag{10.3.7}$$

The bounded region of the independent variables is shaded; the effect of the additional constraints is to increase the number of vertices that must be evaluated.

Exercise 10.3.3

Is vertex 12 in Fig. 10.3.5b one that must be evaluated?

Exercise 10.3.4

Suppose the constraint in Eq. (10.3.5) were to read

$$Ax_1 + Bx_2 = 1$$

rather than

$$Ax_1 + Bx_2 \leq 1$$

Which vertices would be the only valid ones to be used in the evaluation?

A simple example will serve to illustrate the procedures.

Example 10.3.3

Consider an idealized (and largely mythical) ethylamine plant, not unlike the one shown in Fig. 3.5.3, but without the services of a cracking furnace. With proper built-in flexibility, let us say that we have the option of producing marketable monoethylamine and triethylamine, with any diethylamine recycled into the alkylation reaction.

The product split is not limited by plant capacity, but the raw material supply is limited to the following quantities (maximum available supply):

Ethyl chloride: 400 lb · mol/day (limited by capacity of chlorination facility)

Caustic soda (100% basis): 500 lb · mol/day (limited by large caustic demand for enhanced oil recovery)

Ammonia: 300 lb · mol/day (limited by storage and distribution)

The problem is to decide how much of each product to manufacture to maximize the before-tax profit.

Because of the need to recycle the diethyl compound, the maximum ratio of mono- to triethylamine that can be produced is 9 : 1.

The selling prices are:

Monoethylamine: $41.50/(lb · mol)
Triethylamine: $121.20/(lb · mol)

Raw material costs are:

Ethyl chloride: $12.90/(lb · mol)
Sodium hydroxide: $4.00/(lb · mol) (used for neutralization of HCl generated in alkylator)
Ammonia: $1.70/(lb · mol)

The costs of utilities, labor, and the fixed charges are essentially unaffected by variations in feed and product spectrum, except that extra processing imposes an operating charge penalty of $10.00/(lb · mol) of the triethylamine.

SOLUTION The reactions involved are

$$CH_3CH_2Cl + NH_3 \longrightarrow CH_3CH_2NH_2 + HCl$$
$$3CH_3CH_2Cl + NH_3 \longrightarrow (CH_3CH_2)_3N + 3HCl$$

Let

$$x_1 = \text{lb} \cdot \text{mol/day of monoethylamine}$$
$$x_2 = \text{lb} \cdot \text{mol/day of triethylamine}$$

For either product, with stated conditions,

Profit = sales revenue − raw materials cost − operating penalty

1 lb · mol of monoethylamine requires:

1 lb · mol of ethyl chloride
1 lb · mol of NaOH
1 lb · mol of NH_3

1 lb · mol of triethylamine requires:

3 lb · mol of ethyl chloride
3 lb · mol of NaOH
1 lb · mol of NH_3

Thus

$$\text{Profit on monoethylamine (per lb} \cdot \text{mol)} = 41.50 - (12.90 + 4.00 + 1.70)$$
$$= \$22.90/(\text{lb} \cdot \text{mol})$$
$$\text{Profit on triethylamine} = 121.20$$
$$- (3 \times 12.90 + 3 \times 4.00 + 1.70)$$
$$- 10.00$$
$$= \$58.80/(\text{lb} \cdot \text{mol})$$

Objective function:

$$\text{Profit } \phi = 22.90x_1 + 58.80x_2 \qquad (10.3.8)$$

Constraints:

$$\text{Production ratio: } \frac{x_1}{x_2} \leq 9 \qquad (10.3.9)$$

$$\text{Ethyl chloride: } x_1 + 3x_2 \leq 400 \qquad (10.3.10)$$

$$\text{NaOH: } x_1 + 3x_2 \leq 500$$

In view of Eq. (10.3.10), this equation is redundant.

$$NH_3: x_1 + x_2 \leq 300 \qquad (10.3.11)$$

The constraints are plotted in Fig. 10.3.6. The bounded region is shaded; the vertices to be evaluated are 0, 1, 2, and 3.

At 0:

$$x_1 = x_2 = 0 \quad \text{and, from Eq. (10.3.8)} \quad \phi = 0$$

At 1:

$$x_1 = 0 \qquad x_2 = 133.3 \qquad \phi = \$7838/\text{day}$$

At 2:

$$x_1 = 250 \qquad x_2 = 50 \qquad \phi = \$8665/\text{day}$$

At 3:

$$x_1 = 270 \qquad x_2 = 30 \qquad \phi = \$7947/\text{day}$$

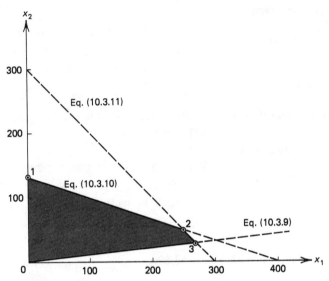

FIGURE 10.3.6
Graphical solution of Example 10.3.3.

Vertex 2 results in highest profit. Strategy therefore is to produce

250 lb · mol/day monoethylamine
50 lb · mol/day triethylamine

In the general case, the objective function in n independent variables is

$$\phi = a_1 x_1 + a_2 x_2 + \cdots + a_n x_n \qquad (10.3.12)$$

and the m linear constraints take the form

$$b_1 x_1 + b_2 x_2 + \cdots + b_n x_n \leq b_0 \quad \text{(inequalities)} \qquad (10.3.13)$$

and

$$c_1 x_1 + c_2 x_2 + \cdots + c_n x_n = c_0 \quad \text{(equalities)} \qquad (10.3.14)$$

(Any of the coefficients may, of course, be zero.)
The procedure calls for changing all inequalities (10.3.13) to equalities by adding "slack" (or "dummy") variables S_i to each inequality:

$$b_1 x_1 + b_2 x_2 + \cdots + b_n x_n + S_1 = b_0 \qquad (10.3.15)$$

Equations (10.3.14) and (10.3.15) constitute a system of m simultaneous equations in $n + j$ unknowns,[1] where, in general, $n + j > m$. We now set $(n + j) - m$ independent variables equal to zero and solve for the remaining m variables, finally substituting their value into Eq. (10.3.12). The procedure is repeated for another combination of $(n + j) - m$ variables; the number of independent trials and computed values of ϕ is given by

$$\frac{(n + j)!}{m! \, (n + j - m)!} \qquad (10.3.16)$$

In general, only positive values of the x_i are stipulated.

Linear programming techniques for the optimum allocation of resources find wide application in the process industries—allocation of feeds and products in the petroleum industry, warehousing and distribution problems, labor allocation (including shift assignments), planning of business strategy. A number of consulting firms specialize in supplying linear programming computer software. Indeed, the computational complexity grows very rapidly with the number of variables to be considered, as can be appreciated from the factorials in expression (10.3.16). The simultaneous equation solution that has been outlined becomes too inefficient for the more complex situations. An algorithm known as the *simplex method* is available which reduces considerably the number of required trials by focusing upon the best path toward an optimum solution. A description of the simplex algorithm is given by Peters and Timmerhaus (1980).

Theoretical developments and the application of linear programming to the allocation of resources form the subject of an eminently readable article by Bland (1981) in the journal *Scientific American*.

10.4 REPLACEMENT ANALYSIS

Concepts

The term "replacement analysis" refers to the economic analysis of the two alternatives:

1 Replace an item of equipment in an operating plant with a new or improved item that will perform the same service.

2 Keep the old item of equipment in service.

Now, there may be intangible reasons for replacing equipment, but most reasons can be translated, at least potentially, into the language of economics. If a particular pump is giving you an awful lot of trouble and you are ready to take a sledgehammer to it, it is likely that your ire may be assuaged by cold economic facts; a new, more expensive pump may, indeed, be justified in light of the burden of maintenance costs and loss of production borne by the troublesome pump.

In replacement analysis parlance, the item which is under consideration for

---

[1] j is the number of inequalities; if equalities only are involved, $j = 0$.

replacement is known as the *defender,* and the candidate to do the replacing is the *challenger.* The technique for making the challenger-defender comparison is the break-even analysis (equivalent annual cost) or any other related method, such as a NPW calculation. To perform a meaningful break-even analysis, all items defining the cash flows associated with the equipment must be known or reasonably estimated. In particular, maintenance and other operating costs characteristic of both the defender and the challenger must be projected, usually on the basis of past experience.

An important part of the analysis is the estimation of the current value of the defender. Writers on engineering economics emphasize the dire error incurred by using the book value of an item as its current value; its true value, as a matter of fact, is what can be currently obtained from buyers of used equipment. For items such as used trucks and machine tools this sales value (or current salvage value) can be ascertained reasonably well; for chemical plant process equipment, the task is more difficult, and in any case the salvage value is likely to be very small after a few years of service. At any rate, past expenditures and depreciation charges bear no relation to the current value—they are sunk costs. They may, however, have an important effect upon tax obligations or credits that the replacement procedure may incur.

The tax obligations or credits are an inseparable component of the replacement economic comparison, and the replacement analysis must therefore be made on an after-tax basis (Stermole, 1974). The tax situation is a matter of the difference between the true value (current salvage value) and the book value. If the sales value is higher than the book value, a situation that is not very likely, particularly in the chemical process industry, the sales gain is a long-term capital gain (assuming the asset has been held for more than a year). The tax rate on long-term capital gains changes with the tides of government philosophy (chemical engineers should read the newspapers); in 1981, the combined federal and state obligations amounted to a tax rate of approximately 30 percent. If the sales value is less than the book value, the sales transaction represents a long-term capital loss. Tax laws permit long-term capital losses to be claimed only up to the extent of long- (and, in some cases, short-) term capital gains (with provisions for carry-over); the transaction therefore generates a tax credit of about 30 percent (i.e., it eliminates a tax payment that would otherwise have to be made on the matching capital gain). There is, however, an important exception to this capital loss rule, and that occurs if the equipment is sold (or thrown out) because it is *obsolete.* Obsolescence ipso facto implies that the original depreciation schedule was wrong, and in such a case (IRS willing) the item may be depreciated down to the salvage value in the last year of use. The effect of this is to increase the tax credit to about 50 percent of the loss.

The equivalent after-tax annual cost of equipment is given by Eq. (10.1.5). For the challenger, the equation can be used as is, provided the service life is known and the quantity $E - \Re$ is constant. If the maintenance costs, for instance, are expected to increase with time, then $E - \Re$ is no longer constant; for comparison purposes it is necessary to compute the *equivalent uniform annual cost* (EUAC). This is done by discounting all the cash flows (expressed as *costs*),

$$(1 - t)(E - \mathcal{R}) - tD$$

to zero time and then computing an equivalent annuity R_E over the service life of the item. The example demonstrates the procedure.

Example 10.4.1

A new inert gas generator is under consideration as a replacement for an old unit. The generator costs $100,000 (installed) and has a projected service life of 12 years, a tax life of 6 years. The annual utilities cost is expected to be $3000; past experience indicates that annual maintenance and repair costs will escalate according to the relationship $(1500 + 600Y)$, where Y is the time in years. Fixed charges are 10 percent of the installed cost; other operating costs (including labor) are $1000/year. The salvage value at the end of 6 years (and thereafter) is $10,000.

Calculate the EUAC for straight-line depreciation, a tax rate of 46 percent, and an after-tax *effective* internal rate of return of $i_0 = 10$ percent. Consider all annual cash flows, including depreciation, to be *continuous* in time.

SOLUTION   In Eq. (10.1.5):

$$E = 3,000 + (1,500 + 600Y) + 0.10(100,000) + 1,000$$
$$= 15,500 + 600Y$$
$$\mathcal{R} = 0 \text{ for this case}$$
$$D = \frac{100,000 - 10,000}{6} = 15,000$$

For continuous discounting of continuous cash flows, use Eq. (2.2.6):

$$P = \int_0^n \phi(Y) e^{-iY} dY$$

The nominal rate of return i can be computed from Eq. (2.1.22):

$$i_0 = e^i - 1 = 0.10 \quad i = 0.0953$$

Therefore for the term $(1 - t)(E - \mathcal{R}) - tD$,

$$P = \int_0^{12} (1 - 0.46)(15,500 + 600Y) e^{-0.0953Y} dY - \int_0^6 0.46 \times 15,000\, e^{-0.0953Y} dY$$
$$= \$39,614*$$

From Eq. (2.3.8),

$$\overline{R}_E \text{(uniform cash flow)} = Pi\, \frac{e^{in}}{e^{in} - 1}$$

$$= 39,614 \times 0.0953\, \frac{e^{0.0953 \times 12}}{e^{0.0953 \times 12} - 1} = \$5541$$

*If so desired, the depreciation tax credits may be handled, and perhaps more properly so, as continuously discounted *discrete* cash flows.

Similarly,

$$\overline{R}_A = 100{,}000 \times 0.0953 \, \frac{e^{1.1436}}{e^{1.1436} - 1} = \$13{,}987$$

and

$$\overline{R}'_A = S_v \frac{i}{e^{in} - 1} = 10{,}000 \times \frac{0.0953}{e^{1.1436} - 1}$$
$$= \$446$$

Thus

$$\text{EUAC} = \overline{R}_E + \overline{R}_A - \overline{R}'_A = \$19{,}082/\text{year}$$

It may happen that the out-of-pocket operating expenses E may be projected to grow, as in the example, but that no definite service life may be projected for the challenger. In that case R_E increases with years of service, R_A decreases, and EUAC reaches a minimum during some year m. It is the minimum EUAC of the challenger that is then compared with the defender's annual cost.

Methodology

Replacement analysis requires no new concepts of economic analysis; the technique is that of the analysis of alternatives. The key steps in the analysis involve a firm, unequivocal definition of the alternative situations and a reasonable projection of salvage and out-of-pocket expenditures associated with equipment, including maintenance costs. In the chemical industry, such projection is, more often than not, fraught with uncertainty.

The definition of alternative situations is, in the final analysis, a matter of common sense, but the problem of proper definition is compounded by conflicting recommendations in the engineering economics literature. The handling of different situations will be illustrated with examples. The general viewpoint taken will be what Stermole (1974) refers to as the *accounting viewpoint:* only actual receipts and disbursements will be recognized as characterizing each alternative. In particular, we will not use the procedure of characterizing the choice of the defender alternative as including a "ghost" investment of the current salvage value (called *opportunity cost* by Stermole, 1974, or *salvage forgone* by Uhl and Hawkins, 1971).

The first common situation that arises in replacement analysis is that of a possible *improvement* in a current operating procedure which involves no defender equipment (or else completely depreciated equipment with no salvage value). The procedure is to be carried out for a specified number of years into the future.

Example 10.4.2

A batch resin reactor is presently unloaded manually into trays; the resin cools and freezes, is broken up into pieces with mallets, and is transferred into Fiberpac drums for shipment. The annual labor cost for this operation is $50,000. It has been proposed that the manual unloading be replaced with an automated system involving a water-cooled apron conveyor, a bin, and a drum loader costing $150,000 installed. The projected out-of-pocket annual expenses associated with the automated system are $15,000, including $5,000 for labor. The reaction system will be operated for another 5 years.

Should the automated system be installed? Assume SL depreciation, 50 percent tax rate, $i_0 = 10$ percent, $S_v = 0$.

SOLUTION

Alternative 1: Keep manual unloading

Annual cost = $(1 - 0.50)\,50,000 = \$25,000$

Alternative 2: Install automated system

Annual cost = $(1 - 0.50)\,15,000 - 0.50 \times \dfrac{150,000}{5} + 150,000\,\dfrac{0.10\,(1.10)^5}{(1.10)^5 - 1}$

= $32,070$

Therefore keep manual unloading.

A variation on the above (with limited remaining years of scheduled operations) is the defender with book value and salvage value.

Example 10.4.3

Compressor A was installed 3 years ago at a cost of $200,000. It has a tax life of 8 years and a salvage value of $20,000 at all times. Out-of-pocket operating charges are $40,000/year (*after* taxes).

Compressor B (the challenger) may be installed for $150,000; it has a tax life of 8 years with a salvage value of $20,000 at all times. Operating charges would be $25,000/year (*after* taxes).

Operations are expected to continue for another 10 years, and either compressor is expected to be serviceable for that length of time. With SL depreciation, $t = 50$ percent, and $i_0 = 10$ percent, should compressor A be replaced?

SOLUTION

Alternative 1: Keep compressor A

$$D = \dfrac{200,000 - 20,000}{8} = \$22,500$$

PW of depreciation tax credits (for five more years) is

$$22,500 \times 0.5\,\dfrac{(1.10)^5 - 1}{0.10(1.10)^5} = \$42,646$$

Thus

$$R_E = -42,646\,\dfrac{0.10(1.10)^{10}}{(1.10)^{10} - 1} + 40,000 = \$33,060$$

(The minus sign converts the credit to a cost.)

$$R_A = 0 \quad \text{(Why?)}$$

$$R'_A = 20,000 \frac{0.10}{(1.10)^{10} - 1} = \$1255$$

Therefore

$$\text{Annual cost} = 33,060 - 1,255 = \$31,805$$

Alternative 2: Install compressor B

PW of new investment = installed cost of compressor B
− salvage value of compressor A
$$= 150,000 - 20,000 = \$130,000$$
Book value of compressor $A = 200,000 - 3 \times 22,500$
$$= \$132,500$$

Thus the long-term capital loss on compressor A is

$$132,500 - 20,000 = \$112,500$$

Assume a tax rate of 30 percent;

$$\text{Tax credit} = 0.30 \times 112,500 = \$33,750$$

Therefore

$$\text{Net PW of new investment} = 130,000 - 33,750$$
$$= \$96,250$$

$$D = \frac{150,000 - 20,000}{8} = \$16,250$$

PW of depreciation tax credits is

$$16,250 \times 0.5 \frac{(1.10)^8 - 1}{0.10(1.10)^8} = \$43,346$$

Thus

$$R_E = 25,000 - 43,346 \frac{0.10(1.10)^{10}}{(1.10)^{10} - 1} = \$17,946$$

$$R_A = 96,250 \frac{0.10(1.10)^{10}}{(1.10)^{10} - 1} = \$15,664$$

R'_A (same as Alternative 1) = \$1255

Therefore

$$\text{Annual cost} = 17{,}946 + 15{,}664 - 1{,}255 = \$32{,}355$$

The annual costs are so close that other considerations would probably enter into the decision.

The effect of tax laws may be appreciated by referring back to the long-term capital loss tax credit in the preceding example. A tax rate of 30 percent was assumed. If however, the switch to compressor B were made because of the *obsolescence* of compressor A, and a tax credit rate of 50 percent were permitted, the annual cost of the compressor B alternative would be reduced to $28,693, an appreciably lower cost than that for alternative 1. An extreme example of the effect of obsolescence is given in the statement of Prob. 6.8.

Exercise 10.4.1

Rework the annual cost of alternative 2 in Example 10.4.3 for the case of zero salvage for compressor A and an allowance for obsolescence.

The most common replacement analysis situation involves the evaluation of existing equipment vis-à-vis a challenger, with no plans to terminate operations at any particular time. The choice then is to operate the defender *for one more year,* or to replace it.

Example 10.4.4

A copper water still supplies distilled water to an industrial laboratory facility. The still is fully depreciated; it has a current salvage value of $35,000, but this is expected to decrease to $25,000 in 1 year. Annual operating costs are $47,000 (after taxes, i.e., tax-adjusted).

A new reverse osmosis unit is available at an installed cost of $250,000; it has an expected service life of 12 years, and the associated annual operating costs (excluding depreciation) are $10,000 after taxes. The salvage value is zero, the tax life 5 years.

Should the copper still be replaced now? Use SL depreciation, $t = 0.48$, and a nominal rate of return $i_0 = 0.12$. Use continuous discounting of continuous cash flows.

SOLUTION

Alternative 1: Keep still one more year

$$\overline{R}'_A = \frac{S_v\, i}{e^i - 1} \quad (S_v = \text{salvage value at end of year})$$

$$= \frac{25{,}000 \times 0.12}{e^{0.12} - 1} = \$23{,}530$$

Therefore

$$\text{Annual cost} = 47{,}000 - 23{,}530 = \$23{,}470$$

Alternative 2: Install RO unit now

$$\text{Investment} = \text{installed cost} - \text{salvage value } now$$
$$= 250{,}000 - 35{,}000 = \$215{,}000$$
$$D = {}^{250{,}000}\!/_5 = \$50{,}000$$
$$\text{PW of tax credits} = \int_0^5 0.48 \times 50{,}000 \, e^{-0.12Y} \, dY$$
$$= \$90{,}238$$

Thus

$$\bar{R}_E = 10{,}000 - 90{,}238 \, \frac{0.12 e^{0.12 \times 12}}{e^{0.12 \times 12} - 1}$$
$$= -\$4{,}191$$
$$\bar{R}_A = 215{,}000 \, \frac{0.12 e^{0.12 \times 12}}{e^{0.12 \times 12} - 1}$$
$$= \$33{,}811$$
$$\text{Annual cost} = 33{,}811 - 4{,}191 = \$29{,}620$$

Thus keep still another year.

Exercise 10.4.2

At the end of each subsequent year, the still salvage value in the above example decreases as follows:

End of year	S_v, $
1 (this year)	25,000
2	15,000
3	5,000
4	0

Should the still be replaced, and if so, at the end of which year?

10.5 PLANT MODIFICATION DECISIONS

Equipment Debottlenecking Strategy

We conclude the discussion of the analysis of alternatives with a few examples of applications involving plant modification decisions. These examples demonstrate how a variety of problems can be formulated to be amenable to solution by using analytical techniques already presented.

A common problem arises when demand for a product reaches the point where it exceeds the original design capacity of a plant. Certain key items of equipment may be the "bottlenecks" which prevent a plant production throughput appreciably above design. Such equipment must then be modified or replaced so as to "debottleneck" the plant.

If the increase in demand can be projected as a step function in the future at the time of the initial plant design, then various strategies can be adopted to handle the future debottlenecking demand:

Strategy I: Build equipment for initial demand. Build new, larger equipment at the time of the demand increase, and scrap the smaller equipment.
Strategy II: Build equipment to meet future demand now.
Strategy III: Build equipment for initial demand. When demand increases, match the increase with a parallel item of equipment.

A simplified approach to the selection of the optimum strategy has been outlined by Resnick (1981)*. The author bases his choice upon the PW of the investments associated with each strategy.

Suppose the initial demand is p_1, and this is expected to escalate to p_2 as a step function at the end of j years. Let the DFC associated with equipment having capacity p_1 be K_1; then, from Eq. (4.2.3), the DFC cost of equipment having capacity p_i is

$$K_i = K_1 \left(\frac{p_i}{p_1}\right)^{0.6} \tag{10.5.1}$$

For strategy I, with continuous discounting and zero salvage value for scrapped equipment,

$$\text{PW}_\text{I} = -K_1 - K_1 \rho^{0.6} e^{-i\tau} \tag{10.5.2}$$

where ρ is the ratio p_2/p_1, i is the nominal after-tax required rate of return, and τ is a break-even time which we will determine by comparison with the other strategy PWs; the idea will be to see whether $\tau > j$.

For strategy II,

$$\text{PW}_\text{II} = -K_1 \rho^{0.6} \tag{10.5.3}$$

Equating Eqs. (10.5.2) and (10.5.3),

$$\tau = \frac{\ln[\rho^{0.6}/(\rho^{0.6} - 1)]}{i} \tag{10.5.4}$$

If $\tau = j$, the projected year for demand increase, then either strategy is acceptable. If $\tau > j$, strategy II is preferred.

*Adapted, with permission.

Exercise 10.5.1

Prove that strategy II is preferred for $\tau > j$.

For strategy III,

$$\begin{aligned}\text{PW}_{\text{III}} &= -K_1 - K_1\left(\frac{p_2 - p_1}{p_1}\right)^{0.6} e^{-i\tau} \\ &= -K_1[1 + (\rho - 1)^{0.6} e^{-i\tau}]\end{aligned} \quad (10.5.5)$$

If the statements of strategies I and III are compared, it is clear that, in the absence of salvage value, strategy III will always be cheaper than I. Comparing strategies III and II,

$$\tau = \frac{\ln[(\rho - 1)^{0.6}/(\rho^{0.6} - 1)]}{i} \quad (10.5.6)$$

If $\tau > j$, strategy II is preferred.

Exercise 10.5.2

Which is the preferred strategy for $\rho = 2.0$, $i = 0.10$, and $j = 5$?

Resnick's analysis may be extended to incorporate operating costs, tax credits, and even salvage value. In order not to complicate the problem unnecessarily, we will make a few simplifying assumptions:

1 Salvage value of scrapped equipment will be taken as zero.
2 Equation (6.1.11) for the after-tax cash flow will be formulated as

$$C_A = tD - (1 - t)E \quad (10.5.7)$$

All components of E (variable charges, regulated charges, and general expenses, all except the fixed charges) will be assumed to be the same regardless of strategy. These therefore need not be considered in the PW computations.

3 The fixed charges (which do not include depreciation in this case) are taken as a percentage of K per annum, say,

$$F = 0.1K \quad (10.5.8)$$

4 Depreciation will be taken as straight-line over the period of j years. $t = 0.5$.

For strategy I, then,

$$\text{NPW}_\text{I} = -K_1 - K_1\rho^{0.6}e^{-i\tau} + \frac{0.5K_1}{j}\int_0^j e^{-i\theta}d\theta - 0.5 \times 0.1\, K_1\int_0^\tau e^{-i\theta}d\theta \quad (10.5.9)$$

We designate

$$\int_0^\theta e^{-i\theta}d\theta = \frac{1}{i}(1 - e^{-i\theta}) \equiv L_\theta \qquad (10.5.10)$$

Then

$$\text{NPW}_\text{I} = -K_1 - K_1\rho^{0.6}e^{-i\tau} + \frac{0.5L_jK_1}{j} - \frac{0.05K_1(1 - e^{-i\tau})}{i} \qquad (10.5.11)$$

Similarly,

$$\text{NPW}_\text{II} = -K_1\rho^{0.6} + \frac{0.5L_jK_1\rho^{0.6}}{j} - \frac{0.05K_1\rho^{0.6}(1 - e^{-i\tau})}{i} \qquad (10.5.12)$$

for $\text{NPW}_\text{I} = \text{NPW}_\text{II}$,

$$\tau = \frac{1}{i}\left\{\ln[\rho^{0.6} + \frac{0.05(\rho^{0.6} - 1)}{i}] - \ln[(\rho^{0.6} - 1)\left(1 - \frac{0.5L_j}{j} + \frac{0.05}{i}\right)]\right\} \qquad (10.5.13)$$

If $\tau > j$, strategy II is preferred.

A similar equation may be derived for the break-even time for comparing strategies II and III. The derivation is left as an exercise (Prob. 10.30).

Plant Expansion Strategy

Many new products on the market experience a growth period, only to plateau out at maturity. A typical life cycle is illustrated in Fig. 3.3.9; the region B represents the growth period, and region C represents maturity. If the life cycle pattern may be initially projected, the question sometimes arises whether a plant should be built to meet the mature demand, or whether it would be better to build two or more smaller plants in stages to match the growing demand. This and several similar problems have been addressed by Rose (1976).

Region B in Fig. 3.3.9 may be approximated with the linear growth shown in Fig. 10.5.1. Three strategies are indicated to match the growth: one, two, or three plants of equal size (only the two-plant strategy is shown on the diagram). The simplest approach to the optimum choice is to compare the PW of the investments, as was the case with the debottlenecking strategy. Using Eq. (5.1.3) (0.7 capacity exponent) for cost scaling, we have, for the three strategies,

Strategy I: One plant

$$\text{PW}_\text{I} = K_1 \qquad (10.5.14)$$

578 CHAPTER 10: THE ANALYSIS OF ALTERNATIVES

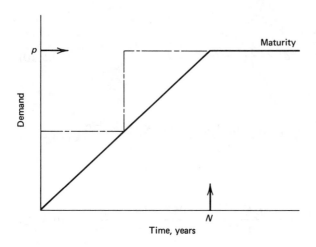

FIGURE 10.5.1
Two-plant strategy to meet growing demand.

Strategy II: Two plants

$$\text{PW}_{\text{II}} = K_1(\tfrac{1}{2})^{0.7} + K_1(\tfrac{1}{2})^{0.7} e^{-iN/2}$$
$$= 0.616 K_1(1 + e^{-iN/2}) \qquad (10.5.15)$$

Strategy III: Three plants

$$\text{PW}_{\text{III}} = 0.463 K_1(1 + e^{-iN/3} + e^{-2iN/3}) \qquad (10.5.16)$$

The best strategy is chosen by numerical evaluation.

Example 10.5.1

A product is expected to grow linearly to maturity in 7 years. If the maturity demand is p, what size of plant should be built initially if $i = 0.15$?
SOLUTION

$$\text{PW}_{\text{I}} = K_1$$
$$\text{PW}_{\text{II}} = 0.980 K_1$$
$$\text{PW}_{\text{III}} = 1.019 K_1$$

Initial construction of a plant having capacity of $p/2$ is indicated, although the differences, in this case, are so small that the full-scale plant would probably be built.

Rose (1976) shows that the two-plant strategy is favored by long maturity times.

Termination of Operations

The shutdown of plant operations has been discussed in connection with break-even charts (Chap. 8). The emphasis of that discussion was sensitivity analysis of a projected operation. What of the ongoing operation? How can one decide whether plant operations should be continued, or whether they should be terminated? The subject of the termination of operations is reviewed by Happel and Jordan (1975).

Let us suppose that we are at the end of year α from start-up and we wish to determine whether the plant should be operated through the end of year β. An important consideration here is the tax life γ, and we consider first the case where $\beta < \gamma$. The criterion of decision will be whether the NPW of remaining operations is larger than zero, using a discount rate i which management may well decide to escalate in the light of attractive competitive projects.

The NPW no longer incorporates any investments (except for possible working capital fluctuations), unless expansions or modifications are contemplated. Five principal components may be identified:

1 PW of the operating margins $(\Re - E)_k$; these are, of course, a part of the after-tax continuous cash flow

$$C_A = (1 - t)(\Re - E) + tD \qquad (6.1.11)$$

2 PW of the depreciation tax credits tD_k [see Eq. (6.1.11)].
3 Discounted tax credit on capital loss if salvage value at the end of year β is less than the book value. For present purposes, we will assume that the loss is fully deductible from income.
4 Discounted working capital credit.
5 Discounted salvage value credit.

Written in the order listed,

$$\text{NPW}_\alpha = \sum_{\alpha+1}^{\beta} \frac{(1-t)(\Re - E)_k}{(1+i)^{k-\alpha}} + \sum_{\alpha+1}^{\beta} \frac{tD_k}{(1+i)^{k-\alpha}}$$

$$+ \frac{t(K_F - \sum_i^\beta D_k - S_v)}{(1+i)^{\beta-\alpha}} + \frac{K_w}{(1+i)^{\beta-\alpha}} + \frac{S_v}{(1+i)^{\beta-\alpha}} \qquad (10.5.17)$$

Note that all components of the NPW are positive, and the NPW will remain positive (implying plant should not be shut down) unless the operating margins $(\Re - E)_k$ become appreciably negative. That is, even if costs exceed sales revenues, the plant may continue to be an economically operable venture.

For the case of $\beta > \gamma$ and $\alpha > \gamma$, the plant has been depreciated and the fixed capital is sunk. For that case the second and third terms in Eq. (10.5.17) disappear; moreover, for sunk capital, S_v is usually zero. The usual decision for old, depreciated plants is: Do we shut down, or do we operate another year?

For shutdown,

$$\text{NPW}_0 = K_w \tag{10.5.18}$$

For one more year ($\beta - \alpha = 1$),

$$\text{NPW}_1 = \frac{(1-t)(\mathcal{R} - E) + K_w}{1 + i} \tag{10.5.19}$$

At parity,

$$\mathcal{R} - E = \frac{iK_w}{1 - t} \tag{10.5.20}$$

NOMENCLATURE

$a, b, c, \ldots$	constants
$A, B, C, \ldots$	constants
A	gauge pressure, atm
b	installation-cost factor
c	equivalent annual cost, \$/year
C	cash flow, \$/year
C_R	replacement cost, \$: see Eq. (2.3.15)
D	annual depreciation charges, \$/year
e	sales increase, \$/\$ of TS&D investment
E	annual out-of-pocket expenditures, \$/year
f	cost-penalty factor
$\mathbf{f}$	annual inflation factor
F	annual fixed charges excluding depreciation, \$/year
$\mathcal{F}_K$	capitalized-cost factor, Eq. (2.3.14)
$\mathcal{F}_R$	capital-recovery factor, Eq. (10.1.3)
$(\mathcal{F}_R)_0$	capital-recovery factor computed with i_0
G	annual general expenses, \$/year
i	nominal interest rate, fractional
$\mathbf{i}$	effective discrete interest rate, fractional; $\mathbf{i}_0$ is the corporate internal interest rate
I	total investment, \$
j	number of inequalities, as in Eq. (10.3.13); also year during which demand increases as a step function
k	index year
K	capital investment, \$; K_F is fixed capital, K_w is working capital; K_R is research investment (which may, however, be handled as an expense)
L	a function: see Eq. (10.5.10)
m	number of years between equipment replacement; also, lead time between start of research and generation of revenue
n	lifetime of project, years; also, indexing limit
N	number of years to demand maturity

p	production, kg/year
p	probability of success
P	principal, or present worth of annuities (no subscript); with subscript, annual profit, \$/year: P_B is profit before taxes, P_A is profit after taxes
q	probability of failure = $1 - $ **p**
Q	permeation rate, gal/(h · ft²)
r	discounted cash rate of return for discrete discounting (effective); $\mathbf{r}_D$ is the differential DCRR
R	annuity; equivalent annual cost of capital, \$/year
R'	equivalent annual credit on salvage, \$/year
$\mathcal{R}$	annual revenue from sales, \$/year
s_v	salvage factor
S	accumulated sum. Also, capitalized cost, \$. S_{00} is capitalized cost of investment plus expenses
S_v	salvage value
$S_1, S_2, \ldots$	dummy variables, Eq. (10.3.15)
t	income tax rate; also, time, years
W	purchased cost, \$
x, y	variables; cartesian coordinates
Y	time index, years
α	time span in years from start-up
β	time span from start-up to shutdown, years
γ	tax life, years
Δ	differential, as in ΔP_B, ΔROIBT, etc.
θ	thickness, in; also, alternate time symbol
ρ	production ratio
τ	break-even time, years
ϕ	objective function
ω	dollars of new capital per dollar of research

Subscripts

A	after taxes; also used for indexing $(A, B, \ldots)$
B	before taxes
D	differential
E	equivalent; C_{AE} is equivalent ("phantom") cash flow after taxes
F	fixed (as in K_F); also, failure
i, j, k	indexing symbols
R	research; replacement
S	success
w	working (as in K_w)

Abbreviations

CFS	capital for sale
COM	cost of manufacture
CSTR	continuous stirred-tank reactor
DCRR	discounted cash rate of return
DFC	direct fixed capital
EUAC	equivalent uniform annual cost

Max (ϕ)   function maximum
Min (ϕ)   function minimum
NPW   net present worth
PWR   present worth ratio
ROI   return on investment
ROIBT   return on investment before taxes
SL   straight line
TS&D   technical service and development

REFERENCES

Allen, D. H.: "A Guide to the Economic Evaluation of Projects", The Institution of Chemical Engineers, London, 1972.

Baasel, William D.: "Preliminary Chemical Engineering Plant Design," Elsevier, New York, 1980.

Bland, Robert G.: The Allocation of Resources by Linear Programming, *Sci. Am.*, **244** (6):126 (1981).

Happel, John, and Donald G. Jordan: "Chemical Process Economics," Marcel Dekker, Inc., New York, 1975.

Kasner, Erick: Break-Even Analysis Evaluates Investment Alternatives, *Chem. Eng.*, **86**:117 (Feb. 26, 1979).

King, C. Judson: "Separation Processes," 2d ed., McGraw-Hill, New York, 1980.

Kroeger, Herbert E.: When Should You Lease?, *Chem. Eng.*, **72**:73 (Feb. 1, 1965).

Newnan, Donald G.:"Engineering Economic Analysis," Engineering Press, San Jose, Calif.,1976.

Peters, Max S., and Klaus D. Timmerhaus: "Plant Design and Economics for Chemical Engineers," 3d ed., McGraw-Hill, New York, 1980.

Resnick, William: "Process Analysis and Design for Chemical Engineers," McGraw-Hill, New York, 1981.

Rose, L. M.: "Engineering Investment Decisions: Planning under Uncertainty," Elsevier, Amsterdam, 1976.

Sarin, R. K., A. Sicherman, and K. Nair: Evaluating Proposals Using Decision Analysis, *IEEE Trans. Syst., Man Cybern.*, **SMC-8** (2):128 (1978).

Stermole, Franklin J.: "Economic Evaluation and Investment Decision Methods," 2d ed., Investment Evaluations Corporation, Golden, Colo., 1974.

Thermal Insulation Manufacturers' Association: "ECON-I: How to Determine Economic Thickness of Thermal Insulation," Mt. Kisco, N.Y., 1973.

Uhl, V. W., and A. W. Hawkins: "Technical Economics for Engineers," AIChE Continuing Education Series, American Institute of Chemical Engineers, New York, 1971.

PROBLEMS

10.1 Consider the choice between installing new rubber lining in a used steel tank to provide process storage for an aqueous pyridine hydrochloride stream, or else purchasing a new FRP* tank for the same job. The total installed cost of the new tank is $10,000, and the tank

*Fiber-reinforced plastic.

will last 10 years (zero salvage). The rubber lining in the steel tank will last 3 years, at which time the lining will have to be replaced. Money is worth 8 percent compounded annually. On the basis of equal capitalized costs, how much can be spent for the rubber lining?

10.2 Using the method of induction, revise Eq. (2.3.13) to account for inflation f. Repeat Prob. 10.1 for the case of $i = 0.17$, $f = 0.09$.

10.3 Basing your judgment upon Eq. (2.3.19) or the results of Prob. 10.2, what do you think is the relationship between i_0, the inflation-adjusted corporate internal rate of return, and i, the rate of return based upon inflated corporate earnings? Which of the two rates do you think is best used in break-even analysis?

10.4 A 5000 gpm stream of C_2Cl_4 (tetrachloroethylene, sp gr 1.62) is circulated to the top of a large scrubbing tower against a 60-ft total dynamic head. A steel centrifugal pump is currently in use for the purpose, but the pump needs to be replaced because of serious corrosion. Past experience has shown that steel pumps must be replaced every 4 years, in spite of the fact that traces of corrosive acid are occasionally removed from the circulating tetrachloroethylene by caustic neutralization at an annual cost of $2000.

It has been proposed that the steel pump be replaced with a pump made from a high-nickel alloy; it is argued that the pump is much more corrosion-resistant and will therefore last longer, and that the caustic neutralization may be omitted. The motor currently in use may be used with either pump and is expected to last indefinitely; it has an energy efficiency of 80 percent.

Estimate the minimum service life that must be expected from the alloy pump to justify its selection, given the following information:

	Steel pump	Alloy pump
Purchased cost, $	12,000	31,500
Replacement installation cost, $	2000	2000
Energy efficiency, %	70	75
Salvage value, $	None	8000

Plant stream factor: 0.85
Delivered power cost: $2\frac{1}{2}$¢/kWh (note 1 hp = 746 W)
Time value of money: 8% (inflation-adjusted, after taxes)
Tax rate: 50%
Routine maintenance costs equal for both pumps

Does installation of the alloy pump appear justified to you?

10.5 Four different reactor choices for a polymerization reaction have been sized and costed out as follows:

Reactor type	Installed cost, $	Annual operating costs (excluding depreciation), $
A	150,000	45,000
B	190,000	35,000
C	210,000	32,000
D	170,000	42,000

All four reactors would have a tax and service life of 5 years, and straight-line depreciation is contemplated. Use a ΔROIBT analysis to choose the best reactor. Neglect salvage value.

10.6 Two types of incinerator are available for burning solid waste generated by a chemical operation:

	Incinerator alpha	Incinerator omega
Installed cost, $	750,000	500,000
Annual variable expenses, $	20,000	29,000
Annual maintenance, insurance, taxes, and factory expense, $	38,000	35,000
Service life, years	20	10

If salvage value for either incinerator is zero, which one should be chosen? Use $i_0 = 0.13$.

10.7 Kasner (1979) explores the effect of the investment tax credit upon break-even analysis. Suppose two systems are available for the removal of particulates from stack gases; both are eligible for a combined first-year investment and energy tax credits of 20 percent.

	System A	System B
Equivalent uniform annual operating cost, $	15,000	12,000
Investment, $	50,000	100,000
Service life, years	5	10

Assume that the tax credits accrue during the first year of operation and may be discounted to zero time at the after-tax rate of return $i_0 = 0.10$. Compute the capital-recovery cost on the basis of the zero-time investment corrected for the tax credit.

Which pollution control system do you recommend?

10.8 Use break-even analysis to determine how many years of operations must be guaranteed before the more expensive of two available catalyst pellet extruder alternatives is accepted. The more expensive extruder costs $50,000 more, but it results in equivalent uniform annual operating cost savings (after taxes) of $10,000. Base the comparison upon a corporate internal rate of return (after taxes, inflation-adjusted) of 9 percent.

10.9 Three different packaged units for supplying a high-temperature heat-transfer medium are available; all are rated at 1×10^8 Btu/h. Any of the three is expected to stay in service for the projected 10-year life of the plant.

	System		
	A	B	C
DFC, $	800,000	973,000	1,175,000
Annual fuel cost, $	4,200,000	3,900,000	3,700,000
Annual power cost, $	100,000	110,000	230,000

Fixed charges (excluding depreciation) are 10 percent of DFC/annum. Salvage at the end of 10 years is negligible.

a Use ΔROIBT (straight-line depreciation) to pick the best choice. Use the criterion ΔROIBT $\geq$ 25 percent.

b Will the annual cost method (break-even analysis) result in the same choice for $i_0 = 14$ percent, $t = 0.46$?

10.10 A mixer-settler extraction train is to be built to extract an aqueous solution of a valuable metal. At this point in the process the metal is valued at 20 cents/lb. Any metal not extracted is lost to a tailings pond. From the information given, specify the optimum economical number of extraction stages, and state your reasoning.

Feed: 1×10^6 lb metal/year
Equipment life: 5 years

	Number of stages			
	2	3	4	5
Capital investment, $	25,000	35,000	44,000	52,000
Recovery, %	75	95	98	99.5
Annual operating cost (excluding depreciation), $	6,000	8,000	10,000	11,000

10.11 Three investment alternatives are listed in the table below, and you are to recommend which one your company should choose, and why.

Company policy sets the following economic evaluation criteria:

1 An acceptable project must show at least 20 percent return on investment before taxes (ROIBT).

2 If a project with a higher capital for sale (CFS) is favored, it must show an incremental ROIBT of at least 30 percent.

	Project number		
	1	2	3
Direct fixed capital, $ $\times 10^3$	5,800	7,600	12,000
CFS, $ $\times 10^3$	6,000	8,200	12,500
Projected net annual cash flow after taxes, $ $\times 10^3$	1,200	1,800	2,300

Each project has a 10-year life, no salvage value. Assume straight-line depreciation.

10.12 A choice of alternatives is occasionally encountered in the course of designing very large and expensive processing units. If a large unit is subjected to unscheduled shutdowns, the loss of production may result in unacceptably large income losses after only a few days. Partial protection against such a catastrophe is afforded by constructing the plant as a two-parallel-train system, each operating at one-half the total rated capacity. If one train must be shut down due to equipment malfunction, production may at least continue at half capacity—provided, of course, that both trains do not suffer an unscheduled shutdown simultaneously, a much less likely occurrence. If maintenance requirements demand a certain number of *scheduled* shutdown days each year (accounted for by a specified "stream factor"), the maintenance jobs may be performed in one train while the other is still operating, and vice versa. However, in this case the two-train scheme does *not* result in any increased annual production (why?). The greater direct fixed capital investment of the two-train design must be justified by an appropriately greater annual profit engendered by the larger production, as confirmed by the ΔROIBT concept.

Suppose you were responsible for the design of a large styrene plant rated at a nominal capacity of 3 million lb/day. The direct fixed capital investment for a single-train plant is $134 million. You must decide whether a two-train scheme is justified. How many unscheduled days would either the one-train unit or one of the two-train units have to be shut down before the two-train unit was justified, based on ΔROIBT = 20 percent? (The premise here is that if the one-train unit is shut down for d unplanned days, *on the average* only one of the two-train units will be shut down for d unplanned days, a premise that is certainly debatable.)

Use the following parameters:

Sales price: 58¢/lb
Variable charges: 34¢/lb
Regulated charges and general expenses: same for either scheme
Working and allocated capital: same for either scheme
Depreciation: straight-line, 10 years, no salvage
Designed stream factor: 0.95

Based on your answer, do you think a two-train scheme is justified?

10.13 A proprietary adhesive has been prepared in equipment centered about a batch reaction kettle. The glass-lined vessel is expected to last no more than 2 years, and at that time the vessel (and some peripherals) is expected to be replaced at a cost of $250,000, with the further expectation that production of the adhesive will continue for another 10 years (that is, 12 years from now).

The R&D department has proposed that during the next 2 years (before the equipment has to be replaced anyway) research be undertaken into a continuous processing arrangement. Continuous equipment is expected to cost quite a bit more ($800,000 fully installed, including controls), but it is claimed that improved quality control, operating discipline, and reduced operating costs will save $250,000/year compared with present operations.

How much annual research money (expended at the start of each year for the next 2 years) can be spent to justify the continuous process? Determine this by equating the NPW of the two alternatives calculated at the specified rate of return of 20 percent after taxes.

In your opinion, is the research justified?

10.14 The vacuum devices listed in the table (top of p. 587) are available for removing 10 lb of N_2/h (water-saturated at 80°F) from equipment that is to be maintained at an absolute pressure of 100 mmHg. Which device is preferred? Use $i_0 = 0.10$.

10.15 Determine the discounted cash rate of return **r** for each of the following two mutually exclusive projects A and B. If the project with the higher investment has the better DCRR, determine the ΔDCRR, and, with the criteria DCRR > 0.1 and ΔDCRR ≥ 0.15, determine which project should be chosen. For each project, assume a 14-year lifetime, invariable

	Liquid jet eductor	Steam ejector	Liquid-ring vacuum pump
In-place installed cost, $	10,000	11,700	11,000
Annual maintenance, fixed charges, $	1,500	800	2,500
Operating requirements:			
Power, kW	7		4
Steam, lb/h		144	
Cooling water, gpm	4.5	10	1
Service life, years	15	20	7

working capital throughout the lifetime, and zero salvage. The corporate internal rate of return is 12 percent. (All numbers in the table are millions of dollars.)

Use of a programmable calculator is recommended.

	Project A	Project B
Investments:		
Fixed capital, time zero	13.3	29.4
Working capital, time zero	2.0	4.0
Fixed capital, end of year 7	6.1	4.8
After-tax cash flows:		
Year 1	1.0	2.0
Year 2	1.2	4.0
Year 3	1.5	6.0
Year 4	2.0	7.0
Year 5	3.0	10.0
Year 6	4.7	14.0
Year 7	5.3	10.2
Year 8	5.0	6.6
Year 9	4.7	4.4
Year 10	4.0	3.4
Year 11	3.0	2.4
Year 12	2.0	2.0
Year 13	1.5	2.0
Year 14	1.0*	2.0*

*Does not include working capital returned.

10.16 The differential DCRR concept may also be used to choose equipment alternatives that are expected to have the same service life.

An organic crystal product may be successfully washed and dewatered by using a semibatch basket centrifuge or a continuous pusher centrifuge. The in-place cost of the basket machine is $73,000; the corresponding cost of the pusher is $108,000, but labor and utilities costs are expected to be $14,000 less per annum. Both units will have a service (and tax) life of 7 years.

Use the criterion of $r = 20$ percent to decide which machine is preferred. Assume straight-line depreciation and a tax rate of 46 percent.

What would be your decision by using the ΔROIBT criterion?

10.17 A small corporation has $250,000 to invest. Its current internal rate of return after taxes is $i_o = 0.08$.

Two proposals have been made for the investment:

a A 7-year project which is expected to generate after-tax annual cash flows of $75,000.

b The money can be loaned for 15 years at an effective before-tax interest rate of 17 percent.

The corporation expects to pay total federal and state income taxes of 40 percent. Which investment should be chosen?

10.18 Either one of two mutually exclusive projects must show an after-tax rate of return of $i = 0.20$. Use a benefit-cost ratio analysis (ΔPWR analysis) to see whether the more expensive project is justified.

	Project A	Project B
Investment, $ $\times 10^6$	17.0	27.0
Project life, years	15	8
Annual after-tax cash flow generated, $ $\times 10^6$	8.3	14.7

10.19 An item of equipment has a capitalized cost of $85,000, computed at an inflation-adjusted after-tax rate of return of 12 percent. Service and tax lives are 10 years, and the corporation uses straight-line depreciation. Cost of installation and salvage value are both 10 percent of the purchased cost.

For a tax rate of 50 percent, what is the EUAC (after taxes)?

10.20 In Chap. 1 a rule was stated that TS&D efforts should generate, over a period of 10 years, about $1/year increased sales per dollar invested in the TS&D effort. The rule may be proved by setting up the following two alternatives:

A. Invest the $1 at the standard corporate rate of return after taxes of 10 percent. Existing facilities will show no gain in sales.

B. Invest the $1 in TS&D activities (a tax-deductible expenditure) which will result in a sales increase of e dollars/year for the basis time of 10 years.

a What is the PW of alternative A?

b Assume that the corporate investments I generate annual sales in conformity to the industry's average turnover ratio (Chap. 5). What are the total annual sales? For $i_o = 0.10$ and an average investment life of 10 years, what is the after-tax corporate cash flow? How much cash flow, then, does $1 of sales generate?

c What is the PW of the sales increases e?

d What is then the PW of alternative B, assuming a 50 percent tax rate?

e What is the value of e for equal PWs of the two alternatives? Do you agree with the rule as stated?

10.21 Use benefit-cost ratio analysis to select the best project among the six listed. The PW of the total generated after-tax cash flows has been computed at the desired rate of return. All amounts shown are in millions of dollars.

	Project					
	1	2	3	4	5	6
Investment	5.33	2.67	8.00	1.33	12.00	13.33
PW of cash flows	9.75	6.25	11.60	1.78	12.00	12.64

10.22 The board of directors must decide how to allocate available investment capital among projects submitted for the next year. In the following list of nonexclusive projects, the project parameters have been simplified to show:

Initial investment (in millions of dollars)
The project lifetime
Anticipated annual after-tax cash flows generated (in millions of dollars)
The board's assigned rate of return (after taxes) required of each project (discrete annual discounting)

Project	Investment	Life, years	Annual cash flow	Rate of return, %
A	50	10	13.8	10.0
B	300	4	173.8	20.0
C	100	10	33.6	25.0
D	50	2	60.0	15.0
E	100	10	13.1	7.5
F	200	10	81.4	10.0
G	100	6	18.0	20.0
H	100	10	41.5	12.5
I	300	10	150.3	20.0

The board does not want to borrow capital this year; only retained earnings of $650 million are available for capital budgeting. Which projects should the board select?

10.23 Three proposals have been submitted for a capital budgeting decision. The proposals are mutually nonexclusive, but two of the proposals consist of two or more mutually exclusive alternatives. Any proposal must meet the minimum criterion of 14 percent rate of return after taxes; $300 million are available for the budget, and an exact match is required. Do the criteria imposed allow an exact match, and if so, what spectrum of proposals and alternatives should be picked?

Proposal	Alternative	Investment, $ \times 10^6$	PW of cash flows, at 14%
1	A	50	70
2	A	100	120
	B	150	140
3	A	100	180
	B	150	200
	C	200	230

10.24 Equipment alternatives which have associated out-of-pocket expenditures may also be compared by using the capitalized-cost concept. The portion of the capitalized cost accounting for the operating expenditures is computed as a perpetuity. Note that a depreciation tax credit is subtracted from the tax-adjusted out-of-pocket expenditures.

a Use the modified capitalized-cost approach to compare the two listed incinerators. Use discrete discounting ($i_0 = 0.10$), a tax rate of 45 percent, straight-line depreciation, and equal tax and service lives. Neglect salvage.

	Incinerator A	Incinerator B
Investment (total), $	120,000	170,000
Life, years	7	10
Out-of-pocket annual expenditures, $	50,000	40,000

b What is the relationship between this capitalized cost, S_{00}, and the EUAC [as in Eq. (10.1.5)]?

10.25 A computer to control a plant packaging a spectrum of health products may be purchased or else leased from the manufacturer. The equipment is projected to have a useful life of 10 years. The purchased cost is $125,000; if leased, the monthly charges will be $1700 for 10 years. The leasing agreement includes a maintenance contract guaranteeing up to 100 h/year of maintenance and repair service, valued at $5000/year. If purchased, the salvage value of the computer at the end of 10 years will be $25,000.

Decide whether the computer should be purchased or leased. As in Example 10.2.5, make the comparison as if purchasing involved borrowing the capital at the prime rate of interest—for computations, assume that the prime rate is 18 percent, the combined income tax rate 50 percent. To simplify some of the computational complexities, make the comparison by calculating the PW of the two alternatives by using an internal rate $i_0 = 12$ percent. For the purchasing alternative, the PW total includes:

a PW of annual payments on "borrowed" capital
b PW of maintenance expenses, assumed to be equal to maximum benefit under maintenance contract
c PW of tax credits on depreciation
d PW of tax credits on interest payments [use Eq. (2.3.23) as basis]
e PW of salvage

10.26 Go through the numerical calculations for all branches of Fig. 10.2.5 and, using the rollback technique, confirm the position value at decision point 1.

10.27 Calculate the position value of decision point X in Fig. 10.2.6. Use required data in the accompanying text.

10.28 In Fig. 10.2.5, suppose the fourth option at decision points 2 to 7 were to keep the semiplant and not to build the main plant. Make a survey of the network to decide whether the possibility of such an option would affect the strategy planning at decision point 1.

10.29 The economic optimization of plant size was discussed in Chap. 8. Figure 8.3.6 illustrates the situation of a maximum in the curve relating ROIBT to plant size. It was pointed out that the maximum is not necessarily the optimum. A reasonable approach is to plot the profit versus capital for sale and to pick the capital (and corresponding plant size) at which the minimum ΔROIBT criterion is met.

Repeat the solution of Prob. 8.16 and select the plant size at which the criterion of $\Delta ROIBT \geq 35$ percent is satisfied.

10.30 Derive an equation analogous to Eq. (10.5.13) for the break-even time τ for comparing debottlenecking strategies II and III when operating costs are considered. Which strategy is preferred for $\rho = 2.0$, $i = 0.15$, $j = 5$?

10.31 (Adapted from Resnick, 1981*). A process intermediate is prepared in a batch reactor. The intermediate is unstable and must be used in the subsequent process immediately after preparation. The preparation is quite "temperamental," and the probability of obtaining an acceptable batch is **p**. The total cost per batch is c; if the batch is unacceptable, it must be discarded, but the subsequent process is delayed, resulting in a cost penalty of fc.

It has been proposed that each time a batch is needed that N batches be prepared in N reactors. Any bad batches or unneeded acceptable batches would be discarded at no extra gain or loss. Show that the number of reactors resulting in minimum total cost is given by

$$N = \frac{\ln \dfrac{1}{f \ln [1/(1-\mathbf{p})]}}{\ln (1 - \mathbf{p})} \tag{P10.1}$$

What is N for $\mathbf{p} = 0.5$, $f = 25$?

10.32 In Example 10.3.2, the optimum number of effects in a multiple-effect evaporator system was found to be two, if the criterion $\Delta ROIBT = 25$ percent was used. How many effects would be indicated if the "traditional" optimum were used,

$$\frac{dP}{dI} = 0 \tag{P10.2}$$

10.33 One thousand gallons per hour of an aqueous solution of biologically active macromolecules is to be processed through an ultrafiltration device. From the data given, calculate the optimum pressure (expressed in atmospheres gauge) for the ultrafiltration.

a The capacity of the ultrafiltration membranes is Q (gallons per hour per square foot of membrane area) $= 0.1A$, where A is the gauge pressure in atmospheres. The complete installed cost of the ultrafiltration device (but not counting the pump cost) is $30/ft^2$ for a device operating at 10 atm. The cost is proportional to $A^{0.6}$.

b A 3-hp pump and motor would be required to forward 1000 gph of the solution from atmospheric pressure into a device operating at 10 atm. The complete installed cost of the 3-hp pumping unit is $10,000, and the cost varies as the 0.33 power of the horsepower rating.

c The only out-of-pocket operating expenses varying with size are fixed charges (10 percent of investment/year not counting depreciation) and power (4.8¢/kWh).

d Stream factor $= 0.90$; salvage $= 0$; tax rate $= 50$ percent; internal rate of return $i_0 = 12$ percent; tax life = service life = 10 years; straight-line depreciation; 1 hp = 0.746 kW.

10.34 Sarin et al. (1978) describe how elements of decision analysis may be used to

---

*Copyright 1981, McGraw-Hill Book Company. Used by permission.

evaluate alternative proposals. Refer to the article and outline how you would go about selecting objectively the most acceptable of several proposals submitted by various research institutions to build and run large-scale solar energy experimental stations!

10.35 In the per-tet process, carbon tetrachloride (CCl_4) and perchloroethylene (C_2Cl_4) are coproduced in any desired ratio by chlorinating hydrocarbon feedstocks. The only important byproduct is anhydrous HCl. A combination of methane and propane is typically used as feed.

Specify the best combination of feedstocks (CH_4, C_3H_8) and products (CCl_4, C_2Cl_4) to maximize plant profits, assuming, of course, that the particular product spectrum can be sold. The chlorine input to the plant is to be kept at a fixed level determined by demands of the chloralkali complex. The costs of utilities, labor, and the fixed charges are unaffected by variations in feed or product spectrum.

Raw material costs:

Cl_2 $4.97/(lb \cdot mol)$
CH_4 21.3¢/(lb · mol)
C_3H_8 75.0¢/(lb · mol)

Product selling prices:

CCl_4 $28.08/(lb · mol)$
C_2Cl_4 $27.80/(lb · mol)$
HCl $2.74/(lb · mol)$

10.36 Your company produces two polymer blends with the trademarks "Supreme A" and "Kotrand." Each is a blend of five polymers as follows:

	Polymer ingredient, % by weight				
	PVA	PVC	ABS	PS	PAN
Supreme A	23	17	31	20	9
Kotrand	9	42	10	3	36

The value added by manufacture (Chap. 7) is $2.50/kg for Supreme A, $1.25/kg for Kotrand.

The purchasing department can locate the following maximum annual supplies of specification-grade polymer ingredients (in metric tons):

PVA 2500
PVC 3000
ABS 7000
PS 6000
PAN 3000

What production strategy do you recommend to maximize value added by manufacture?

10.37 A batch semiplant installation can be used to produce three different products, α, β, and γ. Each batch produced consists of 10,000 lb of product. The following are labor and equipment requirements per batch, expressed as 8-h shifts:

Product	Labor	Equipment
α	1	2
β	3	1
γ	2	1

In the case of product α, the reactor mass must be stirred for one shift without labor attendance. To make products β and γ, additional labor-intensive processing is required away from the semiplant installation.

Nine hundred annual shifts of labor are available, and these must all be used. The semiplant equipment, however, is available for only 600 shifts because of maintenance problems. α sells for $4/lb, β for $5, γ for $6. What production strategy will maximize annual sales?

10.38 A project involving an initial investment of 10×10^6 and a 10-year life has a NPW of zero. What is the EUAC associated with the project?

10.39 In Example 10.4.1, the EUAC was computed for an inert gas generator having a projected service life of 12 years. Suppose the equipment has, in fact, an indefinitely long service life. For what service life value (of m years) will the EUAC be a minimum?

10.40 A company invests $5 million in a manufacturing process which gives positive net annual cash flows after tax of $1 million in each subsequent year. During the third year of operation an alternative process becomes available. The new process would require an investment of $4 million which would give positive net annual cash flows after tax of $1.8 million in each subsequent year. The after-tax value of capital is 10 percent (discrete annual compounding). It is estimated that a market will exist for the product for at least a further 7 years. Should the company continue with the existing process (project A) or should it scrap project A and adopt the new process (project B)? The IRS will not grant an obsolescence allowance. The original plant had a predicted life of 10 years with no salvage value, but the early scrapping would yield a salvage value of $1 million.

10.41 An older plant has a number of field-located instruments which must be routinely checked by operators. It has been estimated that the associated labor cost amounts to some $50,000/year, and that partial automation could potentially save this cost (about half an operator per shift). Installation of automatic instruments logged in the control room will cost $250,000. Annual maintenance costs will increase by $10,000, and the installation will be in use for 10 years. If the corporation's internal rate of return after taxes is 8 percent, is the improvement justified?

10.42 A glass-lined reactor in a completely depreciated CSTR train has developed a crack in the lining, and the vessel will have to be repaired or replaced. The repairs will cost $12,000 and will result in reduced production equivalent to an after-tax cash flow loss of $25,000. The repair is guaranteed for 1 year.

A new vessel may be installed at a total cost of $135,000 and is expected to have a service (and tax) life of 10 years. The replacement will also result in a $25,000 loss of cash flow. The vessel supplier is ready to pay $5000 in trade-in for the damaged vessel.

For $\mathbf{i}_0 = 0.15$, should the vessel be repaired or replaced? Make the comparison on the basis of the anticipated cost of each alternative over the next year; use $t = 0.50$.

10.43 How large an investment may be made to save $1 of energy cost/year for 10 years? Make the following assumptions:

Out-of-pocket expenses to operate new facility: 20% of investment/annum
After-tax required rate of return: 15%
Tax rate: 50%
Tax life: 10 years (straight-line depreciation)
Investment and energy tax credit: 20% in first year

10.44 Modify Eq. (10.5.13) for the case of salvage at the end of j years equal to sK_1. Decide between strategies I and II for $\rho = 1.5$, $j = 7$, $s = 0.2$, $i = 0.12$.

10.45 The plant expansion strategies considered in the text involved construction of plants of equal size.

a Suppose the demand growth shown in Fig. 10.5.1 is known with certainty. Are plants of equal size necessarily the best strategy? What, for example, is the optimum year to build the second plant in a two-plant strategy if $i = 0.10$, $N = 10$?

b Rose (1976, p. 265) argues that for linear demand growth of indefinite length the optimum strategy is, indeed, equal plant increments. Do you agree with this assessment?

10.46 A completely depreciated production unit has $2 million of identifiable working capital. Product demand has decreased to 20×10^6 lb/year; the sales price is 50 cents/lb, the total production expenses are 45 cents/lb. Corporate management feels that the old plant must demonstrate an after-tax rate of return of 20 percent if operations are to continue. Should the plant be shut down?

CHAPTER 11

ENGINEERING MANAGEMENT OF CONSTRUCTION PROJECTS

The various stages of development of a project are climaxed by the construction of facilities which, in turn, reflect the complexity of each stage. The construction effort may involve a modest pilot plant, or, in the final stages of a successful developmental sequence, it may encompass the facilities of a large commercial unit. Occasionally such commercial units are very large, indeed, and the task of construction becomes a project of awesome complexity, with gargantuan demands upon capital expenditure and human organization. Regardless of scale, however, construction projects are characterized by much the same organizational sequence, and their success is to a considerable extent dependent upon the organizational ability of the construction project manager. In the chemical process industry, the duties of construction project manager are frequently assumed by a chemical engineer.

Some familiarity with the sequence of operations and tasks characterizing the construction of facilities is important for the appreciation and proper evaluation of the project that those facilities represent. In what follows, an attempt will be made to impart some insight into the nature of construction-oriented tasks by outlining the needed organizational effort to carry a project forward from the laboratory up to the point of decision whether a full-scale production unit should be built. In fact, it is this sequence that chemical engineers normally have to deal with in the earlier stages of their industrial career—the developmental sequence first outlined in Chap. 3 and involving the construction of miniplants, pilot plants, semiplants. Experience gained as part of these more modest construction efforts is invaluable in helping the chemical engineer to assume and carry out the more burdensome duties of project management at the full-production plant scale.

A review of the organizational effort required for carrying out the various pilot-plant-scale construction tasks is thus a useful introduction to large-scale project management. The proper organization and control of large-scale projects is aided by the use of checklists and forms which act as reminders and reporting formats for the project manager; each corporate construction division or construction company has its own collection of such construction aids. We will not concern ourselves with the detail of these aids; publications are available that do describe such lists and forms (Landau, 1966; Ludwig, 1974; Hajek, 1977; Roth, 1979). We will, nevertheless, review the general nature of the tasks required of the manager of a large construction project to give the neophyte some concept of the magnitude of the organizational effort that must be undertaken. We will concern ourselves more specifically with the methods, used as part of the plant construction sequence, to answer the two questions:

Will the plant be finished within the scheduled time?
Will the costs be confined to the budgeted amount?

In addition to techniques that are used to monitor these fundamental project evaluation questions, techniques will also be outlined which are commonly used to monitor the safety, environmental health, and loss prevention aspects of the facilities under construction.

The material in this chapter really does little more than scratch the surface of a complex discipline—it is but a brief introduction. But then construction project management is hardly the discipline to be acquired from textbooks; it can be outlined and systematized to some extent, but its successful practice is ultimately dependent upon a background of personal experience, innate organizational ability, and good judgment.

11.1 THE MANAGEMENT OF THE PROCESS DEVELOPMENT SEQUENCE

The Substance of Process Development

In Chap. 3 the process development sequence was first described in terms of a number of developmental "landmarks." These were briefly outlined to emphasize the need for expanded and more thorough project evaluation techniques to reflect the increased complexity of succeeding stages of development. Our present purpose is to see how one goes about organizing the various steps of the process development sequence. We will attempt to demonstrate this by means of a step-by-step guide which lists some of the detail of the tasks that the project manager must tackle at various stages of the sequence. It is too much to expect a short written guide to be all-inclusive; our purpose will be to impart some degree of insight into the nature of the organizational effort, rather than to outline a guide that is to be followed religiously. Some aspect of the developmental sequence often constitutes one of the first assignments of chemical engineers in industry; nevertheless, it is an important assignment, perhaps the most challenging one in an individual's career.

SECTION 11.1: THE MANAGEMENT OF THE PROCESS DEVELOPMENT SEQUENCE

The developmental landmarks of a project include several potential phases of construction beyond the laboratory stage, but preceding the decision to build a full-scale production unit. These intermediate facilities have various appellations, but we will confine our discussion to three readily distinguishable categories: miniplants, pilot plants, and semiplants. Not all these are necessarily built as part of a single process development sequence, but frequently they are. What, then, is the purpose of this costly, time-consuming construction and operating sequence?

In spite of our advanced level of systematic knowledge and experience in process synthesis, equipment design, and process scale-up, experimental work at various scale-up levels is needed to prove out all aspects of a process, to avoid surprises in the full-scale plant. It is necessary to define the full spectrum of the interrelated processing steps in all their important detail to assure smooth, efficient, economical, and safe operations in the production unit. We often accomplish this by trying out the visualized process in equipment of increasing size and complexity; we may not be too sure of ourselves, so, just like the timid swimmer, we get in a little deeper a step at a time, and we learn with each step. Many processing steps are regrettably too complex to be described in simple enough mathematical terms to permit unequivocal scale-up to the production scale from the laboratory. More often than not it is difficult during early stages of process development *to define all the important parameters that fully describe each processing step*; many such parameters are overlooked, not really thought of, or unnoticed on a small scale, but their importance is gradually revealed by increasing the scale of operations and gaining a better understanding.

Example 11.1.1

A bench chemist places 250 ml of liquid A [sp gr = 1.20, specific heat = 0.45 cal/(g · K)] into an uninsulated 500-ml round bottom flask. The liquid is heated to 100°C, and then over a period of 30 min gaseous reagent B is added with gentle stirring. At this time the reaction is completed, and the chemist reports "no exotherms"—the reaction temperature remained at about 100°C with no additional heating. The conclusion is made that over 30 min the heat of reaction is mild enough to avoid heating the product to its decomposition temperature of 150°C.

What, in fact, would happen if the reaction were undertaken with 1500 gal of liquid A, preheated to 100°C, placed in a 6-ft-diameter by 10-ft glass-lined jacketed vessel (with cooling water in jacket), and sparged with gas B over 30 min?

SOLUTION The surface area of a sphere with volume = 500 ml is 0.330 ft^2; for 250 ml of liquid, the area in contact with glass (neglecting "swirl" due to stirring) is 0.165 ft^2.

For uninsulated glass flasks, the overall heat-transfer coefficient from the stirred liquid to air is about 5 Btu/(h · ft^2 · °F).[*] Assume air temperature of 25°C in laboratory; then, for experiment, $\Delta T = 75°C$ or 135°F.

Therefore, in laboratory experiment,

$$Q = UA\,\Delta T = 5 \times 0.165 \times 135 = 111.4 \text{ Btu/h}$$

---
[*]It is much smaller in the vapor space of the flask.

and, during the 30 min of addition of gas B, heat generated per gallon (250 ml ≡ 0.0661 gal) is

$$111.4 \times 0.5/0.0661 = 842.7 \text{ Btu/gal}$$

In 1500 gal, heat generated in 30 min is

$$842.7 \times 1500 = 1.264 \times 10^6 \text{ Btu}$$

$$\text{Height of liquid in vessel} = \frac{1500}{7.48 \times \frac{\pi}{4}(6)^2} = 7.09 \text{ ft}$$

Assume jacket on large vessel includes base of vessel. Then jacket area in contact with liquid is

$$\pi \times 6 \times 7.09 + \frac{\pi}{4}(6)^2 = 161.9 \text{ ft}^2$$

For jacketed glass-lined vessels with agitation, $U \approx 30$ Btu/(h·ft²·°F). Let final temperature of contents be T°F, and assume temperature increases approximately linearly during the 30-min addition. Then average temperature of liquid is $(212 + T)/2$. Take cooling water temperature of 77°F. Then total heat transferred to cooling water in the large vessel is

$$30/60 \times 30 \times 161.9 \left(\frac{212 + T}{2} - 77\right) = 1214.2T + 0.0704 \times 10^6 \text{ Btu}$$

Heat absorbed by reactor contents (initially at 100°C = 212°F) is

$$1500 \times 8.34 \times 1.20 \times 0.45(T - 212) = 6750T - 1.431 \times 10^6 \text{ Btu}$$

Heat generated = heat absorbed + heat transferred

$$1.264 \times 10^6 = (6750T - 1.431 \times 10^6) + (1214.2T + 0.0704 \times 10^6)$$

whence

$$T = 330°F \quad \text{or} \quad 165°C$$

that is, on the large scale, in spite of deliberate cooling, the reactor mass temperature would exceed the decomposition temperature of 150°C!

The example serves to illustrate the danger of failing to define and understand all aspects of a processing step. Chemical engineers, by virtue of their training, are adept at scaling up processing steps to conform to the increasing size of facilities in the process development sequence, provided that the processing steps are adequately defined. The technique of scale-up has been, to a large degree, adequately systematized (see, for instance, the book of Johnstone and Thring, 1957). A number of excellent publications are also available to remind the chemical engineer what elements to look for so as to define adequately all aspects of a processing step in the first place, and, in particular, what some of the important associated safety aspects are. Marinak (1967), Hudson and Marinak (1972), and Kline et al. (1974) are useful sources of such safety-oriented information.

Returning to the substance of Example 11.1.1, we note that the result indicates the possibility of attaining a temperature of decomposition. An important aspect of the process development sequence is to check for dangerous instability of reactants, products, and even intermediate mixtures, to see whether an unanticipated reaction, perhaps well above normal operating temperatures, results in a self-sustaining exotherm with dangers of rapid vessel overpressuring. The decomposition reaction above 150°C, mentioned as part of Example 11.1.1, could turn out to be exothermic and even gas-producing, and under the proper circumstances such a reaction could "run away" in spite of the cooling available. Various calorimetric screening techniques are available for detecting such instability (see Prob. 3.5), and it is virtually mandatory that instability information be obtained for compounds and mixtures not previously subjected to similar testing.

The course of runaway reactions may be understood by reference to the graph in Fig. 11.1.1. Consider a batch reactor with a cooling jacket or cooling coils; add all the reactants to the vessel and preheat to the desired operating temperature, say, T_A. If the reaction is exothermic, cooling is required to maintain the temperature at T_A. The exothermic heat generation rate increases with temperature above T_A in the manner shown, since the reaction rate constant increases with temperature in an exponential manner, in accordance with the Arrhenius equation. (We neglect, for the moment, the effect of decreasing concentration with time.) The heat transferred to the coolant is given by

$$Q = UA(T - T_C) \qquad (11.1.1)$$

where U, the overall heat-transfer coefficient, A, the heat-transfer area, and T_C, the

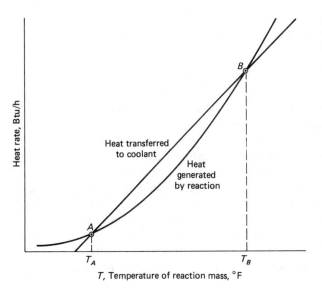

FIGURE 11.1.1
Characterization of reaction temperature control.

coolant temperature, are essentially invariable. As a result, the heat transferred to the coolant is linear in T, as shown in the graph. The two lines intersect at point A, and T_A is the stable equilibrium temperature. If, for some reason, the temperature goes above T_A, the heat transferred exceeds the heat generated, and the reactor mass cools back down toward T_A. If the temperature should drop below T_A, the generated heat exceeds the heat transferred, and the reactor mass temperature moves back up toward T_A.

The intersection at B is the runaway point. If the reactor temperature, for any reason, approaches T_B from a lower temperature, the tendency continues to be for the reactor mass to cool down from T_B to T_A, since the heat transferred exceeds the heat generated. If, for any reason, the reactor temperature exceeds T_B, we are in trouble, for the generated heat exceeds the heat transferred, and the reaction mass will heat up in increasingly rapid fashion—it will "run away."

If coolant enters the jacket or coil at T_C, and the equilibrium temperature T_A is above the desired operating temperature, little can usually be accomplished by increasing the coolant flow; T_C is the limit of the *average* coolant temperature. The only practical recourse is to supply a colder coolant. To make sure that the decomposition temperature is not reached and that dangerous overpressures are not obtained, several safeguards may be undertaken:

1 Provide effective cooling so that T_A, the equilibrium temperature, will be well below the decomposition temperature.

2 Feed a key reactant into the batch vessel at a controlled rate, rather than premixing all reactants. (In case of a gaseous reactant such as in Example 11.1.1, this is almost the only way to add it.) Provide automatic shutoff on reactant if temperature exceeds an upper limit.

3 Provide reactor venting capability to vent any generated gas at its maximum possible generation rate, at a pressure well below the reactor's rated pressure.

4 If temperature exceeds an upper limit, dump reactor contents (perhaps into a pool of water).

5 If temperature exceeds an upper limit, dump cold material into reactor to quench reaction.

Kline et al. (1974) recommend that *two* safeguards be used to prevent a runaway of the intended reaction. The first two, in fact, are commonly employed. If, however, secondary exothermic reactions or decomposition reactions have not been properly identified by calorimetric screening as part of process development, a dangerous situation may arise. The relationship in Fig. 11.1.1 may be developed on the basis of known information about the primary reaction of interest, but the unexpected secondary reactions could inadvertently heat the reactor mass well above T_A. Cutting off the flow of a primary reactant could now prove ineffective, for *all the reactants to feed the secondary reaction* are already in the vessel! The result of this lack of adequate knowledge could be a temperature rise beyond the runaway point of the secondary reaction and a potential overpressure catastrophe.

The vigorous pursuit of chemical stability information is only one of many aspects of the process development effort which the project manager must keep in mind during progress through the stages of miniplants, pilot plant, semiplant. These intermediate facilities, described in Chap. 3, progressively impose more demand upon the ingenuity and organizational ability of the individual who is called upon to manage them, because of their increasing size and complexity—and yet, in many ways, these facilities do resemble the full-scale production units. The differences are primarily the obvious ones of size, and of built-in flexibility in the small-scale facilities, permanence in the large ones. There is often a certain tendency to extend the "pots and pans" development concept from the lab bench to the pilot plant and even beyond, and the chemical engineer must insist on diverting the process development effort from this path of least resistance toward the path of a rational engineering process design. Moreover, the chemical engineer must continue to insist that cost-conscious management recognize the difference and the need for both scale-up–oriented pilot plants and market development–oriented semiplants; the two are *not* the same.

Exercise 11.1.1

Continuous processing steps are usually preferred over batch processing steps for reasons of easier process control, better quality control, less labor demand, and simpler equipment arrangement. The preference extends to the use of continuous stirred reactors, even though they require a larger liquid volume than batch reactors for equal productivity and conversion, and even though laboratory development is normally effected in batches.

A first-order irreversible reaction is to be carried out to a reactant conversion level of 90 percent. In the batch reactor, the turnaround time (i.e., time from start of one batch to start of the next) is twice the reaction time (due to time demand for loading and unloading). What is the cost of a single-stage continuous stirred reactor relative to that of a batch stirred reactor?

A job assignment within the context of a process development sequence is, indeed, quite a challenge to the neophyte chemical engineer, not only because of the organizational complexity but also because of the need to work with a bewildering array of unfamiliar equipment items—pumps, instruments, valves, an endless variety of fittings. Most of us eventually learn to recognize and live with this Niagara of hardware.

The Organization of Pilot Plant Operations

Exhibit 11.1A is an outline of the sequence of activities characterizing a pilot plant project. The intended purpose of the outline is to serve as a guide to the project engineer—a reminder of what should be done, in approximate sequence, to carry a project toward its intended goal, from the bench scale to the full-scale plant.

Strictly speaking, it is not a "checkoff" list, because the total effort of converting a laboratory experiment into a commercial process is much too complex to be amenable to a shopping list approach. Pilot plants by their very nature are instruments of unpredictable change, and technological as well as business prospects are in a constant state of flux. Nevertheless, a systematic approach to project manage-

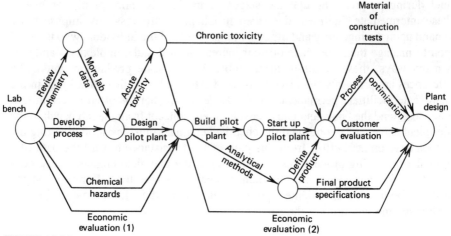

FIGURE 11.1.2
A process development network.

ment and engineering is possible, and hopefully the guide will serve to emphasize such systematization.

The problem of process development and scale-up is one that has received a great deal of attention—in the published literature, in management conferences, in seminars, and advanced education courses. Frequently the focus of interest has been the primarily technical scale-up problem, and technical guides and checkoff lists are plentiful. The emphasis in Exhibit 11.1A is on *organizational* problems, the sequence required to organize a successful process development effort.

Users are encouraged to modify and expand the guide to fit their own experience; on the other hand, many listed steps may be redundant for the situation on hand. It must be understood that the sequential listing does not imply a sequential operation in actual practice; a properly undertaken development process consists of a network of series and parallel operations, as shown in Fig. 11.1.2.

EXHIBIT 11.1A

PROJECT ENGINEERING IN THE PILOT PLANT

I Review of bench scale work

The first step in pilot plant development work is the assembly of all reports, pertinent memoranda, and other written sources. If a formal report has not been written summing up the laboratory work, it is best that time be taken to write such a report before serious pilot plant design work is undertaken; the very effort of collating information for a report frequently results in a better understanding of the process and a more thorough evaluation and interpretation of data. A collection of scraps of paper, monthly progress reports, and scattered notebook references by itself is *not accept-*

able; on the other hand, *all* pertinent information (including oral information) should be available to the project engineer to supplement the formal reports.

A thorough project review should impart to the project manager a good, quantitative grounding in the following areas:

Product
   Product formula or composition, general description
   Existing or contemplated uses; proposed production rate
   Preferred physical form, proposed specifications

Reactions and stoichiometry
   Raw materials—availability, quality requirements
   Primary reactions and coproducts
   Effects of impurities upon reaction
   Side reactions, by-products

Material and energy balances
   Preliminary block flow diagrams
   Preliminary process flow sheet
   Process research material balances
   Projected flow table, including operating conditions
   Thermodynamic data (if available), thermal data
   Energy balances

Unit operations and processes
   Results of separate studies on reactions, separation processes, etc.; investigation of effect of operating variables, including agitation intensity
   Specifically, extent of current knowledge of reaction kinetics and reaction equilibria
   Results of other rate process studies (crystallization, electrolysis, drying, etc.)
   Projected special equipment requirements (reactor configurations, cell configurations, centrifuges, vacuum evaporators, etc.)

Unusual processing effects
   Proven or suspected effect of impurities
   Foaming, bumping
   Catalytic effects of materials of construction
   Tar formation, coking, product discoloration
   Any "mysterious" phenomena (unexpected inhibition, sudden appearance of mushy crystals, "popping" sounds, etc.)

Mechanical systems
   Crushing, spinning, extrusion, special pumping, etc.

Engineering properties
   Vapor-liquid equilibria, phase equilibria, distribution constants, transport properties, etc.
   Studies performed, literature references, estimates

Safety and toxicology
   Danger of runaway reactions due to improper addition sequence, inadequate cooling, overheating, light sensitivity, poor mixing, improper inert padding, etc.

(Continued)

Have DTAs* or ARCs* been run on all reaction materials, *including intermediates?* Mechanical shock tests? Have thermal labs been consulted? Reactive chemicals committees?

Are flammability data available? NFPA ratings?

Have safety standards been reviewed?

Toxicity data on all raw materials, products, and intermediates. If not readily available, assemble toxicity information on spectrum of similar materials. Chronic and acute toxicity studies

Safe handling methods, personnel protection

Analytical methods

Methods of monitoring reactions, separations efficiency

Critical evaluation of reliability of analytical methods—choices available, automation

Summary of best methods to determine product specifications

Materials of construction

Bench scale corrosion studies

Catalytic or inhibition effects of materials of construction

Literature recommendations

Stray electrolytic effects

Environmental impact

Waste gas streams, venting problems, odors

Aqueous waste streams, BOD and COD, known or projected methods of treatment

Projected cooling water loads

Problems associated with spills, vapor emissions

Disposal of spent catalyst, waste tars, samples, etc.

Economic status

Extent of marketing analysis

Projected production rate, distributions methods, pricing structure

Preliminary capital investment estimates, estimates of cost of manufacture, return on investment

Market development requirements

Business and politics

Patent status

Degree of support for project by division management; product department, business management team; other divisions, top management

Status of financial support

Integration of project into corporate plans

Have contacts been established with sales, TS&D?

II Project status decision

The decision as to the future course of a given project is primarily a managerial one, but the project engineering manager has the duty to impart to management information bearing on such decisions, including personal value judgments and recommen-

*Differential thermal analysis; accelerated rate calorimeter.

dations. If a project is at the stage where most of the information in part I has been acquired, it is generally considered to be a promising one, and scale-up to at least a miniplant scale may be anticipated.

Nevertheless, if project managers find good reasons why projects should be dropped at this stage, they should not hesitate to voice their convictions; on the other hand, they may be able to acquire the information and insight required to defend and promote a project which is not viewed favorably by management.

The decision on how to advance a particular project is rarely a one-shot affair and is subject to continuous review. In some instances a market development semiplant may be projected while bench scale work is still in progress; nevertheless, the specific design of a scaled-up unit should be delayed until the information in part I has been substantially accumulated.

The project advancement decision network is sketched out in Fig. 11.1.3. Some of the end functions may be combined; for example, a market development semiplant may double up as a scale-up pilot plant. In exceptional cases the decision may be made to build a full-scale plant with bench scale information alone.

The decision to proceed with scale-up bench scale work and whether to bypass some of the stages of scale-up depends upon a number of criteria:

Scale-up problems
Recycle problems, other long-range effects

FIGURE 11.1.3
Project advancement decision network.

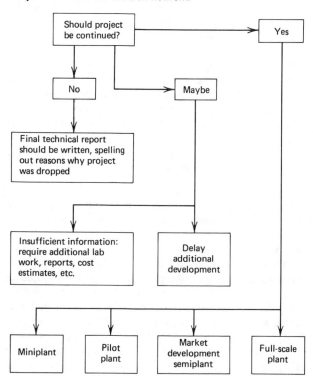

(*Continued*)
   Inherent size limitations (tube polymerizers, regenerative furnaces, etc.)
   Budgetary limitations
   Market status, market development needs
   Technological status

III Pilot plant design

The following design sequence applies to any of the scale-up alternatives: miniplant, pilot plant, or semiplant.

A. Preliminary procedures

   1 Define the principal problem areas—scale-up uncertainties, waste disposal problems; outline the studies required to solve the problem areas, the data and measurements to be taken.

   2 Define the criteria to establish pilot plant throughput. The key criterion may be the estimated size of a reactor or other piece of processing equipment judged to be adequate to undertake meaningful processing studies; it may be the estimated production rate to satisfy market development requirements.

   3 Decide whether process is to be batch or continuous, and reformulate a block flow diagram corresponding to the proposed pilot plant (not all processing steps may need to be piloted).

   4 With the help of management, define personnel requirements for the design and pre-start-up stage, as well as the areas of responsibility.

   5 At an early stage, define the major safety problems. Pin down reaction energetics, and continue to explore possibilities of unexpected side reactions, heat- and mass-transfer problems associated with scale-up, and the possible detrimental effect of operational upsets.

   6 At this point, it is the project manager's responsibility to start a *job book*. Eventually such a book may consist of several volumes. The job book should constitute an assemblage of all pertinent background material, calculations, correspondence, purchase orders, drawings, and any other material bearing upon the project, properly annotated and *indexed*. The job book should be a document that can be picked up and understood by the project manager's colleagues. Eventually it will be supplemented with other documents—large layout diagrams, operating logbooks, reports—but all other "loose" material should be incorporated.

B. Process design

   1 Recalculate and adapt material and energy balances, process flow sheets to the selected pilot plant situation. Outline preliminary operational sequences to firm up process design.

   2 Equipment design and selection. Design and size needed equipment. Sum up results (including design methods if applicable) on specification sheets. Keep separate spec sheets for each item of equipment, with the possible exception of spared or multiple items (pumps, ejectors, etc.). Identify equipment in conformity with accepted symbolism. Include equipment sketches as required (vessels, special designs).

   It must be understood that at this point pilot plant equipment design is an evolutionary process. The final design may be a matter of availability of equipment in pilot plant stores. Nozzle and manhole layouts may need modification as ideas on process requirements and piping layouts crystallize. Accessibility of equipment "innards" (impellers, seals, demisters, packing supports) should be an early consideration as well as an adequate supply of bottom drains, extra nozzles, and sampling cocks.

In many cases complex process design problems may have to be resolved at this stage—inasmuch as possible. The inherent uncertainties at the pilot plant stage require the resolution of processing and equipment alternatives; in some cases alternate equipment may have to be provided to resolve the uncertainties experimentally.

The project manager may seek help from various sources:

Computer center and systems: optimization studies, staged process computations.
Vendors: sales representatives may have valuable advice, technical literature.
Resource people: Individuals with specialized knowledge, design groups, auxiliary functional groups are ready to help.

In some cases, special laboratory studies may be required to firm up pilot plant designs—flame velocity studies, mixing simulations, pressure drop measurements in unusual configurations, unavailable engineering properties, ultraviolet light absorption studies.

   3 Instrumentation and control

      a As operating procedures are firmed up, specify measuring and control instruments required to control and monitor the process. Incorporate information on flow sheet.

      b Undertake preliminary exploration of computer control of operations—particularly routine semiplant operations, data acquisition and processing.

   4 Process safety

      a Analyze process operations to establish requirements for pressure-relief devices, alarms, emergency dumps, other emergency provisions. Account for utilities upsets, reasonable human error to design a "fail-safe" system.

      b Review equipment specifications with resource people and specialists (safety, high-pressure, mechanical, electrical). Consult plant safety standards.

   5 Supply and distribution

      a Raw materials: availability, form, storage requirements on site, transfer method from packages. May have to be ordered far in advance of start-up.

      b Product: packaging, on-site storage, distribution for market development. Experimental designation, product department clearance for distribution.

   6 Economic analysis

      a Review process economics to conform with developing process design.

      b Monitor existing and projected pilot plant costs to avoid cost overruns.

C. *Plant design*

   1 Utilities

      a Water requirements (cooling, process, distilled), aqueous discharges, water cooling loads, water recycle, water treatment systems, etc. Seal flush requirements.

      b Steam: demand, trap types, condensate handling and distribution, insulation requirements.

      c Electrical: power demand schedule, ac or dc, voltage requirements.

      d Natural gas: distribution system, burners and furnace systems.

      e Instrument air: amount required, availability, drying and filtering.

(*Continued*)

    f Nitrogen for inert pads.
    g Hot oil systems: furnace, storage, distribution system.
    h Refrigeration: peak demand, refrigeration level, refrigerant, brine systems, equipment requirements.
2 Vent systems
    a Review filling and emptying sequences to check proper venting, possible vapor mixing problems in common header tie-ins. Check operation of check valves, control valves on vent system.
    b Design vent headers; pay particular attention to possible trapping of condensate, plugging by sublimed solids, rain penetration through exhausts, noise control.
    c Vent gas scrubbing systems, knock-out pots.
    d Flares and flame arresters.
3 Waste disposal
    a Treatment of waste gas streams (if not part of process flow sheet).
    b Treatment and disposal of waste aqueous streams, including lagoons, aeration basins, chemical precipitation, etc.
    c Treatment and disposal of solid wastes, tars, and organic liquids: combustion, land burial problems.
    d Disposal of samples, unmarketable products, spills, emergency reactor dumps.
4 Instrumentation and control
    a Specify measuring and control instruments, including make, modes. Avoid glass.
    b Specify instrument mounting type and location: control panels, thermowells, chemical gauges, etc.
    c Specify alarms, including type, location, etc.
    d Specify safety relief valves, bursting disks, overload devices.
    e Interface with computer systems group:

        Check complex control systems for "bucking," interlock problems, critical frequency response.
        Design process computer control system if initial investigation justifies it.

5 Operations sequence
    a Complete final analysis of projected operations to establish possible additional needs in equipment, piping, controls. Be sure to account for transfer procedures that could lead to cross-contamination; emergency storage needs, sampling and purge needs; again review the effect of operational upsets and operators' errors.
    b Outline operating instructions (preliminary draft):

        Start-up
        Steady state
        Normal shutdown
        Emergency shutdown

    c Devise emergency instruction posters for mounting close to equipment (if needed).

SECTION 11.1: THE MANAGEMENT OF THE PROCESS DEVELOPMENT SEQUENCE

    d Review required data and measurements to make sure the proper instruments, sampling cocks, etc., are available.

6 P&IDs. At this point all the previous design information should be assembled in piping and instrumentation diagrams. For miniplants and "single pot" experiments, the P&ID is likely to be little more than a detailed sketch of the setup; in particular piping may consist of rubber or plastic tubing, and much of the peripheral equipment will be portable and interchangeable. For normal pilot plant and semiplant design purposes, a P&ID drawn along engineering department specifications is required. The project manager should have drafting help to assemble the P&IDs.

Completion of P&IDs requires the design of piping systems at this point:

Pipeline sizing and material specifications
Sizing of all valves, including control valves
Piped-in utilities, but refer to utility drawings to designate header connections

7 Layout and siting
  a Site selection

    Indoor or outdoor
    Existing versus new buildings
    Process and process hazard considerations: need for bulkheads, waste ponds, etc., hazard of possible explosions, accidental emissions, noise nuisance
    Availability of utilities, roads, storage areas
    Personnel considerations: change rooms, lunch rooms, toilets, heating and ventilation, office space

  b Equipment layout

    Plot plans
    Building layouts
    Equipment layout plans and elevations
    Control panel layouts

    In case of particularly complex pilot plant layouts, the construction of models may prove to be time saving.

  c Building requirements

    Repairs to existing buildings; inspection to guarantee structural integrity, conformity to codes
    Heating, ventilation, air conditioning, lighting
    Control rooms, offices, laboratory space
    Fire control, sprinkler floor requirements, CO_2 floods
    Personnel protection: showers, eye baths, protective bulkheads, blowout panels, emergency exits
    Sewers, waste lines, stacks
    Electrical distribution system housing
    Doorways, hoists, special requirements

  d Site process requirements

    Concrete pads
    Equipment supports

(Continued)
> Stairways, platforms, catwalks, ladders
> Furniture
> Insulation, painting
> Protective shields
> Hoods, lab benches, work areas, equipment storage

> **8** Interface with other engineering disciplines and groups
>   **a** Buildings and structures, building code requirements.
>   **b** Piping layouts. For relatively simple installations, the P&IDs may be sufficient. In particularly complex systems, models may prove helpful for both parts procurement and during construction.
>   **c** Electrical distribution: code requirements, distribution layouts, grounding requirements.
>   **d** Plant instrumentation and analytical group: continuous monitoring instruments.
>   **e** Computer systems: computer control. Hardware, programming.
>   **f** Environmental control: waste disposal systems.
>   **g** Safety department: safety engineer and industrial hygiene OK, safety checkoff. Input from related groups: fire department, pressure vessel committee.
>   **h** Economic evaluation: final construction cost estimates.
>   **i** Building services: tie in with pilot plant block services.
> **9** Go through scale-up checkoff lists for final revisions.

IV Pilot plant construction

The project manager has the responsibility of supervising the plant construction and of coordinating the activities of the various support groups. It will be understood that many of the duties outlined below will actually be initiated at a much earlier stage, usually during the design period; for example, delays in placing orders with the purchasing department could result in serious delays in completion of plant construction.

> **1** Interface with other pilot plant services groups
>   **a** Purchasing and equipment procurement. Includes borrowing and scrounging, renovation of used equipment, locating used equipment for purchase.
>   **b** Construction and maintenance. Arrange for contractors, external engineering support, corporate construction and maintenance support. Help draw up construction schedules, cost accounting schedules, maintenance projections. Draw up special supplies requirements (not regularly stored).
>   **c** Operating staff. Project pilot plant operating personnel requirements, time scheduling.
>   **d** Management. Line management acts as liaison, "protective shield" from operations management.
> **2** Construction monitoring. Project manager checks progress of construction with help of construction schedules, P&IDs and layout drawings. Cost control.
> **3** Individual personnel protection. Procure special clothing, gloves, face shields. Assemble emergency first-aid kits (get help from safety department). Arrange for air pacs, other breathing apparatus, fire blankets. Obtain safety signs, warning signs.
> **4** Draft operating instructions. By this time operating sequences should be well-

established—or as well-established as expected in view of pilot plant uncertainties. These should go through a final review, and a *detailed, step-by-step* outline is written to incorporate (in separate sections):

Start-up procedures (process)
Steady-state operations
Normal shutdown procedures
Emergency shutdown procedures

In addition, instructions should be written on how to operate peripheral equipment—furnaces, refrigeration systems, utilities, fire fighting equipment, etc. It will be understood that operating instructions will be continually revised during the lifetime of the pilot plant.

Operating instructions should be incorporated into *an operating book,* which may be considered to be one volume of the job book. The operations book also should contain specific equipment operational instructions, maintenance and repair instructions, and troubleshooting checkoff lists, all of which are frequently supplied by the vendor. More detailed equipment brochures may be kept in a separate readily accessible file.

In addition to detailed instructions, simplified instructions relating to a particular piece of equipment or operating step may be drafted for operators' benefit, particularly if an unthinking error leads to disaster. ("Pull handle *down* to *close* valve.")

5 Operator training. Start before construction is completed (in small installations, operators are likely to be in on the construction). Include safety drills. Computer indoctrination.

With help of management, establish technical personnel requirements for pilot plant operation, particularly during start-up; get these people involved well before start-up.

6 Final review of required data and measurements. Prepare necessary data sheets, logbooks.

7 Laboratory arrangements. Procure pH meters, temperature recorders, O_2 analyzers, chromatographs, titration equipment, standard solutions, and any other materials not routinely stored in research or the plant storeroom. Also get supply of basic tools, appropriately marked and locked in tool cabinets or drawers.

8 Equipment checkout

a Check over equipment after delivery; contact vendor representatives to settle problems (e.g., broken Pfaudlers).

b Start equipment card file for plant records.

c Check for leaks, preferably by pressuring with air and using Leak-Tek. Glass vacuum systems may be checked with an induction-coil leak detector.

d Fill equipment with water (or solvents); check for leaks, pump operation. Steam up exchangers, reboilers, jackets.

e Calibrate instruments, rotameters, variable-speed drives, etc.; prepare calibration charts, post in appropriate locations.

f Check for proper grounding of all equipment.

g Check direction of rotation of all rotating machinery. Check seals, gland flushes.

h Check controller action. Preset proportional bands, other modes. Inasmuch as possible, check out action of all remote indicating instruments: level indica-

(*Continued*)

tors, multiple-point temperature recorders, alarms. Check out computer integrity.

i Make sure special equipment is thoroughly preserviced—for example, air purge from Pulsafeeder oil system, all Swagelock fittings tight, electrode spacing adjusted, etc.

j Check out action of peripheral equipment: refrigeration units, compressors, cells, generators, Dowtherm units. Procure supply of refrigerants, heat-transfer fluids, vacuum pump oils, etc.

k Flush out piping systems with water (or solvents) to remove contaminants; check for slag, forgotten tools, plugs, fouled strainers, etc.

l Check out tracing systems.

9 Prepare raw materials for start-ups; provide safe storage.

10 Prepare and submit preoperational checklist. Arrange for preoperational safety committee checkup.

V Pilot plant operations

The project manager now has the most difficult task of all—the creation of a commercially successful process through pilot plant operations. Duties include:

Pilot plant start-up. Troubleshooting, revision of operations, equipment modifications.
Analysis of operational problems.
Data analysis and correlation.
Equipment revision and expansion.
Periodic and final project analysis.
Monitor process to discover new areas of information needed for full-scale plant. Write reports and summaries.
Product evaluation, updating of product specs.
Economic analysis updating.
Project decision:

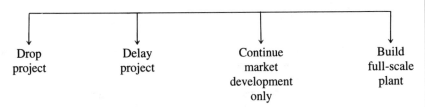

Criteria for the decision:
 Technological success
 Marketing success
 Economic success

VI Preliminary package, full-scale plant

If the decision is made to proceed with construction of the full-scale plant, the project manager may expect to be involved in all phases of design (perhaps as a "manufacturing representative") as well as start-up operations. The project manager also will have the responsibility of preparing a preliminary design package to be used by engineering as source material for the plant design. The preliminary package will include:

Mass and energy balances
Flow sheets and flow tables
Preliminary equipment design
Details of special equipment design
Preliminary instrumentation
Reports and summaries, including engineering property data

11.2 COST CONTROL OF CONSTRUCTION PROJECTS

The Requisites of a Successful Construction Project

Let us now turn our attention toward the ultimate phase of the project development sequence, the construction of a full-scale production plant. We will concern ourselves primarily with two aspects of the evaluation of *how well the construction effort is doing:*

Project cost control
Project timing (scheduling) control

Neither subject can be discussed without some understanding of how the construction effort is organized in the first place, and a brief review of this complex task of organization will therefore be undertaken. It should be understood that the comments which follow apply to a large extent to *any* construction during *any* phase of process development, but our comments will be tailored to the large construction project. The difference is one of scale and organizational complexity only. For this reason, we will forgo the enumeration of task details as in Exhibit 11.1A and concentrate upon an overview of the nature of the tasks that construction project management entails.

A successful construction project has four elements which contribute to its success:

1 A Proven Process A formula for disaster is to initiate a construction project without the complete and unequivocal understanding of each and every step in the proposed process, backed up by hard data and process experience in smaller equipment. It simply will not do to say, "We'll catch up on this step later without any trouble; let's go on with the construction preliminaries." If there is any doubt at the time the decision is made to build, the time to resolve that doubt is then and there—in the laboratory or in the pilot plant. Even the accumulation of physical property information (such as the packing behavior of powdered products, for example) obviously cannot be delayed beyond the scheduled time of designing and specifying equipment.

Once a plant is built and started up, problems with equipment are commonplace and certainly annoying, but such problems can almost always be resolved. Problems with *chemistry* are much more serious and usually result in catastrophic production delays, major process revisions, even project abandonment. The production unit is no place to prove out the chemistry. Pilot plants should really be used even for demonstrating the adequacy of equipment that is to be used in the production unit—for instance, pump seals for high-temperature organic fluid service.

2 Adequate Scope Definition An obvious prerequisite of any construction project is that the scope of the project be properly defined before capital expenditure is authorized. By scope we mean exactly what it is that is going to be built—how large (in terms of rated capacity), and with what auxiliary facilities. And yet, in practice, inadequate scope definition is perhaps the primary reason why some construction projects incur cost overruns and start-up delays. Failure to identify and properly account for those off-paper arrows on flow sheets constitute one of the precursors of poor scope definition.

3 Realistic Schedules A construction project must be carried out within a realistic time frame, one based upon reasonable projections of the time required to accomplish each of the network of tasks that constitutes the project. All too often project scheduling is done in much too cavalier a fashion, with little regard for the timing nuances of sequential or overlapping jobs, and with timing projections based more upon wishful thinking than experience and fact. It is unlikely that such a casually assembled schedule will be attained in practice.

Provided that reasonable estimates of the time requirements for the individual components of the construction sequence can be made (time for completing piping diagrams, for example, or equipment delivery times), systematic methods are available to formulate a realistic schedule. Such methods need not be complex; simple bar charts, for instance, are often adequate for purposes of time control of smaller-scale projects. More complex projects may require the use of scheduling techniques such as CPM (critical path method) and PERT (program evaluation and review technique). These methods are intended for handling the complex task networks characteristic of complex projects; they are readily adaptable to computer manipulations. The methods will be further explained as part of the subsequent discussion of time control of construction projects (Sec. 11.3).

4 Reliable Cost Estimate The cost estimate is the basis for the capital budget authorization. If the estimate is too low, the project will result in cost overruns above the amount authorized. If the estimate is too high, this may jeopardize the decision to proceed with the project in the first place, and it will certainly jeopardize the reputation of the estimate originators. Actually, provided that the project scope has been properly defined, a reliable cost estimate may be generated by experienced cost estimators even in the absence of detailed design. This is particularly important, since the authorized budget is based upon an *authorization estimate* (Chap. 1), made from flow sheets and preliminary equipment specifications only. The subsequent *project control estimate* is based upon firm equipment bids and detailed design; if it varies appreciably from the authorization estimate, additional requests for capital may have to be made. Any such additional requests, needless to say, are a source of intense discomfort to the responsible requestors; nobody relishes the job of carrying the message, "We goofed, please send more money."

Exercise 11.2.1

The $10 million DFC estimate of a plant includes a contingency of 5 percent. Following authorization of the construction funds, the project manager is informed that the existing cooling towers in the integrated complex will not handle the large cooling water requirement of 8000 gpm; the new plant will have to be built with its own cooling towers (a change in scope). Will the contingency accommodate the additional cost? (See Fig. P5.6.) What other possible cost variances must the contingency normally accommodate?

The Organization of a Construction Project

Suppose the decision has been made at the appropriate level to build a plant. What happens now?

We will assume that at this point the project has been appropriately defined within the corporate structure. Process development, economic, and marketing studies have presumably advanced to the stage where the project has been judged commercially attractive, and the decision to build a plant with a specified capacity has been made. This decision has been made with the concurrence of key corporate authorities, up to the board of directors in the case of projects involving projected expenditures above some stipulated minimum. The project definition usually (but not always) incorporates a geographical siting and identifies a *sponsoring department manager,* an individual who will have the task of steering the project through the top echelons of the corporate financial management. In the case of process development construction projects, this is usually the divisional research director; in the case of production plants, the sponsoring department manager is likely to be the divisional manager.

The first task facing the sponsoring manager is the appointment of a *project manager* (sometimes called the *manufacturing representative*). The project manager's job is just what the title implies: to manage the total organization of the construction project from definition to plant start-up, and to assume responsibility for the safety, operability, and cost of the facility. At this time, the *project engineer* may also be identified; the project engineer's responsibility is to oversee and coordinate all those engineering functions and disciplines required to design and construct the project. The project engineer is the representative of the divisional engineering department and will work closely with corresponding representatives from the construction company which is eventually awarded the construction bid (or else the corporate construction division, if one exists). The appointed project manager often is the same individual who managed the advanced phases of the process development sequence. In the case of process development construction projects, the project manager and project engineer may be the same individual.

The project sequence is characterized by two phases, one prior to the authorization of funds, one following the authorization. Let us first enumerate some of the project manager's tasks prior to the authorization request:

1 Assembly of pertinent technical information; initiation of any required R&D tasks to establish a "proven process"
2 Definition of project scope
3 Development of a preliminary design and alternates
4 Preparation of authorization estimate for the preliminary design
5 Solicitation, with the sponsoring manager's help, of approval of the design package by key business and technological centers in the corporation
6 Preparation of authorization request documents

The authorization request is the formal medium for obtaining the money to build the project. The request package need not be voluminous. Of course, it must contain the total capital amount requested (with backup documentation), along with a brief outline of the project and its business potential, and a summary of environmental, safety, and loss prevention evaluations of the project. For projects that will incur expenditures above some arbitrary minimum, the authority that grants the fund request is the board of directors; small projects may be authorized at a lower level of authority (for instance, by the divisional manager). The package is submitted to the board which, more often than not, requests the presence of the project manager to present the package and answer questions. The attendance of the project manager also serves to emphasize the heavy responsibility that the manager is requesting to assume, the responsibility of controlling the money granted—for only the project manager has authority to commit the funds once they have been granted!

If the granting authority does release the requested funds, the actual job of construction can now commence. The approximate sequence of tasks looks something like this:

1 Preparation of a "design package" (i.e., the sum total of material needed for the detailed design). This is submitted to construction companies for bids or else to the corporate engineering and construction division. In some cases the engineering and construction functions may be separated, with the divisional engineering department assuming total responsibility for the engineering job, and the construction job awarded partly or totally to outside contractors; engineering, however, takes on the job of overall construction coordination.
2 Establishment of accounting and cost control procedures.
3 Organization of project personnel. By this time the plant superintendent will have been identified to take on the job of preparing for operations. In the case of complex projects, responsibilities may be subdivided among a number of functional managers, as shown in Fig. 11.2.1.
4 Establishment of a time schedule and time control procedures (concurrently with some subsequent tasks).
5 Completion of the engineering design. A few details of the design task have been outlined in Chap. 5. The design is often accomplished with the active partici-

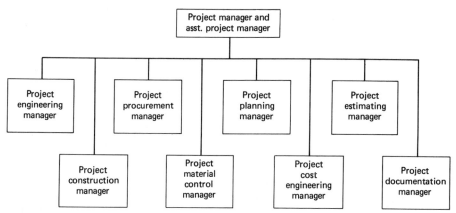

FIGURE 11.2.1
An organization chart for complex projects. *(Adapted from Datz, 1981, by permission.)*

pation of R&D personnel; on the other hand, much of the "nut and bolt" mechanical design will be left up to the equipment vendor or contractor with the winning bid. The normal engineering design is summed up in specification sheets; an example of a heat-exchanger specification sheet is given in Fig. 11.2.2. Details of items such as shell thickness or support brackets are left up to the vendor, although the final design must, of course, be approved by the project engineer. The process design package contains, as a minimum, the items listed in Table 11.2.1.

6 Initiation and execution of equipment procurement. Winning bids by vendors and contractors are awarded with the help of the purchasing department.

7 Organization of project reviews with functional groups such as environmental control, safety, and fire departments. Environmental control, for example, has the responsibility of ensuring conformity with environmental regulations and interacting with enforcing agencies.

8 Completion of the project control estimate.

9 Initiation and execution of the construction phase.

10 Monitoring and reporting construction progress.

11 Preparations for start-up, including operating personnel requisitioning and organization, ordering of spares and raw materials, assembly of operating procedures and computer programs, disaster planning, and initiation of personnel training. These chores are often taken on by the prospective plant superintendent, who, in a few instances, is the same individual as the manufacturing representative.

12 Final inspection and testing.

13 Start-up.

14 Project closeout.

Clearly, the project manager requires a great deal of help in carrying out this formidable array of tasks; the functional groups listed in Table 11.2.2 are available to give such support.

CHAPTER 11: ENGINEERING MANAGEMENT OF CONSTRUCTION PROJECTS

PLANT						FILE/JOB NO.		
LOCATION			BLDG. NO.			CHARGE NO.		
MANUFACTURER			NO. UNITS			B/M NO.		
						P. O. NO.		
DUTY/UNIT		BTU/HR	NO. UNITS		SHELLS/UNIT	TEMA SIZE/TYPE		

1	TYPE	Horiz. Vert. Sloped ° to horiz.		FTS-with Exp. Joint	U-Tube	Kettle	Coil	Hairpin	Box
2		Floating Hd.-Pull Thru-Clamp Ring-Packed		Thermosyphon	Fintube	Other:			
3		Additional Proc. Data on Sheet No.		Inlet	SHELL SIDE - Outlet		Inlet -	TUBE SIDE -	Outlet

#		Parameter	Units					
4		Fluid						
5		Total Flow	lb/hr					
6		Liquid	lb/hr					
7		Density	lb/cu ft					
8		Viscosity	CS-CP					
9		Specific Heat	Btu/lb·°F					
10	PERFORMANCE OF ONE UNIT	Thermal Conductivity	Btu/hr·sq ft·°F/ft					
11		Vapor	lb/hr					
12		Mol Wt Density	lb/cu ft					
13		Viscosity	CS-CP					
14		Specific Heat	Btu/lb·°F					
15		Thermal Conductivity	Btu/hr·sq ft·°F/ft					
16		Latent Heat	Btu/lb					
17		Non-Condensables	lb/hr		M.W. =		M.W. =	
18		Velocity Max./Min.	ft/sec					
19		Norm. Max. Operating Temp.	°F					
20		Operating Press. (Inlet)	psig					
21		Pressure Drop	psi	Allow.	Calc.	Allow	Calc.	
22		Film Coeff.	BTU/Hr. Sq. Ft.					
23		Fouling Resistance,	hr·sq ft·°F/Btu					
24		Over-all Coefficient "U"	Btu/hr·sq ft·°F; Clean	Service	LMTD (Corrected)			°F
25		Installed Area/Unit	sq ft (Outside) Including-Excluding Area in tube sheets.					

#		Parameter	SHELL SIDE	TUBE SIDE				
26					Shell ID (Approximate)			in.
27		Design Temperature	°F		No. Tubes (Approximate)			
28		Design Pressure	psig		Tube OD	in.	Tube Gage	BWG
29		Hydrostatic Test	psig		Tube Length			ft
30		Corrosion Allow./Lining	in.		Tube Pitch △ □ ◇			in.
31	DESIGN DATA PER SHELL	Number of Passes			Joint			
32		Insulation			Make/Type Fintubes			
33		Cross Baffles; Type	; Segment Cut	%	No. Fins	per tube;		per in.
34		Number	; Spacing Approx. Equal - See Sheet No.		Fin Height	in.	Fin Thick.	in.
35		Provide	in. Horizontal Cut on Bottom for Condensate Drain		Vessel Supports Saddles - Lugs Other;			
36		Long Baffles: Type	; Number		Weir Height	in.; Shell After Weir		in.
37		Impingement Baffle Yes-No	Condensate Lift Yes-No	Removable Bundle Yes-No	Cathodic Protection Yes-No			
38		TEMA Class	Lethal Yes-No	Code ASME	National Board	Other:	Stamp	Yes-No
39		Spot Radiograph: Shell	Shell Cover	Channel	Stress Relieve: Shell	Shell Cover	Channel	Floating Head
40		Weight Complete Empty/Full of Water	/		lb	Weight Bundle Only		lb
41		Sandblast;	Paint:	Maintain	Shell ID	Tube Count	Installed Area+	% - %

#			SHELL SIDE				TUBE SIDE						
42											Tubes	Chan. Noz. Necks	
43		Service	Mk	Size	Rtg	Face	Type	Mk	Size	Rtg	Face	Type	Stat Tube Sheet	Chan. Noz. Flanges
44	NOZZLES PER SHELL	Inlet										MATERIAL SPECIFICATIONS	Fltg. Tube Sheet	Channel Flanges
45		Outlet											Cross Baf./Tube Sup.	Fltg. Head
46		Drain											Long Baffles	Fltg. Hd. Flange
47		Vent											Impinge. Baffle	Clamp Ring
48													Weir/Lift	Bolts/Studs
49													Tie Rods & Spacers	Nuts
50													Shell	Vessel Supports
51													Shell Cover	GASKETS & PACKING
52													Channel	Shell Cover
53		See Nozzle Sketch on Sheet No.											Ch. Cover/Bonnet	Shell-Chan. Side
54	CONN.	Shell Side:		Parallel Banks of			Shells In Series						Shell Noz. Necks	Channel
55		Tube Side:		Parallel Banks of			Shells in Series						Shell Noz. Flanges	Channel Cover
56		Stacked:		Wide			High						Shell Flanges	Fltg. Hd.

						EQUIP. NO.
SPEC. BY						
CHECKED:		SERVICE				**HEAT EXCHANGER SPECIFICATIONS**
APP'D:						
DATE:		REVISION DATE	A	B	C	SPEC. NO.
VENDOR TO COMPLETE ALL INFORMATION MARKED—				SHEET OF		

FIGURE 11.2.2
An engineering specification sheet.

TABLE 11.2.1
PROCESS DESIGN PACKAGE INFORMATION

1. Detailed flow sheets, with unique requirements necessary for effective operation noted
2. General plot plan showing buildings, process towers, cooling towers, pipeways, furnaces, sewer lines, etc.
3. Equipment specification sheets
4. Listing of vendors for critical equipment
5. P&IDs and piping diagrams indicating pipe size, material, insulation, etc.
6. Layouts and models
7. Instrument specifications, instrument panel layouts, computer controls
8. Pump and control valve schedules
9. Specifications other than equipment

The Significance of Cost Control

What is meant by cost control, and why is it needed? The answer is that whether we are talking of a private household or a giant process industry, to stay within one's means, one must work at it. Once the board of directors has authorized the expenditure of a fixed amount of capital to build a plant, it is most important that the project expenditures remain within the limits of the budgeted amount. The task of avoiding overruns is considered to be the primary responsibility of the project manager. If overruns do occur, the economic fate of the project may be seriously jeopardized, and an unwanted extra burden may be imposed upon the corporation's available capital pool. In any case, close adherence to budgets is a manifestation of rational financial management; a cavalier attitude and sloppy control tend to be infectious, and the resulting overruns are cumulative and potentially catastrophic to the corporation.

What are some of the causes of cost overruns? Perhaps the most serious cause is *change in scope*. Clearly, if the process is changed in the midst of construction, or if unanticipated construction of auxiliary facilities is needed, the project manager is quite likely to lose cost control. A judicious evaluation of the consequences must be made; if the impact is minor, the contingency fund may absorb the change, but in case of doubt, the project manager has no choice but to make a request of the board for additional funds.

TABLE 11.2.2
CONSTRUCTION SUPPORT FUNCTIONS

Research	Purchasing
Environmental control	Maintenance
Industrial hygiene	Accounting
Safety	Medical
Process engineering	Legal
Plant engineering	

Design modifications and *extras* (items not thought of at the time of fund authorization) are related to change in scope. They are hopefully accounted for by contingency funds, but it is surprising how minor-appearing modifications can increase costs out of proportion to their relative magnitude. Minor modifications are an inevitable concomitant of the construction process, but every effort should be expended to minimize them through firm design backed by pilot plant (or similar) experience, and through expert help on items such as pilings and foundations, items perhaps a bit foreign to the chemical engineering manager.

A *poor authorization estimate* is, of course, a potential source of cost overruns. Given proper scope definition, however, this is not a likely cause, unless inexperienced estimators have been used. A more likely source of trouble is the failure to emphasize or to heed admonitions regarding inflation-caused *cost escalation*. Unrealistic time scheduling may delay projects so that cost escalation is more serious than originally anticipated.

People problems are common, but uncommonly difficult to solve. They are manifested by poor worker productivity, or by unusual obstreperousness of middle-level supervisors and managers. One frustrating aspect is the subcontractor who escalates costs unreasonably and repetitively with every minor change, a possible indication of a mistakenly low bid to start with. People problems are frequently the result of the project manager's failure to exercise close control and surveillance.

What, then, can be done about some of these potential problems? There are three principal avenues of recourse:

Anticipation
Monitoring
People control

Anticipation of the effect of possible extras, design modifications, and cost escalation is a matter of experience with previous projects of a similar nature. On the basis of such experience, knowledgeable estimators adjust the estimated costs of items such as piping or structures upward from the totals which are based upon design diagrams and specifications.

Monitoring refers to the use of cost control and scheduling tools, which will be described presently, to keep tight reins upon the project.

People control is a matter of establishing "who is boss" at an early stage of the construction project. Hopefully, this can be done by reason and persuasion, by gaining the trust and respect of the project managerial hierarchy through effective communications and intuitive fairness, and by using a systematic control methodology that is well understood. A good project manager must learn to ferret out human conflicts (they are always there) and to resolve them calmly, fairly, and firmly, with the leverage of clout when needed. The author witnessed a scene aboard a supertanker, dry-docked in Singapore for overhaul, involving the project manager and the ship's captain screaming at each other, toe to toe and nose to nose, regarding the cost of shore accommodations for the ship's crew. This is not a recommended cost control procedure, but the project manager must establish who is boss. (In the Singapore incident, the project manager "won," after yielding a little on the question of the luxury of accommodations.)

Cost Control Tools

The "tools" that one uses for properly controlling and monitoring the costs of construction are not particularly unique; they are much the same tools that any reasonably capable organizer would think of anyway, not unlike the tools one would use, for example, to control household expenditures.

The first such tool has already been alluded to as part of the proposition that proper care be taken to anticipate small changes in scope. In fact, the tool of *allowance allocation to estimates* is employed even before the authorization request; an example, taken from the work of Roth (1979), is shown in Table 11.2.3. The estimated material and labor costs of various coded categories are first supplemented with allowances for cost escalation, extras, and design modifications, based upon recent experiences with those categories in similar plants. Finally, a contingency factor is added to the grand total to account for the "level" of the estimate; since preauthorization estimates are not based upon detailed design, the contingency accounts for the lack of completeness of design as well as unanticipated

TABLE 11.2.3
ALLOWANCE ALLOCATIONS TO ESTIMATE
(In Thousands of Dollars)

A. Base estimate

Construction code no.	Category	Estimate Material	Estimate Labor	Estimate Totals
100	Process equipment	100	78	178
500	Process accessories	65	50	115
850	Engineering		50	50
		165	178	343

B. Allowances

Construction code no.	Escalation Material	Escalation Labor	Extras Material	Extras Labor	Design modifications
100	12	7.8	10	7.8	5
500	7.8	5	6.5	5	4.3
850		5		5	

C. Subtotals plus contingency

Construction code no.	Base totals	Allowances	Totals
100	178	42.6	220.6
500	115	28.6	143.6
850	50	10	60
		Sum of totals	424.2
		Contingency of 11%	46.7
		Total	471

problems reviewed in Chap. 5. Such informed allowance allocation is similar to the establishment of budget goals for a private household; expenditures in each category are reviewed over the past few years, and a new budgeted amount is obtained by extrapolation and adjustment to account for anticipated inflation trends and expectations of changes in living style.

Table 11.2.3 also incorporates the important concept of a *code of accounts*. Just as it is done in a simple household budget, the sources of expenditure are judiciously categorized; in the case of industrial construction projects, the categories are identified by means of a numerical code. There is no hard and fast rule as to how this is to be done, and each corporation has its own standard code of accounts which can then be adapted to the needs of each project. A portion of a code of accounts is illustrated in Table 11.2.4. In this particular case, the 700 code series is reserved for the category of instrumentation. The second and third digits are used for identifying the principal category subdivisions; thus 700 identifies general documentation, 722 control valves, and so forth. Further numerical coding is at the project manager's discretion and may be used to identify sections of the plant (722–03: control valves in tank farm), individual items (722–03–15), even vendors.

Numerical coding is first used to identify principal categories and subdivisions in the preauthorization and cost control estimates. The code is subsequently used as a flexible cost-monitoring tool to keep track of expenditures within any category or subdivision desired; in this way problem sources may be quickly identified. For example, the total 700 code account may be monitored, or the 721 account (relief valves), or, with proper coding, the total account of a particular vendor, the total account ascribable to a particular material of construction (say, nickel), and others. It will be appreciated that a flexible numerical code facilitates access to computer-stored data.

The utility of the code of accounts is not restricted to cost monitoring. The same code is used for proper identification in any construction-oriented endeavor: filing, purchase order and other document identification, status reporting. It should be emphasized, however, that the code of accounts is distinct from the equipment labeling appearing on flow sheets and P&IDs (Exhibit 3.5B); thus distillation column B-16 may wind up with an account code of 506-17-09-33.

The total project cost status is monitored best by using the project *cash outflow curve (S-curve)*. A typical curve is shown in Fig. 9.1.3; note that the coordinates are normalized as cumulative percent of budgeted cash expended versus cumulative percent of scheduled time elapsed. For each project, a cash outflow curve shape may be projected, provided a realistic time schedule has been formulated. Construction costs may then be tracked on the curve to indicate general adherence to budgets and schedules.

A simple example will serve to illustrate the method of constructing a cash outflow curve. The plan of construction for installing a new batch drying furnace is given in Fig. 11.2.3. The plan is to have the job completed in 5 weeks; delivery of the furnace is anticipated in the middle of the third week ($30,000 delivered cost), and a total of $15,000 worth of accessories will be delivered (or obtained from the plant storeroom) during the fourth week.

TABLE 11.2.4
AN EXCERPT FROM A TYPICAL CODE OF ACCOUNTS

	700-00 Instrumentation
Acct. no.	Description
700-10	Design calculations
700-20	Specifications—general
700-30	Vendor correspondence
700-40	Inquiries
700-50	Operating instructions and start-up data
700-60	Correspondence—general
701-00	Indicators—gauges
702-00	Controllers—recorders
703-00	Transmitters
704-00	Thermocouples—RTD
705-00	Analyzers
706-00	Transducers
707-00	Switches
708-00	Alarms—annunciators
709-00	Intercom
710-00	Panel—enclosures—mounting shelves
711-00	Graphics
712-00	Control computer
713-00	Management information computer
714-00	Input terminal
715-00	Printer
716-00	Video terminal
717-00	Modems
718-00	Air supply set
719-00	Orifice meter run
720-00	Valve manifolds
721-00	Relief valves
722-00	Control valves
723-00	Regulators
724-00	Piping
725-00	Wiring
726-00	Supports
727-00	Installation
728-00	Painting
729-00	Testing

For the sake of simplicity, assume:

1 All labor, including design and supervision, is $25/h.
2 All labor and material costs are disbursed at the end of the week in which incurred.
3 Personnel requirements include:
  Project manager: one person, all 5 weeks
  Design and specifications: one person
  Site preparation: one person
  Installation: three persons

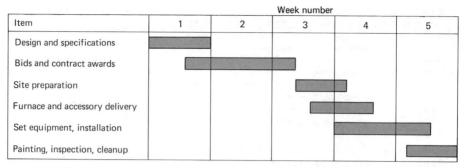

FIGURE 11.2.3
Installation of a drying furnace: plan of construction.

>Painting, etc.: two persons
>Bids and contracts: total of 9 person-hours

Total costs for each week are now generated. For example, for week 3,

$$\begin{aligned}
\text{Project manager cost} &= 40 \text{ h} \times 1 \text{ person} \times \$25/\text{person-h} \\
&= \$1{,}000 \\
\text{Furnace delivery} &= \$30{,}000 \\
\text{Site preparation} &= 3 \text{ days} \times 8 \text{ h/day} \times 1 \text{ person} \times \$25/\text{person-h} \\
&= \$600 \\
\text{Purchasing functions} &\approx 2 \text{ person-h} \times \$25/\text{person-h} \\
&= \$50 \\
\text{Total cost for week 3} &= \$31{,}650
\end{aligned}$$

The total of all budgeted costs is $57,425, of which the third week represents 55.1 percent. In this way, a cash outflow curve may be plotted, as in Fig. 11.2.4. The shape of the curve, in fact, is quite typical of many projects, with very slow cash disbursement until the sudden spurt at the 50–60 time percent point, when equipment delivery commences.

Clearly, the job of constructing a cash outflow curve for a construction project becomes enormously complex for large-scale projects, and it is customary to employ, even for cost control purposes, standard curves based upon past experience with similar plants, curves not unlike Fig. 9.1.3. Often separate curves are used for the engineering design and the construction phase. Start-up costs are normally not included as part of the construction cost control effort; start-up costs may be estimated separately (see, for example, Malina, 1980).

Cost control during the construction project is characterized by *cost tracking*. For instance, progress of the project may be monitored as in Fig. 11.2.5; actual disbursements are superimposed upon the projected cash outflow curve, and deviations as well as trends are seen at a glance.

SECTION 11.2: COST CONTROL OF CONSTRUCTION PROJECTS **625**

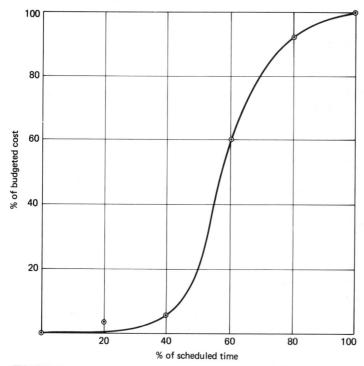

FIGURE 11.2.4
Installation of a drying furnace: cash outflow curve.

FIGURE 11.2.5
Cost tracking of cash outflow curve.

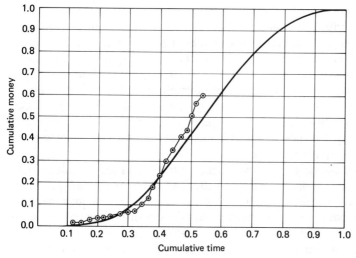

Exercise 11.2.2

Does the trend in Fig. 11.2.5 necessarily indicate a danger of cost overruns? What other explanation could there be for the trend?

Effective cost control, of course, involves a great deal more than graphical tracking. For example, *cost control logs* are recommended as a means of monitoring the cost performance of various code-of-accounts classifications, of contractors, of vendors, of in-house design and construction groups. If deviations occur (for example, 70 percent of budgeted funds expended with 50 percent of job completed or material delivered), the project manager must determine the cause and somehow get costs back in line. Finally, corporate management is kept abreast of the progress of construction by means of periodic *status reports* and *review conferences;* these serve to remind the project manager of the importance of keeping tight reins upon the project at all times.

Exercise 11.2.3

Curves similar to Fig. 11.2.5, but with different coordinates, are also useful for project monitoring and cost control. One such curve may represent percent project completion *versus* percent of scheduled time. How do you think that the percent completion could be characterized and computed?

11.3 TIME CONTROL OF CONSTRUCTION PROJECTS: SCHEDULING

Bar Charts

The other important aspect of construction project evaluation is *time control*. The purpose of time control is to make sure that a construction project will be completed within the allocated time; time overruns delay plant start-up and may result in costly production delays and failure to live up to product delivery schedules to customers.

As was the case with cost control, a number of planning and monitoring tools is available to the project manager. Some of these again are rather self-evident to the person with good organizational perception, but some require a certain degree of elaboration to be properly understood. In particular, project *planning* methods involve network analysis not unlike the decision tree network analysis described in Chap. 10. These network planning methods (and the resource *scheduling* based upon the plans) are readily adaptable to computer usage. In fact, the complexity of the networks normally demands the use of computers, and a thriving business exists in the appropriate computer software.

The simplest time planning tool is the ordinary *bar chart*. A very simple example is shown in Fig. 11.2.3; perhaps a more realistic schedule is the one in Fig. 11.3.1. For small jobs (say, below $250,000), the bar chart is an adequate resource allocation and time control tool. No matter what the job size or the complexity of the scheduling tool, the resulting schedule is only as reliable as the projected timing of the individual component tasks. If, for instance, the reactor procurement and

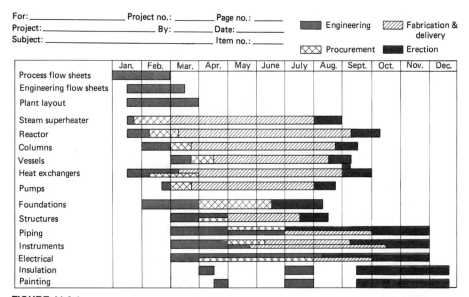

FIGURE 11.3.1
A preliminary project schedule. *(Landau, 1966. Reproduced by permission of Van Nostrand Reinhold Company, copyright 1966.)*

delivery time in Fig. 11.3.1 is missed by 2 months, the reactor may, indeed, still be installed by the scheduled completion date (end of December), but other jobs which depend upon the reactor installation will be correspondingly delayed, and so will the completion of the project. No matter how sophisticated the scheduling methodology may be, its effectiveness depends upon the realism of the time scheduling of the individual tasks. Datz (1981) refers to this as GIGO, garbage in, garbage out. The self-delusion schedule in Exhibit 11.3A illustrates the perils of GIGO.

Reference to any bar chart schedule, such as the simple one in Fig. 11.2.3, reveals a number of characteristics of the component tasks:

1 Some jobs are characterized by a *precedence hierarchy*. For instance, bids, in effect, cannot be sent out until the equipment specifications have been completed. Equipment cannot be set in place until the foundations have been finished.

2 Other jobs do not have the burden of restraint of a precedent job, or else there is "plenty of time" to complete the job within a relatively wide time span. Such jobs are called *floaters*. The site preparation job in Fig. 11.2.3, for example, could have been scheduled for the second week (although it must, of course, follow the design phase). Floaters are important in the appropriate scheduling of resources such as labor to level out excessive demand fluctuations.

3 Other *restraints*, in addition to the precedence hierarchy restraint, may occur. The site preparation job mentioned may, in fact, be restrained by the requirement

> **EXHIBIT 11.3A**
>
> AN EXAMPLE OF AN UNREALISTIC PROCUREMENT TIME PROJECTION
>
> *Problem*
> Purchase a tank, 6 ft in diameter by 36 ft long, 150 psi rating, SS 316, and tell the construction department when it will be on hand. All specifications have been written and checked.
>
> | Activity | Self-delusion | Reality |
> |---|---|---|
> | A. Bid period | 14 days | 18 days |
> | B. Bid award | 1 day | 3 days |
> | C. Vendor delivery statement, 6 to 8 weeks | 42 days | 56 days |
> | D. Vendor drawing review and approval | Included in item C | 7 days |
> | E. Shipping time | 5 days | 12 days |
> | F. Unload and check | Not thought of | 2 days |
> | | 62 days | 98 days |
>
> *Source*: Courtesy of William G. Clark, Manager of Engineering at Dow's Western Division.

of minimizing the total time of construction-caused interruptions upon existing production.

4 Jobs may *overlap* in time. Some jobs that are not interdependent may be carried out in parallel. Even jobs that have precedence restraint may be *blended* to some extent; for example, painting need not be delayed until every last scrap of insulation has been installed.

A consideration of these characteristics makes it clear that to minimize the time required to complete a construction project, some optimum arrangement of the component tasks exists. In fact, there is a singular sequence of a portion of the total ensemble of tasks that defines the so-called *critical path* through the project. The tasks that lie on the critical path are the *critical jobs,* and the sum total of the times projected for the sequence of critical jobs is the minimum project time.

Bar charts may be used to illustrate the nature and method of defining the critical path, as shown by Lowe (1980). We will take Lowe's example to demonstrate the application of critical path concepts. The first and most important step is to list the required project jobs, to estimate the job duration, and to define in an unequivocal manner the precedence hierarchy. Lowe refers to this listing as a *job sheet,* and his example is reproduced in Table 11.3.1. Never mind what the actual project represents; accept, for the moment, that the list of jobs is complete.

Each job, then, is assigned a number, and the job duration is estimated for some *standard complement of workers and equipment.* For example, pipework insulation (job 19) might involve eight insulators using standard tools. The precedence hierarchy (*job sequence*) is then defined. Job 1 must be the first one; job 2 follows

TABLE 11.3.1
A JOB SHEET

Job no./job	Job sequence	Job duration, days
1 Clear site	0:1	4
2 Excavations and foundations	1:2	9
3 Install support steelwork	2:3	9
4 Install stairways	3:4	3
5 Install platforms	3:5	8
6 Install handrails and platform angles	3:6	7
7 Install equipment	4, 5, 6:7	4
8 Calibrate equipment	7:8	7
9 Install prefabricated ductwork	7:9	21
10 Install process pipework	4, 5, 6:10	10
11 Connect pipework to equipment	8, 10:11	4
12 Install service pipework	4, 5, 6:12	7
13 Connect service pipework to equipment	8, 12:13	3
14 Install instrumentation	4, 5, 6:14	14
15 Connect instrumentation to equipment	8, 14:15	4
16 Install electric apparatus	4, 5, 6:16	9
17 Connect electrics to equipment	8, 16:17	3
18 Insulate equipment	8:18	10
19 Insulate pipework	11, 13:19	14
20 Insulate ductwork	9:20	6
21 Testing equipment	1–17:21	4
22 Painting	1–13:22	7
23 Cleanup	1–20:23	1

Source: Lowe (1980); by permission of *Chemical Engineering*.

job 1, and no other job can be performed directly following job 1 until job 2 has been completed. When job 3 has been completed, *three* independent jobs (4, 5, and 6) can be started in parallel. Job 7, however, cannot be started until *all three* have been completed; therefore the start of job 7 is dictated by the completion of the job of the longest duration (number 5).

Note that not all precedent jobs need to be listed for clarity. For instance, job 17 has only two antecedents listed, 8 and 16. But 16 has three antecedents listed—4, 5, and 6, which then, of course, become antecedents of 17 as well. The hierarchy listing must be unequivocal, but the person assembling the list need not worry about the *direct* antecedent of a particular job at this point. Thus job 23 (cleanup) must await the completion of all jobs from 1 to 20; later we find that job 20 is the direct antecedent, but this need not be known at the time the job list is assembled.

It should be emphasized that there is nothing unique or sacred about the specific jobs listed in Table 11.3.1. Two different persons could come up with different listings and a different number of jobs. For instance jobs 11 (connect pipework to equipment) and 13 (connect service pipework to equipment) could possibly be combined, although the precedence hierarchy would then demand completion of jobs 10 and 12 before proceeding with the combined job. A different list might include job 22 (painting) as an antecedent of 23 (cleanup).

The bar chart constructed by Lowe on the basis of the job sheet in Table 11.3.1 is shown in Fig. 11.3.2. If desired, a horizontal format as in Fig. 11.3.1 may be used. Each job is numbered at the top of the bar (line segment) representing it. Jobs 1, 2, and 3 are in direct sequence. Jobs 4, 5, and 6 are parallel, but job 5 is the longest. We can ask, "Is there a job that can be started when job 4 has been completed, but before 5 or 6?" A scan of the list shows that there is no such job; thus 7 depends upon completion of 4, 5, and 6, and effectively upon the completion of the longest of the three, number 5.

In this fashion we continue to develop the bar chart. Can other jobs be started in parallel with 7? Certainly not 8 or 9, but we soon discover that 10, 12, 14, and 16 can all be run in parallel with 7! Why can't job 13 be started right after 12 has been completed? Because job 8 is another required antecedent, and, as seen from the chart, job 13 is inevitably delayed by it.

After completing the bar chart, we note the path from start to finish which has no gaps in it. This is the critical path, for it defines the minimum time in which the

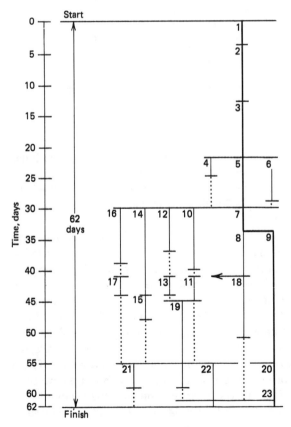

FIGURE 11.3.2
Bar chart with critical path indicated. *(From Lowe, 1980. By permission of Chemical Engineering.)*

project can be completed. In Fig. 11.3.2, the critical jobs are 1–2–3–5–7–9–20–23 (job 22 can substitute for 20 plus 23 on the critical path). The minimum project time is thus 62 days, a considerable reduction from the sum total of all jobs, 168 days. Clearly, nobody in their right mind would run all the jobs sequentially, but it would be difficult to guess the 62-day minimum sequence with a more haphazard approach.

Exercise 11.3.1

Suppose painting had to be completed before cleanup in Table 11.3.1. What would be the minimum project duration?

Floaters are readily spotted on a bar chart such as Fig. 11.3.2. Job 18, for instance, could be delayed for as long as 10 days without affecting the project duration.

Exercise 11.3.2

Can job 11 be floated?

The bar chart is a particularly convenient time control tool, for it can be transferred onto a calendar date scale, and the progress of various scheduled jobs can be monitored at a glance. Often it is useful to update the bar chart to reflect the reality of actual completion dates.

Exercise 11.3.3

Job 9 has been unexpectedly completed 5 days early, and the workers on jobs 20, 21, 22, and 23 are "ready to go." How many days can be knocked off the scheduled 62-day project duration in Fig. 11.3.2?

CPM (Critical Path Method)

Large construction projects involve many more jobs than the 23 listed in Table 11.3.1. The job described as "excavations and foundations" may in itself involve a complex network of operations that needs to be incorporated into the critical path definition. Many thousands of jobs may thus be involved, and a simple bar chart rapidly becomes unwieldy and difficult to construct.

The planning methodology which led to the construction of Fig. 11.3.2 has been formalized as the *critical path method*. Algorithms have been devised (dynamic programming) which permit the computer generation of the CPM plan, often with the input of network modules representing standard tasks such as piling installation (Datz, 1981). Programs are also available to facilitate resource scheduling based upon the CPM plan (for instance, workforce demand leveling), and to help monitor the progress of construction, pinpoint timing "pinch points," and update the plan.

CPM diagrams, used primarily to illustrate the method, are not unlike the bar chart of Fig. 11.3.2. The symbolism of CPM diagrams, in some ways analogous to the symbolism of decision trees (Chap. 10), is shown in Fig. 11.3.3. The jobs (or *activities*) are symbolized by arrows which point in the general direction of the project time flow, from start to end. Each arrow starts at a preceding *node* P and ends on a succeeding node S. The nodes are usually symbolized by small labeled circles; their exact position, and the arrow lengths, are immaterial. If an activity ends at a particular node, that activity must be completed before any activities originating at that node can be started. For instance, none of the activities marked B, C, and D can be started until A has been completed. Note that there is no requirement or implication that B, C, and D need to be started simultaneously!

The activities A, B, . . . are labeled, for purposes of computer manipulation, on the basis of their numbered P and S nodes, as (P, S); thus activity F is labeled (40, 50). The node numbering is arbitrary, but for any activity the S number must be higher than the P number. An obvious rule is that *no two activities can have the same nodes*, for otherwise they would have the same label and thus be indistinguishable by the computer. Nevertheless, two activities can certainly have a common antecedent and consequence; for example, jobs B and C cannot be started until A has been completed, and they both must be finished before G can be started. To avoid an anomaly, job B, for instance, is provided with its own S node, and a *dummy activity E* (30, 50), having zero duration, is added to connect job B to the desired P node of job G. One might ask, "Why was the dummy added to B? Why not C?" It turns out that it makes no difference, but for the sake of internal consistency it is usually best to add the dummy to the shorter duration job. Thus job B presumably takes less time than job C. Dummies also indicate required precedence.

The basis for the CPM diagram is the job sheet with its precedence hierarchy. The initial job sheet for Fig. 11.3.3 might look something like this:

Job	Sequence	Duration
A	A	4
B	A:B	4
C	A:C	12
D	A:D	6
F	D:F	4
G	B, C, F:G	6

(There is no E listing initially; the existence of a dummy activity is not known a priori.)

This may be converted to a job sheet which displays the (P, S) notation. The conversion may be made, in our case, with the help of the CPM diagram, or else by "asking the right kind of questions":

Start at node 10.
End shortest job at node 20.
Are there any other jobs with P = 10? No.
End shortest job (B) starting at P = 20 at node 30.
Are there any other jobs with P = 20? Yes.

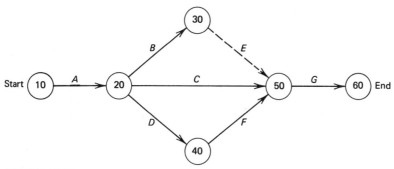

FIGURE 11.3.3
A simple CPM diagram.

End the next shortest job (*D*) at node 40.
Any more? Yes.
End the next shortest job (*C*) at node 50.
Any more? No.
Therefore go to next numbered node, 30.
End shortest job (*G*) at node 60.
Are there any other jobs with P = 30? No.
Therefore go to next numbered node, 40.
Etc.

The need for the dummy *E* is soon revealed when it is discovered that nodes 30 and 50 would both be node P for job *G*.

The modified job sheet is shown in Table 11.3.2.

The critical path and its duration are determined by computing the duration of all possible paths between nodes 10 and 60 and picking the *longest* one. For the simple case considered, three paths are possible:

10–20–30–50–60*   Σ time = 14

---

*Dummy is part of the path; in fact, dummies may be part of the critical path if they are not consistently associated with jobs of shorter duration.

TABLE 11.3.2
CPM JOB SHEET FOR FIG. 11.3.3

Job	Duration
(10, 20)	4
(20, 30)	4
(20, 40)	6
(20, 50)	12
(30, 50)	0
(40, 50)	4
(50, 60)	6

10–20–50–60   Σ time = 22
10–20–40–50–60   Σ time = 20

Therefore the critical path is the second one listed, and the project minimum time is 22.

A CPM diagram equivalent to the bar chart in Fig. 11.3.2 is shown in Fig. 11.3.4. The bar chart is decidedly easier to draw, and the project timing is more easily visualized, since the bar chart includes a time scale, and parallel jobs are more obviously juxtaposed. In the CPM diagram, the activities are labeled with their number assignments from Table 11.3.1; the dummy activities have been arbitrarily labeled with letters. Most of the dummies are incorporated to avoid duplication of activities between two specific nodes. Examples include A and B; note that dummies are associated with the shorter duration activities. Dummy E is noteworthy in that it is not used to avoid activity duplication; it only indicates the precedence of job 8 over job 15.

Exercise 11.3.4

Pencil in node numbers on the CPM diagram, Fig. 11.3.4, which will conform to the requirement that in each activity label (P, S), S > P.

FIGURE 11.3.4
CPM diagram of example in Table 11.3.1.

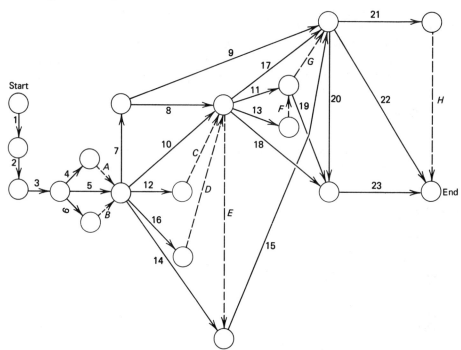

SECTION 11.3: TIME CONTROL OF CONSTRUCTION PROJECTS: SCHEDULING

The final format of the CPM diagram is decidedly a function of the relative job durations, as illustrated by the following example.

Example 11.3.1

Suppose job 17 in Table 11.3.1 were 19 days long instead of 3. How would the bar chart and CPM diagram be affected? Would the critical path be any different?

SOLUTION See Fig. 11.3.5. The bar chart (only the lower, modified portion is shown) quickly reveals that the critical jobs now are 1–2–3–5–7–8–17–21, with a total duration of 64 days. The 16-day increase in job 17 has resulted in only a 2-day increase in project duration, since job 17 had not previously been critical.

The CPM diagram (terminal portion only) is modified by adding the new node and the dummy J. Note that the direction of dummy activity H is reversed.

We saw that the job durations in Table 11.3.1 were based upon a standard complement of workers and equipment, and the project duration turned out to be 62 days. Suppose this much time is unacceptable; what can be done about it? The answer is that some of the project's component jobs can be *crashed;* that is, extra workers, equipment, and overtime can be thrown on jobs to shorten the job duration. Some jobs cannot be crashed, and other jobs need not be crashed, as we will see.

We return to Lowe's example and assume that the various jobs, first listed in Table 11.3.1, can be crashed as shown in the following table:

FIGURE 11.3.5
Modified diagrams, Example 11.3.1. (a) Bar chart; (b) CPM diagram.

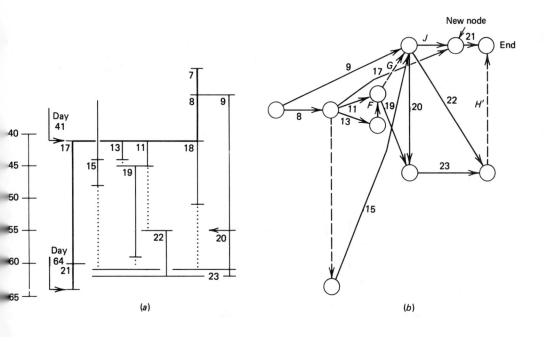

Job no.	Crashed time, days	Job no.	Crashed time, days
1	2	13	2
2	6	14	7
3	7	15	2
4	2	16	6
5	5	17	2
6	5	18	5
7	3	19	7
8*	7	20	3
9	14	21*	4
10	7	22*	7
11	2	23*	1
12	6		

Comparison with Table 11.3.1 shows that the jobs marked with an asterisk cannot be crashed.

The results of maximum crashing are shown in Fig. 11.3.6. The project duration has been shortened by 18 days to 44 days. For maximum crashing, even some jobs that are not on the critical path may have to be crashed; for instance, job 6 has to be crashed to accommodate the crashing of job 5. In fact, crashing may change the critical path; if job 6 were to have a crash time of 6 days instead of 5, it would enter the critical path. Note in particular that there is no point in crashing some jobs; for instance, crashing job 14 even by 7 days would not shorten the project one iota.

In the above analysis, we have assumed that joint crashing of jobs is possible, that the project will not be choked with people getting in each other's way. We have

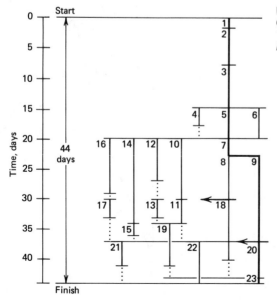

FIGURE 11.3.6
Crashed project bar chart. *(From Lowe, 1980. By permission of Chemical Engineering.)*

also not considered economic factors which may modulate our desire to maximize crashing. Put simply, crashing costs a lot of money, and this must be more than balanced by economic benefits of an earlier start-up. Provided that reasonably reliable data can be assembled on the cost of crashing any one job an additional day, it stands to reason that the first job to be crashed is the one that costs the least. Even then, the additional cost must be more than offset by increased profits (on a pretax basis) generated as the result of an earlier start-up.

For example, suppose each additional day of operation of a new process modification will result in an increased profit before taxes of $300. The critical jobs that can be crashed are shown:

Job	Max. number of days that can be crashed	Cost per day crashed, $
A	2	150
B	3	250
C	4	400
D	2	450

Job A is clearly the one to start with in the crashing analysis; crashing 1 day will cost $150 but save $300. In this way, the following table can be developed:

Total crash time, days	Total increased profit, $	Jobs and days crashed	Cost of crashing, $	Net saving, $
1	300	A1	150	150
2	600	A2	300	300
3	900	A2, B1	550	350
4	1200	A2, B2	800	400
→ 5	1500	A2, B3	1050	450
6	1800	A2, B3, C1	1450	350
7	2100	A2, B3, C2	1850	250
8	2400	A2, B3, C3	2250	150
9	2700	A2, B3, C4	2650	50
10	3000	A2, B3, C4, D1	3100	−100
11	3300	A2, B3, C4, D2	3550	−250

The maximum saving is attained with 5-day crashing; maximum crashing would, in fact, result in a net loss. A better estimate of the optimum could be obtained by using considerations of timing and the time value of money to compute a NPW balance.

Crashing is one of the considerations that enters into the scheduling of people, equipment, and other resources. CPM results may be used for scheduling tasks such as workforce leveling. A simple example will serve to illustrate the idea.

Suppose a network of construction tasks includes a portion wherein the following five jobs must be completed within a span of 10 days, as demanded by other critical jobs:

Job	Normal job duration (no crashing), days	Normal worker requirement
A	6	3
B	5	5
C	4	7
D	3	10
E	2	14

If the jobs were all to be completed as soon as possible, the daily worker requirement would be as shown in Fig. 11.3.7. In most cases such a demand schedule would be intolerable; 39 workers would be employed at the start, and a few days later most of them would be sent home. By recognizing that the jobs are floaters, a better workforce scheduling job can be done, as in Fig. 11.3.8. The leveling is not perfect, but there is a much better chance of moving individuals from and into other jobs outside the scope of the five jobs considered. The distribution in Fig. 11.3.8 was obtained by trial and error; computer-oriented optimization methods can be devised.

FIGURE 11.3.7
Workforce demand in absence of leveling.

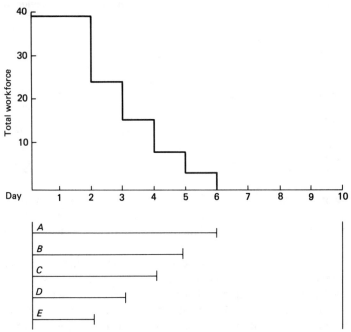

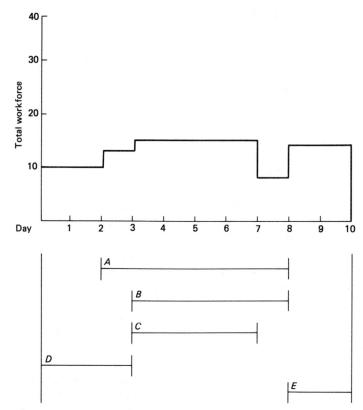

FIGURE 11.3.8
Workforce demand with workforce leveling.

PERT (Program Evaluation and Review Technique)

The program evaluation and review technique is a variant of CPM which incorporates the element of risk into the computations. It is recognized that the estimation of even a normal job duration is not terribly reliable, and an attempt is made to compensate for this uncertainty by defining an *expected time* t_e given by

$$t_e = \frac{t_l + 4t_m + t_h}{6} \tag{11.3.1}$$

Here t_m is the most likely job duration, and t_l and t_h are, respectively, low and high estimates so that the job will take less time than t_l 5 percent of the time, and 5 percent of the time it will take longer than t_h. Equation (11.3.1) is based upon the hypothesis that individual job durations will follow a beta distribution.

Assuming that values of t_m, t_l, and t_h can be reasonably estimated, we can now calculate values of the time T_E, *the earliest expected completion time of each job* in the project network. Let us return to the example of Fig. 11.3.3 to see how this is done; the estimated t's and the calculated T_E's are given in Table 11.3.3. We start by calculating t_e for each job by using Eq. (11.3.1). Note that for dummies all t's are zero. For some jobs there is an invariant completion time, and $t_l = t_m = t_h$.

We now start at node 10; for job (10, 20), $T_E = t_e$. Since no other jobs have P = 10, we move on to node 20 and note that three jobs originate here. Node 30 is reached with jobs (10, 20) and (20, 30), and T_E is the sum of the t_e's for those two jobs; $T_E = 4.0 + 3.8 = 7.8$. Similarly, the T_E's for jobs (20, 40) and (20, 50) are computed as 10.2 and 16.2, respectively. Continuing with the next node in numerical order, we note that t_e for (30, 50), a dummy, is zero, so that T_E for (30, 50) is the same as T_E for (20, 30), or 7.8. In like manner, T_E for (40, 50) is $4.0 + 6.2 + 4.0 = 14.2$.

No other activities originate from nodes 30 or 40, so we can proceed to 50. We have now determined that three jobs terminate at 50; the job that originates at 50 cannot start until *all* the jobs terminating at 50 have been completed. The one that has the longest T_E is (20, 50), whence the T_E for (50, 60) is $16.2 + 6.0 = 22.2$. Note how the whole developmental sequence proceeded in a logical and systematic fashion. Once the individual responsible for the planning has properly established a precedence hierarchy represented by the (P, S) terminology, the follow-up PERT computations are automatic and readily adaptable to programming.

The final value of 22.2 days for the earliest expected completion time for the whole project represents much the same kind of information that was obtained with CPM, except that the elements of risk have been recognized. With PERT, we can now go one step further and ask, "What is the probability that the project can be finished in less than X days?" For example, for the project represented by Table 11.3.3, what is the probability that the project can be finished in 20 days or less?

We start by listing values of T_L, the *latest time* any activity must be completed to meet the demand of 20 days. Thus, for (50, 60), the only job with S = 60, $T_L = 20$. Three activities end at node 50, and therefore all three must end, at the latest, in 20 minus t_e of (50, 60), or $20 - 6.0 = 14.0$ days. (30, 50) is a dummy, therefore (20, 30) must also have $T_L = 14.0$. (40, 50) has $t_e = 4.0$, so that T_L for (20, 40) is $14.0 - 4.0 = 10.0$. Finally, T_L for (10, 20) must be the *smallest* (why?) of the three possibilities (there are three activities originating at node 20),

TABLE 11.3.3
PERT COMPUTATIONS FOR FIG. 11.3.3

Job	t_l	t_m	t_h	t_e	T_E	T_L	Slack, $T_L - T_E$
(10, 20)	3	4	5	4.0	4.0	1.8	−2.2*
(20, 30)	2	4	5	3.8	7.8	14.0	6.2
(20, 40)	5	6	8	6.2	10.2	10.0	−0.2
(20, 50)	10	12	15	12.2	16.2	14.0	−2.2*
(30, 50)	0	0	0	0	7.8	14.0	6.2
(40, 50)	4	4	4	4.0	14.2	14.0	−0.2
(50, 60)	5	6	7	6.0	22.2	20.0	−2.2*

$$T_L - t_e \text{ of } (20, 30) = 14.0 - 3.8 = 10.2$$
$$T_L - t_e \text{ of } (20, 40) = 10.0 - 6.2 = 3.8$$
$$T_L - t_e \text{ of } (20, 50) = 14.0 - 12.2 = 1.8$$

and the answer is 1.8.

The next step is to calculate the so-called *slack* of each job, the value of $T_L - T_E$. Positive values of slack mean that, under the imposed project time limit, the *latest* that a job must be finished is beyond the time that *it is expected* to be completed; floaters are more likely to have a positive slack. Negative values of slack indicate a "pinch point" situation—the *expected* time (but not necessarily the low estimate) goes beyond the latest compulsory time. The activities on the critical path will, by necessity, *all have the same slack and the lowest value of slack*. Note that, in this manner, *the activities on the critical path are automatically identified* (they are marked with asterisks in Table 11.3.3). Computations and the critical path identification may be performed without the aid of any diagrams, although these do promote easy visualization.

The variance of the individual activity durations having a beta distribution is given approximately by (Fuller, 1966)

$$\sigma_i^2 = \left(\frac{t_h - t_l}{6}\right)_i^2 \tag{11.3.2}$$

If we add the duration variances of all the critical activities, we obtain the duration variance of the critical path, or, in other words, the duration variance of the project [see also Eq. (4.5.12)]:

$$\sigma_{CP}^2 = \sum_1^n \left(\frac{t_h - t_l}{6}\right)_i^2 \tag{11.3.3}$$

Furthermore, we assume that the distribution of the project durations will be normal, and the standardized variable y is defined,

$$y = \frac{(T_L - T_E)_{CP}}{(\sigma_{CP}^2)^{1/2}} \tag{11.3.4}$$

for which the mean is zero and variance is 1. Tables of normal distribution functions can then be used to calculate the probability that T_L (normally distributed about T_E) will be less than some specified amount.

For the critical path in Table 11.3.3,

$$\sigma_{CP}^2 = \tfrac{1}{36}[(5-3)^2 + (15-10)^2 + (7-5)^2] = 0.917$$

$(T_L - T_E)_{CP} = T_L - T_E$ for activity (50, 60) (or *any* critical activity). Therefore

$$y = \frac{-2.2}{\sqrt{0.917}} = -2.30$$

From tables, for $y = -2.30$, $1 - F(y) = 0.0107$; that is, there is a 1.07 percent probability that the project in Table 11.3.3 can be finished in 20 days or less. In plain words, completion within 20 days is very unlikely.

Exercise 11.3.5

What is the probability of completing the project in Table 11.3.3 in 24 days or less?

A number of corporations are in the business of providing software for computer control of construction projects by use of CPM and PERT. It will be appreciated, for example, that workforce or cost control may be exercised by proper scheduling on the basis of the PERT time network and subsequent monitoring by comparing actual numbers with the plan.

11.4 SAFETY AND LOSS PREVENTION

Preoperational Reviews

The safety and loss prevention aspects of a project are vital components of the noneconomic evaluation of the project; in fact, modifications which arise from such evaluation may have an important economic impact. Considerations of *safety* incorporate personal safety and hygiene of workers in the proposed process area as well as safety and integrity of the human and natural environment. *Loss prevention* considerations refer to equipment and procedures to be planned out before plant start-up so that catastrophic accidents involving capital loss and business interruption will be avoided or, at worst, limited to an economically acceptable level.

An important tool in the safety and loss prevention effort is the *preoperational review*. The review involves the assembly of all pertinent safety and loss prevention information regarding a particular facility so that such information may be evaluated by a group of knowledgeable individuals who are not directly associated with the project but who, by virtue of their experience, can discover unanticipated problems and make appropriate recommendations. Preoperational reviews are mandatory for full-scale plants as well as semiplants and pilot plants. They are recommended for miniplants, and even the more ambitious or hazardous laboratory experiments can benefit from perhaps more simply structured reviews. The timing of the reviews depends upon the nature and scale of the project. The construction of full-scale plants may involve several such reviews, from the initial planning stage to the stage just before start-up of the completed facilities.

The individuals who make such reviews may be members of permanent committees on the integrated complex site (such as a divisional safety committee), or they may be ad hoc members of the immediate review committee. However, departments that normally are responsible for monitoring various safety and loss prevention aspects in the plant should be represented. Such departments may include industrial hygiene, medical, environmental control, safety, plant protection, and reactive chemicals (the last a department or committee entrusted with the monitoring of hazards arising from the reactivity of process chemicals). The multifaceted expertise of the preoperational review committee is there to help the project manager achieve the goal of a successful and safe project. Nevertheless, the committee must be given the clout to enforce its recommendations; manifestations of that clout include the requirement of sign-off by key departmental representatives and the rule that no facility may be started up until all committee recommendations have been carried out.

In order that the committee be able to carry out its duties, the safety and loss prevention aspects of the project must be properly documented, a responsibility of the project manager. The documents are assembled in a *pre-op package* which, first of all, must have information that will give the committee members a thorough understanding of the nature of the process and facility:

Process purpose and description
Reaction scheme
Flow sheets and P&IDs, material balance
Equipment description and layout
Operating instructions (including control computer programs)

The assembly of this material and other information that follows is a job of considerable magnitude; however, it is information that the project manager must have anyway, and the pre-op package serves the purpose of centralizing and systematizing information for subsequent use as backup documentation for the completed facility.

The following are some of the many potentially useful items that are frequently incorporated into the pre-op package:

Chemical stability information. Results of calorimetric screening methods to detect hazardous reactivity or instability of all process chemical streams. Results of shock sensitivity tests, dust explosion tests, etc. NFPA ratings (Chap. 3), flammability data.

Toxicological data. Assessment of all compounds and mixtures that could potentially contact workers for hazard to eyes and skin and toxicity by ingestion or inhalation. Allowable workplace concentrations (TLVs—see Prob. 3.2). Long-range effects (carcinogenicity, chronic toxicity).

Thermodynamic and thermochemical analysis. Heats of reaction and energy balances. Thermochemical analysis of operator errors and runaway reactions, both primary and secondary (decompositions)—maximum attainable temperatures, pressures.

Reactive chemicals and hazard review. The project manager's own assessment of the process hazards and what has been done to overcome them. An example might be an assessment of the hazards of flammable vapor leaks, the damage potentially caused by vapor cloud detonations (calculated as overpressures upon buildings and structures), and preventive measures (quench sprays, barriers, inerting gases).

Emergency procedures. Automatic and operator-initiated actions to deal with emergencies: power loss, cooling water loss, massive spills, fires, and many others. Quick emergency shutdown procedures.

Job safety analysis. Step-by-step analysis of operating jobs which pose a hazard to the operators, and the precautions to be taken, including protective clothing and equipment.

Emissions and waste disposal. Identification of all vents, assessment of fugitive emissions. Compliance with enforcement agencies' regulations. Methods of waste disposal. Assessment of environmental hazards.

Relief devices. A listing and documented design justification of all pressure-relief devices.

Fire and explosion index. Computation of this index, which is described in what follows, facilitates the assessment of the risk of fire and explosions and the maximum probable property damage caused by such catastrophes.

Special permits. These permits may be granted by the managers of the integrated complex so that normally proscribed activities or equipment may be used. An example might be the use of a sparking device in an area with a flammable vapor potential. Proper safety precautions to negate the disadvantages must be outlined.

Other documentation might include, for example, evidence that neighboring operations have been consulted to make sure that no hazardous interactions will occur and that emergency information is exchanged. Experienced preoperational reviewers are adept at analyzing pre-op packages and interviewing project participants to spot potential problems. Operating instructions, even computer programs, must be scanned in detail, a tedious job for individuals not closely associated with a project. Yet it is the only way to discover potentially disastrous problem nuances such as the operator error of not turning on an agitator on time. Serious industrial accidents have occurred when operators, suddenly realizing their error, have rushed to push the reactor agitator button—*after* adding all reactants—to initiate an instantaneous energy release when the layered reactants are suddenly blended. Proper instructions or controls must anticipate such possibilities and indicate means of obviating the problems.

Fire and Explosion Index

The fire and explosion index (F&EI) represents a systematic approach to the identification of significant areas of loss potential in chemical plants. The index serves as a basis for estimating the maximum probable property damage (MPPD) in a plant if a fire or explosion originates in one part of it; the business interruption (BI) loss can also be estimated on the basis of the maximum probable days of outage (MPDO) following a catastrophe.

Computations required to generate an index number are outlined in the booklet, "Fire and Explosion Index Hazard Classification Guide," 5th edition, published by the American Institute of Chemical Engineers (1981). The number that is obtained for a particular plant (or isolated portion of a plant) is quite arbitrary, but comparison with the numbers that are computed for plants having a perceived range of operating risks results in some feel for the hazard associated with the operation of the plant in question.

The computation procedure starts with the highest *material factor* (MF) of any chemical component or mixture in the plant. The higher the MF, the more reactive and flammable is the component. For example, if the facility in question is a distillation unit for separating pyridine and octane with MF = 24 and 16, respec-

---

EXHIBIT 11.4A

LOSS CONTROL CREDIT FACTORS

1 Process control (C_1)

a Emergency power	.97	e Computer control	.89 to .98
b Cooling	.95 to .98	f Inert gas	.90 to .94
c Explosion control	.75 to .96	g Operating procedures	.86 to .99
d Emergency shutdown	.94 to .98	h Reactive chemical review	.85 to .96

C_1 total _____*

2 Material isolation (C_2)

a Remote control valves	.94	c Drainage	.85 to .95
b Dump/blowdown	.94 to .96	d Interlock	.96

C_2 total _____*

3 Fire protection (C_3)

a Leak detection	.90 to .97	f Sprinkler systems	.60 to .96
b Structural steel	.92 to .97	g Water curtains	.95 to .97
c Buried tanks	.75 to .85	h Foam	.87 to .98
d Water supply	.90 to .95	j Hand extinguishers	.92 to .97
e Special systems	.85	k Cable protection	.90 to .96

C_3 total _____*

$$C_1 \times C_2 \times C_3 = \underline{\qquad}$$
Actual credit factor† (to line D below)

Risk analysis summary

A1. F&EI _____
A2. Radius of exposure _____ ft
A3. Value of area of exposure $ × 10⁶ _____
 B. Damage factor _____
 C. Base MPPD (A3 × B) $ × 10⁶ _____
 D. Credit factor, actual _____
 E. Actual MPPD (C × D) $ × 10⁶ _____
 F. Days outage (MPDO)   _____ days

*Product of all factors used.
†Obtained from correlating graph in the Guide.

tively, the higher MF is used for the computation. The MF is tabulated for many compounds and mixtures; it can also be estimated from NFPA ratings or from stability and flammability data.

The F&EI is generated by multiplying the material factor by a *hazard factor;* the latter is obtained as a sum of penalties for each of a number of listed potentially hazardous situations—enclosed process units, operation above autoignition temperature, corrosive fluids, and many others. Typically, a light hydrocarbon distillation facility might have an index of 100, an epoxy resin semiplant 60, a warehouse for a bagged powder product 15. An index above 100 indicates an appreciable hazard; above 135 the hazard is considered very severe, indeed.

A more meaningful correlation is the one between the F&EI and the MPPD. The procedure is outlined in Exhibit 11.4A. A *credit factor* is developed which depends upon the degree of fire protection and the use of other devices to moderate the probability of disaster. Graphical correlations are available in the reference so that a *radius of exposure* may be estimated from the F&EI (based upon analysis of actual industrial accidents). The DFC within the radius of exposure is then estimated and multiplied by the credit factors to generate the MPPD. This kind of computation gives management (and insurance carriers) a great deal of insight about the risks associated with a hazardous operation. Moreover, the effect of additional disaster control investments upon MPPD can be computed (by means of adjusted values of the credit factors); the method is thus a quantitative loss prevention control tool.

A graph of MPDO versus MPPD, based upon actual incidents, is also available. The business interruption loss can be calculated from the MPDO and the daily VPM (value of product manufactured).

NOMENCLATURE

A area, m^2
C cash flow
i_0 after-tax, inflation-adjusted internal rate of return
k_1 first-order reaction rate constant, h^{-1}
K capital investment
m construction period, years
n plant useful life, years
Q heat transfer rate, W
r discounted cash rate of return
t activity duration
T duration of a sequence of activities; also, temperature
U overall heat transfer coefficient, $W/(m^2 \cdot K)$
y standardized variable, normal distribution
Δ difference
σ^2 variance

Subscripts

C coolant
CP critical path

e   expected
E   earliest expected
h   highest expected
l   lowest expected
L   latest
m   medium expected

Abbreviations

ARC   accelerated rate calorimetry
BI   business interruption
BOD   biochemical oxygen demand
COD   chemical oxygen demand
CPM   critical path method
DCRR discounted cash rate of return
DFC   direct fixed capital
DTA   differential thermal analysis
F&EI fire and explosion index
IR   infrared
MF   material factor
MPDO maximum probable days outage
MPPD maximum probable property damage
NFPA National Fire Protection Association
NPW   net present worth
PERT program evaluation and review technique
P&ID piping and instrumentation diagram
(P, S) (preceding, succeeding) node notation
TLV   threshold limit value
VPM   value of product manufactured

REFERENCES

Baasel, William D.: "Preliminary Chemical Engineering Plant Design," 4th printing, Elsevier, New York, 1980.

Datz, Marvin: Develop Project Scope Early, *Hydrocarbon Process.*, **60**(9):161 (1981).

Fuller, Don: "Organizing, Planning, and Scheduling for Engineering Operations," Industrial Education Institute, Boston, 1966.

Hajek, Victor G.: "Management of Engineering Projects," McGraw-Hill, New York, 1977.

Horwitz, Benjamin A.: How Does Construction Time Affect Return?, *Chem. Eng.*, **88**(19):158 (Sept. 21, 1981).

Hudson, W. G., and M. J. Marinak: "AIChE Pilot Plant Safety Manual," American Institute of Chemical Engineers, New York, 1972.

Johnstone, Robert Edgeworth, and Meredith Wooldridge Thring: "Pilot Plants, Models, and Scale-up Methods in Chemical Engineering," McGraw-Hill, New York, 1957.

Klimpel, Richard R.: Operations Research: Decision-Making Tool - II, *Chem. Eng.*, **80**:87 (Apr. 30, 1973).

Kline, P. E., et al.: Guidelines for Process Scale-up, *Chem. Eng. Prog.*, **70**(10):67 (1974).

Landau, Ralph: "The Chemical Plant: From Process Selection to Commercial Operation," Reinhold, New York, 1966.

Lowe, C. W.: Get More Control over any Project with Bar Charts, *Chem. Eng.*, **87**(14):139 (July 14, 1980).

Ludwig, Ernest E.: "Applied Project Management for the Process Industries," Gulf, Houston, 1974.

Malina, Marshall A.: Upgrading Predictions of Startup Costs, *Chem. Eng.*, **87**(16):167 (Aug. 11, 1980).

Marinak, M. J.: Pilot Plant Pre-Start Safety Checklist, *Chem. Eng. Prog.*, **63**(11):58 (1967).

Martino, Robert L.: Plain Talk on Critical Path Method, *Chem. Eng.*, **70**:221 (June 10, 1963).

Roth, Joanne E.: Controlling Construction Costs, *Chem. Eng.*, **86**(21):88 (Oct. 8, 1979).

PROBLEMS

11.1 Methylene chloride, CH_2Cl_2, is a commonly used industrial solvent because of its relatively low toxicity, essential nonflammability, and acceptable degree of chemical inertness and stability. It is commonly used as a solvent carrier for organic reactions. Given a sufficiently high activation energy, however, it can undergo a rapid decomposition reaction which releases a large amount of heat. The high activation energy may be imparted, for instance, by very energetic and rapid reactions occurring in the bulk of the solvent.

a Write a heat-producing decomposition reaction for CH_2Cl_2.

b Calculate the heat of reaction (starting with liquid CH_2Cl_2) by using handbook data.

c One cubic meter of liquid CH_2Cl_2 at its atmospheric boiling point (40°C) is contained in a 2-m^3 closed vessel; the vapor space has CH_2Cl_2 only. Calculate the *approximate* pressure (in atmospheres) potentially developed in the vessel if the solvent decomposes (use estimated average heat capacities, generalized compressibility correlations, etc.).

11.2 In Exercise 11.1.1 the relative costs of a batch and a single-stage continuous reactor were asked for. Would any cost saving be achieved by using multiple-stage continuous reactors? If so, how many stages would give the lowest relative cost?

11.3 An organic peroxide dissolved in excess benzene oxidizes the benzene to phenol in the presence of a suspended catalyst. The reaction is second-order, but in the presence of a large excess of benzene it is pseudo-first-order, with the reaction rate constant given by

$$k_1 = 8.583 \exp \frac{-1706}{T} \, h^{-1}$$

where T is the absolute temperature (K).

The reaction is exothermic with a heat release of -1.67×10^5 J/(g · mol) of peroxide reacted.

Suppose 3.80 m^3 of the reaction mixture is introduced into a stirred reaction vessel which is cooled with 27°C water circulated through an externally clamped cooling panel coil. The manufacturer claims that the effective heat transfer area is 4.5 m^2 and that the effective heat-transfer coefficient is 75 W/(m^2 · K).

a If the reaction mixture has an initial peroxide concentration of 1500 g · mol/m^3, what is the "temperature of no return"—i.e., runaway point B in Fig. 11.1.1—after the catalyst has been introduced?

b Suppose, because of a poorly fitted panel coil, the heat transfer coefficient is only 70 W/(m^2 · K). What is the effect upon the temperature of no return?

11.4 In the laboratory experiment of Example 11.1.1, liquid A was hexyl bromide solvent containing 12.5 g of dissolved 2,4-heptadiene. The gaseous reagent B was bromine vapor introduced at 100°C. The chemist performing the experiment fed excess bromine into the flask, but did not analyze the off-gases. The desired reaction was

$$CH_3(CH=CH)_2C_2H_5 + 2Br_2 \longrightarrow CH_3(CHBr-CHBr)_2C_2H_5$$

for which the calculated heat of reaction is -53.4 kcal/(g·mol) of heptadiene. A small quantity of LiBr acted as catalyst. Reaction completion was judged by absence of the characteristic double bond band on IR spectrum.

Is the observed temperature behavior compatible with the thermochemistry of the desired reaction? If not, what energetic side reactions would you anticipate?

11.5 Laboratory experiments have demonstrated that an acceptable grade of industrial calcium chloride solution may be prepared by reacting a waste 35% HCl stream with crushed limestone. In small lab batches, 99 percent of the acid reacted with excess limestone in 10 min to produce a 42% $CaCl_2$ solution according to the reaction

$$CaCO_3(s) + 2HCl(aq) \longrightarrow CaCl_2(aq) + CO_2(g) + H_2O(l)$$

The last of the residual acid is slow to react with the limestone and is preferably neutralized with 50% NaOH.

Sketch a reactor configuration that you would recommend for the pilot plant study of the $CaCl_2$ process, using 1-in limestone lumps as feed. Draw a preliminary flow sheet indicating the pilot plant equipment that might be required for the studies. For present purposes, the equipment need not be sized.

11.6 A waste tar stream is presently burned in a furnace. The tar contains about 10 wt % of a valuable by-product, benzoic acid. Preliminary experiments have shown that at 150°C (under pressure) the benzoic acid may be extracted with liquid water; the other tar components are not water-soluble to any appreciable extent. The distribution constant (i.e., the ratio of concentrations in two equilibrated phases) favors the aqueous phase. Water and benzoic acid are miscible in all proportions above the melting point of the acid (122°C). The high temperature is required to keep the viscosity of the extracted tar to a manageable level (about 1000 cP); the tar freezes at 80°C. Substantially all the benzoic acid may be recovered by cooling the extract and precipitating out the benzoic acid.

Sketch a pilot plant flow sheet showing equipment that might reasonably be used for studying a continuous extraction and dry product crystal recovery process.

11.7 Thionyl chloride may be prepared at 200°C by passing mixtures of S_2Cl_2 (sulfur monochloride), SO_2, and Cl_2 over activated charcoal:

$$S_2Cl_2 + 2SO_2 + 3Cl_2 \rightleftharpoons 4SOCl_2$$

In continuous-flow tubular nickel reactors, equilibrium is reached within a few seconds.

A miniplant facility is required to check the effectiveness and lifetime of a number of commercially available charcoal catalysts. With the aid of sketches, explain what you would do to come up with the required facility to carry out a long-range testing program. Be sure to specify (in general terms) the instrumentation needed for process control and data gathering. Chlorine and sulfur dioxide are available in compressed gas cylinders; sulfur monochloride may be purchased in small 5-gal drums fitted with screwed (standard pipe threads) bung holes. What would you do with the reaction products?

11.8 One hundred gram mols per hour of methane is cracked in a pilot plant high-temperature furnace. The product gas is cooled in an oil quench tower where tars are scrubbed out. The gas stream is then passed through a low-temperature condenser to remove condensable fractions and finally into a gas separation system.

The low-temperature condensate is essentially pure benzene (59 ml/h, sp gr = 0.879). Analysis of the tars in the quench oil indicates 4.03 wt % hydrogen, the rest carbon. The analysis of the gas out of the condenser is:

  0.85% C_2H_4 (by volume)
  5.12% CH_4
  10.52% C_2H_2
  83.51% H_2

A full-scale plant is to be built to produce 10 metric tons/day of acetylene (C_2H_2). How much tar will be produced, and what would you propose to do with it?

11.9 The problem of reduced project profitability due to construction delays has been addressed by Horwitz (1981).

Consider a project which will generate equal after-tax cash flows C_A each year for a period of n years after the completion of construction. Let the C_A's be discrete events at the ends of years 1, 2, ... , n, counting from construction completion. Suppose, furthermore, that the investment K is expended in equal portions at the end of each one of m years of construction. Thus, if $m = 1$, the total of K is expended at the end of year -1 (or start of year 1).

a Derive a general expression for calculating the DCRR, **r**, in terms of K, C, m, and n, for the case where both K and C are discretely compounded (or discounted) at the same rate **r**. This is the case considered by Horwitz.

b If the construction takes place over a period of time $m > 1$, the investment is time-distributed, and, as we saw in Chap. 9, the investments should be compounded at the inflation-adjusted after-tax internal rate of return i_0. Modify the expression derived in part (a) to account for this. (Take "zero time" at the end of construction.)

c A $5 million construction project, scheduled to be completed at the end of 1 year, was estimated to have a DCRR of 12.4 percent, with a useful life of 15 years. Because of tight capital, the decision has been made to slow the construction effort so as to span 3 years. What will the new projected value of DCRR be, with and without considering the corporation's internal rate of return of 10 percent?

d Modify the expression in part (b) for the case of continuous discounting at the nominal rate r, a nominal internal rate of return i_0, uniform cash flows $\overline{C_A}$ over each of n years of operation, and uniform construction disbursements $\overline{K}/m$ over the m-year span of construction ending at time zero.

11.10 Malina (1980) proposes guidelines for estimating plant start-up costs. He finds that a reasonably reliable projection can be based upon a certain percentage of "normal" raw materials costs plus a percentage of "normal" overhead costs (maintenance, supplies, utilities, fixed charges except depreciation, all labor-related costs).

Consult Malina's work and estimate the total start-up cost for the new facility described in the statement of Prob. 7.28.

11.11 Construct an S-curve (cash outflow curve) for labor-related expenditures for the project outlined in Table 11.3.1 and Fig. 11.3.2. Labor requirements are given in Table P11.1; the construction work will be done on day shift only, 5 days/week, with no overtime.

Use the following pay scales:

TABLE P11.1
LABOR REQUIREMENTS, PROB. 11.11

Job no.	Supervisors	Construction workers, level				Job no.	Supervisors	Construction workers, level			
		I	II	III	IV			I	II	III	IV
1	1	3		1		13	1			3	
2	2	6	2	2		14	1		2		5
3	2	2	2	7	1	15	1		2		5
4	1		4			16	2		2	3	3
5	1		2	6		17	1		1	2	1
6	1		1	3		18	1	3	4		
7	1	1	2	6	2	19	1	3	4		
8					2	20	1	3	4		
9	2	3	4	2		21				2	2
10	1			4	1	22	1	4	2		
11	1			4	1	23	1	3			
12	1			3							

Construction workers, level I $12/h
Construction workers, level II $16/h
Construction workers, level III $20/h
Construction workers, level IV $25/h
Supervisors $28/h

Rules:

1 Base the curve on the basic pay scales only.

2 Include services of one project manager and one engineer ($30,000/annum each) over full 62-day project span.

3 Actual disbursement is to take place at the end of each 5-day period for work performed during that period (the last period is only 2 days).

11.12 Baasel (1980) uses the example of changing a flat tire to illustrate CPM. Assume a passenger automobile has just experienced a flat tire and has been pulled off to a safe spot, off the road.

a Generate a job sheet listing each activity required to complete the job of tire changing, starting with setting the hand brake and ending with the hand brake release just before restarting the car.
b Define the precedence hierarchy, using the notation of Table 11.3.1.
c Estimate the time (in seconds) required to complete each activity.
d How long will the project take if only one person is available to do the work?
e Draw a CPM diagram and define the critical path.
f What is the critical path duration?
g Can two persons finish the project within the time span of the critical path? Three?

11.13 Draw a CPM diagram for Fig. 11.3.6.

11.14 Table P11.2 is a job list for a project to install a scrubber in an existing process.
a Estimate the precedence hierarchy of the activities.

TABLE P11.2
JOB SHEET, SCRUBBER INSTALLATION

Code	Activity	Normal time, days
A	Process design	14
B	Project design	14
C	Procure scrubber	85
D	Procure pump	68
E	Procure piping	45
F	Prefabricate pipe	5
G	Erect scaffolding	2
H	Site preparation	40
J	Pour concrete and allow to set	40
K	Install pump	3
L	Install scrubber	4
M	Fit up pipe	4
N	Weld pipe	3
P	Shut down process	14
Q	Remove some existing piping and make changes in existing equipment	16
R	Pressure test, flush, and make dry run	7
S	Start up process and test scrubber	5
T	Remove scaffolding	1
U	Cleanup	3

Source: Baasel (1980); reproduced by permission.

b Draw a bar chart of the project, analogous to Fig. 11.3.2.
c Draw a CPM diagram.

11.15 Figure 11.3.6 represents the maximum crashing of the project outlined in Table 11.3.1. We have seen that maximum crashing is not necessarily the most economical. Suppose each day saved by crashing results in a profit before taxes of $300. The schedule of costs incurred by crashing is as follows:

Job no.	Maximum days crashed	Cost per day crashed, $
1	2	175
2	3	200
3	2	310
5	3	290
6	2	350
7	1	300
9	7	230
15	2	410
19	7	280

What is the optimum crashing schedule, and how many days of construction will it save?

11.16 A project has four sequential activities with the following characteristics:

Activity	Normal duration, days	Normal total cost, $	Max. number of days that can be crashed	Premium cost per day crashed, $
A	6	8,500	2	800
B	4	7,000	3	500
C	12	22,100	4	1000
D	7	13,000	2	700

Each day saved results in an additional profit (before taxes) of $800. Use the NPW concept as a criterion of optimum crashing. Use daily discrete discounting or compounding based upon value of money (before taxes) of 25 percent/annum. Assume each activity is paid off completely at its completion, but take additional profit credit for each day saved.

11.17 (Adapted from Baasel, 1980.) Table P11.3 is a job sheet for a pipeline renewal project. The job sequential code is in the (P, S) notation. Activity R represents the required time span from the present before the pipeline may be deactivated for construction work. "Lead time" (activity Q) is the time necessary for organizing the whole effort.

TABLE P11.3
ACTIVITIES, COSTS, AND TIMES FOR RENEWING A PIPELINE

Job code/job		Sequential code	Elapsed time, days (max.)	Cost, $	Elapsed time, days (min.)	Total crash cost, $
Q	Lead time	1, 2	10		5	
R	Line available	1, 5	44		28	
A	Measure and sketch	2, 3	2	300	1	400
B	Develop materials list	3, 4	1	100	1	100
C	Procure pipe	4, 7	30	850	20	1100
D	Procure valves	4, 8	45	300	30	600
E	Prefabricate sections	7, 9	5	1200	3	2000
F	Deactivate line	5, 6	1	100	1	100
G	Erect scaffold	4, 6	2	300	1	500
H	Remove old pipe	6, 9	6	400	3	1000
I	Place new pipe	9, 10	6	800	2	2000
J	Weld pipe	10, 11	2	100	1	300
K	Place valves	8, 11	1	100	0.5	250
L	Fit up	11, 12	1	100	0.5	250
M	Pressure test	12, 14	1	50	0.5	100
N	Insulate	11, 13	4	300	2	700
O	Remove scaffold	13, 14	1	100	0.5	200
P	Cleanup	14, 15	1	100	0.5	200
				$5200		$9800

Source: Baasel (1980); by permission. (Table originally appeared in monograph, "Arrow Diagram Planning," copyright 1962, E. I. du Pont de Nemours & Company.)

a Draw a CPM diagram of the "normal" (maximum) time sequence.
b Identify the critical path and determine the minimum time to complete the project.
c Determine the minimum crash cost, no time value of money.
d Determine the minimum crash time, no time value of money.
e Determine the optimum crash time if each day of construction saved results in an extra $300 profit. Assume that each activity must be crashed to its maximum extent, or not at all.

11.18 For the project characterized by the following requirements (Martino, 1963), develop a reasonable workforce schedule that will allow the project to be completed in minimum time.

Job code	Time, days	No. of workers of specific craft	
		A	B
1, 2	2	2	
1, 3	1		2
2, 3	4	2	
2, 4	2	2	
3, 4	2	2	
3, 5	5		2
4, 5	1	2	
4, 6	2		2
5, 6	3	2	

11.19 (Adapted from Baasel, 1980.) A project has the CPM diagram in Fig. P11.1a and workforce requirements in Fig. P11.1b. Devise a reasonable schedule for optimum workforce leveling.

11.20 (Adapted from Hajek, 1977.) A unique process control electronic device is to be built according to the activity sequence outlined in Table P11.4. The activities are coded using the (P, S) notation for nodes; dummies are included. The "elapsed times" (in days) are the best estimates of t_l, t_m, and t_h.
  a What is the earliest *expected* completion time for the project?
  b What is the probability that the manufacturing operation can be started at the end of 32 days?
  c What is the probability that the project will be completed in 48 days?

11.21 A PERT activity sequence for introducing a new chemical product has been assembled by Klimpel (1973); it is reproduced in Table P11.5.
  a What is the earliest expected completion time?
  b What is the probability of a completion time 10 percent smaller than the earliest expected one?

11.22 Table P11.6 lists the expected times (including low, medium, and high times) for each of the activities for the scrubber installation project of Prob. 11.14. Assign (P, S) notation to each activity, and estimate the earliest expected completion time.

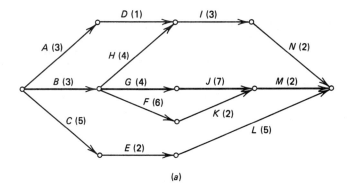

(a)

Activity designation	A	B	C	D	E	F	G	H	I	J	K	L	M	N
Time to complete activity, days	3	3	5	1	2	6	4	4	3	7	2	5	2	2
Number of workers required	14	16	6	13	15	6	3	3	12	5	13	7	12	6

(b)

FIGURE P11.1
(a) CPM diagram, Prob. 11.19. (b) Workforce requirements, Prob. 11.19. *(Baasel, 1980. By permission.)*

TABLE P11.4
JOB SHEET, ELECTRONIC CONTROL DEVICE

Activity	Identity	Elapsed times, days
(1, 2)	Data gathering	8–10–12
(1, 3)	Design analysis	9–11–12
(2, 3)	Dummy	
(3, 4)	Breadboard analysis	4– 5– 8
(4, 5)	Dummy	
(3, 5)	Electronic design	5– 6– 8
(3, 6)	Mechanical design	6– 7– 9
(3, 7)	Reliability engineering	4– 5– 6
(5, 7)	Dummy	
(6, 7)	Dummy	
(7, 8)	Prepare purchase orders	1– 2– 3
(7, 10)	Integrate design	8–10–12
(8, 9)	Purchasing, delivery	12–13–15
(9, 11)	Dummy	
(10, 11)	Drafting	6– 7– 9
(11, 12)	Manufacturing and assembly	5– 6– 9
(12, 13)	Final testing	2– 2– 5

Source: Based on PERT example in Hajek (1977), with permission of McGraw-Hill Book Co.

TABLE P11.5
ACTIVITIES AND TIMES FOR INTRODUCING A NEW PRODUCT

Activity	Number	Optimistic, weeks	Most likely, weeks	Pessimistic, weeks
Large-scale engineering feasibility studies	1–3	8	9	10
Prepare miniplant quantity of product	1–2	6	7	9
Conduct lab tests for government approval	2–4	5	5	5
Train sales force	2–5	8	10	11
Plant construction	3–6	9	12	15
Analyze shipping, storage, warehouse, etc.	2–6	11	15	20
Time for government to approve submitted tests	4–5	3	4	6
Install quality control and special safeguards	5–6	5	6	8
Plant start-up	6–7	4	5	10
Advertising campaign and initial customer contacts	5–7	8	10	12

Source: Klimpel (1973), by permission.

TABLE P11.6
TIME ESTIMATES (DAYS) FOR THE INSTALLATION OF A SCRUBBER IN AN EXISTING PROCESS

Activity	t_l	t_m	t_h	t_e
Process design (A)	11	14	16	13.8
Project design (B)	12	14	17	14.2
Procure scrubber (C)	75	85	100	85.8
Procure pump (D)	58	68	78	68.0
Procure piping (E)	40	45	55	46.7
Prefabricate pipe (F)	4	5	6	5.0
Set up scaffolding (G)	2	2	2	2.0
Site preparation (H)	30	40	60	41.7
Pour concrete and allow to set (J)	40	40	41	40.2
Install pump (K)	3	3	4	3.2
Install scrubber (L)	3	4	5	4.0
Fit up pipes and valves (M)	3	4	5	4.0
Weld pipe (N)	3	3	4	3.2

TABLE P11.6
TIME ESTIMATES (DAYS) FOR THE INSTALLATION OF A SCRUBBER IN AN EXISTING PROCESS (*Continued*)

Activity	t_l	t_m	t_h	t_e
Preparation for and shutdown of process (*P*)	12	14	16	14.0
Remove some existing piping and make changes in existing equipment (*Q*)	12	16	20	16.0
Pressure test, flush, and dry run (*R*)	6	7	10	7.3
Start up process and test scrubber (*S*)	4	5	8	5.3
Remove scaffolding (*T*)	1	1	1	1.0
Cleanup (*U*)	3	3	3	3.0

Source: Baasel (1980), by permission.

CHAPTER 12

CORPORATE PERFORMANCE ANALYSIS

In previous chapters we dwelt upon the evaluation of individual projects, their profitability, and the assessment of their commercial acceptability. A corporation can be thought of as an assembly of projects. Let us explore, then, how well our stipulated criteria of economic performance are realized in practice by the assembly of projects. By analyzing the economic performance of corporations in the chemical and allied process industries, we will try to establish whether the assembly of projects meets the expectations of the individual projects.

The input data required for corporate performance analysis are contained in the corporation's annual financial report. The typical report contains a great deal of information, but for our purposes we are interested in two documents that constitute the substance of a financial report:

1 Balance sheets
2 The income statement

The annual business transactions of the corporation are condensed within these documents; therein we will find the raw materials for our performance analysis. We will see that to gauge the performance trend of a particular corporation, or to compare the performance of different corporations, the absolute values which constitute the financial reports must in some ways be modified. That is, if one company, on an absolute basis, has 10 times the gross sales of another one, the former is not necessarily the better performer.

To facilitate comparison, including comparison with some of the profitability standards of individual projects that we have discussed, we will define certain *dimensionless ratios* which may be computed from the raw data in the annual report. In this way corporations of different size and structure may be characterized in a unified manner, in much the same way that physical systems of different size and nature may be characterized with the dimensionless ratios so familiar to chemical engineers. The corporate annual reports often list the year-by-year trend of some of the dimensionless ratios. One such ratio that is familiar to us is the return on investment; for the individual project, it is the ratio of annual profit to capital for sale. We will learn how to compute that ratio for a corporation.

The methods of analyzing the performance of securities will be briefly outlined, not for the purpose of promoting or aiding private investments, but to allow an evaluation of the economic significance of the securities to the corporation. The chapter and book conclude with an outline of the performance of a few specific corporations in the chemical process industries and an assessment of the role that the chemical process industries play in the national economy of the United States.

12.1 PERFORMANCE DOCUMENTATION

The Annual Report

Perhaps the most convenient source of financial information about a particular corporation is the corporation's annual report. There is no legal obligation that such a report be regularly published, but it certainly is customary. The purpose of the annual report is to outline the company's financial status for the benefit of existing and potential investors. Perhaps another purpose is an internal, introspective one—to let the employees see how well their corporation is doing, to provide a vehicle for documenting its current financial status. Since the primary focus is an external one, the appearance format is considered to be important, and annual reports are often characterized by printed opulence—slick paper and a great deal of bright coloring. The pages abound with pictures and easily understood bar charts and graphs. Photographs represent key members of the board of directors as stern, conservative individuals, not likely to squander the investors' money foolishly, but also as flexible people who can bend with the winds of change and who understand all levels of their business venture very well (often they are pictured, symbolically, wearing hard hats). The thrust of the text is ebullient and optimistic, in conformity with the purpose of attracting investors; the wise reader learns to separate promotional chaff from reality.

Viewed in realistic context, the annual report, provided it is reasonably well-written, represents a good outline sketch of the corporation. Most reports incorporate a great deal of useful information:

An outline of the nature of the corporation's business
New developments (R&D, for example) that promise to expand current business
New significant acquisitions, partnerships, etc.
Discussion of critical public issues (such as environmental concerns) that affect the business
Employee profiles

The board's own appraisal of the company's financial status
Auditing reports by independent accountants certifying that financial data meet accepted accounting principles
Lists of directors, officers, subsidiaries

For our present purposes, however, we are primarily interested in the portions that contain all the numbers:

1 The consolidated balance sheet, which lists the corporation's assets and its liabilities
2 The consolidated statement of income, a summary of the corporation's revenues and expenditures over the past year
3 The consolidated statement of retained earnings, an annual update on the total amount of money ploughed back into the corporation
4 Notes to the financial statements, an explanation of how the consolidated statements were assembled, including the accounting principles employed

Many annual reports have additional sections of considerable value to the performance analyst:

1 Statements of annual changes in the company's financial position—a cash flow analysis.
2 Comparative statements outlining the historical trend over a period of perhaps a decade of the corporation's financial parameters.
3 In recent years a welcome addition to many reports has been the adjustment of certain selected financial data to account for the effect of inflation. The adjustment guide is the Consumer Price Index (Chap. 1); the adjusted data reflect the true corporate growth more faithfully.
4 Sales analysis of various product groupings.
5 The management's own analysis of the corporation's financial performance.

Balance Sheets

The corporation's balance sheet is, in some respects, analogous to a household balance sheet. The total value of what you own (*assets*) minus the total value of what you owe (*liabilities*) is equal to the household's net worth (*equity*):

$$\text{Assets} - \text{liabilities} = \text{equity} \qquad (12.1.1)$$

The balance represented by Eq. (12.1.1) is established at *one point in time;* the balance sheets in a corporate annual report are prefaced by the date that represents that point in time, usually December 31 of the year in question.
To illustrate the makeup of a balance sheet, almost any annual report will do, but we will use the balance sheet of an entirely mythical company, outlined in the brochure, "How to Read a Financial Report," published by the investment brokerage firm of Merrill Lynch Pierce Fenner & Smith Inc. (4th edition, May 1979). In

conformity with Eq. (12.1.1), the balance sheet has three sections, the first one of which lists the corporate *assets;* this section is shown in Table 12.1.1.

This table (and the others which constitute a balance sheet) has perhaps fewer listed items in it than most actual annual reports; nevertheless, the important items are all there, and the resulting simplicity is one reason for choosing a mythical company for the illustration. First of all, note the two subsections marked *current assets* and *fixed assets*. Current assets are items which have been identified, in the context of individual projects, with *working capital* (Chap. 8):

Cash on hand
Marketable securities (at *cost*)
Accounts receivable (adjusted for bad debts)
Inventories (raw materials, products, supplies, in-process materials)

Exercise 12.1.1

Why are marketable securities not listed at their current market value?

The fixed assets—property, plant, and equipment—are listed at their total acquisition cost (direct fixed capital) but are corrected for accumulated depreciation. The

TABLE 12.1.1
Balance sheet: Assets
December 31, 19__

Assets

Current assets			
Cash			$ 950,000
Marketable securities at cost (market value $1,570,000)			1,550,000
Accounts receivable		$2,100,000	
Less: Provision for bad debts		100,000	2,000,000
Inventories			1,500,000
Total current assets			$6,000,000
Fixed assets (property, plant, and equipment):			
Land		$ 450,000	
Buildings		3,800,000	
Machinery		950,000	
Office equipment		100,000	
		$5,300,000	
Less: Accumulated depreciation		1,800,000	
Net fixed assets			3,500,000
Prepayments and deferred charges			100,000
Intangibles (goodwill, patent, trademarks)			100,000
Total assets			$9,700,000

Source: Reproduced by special permission, Merrill Lynch Pierce Fenner & Smith Inc., from "How to Read a Financial Report," copyright 1979.

net fixed assets are therefore identifiable with the *book value* of the corporation's total fixed assets (or, if different depreciation schedules are used for tax and book purposes, the "book" book value; see Chap. 6). Assets are kept on the books as long as they are in use, even though completely depreciated; if they are discarded, both gross assets and accumulated depreciation are correspondingly reduced.

The other items on the credit side of the balance sheet merit further explanation. *Prepayments* refer to payments for materials or services for which full value has not yet been received. Examples are prepayments on insurance premiums, down payments on construction contracts, or prepayments on future delivery of scarce supplies (Chap. 9). The idea is that the prepayment is a true potential asset; thus the scarce material, once delivered, will be transferred to the inventory category, or to some other asset category as a result of its contribution to manufacture. *Deferred charges* are, in some ways, similar to prepayments, in that an expenditure has been made from which benefits will accrue over a number of years. If the expenditure happens to be large (for example, R&D expenditures), the corporation may feel that it is more reasonable to consider the expenditure as if it were spread out over several years. The deferred charges specific to a particular expenditure then decrease annually over the full period of deferment.

The *intangibles* are a particularly puzzling category. The term refers to assets which have no physical existence but which nevertheless have assignable value. Patent rights are an example of an intangible asset; the value of such an asset may be determined by the purchase cost of the rights, or by the cost of developing the patent. *Goodwill* is the assigned value of those qualities which bring patronage to a business. For instance, the exclusive right to distribute a product in a particular area is a goodwill item having some inherent value. The assignment of value must not be capricious; preferably, it is based upon monetary transactions which first involved the value of the goodwill, for instance, the cost of a franchise. The assessed value of intangibles must be such as to be appropriately reflected in the income statement when the intangible assets are sold, depreciated, or otherwise used up, in accordance with acceptable accounting principles.

The debit side of the balance sheet (the *liabilities*) is illustrated in Table 12.1.2. There are two categories of liabilities: current ones and long-term ones. The *current liabilities,* just like the current assets, are identified with working capital. They are short-term (1-year) obligations which diminish the working capital total. *Accounts payable* represents money that is owed for materials and recurring services that have already been delivered (raw materials, equipment, and so forth). The *notes payable* item constitutes the total owed upon short-term (1-year) loans from banks and other lending institutions. *Accrued expenses payable* are salaries, interest payments, insurance premiums, and other recurring expenses which remain unpaid at the date of the balance sheet; unpaid income taxes are also listed.

Long-term liabilities are debts (promissory notes, debentures, bonds), due after more than 1 year from the date of the balance sheet. In Table 12.1.2 the long-term liability consists of first mortgage bonds—promissory notes to the bond purchasers which guarantee that the purchasers will have top priority claims upon mortgaged assets of the corporation if, for some reason, the corporation cannot pay off the

TABLE 12.1.2
Balance Sheet: Liabilities and Stockholders' Equity

Liabilities

Current liabilities:		
Accounts payable	$1,000,000	
Notes payable	850,000	
Accrued expenses payable	330,000	
Federal income tax payable	320,000	
Total current liabilities		$2,500,000
Long-term liabilities:		
First mortgage bonds, 5% interest, due 19—		2,700,000
Total liabilities		$5,200,000

Stockholders' Equity

Capital stock:		
Preferred stock, 5% cumulative, $100 par value each; authorized, issued, and outstanding 6000 shares	$ 600,000	
Common stock, $5 par value each; authorized, issued, and outstanding 300,000 shares	1,500,000	
	2,100,000	
Capital surplus	700,000	
Accumulated retained earnings	1,700,000	
Total stockholders' equity		4,500,000
Total liabilities and stockholders' equity		$9,700,000

Source: Reproduced by special permission, Merrill Lynch Pierce Fenner & Smith Inc., from "How to Read a Financial Report," copyright 1979.

value of the bonds at the maturity date. The table entry also identifies the annual "interest" payments which accrue to the bondholders.

Table 12.1.2 also lists the typical items that go to make up the *stockholders' equity*, which must satisfy the balance indicated in Eq. (12.1.1). Capital stock, both preferred and common (see Chap. 1), is listed at its par value. It should be noted that par value has no great significance, and some corporations issue stock with no stated par value. Stock is almost always sold by the corporation above par value; the excess that is received from shareholders above the par value is listed as *capital surplus*.

Exercise 12.1.2

Why is capital stock not listed on the balance sheet at its market value?

The *accumulated retained earnings* category is the other important component of stockholders' equity. The retained earnings, as we will see, represent all the profits that have cumulatively been ploughed back into the corporation. Each year's net profit (after subtracting all costs and taxes), diminished by the amount of stock dividends, contributes to the accumulated retained earnings total. The retained earnings are *not* a pot full of money that is available, for instance, for additional investments; however, retained earnings increases are reflected in increasing assets, and investments (fixed assets) may, indeed, be made by converting some of the current assets.

The accumulated retained earnings category is thus a running balance which serves to satisfy, year after year, the rigid requirements of Eq. (12.1.1). The rigidity of this requirement is a source of puzzlement to many. Why is it that, based upon apparently disparate and even vague elements such as goodwill, the equity *always* equals assets minus liabilities, down to the last cent? One is tempted to say that that is what accountants are for, and that is why each corporation has so many of them. In point of fact, consistent accounting principles are, indeed, necessary to come up with a credible balance sheet; a bit of reflection will satisfy even those not well-versed in accountancy how the required balance is maintained.

Example 12.1.1

Suppose, in Table 12.1.1, the securities are sold at their market value ($1,570,000) to settle all accounts payable ($1,000,000, Table 12.1.2) and to buy $300,000 worth of raw materials. Assuming no other business transactions have been made, trace the effect of the sale upon the balance sheet.

SOLUTION The sale represents a long-range capital gain of $20,000 over the acquisition cost. The tax is assumed to be 30 percent, so that net profit on the sale, after tax payment, is $14,000. This amount increases the accumulated retained earnings to $1,714,000, and the stockholders' equity becomes

$$\text{Stockholders' equity} = \$4,514,000$$

The liabilities are reduced by $1,000,000, so that

$$\text{Total liabilities} = \$4,200,000$$

The total cash generated by the sale is the sales income diminished by the long-term capital gains tax, or $1,564,000.

One million dollars is used to reduce the liabilities, and $564,000 is therefore left to increase the assets ($300,000 into inventories, the balance of $264,000 presumably as cash). The net decrease in assets is this $564,000 minus the sold securities at cost, or

$$564,000 - 1,550,000 = -986,000$$

and the total assets equal 9,700,000 − 986,000; that is,

$$\text{Total assets} = \$8,714,000$$

The requirement of Eq. (12.1.1) is maintained:

$$8,714,000 - 4,200,000 = 4,514,000$$

Exercise 12.1.3

"Stockholders' equity" implies that that is how much the stockholders own. Do stockholders own the working capital?

Exercise 12.1.4

With the so-called *first-in first-out* (FIFO) method of inventory control, the accounting procedure assumes that the raw material or purchased product produced first is the first to be consumed or sold. The *last-in first-out* (LIFO) method, on the other hand, assumes that the last inventory acquired is the first disposed of. The choice of accounting method becomes significant in time of inflation.

Suppose a corporation has been using the FIFO method and decides to adopt the LIFO method. In time of inflation, what will be the effect of the switch upon the magnitude of the inventories item on the balance sheet?

A change in the magnitude of the inventories will affect, in turn, the size of the total assets. Trace out how the balance demanded by Eq. (12.1.1) is maintained.

Income Statements

Whereas balance sheets show the corporate position at one point in time, the *income statement* reflects the results of operations throughout 1 year. An income statement is illustrated in Table 12.1.3.

The statement starts with the total sales revenue for the year; this total is then diminished by items (cost of goods sold, depreciation, selling and administrative expenses) which have been termed, in context of the individual project, as *cost of*

TABLE 12.1.3
Income Statement Example
Year 19 __

Net sales		$11,000,000
Cost of sales and operating expenses:		
Cost of goods sold	$8,200,000	
Depreciation	300,000	
Selling and administrative expenses	1,400,000	9,900,000
Operating profit		$1,100,000
Other income:		
Dividends and interest		50,000
Total income		$1,150,000
Less: interest on bonds		135,000
Income before provision for income tax		$1,015,000
Provision for income tax		480,000
Net profit for the year		$ 535,000

Source: Reproduced by special permission, Merrill Lynch Pierce Fenner & Smith Inc., from "How to Read a Financial Report," copyright 1979.

manufacture and *general expenses* (or out-of-pocket expenditures plus depreciation). The balance, the *operating profit*, is augmented with interest and dividend income from securities held by the corporation.

The *total income* is diminished by interest on bonds and other moneys borrowed. Here, at last, we account for interest, something we agreed we would not consider in evaluating the profitability of individual projects. The balance is the *gross profit* upon which income tax is paid. The *net profit* is computed in the same manner as that for individual projects—the gross profit minus the income tax.

Exercise 12.1.5

What is the corporation's continuous cash flow for the year, before and after taxes?

A useful adjunct to the income statement, one that appears in most annual reports, is the *accumulated retained earnings statement;* an example is reproduced in Table 12.1.4. The statement is intended to show how the accumulated retained earnings—the accumulated total ploughed back into the corporation—has changed during a particular year. The change is the result of the net profit accruing to the corporation, diminished by the year's stock dividends.

Footnotes

Corporations are complex financial entities, and the consolidated statements illustrated in Tables 12.1.1 to 12.1.4 cannot be expected to convey all the nuances of that financial complexity. For this reason corporate annual reports incorporate a section entitled *notes to financial statements,* footnotes that explain the methodology and data used in formulating the statements. In fact, such footnotes may contain a great deal of important supplemental information, and occasionally they may constitute fascinating reading, indeed, revealing some of the financial "skeletons in the closet."

TABLE 12.1.4
Accumulated Retained Earnings Statement
(Earned Surplus)
Year 19 __

Balance Jan. 1, 19 —			$1,315,000
Net profit for the year			535,000
Total			$1,850,000
Less dividends paid on:			
Preferred stock	$	30,000	
Common stock		120,000	150,000
Balance Dec. 31, 19 —			$1,700,000

Source: Reproduced by special permission, Merrill Lynch Pierce Fenner & Smith, Inc., from "How to Read a Financial Report," copyright 1979.

There is no specific requirement regarding the substance of the footnotes. A cursory survey of a typical annual report will come up with items such as:

The effective income tax rate paid by the corporation, including a subdivision among federal, state, local, and foreign taxes

Explanation of the effect of using different depreciation schedules for tax and book purposes

A breakdown of plant properties

Statistics on mineral and energy reserves

Retirement plan statistics

Details of long-term debts

Stock options and awards

Product group sales analysis

Exercise 12.1.6

In 1979, Standard Oil Company of California reported an effective income tax rate of 39.2 percent, at a time when the federal statutory maximum rate alone was 46 percent. Give some possible reasons for this apparent discrepancy.

Other useful information will be found scattered throughout the annual report, not necessarily in the footnotes section. Table 12.1.5, for example, is a statement which outlines the disposition of the total corporate annual cash flow and the effect upon working capital distribution; the data in this particular table are related to the data in Tables 12.1.1 to 12.1.4. Other useful tables may focus upon a historical financial summary covering several years, or the effects of inflation upon financial performance criteria.

12.2 RATIO ANALYSIS

The Significance of Ratio Analysis

Ratio analysis is a system used for the analysis of the financial performance of corporations. Ratio analysis facilitates the year-by-year monitoring of the performance of a particular company, and it also permits a more rational comparison of performance of different companies. We have seen that the net profit is the "bottom line" of the income statement, and let us suppose that a corporation boasts a 10 percent annual increase in net profit over the past 5 years. Does this trend indicate a "good" performance? Much of that apparent growth could be due to inflation, but even when corrected for that effect, the growth could be partly ascribable, for instance, to the reduction of inventories. We need to compare the profit with some other financial parameter, to ratio it to obtain a "normalized" value that can be monitored more rationally. Similarly, if two corporations are compared and one has twice the net profit of the other one, is the one with the higher profit the better performer? Again, we can see intuitively that this is not necessarily so, but we need to normalize the profit values for rational comparison.

TABLE 12.1.5
Statement of Source and Application of Funds
19 __

Funds were provided by:		
Net income	$ 535,000	
Depreciation	300,000	
Total		$ 835,000
Funds were used for:		
Dividends on preferred stock	$ 30,000	
Dividends on common stock	120,000	
Plant and equipment	305,000	
Sundry assets	10,000	
Total		465,000
Increase in working capital		$ 370,000

Analysis of Changes in Working Capital: 19—

Changes in current assets:		
Cash	$ 150,000	
Marketable securities	390,000	
Accounts receivable	100,000	
Inventories	(300,000)	
Total		$ 340,000
Changes in current liabilities:		
Accounts payable	$ 60,000	
Notes payable	(150,000)	
Accrued expenses payable	30,000	
Income tax payable	30,000	
Total		$ (30,000)

Source: Reproduced by special permission, Merrill Lynch Pierce Fenner & Smith, Inc., from "How to Read a Financial Report," copyright 1979.

As a matter of fact, we have resorted to a form of ratio analysis in establishing the performance of individual projects and in the course of the analysis of project alternatives. The return on investment (ROI) criterion of economic performance is a ratio of profit to capital for sale, the two separate results of the investment and net income analyses, respectively. The benefit-cost (present worth) ratio analysis was judged to be the best for comparing project alternatives. We will see that similar ratios can be used to characterize the grand ensemble of projects that constitutes the corporation. A great many ratios may be defined, and each represents some perceived financial attribute of the corporation; we will look at some of the commonly used ratios, but by no means all of them.

Balance Sheet Ratios

We start with the examination of some ratios which are based upon data in the balance sheet only.

Current Ratio The current ratio is defined as

$$\text{Current ratio} = \frac{\text{current assets}}{\text{current liabilities}} \quad (12.2.1)$$

We will use data in Tables 12.1.1 and 12.1.2 to illustrate the ratios. Thus

$$\text{Current ratio} = \frac{6.0 \times 10^6}{2.5 \times 10^6} = 2.4$$

The current ratio is a measure of the corporation's working capital position, its ability to meet current obligations and still have sufficient operating funds. A "safe" current ratio is thought to be about 2.0. Chemical companies tend to have a low current ratio, for chemical inventories are generally quite low. On the other hand, the nature of the chemical business is such that the accounts receivable are more readily collectible, a hidden "plus" to the assets.

Acid Test Ratio The acid test ratio, often called the *quick assets ratio,* is given by

$$\text{Acid test ratio} = \frac{\text{cash + marketable securities + receivables}}{\text{current liabilities}} \quad (12.2.2)$$

where the sum in the numerator is collectively termed *quick assets*. For our balance sheet,

$$\text{Acid test ratio} = \frac{(0.95 + 1.55 + 2.00) \times 10^6}{2.50 \times 10^6} = 1.8$$

Some analysts feel that the acid test ratio is a more valid measure of the ability of a company to meet its obligations. The quick assets are the current assets without the inventories. Quick assets are readily convertible into cash, whereas inventories must first be sold for their book value, something that takes time and is by no means guaranteed. A ratio of 1 or more is generally considered desirable.

Exercise 12.2.1

What is wrong with having an acid test ratio of less than 1?

Equity Ratio The equity ratio is defined as

$$\text{Equity ratio} = \frac{\text{stockholders' equity}}{\text{total assets}} \quad (12.2.3)$$

which, for the balance sheet studied, is

$$\frac{4.5 \times 10^6}{9.7 \times 10^6} = 0.46$$

or 46 percent (usually quoted as a percentage).

The equity ratio is a measure of the credit worthiness of a corporation, its long-term financial strength and its ability to weather business declines. A ratio of about 50 percent is considered acceptable. A higher ratio implies that the corporate management does not believe in heavy indebtedness as a vehicle for corporate growth, but it may also mean that the corporation is not taking advantage of *capital leverage,* discussed in what follows.

Interstatement Ratios

Other commonly used ratios are based upon data in the income statement alone, and still others require data input from both the balance sheet and the income statement. One such interstatement ratio is the *return on stockholders' equity,*

$$\text{Return on stockholders' equity} = \frac{\text{net profit}}{\text{stockholders' equity}} \qquad (12.2.4)$$

which, for the numerical example, is

$$\frac{0.535 \times 10^6}{4.5 \times 10^6} = 0.119 \quad \text{or} \quad 11.9\%$$

The return on stockholders' equity is a measure of the corporate earning power from the stockholders' point of view. For the 15 largest chemical companies, the average ratio was 11.0 percent in 1980.

A whole family of interstatement ratios is based upon the ROI concept, and we will return to these types of ratios presently. First, let us examine two ratios from a group that are collectively called *operating ratios* and are based upon data in the income statement alone.

Operating Margin of Profit This ratio, sometimes called simply the *operating margin,* is given by

$$\text{Operating margin} = \frac{\text{operating profit}}{\text{net sales}} \qquad (12.2.5)$$

$$= \frac{1.10 \times 10^6}{11.00 \times 10^6} = 0.10 \quad \text{or} \quad 10.0\%$$

The operating profit excludes income from dividends and interest and is computed before income tax payments and interest payments on debt. The operating margin is clearly a measure of the operating efficiency of the corporate production facilities. The ratio is useful in monitoring the nature of the corporation's year-by-year growth; perhaps there is a steady growth of sales, but is the firm becoming more profitable? The operating margin is also a useful index to compare performance with other companies, or with the whole chemical industry.

Net Profit Ratio This ratio, which is also called the *return on sales* (Chap. 8) or the *profit margin,* is conceptually related to the operating margin:

$$\text{Net profit ratio} = \frac{\text{net profit}}{\text{net sales}} \quad (12.2.6)$$

$$= \frac{0.535 \times 10^6}{11.00 \times 10^6} = 0.049 \quad \text{or} \quad 4.9\%$$

The applications of the net profit ratio are similar to those of the operating margin. In 1980, the profit margin of chemical companies having sales over $1 billion was 5.1 percent. Smaller companies appeared to have a significantly higher profit margin.

Exercise 12.2.2

Does inflation per se have an impact upon the year-by-year trend of a corporation's profit margin?

The Return on Investment

We have seen that the return on investment concept and its variants constitute one of the most commonly used criteria of the economic performance of individual projects (Chap. 8). Indeed, certain interstatement ratios related to the ROI concept are commonly used for the performance analysis of corporations. Two such ratios are

$$\text{Return on total assets} = \frac{(\text{total income})100}{\text{total assets}} \quad (12.2.7)$$

and

$$\text{Return on investment} = \frac{(\text{net profit})100}{\text{current assets} + \text{gross fixed assets}} \quad (12.2.8)$$

The total income in Eq. (12.2.7) is taken before taxes and interest payments, and the ratio is therefore related to the ROIBT concept of individual projects. For the corporation represented by the tables in this chapter,

$$\text{Return on total assets} = \frac{100 \times 1.15 \times 10^6}{9.70 \times 10^6} = 11.9\%$$

Exercise 12.2.3

The return on total assets is perhaps more related to ROAIBT of individual projects. Why? In what respects does it differ from ROAIBT?

The gross fixed assets in Eq. (12.2.8) are the *undepreciated* fixed assets, i.e., net fixed assets plus accumulated depreciation. The return on investment is related to the ROIAT of individual projects, except that interest payments have been subtracted to obtain the net profit. For the corporation discussed,

$$\text{Return on investment} = \frac{1.00 \times 0.535 \times 10^6}{(6.00 + 3.50 + 1.80)10^6} = 4.7\%$$

The problem with the ratios in Eqs. (12.2.7) and (12.2.8) is that they do not relate directly to the ROI variants as they were defined for individual projects. We have characterized a corporation as an assembly of projects, and we would therefore like to define the corporate ROI ratios in exactly the same way as they were defined for individual projects so that a direct comparison could be made. The question is, "How closely does the assembly of projects approach the performance criteria stipulated for acceptable new individual projects?"

To answer this question, let us examine the significance of the terms in the various project ROI ratios, and let us start with the most commonly used one, the ROIBT:

$$\text{ROIBT} = \frac{100 P_B}{\text{CFS}} \qquad (8.2.1)$$

It will be recalled that P_B, the project's profit before taxes, is, indeed, the profit before tax payments; moreover, interest payments were *not* considered in computing the project's profit. One is therefore tempted to equate the corporate total income (before taxes and interest) with P_B. There is, however, another complication. Some of the corporate income is obtained from dividends on securities and interest on bank deposits. Such income is not the result of investing capital into production facilities, the sole income source when projects are evaluated. P_B is therefore more properly identified with the *operating profit* (Table 12.1.3).

The denominator in Eq. (8.2.1) is the capital for sale associated with a particular project. It will be recalled that CFS consists of fixed capital and working capital. Let us consider the working capital first. On the corporate balance sheet (Tables 12.1.1 and 12.1.2), the working capital incorporates the current assets and the current liabilities categories; the long-term liabilities (i.e., long-term debt) were not considered a liability upon an individual project's working capital. As a first approximation, the working capital is equal to the total current assets minus the total current liabilities, but we do need to examine some of the balance sheet items in the light of the project working capital definition.

We have seen that some cash on hand is associated with every project. On the corporate balance sheet the cash assets on deposit may represent considerably more than immediate cash required for current operations. Since interest on bank deposits has been specifically excluded from P_B, we make the approximation of excluding cash from current assets. Similarly, we exclude marketable securities from current assets, since these do not represent an investment in production facilities. On the other hand, we will consider prepayments and deferred charges as part of the working capital; those prepayments *are* project-oriented. The current liabilities should logically *all* be subtracted, so that

Working capital = current assets + prepayments and deferred charges
$$- \text{current liabilities} - \text{cash} - \text{securities} \quad (12.2.9)$$

The fixed capital component of CFS is the total of the *gross*, i.e., *undepreciated*, assets (the initial investment). Thus

$$\text{Gross assets} = \text{net fixed assets} + \text{accumulated depreciation} \quad (12.2.10)$$

We will not include the intangibles in any of the assets, since these are not a legitimate component of the project's CFS.

Exercise 12.2.4

Does the book method of depreciating fixed capital have an effect upon the proper computation of gross assets?

If the properly modified definitions of P_B and CFS are substituted into Eq. (8.2.1), the result is

$$\text{ROIBT} = \frac{100(\text{operating profit})}{[\text{net fixed assets} + \text{accumulated depreciation}]\phi + [\text{current assets} + \text{prepayments} - \text{current liabilities} - \text{cash} - \text{securities}]}$$

$$(12.2.11)$$

The significance of the factor ϕ will be pointed out presently. Note that the format of the ROIBT expression involves the comparison of a particular year's actual profit relative to the total initially invested many years before. When the ROIBT of individual projects was discussed in Chap. 8, the focus was upon the *prediction* of profitability using constant-value money. The ROIBT concept, of course, can be used to evaluate the actual performance of individual projects, using data on profit obtained at some time following the initial investment; however, a factor ϕ must be used to account for inflation, as in Eq. (12.2.11).

Exercise 12.2.5

Formulate an equation analogous to Eq. (12.2.11) which could be used to compute the corporate ROIBT for *all* the corporation's investments.

It will be recalled that predicted ROIBT was assessed to be independent of the vagaries of inflation since revenues and expenses (except depreciation) could be expected to increase with time proportionally, so that the effective profit would remain invariant in terms of constant-value money. However, if ROIBT is based upon the actual profit during some year (in cocurrent dollars) relative to an investment made a number of years before, Eq. (8.2.1) and its analogs must be corrected for inflation either by inflation-discounting the profit to the time of investment or, as we will do, by obtaining the current, inflation-adjusted value of the past investment. The *inflation adjustment factor* ϕ in Eq. (12.2.11) brings the sum total of past investments to their value at the time of the operating profit under consideration. The computation of ϕ turns out to be a bit sticky.

If we had access to all the corporation's books, we could laboriously calculate ϕ by using a project-by-project analysis. Since annual reports do not have information that detailed, we must make approximations to estimate the temporal distribution of the fixed asset investments in the denominator of Eq. (12.2.11).

We start by assuming that projects have an average lifetime of n years (i.e., time between fixed investment and the removal of the investment from the books). The average annual inflation factor over the n years is taken as **f** (Chap. 1), and the average annual increase of the gross fixed investments in the denominator of Eq. (12.2.11) is designated as **k**. Furthermore, let G be the balance sheet value of those gross plant property items, and P be the current value of G adjusted for inflation; that is,

$$P = G\phi \qquad (12.2.12)$$

Let R be the fixed investment at the start of year $-n$. Its value in current dollars is

$$P_{-n} = R(1 + \mathbf{f})^n \qquad (12.2.13)$$

The fixed investment at the start of year $(-n + 1)$ is $R(1 + \mathbf{k})$, and

$$P_{-n+1} = R(1 + \mathbf{k})(1 + \mathbf{f})^{n-1} \qquad (12.2.14)$$

Finally, for year -1, the year of the balance sheet, the fixed investment is $R(1 + \mathbf{k})^{n-1}$, and

$$P_{-1} = R(1 + \mathbf{k})^{n-1}(1 + \mathbf{f}) \qquad (12.2.15)$$

In the above development, we have assumed that each year's investment is made

as a lump sum at the start of each year; perhaps more reasonable assumptions may be made, but they will not change the thrust of the arguments.
Now,

$$P = \sum_{1}^{n} P_{-j}$$
$$= R \sum_{1}^{n} (1 + f)^j (1 + k)^{n-j}$$

Evaluation of the series gives

$$P = R(1 + k)^{n-1}(1 + f) \frac{1 - \left(\dfrac{1 + f}{1 + k}\right)^n}{1 - \dfrac{1 + f}{1 + k}} \qquad (12.2.16)$$

But

$$G = R \sum_{1}^{n} (1 + k)^{j-1} = R \frac{(1 + k)^n - 1}{k} \qquad (12.2.17)$$

Combining Eqs. (12.2.16) and (12.2.17),

$$\frac{P}{G} = \phi = k \frac{(1 + k)^{n-1}(1 + f)}{(1 + k)^n - 1} \frac{1 - \left(\dfrac{1 + f}{1 + k}\right)^n}{1 - \dfrac{1 + f}{1 + k}} \qquad (12.2.18)$$

where, normally, $k > f$.

Thus $\phi = \phi(f, k, n)$ only. The average f may be computed from CPI data such as those in Table 1.3.1. Many annual reports contain comparative statements which list additions to plant properties over a period of several years; k may be approximated from such data, as illustrated in the example that follows. Admittedly these data apply only to the fixed assets, whereas the CFS in the denominator of Eq. (12.2.11) must also include the working capital. The annual changes in working capital may be computed from the comparative statements if such data happen to be available; otherwise, we may assume that the working capital remains at a relatively fixed percentage of the fixed assets, as suggested in Chap. 8. We will adopt this simplification here. We use the current balance sheet value of fixed capital as an indication of that percentage; since the value is already expressed in current currency, it need not be corrected with an inflation adjustment factor.

n can be taken as the average for the chemical industry (about 10 years), or it may be computed from the value of **k**, the current gross assets, and the additions to plant properties for the year currently ended, as illustrated in Example 12.2.1.

Exercise 12.2.6

What format does Eq. (12.2.18) assume for the case of **k** = **f**?

Example 12.2.1

The data in Tables 12.2.1 and 12.2.2 are extracted from the 1980 annual report of The Dow Chemical Company. What is the ROIBT of Dow's projects for 1980?

SOLUTION The additions to property are plotted in Fig. 12.2.1. The best straight line through the data on semilog paper (eyeballed, or fitted with least-squares method) represents the relationship

$$R_t = R_0(1 + k)^t \qquad (12.2.19)$$

where R_0 represents the additions during some arbitrary year zero and the time t is measured in years. On semilog paper, the slope is log $(1 + k)$; measurement of the slope yields

$$k = 0.078$$

TABLE 12.2.1
1980 ANNUAL REPORT, THE DOW CHEMICAL COMPANY: BALANCE SHEET AND INCOME STATEMENT ITEMS

(All Figures in Millions of Dollars)	
1. Total assets	11,538
2. Gross plant properties	9,873
3. Accumulated depreciation	4,201
4. Cash	25
5. Securities and investments	1,370
6. Current assets	4,390
7. Current liabilities	2,803
8. Prepayments and deferred charges	269
10. Sales	10,626
11. Operating profit	1,212
12. Other income	411
13. Interest payments	385
14. Income before taxes	1,238
15. Income tax	424
16. Net income*	814
17. Depreciation	728

*Includes minority interests' share in income.

TABLE 12.2.2
1980 ANNUAL REPORT, THE DOW CHEMICAL COMPANY: ANNUAL ADDITIONS TO PLANT PROPERTIES

(All Figures in Millions of Dollars)

Year	Addition
1974	890
1975	935
1976	1200
1977	1163
1978	1075
1979	1268
1980	1184

FIGURE 12.2.1
Annual additions to property, The Dow Chemical Company. *(1980 Annual Report, The Dow Chemical Company.)*

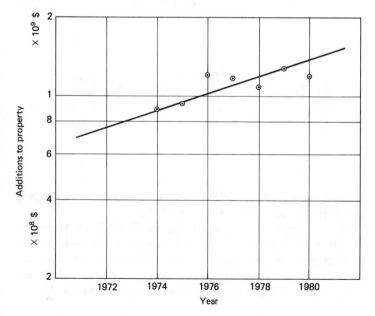

that is, a 7.8 percent average annual increase in plant properties investments. (Do you agree with that result?)

Now, suppose G is the current value of the gross plant properties; let R_{-j} be the addition to plant properties during year $-j$, measured from time zero (end of year -1). Then

$$G = R_{-1}[1 + (1 + k)^{-1} + (1 + k)^{-2} + \cdots + (1 + k)^{-n+1}]$$
$$= R_{-1} \sum_{1}^{n} (1 + k)^{1-j} \qquad (12.2.20)$$

where R_{-1} is the addition during the current year ending.

It can be shown (Exercise 12.2.7) that n may be computed from the expression

$$n = \frac{-\log\left[1 - \dfrac{k(G/R_{-1})}{1 + k}\right]}{\log(1 + k)} \qquad (12.2.21)$$

Since, however, k is obtained from a graph such as Fig. 12.2.1, R_{-1} must be taken from the graph, rather than the actual value in the annual report. In our case,

$$R_{-1} = 1360 \quad \text{and} \quad G = 9873$$

whence $n = 9.92 \approx 10$ years, in good agreement with the industrywide rule of thumb.

During the 1970–1980 period, f (from data in Table 1.3.1) averaged out to be 0.078, the same value as that derived for k (the annual increases in additions managed to match inflation). For the case of $f = k$, Eq. (12.2.18) simplifies to

$$\phi = \frac{nk(1 + k)^n}{(1 + k)^n - 1} = 1.477 \qquad (12.2.22)$$

and, from Eq. (12.2.11),

$$\text{ROIBT} = \frac{100(1212)}{[9873]1.477 + [4390 + 269 - 2803 - 25 - 1370]}$$
$$= 8.1\%$$

Exercise 12.2.7

Derive Eq. (12.2.21).

Perhaps a more realistic assessment of ROIBT would be obtained by averaging the ratio over several years, in conformity with Eq. (8.4.1).

A corporate ROIAT, analogous to a project-oriented ROIAT, may also be defined. Since the interest on debt is not included as part of the cash flow of individual projects, the interest must be added to the net corporate profit to calculate the equivalent ROIAT. Moreover, the net profit must be adjusted to include the net profit from operations only, not including profit from other investments. We assume that the adjustment may reasonably be made by proportioning income tax payments equitably between the operating profit and the profit from other investments. This procedure neglects the effect of various tax credits. Thus

$$\text{Adjusted income tax} = \frac{\text{income tax} \times \text{operating profit}}{\text{operating profit} + \text{interest income}} \quad (12.2.23)$$

and

$$\text{Net income} = \text{operating profit} - \text{adjusted income tax} \quad (12.2.24)$$

so that

$$\text{ROIAT} = \frac{100(\text{operating profit} - \text{adjusted income tax})}{[\text{net fixed assets} + \text{accumulated depreciation}]\phi + [\text{current assets} + \text{prepayments} - \text{current liabilities} - \text{cash} - \text{securities}]} \quad (12.2.25)$$

For the data in Example 12.2.1,

$$\text{ROIAT} = \frac{100\left(1212 - \dfrac{424 \times 1212}{1212 + 411}\right)}{[9873]1.477 + [4390 + 269 - 2803 - 25 - 1370]}$$
$$= 6.0\%$$

Exercise 12.2.8

We saw in Chap. 8 that for individual projects

$$\text{ROIAT} = (1 - t)\text{ROIBT}$$

where t is the tax rate. From the data in the above example, what tax rate t do you compute? Does that value of t yield the anticipated relationship between ROIBT and ROIAT? Explain any anomalies.

The Internal Rate of Return

In previous chapters we have repeatedly referred to i_0, the corporate internal rate of return on the total of the corporation's investments—after taxes, and adjusted for inflation. We will now explore the possibility of deriving i_0 from corporate data normally available in an annual report.

We will find that, just as was true of the ROIBT computation, many simplifying assumptions must be made; some of these are the same as those used in the ROIBT method. The i_0 concept differs from the ROIBT concept in two important respects:

1 The corporation i_0 is a measure of the profitability on *all* the corporation's investments, including securities and deposits. In this way i_0 is a more faithful criterion of total profitability, to be used as a standard against which all projects are to be gauged.

2 Neglecting the effects of inflation, ROIBT can be computed on the basis of data for one fiscal year, but i_0 depends upon the complete past *and future* performance of the investments.

To appreciate the significance of the second point in particular, let us review the DCRR concept for the individual project; we will then see how the concept may possibly be extended to the sum total of the corporation's projects. For the sake of simplicity, consider a project with the capital investment $-K$ at time zero and equal after-tax annual cash flows C_A over the project's lifetime of n years. If this were the only corporate project, then i_0 would be related to n as shown:

$$n = \frac{-\log(1 - i_0 K/C_A)}{\log(1 + i_0)} \qquad (12.2.26)$$

Exercise 12.2.9

Derive Eq. (12.2.26).

The relationships expressed by the equation are illustrated graphically in Fig. 12.2.2. Thus, for example, if $K/C_A = 2.0$, the project must last at least 2 years to show any return at all; beyond about 8 years, the project's DCRR is no longer dependent upon the project length. The problem we are faced with is to determine i_0 if we know K/C_A for some year m, for this is what the annual report effectively gives us, the cash flow and investment for a particular year. For the specific case of a single project and $K/C_A = 2.0$, if $m \geq 8$, we have no problem; $i_0 \approx 0.5$, an attractive DCRR. But what if $m = 3$? The curve tells us that if this were the project's terminal year, i_0 would be 0.24, but we really cannot tell how much longer the project will last and, therefore, what its DCRR is eventually going to be.

Fortunately, a corporation is an assembly of many projects, and the performance of such an assembly tends to average out. For instance, no real project is characterized by equal annual cash flows, yet the ups and downs of individual project cash flows tend to compensate each other in a large assembly, so that the total corporate cash flow is relatively uniform (or else changes annually in a relatively uniform manner). Similarly, projects vary considerably in their effective lives, but their ensemble may be characterized by an average life of n years; the value of n may be computed from Eq. (12.2.21). Admittedly, this value of n applies to plant properties only, but we will assume that the same value holds for securities, intangibles, and other such investments.

As another simplification, consider a corporation which happens to consist of just n projects, each with a life of n years, each representing an equal capital investment and rate of return; no two projects are started in the same year. If we ignore the effects of inflation, the total corporate K and the annual C_A remain the same year after year, for each year one project is terminated and an equal one is initiated. Since we know n, and K/C_A is invariant, i_0 may be obtained from the relationship in Eq. (12.2.26).

Suppose, now, the annual capital investment for each new project varies in some fashion; perhaps it increases from year to year by the ratio $1 + k$. The fact is that the K reported in the balance sheet still represents the sum total of the investments over the past n years, and C_A is still the total cash flow generated by *those* investments. If the average life continues to be n years, then Eq. (12.2.26) still applies. Thus a corporation may be looked upon during any particular year as a single giant project with investment K and cash flow C_A; we know the total life of the project is n years, and we can therefore calculate i_0.

To make the i_0 thus computed compatible with the definition of DCRR for individual projects, we must examine the meaning of K and C_A more closely. We

FIGURE 12.2.2
DCRR as a function of project lifetime.

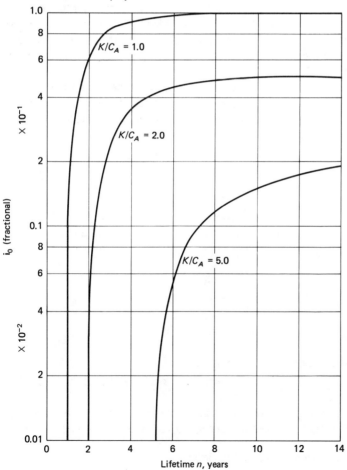

start our examination by neglecting inflation. We note that:

1 The initial investment ($-K$) used in cash flow analysis is the total *undepreciated* amount.
2 Debt considerations do not normally enter into cash flow analysis.

Since the rate of return on *all* the corporation's investments is to be determined, K is then defined as

$$K = \text{total assets} + \text{accumulated depreciation} - \text{current liabilities} \quad (12.2.27)$$

We saw in Eq. (6.1.1) that cash flow after taxes is the sum of the profit after taxes (net profit) and the year's depreciation charges,

$$C_A = P_A + D \quad (12.2.28)$$

However, we do not wish to have long-term debt considerations enter into our definition of C_A, and we therefore also add *interest on long-term debt* to the terms making up C_A. And still another term must be added, a term which will be understood when we recall that DCRR calculations include the working capital, recovered at the termination point of a project, as part of the overall cash flow. How can this term be estimated?

Again, we must make some simplifying assumptions. One is that the ratio of total working capital (= current assets minus current liabilities) to gross plant properties is invariable from year to year and is typified by data in the latest annual report. The working capital recovered this year is then given by

$$(\text{WC})_R = \left(\frac{\text{WC}}{G}\right)_{-1} \cdot R_{-n} \quad (12.2.29)$$

where $(\text{WC}/G)_{-1}$ is the ratio calculated from the latest annual report, and R_{-n} is the plant properties investment at the start of year $-n$. R_{-n} may be obtained from Eq. (12.2.17):

$$R_{-n} = \frac{Gk}{(1+k)^n - 1} \quad (12.2.30)$$

The value of **k** is obtained as in Example 12.2.1. Thus C_A is given by

$$C_A = \text{net profit} + \text{depreciation*} + \text{interest payments} + \text{recovered working capital} \quad (12.2.31)$$

where the recovered working capital is obtained from Eqs. (12.2.29) and (12.2.30). It should be noted that such recovered working capital does not appear anywhere

---
*The year's depreciation, *not* accumulated depreciation.

in the annual report, although it affects the working capital distribution shown in Table 12.1.5.

The next step in developing an acceptable method for computing the corporation i_0 is to adjust the various terms for inflation. We will use the same device as was used in the ROI computations—that of inflation-adjusting the total investments to match the current cash flow. The only components of K in Eq. (12.2.27) that need to be corrected for inflation are the gross fixed assets; the inflation adjustment factor ϕ is the same one as in Eq. (12.2.18).

$$K = \text{(net fixed assets + accumulated depreciation)}\phi \\ + \text{ total assets } - \text{ net fixed assets } - \text{ current liabilities}$$

(12.2.32)

In particular, the purchased cost of marketable securities, implicitly included in Eq. (12.2.32), need *not* be corrected for inflation. Any eventual sales price above the purchased cost is reported as income.

In the expression for C_A, Eq. (12.2.31), the only term that must be corrected for inflation is the recovered working capital; the first three terms are already expressed in current dollars. The reason that the $(WC)_R$ term must be inflation-adjusted with the factor $(1 + \mathbf{f})^n$ is that the R_{-n} calculated from Eq. (12.2.30) is expressed in uninflated dollars, n years ago; the working capital recovered at the end of the nth year has been inflated. Thus

$$C_A = \text{net profit + depreciation + interest payments} \\ + \text{ recovered working capital}(1 + \mathbf{f})^n$$

(12.2.33)

The procedure for estimating i_0 may be summed up as follows:

1 Estimate k, the average annual rate of increase of the plant properties (gross fixed investments), using historical data in the annual report. See Fig. 12.2.1 for example.

2 Estimate n, the average project life, from Eq. (12.2.21), or else take n as the average for the chemical process industry (10 years).

3 Calculate the average inflation factor f over the n years, using data in Table 1.3.1 (or similar CPI data sources).

4 Calculate the inflation factor ϕ from Eq. (12.2.18).

5 Calculate K and C_A [Eqs. (12.2.32) and (12.2.33), respectively], and form the ratio K/C_A.

6 Compute i_0 from Eq. (12.2.26), or use Fig. 12.2.3 for the case of $n = 10$.

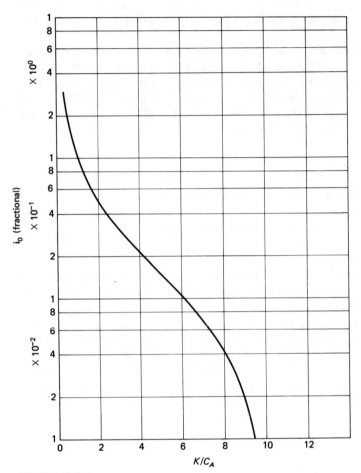

FIGURE 12.2.3
Internal interest rate for 10-year project life.

Example 12.2.2

Determine i_0 for The Dow Chemical Company for 1980 from data and results in Example 12.2.1.

SOLUTION In Example 12.2.1, we found that $n = 10$, $\mathbf{k} = \mathbf{f} = 0.078$, $\phi = 1.477$. From Table 12.2.1,

$$\text{Net fixed assets} = \text{gross plant properties} - \text{accumulated depreciation}$$
$$= 9873 - 4201 = 5672$$

From Eq. (12.2.32),

$$K = 9{,}873 \times 1.477 + 11{,}538 - 5{,}672 - 2{,}803 = 17{,}645$$

In Eq. (12.2.30),

$$R_{-10} = \frac{9873 \times 0.078}{(1.078)^{10} - 1} = 688$$

$$\left(\frac{WC}{G}\right)_{-1} = \frac{\text{current assets} - \text{current liabilities}}{\text{gross plant properties}}$$

$$= \frac{4390 - 2803}{9873} = 0.161$$

From Eq. (12.2.29),

$$(WC)_R = 0.161 \times 688 = 110.8$$

From Eq. (12.2.33),

$$C_A = 814 + 728 + 385 + 110.8(1.078)^{10} = 2{,}162$$

$$\frac{K}{C_A} = \frac{17{,}645}{2{,}162} = 8.16$$

From Fig. 12.2.3, $i_0 = 0.039$.

An average i_0 over several years is, of course, a better indication of a corporation's long-range performance.

12.3 ANALYSIS OF PERFORMANCE OF SECURITIES

Corporate Performance and Market Value of Securities

In Chap. 1, three categories of corporate securities were described:

Bonds
Preferred stock
Common stock

The bonds are instruments of debt; shares of stock represent evidence of ownership. The market value of stock, and common stock in particular, is considered to be an indicator of the performance of a corporation. Our purpose here will be to see why this should be so and how, in fact, the performance of securities mirrors the performance of the assembly of projects that constitutes the corporation.

When a corporation, with the approval of appropriate governmental agencies, floats a new issue of stock, it receives, as a result of sales to the public, the par value, plus any amount that the public is willing to pay to the corporation above the par value (*capital surplus*). Once the stock has been sold, however, the corporation no longer has any direct control over the financial fate of the stock. The stock is now on the open market and is traded just like any commodity that is beyond the direct control of the originator. If the market price goes above or below the price that originally accrued to the corporation, this fact alone is in no manner reflected on the balance sheets or in the income statements. And yet the corporation's management continues to view the market performance of its stock with an almost passionate degree of interest and attention. Market advances are a cause for joy and exuberance, declines cause gloom and self-doubt. Why should this be so?

Among the various reasons for this attitude is the belief, both by corporate managements and the public, that stock performance in the market somehow reflects the performance of the corporation, that the market price variations constitute an internal as well as an external monitoring basis. There is certainly some justification for this belief; after all, to a considerable extent the market price is, indeed, determined by the public's evaluation of the corporate performance potential and trust in the competence of management, both of which establish the demand for the stock. However, stock market prices are affected by more than corporate performance; all the forces of the marketplace—economic, political, and psychological—combine to induce price fluctuations that are unpredictable and often apparently irrational. The market value of a stock should reflect its *book value* (which we will learn to calculate) and its earning potential, but often it does not.

We have seen in Chap. 1 that the common stockholders are the ultimate "boss" of the corporation. They certainly wish to see adequate and growing earnings per share, but they also wish to have the market value of their shares grow, and a declining market value may well convince them that there is something wrong with the company they own. Here is perhaps another reason why top corporate management keeps a wary eye upon stock market prices, for stockholders who perceive that their company is going wrong can prove to be a most uncomfortable influence. If they happen to organize themselves into a unified body, they can be a source of unwelcome harassment, and they may even drive the corporate board of directors out of their jobs, using their legal right to vote for members of the board at the annual stockholders' meeting. In some cases the combination of low stock price and dissatisfied stockholders may act as an incentive for some other corporation to gain control by buying up a majority of the shares of stock.

There is also the argument that good market performance of a stock encourages new investors to buy stocks (and bonds, for that matter) in the future, to supply corporate investment capital needs. This is no longer an important consideration with established corporations in the chemical process industries, for very little of their capital is raised from new issues of stock (Table 1.3.3). What little is raised is primarily for the purpose of employee stock purchase plans; employees are encouraged to purchase stocks at well below the market value. Quite often, however, the corporation will purchase its own stock on the open market for this purpose.

Stock ownership by employees is considered a strong incentive to identify more closely with the corporation, to "work harder" to make the corporation and therefore its stock more successful. Corporate directors and officers often own large blocks of shares of their own company's stock, and members of the higher managerial and technical hierarchy are given options to purchase stock at attractively low prices. Perhaps this explains management's passionate interest in the stock market.

The current market value of all the corporation's outstanding stock is viewed by some economists as the most faithful indication of the true net present worth of all the corporation's investments.

Book Value of Securities

How much of the corporation does the stockholder really own? The answer lies in the *book value* of the various securities, which can be computed from balance sheet data. The procedure involves a sequential payoff of financial obligations in the order of priorities that would prevail if the corporation were undergoing liquidation. The order of priorities is:

1 Current liabilities (such as wages owed)
2 Long-term liabilities (bonds)
3 Preferred stock
4 Common stock

The current liabilities should be met by the available cash, marketable securities, and receivables; i.e., the *acid test ratio* should exceed 1. Regardless, the book value of the securities (stocks and bonds) is based upon the net assets after the current liabilities have been liquidated. Moreover, intangibles are not normally included in the net asset basis for calculating book value; the net assets, reduced by intangibles, are sometimes called *tangible assets:*

$$\text{Tangible assets} = \text{total assets} - \text{intangibles} \qquad (12.3.1)$$

The procedure for calculating book value is best illustrated by example.

Example 12.3.1

For the corporation described in Tables 12.1.1 and 12.1.2, calculate the book value of bonds, preferred stock, and common stock.
SOLUTION There are 2700 bonds outstanding, each with a redemption value of $1000, for a long-term liabilities total of $2,700,000.

$$\begin{aligned}\text{Book value per bond} &= \frac{\text{tangible assets} - \text{current liabilities}}{\text{bonds outstanding}} \qquad (12.3.2)\\ &= \frac{9,700,000 - 100,000 - 2,500,000}{2,700}\\ &= \$2629/\$1000 \text{ bond}\end{aligned}$$

The bond book value thus exceeds substantially the redemption value; the bond holders may rest assured that their investment is well protected in emergencies.

In Table 12.1.2, the outstanding shares of preferred stock ($100 par value) are listed at 6000. After the bond holders have been paid off,

$$\text{Book value per share of preferred stock} = \frac{\text{tangible assets} - \text{total liabilities}}{\text{shares of preferred stock}} \quad (12.3.3)$$

$$= \frac{9,600,000 - 5,200,000}{6,000}$$

$$= \$733/\text{share}$$

Finally, for the common stock, if the remaining assets are portioned out among the 300,000 shares outstanding ($5 par value),

$$\text{Book value per share of common stock} = \frac{\text{tangible assets} - \text{total liabilities} - \text{preferred stock par value}}{\text{shares of common stock}} \quad (12.3.4)$$

$$= \frac{9,600,000 - 5,200,000 - 600,000}{300,000}$$

$$= \$12.67/\text{share}$$

The book value of both the preferred and the common stock shares is above par value in the above example. The par value, however, has little meaning once shares are traded on the open market. The market value is a much more significant basis of comparison; clearly, the market value of the stock of a successful company should be above the book value. Such price premium indicates confidence in the future performance of the corporation, and confidence in the current performance as measured by the earnings per share (see below).

Exercise 12.3.1

In 1980, The Dow Chemical Company had an average of 182,162,000 shares of common stock outstanding, no preferred stock. The closing market price of its stock on December 31 was $32, at which time the stockholders' equity amounted to $4.440 billion, with intangibles worth $109 million. How did the book value of the common stock compare with the market value?

Characterization of Security Performance

Shares of stock generate two types of income to the stockholder:

1 Income from dividends (usually on a quarterly basis)
2 Income from trading the securities on the market

A steady annual growth of dividends on common stock is considered by most corporate boards to be a particularly understandable indication of good corporate performance, and many companies will go to extraordinary lengths to maintain such annual growth. Indeed, the investor considers dividends to be an important component of stock-generated income and welcomes an unflinching growth trend, provided it is not all inflation-caused.

Exercise 12.3.2

Standard Oil Company of California has paid its common stockholders the following annual cash dividend per share during the 1976–1980 period:

	1976	1977	1978	1979	1980
Dividend, $	1.08	1.18	1.28	1.45	1.80
Year-end market price of stock, $	30	26	29	32	50

From data in Table 1.3.1, decide whether the dividend growth rate has kept up with inflation. What average rate of return is the stockholder getting in dividends, relative to the value of the annual stock investment?

Perhaps a more faithful, if less immediately tangible, indication of performance is the ratio of *earnings per common share*. To compute this ratio properly, it is necessary to subtract preferred stock dividends, if any, from the year's net profit:

$$\text{Earnings per common share} = \frac{\text{net profit} - \text{preferred stock dividend}}{\text{outstanding shares of common stock}} \quad (12.3.5)$$

For the corporation represented by Tables 12.1.1 to 12.1.3,

$$\text{Earnings per common share} = \frac{535{,}000 - 6{,}000 \times 100 \times 0.05}{300{,}000}$$
$$= \$1.68 \text{ per share}$$

The earnings per common share, prominently displayed in corporate annual reports, are a well-recognized criterion of the year-by-year assessment of corporate performance. The ratio is used, in turn, to generate another commonly used entity, one that combines dividend performance and the market performance of the stock—the *price-earnings (P/E) ratio*:

$$P/E = \frac{\text{market price of stock at year's end}}{\text{annual earnings per share}} \quad (12.3.6)$$

In 1980, the P/E ratios for the 10 largest U.S. chemical corporations averaged at 7.6, a considerable drop from the decade of the seventies, when 15.0 was considered to be more of a standard. The ratio is still another monitoring device of year-by-year corporate performance; the preferred behavior is one of constancy at a fairly high value—but not too high, for this would indicate an unrealistically overpriced stock, or even worse, a decrease in annual earnings. A low ratio may occur during periods of rapidly expanding prices, but a low value such as 7.6 usually indicates underpriced stock.

Leverage

The financial practices of a corporation can contribute directly to the market attractiveness of its stock. In considering the sources of capital in Chap. 1, we saw that there are those corporations that prefer to finance their investments with external loans, that maintain a high debt-to-equity ratio. Heavy borrowing is justified on the basis of the improvement of opportunities for the aggressive penetration of new markets and, during periods of inflation, of paying off debts with cheaper, inflated currency. Another more subtle effect of borrowing has to do with the relationship between debt payments and earnings per share; the effect is commonly called *leverage*.

We have seen that a steady year-by-year increase in earnings per share is considered to be a particularly understandable criterion of a corporation's financial success, one that is likely to have a beneficial effect upon the market price and investment attractiveness of the corporation's securities. Corporate earnings (or earnings per share) are calculated after taxes and *after interest payments*; the interest payments, needless to say, may have a profound effect upon the reported earnings. Particularly during periods of steady business growth and expansion of markets, interest payments may have the arithmetical effect of magnifying the rate of growth of earnings, to bedazzle the delighted investor.

For the sake of illustration, let us suppose that a corporation has $10 million worth of outstanding bonds that pay 8 percent annual interest. Now consider the corporation's hypothetical total income, after taxes but before interest payments, for two successive years during a period of modest business growth:

	1982	1983
Total income	$880,000	$970,000
Interest	$800,000	$800,000
Profit	$ 80,000	$170,000

In the 2 years shown, the income has increased by only 10.2 percent, but the reported earnings have increased by 112.5 percent!

Clearly, the magnifying effect will also work in reverse; during periods of sluggish business and inflexible pricing, a modest decrease in income may result in a most uncomfortable decrease in net profit. Considerable managerial skills are required to manipulate leverage to optimize the corporate financial image.

Stock Performance History

If we agree that the performance of stocks is a reflection of the performance of the corporation, we must ask how best to gauge the performance of the stock. A number of criteria have been examined, but each of these is restricted in scope; we need a criterion that will combine the market and dividend performances, and do so to account for those performances over a period of several years. Indeed, a historical perspective is needed, for any human activity such as the corporation represents will be characterized by "bad years," by temporary setbacks. What is of interest is to find out whether, over the years, a corporation has been able to cope with the ups and downs of the general economy and its own fortunes to exhibit a steady pattern of growth and financial strength. If such a pattern does exist, the principle of historical continuity (Sec. 3.3) suggests that the pattern may be extrapolated into the future.

A criterion to satisfy these requirements is incorporated in the answer to the question, "If I had invested $1 in the corporation's stock, say, 10 years ago, and I ploughed all dividends back into more stock, how much would I have today—in constant-value dollars?" The following example illustrates the method of computing the current value of the investment.

Example 12.3.2

Consider $1 invested into Standard Oil of California common stock at the end of 1976 (CPI = 170.5). Assuming dividends are reinvested at the end of each year, what is the value of the investment at the end of 1980 (CPI = 246.8) in 1976 dollars?

SOLUTION The following table may be generated in terms of inflated currency. To make the numbers more manageable, take as basis $100 invested at the end of 1976; this will buy $100/30 = 3.333$ shares.

Year	Shares at end of year	Value of shares, $	Total dividend at end of year, $	Shares purchased with dividend
1976	3.333	100.00	0.00	0.000
1977	3.484	90.58	3.93	0.151
1978	3.638	105.50	4.46	0.154
1979	3.803	121.70	5.28	0.165
1980	3.940	197.00	6.85	0.137

The value at the end of 1980, in 1976 dollars, is

$$\frac{197.0 \times 170.5}{246.8} = 136.1$$

That is, the $1 has grown to 1.361 constant-value dollars, a compounded growth of about 8.0 percent/year, inflation-adjusted.

Stock splits, capital gains distributions, and other financial phenomena complicate the computation somewhat.

Is the growth revealed by a calculation such as in Example 12.3.2 "good enough"? In the final analysis, as was the case with individual projects, the decision is a matter of comparison with other potential investments, with similar performance criteria of other corporations. For the specific case in Example 12.3.2, most investors would look at the result with a great deal of favor, although it must be remembered that the computation does not consider the effect of income taxes.

Publications exist which contain a great deal of information on the financial history and current status of corporations in the chemical process industries. Some are listed in Exhibit 12.3A; the journal *Value Line* is a particularly good source of historical data and ratio analysis information.

12.4 PERFORMANCE OF THE CHEMICAL INDUSTRIES

Some Corporate Statistics

The discussion of corporate performance analysis will be concluded with a few individual corporation statistics. The purpose is to impart some feeling as to typical financial data ranges that characterize the chemical process industries. For more detail, the references in Exhibit 12.3A may be consulted, although corporate annual reports are unquestionably the best source of raw data.

The 10 largest corporations in the world that are primarily involved in chemical manufacture are listed in Table 12.4.1. The ranking is on the basis of total 1980 sales (although not all the sales are necessarily chemicals). The world's leaders are clearly formidable economic giants with multibillion dollar sales turnovers that challenge and exceed the gross national product of many a nation. The list is dominated by Western Europe, but some members of the Japanese chemical industry are not far behind. Three giants (Mitsubishi, Sumitomo, and Asahi) boast annual sales in the $3 billion range.

EXHIBIT 12.3A

SOURCES OF FINANCIAL DATA OF CORPORATIONS IN THE CHEMICAL PROCESS INDUSTRIES

1 "Moody's Industrial Manual," Moody's Investors Service, Inc., New York. An annual detailed review of the financial profile of individual corporations; includes credit risk ratings.

2 "The Kline Guide to the Chemical Industry," 4th ed., Susan Curry and Susan Rich (eds.), Charles H. Kline and Co., Inc., Fairfield, N.J., 1980. A multifaceted overview of the chemical industry business.

3 Wei, J., T. W. F. Russell, and M. W. Swartzlander, "The Structure of the Chemical Processing Industries," McGraw-Hill, New York, 1979. The book incorporates thumbnail sketches of some of the larger corporations and an outline of the general characteristics of the industry.

4 *Value Line Investment Survey*, Arnold Bernhard and Co., 5 East 44th St., New York, N.Y. 10017. Weekly journal; most major corporations are profiled at least once per year with a wealth of graphs, historical performance data, and assessments of projected performance.

5 The weekly journals *Chemical Week* and *Chemical and Engineering News* abound with corporate financial data and industry reviews.

TABLE 12.4.1
TEN TOP-RANKED CHEMICAL CORPORATIONS, RANKED BY 1980 TOTAL SALES

Rank	Corporation	Country	Total 1980 sales, millions of dollars
1	Hoechst	West Germany	16,437
2	Bayer	West Germany	15,838
3	BASF	West Germany	15,237
4	Du Pont	United States	13,652
5	ICI	United Kingdom	13,291
6	Dow	United States	10,626
7	Union Carbide	United States	9,994
8	Montedison	Italy	9,090
9	DSM	Netherlands	7,497
10	Rhône-Poulenc	France	7,139

Source: Data extracted, with the permission of *Chemical and Engineering News*, primarily from the June 8, 1981, issue. Copyright 1981, American Chemical Society.

A similar list for the United States alone is shown in Table 12.4.2. The ranking is based upon chemical sales exclusively, including sales abroad, of all corporations, not necessarily just those primarily engaged in chemical manufacture. The petroleum corporations clearly perform very well in the chemical area, even though for a giant such as Exxon the chemical business constitutes only about 7 percent of its 1980 annual sales of over $100 billion.

The 1980 financial performance of a number of U.S. chemical companies is summed up in Table 12.4.3. In general, the ratios tend to be confined within rather well-defined ranges, but some trends are detectable even with this very limited sample. For instance, the net profit ratio [Eq. (12.2.6)] is higher for corporations with predominantly specialty chemicals sales, particularly pharmaceuticals. Simi-

TABLE 12.4.2
TEN TOP-RANKED U.S. CORPORATIONS, RANKED BY 1980 CHEMICAL SALES

Rank	Corporation	Industry Classification	1980 chemical sales, millions of dollars
1	Du Pont	Basic chemicals	10,250
2	Dow	Basic chemicals	7,217
3	Exxon	Petroleum	6,936
4	Union Carbide	Basic chemicals	5,650
5	Monsanto	Basic chemicals	5,453
6	Celanese	Basic chemicals	3,200
7	Shell Oil	Petroleum	3,089
8	W. R. Grace	Specialty chemicals	2,733
9	Gulf Oil	Petroleum	2,569
10	Occidental Petroleum	Petroleum	2,458

Source: Data extracted, with the permission of *Chemical and Engineering News*, primarily from the June 8, 1981, issue. Copyright 1981, American Chemical Society.

TABLE 12.4.3
COMPARATIVE FINANCIAL PERFORMANCE (1980) OF SELECTED U.S. CHEMICAL COMPANIES

Company	Current ratio	Equity ratio	Net profit ratio, %	Return on stockholders' equity, %	Sales as % of assets	P/E
Predominantly basic chemicals						
Allied Corp.	1.3	0.42	5.5	16.3	123	5.9
Celanese	2.0	0.42	3.6	11.1	127	6.2
Du Pont	2.3	0.60	5.2	12.6	143	8.3
Union Carbide	2.1	0.49	6.7	14.3	104	4.4
Predominantly specialty chemicals						
Int. Minerals	2.0	0.48	8.2	16.6	97	9.0
Lubrizol	2.5	0.70	12.4	24.6	139	11.6
Nalco	2.7	0.74	11.7	23.6	148	11.2
Williams	1.8	0.41	6.7	14.4	87	7.5
Mixed commodities and specialties						
Dow	1.5	0.38	7.1	17.4	93	8.2
Ferro	2.0	0.51	4.2	13.2	162	5.5
W. R. Grace	1.5	0.40	4.7	16.6	142	7.9
Stauffer	1.4	0.44	7.2	13.4	81	7.0
Pharmaceuticals						
Eli Lilly	2.0	0.67	13.4	19.7	98	12.3
Merck	2.3	0.65	15.2	22.3	95	13.0
Pfizer	1.9	0.49	8.4	17.5	93	12.3
Upjohn	1.9	0.52	9.7	19.7	105	9.6

Source: Data extracted, with the permission of *Chemical and Engineering News*, primarily from the June 8, 1981, issue. Copyright 1981, American Chemical Society.

larly, the P/E ratio is higher for the same category of corporation, perhaps reflecting the investor's higher confidence in the performance of specialty companies.

On the other hand, the apparently better performance of the specialty-oriented corporations may be a function of their size—they tend to be smaller. Differences in the performance of corporations, based upon size, are shown in Table 12.4.4. The differences exhibited are rather remarkable. Smaller corporations tend to be characterized by a significantly higher return on investment [Eq. (12.2.8)], higher employee productivity as measured by sales, considerably less capital spending to generate a dollar of sales, and a solidly smaller reliance upon debt financing. The statistics constitute food for thought as to what constitutes an economically optimum size of an independent corporate unit.

TABLE 12.4.4
1980 MEDIAN PROFITABILITY OF CHEMICAL COMPANIES

Return on Investment, %	Sales per employee	Capital spending, % of sales	Sales, % of assets	Debt as % of debt plus equity
Sales over $1 billion; median of 23 companies				
4.7	$104	9.3	119.5	32.8
Sales from $80 million to $1 billion; median of 17 companies				
9.8	$148	5.8	123.8	13.2

Source: Data extracted, with the permission of *Chemical and Engineering News*, primarily from the June 8, 1981, issue. Copyright 1981, American Chemical Society.

Exercise 12.4.1

The ratio of sales to total assets in Table 12.4.3 is somewhat analogous to the turnover ratio for individual plants [Eq. (5.1.5)]. In what ways do the ratios differ? Would you expect the "turnover ratio" for a corporation to be more or less than the turnover ratio for plants?

Position of the Chemical Industry in the U.S. Economy

The relative position of the chemical industry in the U.S. economy is illustrated in Fig. 12.4.1. The value added by manufacture by the total U.S. industry exceeded the $500 billion mark toward the end of the decade of the seventies; of this total, 10 percent was contributed by chemicals and allied products, close on the heels of the food processing industry, which contributed 12 percent. The relative contribution of some of the other processing industries is shown in the graph. Indeed, the chemical industry is a strong and vital component of the U.S. and world economy; its continued growth and vitality will in no small way be enhanced by the adequate evaluation of the myriad of new projects that are intended to guarantee future growth.

EPILOGUE

With these words, our survey of the various aspects of project evaluation comes to an end.

It would be presumptuous and, clearly, quite incorrect to say that the subject areas important to adequate project evaluation have been exhausted in the course of the discussion contained within these pages. In fact, most subjects have been treated in what experienced practitioners of project evaluation would characterize as a preliminary, an introductory, manner. Some subjects which can be legitimately considered as part of the milieu of project evaluation have not even been men-

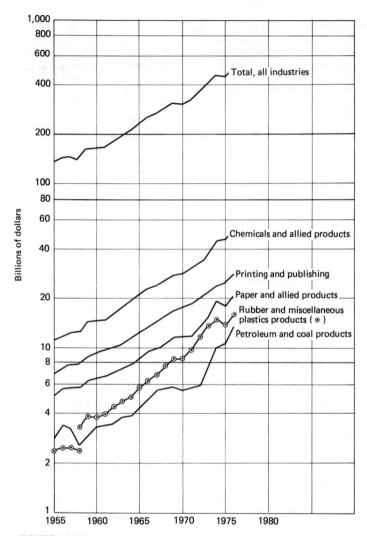

FIGURE 12.4.1
Value added by manufacture, by industry. *(Reproduced, with permission, from "Chemical Economics Handbook," Stanford Research Institute, Menlo Park, Calif.)*

tioned. In particular, except for a brief discussion of technological forecasting, little has been said about attempts to evaluate projects and assemblies of projects within a context beyond the boundaries of the sponsoring corporation. What effect do new projects have upon other industries? Upon the society that spawns and shelters the chemical corporations? Economists, engineers, and business people have worked together to create models to quantify and study such interactions. Econometric models describe the interactions within project assemblies, within a whole industry, within a whole economy. The limitations of energy resources have forced a more detailed accounting of project energy inputs and outputs, including energy that is inherently contained in raw materials, products, even the production equipment. Product or process costs are frequently evaluated in terms of *life cycle costs,* costs that include expenditures for ultimate disposal, as well as other peripheral expenditures inevitably incurred through the use of the product or process, beyond the immediate production and distribution context.

Risk analysis has been highlighted herein as an important aspect of project evaluation, but methods of generating probability data (the Delphi method, for instance) have not been emphasized. Safety aspects of project evaluation have also been stressed, but nothing has been said about fault tree analysis, a method of increasing the reliability and safety of operations. Many other subjects lie upon the common ground between project evaluation and process design; again, many such subjects have been ignored—evolutionary operations (EVOP), or the economic evaluation components of the process computer-simulation program ASPEN.

Nevertheless, the material that has been presented should serve as an adequate basis for the evaluation of a great majority of process development projects. Hopefully, it also will intrigue the practicing chemical engineer sufficiently to investigate, learn, and use the more advanced techniques which spatial considerations alone have excluded from these pages.

NOMENCLATURE

C_A cash flow after taxes, $/year
D annual depreciation charge, $
f inflation factor, fractional
G gross plant properties on balance sheet, $
i_0 discrete annual internal interest rate on all the corporation's investments, adjusted for taxes and inflation; i_0 is the nominal continuous interest rate
k average annual increase in gross fixed assets, fractional
K capital investment, $
m year number
n total number of years
P present value
P_B profit before taxes
R annual plant properties investment, $
t time, years
ϕ inflation adjustment factor

Subscripts

$-j$ for year $-j$

Abbreviations

CFS   capital for sale
CPI   Consumer Price Index
DDB + SL   double-declining-balance and straight-line (depreciation)

DCRR   discounted cash rate of return
EVOP   evolutionary operations
FIFO   first in, first out
LIFO   last in, first out
P/E   price-earnings ratio
ROI   return on investment
ROAIBT return on average investment before taxes
ROIAT   return on investment after taxes
ROIBT   return on investment before taxes
WC   working capital
$(WC)_R$ recovered working capital

PROBLEMS

12.1 Changes in retained earnings (Table 12.1.4) are computed from the annual net profit. Why is cash flow after taxes not used for the computation? If it were, how would the elements of the balance sheet have to be reconstituted?

12.2 Use arguments such as those in Example 12.1.1 to trace the effect of the following actions upon the corporate balance sheet:
  a Some raw materials inventories are sold.
  b A plant with all capital depreciated a few years ago is terminated.
  c A customer who has not yet paid for some shipments goes bankrupt.
  d Product distribution rights in the west are sold to a competitor below the asset value.
  e Warehouses in a distribution center are sold at book value.

12.3 In Chap. 8 the phenomenon of uninflated depreciation charges was discussed (see, for instance, Exercise 8.3.1). The Financial Accounting Standards Board (FASB) has suggested the following procedure for calculating the effective tax rate on income in constant dollars:
  a For the income statement in Table 12.1.3, what is the effective tax rate "as reported"?
  b Suppose, on the average, the corporate assets are depreciated in n years, the rate of growth of annual investments in new plant properties is k (in cocurrent dollars), and the average inflation rate is f; the useful life of the plant properties is also n years. Show that if depreciation charges properly reflected inflation, the total annual depreciation charge D' would be related to the uninflated depreciation charge D reported on the income statement as

$$\frac{D'}{D} = k \frac{(1+k)^n - (1+f)^n}{(k-f)[(1+k)^n - 1]} \qquad (P\,12.1)$$

  c What would be the inflated depreciation charge in Table 12.1.3 for the case of $n = 10$, $k = 0.12$, $f = 0.08$?
  d What is the net profit, adjusted for inflation?
  e What, in fact, is the effective tax rate on the inflation-adjusted income before tax? How does this effective tax rate in constant dollars compare with the reported tax rate?

12.4 A simplified version of the 1980 Annual Report of Standard Oil Company of California is reproduced in Table P12.1 (p. 700). From the data, calculate the following ratios:
  a Current ratio   **c** Equity ratio
  b Acid test ratio   **d** Return on stockholders' equity

e Operating Margin
f Profit margin
g Return on total assets
h Return on investment
i Sales as percent of assets
j Debt as percent of debt plus equity

Use the results to prepare an assessment of the financial status of the company.

12.5 A simple corporation is represented by four projects described by the following data:

Project no.	Year started	Fixed assets investment, millions of dollars	Tax life, years	Book life, years	Relative value of Consumer Price Index
1	1975	5.00	8	15	1.00
2	1977	3.00	7	7	1.13
3	1980	7.00	10	10	1.53
4	1983	4.00	6	10	1.86

The book life is equal to the useful life of each project, with zero salvage value at the project's termination. Book depreciation is taken as straight-line over the book life; tax depreciation is computed using the DDB + SL method over the tax life. Each investment is made at the start of the stated year; the investment shown is expressed in cocurrent dollars, i.e., dollars actually paid out at the start of the year in question.

    a In the 1985 annual report (relative CPI = 2.25), what value of the gross fixed assets is reported?
    b What is reported as the accumulated depreciation?
    c What are the reported net plant assets?
    d What is the value of the inflation adjustment factor ϕ?

12.6 As a result of differences in book and tax depreciation methods (Chap. 6), a corporation may defer part of its income tax payments to later years. This, of course, is an advantage, for extra funds are available for investment. Deferred income taxes are listed on the balance sheet as an asset.

    a Justify the listing of deferred income taxes as an asset. If this asset is partly or totally eliminated during any one year, how is a financial balance maintained on the balance sheet?
    b Suppose the "simple corporation" in Prob. 12.5 pays an effective income tax rate of 40 percent. What is the total of the deferred income tax listed in the 1985 annual report?
    c What would the reported net plant assets be if the tax depreciation methods were to be used on the books as well?

12.7 The second bracketed term in the denominator of Eq. (12.2.11) represents the working capital equivalent of individual projects. Do you agree with the format developed? For example, do you think all the current liabilities should be subtracted to maintain the analogy to individual projects? Justify the format, or propose a better one.

12.8 In Sec. 12.2 a method was described for deriving the value of i_0, the *discrete* (effective) corporate internal rate of return adjusted for taxes and inflation, from data in the annual report. How would you modify the method to derive i_0, the nominal *continuous* rate of return?

12.9 The balance sheet and various financial data for a corporation are reproduced in Tables 12.1.1 to 12.1.5. For the latest year, additions to plant properties, R_{-1}, are $305,000 (Table 12.1.5). If n is known to be 22 years, what is the likely value of k, the average annual increase of the gross fixed investments? If the average value of f over the past 22 years is 0.066, what is the value of i_0 for the corporation?

12.10 Use the data in Table P12.1 to calculate the book value of common stock of Standard Oil Company of California, and compare that value with the market value of the stock at the end of 1980 (see Exercise 12.3.2).

12.11 The historical performance of common stock of The Dow Chemical Company is shown in the following table:

Year	Year-end market price of stock, $	Cash dividends paid per share, $
1971	19.72	0.44
1972	25.38	0.45
1973	28.75	0.48
1974	27.50	0.55
1975	45.81	0.73
1976	43.38	0.90
1977	26.75	1.10
1978	24.88	1.25
1979	32.13	1.45
1980	32.13	1.60

Values in the table have been adjusted for stock splits.

Consider $1 invested into Dow common stock at the end of 1970 (market price = $12.27 per share; Consumer Price Index = 116.3). Assuming dividends are all reinvested at the end of each year, what is the value of the investment at the end of 1980 (CPI = 246.8) in 1970 dollars?

12.12 Obtain annual reports of about six major chemical corporations for the past year. Using criteria of your choice, rank the corporations in order of performance from the investor's point of view.

TABLE P12.1
1980 Annual Report,
Standard Oil Company of California
Simplified Balance Sheet and Income Statement
(Millions of Dollars)

Assets

Current assets:		
Cash	$ 253	
Marketable securities, cash investments	2,789	
Accounts receivable	5,162	
Inventories	2,052	
Total current assets		10,256
Fixed assets	15,829	
Less: Accumulated depreciation	(7,049)	
Net fixed assets		8,780
Miscellaneous prepayments and deferred charges		3,126
Total assets		$ 22,162

Liabilities

Current liabilities:		
Accounts payable	$ 4,994	
Notes and expenses payable	246	
Income taxes payable	1,745	
Total current liabilities		6,985
Long-term liabilities:		
Long-term debt	1,109	
Other long-term obligations	2,991	
Total long-term liabilities		4,100
Total liabilities		$ 11,085

Stockholders' Equity

Common stock, $3 par value, 342,109,258 shares outstanding	$ 1,026
Capital surplus	881
Accumulated retained earnings	9,170
Total stockholders' equity	11,077
Total liabilities and stockholders' equity	$ 22,162

Income Statement, 1980

Net sales		$ 41,553
Costs and operating expenses:		
Operating expenses	36,742	
Depreciation	767	
Selling and administrative expenses	954	(38,463)
Operating profit		3,090
Other income		1,366
Total income		4,456
Interest and debt expense		(158)
Income before provision for income tax		4,298
Provision for income tax		(1,897)
Net profit for the year		$ 2,401

APPENDIX A

THE SUMMATION OF SERIES USING THE CALCULUS OF FINITE DIFFERENCES

The purpose of this short review is to give the student a working, nonrigorous exposition to methods of summing series by using the calculus of finite differences. The summation of series is only one of a multitude of applications of this calculus, and only those aspects which have immediate application will be touched upon here. No previous exposure of the student is assumed, and the interested student is certainly urged to delve into more rigorous and wide-ranging expositions in books such as the treatise of C. H. Richardson, "An Introduction to the Calculus of Finite Differences," Van Nostrand, New York, 1954.

The specific problem we wish to address is this: suppose we have a series such as

$$\sum_{1}^{n} x^2 a^{-x}$$

where the variable x takes on discrete values between 1 and n, and these values differ by a unit amount. How can we find an expression for the sum indicated? Before we can attack the problem directly, we need to explore some of the fundamental concepts of the calculus of finite differences.

A.1 FINITE DIFFERENCES

Let the general term of a series be U_x; in the series above, for example,

$$U_x = x^2 a^{-x}$$

We define the *first difference* of U_x as

$$\Delta U_x = U_{x+1} - U_x \qquad (A.1.1)$$

where U_{x+1} is evaluated by substituting $x + 1$ for x in the general term U_x. For example, for $U_x = \log x$,

$$\Delta U_x = \log (x + 1) - \log x = \log \frac{x + 1}{x}$$

Also,

$$\Delta a^{bx} = a^{b(x+1)} - a^{bx} = (a^b - 1)a^{bx}$$

and for the series first mentioned,

$$\Delta x^2 a^{-x} = (x + 1)^2 a^{-(x+1)} - x^2 a^{-x}$$
$$= a^{-x} \left[\frac{(x + 1)^2}{a} - x^2 \right]$$

The difference in the calculus of finite differences corresponds to the differential in the calculus of continuous functions. Higher-order differentials similarly correspond to higher-order differences. For example, the second difference is the difference between first differences, but this need not concern us for the present.

A.2 FACTORIAL NOTATION

There are two factorial forms which need to be defined:

$$(a + bx)^{(n)} = (a + bx)[a + b(x - 1)] \cdots [a + b(x - n + 1)] \quad (A.2.1)$$

and

$$(a + bx)^{|n|} = (a + bx)[a + b(x + 1)] \cdots [a + b(x + n - 1)] \quad (A.2.2)$$

These should be distinguished from the standard factorial notation

$$(a + bx)^n$$

We introduce these two forms because they are particularly easy to difference or sum (integrate), and the form of the differences (or sums) is similar to that of conventional differentials (or integrals) in the infinitesimal calculus. Thus

$$\Delta(a + bx)^{(n)} = bn(a + bx)^{(n-1)} \quad (A.2.3)$$

This theorem can be easily proved by the student. Similarly,

$$\Delta \frac{1}{(a + bx)^{|n|}} = \frac{-bn}{(a + bx)^{|n+1|}} \quad (A.2.4)$$

where $(a + bx)^{(0)}$ and $(a + bx)^{|0|}$ are both defined as 1.

A.3 CHANGING POLYNOMIALS INTO FACTORIALS

Polynomials are not difficult to difference, but they are not easy to sum. They are best converted to one of the two factorial forms previously defined. The method is best illustrated by example.

It stands to reason that if an nth-degree polynomial has a factorial form equivalent, the factorial will also be nth degree. Consider, for example, the third-degree polynomial

$$U_x = 3x^3 + 5x^2 - 4x + 2$$

Assume a factorial equivalent,

$$U_x = Ax^{(3)} + Bx^{(2)} + Cx^{(1)} + 2$$

Expanding,

$$3x^3 + 5x^2 - 4x + 2 = Ax(x - 1)(x - 2) + Bx(x - 1) + Cx + D$$

The equality must hold for all values of x. To determine the values of the coefficients A, B, C, and D, the equality is evaluated, in turn, for $x = 0$, $x = 1$, and $x = 2$ (to make x, $x - 1$, $x - 2$ equal to zero in turn).

For $x = 0$, we immediately obtain $D = 2$. Since the constant terms are equal, we can now simplify by dividing by x:

$$3x^2 + 5x - 4 = A(x - 1)(x - 2) + B(x - 1) + C$$

For $x = 1$,

$$3 + 5 - 4 = C = 4$$

Therefore

$$3x^2 + 5x - 4 = A(x - 1)(x - 2) + B(x - 1) + 4$$

We now divide both sides by $x - 1$; the left side will have a remainder just exactly equal to C (=4), a consequence of the *remainder theorem* which states that if a polynomial $f(x)$ is divided by $x - a$, the remainder is $f(a)$. Thus, using long division,

$$\begin{array}{r} 3x + 8 \\ x - 1 \overline{\smash{)}3x^2 + 5x - 4} \\ \underline{3x^2 - 3x} \\ 8x - 4 \\ \underline{8x - 8} \\ + 4 \text{ remainder} \end{array}$$

that is,

$$\frac{3x^2 + 5x - 4}{x - 1} = 3x + 8 + \frac{4}{x - 1}$$

Therefore

$$3x + 8 + \frac{4}{x-1} = A(x-2) + B + \frac{4}{x-1}$$

or

$$3x + 8 = A(x-2) + B$$

For $x = 2$, $B = 6 + 8 = 14$. Therefore

$$A(x-2) = 3x - 6 = 3(x-2) \quad A = 3$$

Thus

$$U_x = 3x^3 + 5x^2 - 4x + 2 = 3x^{(3)} + 14x^{(2)} + 4x^{(1)} + 2$$

The student may wonder why things "worked out just right." Newton's theorem shows that an nth-degree polynomial can, in fact, be expressed as an nth-degree factorial *exactly*; for detail, the student is referred to the book by Richardson.

We will see that U_x in factorial notation is more readily integrated.

A.4 FINITE INTEGRATION

Finite integration denotes the process of finding a function V_x which, when differenced, yields the function U_x. That is,

$$\Delta V_x = U_x$$

and using notation to indicate integration,

$$\Delta^{-1} U_x = V_x$$

By analogy with infinitesimal calculus integration, the indefinite integration is better written

$$\Delta^{-1} U_x = V_x + C \qquad (A.4.1)$$

since $\Delta C = 0$ (why?).

Some important finite integrals and operations are listed in Table A.4.1; most can be readily derived by the student or proved by redifferencing.

Example A.4.1

Find $\Delta^{-1}(3x^3 + 5x^2 - 4x + 2)$.

SOLUTION We have seen that this polynomial may be written as

TABLE A.4.1
SOME COMMON FINITE INTEGRALS AND OPERATIONS

1. $\Delta^{-1}(U_x + V_x - W_x) = \Delta^{-1}U_x + \Delta^{-1}V_x - \Delta^{-1}W_x$
2. $\Delta^{-1}kU_x = k\Delta^{-1}U_x$
3. $\Delta^{-1}(U_x \Delta V_x) = U_x V_x - \Delta^{-1}(V_{x+1} \Delta U_x)$
4. $\Delta^{-1}a^{bx} = \dfrac{a^{bx}}{a^b - 1}$ $(a \neq 1)$
5. $\Delta^{-1}C = Cx$
6. $\Delta^{-1}\log\dfrac{x+1}{x} = \log x$
7. $\Delta^{-1}(a + bx)^{(n)} = \dfrac{(a+bx)^{(n+1)}}{b(n+1)}$
8. $\Delta^{-1}\dfrac{1}{(a+bx)^{|n|}} = \dfrac{1}{b(1-n)(a+bx)^{|n-1|}}$
9. $\Delta^{-1}x(x!) = x!$
10. $\Delta^{-1}\cos(a+bx) = \dfrac{1}{2\sin(b/2)}\sin\left(a - \dfrac{b}{2} + bx\right)$
11. $\Delta^{-1}\sin(a+bx) = \dfrac{1}{2\sin(b/2)}\cos\left(a - \dfrac{b}{2} + bx\right)$

$$U_x = 3x^{(3)} + 14x^{(2)} + 4x^{(1)} + 2$$
$$\Delta^{-1}U_x = 3\cdot\dfrac{x^{(4)}}{4} + 14\cdot\dfrac{x^{(3)}}{3} + 2x^{(2)} + 2x^{(1)} + C$$

This answer may be converted back to a polynomial by multiplying out the factorials, or, more easily, by using the *Stirling numbers of the first kind* listed in Table A.4.2. The Stirling numbers are the coefficients in the polynomial corresponding to each factorial. The notation

TABLE A.4.2
STIRLING NUMBERS OF THE FIRST KIND*
(Up to $n = 7$)

n	$\mathscr{S}_1^n$	$\mathscr{S}_2^n$	$\mathscr{S}_3^n$	$\mathscr{S}_4^n$	$\mathscr{S}_5^n$	$\mathscr{S}_6^n$	$\mathscr{S}_7^n$
1	1						
2	-1	1					
3	2	-3	1				
4	-6	11	-6	1			
5	24	-50	35	-10	1		
6	-120	274	-225	85	-15	1	
7	720	-1764	1624	-735	175	-21	1

*See C. R. Richardson, "An Introduction to the Calculus of Finite Differences," Van Nostrand, New York, 1954, for methods of expanding this table.

means this: n is the degree of the factorial being converted (or the resulting polynomial), and j is the exponent of a particular x term in the polynomial expansion. For example, from row $n = 4$,

$$x^{(4)} = x^4 - 6x^3 + 11x^2 - 6x$$

To convert the previous answer into a polynomial, we proceed as follows:

Coefficients of	x^4	x^3	x^2	x
$¾x^{(4)}$	¾	-9/2	33/4	-9/2
$1⅓x^{(3)}$		1⅓	-14	28/3
$2x^{(2)}$			2	-2
$2x^{(1)}$				2
Sum:	¾	⅙	-15/4	29/6

$$\Delta^{-1} U_x = ¾x^4 + ⅙x^3 - 15/4 x^2 + 29/6 x + C$$

One word of caution in using Table A.4.1. For example, in expression number 6, one is tempted to substitute

$$y = \frac{x + 1}{x} \quad \text{or} \quad x = \frac{1}{y - 1}$$

so that $\Delta^{-1} \log y = -\log(y - 1)$.

Right? Wrong! Remember, the calculus of finite differences is based on unit differences in the independent variable. Expression 6 holds for discrete values of x differing by unity, but in the derived expression y does not assume discrete values differing by unity, and the expression is incorrect.

A.5 THE DEFINITE INTEGRAL

We now return to the problem of the summing of series, which turns out to be a problem in finite integration. It can be readily shown that if

$$V_x = \Delta^{-1} U_x$$

then

$$U_a + U_{a+1} + \cdots + U_n = \sum_{a}^{n} U_x = V_{n+1} - V_a$$

or,

$$\sum_{a}^{n} U_x = V_x \Big|_{a}^{n+1} \quad (A.5.1)$$

that is, the sum of the series is the *definite integral* between the limits a and $n + 1$. Note that the limits of the integral are somewhat different than in the infinitesimal calculus.

Example A.5.1

Find the value of $\sum_0^5 \cos 2x$.

SOLUTION From Table A.4.1,

$$\Delta^{-1} \cos 2x = \frac{1}{2 \sin 1} \sin (2x - 1)$$

$$\sum_0^5 \cos 2x = \frac{1}{2 \sin 1} \sin (2x - 1) \Big|_0^6$$

$$= \frac{1}{2 \sin 1} [\sin 11 - \sin (-1)]$$

$$= -0.0942$$

Example A.5.2

Evaluate the sum of the first n terms of the series

$$1 + 8 + 27 + 64 + 125 + \cdots$$

SOLUTION By inspection, the general term is x^3 (the general term of a series of numbers is not always easy to deduce). Let

$$x^3 = Ax^{(3)} + Bx^{(2)} + Cx^{(1)} + D$$
$$= Ax(x - 1)(x - 2) + Bx(x - 1) + Cx + D$$

For $x = 0$, $0 = 0 + D$, and $D = 0$. Dividing by x,

$$x^2 = A(x - 1)(x - 2) + B(x - 1) + C$$

For $x = 1$, $C = 1$. Dividing by $x - 1$,

```
            x + 1
   x - 1 / x²
           x² - x
           ─────
              + x
              + x - 1
              ───────
                  + 1 = remainder
```

Therefore

$$x + 1 + \frac{1}{x-1} = A(x-2) + B + \frac{1}{x-1}$$

$$x + 1 = A(x-2) + B$$

For $x = 2$, $2 + 1 = B = 3$.

$$x + 1 = A(x-2) + 3$$

hence $A = 1$. Therefore

$$x^3 = x^{(3)} + 3x^{(2)} + x^{(1)} = U_x$$
$$V_x = \tfrac{1}{4}x^{(4)} + x^{(3)} + \tfrac{1}{2}x^{(2)}$$

and by (A.5.1),

$$\sum_1^n x^3 = \left| \tfrac{1}{4}x^{(4)} + x^{(3)} + \tfrac{1}{2}x^{(2)} \right|_1^{n+1}$$
$$= \tfrac{1}{4}(n+1)^{(4)} + (n+1)^{(3)} + \tfrac{1}{2}(n+1)^{(2)}$$

(since $1^{(2)} = 0$—why?)

Converting to polynomial form by using Stirling numbers,

$$\sum_1^n x^3 = \tfrac{1}{4}(n+1)^4 - \tfrac{1}{2}(n+1)^3 + \tfrac{1}{4}(n+1)^2$$

A.6 INTEGRATION BY PARTS

The integral forms listed in Table A.4.1 are clearly not sufficient to solve any but a few of the possible series formats. Two auxiliary operations are available to expand greatly the scope of finite integration. One such operation is integration by parts, analogous to the same operation in the infinitesimal calculus. The operation is listed as item 3 in Table A.4.1. It is worth exploring as a potential simplification whenever the general term U_x is a multiple of two or more separable functions of x:

$$U_x = f_1(x) f_2(x)$$

An example will serve to illustrate its use: Find $\Delta^{-1} x a^{-x}$. Let

$$U_x = x \qquad \Delta V_x = a^{-x}$$

Then

$$\Delta U_x = (x+1) - x = 1$$

$$V_x \text{ (from Table A.4.1)} = \frac{a^{-x}}{a^{-1} - 1} = \frac{a^{-x+1}}{1-a}$$

and

$$V_{x+1} = \frac{a^{-x}}{1-a}$$

Thus

$$\begin{aligned}\Delta^{-1} x a^{-x} &= x \cdot \frac{a^{-x+1}}{1-a} - \Delta^{-1} \frac{a^{-x}}{1-a} \\ &= x \cdot \frac{a^{-x+1}}{1-a} - \frac{a^{-x+1}}{(1-a)^2} \\ &= \frac{1}{a^{x-1}(1-a)^2} [x(1-a) - 1]\end{aligned}$$

The other operation promised is described next.

A.7 METHOD OF UNDETERMINED COEFFICIENTS AND FUNCTIONS

This method of long-winded appellation, familiar to those who are into differential equations, is best illustrated by example. We will choose the series which introduced Appendix A:

$$\sum_1^n x^2 a^{-x}$$

We note first that, from the previous example,

$$\Delta^{-1} x a^{-x} = a^{-x} \cdot g(x) = V_x$$

We assume that $\Delta^{-1} x^2 a^{-x}$ has a solution of similar form:

$$\Delta^{-1} x^2 a^{-x} = a^{-x} \cdot f(x) = V_x \qquad (A.7.1)$$

This is the core of the method; we recognize that the operator Δ does not alter the basic nature of the function upon which it operates, and we look for portions which, by experience or intuition, will remain unchanged in form. For example, we see from item 4, Table A.4.1, that the portion a^x remains in identifiable form. It is then assumed that the solution is the unchanged form multiplied by some other unknown function, $f(x)$.

We have no guarantee that the chosen format will result in a viable solution; for example, we might have chosen

$$V_x = a^{-x} x \cdot g(x)$$

and perhaps done better. Nevertheless, let us stick with our first choice. We have

$$\Delta^{-1} x^2 a^{-x} = a^{-x} \cdot f(x)$$

or
$$x^2 a^{-x} = \Delta[a^{-x} \cdot f(x)] = f(x+1)a^{-(x+1)} - f(x)a^{-x}$$

Simplifying,
$$x^2 = \frac{1}{a}f(x+1) - f(x) \qquad (A.7.2)$$

To satisfy this equation, $f(x)$ must evidently be a second-degree polynomial in x. Let
$$f(x) = Ax^2 + Bx + C$$

so that
$$f(x+1) = Ax^2 + (2A + B)x + (A + B + C)$$

Substituting into (A.7.2),
$$x^2 = \frac{A}{a}x^2 + \frac{2A + B}{a} + \frac{A + B + C}{a} - Ax^2 - Bx - C$$
$$= x^2\left(\frac{1-a}{a}\right)A + x\left(\frac{2A+B}{a} - B\right) + \left(\frac{A+B+C}{a} - C\right)$$

Equating coefficients of like powers of x,
$$A\left(\frac{1-a}{a}\right) = 1 \qquad A = \frac{a}{1-a}$$
$$\frac{1}{a}(2A + B) - B = 0$$

Substituting for A,
$$B = \frac{-2a}{(1-a)^2}$$
$$\frac{1}{a}(A + B + C) - C = 0$$

Substituting for A and B,
$$C = \frac{a}{(1-a)^2}\left(\frac{2}{1-a} - 1\right)$$
$$f(x) = \left(\frac{a}{1-a}\right)x^2 - \left[\frac{2a}{(1-a)^2}\right]x + \frac{a}{(1-a)^2}\left(\frac{2}{1-a} - 1\right)$$

and

$$V_x = a^{-x} \cdot f(x)$$
$$= \left(\frac{a^{-(x-1)}}{1-a}\right)x^2 - \left[\frac{2a^{-(x-1)}}{(1-a)^3}\right][x(1-a) - 1] - \frac{a^{-(x-1)}}{(1-a)^2}$$

Finally,

$$\sum_1^n x^2 a^{-x} = V_x \bigg|_1^{n+1}$$
$$= \frac{a^{-n}}{1-a}(n+1)^2 - \frac{2a^{-n}}{(1-a)^3}[(n+1)(1-a) - 1] - \frac{a^{-n}}{(1-a)^2}$$
$$- \frac{1}{1-a} - \frac{2a}{(1-a)^3} + \frac{1}{(1-a)^2} \quad (A.7.3)$$

It should be evident to the student by this time that even series with algebraically simple general terms yield expressions for their sum of extraordinary complexity. The opportunities for error in the derivations are manifold, and it would seem wise that for series with not too many terms, term-by-term numerical addition might be the least time-consuming approach. Such numerical integration is easily programmed for hand-held computers.

PROBLEMS

A.1 Derive the expression for the sum of an arithmetic progression, Eq. (2.2.18).
A.2 Derive the expression for the sum of a geometric progression, Eq. (2.2.20).

Deduce the general term of the following series and find the sum of the first 10 terms:

A.3 $1 + 4 + 9 + 16 + \cdots$
A.4 $\dfrac{1}{1} + \dfrac{1}{1+2} + \dfrac{1}{1+2+3} + \cdots + \dfrac{1}{1+2+3\cdots+n}$
A.5 $1 \times 3 + 2 \times 4 + 3 \times 5 + \cdots$
A.6 $2 \cdot 2 + 6 \cdot 2^2 + 12 \cdot 2^3 + 20 \cdot 2^4 + 30 \cdot 2^5 + \cdots$
A.7 Evaluate Eq. (A.7.3) for $a = 2$ and $n = 5$, and compare your result with the sum obtained by evaluating the first five terms numerically.

Derive expressions for the following:

A.8 $\sum_1^n x(x!)$

A.9 $\sum_1^{n-1} \sin\dfrac{\Pi}{n}x$

A.10 $\sum_1^n (x^2 + 1)x!$

A.11 $\sum_0^k \sin^2(a + bx)$

A.12 $\sum_{3}^{n} 8x^2 + 3x + 2$

Integrate the following expressions:

A.13 $\Delta^{-1} x \cdot \sin x$

A.14 $\Delta^{-1} e^x x^{(3)}$

A.15 $\Delta^{-1} \dfrac{x \cdot 2^x}{(x+1)^{|2|}}$

APPENDIX B

THE INTERCONVERSION OF ENGINEERING AND SI UNITS

The technical and scientific community in the United States is in a state of transition regarding the convention on units. The International System of Units (SI) is, without question, the most rational and easiest to use of all systems, and yet usage of the entrenched "engineering (fps) units" has by no means disappeared. Old habits die hard, and anachronistic units continue to permeate the fabric of technology, particularly at its interface with business. Petroleum people still speak of barrels of oil; chemical prices are still quoted on a pound or short ton basis; pipe diameters are still referred to in inches and pipe lengths in feet; and experienced chemical engineers continue to rate production unit heat exchangers in terms of $Btu/(h \cdot ft^2 \cdot °F)$. The fact is that emphasis upon SI unit usage has confronted practicing engineers with the need to be versatile in either system of units—in fact, we must not even forget the cgs unit system, with its ergs and dynes, still solidly entrenched with the scientific community. In this textbook the discussion, examples, and problems are predominantly characterized by engineering fps units, to reflect the industrial world as it is, and not perhaps the way it should be. The symbols in the Nomenclature, however, are expressed in SI units.

Interconversion among the various systems of units poses no particular problem; it is aided by use of well organized conversion tables such as the one in the back of the fifth edition of Perry's handbook. Peters and Timmerhaus (1980, Exhibit 4.2A) present a very thorough review of the SI system in their appendix A. The alternative to the use of tables, and probably the better one, is to decompose all derived units into their fundamental components of mass, length, time, and temperature and to perform the conversion of the fundamental units with a few easily remembered numerical relationships.

If there is an interconversion problem, it is ascribable to the abomination of mixed usage of pound-force and pound-mass units in the engineering fps unit system. The purpose of this short review is to present a rule of thumb which helps interconversion whenever mixed mass-force units must be dealt with. The only requirement is that one must know a priori whether the unit of "pound" is a mass unit or a force unit, a fact regrettably not always clarified in published nomenclatures.

Fundamental SI Units

Mass	kilogram	kg
Length	meter	m
Time	second	s
Temperature	kelvin	K
Amount of material	mole	g·mol or mol

Derived and Frequently Used Units

Force	newton	kg·m/s²	N
Work, energy	joule	N·m	J
Power	watt	J/s	W
°C	(OK to use, but called degrees Celsius)		
Gravitational constant		9.807 m/s²	g_n
Pressure	pascal	N/m²	Pa
	bar	10⁵ Pa	
	1 atm	1.013 bars	
	(1 torr ≡ 1 mmHg = 1/760 atm; torr not used)		
Heat	Use J (Cal$_{IT}$ = 4.187 J; calories not used)		

Interconversion, FPS and SI Units

There is no fundamental problem, but remember that fps units are *mixed* mass-force units. For example,

Density is expressed as lbm/ft³ (lbm ≡ pound-mass)
Pressure is expressed as lbf/ft² (lbf ≡ pound-force)

The *defining relationship* is

$$1 \text{ lbf} \equiv 32.17 \text{ lbm} \cdot \text{ft/s}^2$$

Example

In fps units, what is the pressure exerted by 1 ft of water at elevation where $g = 30$ ft/s²?

$$P = \frac{V \cdot \rho \cdot g}{A} = \rho \cdot g \cdot h$$

$\rho = 62.4 \text{ lbm/ft}^3$
$g = 30.0 \text{ ft/s}^2$
$h = 1 \text{ ft}$

Thus

$$P = 30 \times 62.4 \times 1 \frac{\text{lbm}}{\text{ft} \cdot \text{s}^2}$$

From defining relationship,

$$\text{lbm} = 1/32.17 \; \text{lbf} \cdot \frac{\text{s}^2}{\text{ft}}$$

Thus

$$P = \frac{30 \times 62.4}{32.17} \text{lbf} \cdot \frac{\text{s}^2}{\text{ft}} \cdot \frac{1}{\text{ft} \cdot \text{s}^2}$$
$$= 58.2 \; \text{lbf}/\text{ft}^2 \quad (0.404 \text{ psi})$$

The rule of thumb is that if a functional relationship exists such that

$$\phi_m = \phi(\text{lbm})$$

that is, units of ϕ_m have lbm, and it is required to obtain ϕ_f in units of lbf, then

$$\phi_f = \phi\left(\frac{\text{lbm}}{g_c}\right) \quad \text{and} \quad \phi_m = \phi(\text{lbf} \cdot g_c)$$

where

$$g_c = 32.17 \frac{\text{lbm} \cdot \text{ft}}{\text{lbf} \cdot \text{s}^2}$$

Example

Give previous answer in SI units.

$$P = 58.2 \frac{\text{lbf}}{\text{ft}^2} = 58.2 \times 32.17 \frac{\text{lbm} \cdot \text{ft}}{\text{ft}^2 \cdot \text{s}^2}$$
$$= 58.2 \times 32.17 \frac{\text{lbm}}{\text{ft} \cdot \text{s}^2} \times 0.4536 \frac{\text{kgm}}{\text{lbm}}$$
$$= 58.2 \times 32.17 \times 0.4536 \frac{\text{kgm}}{\text{ft} \cdot \text{s}^2} \times 3.281 \frac{\text{ft}}{\text{m}}$$
$$= 2786 \; \text{kgm}/(\text{m} \cdot \text{s}^2)$$
$$= 2786 \; \text{kgm} \cdot \text{m}/(\text{s}^2 \cdot \text{m}^2)$$
$$= 2786 \; \text{N/m}^2 = 2786 \; \text{Pa}$$
$$= 0.02786 \; \text{bars} = 0.02751 \; \text{atm}$$

Examples of Mixed Units, FPS System

1 All quantities involving the concept of work, heat, energy, and power are commonly expressed in terms of lbf. Specifically, the units of work are ft · lbf.

2 On the other hand, many physical properties are commonly expressed in terms of lbm. For example,

Units of viscosity are lbm/(ft · s).
Units of mass-transfer coefficient are lbm/(ft² · s).

3 Some physical properties, such as surface tension, are expressed in lbf units (lbf/ft). This results in considerable confusion; for example, in calculating dimensionless groups such as the Weber number,

$$\mathrm{We} = \nu^2 \frac{\rho L}{\sigma}$$

it is important to account for mixed lbm–lbf units, and using the rule of thumb, We is often written as

$$\mathrm{We} = \nu^2 \frac{\rho L}{\sigma g_c}$$

No such problems arise in the use of the SI system. However, in converting from the *fps* to the SI system, *it is important to be aware whether the fps units are in lbf or lbm.*

INDEX

Page numbers in *italic* indicate illustrations and tables.

Accounting:
   depreciation, role of, 347–351
   inventory control methods, (exercise) 665
Accounts payable, 662, *663*
Accounts receivable, 428, *429*, 430, 661
Accumulated sum, 78
Acid test ratio, 669
Administrative costs, 429–430
Agitation:
   preliminary equipment sizing, 203
   standard equipment geometry, *204*
Allocated capital (*see* Capital, fixed, allocated)
Allocation of resources, 560–567
Allowance allocation, 621–622
Alternative investment analysis (*see* Analysis of alternatives)
American Institute of Chemical Engineers, 48–51, 53, 54
   code of ethics, 54
   continuing education, 53
   dynamic objectives, 49–51
   professional guidelines, 49
Amortization (*see* Depreciation)
Analysis of alternatives, 63–64, 521–522
   break-even analysis, 529–531
   capitalized cost in, 527–529
   choice of criteria for, 531–532, 538
   differential methods: differential return on investment, 523–525, 532–535
      effect of service life, 525, 535
   effect of inflation, 535
   equipment selection methods, 522–531, 538
   incremental methods (*see* differential methods, *above*)
   investment selection methods, 531–532, 538
   process selection methods, 538
   project selection methods, 531–532, 538
   risk analysis in, 544–546

Analysis of alternatives (*Cont.*):
   risk associated with choices, 497
   (*See also* Replacement analysis)
Annual report, 658–660
   footnotes, 666–667
Annuity, 102–107
   nonuniform, 105–106
Arithmetic progression, 96–97
ASPEN (program), 697
Assets, 660–662
   current, 661
   fixed, 661
      average annual increase, 674–678
   tangible, 687
Authorization request, 616
Auxiliary facilities, 257–259, 285–286
   and allocated capital, 267
   categories of, 268, *286*
   cost of, *262*, 268, 288–290
   dedicated raw materials, 290
   definition of, 267
   evaporation ponds, 288–290
   flow sheet convention, 162–165
   in grass roots plants, 267
   offsite facility, 257
   oxygen plants, 290, *291*
Azeotropes:
   heterogeneous, 191–192
   prediction of, 191

Backup capital (*see* Capital, fixed, allocated)
Balance sheet, 658, 660–665
   ratios, 668–670
Bar chart in scheduling, 626–631
Bare cost (*see* Purchased cost)
Bare module factor (*see* Modular factor)
Battery limits, 257, *258*, 260n.

719

Benefit-cost ratio analysis (*see* Present worth ratio analysis)
Biochemical oxygen demand (BOD), (prob.) 412
Block flow diagram, 155–157
  graphics, 155–156
  as precursor of flow sheet, 160–162
Block operations, 373
Bonds, 69, 685
  book value of, 687
  (*See also* Securities)
Book value:
  in annual report, 661–662
  "book," 348
  definition of, 340
  of securities, 687–688
  "tax," 348
Break-even analysis, 439–442
  in analysis of alternatives, 529–531
  in economic optimization, 555, (example) 556–557
Break-even chart, 439–442
  break-even point, *440*, 441
  construction method, 440–441
  shutdown point, *440*, 441
Break-even point, 439, *440*, 441, *442*
Buildings:
  cost of, *262*, 266, 271–272
  process requirements, 266
Bulk cost, 371, 398
Business interruption loss, 644
By-products:
  in cost of manufacture, 372–373
  as criterion of process choice, 153

Capacity exponent, 215, 218, 273–274
  processing plants, *274*
Capital:
  administrative, *429*, 430
  analysis of investment alternatives, 531–551
  average undepreciated, 448
  backup (*see* fixed, allocated, *below*)
  borrowed: cost of, 71, 666
    in DCRR calculations, 490–493
    effect on DCRR, 490–493
  budgeting of, 537–539
  in cash flow analysis, *468*, 469
  cost of, 70–72
  definition of, 259*n*.
  differential, 524
  expensed, 351
  fixed: allocated, 259, 260, 374–376, *382*, 421–425
    depreciation on, 472–473

Capital, fixed (*Cont.*):
  direct, 256, 259–260
    (*See also* Direct fixed capital)
  general and administrative, (fig.) 429, 430
  incremental (*see* differential, *above*)
  leverage, 670, 690
  market, 422, 430–431
  recoverable, 337, 425, 428
    in profitability calculations, 446
  research, *429*, 430
  selling, *429*, 430
  sources of, 69–70
  sunk, 349, 579
  time-distributed investments, 486–490
  working, 259, 337, 421–422, *423*, 425–428
    accounts receivable, 428, 430
    on balance sheets, 661
    in cash flow analysis, 468–469
    cash on hand, 428, 430
    categories of, *423*, 425–428
    computed from annual report, 682
    and depreciation, 425
    estimation of, 425–428
Capital budgeting, 537–539
Capital gains and losses:
  long-term, 349, 568
  short-term, 350
Capital investment, 61
  estimation of (*see* Cost estimates)
Capital-recovery cost, 529
  and depreciation, 530
  salvage, 529
Capital-recovery factor, 529
Capital for sale, 422, 430–431
  estimation of, *429*, 430, (example) 431–432
Capital surplus, 663
Capital for transfer:
  categories of, 422
  definition of, 422, *423*, 430
  estimation of, 422–428
Capitalized cost, 107–111, 526–527
  in analysis of alternatives, 527–529
  and depreciation, 528
  factor, 108, 526
  inflation, effect of, 111, 528
  interest rate in, 527–528
  in plant with finite life, 109–111
Careers and career planning, 26–54
  advancement, 41–48
  career guidance, 44–48
  career objectives, 26, 28–29
  dual ladder system, 41–43
  job functions, 39–40, 42
  job interview, 27–34
  job performance review, 45

Careers and career planning (*Cont.*):
  politics, corporate, 47
  résumé, 30–33
    letter of transmittal, 31
  salaries, 30, 34–36
  starting assignments, 36–39
  triangle of success, 46–48
  (*See also* Professional development)
Cash flow, 86–87, 466
  analytical distributions, 94–96
  categories of, 467–469
  computation methods, 345–347, *468*
  continuous, 87, 338–339, 345–346, 467
  and depreciation, 338
  diagram, 469–472
  discrete, 87, 467–469
  equation of continuity, 468
  net, 87, 467
  normally distributed, 94–95, 482–483, (prob.) 514–515
  ramp, 92–93
  relation to profit, 339–340, 346, 446
  sign convention, 87, 467
  after taxes, 345–346
Cash on hand, 428, *429*, 430, 661
Cash outflow curve, 470, *471*, (prob.) 512, 622–625
  cost tracking, 624–625
Cash position, cumulative, 469–470
Centrifuges, (example) 206–207
Certification, 52–53
Challenger, 568
Chance event point, 546
Chart:
  bar (in scheduling), 626–631
  break-even, 439–442
  corporate organizational, 10–11
  product profile, 18–21
Chemical Engineering Plant Cost Index, 224
Chemical process industries:
  corporate ranking, *693*
  functional groups, 5–9
  job environment, 2–5
  objective, 3–4
  performance analysis, 692–695
  ranking in U. S. economy, 695, *696*
Chemicals:
  basic, 16
  commodity, 16
  fine, 16
  heavy, 16
  stability of, in scale-up, 599
  unit value of, 372
Code of accounts, 622, *623*
Cogeneration, 383–384

Communications, 56–58
Component build-up method, 212–214, (example) 219–223
Compounding, 78–81
  annual, 79
  continuous, 81
  discrete, 79–81
  factors, 86, 92
  time interval, 79–80
  (*See also* Discounting)
Computer-aided economic analysis, 481, (prob.) 510–512
Construction (*see* Plant construction)
Consumer demand curve (*see* Demand-price curve)
Consumer Price Index, 66–67, 223–224, *225*
Contingency factors, 270
Contractor fees, 230, *262*, 270
Coproducts:
  in cost of manufacture, 371–372
  as criterion of process choice, 153
Corporation:
  annual report, 658–660
  financial data sources, *692*
  life-style, 43–48, 54
  long-range planning, 148–149
  organization, 10–13
  performance analysis, 658–659, 692–695
  politics, 47
Cost (*see specific designation*)
Cost control, 613, 619–626
  cash outflow curve, 622–625
  code of accounts, 622, *623*
  cost tracking, 624–625
Cost data, sources, 215–217
Cost estimates:
  authorization, 64–65, 614
  categories of, 64–65
  common errors in, 312–314
  detailed, 271
  factored, 65, 229, 261–271
  functional step scoring, 278–285
  order-of-magnitude, 64
  piping, *262*, 305–311
    shortcut methods, 261–263, 306, 308–311
  project control, 65, 614–615
  reliability of, 232–240, 311–314
  shortcut methods, 271–285
  statistical analysis, 232–240
  study, 64
  total plant cost, *262*, 270
  total plant direct cost, 261, *262*, 268
  total plant indirect cost, *262*, 268
Cost index, 223–227
  Chemical Engineering Plant Cost Index, 224

Cost index (Cont.):
  Consumer Price Index, 66–67, 223–224, 225
  data sources, 224
  Engineering News Record (ENR) Index, (prob.) 316
  limitations, 225–227
  Marshall and Swift (M&S) equipment cost index, 224, 225
  projection, 225
Cost of manufacture (COM), 367–373
  bulk cost, 370, 371, 398
  by-products, effect on, 372–373
  components of, 337, 367
  coproducts, effect on, 371–372
  cost for transfer, 370, 371
  definition of, 334, 367, 368
  depreciation in, 334
  estimation of, 370–373
  fixed charges, 371, 395–397, 399
  regulated charges, 371, 392–395, 399
  scope of, 368–369
  shortcut methods, 398, 399
  variable charges, 371, 399
Cost for sale (see Cost of manufacture)
Cost tracking, 624–625
Cost for transfer, 370, 371
Crashing, 635–639
Criterion of economic performance, 62–63, 89, 124, 421
  choice of, 434, 466–467, 496–497, 531, 538
  in plant sizing, 143
  use of, in pricing, 436–437
Critical path, 628
Critical path method (CPM), 631–637
  crashing, 635–639
  in workforce leveling, 637–639
Cumulative cash position, 469–470
Current ratio, 669

Debentures, (prob.) 117
Debottlenecking, 574–577
Debt (see Capital, borrowed)
Debt ratio, 69, 695
Decision networks (see specific designation)
Decision point, 546
Decision trees, 544–551
  "roll-back" technique, 547–549
  strategy deduction from, 549
Defender, 568
Deferred charges, 661, 662
Delivered cost, 209–210, 229
Delphi method, 697
Demand:
  versus price, 133–135

Demand: (Cont.):
  price elasticity of, 133–135
  projection of, 128–149
  regional, estimation of, 140–146
  relation to production, 129
  utilities, 286, (example) 287–288
Demand-price curve, 133, 134, 443
  analytical correlation, 134
  in economic optimization of plant size, 443
Depletion, 337
Depreciation, 334–337, 395
  accountancy, 347–351
  on allocated capital, 472–473
  and cash flow, 338, 345–347
  effect of inflation, 435–436
  effect upon profit, 337–340, 345
  methods: composite, 357–358, 359
    double-declining-balance (DDB), 353–357
    effect on assets value, 348
    Matheson formula, (prob.) 366
    preferred, 347, 360
    sinking fund, (prob.) 366
    straight-line (SL), 341, 351–353
    sum-of-years-digits (SOYD), 358–360, 361
  as operating cost, 335–336
  in preliminary estimates, 360
  tax credit, 360, 530
  and working capital, 425
Depreciation tax credit, 360, 530
Design:
  economic optimization in, 551–553
  of pilot plants, criteria for, 606–610
  of piping systems, 290–304, 552
  preliminary, 188–209
    sources of methods, 190
  safety factors in, 207, 209
Differential return on investment (see Analysis of alternatives, differential methods)
Differential thermal analysis (DTA), 128
Dimensionless ratios (see Ratio analysis)
Direct fixed capital (DFC):
  in cash flow analysis, 468, 469
  definition of, 256, 259
  elements of, 261–271
  estimation of (see Cost estimates)
Discounted cash flow (see Discounted cash rate of return)
Discounted cash rate of return (DCRR), 478–481
  differential, 533–535, 538
    with nonuniform cash flow, 534–535
  effect of borrowed capital, 490–493
  for normally distributed production, 482–484, (prob.) 514–515
  probability distribution, 502–507

INDEX   **723**

Discounted cash rate of return (*Cont.*):
　relation to return on investment, 494–496
　in research project evaluation, 493
　with time-distributed investments, 486–490
Discounting, 87
　continuous, 87
　　of continuous cash flows, 90–91
　　of discrete cash flows, 99–100
　discrete, 87
　　of discrete cash flows, 96–99
　factors, 92
Distillation:
　flow sheet, 161–162
　preliminary design, 190–199
　　column dimensions, 194–195, (example) 195–199
　　packed columns, 194–195
　　stage calculations, 192–194, (example) 195–199, (prob.) 251
　　vapor-liquid equilibria, estimation of, 190–192
Distribution:
　barges, 454–455
　cost of, 141, 369, 377, 452–455
　emergency response teams, 451
　as functional group, 9
　loading facilities, 398–400, 452
　marine transport, 454–455
　by railroad, 141, 451–453
　　car categories, 451–452
　　tank car sizes, 452
　by truck, 453
　warehousing strategy, (prob.) 410–412
Distribution (mathematical):
　cumulative, 501
　normal, in life cycle theory, 482–483
　probability: of economic parameters, 499–502
　　trapezoidal, 500–501
　of profitability index, 502–507
Double-declining balance (DDB) depreciation, 353–357
　salvage value, effect of, 354, 356–357
　straight-line, combination with, 357–358, *359*
Dual ladder system, 41–43
"Dummy" activities (CPM), 632
"Dummy" variables, 566

Earnings per share, 689
Economic analysis:
　categories of, 63–64
　computer-aided, 481, (prob.) 510–512

Economic analysis (*Cont.*):
　criterion of economic performance, 62–63, 124, 466
　effect of inflation, 67
　investment analysis, 62–63
　net income analysis, 62–63
　optimization (see Economic optimization)
　of process improvements, 570–571
　profitability, 124
　profitability index, 63
　of research investments, 541–544
　scope of, 61–66
Economic evaluation, as industrial functional group, 8–9
Economic optimization, 63, 551–553
　allocation of resources, 560–567
　with break-even analysis, 555, (example) 556–557
　of continuous objective functions, 553–557
　with differential ROIBT, 554–555, (example) 556, 557–560
　of discrete objective functions, 557–560
　of plant size, *438*, 443–444
　strategy of, 549
Economic pipe diameter (see Piping systems, sizing of)
Energy balance, preliminary, 159–160
Energy tax credit, 472
Engineering:
　construction, 7, 269–270
　as industrial functional group, 6–7
　plant, 7
　process, 6–7
Engineering registration, 51–53
　examinations, 52
　practice act, 51
　title act, 51
Environmental control, 2–3, 400–402
　esthetics, 402
　as industrial functional group, 9
　in process choice, 154
　regulations, 401
Equipment:
　analysis of alternatives, 522–531
　　acceptable differential ROIBT, 524, 558
　　break-even analysis, 529–531
　　use of capitalized cost, 527–529
　characterization of, for costing purposes, 189
　debottlenecking strategy, 574–577
　delivered cost, 209–210
　depreciation of, 335
　equipment list conventions, *208*, 209, 379
　field fabrication, 228

724 INDEX

Equipment (*Cont.*):
  flow sheet labels, 162, *164*
  installation of, 227–228
  installed cost, 210, 228–232, 261, *262*
  leasing of, 539–541
  life, 147, 340–342
  need for duplication, 169
  pictographs (symbols), 162, *163*
  for pilot plants, 606–607
  preliminary design of, 188–209
  purchased cost: adjustment factors, 213, *214*
    capacity exponent, 215, 218
    component build-up method, 212–214,
      (example) 219–223
    correlation methods, 210–215
    data sources, 215–217
    estimation of, 209–223
      by analogy, 219–223
    inflation adjustment, 223–227
    reliability of estimates, 232–240
    size scaling, 217–219
    variability of data, 232–235
  safety factors in design of, 207, 209
  size criterion, 189
  for storage, 205–206
  vendors, information sources, 245
Equity, 660, 663
Equity ratio, 669–670
Equivalent annual cost, 529–531
Equivalent maximum investment period
  (EMIP), 473–474
Equivalent uniform annual cost (EUAC),
  568–569
Esthetics, 402
Ethics, 47, 54–56
Evaporation ponds, 288–290
Evolutionary operations (EVOP), 697
Expectation (mathematical), 498, 507, 545
Expected value, 498, 507, 545

Factor(s):
  adjustment, 213, *214*
  capacity (valves), (prob.) 328
  capital-recovery, 529
  capitalized-cost, 108, 526
  complexity, 280
  compounding, 86, 92
  contingency, 270
  discounting, 92
  inflation adjustment, 673–674
  instrumentation cost, *262*, *264*
  Lang, 272–273
  modular, 277–278, *279*
  power, 273–274

Factor(s) (*Cont.*):
  purchased cost, 261–271
  safety, 207, 209
  seven-tenths factor rule, 273–274
  severity, 280, *282*
  six-tenths factor rule, 218
  size, 270
  stream, 373
Factory expense, 397
  allocated capital, *423*, 424–425
Fault tree analysis, 697
FIFO (inventory accounting), (exercise) 665
Filtration, typical design rates, (prob.) 249
Finite differences, 703–704
Finite integration, 706–708
  definite integral, 708–710
  finite integrals, *707*
  integration by parts, 710–711
Fire and explosion index, 644–646
Fixed charges, 371, 395–397, 399
Floater, 627
Flow sheet, 160–170
  construction methodology, 162–167
  conventions, 162–168
  equipment labeling, 162, *164*
  errors, 169–170
  pictographs (symbols), 162, *163*
  translation from block flow diagram,
    160–162
Flow table, 158–159
Free on board (FOB), 210, 451
Freight costs, 141, 210, 377
  effect on pricing, 451
  freight equalized costs, 451
Fringe benefits, 230, 391
Functional groups in industry, 5–9
Functional step scoring, 275, 278–285
  complexity factor, 280
  functional step definition, 278, 280–281
  severity factor, 280, *282*
Future worth, 86

General and administrative costs, 429–430
General expenses, 422
  categories of, 429–430
  definition of, 338, 369, 428
  estimation of, 428–430
Geometric progression, 97
Gompertz curve, 94, *95*, 130, 132
Goodwill, *661*, 662
Gradient uniform series, (prob.) 116
Grass roots plant, 142, 257
  auxiliary facilities, 267–268
Gross margin, (prob.) 465

INDEX   **725**

Hay points system, 41
Heat exchangers:
   condensers, 201
   fouling, (exercise) 202
   heat transfer coefficients, typical, 200
   preliminary design of, 199–202
   reboilers, 201
Heat of reaction, estimation of, 159–160
Heat transfer:
   approach temperatures, 200–201
   coefficients, typical, 200
   coolant temperature rise, 200
   equipment design, 199–202
   thermal pinch point, 200–201
Heuristics, 188–189
Historical analogy, demand projection from, 139–141
Historical continuity, principle of, 129

Income statement, 658, 660, 665–666
Income taxes (*see* Taxes, income)
Induction, 100–102
Inflation, 66–68, 83–85
   in analysis of alternatives, 528, 535
   in corporate performance analysis, 673–678
   effect on capitalized cost, 111, 528
   effect on equipment costs, 223–227
   effect upon profitability, 435–436
   rate of, 67–68, *71*, 83
   relation to interest rates, 436
Inflation adjustment factor, 673–674
Installation:
   cost of, 210
   estimation of costs of, 228–232
   labor costs, 230–232
   labor requirements, 230–232
   significance of, 227–228
Installed cost, 210, 228–232
   estimation of, 228–232, 261, *262*
   labor component, 230
Instrumentation:
   cost of, *262*, 263–265
   in pilot plants, 608
Insulation:
   cost of, *262*, 265
   economic thickness, 302, 552
Insurance, cost of, 395–397
Intangibles, *661*, 662
Integrated complex, 142, 257
Interest:
   compounding, 78–81
   definition of, 70, 78
   as operating expense, 70, 72, 397, *665*, 666, 682

Interest (*Cont.*):
   payments: present worth of, 111–113
     tax credit on, 491
Interest rate:
   categories of, 71
   compound, 78–81
   continuously compounded, 81, *83*
   definition of, 78
   discounted cash rate of return, 478–481
   discrete compound, 79–81, *83*
   effective, 81–82, *83*
   internal, 71
     in capitalized cost, 527–528
     computation of, 679–685
     inflation-adjusted, 528, 679–685
   on new investments, 71, 476
   nominal, 80, *83*
   in present worth calculations, 476, 478
   prime, 71
   relation to inflation, 436
   simple, 79, *83*
   on time-distributed investments, 488–489
Interstatement ratios, 670–671
Inventories:
   on balance sheet, 661
   control of, (exercise) 665
   in-process, 427
   maintenance and stores, 425
   product, 426–427
   raw materials, 425–426
Investment (*see* Capital)
Investment, return on (*see* Return on investment)
Investment tax credit, 472
Investment per unit capacity, 275–276
Isometric drawings, 303–304

Job performance review (JPR), 45

Labor:
   clerical, 391–392
   construction, pay rate, 230, 309
   cost of, 230, 390–392
   fringe benefits, 230, 391
   operating, 384–392
     annual working hours, 389
     cost of, 390–391
     estimation of requirements for, 384–390
     job function, 384–385
     plant total, 389–390
     for piping installation, 306, 308–310
     supervisory, 391–392
Laboratory, cost of, in quality control, 393–394

726 INDEX

Land:
  cost of, 337, 428
    in cash flow analysis, 468–469
    as recoverable capital, 428
Lang factors, 272–273
Leading indicators, 127, 135
  sources, 139
Learning curve, 442
Leasing, 539–541
Leverage, 670, 690
Liabilities, 660, 662–663
  current, 662, 663
  long-term, 662–663
Life:
  equipment, 147, 340–342
    in analysis of alternatives, 525, 529, 538
  plant, 109, 147, 341–342, 349–350
    in analysis of alternatives, 535–537
    average, from annual report, 674, 678
Life cycle:
  normal distribution approximation, 482–483
  plant, 147, 481–485
  product, 94, 146–147, 485
  production, 147, 481–485
  theory, 94, 146–147, 481–485
Life cycle cost, 697
LIFO (inventory accounting), (exercise) 665
Linear programming, 561
Loading facilities, 398–400, 452
Logistic curve, 94, 95, 130
Loss prevention, 642
Low equation, 494–496

Maintenance:
  contrasts, in equipment leasing, 540
  cost of, 392–393
  scheduling of, 552
Management:
  of construction projects, 595–596, 610–617
  corporate, 7, 10–11, 40, 42
  line, 11
  of pilot plant operations, 601–613
  of process development, 596–613
  project, 1–2, 12–26, 65, 269, 595–596, 601–613
    of construction projects, 269, 595, 610–619
    status advancement decisions, 604–606, 612
  staff, 11
Manufacturing cost (see Cost of manufacture)
Manufacturing representative, 13, 615
Margin:
  operating, 670–671
  profit, 671

Marginal cost, (prob.) 460
Market capital, 422, 430–431
Market development, 125, 127
Market penetration, 140–146
Market value of stocks, 69, 685
Marketing (see Sales)
Marketing research, 15, 124, 126–128
  field survey, (prob.) 174
  methodology of, 126–127
  scope of, 126–127
Marshall and Swift (M&S) equipment cost index, 224, 225
Materials of construction, economic choice, 527
Matheson formula, (prob.) 366
Maximum probable days of outage (MPDO), 644–646
Maximum probable property damage (MPPD), 644–646
Mid-square method, 503
Miniplant, 125, 601
Mixing (see Agitation)
Modular factor, 277–278, 279
Money:
  constant-value, 83
  time value of, 70, 84–86
  unit value of, effect of inflation upon, 66–67, 83
  as yardstick of profitability, 66
Monte Carlo simulation in risk analysis, 498, 502–504

National Fire Protection Association (NFPA) ratings, 128
Natural gas, cost of, 381–382
Net positive suction head (NPSH), 300
Net present worth, 87–90
  choice of interest rate, 476
  as criterion of economic performance, 89, 474, 476–478
  as criterion of profitability, 476–477
  probability distribution, 505–507
  in research project evaluation, 493
  in scheduling analysis, 637
Net profit ratio, 671
Normal distribution:
  of economic parameters, 504–507
  in life cycle simulation, 482–483

Objective function, 551, 560
Obsolescence in replacement analysis, 568, 573
Offsite facility, 257
Oil, cost of, 379–380

INDEX   **727**

Operating block, 388
Operating book, 611
Operating cost (see Cost of manufacture)
Operating discipline, 552
Operating margin, 670–671
Operating profit, *665*, 666
Operating ratios, 670–671
Operating supplies, 379, 395
Opportunity cost, 570
Out-of-pocket expenses, 338, 345–346, 467
  in cash flow analysis, 468
Oxygen, standard cost of, 378

Packaging:
  allocated capital for, *423*, 424
  containers, 400
  cost of, 398–400
  loading facilities, 398–400
Par value of stocks, 69, 663
Patents, 58–61
  classification, 59
  criteria of patentability, 59–60
  as criterion of process choice, 153
  as criterion of project success, 17
  licensing, 61
  royalties, 61, 397–398
  trademarks, 59
  written format, 60–61
Payment on loan (see Annuity)
Payout time (payback period), 449–450
  with interest, 474–476
  without interest, 449–450
Perpetuity, 107
PERT (see Program Evaluation and Review Technique)
Petroleum (see Oil)
Physical cost (see Total plant direct cost)
Pilot plant, 125, 601
  construction of, 610–612
  design criteria, 606–610
  design sequence, 606–610
  organization of operations, 601–613
Piping and instrumentation diagram (P&ID), 166–168, 609
Piping systems:
  control valve piping, *295*, 296
  cost of, 261–263, 305–311
    data sources, *306*
    shortcut methods, 306, 308–310
  design of, 290–304
  insulation, 302
  isometric drawings, 303–304
  manifolds, 301–302
  piping standards, 293, *294*
  relief device piping, *295*, 296–297

Piping systems (*Cont.*):
  sizing of, 297–301
    for avoiding siphons, (prob.) 329
    economic pipe diameter, 297–298, 552
    minimum critical velocity, (prob.) 327–328
    pipe velocity criterion, 298–299
    for slurry flow, (prob.) 331
  standard configurations, 293–297
  supports, 302–303
  tracing, 302, 303
  winterizing, 302
Plant:
  battery limits, 257, *258*
  construction of (see Plant construction)
  expansion strategy, 577–578
  grass roots, 142, 257
  investment, estimation of (see Cost estimates)
  layout, (prob.) 324–325
  life, 109, 147, 341–342, 349–350
  life cycle, 147, 481–485
  maximum sustained capacity, 482–483
  modifications, analysis of, 570–571, 574–578
  overcapacity, 207
  parallel train, (prob.) 585–586
  pilot (see Pilot plant)
  production for prescribed DCRR, 485
  rated capacity, 482–483
  shutdown, *440*, 441, 579–580
  sizing of, 143–145, *438*, 443–444
  start-up, 125, 431
    equipment for, 169
    preoperational review, 642–644
  turnkey, 153
Plant construction, 125
  authorization study, 125
  cash outflow curve, 470, *471*
  contractor fees, 230, *262*, 270
  cost control, 613, 619–626
  elements of success in, 613–615
  engineering and construction fees, *262*, 268–270
  labor rates, 230
  labor requirements, 230–232
  management of, 595–596, 610–617
    authorization request, 616
    project engineer, 615
    project manager, 615
  penalty due to delays, (prob.) 650
  pilot plants, 610–612
  preliminary design package, 612–613
  preoperational review, 642–644
  scheduling, 613, 614, 626–642
    critical path method, 631–637
    PERT, 639–642

Plant gate capital (*see* Capital for transfer)
Power:
   cogenerated, 383–384
   cost of, 381–384
Power factor, 273–274
Precedence hierarchy in scheduling, 627
Preliminary design, 188–209
   sources of methods, *190*
Preoperational review, 642–644
Prepayments, *661*, 662
Present worth, 85–86
   interest payments, 111–113
   net (*see* Net present worth)
   of normally distributed cash flows, 483–484
   ratio (*see* Present worth ratio)
Present worth ratio, 476–477
   in analysis of alternatives, 535–538
   in economic ranking of projects, 537–539
Present worth ratio analysis, 535–538
   effect of project life, 536–537
Price-earnings ratio, 689
Price elasticity, 133–135
   demand-price curve, *133*, 134
   product classification by, 134
   unitary, 134
Prices and pricing:
   criteria of, 151–152, 436–437
   effect upon demand, 133–135
   effect of freight costs, 149–150, 451
   established products, 149–151
   historical, 150
   sources, 149, 150
   strategy, 150–151, 436–437
Prime interest rate, 71
   and inflation, 71, 436
Principal, 78
Principle of minimum investment, 522–523
Probability distribution:
   cumulative, 501
   of economic parameters, 499–502
      correlation between events, 547
   normal, 504
      of net present worth, 504–507
   of profitability index, 502–507
   trapezoidal, 500–501
Process(es):
   analysis of alternatives, 531–537
      acceptable differential ROIBT, 533
   criteria of choice, 152–154
   definition of, for new products, 154
   description, sources, *152*
   energy balance, preliminary, 159–160
   improvements, analysis of, 570–571
   mass balance, 158
   prototypic, 154–155
   synthesis, 154–155, 158

Process development:
   chemical stability testing, 599
   management of, 596–613
   pilot plant operations, 601–613
   sequence of, 596–597, *602–613*
Product profile, 124, 127–128
Product profile chart, 18–21
Product stewardship, 7, 402, 451
Production, 6
   level of: effect on profitability, *438*, 440–442
      at minimum acceptable return, 441–442
   life cycle, 146–147, 481–485
   statistics: analytical correlation of, 129–131
      demand projection from, 128–146
      graphical correlation of, 131
      sources, *129*, *135–137*
      trend analysis of, 128–132
Professional development, 36–39, 43–46, 48–58
   certification, 52–53
   communications, 56–58
   continuing education, 53
   engineering registration, 51–52
   initial assignments, 36–39
   professional project program, 36
   professional society activities, 48–51, 53
   professionalism, 44
   training, 36, 39, 45–46
   whole job concept, 43
Profit, 4
   gross, 338, 666
      (*See also* before taxes, below)
   net, *665*, 666
   operating, *665*, 666
   relation to cash flow, 339–340, 346
   after taxes, 72, 338–339, 345–346
   before taxes, 72, 338, 339, 345–346, 422, 428–430, 524
   differential, 524
Profit margin, 671
Profitability, 63, 124
   criterion of (*see specific designation*)
   index of (*see specific designation*)
Program Evaluation and Review Technique (PERT), 639–642
   earliest expected completion time, 639–640
   probability techniques, 641–642
   slack, 641
Progression:
   arithmetic, 96–97
   geometric, 97
Project:
   analysis of alternatives, 531–539
   classification, 123
   construction (*see* Plant construction)
   cost control of, 613, 619–626

Project (*Cont.*):
  criteria of success, 14–19
  definition of, 13
  development stages, 64, 122–126
  economic basis of, 61–72
  evaluation of (see Project evaluation)
  mortality rate of, 25–26
  mutually exclusive, 522
    ranking of, 531–532
  nonexclusive alternatives, 537
  objectives of, 14
  optimum strategy, 549
  origin of, 13–14
  professional project program, 36
  risk analysis, 497–507, 544–551
  scheduling of, 613, 614, 626–642
  scope of, 313, 614
  scoring, 19
  typical, for neophyte engineers, *37–39*
Project evaluation:
  preliminary, sequence of, 123–124
  of research projects, (prob.) 414–415, 493–494, (prob.) 510–512, 541–544
Project management (see Management, project)
Propagation of errors, 235–240
Property taxes, 344, 395–397
Pumps:
  motor drives, 203
  net positive suction head, 300
  performance characteristics, 300–301
  piping configuration, 295–296
  sizing, 202–203, (example) 203–205
Purchase orders as sources of cost data, 217, *218*
Purchased cost:
  adjustment factors, 213, *214*
  capacity exponent, 215, 218
  component build-up method, 212–214, (example) 219–223
  correlation methods, 210–215
  data sources, 215–217
  definition of, 209–210
  estimation of, 209–223
    by analogy, 219–223
  factors, 229
  inflation adjustment, 223–227
  relationship to delivered cost, 209–210
  reliability of estimates, 232–240
  size scaling, 217–219
  variability of data, 232–235
Purchased cost factors, method of, 229, 261–271
Purchasing:
  as industrial functional group, 9
  as source of cost data, 216, 217, *218*

Quality control, cost of, 393–394
Quick assets, 669

Ramp cash flow, 92–93
Random number generation, 503
Rate of return (see Interest rate)
Ratio analysis, 658–659, 667–672
Raw materials:
  allocated capital for, *423*, 424
  catalysts, cost accounting of, 378–379
  categories of, 16
  cost of: in COM estimation, 371, 374–376
    contract prices, 377
    in-plant transfer costs, 378
    sources, 149–150, 377
    as criterion of process choice, 153
    standard cost of, 374–376
  ultimate, 16
  unit ratio, 377
Reactors, 203
  control of, 599–600
  runaway reactions in, 599–600
  scale-up of, 599–600
Regulated charges, 371, 390, 392–395, 399
  on break-even charts, 439–440
Reliability analysis:
  cost estimates, 311–314
  equipment costs, 232–240, 311–314
Replacement analysis, 64, 567–574
  challenger and defender concepts, 568
  effect of tax structure, 568–570, 573
  equivalent uniform annual cost, 568–569
  methodology of, 570–574
Research and development, 5–6, 19–26
  applied research, 19, 22
  capital costs, *429*, 430
  defensive research, 23
  directed research, 22
  employment statistics, *24*
  environmental research, 23
  evaluation of research projects, (prob.) 414–415, 493–494, (prob.) 510–512, 541–544
  expenditures, 23–25, 398, 429–430
  fundamental research, 19, 22
  long-term research, 22
  measure of effectiveness, 25–26, 544
  mortality of projects, 25–26
  offensive research, 23
  organization, 11–13, *42*
Résumé, 30–33
  letter of transmittal, 31
Retained earnings, 663–664
  statement, 660, 666
Return on average investment, 448–449

Return on investment (ROI), 84, 422, 428, *429*, 659, 668, 671–679
   acceptable values, 432–434
   average DFC basis (ROAIBT and ROAIAT), 448–449
   corporate, 671–679
   as criterion of pricing, 436–437
   effect of inflation, 435–456
   effect of variable cash flow, 446–447
   price dependence, 437–439
   recoverable capital convention, 446
   relation to DCRR, 494–496
   in service facilities, 376
   after taxes (ROIAT), 447, 678–679
      computation from annual report, 678–679
   before taxes, 428, *429*, (example) 431–432, 523–525, 672
      computation from annual report, 672–678
      differential, 523–525, 532–533, 535, 538, 554–555
Return on sales, 450, 671
Return on stockholders' equity, 670
Return on total assets, 671
Risk analysis, 422, 432–434, 497–507, 544–551
   decision trees, 544–551
   quantification of risk, 497–507
      probability techniques, 498–502, 504–507
"Roll-back" technique, 547–549
Royalties, 61, 397–398
Rule of signs, 486–487
Runaway reactions, 599–600
   runaway point, 600
   safeguards against, 600

S-shaped curve (see Cash outflow curve)
Safety factors, 207, 209
Safety and health, 3
   as criterion of design, 402
   fire and explosion index, 644–646
   preoperational review, 642–644
   in process choice, 154
   relief devices, *295*, 296–297, (prob.) 330–331
Salaries:
   of chemical engineers, 30, 34–36
   (*See also* Careers and career planning)
Sales, as industrial functional group, 7–8
Sales taxes, 210
"Salvage foregone," 570
Salvage value, 108, 337, 342–343, 360
   equivalent annuity credit, 529
Scale-up:
   decision criteria, 606
   in process development, 597–598

Scale-up exponent, 215, 218, 273–274
   labor, 386
Scheduling, 613, 614, 626–642
   bar charts, 626–631
   crashing, 635–639
   critical path, 628
   critical path method, 631–637
   floater, 627
   job sheet, 628–629
   PERT, 639–642
   precedence hierarchy, 627
   workforce leveling, 637–639
Securities, 659, 661
   book value, 687–688
   performance analysis, 685–692
   (*See also* Stocks)
Selling cost, 429–430
Semiplant, 125, 601
   economic justification, 547–549
Sensitivity analysis, 147, 437, 444–445
   probability techniques, 498, 502–507
   sensitivity diagram, 445
Series, summation of, 97, 98, 703–713
Seven-tenths factor rule, 273–274
Shutdown point, *440*, 441
Simplex algorithm, 567
Sinking fund, 103, (prob.) 117
Six-tenths factor rule, 218
Slack, 641
"Slack" variables, 566
Specifications, product, 127
Standard cost, 374–376
Statistical analysis:
   of equipment cost data, 232–240
   in establishing probability distribution, 499–502
   in profitability computations, 504–507
   propagation of errors, 235–240
Steam, cost of, *382*
Steam plant (see Auxiliary facilities)
Stirling numbers, 707
Stockholders' equity (see Equity)
Stocks, 69
   book value of, 687, 688
   common, 69, 685
   dividends, 69, 72, 688–689
   earnings per common share, 689
   market value, 69, 685
   par value, 69, 663
   preferred, 69, 685
   price-earnings ratio, 689
Storage facilities, 169, 205–206
   for products, 426–427
   for raw materials, 426
   sizing of, (prob.) 458
   warehouses (see Buildings)

Straight-line (SL) depreciation, 341, 351–353
Stream factor, 373
Sum-of-years-digit (SOYD) depreciation, 358–360, *361*
Sunk costs, 568
Superproductivity, 207, 442, 483
Supervision, cost of, 391–392
Supplies, operating, cost of, 395

Tanks (*see* Vessels)
Taxes:
   capital gains, 349–350, 568
   effect on cash flow, 345–347
   effect of obsolescence, 568, 573
   effect on profit, 338, 339, 345–346
   income: effect of depreciation, 343, 345–347, 360
     laws for, 72, 343–344, 347
     rates, 343–344
     state, 343–344
   property, 344, 395–397
   in replacement analysis, 568–570, 573
   sales, 210
   tax credits, 472
     depreciation, 360, 530
     on interest payments, 491
     on investment, 472
Technical service and development (TS&D), 8
   measure of effectiveness, 25
Technological forecasting, 148–149
Threshold limit value (TLV), 127
Time control (*see* Scheduling)
Time-distributed investments, 486–490
   rule of signs, 486–487
Time value of money, 70, 84–86
Total plant cost (TPC), *262*, 270
Total plant direct cost (TPDC), 261, *262*, 268
Total plant indirect cost (TPIC), *262*, 268
Traffic and distribution, 9
   (*See also* Distribution)
Transportation (*see* Distribution)
Trapezoidal distribution, 500–501
   cumulative, 501
Trend analysis, 128–132
   analytical correlation, 129–131
   graphical, 131
Triangle of success, 46–48
Turnkey plant, 153
Turnover ratio, 276–277
   average for chemical industry, 277
   processing plants, *274*

Unit cost, 369
Unit ratio, 377

Unit value, 372
Units:
   conventions in industry, 715
   interconversion of, 715–718
Utilities:
   allocated capital for, *423*, 424
   auxiliary facilities for, 285, *286*
   cost of, 379–384
     fuel oil, 379–380, *382*
     natural gas, 381, *382*
     power, *382*, 383–384
     steam, *382*
     water, *382*
   demand estimation, 286, (example) 287–288, 379
   flow sheet convention, 162–165, 170
   pilot plant requirements, 607–608
   standard cost of, 374–376

Value added by manufacture, 372
Vapor-liquid equilibria:
   azeotropic behavior, 191–192
   data sources, *191*
   estimation of, 190–192
Variable charges, 371, 399
Venture worth, 478
   (*See also* Present worth)
Very large crude carrier (VLCC), 454
Vessels:
   storage, 205–206
   surge, 205–206

Warehouses (*see* Buildings)
Wastes and waste treatment, 394–395
   biochemical oxygen demand, (prob.) 412
   cost of aqueous waste treatment, 290, 394, *396*
   as criterion of process choice, 154
   evaporation ponds, 288–290
   in pilot plants, 608
   on process flow sheets, 170
Water:
   categories of, *382*
   cost of, *382*
Weibull function, 95
Whole job concept, 43
Workforce leveling, 637–639
Working capital (*see* Capital, working)
Write-off (*see* Depreciation)

Yard improvement costs, *262*, 267

LIBRARY
ROWAN COLLEGE OF NJ
201 MULLICA HILL RD.
GLASSBORO, NJ 08028-1701